The Physics of Sports

The Physics of Sports

Edited by
Angelo Armenti, Jr.
Villanova University

American Institute of Physics
Woodbury, New York

In recognition of the importance of preserving what has been written, it is a policy of the American Institute of Physics to have books published in the United States printed on acid-free paper.

AIP Press
American Institute of Physics
500 Sunnyside Boulevard
Woodbury, NY 11797-2999

Library of Congress Cataloging-in-Publication Data
The Physics of sports / edited by Angelo Armeni, Jr.
p. cm.
Includes bibliographical references.
ISBN 0-88318-946-1 (v. 1)
1. Physics. 2. Sports. 3. Force and energy. I. Armenti, Angelo
QC26.P49 1991 91-30634
796'.01'53–dc20 CIP

10 9 8 7 6 5 4 3 2

DEDICATION

This volume is dedicated to Paul Kirkpatrick, sportsman and scientist, whose early interests, keen insights, and seminal publications on scientific aspects of athletic competition established for all time the avocation which has come to be known as the physics of sports.

ACKNOWLEDGMENTS

A number of people played significant roles in making this book possible, and I would like to thank all of them publicly for their contributions and assistance. Special thanks go to: Rita Lerner of AIP, who first proposed to me the idea for this book; the authors of the 57 articles that are included in this volume; Carolyn Lea, who provided invaluable research assistance; Howard Poss, who introduced me to the work of Vitruvius; Howard Brody, Peter Brancazio, and Peter Hinrichsen, who reviewed and helped develop the Bibliography; John Zumerchik, my energetic and inspiring editor at AIP, who guided the project to fruition; and my wife Barbara for her unfailing love, support, and encouragement.

TABLE OF CONTENTS

CHAPTER 6: THE BICYCLE

CHAPTER 7: THE HUMAN ELEMENT

CHAPTER 8: FLUIDS AND SPORTS

CHAPTER 9: SPORTS ASSORTMENT

CHAPTER 10: BIBLIOGRAPHY

APPENDICES

FOREWORD

A Modern Definition of Sport

The following definition of "sport" is taken from James A. Michener's *Sports in America* (Random House, New York, 1976), and was adopted to help clarify the proper scope of sports activities to be included in this volume:

1. an athletic activity requiring physical prowess or skill and usually of a competitive nature (baseball, football, tennis, Olympic field sports, fencing, boxing, etc.);
2. any form of activity carried on out of doors, often not of a competitive nature (hunting, fishing, horseback riding, sailing, birdwatching, etc.).

Statement of Purpose

The primary purpose of this book is to collect and present in one place interesting examples of the physics of sports for the benefit of physicists, physics teachers, students, and other sports-minded and scientifically curious readers.

If supplemented properly with a "standard" physics text, this volume can be used to *teach* courses in the Physics of Sports. It is also intended to serve as a *reader* for students and others with technical or scientific interests. By combining *enjoyable reading* with *scientific explanations* for fascinating sports questions, this volume is designed to foster greater interest and understanding of sports and the physical principles which govern them.

How This Book is Structured

This book is organized into ten chapters with each chapter devoted either to a *single sport* (e.g., Baseball, Chap. 2; Tennis, Chap. 5; Bicycle, Chap. 6), to a *selection of sports* (e.g., The Golf Ball and Other Spherical Objects, Chap. 3; Running and Jumping, Chap. 4; Sports Assortment, Chap. 9), or to a *sports theme* (The Pioneers, Chap. 1; The Human Element, Chap. 7; Fluids and Sports, Chap. 8; Bibliography, Chap. 10).

This structure was chosen because the physics of sports literature is uneven in two very important respects: First, some sports have received much more scientific attention than others, and second, the mathematical sophistication required for different articles varies considerably with author, subject matter, and intended audience.

In an effort to deal positively with these constraints, the articles included in this volume were selected on the basis of their ability to contribute to breadth *and* depth of presentation, by combining a well-rounded sampling of physics articles on *many* different sports with more intensive study into a selected *few* other sports. The intended result is a book that includes the physics of many sports but does not attempt to be encyclopedic.

Sports Questions Answered

A key goal of this book is to provide answers to some of the most fascinating questions in the world of sport, e.g., What makes a curve ball curve? What allows a sailboat to sail almost directly into the wind? Why does a golf ball have dimples? etc. A more complete selection of the major sports questions answered in this volume, chapter by chapter, is presented below:

Chapter 1: The Pioneers

Shot Put. In throwing the shot put, a 54-ft, 11-in. toss in Boston is actually a better performance than a 55-ft toss in Mexico City. Why?

Discus. Why is it that the range of a well-thrown discus thrown in an eastward direction exceeds that of an otherwise identical westward throw by about one inch (an amount which is even larger than the ostensible precision of such measurements in athletic competition)?

Archery. The initial acceleration of an arrow launched from a competition bow approaches 100g's (i.e., 100 times the acceleration due to gravity!). How can a string be strong enough to deliver such an enormous acceleration to the arrow?

Archery. When a bow is drawn back by a right-handed archer, the arrow is usually aimed directly at the target. But when the string is released, moving forward in the median plane of the bow, the arrow begins to point more and more to the *left* as the string gets closer and closer to its equilibrium point. Despite this change in direction, the arrow nevertheless hits the mark! Why?

Chapter 2: The National Pastime

Hitting for Distance. In hitting a pitched baseball for distance, where is the optimal location ("sweet spot") for bat–ball contact?

Bat Weight. All other things being equal, heavy bats are capable of imparting more energy, velocity, and hence distance to a pitched baseball than lighter ones. Why then do so many major league batters use lighter bats?

Fly Ball. How is it possible that major league outfielders can take a brief look at a fly ball, run to exactly the right spot, and then get there just in time to catch it?

Curve Ball. Why do baseballs, tennis balls, and other "rough" balls curve in the "normal" way, as described by the Magnus effect, while smooth balls (spitballs?) curve in the opposite direction?

Curve Ball. How does the degree of lateral deflection of a curve ball depend on the velocity and spin rate imparted to the ball by the pitcher?

Knuckle Ball. What role do the seams of a baseball play in the erratic aerodynamic flight of a knuckle ball?

Aluminum Bat. What advantages do aluminum bats have over wooden bats? What disadvantages?

Backspin. Why is it that balls hit with backspin (whether in baseball, golf, or tennis) tend to travel farther than balls hit with either little spin or top spin?

Coefficient of Restitution. The coefficient of restitution of major league baseballs is required to be 0.546 ± 0.032. How much variation in the range of a well-hit ball does the allowed variation in coefficient of restitution provide?

Chapter 3: The Golf Ball and Other Spherical Objects

Loft Angle. The angle of launch of a golf ball struck by a golf club is approximately equal to the angle of loft of the club used (approximately 11° for a typical driver golf club). Why then is the optimum angle of launch for a driven golf ball ($\approx$11°) so much smaller than the optimum angle of launch for a projectile in a vacuum (45°)?

Golf Drive. Unlike the trajectory of a projectile in a vacuum which is everywhere convex downward, the trajectory of a well-driven golf ball is initially concave upward. Why is this so?

Bowling. Calculations show that a bowling ball launched with side spin initially moves in a parabolic path and then a straight line as it begins to roll without slipping. Why then do bowlers call it a "hook"?

Superball. Why does a "superball" return to the hand of the person launching it after three collisions with the floor, the underside of a table, and the floor?

Basketball. There is an optimum shooting angle in basketball which requires the smallest launching force and provides the greatest margin of error. What is the relationship between this angle and the familiar optimum angle of launch for an ideal projectile (45°)?

Foul Shots. Rick Barry, the man who holds the NBA record for highest career foul-shooting percentage (90%), used the *underhand* delivery for shooting fouls rather than the almost universal overhand push shot. What, if any, are the relative advantages of the underhand shot?

Shot Put. Although air resistance plays little role in the motion of the shot, the optimum launch angle for maximum distance in the shot put is closer to 42° than to 45°. Why?

Chapter 4: Running and Jumping

Sprinting. When sprinting on a curved track, athletes prefer the middle rather than the extreme inside or outside lanes. What major factors contribute to this preference?

Sprints and Runs. When races of different lengths are run, from the 50-yard dash to the 10,000-m run, the results show that the highest *average* velocity in world class competition is achieved in the 200-m run (10.39 m/s). What major factors contribute to this result?

Endurance Events. For world-class competition at long distances in running, walking, swimming, as well as skating, there is a strong correlation between the distance of the event and the average velocity achieved over that distance. Why should this be so?

Walking. In walking at different speeds, there is a given pace length that minimizes the power expended. How does this optimum pace length vary with walking speed?

Wind Effects. The characteristics of the air in which sporting events take place can affect certain athletic performances. For example, wind has been known to aid or hamper a sprinter, and variations in air density due to changes in temperature or altitude could have similar effects. What effects would these variations have on track events such as the pole vault and the long jump?

Loop-the-Loop. Could an athlete run a 3-m radius "loop-the-loop"?

Track Characteristics. Runners can run faster on a springy track than on a hard track. Why is this so?

Chapter 5: Tennis

Oversize Rackets. What makes oversize tennis rackets so much better than standard rackets?

Sweet Spot. Tennis rackets have been described as having more than one "sweet spot." How can this be so?

Perimeter Weighting. What are the advantages and disadvantages of perimeter weighting system tennis rackets?

Court Speed. The "speed" of a tennis court is related to the coefficient of friction between the surface of a tennis ball and the surface of the court. How do different court speeds determine the fraction of the ball's velocity lost during the first bounce of a well-hit tennis shot?

Top-Spin Lob. How much spin is required to produce a world-class attacking top-spin lob in tennis?

Bounce Characteristics. How much variation is allowed in the bounce characteristics of different tennis balls?

Chapter 6: The Bicycle

Front Fork Geometry. What aspects of the front fork geometry of a bicycle are most responsible for its dynamic stability? In other words, why are some bicycles easier to ride than others?

Stability and Speed. Why are bicycles easier to ride "at speed" than at very low velocities?

Hill Climbing. What limits the speed with which one can ride a bicycle up a hill of constant incline?

Pedal Motion. Why is the circular pedal motion of a bicycle not the most efficient shape?

Chapter 7: The Human Element

Speed and Distance. Are there fundamental limits on how *fast* or how *far* a human can run? What are they?

High Jump and Pole Vault. Are there fundamental limits on human performance in the high jump or pole vault? What are they?

Jumping. Are there limits on how far a human can jump either vertically or horizontally? What are they?

Bicycle Speed. What is the maximum speed at which a human could power a bicycle on level ground?

Running Efficiency. Calculations show that more massive athletes are more energy efficient when running than smaller athletes. Why then do successful distance runners (for whom efficiency would seem to be critical) tend to be rather slender and not very massive?

Karate. What are the origins of the huge forces generated in karate blows and kicks?

Karate. Is the force of a potential karate blow limited in any way by the size and weight (or lack thereof) of the person delivering the blow?

Downhill Skier and Sky Diver. The world speed record for downhill skiing has been reported to be greater than the speed of a sky diver. How can this be possible?

Pre-Jumping. Why do downhill ski racers pre-jump, i.e., jump *before* reaching a steeper section of a slope?

Downhill Skiing. Is it possible to ski 234,000 vertical ft (44 miles!) in a 24-h period?

Chapter 8: Fluids and Sports

Discus. A well-thrown discus will travel farther when thrown *into* the wind than when thrown along the direction of the wind. How is this possible?

Sailboat. A sailboat can sail almost directly into the wind. How is this possible?

Iceboat. Iceboats are capable of speeds greater than the wind speeds that propel them. How is this possible?

Swimming. What are the major factors that limit the speed of a swimmer?

Chapter 9: Sports Assortment

Football Kick. In punting a football, there are *two* different launch angles corresponding to a given launch velocity and desired horizontal range. Why do coaches demand and punters usually (but not always) deliver the steeper of the two launch angles?

Football Pass. Why does a well-thrown football keep its axis pointed along its trajectory?

World Series. Why is a seven-game World Series more than twice as likely to occur as a four-game Series?

Tennis. If a tennis player has a *slightly* greater than 50–50 chance of winning a point, how much greater than 50–50 will be his chance of winning a match?

Drag Racing. In world-class drag racing, the product of [ET] (elapsed time in seconds) and [MPH] (speed in miles per hour at the end of a quarter-mile track) tends to be a constant. Why should this be so?

Drag Racing. World-class drag racers achieve 80% of their ultimate (quarter-mile) speed at the one-eighth mile mark, and one-third of their ultimate speed after only the first 60 ft of travel! What kind of acceleration profile is needed to produce these speeds? In particular, how many g's of initial acceleration are involved?

Sailboat Weight Distribution. How do Olympic measurers infer the weight distribution of racing sailboats without drilling or cutting into the hulls?

Racing Shells. Eight-oared racing shells rowed by heavyweight athletes are typically 5% faster than similar shells rowed by lightweight (155 lb or less) athletes. Why should this be so?

Canoes. Why is it harder to paddle a canoe in shallow water?

Badminton. What is the terminal velocity of a badminton shuttlecock?

Fly Casting. What are the major factors involved in successful fly casting?

Springboard Diving. Do springboard divers violate angular momentum conservation?

INTRODUCTION

As I was growing up in the 1950s in Bridgeport, Pennsylvania, a small blue-collar town near Philadelphia, sports occupied a large part of my life and the lives of my friends. Our span of sports interests was great, ranging from avid spectatorship of big-time professional and college sports to active participation in leagues and "pickup" games.

The seasons were marked by the sports that were played: baseball in the spring and summer, football in the fall, and basketball in the winter. So powerful and unquestioned was the seasonal connection that it was virtually unheard of, for example, either to toss a football around during the summer, i.e., "baseball season," or to have a baseball catch after the World Series had ended and the "football season" had begun in earnest. Of course, some identification of sports with seasons continues to this day despite the overlap created by preseason and postseason expansion of existing sports and the introduction of major new sports to the spectating menu. In the America of the 50s, soccer existed primarily in high school gym classes, and professional ice hockey had not yet caught on in a big way.

In those days, Philadelphia had two professional baseball teams, the Athletics of the American League and the Phillies of the National League, and while each of us naturally had a favorite, we nevertheless followed both teams all season long, from their annual preseason exhibition game to the longed-for World Series clash that never materialized. Although Little League had not yet been formally instituted, baseball leagues were plentiful and opportunities as either a participant or a spectator were great.

Football, marching bands, the return to school, and the crisp autumn air all went together. We watched the Philadelphia Eagles on television or listened to their games on the radio and, if we were fortunate enough to have a relative with a season ticket, we would occasionally attend a game in person at Franklin Field. We attended all our high-school football games and looked forward to the traditional season-ending Thanksgiving Day rivalry with the team from the adjoining town.

A Man Named Wilt Chamberlain

In winter, we played basketball on school teams or in town leagues, followed the Philadelphia Big-Five college basketball season, and rooted for the Philadelphia Warriors of the National Basketball Association. Our interest in the Warriors was heightened by the presence of a former local high-school standout by the name of Wilt Chamberlain.

Chamberlain had been a star at Overbrook High School in Philadelphia and went on to become the premier "big man" of his era in professional basketball. Chamberlain's career spanned 14 professional seasons (1959–60 to 1972–73) and his exploits are truly legendary. Consider first his career statistics: 31,419 career points and 23,924 rebounds in 1045 games for a lifetime average of 30.1 points/game and 22.9 rebounds/game. In addition to that, in a sport for which the regulation time for a complete game is 48 min, Chamberlain averaged 45.8 min/game over his career.

And then there was March 2, 1962.

On that evening, Wilt Chamberlain scored 100 points in a single game of basketball. He played the full 48 min, shot 36 for 63 from the floor and 28 for 32 from the foul line. As amazing as those achievements were, they are probably less impressive than the fact that he *averaged* 50.4 points/game for the entire 1961–62 season! Other notable achievements for that season included averages of 25.6 rebounds/game and 48.5 min/game. (His team played 10 overtime periods that season.)

There was one other aspect of Chamberlain's game that was equally incredible—his phenomenally poor foul-shooting ability. Despite his heroic performance from the foul line during his 100-point game, Chamberlain's career foul-shooting percentage (51.1%) was actually lower than his career field goal percentage (54.0%). In view of his undisputed excellence in every other aspect of the game, Chamberlain's career-long foul-shooting futility was a source of fascination and bemusement to fans and sportswriters alike. It was also a perplexing and constant source of great personal embarrassment to this proud and sensitive man.

At one point in his career, the 7-ft, 2-in. Chamberlain resorted to an underhand delivery from the foul line, a style that was familiar from girls' basketball but virtually unheard of in men's basketball.[1] As I thought at the time, here is the tallest man in basketball trying to become a better foul shooter by, in effect, making himself shorter!

My scientific curiosity with sports began then as I found myself wondering whether Chamberlain was such a bad foul shooter *because* he was so tall.[2] I wondered whether there was something in the kinematics of projectile motion, in particular, the height of the release point above the floor, that preordained not only his failure but, to a somewhat

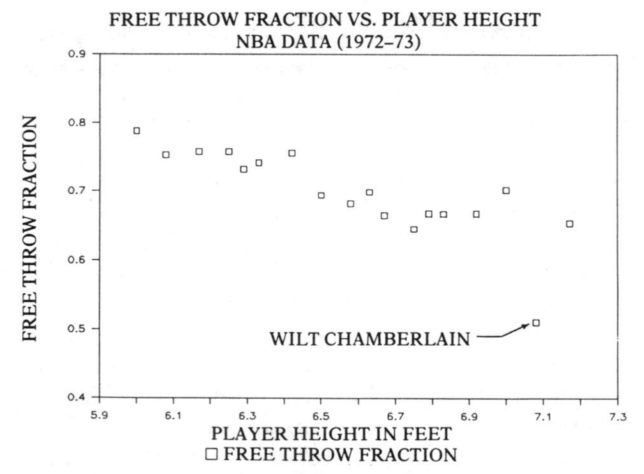

FIGURE 1. With the exception of the strikingly anomalous data point corresponding to Wilt Chamberlain, all other data points are averages representing more than one player. The Official 1972–73 National Basketball Association (NBA) Guide lists career statistics for 216 players. The career foul-shooting percentage of each player was recorded against his height and an average percentage was calculated for each category (25 in all, ranging from 5 ft, 9 in. to 7 ft, 2 in.). Due to some nonuniformity in reporting (most heights are recorded to the nearest inch, a few to the nearest half-inch, and one to the nearest quarter-inch), a number of height values were unique to a single player and hence the corresponding data points were not included above. The raw data for individual players is presented in Appendix 1.

lesser extent, the failure of many of the other big men in the sport at that time (see Fig. 1).

Although a precise explanation for the apparent correlation between player height and foul-shooting futility has yet to be formulated, many important insights into the physics of basketball have been documented. (See, for example, the articles by Brancazio, and Tan and Miller in Chap. 3.)

Seven-Game World Series Occur Twice as Often as Six-, Five-, or Four-Game World Series

There was another puzzling situation involving sports that cried out for a "scientific" explanation, namely, the fact that so many World Series went the full seven games.[3] A look at the data reveals that between 1903 and 1990, seven-game series occurred twice as often as six-game series, and almost twice as often as either five-game or four-game series. (See Table 1 for a summary and Appendix 2 for more detail.)

A friend suggested, somewhat cynically perhaps, that the players probably "extended" the World Series artificially in an effort to maximize gate receipts, television revenues, and player paychecks.

TABLE 1

World Series Data (1903–1990)

Summary	Number	Percent of Total
7-game series:	35*	40%
6-game series:	17	20%
5-game series:	20	23%
4-game series:	15	17%
TOTAL NUMBER OF SERIES	87	100%

*Between 1903 and 1990, there have been four eight-game and 31 seven-game series, for an equivalent total of 35. Also, the game totals shown above have been adjusted by the removal of *tie* games, three of which were recorded between 1907 and 1922.

It is easy to discredit that hypothesis, however. Some checking reveals that players receive the same pay whether the series goes four, five, six, or seven games. And since the players on the winning team do receive more money, and since unnecessary games only increase the likelihood of injury, there is great incentive to sweep the other team, and little incentive to risk ultimate victory by failing to put the other team away as soon as possible.

Actually, the real reason for the great number of seven-game World Series involves elementary probability and statistics. If the two world series contenders are evenly matched, and one expects more often than not that they will be, the probability of either team winning a single game is 0.50 (i.e., 50–50). Then since there are only two possible outcomes for each game (*either* team A wins *or* team B wins), the probability of an n-game series will be proportional to the number of ways a World Series can be concluded in n games, and may be readily calculated using the binomial probability distribution. See the articles by Krane and Fischer (Chap. 9) for elegant discussion and calculation of these and other sports-related probabilities.

Physics and Sport: A New Course for Nonscience Majors

With questions about the poor foul shooting of Wilt Chamberlain and the abundance of seven-game world series some 20 years ago, my interest in the physics of sports began. My first formal experience with the physics of sports took place in 1971 when I designed and taught a course entitled "Physics and Sport" at Temple University. Several years later, after having taught a similar course at Villanova, I described my experiences in an article entitled, "Physics and sport: A new course for nonscience majors" [Phys. Teach. **12**, 349 (1974)].

The success of that course was based on something every physics teacher knows: (a) A great way to increase a student's interest in physics is to relate the material to phenomena of direct personal interest to the student; and (b) a good time to provide an explanation to a student is when that student happens to be very curious about the subject.

Twenty years ago, public and student interest in sports was strong, and hence the use of sports as a vehicle to motivate student interest in physics was great. If anything has changed since then it is that interest in sports has grown even more, as has the ability to motivate student interest in physics through sports.

The 57 articles contained in this reprint volume represent a fairly broad introduction to the literature on the physics of sports and should be of special interest to most physics teachers. However, these articles represent only a tiny fraction of the work which has been done to date. Many excellent articles had to be omitted from this volume due to their length and/or complexity. For that reason, an extensive bibliography with over 300 entries has been included in Chap. 10.

Because of the tremendous increase in research on the physics of sports, a number of sports topics are now being treated in second- and third-generation articles, with greater depth and sophistication the typical result. For example, Chapman's paper (1968) on the geometry of a fly ball trajectory neglected crucial air resistance effects.

Brancazio's paper (1985) both complemented and superseded Chapman's paper. However, as a pedagogical device, many teachers will want to introduce the problem the way Chapman presented it and then proceed to include the substantial corrections imposed by the realities of air resistance.

Multiple-generation articles can now also be found in the literature on a wide variety of sports, from golf and tennis to archery, running, and the shot put. This development, as well as the quality and quantity of recent books and articles, suggest that interest in the physics of sports will continue to grow well into the future.

While the articles reprinted in this volume were selected so as to appeal to a wide variety of students, scientists, the scientifically curious, sports buffs, coaches, etc., the particular selection and mix of articles, together with the material found in the chapter headings, are intended to be of maximal use to the teacher or student who wishes to teach or take a course in the physics of sports.

Although my interest in the physics of sports began more than 20 years ago, it continues undiminished to this day. The hope of this book is to stimulate others to take an interest in the physics of sports, and the goal of this book is to demonstrate just how rewarding that interest can be.

The Ancient Origins of Sports and of Physics

The first organized sports activity for which a written record survives dates to the First Olympic Games in 776 B.C. A youth from Elis (in southwestern Greece) by the name of Coroebus won a footrace at Olympia and became the first Olympic victor whose exploits made written history.[4] Sports almost certainly began many centuries before that event, although the precise date is lost in early human history.

Physics on the other hand traces its origins to the first time measurements by the ancient Babylonians in about 3000 B.C. Also, the use of iron for tools and machines occurred between about 2000 and 1000 B.C. The first description of magnetism together with major advances in geometry and astronomy date to about 600 B.C., while the work of Pythagoras and Zeno appeared around 500 B.C. A primitive atomic theory was proposed by Democritus in about 400 B.C., while at roughly the same time medicine was being developed by Hippocrates and the theory of "forms" (thought to be a precursor of modern theoretical physics) was being developed by Plato. During the reign of Alexander the Great, Aristotle wrote what many today consider to be the very first "textbook" of physics.[5] His writings also included examples of what one would now call the physics of sports!

Aristotle (384–322 B.C.) and the Physics of Sports

That Aristotle was attempting to grasp the significance of Newton's third law, some 2000 years before Newton formulated it, is clear from the following quotation taken from his treatise on the *Progression of Animals* [6]:

> ...the animal that moves makes its change of position by pressing against that which is beneath it; and so, if the latter slips away too quickly to allow that which is setting itself in motion upon it to press against it, or if it offers no resistance at all to that which is moving, the animal cannot move itself at all upon it.

He goes on to say,

> For that which jumps performs that movement by pressing both on its own upper part and on that which is beneath its feet; for the parts in a way lean upon one another at their joints, and, in

general, that which presses leans on that which is pressed.

He then concludes with the following observation:

> Hence athletes jump farther if they have the weights in their hands than if they have not,[a] and runners run faster if they swing their arms;[b] for in the extension of the arms there is a kind of leaning upon the hands and wrists.

Vitruvius (1st century B.C.) Decries Excessive Recognition and Financial Reward for Athletes!

Two hundred years after Aristotle in the first century B.C., Marcus Vitruvius Pollio, a Roman architect and military engineer under Augustus, wrote his famous treatise (in ten books) on Greek and Roman architecture.[7] In his Introduction to Book Nine, *Vitruvius complains that athletes receive more recognition than scholars*:

> The ancestors of the Greeks held the celebrated wrestlers who were victors in the Olympic, Pythian, Isthmian, and Nemean games in such esteem, that, decorated with the palm and crown, they were not only publicly thanked, but were also, in their triumphant return to their respective homes, borne to their cities and countries in four horse chariots, and were allowed pensions for life from the public revenue.
>
> When I consider these circumstances, I cannot help thinking it strange that similar honours, or even greater, are not decreed to those authors who are of lasting service to mankind. Such certainly ought to be the case; for the wrestler, by training, merely hardens his own body for the conflict; a writer, however, not only cultivates his own mind, but affords every one else the same opportunity, by laying down precepts for acquiring knowledge, and exciting the talents of his reader.
>
> What does it signify to mankind, that Milo of Crotona, and others of his class, should have been invincible, except that whilst living they were ennobled by their fellow countrymen? On the other hand the doctrines of Pythagoras, Democritus, Plato, Aristotle, and other sages, the result of their daily application, and undeviating industry, still continue to yield, not only to their own country, but to all nations, fresh and luscious fruit, and they, who from an early age are satiated therewith, acquire the knowledge of true science, civilize mankind, and introduce laws and justice, without which no state can long exist. Since, therefore, individuals as well as the public are so indebted to these writers for the benefits they enjoy, I think them not only entitled to the honour of palms and crowns, but even to be numbered among the gods.

Two-thousand years after Vitruvius wrote these words, debate continues on the proper role of athletics both in colleges and universities and in society as a whole. No doubt, some faculty members today would argue that the 2000-year-old concern of Vitruvius is as valid today (if not more so) as it was then. In any event, whether the specific debate concerns Proposition 48, or the role of colleges and universities as "farm systems" for professional sports, or the multi-million dollar professional contracts offered to "mediocre" players, we are seeing the continuation of a debate that almost certainly did not begin with Vitruvius' generation, and will certainly not end with ours.

Many books have been written on the central role of sports in human societies[8] and, while conclusions differ as to why sports play such a major role, there is no disagreement on the fact that they do play such a role, have played such a role in the past, and will almost certainly continue to play an important and pervasive role in the future.

Some have suggested that the human fascination with sports is both deeply ingrained and genetically compelled by the harsh environment which our early ancestors faced and, in so many ways, successfully overcame. Viewed in this way, sports, with their focus on physical prowess and keen competition, are vestigial and symbolic reenactments of the life and death struggles, challenges, and achievements which form the very fabric of early human evolutionary history.

Physics of Sports in Modern Times

In the early 1850s, a german professor by the name of Heinrich Magnus conducted studies on behalf of the Prussian Artillery Commission in which the sideways deflection of rotating artillery shells was measured and explained.[9] This deflection of spinning projectiles, now known as the "Magnus effect," was first cited in a physics of sports context in 1877 by Lord Rayleigh in a paper which described the swerve of a spinning tennis ball.[10]

Much later, in 1959, Lyman Briggs[11] wrote an historic paper on the lateral deflection (curve) of a spinning baseball in which he cited the Magnus ef-

---

[a]Special weights or sometimes stones were held in the hands and thrown backwards by jumpers while in the air to add to their impetus; cf. Norman Gardiner, *Greek Athletic Sports and Festivals*, pp. 298ff, who proves by experiment the truth of the statement made in the present passage.

[b]On the importance attached by the Greeks to arm-action in running, especially in short races, cf. Norman Gardiner, *op. cit.* p. 282.

[Footnotes a and b are taken from E. S. Forster's English translation of Aristotle's original work.]

fect as the cause of the substantial deflections that were experimentally observed. Although his explanation for the curve ball is routinely accepted today, many people, prior to Briggs' paper, *believed that the curve ball did not deflect at all but was instead an optical illusion!* Additional discussion on this point is found in Chap. 3.

Magnus' work on the deflection of spinning projectiles also influenced P.G. Tait, who between 1893 and 1896 produced the first accurate calculations of the trajectories of spinning golf balls.[12] Tait's work, in turn, set the stage for some major contributions to the physics of golf by J. J. Thomson[13] in 1910 and J. W. Maccoll[14] in 1928.

In 1937, Paul Kirkpatrick[15] linked physics not just to a single sport but to a wide variety of sports, thereby effectively establishing for all time the avocation that would come to be known as the physics of sports. An updated (1944) version of his 1937 paper is reprinted in Chap. 1 and is both a keystone for this volume as well as must reading for anyone with a serious interest in the physics of sports. Kirkpatrick's interest endured a lifetime and his publications on the subject spanned more than a quarter of a century (1937–63). Paul Kirkpatrick can legitimately be called the father of the physics of sports, and this volume is gratefully dedicated to him.

In the 70s and 80s, a veritable explosion of work on the physics of sports took place. The most prolific authors on the subject during this period included P. Brancazio, H. Brody, C. Frohlich, T. McMahon, A. Tan, J. Walker, and R. Watts. A number of their papers are included in this volume and a more extensive listing of their contributions and those of many other authors is found in Chap. 10.

---

[1]There was one notable exception to the all-female underhand delivery rule, namely, Rick Barry, the man who to this day owns the record for the highest (0.900) career foul-shooting percentage in professional basketball. [See *The 1991 Complete Handbook of Pro Basketball* (17th ed.), edited by Zander Hollander (Signet, New York, 1990).]

In addition to the former widespread use of the underhand delivery, girls' basketball in the 50s (there was no "women's" basketball in those days) was also different from boys' basketball for another distinctive reason: it was played with six on a side, three on offense and three on defense, somewhat like ice hockey is played today.

[2]Chamberlain's height has been variously reported over the years to be somewhere between 7 ft, 1 in. and 7 ft, 2 in. Many accounts cite the 7-ft, 2-in. figure, while the Official NBA Guide [*Sporting News Publishing*, St. Louis (1972)] in what was to be his last professional season (1972–73), lists him at 7 ft, 1 in.

[3]The format of World Series play has changed since the early years. In 1903 (the first official World Series) as well as in 1919, 1920, and 1921, five (rather than four) wins were required for victory. Also, tie games were once allowed and, in fact, were recorded in the years 1907, 1912, and 1922. The combination of these two anomalous factors resulted in four 8-game World Series! These occurred in 1903, 1912, 1919, and 1921.

[4]See, for example, John Kieran and Arthur Daley *The Story of the Olympic Games: 776 B.C. to 1972*, (Lippincott, Philadelphia and New York, 1973).

[5]See R. B. Lindsay, *Basic Concepts of Physics* (Van Nostrand Reinhold, New York, 1971), Appendix 1, for a description of the major developments and discoveries in physics and other sciences, together with contemporary political and social events.

[6]See Aristotle, *Progression of Animals*, translated by E. S. Forster (Harvard University, Cambridge, 1945), p. 489.

[7]See *The Architecture of Marcus Vitruvius Pollio, in Ten Books*, translated by J. Gwilt (Priestley and Weale, London, 1826), p. 259.

[8]See, for example, Richard Lipsky, *How We Play the Game: Why Sports Dominate American Life* (Beacon, Boston, 1981); Richard E. Lapchick, *Fractured Focus: Sports as a Reflection of Society* (Heath, Lexington, MA, 1986).

[9]See H. G. Magnus, "On the deviation of projectiles; and on a remarkable phenomenon of rotating bodies," *Memoirs of the Royal Academy*, Berlin (1852). English translation in *Taylor's Foreign Scientific Memoirs* (1853), p. 210.

[10]See Lord Rayleigh, "On the irregular flight of a tennis-ball," *Messenger of Mathematics* **VII**, 14 (1877). Much recent work has been done on the physics of tennis, primarily by Howard Brody. See Chaps. 5 and 10 for more details and additional references.

[11]Lyman J. Briggs, "Effect of spin and speed on the lateral deflection (curve) of a baseball; and the Magnus effect for smooth spheres," *Am. J. Phys.* **27**, 589 (1959).

[12]See, for example, P. G. Tait, "Long Driving," *Badminton Mag.* (March, 1896). In this paper, Tait acknowledges the invaluable assistance provided by his "computer," which, as it turns out, was not a machine but rather an unnamed *person* (graduate student?) who performed many of the tedious calculations on which his results are based.

[13]J. J. Thomson, "The dynamics of a golf ball," *Nature* **85**, 2147 (1910).

[14]John W. Maccoll, "Aerodynamics of a spinning sphere," *J. R. Aeronaut. Soc.* **32**, 777 (1928).

[15]Paul Kirkpatrick, "Bad physics in athletic measurements," *Am. J. Phys.* **12**, 7 (1944).

This photo of Michael ("Air") Jordan is from a recent Nike "Astronautics" commercial, a popularization of the physics of sports which included the following dialogue between Mars Blackman and a real-life Professor:

MARS: Yo Professor, how does Mike defy gravity, do you know, do you know, do you know, do you know?

PROF: Michael Jordan overcomes the acceleration of gravity by the application of his muscle power in the vertical plane thus producing a low altitude earth orbit.

MARS: A what?

PROF: Do you know what I mean? Do you know, do you know, do you know?

MARS: Money, check him out!

As strange as it might seem at first glance, the Professor's description of an "Air" Jordan trajectory as a low altitude earth orbit is, in fact, quite correct. The explanation familiar from high school physics involving parabolic motion is a mere, if convenient, approximation to the *elliptical* shape which every near-earth trajectory, whether that of a leaping athlete or an orbiting space station, follows.

Objects with horizontal speeds larger than about five miles per second (18,500 mph) achieve elliptical earth orbit, i.e., defy earth's gravity sufficiently to attain a radius of curvature at least equal to that of the earth. Such objects achieve orbit by, in effect, taking advantage of that curvature to "fall around" rather than into the earth. All objects with horizontal speeds smaller than that value, including at least for the moment Michael Jordan, attain elliptical trajectories of smaller curvature and, inevitably, fall to—rather than around—the earth.

Because Michael Jordan's "flying" is largely two-dimensional, i.e., involves significant horizontal as well as vertical displacements and, because the energy expended in the one direction is not available for the other, the vertical height and horizontal range which he achieves during his spectacular dunks are each smaller than the respective record performances achieved in the more one-dimensional efforts of the high jumper (about 2 m, vertically) and the long jumper (about 9 m, horizontally.) However, the combination of his vertical and horizontal displacements is clearly prodigious, awesome, and beautiful to watch. (Courtesy of Wieden & Kennedy, Spike Lee Productions, and Nike, Inc.)

Chapter 1
THE PIONEERS

CONTENTS

We are told that a pioneer is one who ventures into unknown or unclaimed territory. On the basis of that definition, the authors of the two articles reprinted in this chapter, Paul Kirkpatrick and Paul Klopsteg, clearly deserve to be considered the modern pioneers in the physics of sports.[1]

Kirkpatrick's paper, "Bad physics in athletic measurements," is both a celebration of and an invitation to the physics of sports. Although it is practical in its outlook, noting that "unscientific" measurement procedures in organized athletics could result in a gold medal being given to the wrong athlete, it also employs fundamental physics such as the Coriolis effect and Newtonian gravitation to predict significant variations in the range of athletic projectiles (shot, discus, javelin, etc.) due to East/West differences in launch direction or small variations in the local value of the acceleration due to gravity.

While Kirkpatrick's paper takes the form of a survey article by including examples from a variety of individual sports, Klopsteg provides a monograph that focuses in depth on the physics of a single sport, archery, in what he describes as the "...joining of his vocation with his avocation."

Klopsteg's article is also a *tour de force* that employs fundamental physics including hysteresis, strength of materials, and vibrations of solids, to relate a number of anecdotal characteristics of bows and arrows (e.g., "cast," "spine," etc.) to experimentally inferred characteristics (e.g., efficiency, figure of merit, etc.) The article also pioneers the use of new physical concepts such as the virtual mass of a bow, to explain some of the phenomenology and accompanying folklore that have evolved during literally thousands of years of human development of the bow and arrow.[2] The article also contributed to further development of the bow and arrow through its analytical discussion on various aspects of scientific bow design.[3]

All in all, these first two articles set a high standard for the balance of the collection included in this volume. They are compelling and thorough, and provide numerous physical insights to and clear explanations for the various sports considered. They also exhibit a similar spirit, a sense of satisfaction, if not outright pleasure, at being able to merge two things the authors clearly care about very deeply: physics and sports.

---

[1]Although, technically speaking, these authors were not the first to link physics with sports, they were the first to do so in a way that made the results of their efforts accessible to a wide audience.

[2]Klopsteg reports, for example, that "The bows of Genghis Khan [1162?–1227] and his hordes were of the *composite* type" (emphasis added). That such a sophisticated design would have emerged before the end of the Middle Ages suggests that development of the bow and arrow must have begun very early. According to one source, "The invention of the bow is lost in the fog of prehistory, but the records and monuments of the earliest peoples and the stone rubbings of men of the Mesolithic Age tell us that for a period of at least eight thousand years the bow has been man's greatest basic offensive weapon." See Vic Hurley, *Arrows Against Steel: The History of the Bow* (Mason/Charter, New York, 1975).

[3]For more recent discussions of bow design and related topics see, for example, Burton G. Shuster, "Ballistics of the modern-working recurve bow and arrow," *Am. J. Phys.* **37**, 364 (1969); and W. C. Marlow, "Bow and arrow dynamics," *Am. J. Phys.* **49**, 320 (1981).

Bad Physics in Athletic Measurements

PAUL KIRKPATRICK
Stanford University, Stanford University, California

THE physics teacher has been accustomed to find in athletic activities excellent problems involving velocities, accelerations, projectiles and impacts. He has at the same time overlooked a rich source of illustrations of fictitious precision and bad metrology. When the student is told that the height of a tree should not be expressed as 144.632 ft if the length of its shadow has been measured only to the nearest foot, the student may see the point at once and yet ask, "What difference does it make?" But when shown that common procedures in measuring the achievements of a discus thrower could easily award a world's record to the wrong man, the student agrees that good technic in measurement is something more than an academic ideal. The present discussion[1] has been prepared partly to give the physics teacher something to talk about, but also to start a chain of publicity which may ultimately make athletic administrators better physicists and so make their awards more just.

If physicists were given charge of the measurements of sport, one may feel sure that they would frown upon the practice of announcing the speed of a racing automobile in six or seven digits—see, for example, the *World Almanac* for any year—when neither the length of the course nor the elapsed time is known one-tenth so precisely. They could and would point out such inconsistencies as that observed in some of the events of the 1932 Olympic games when races were electrically and photographically timed to 0.01 sec, but with the starting gun fired from such a position that its report could not reach the ears of the waiting runners until perhaps 0.03 to 0.04 sec after the official start of the race. In this case, electric timing was used only as an unofficial or semi-official supplement to 0.1-sec hand timing; but it is easy to see that a systematic error of a few hundredths of a second will frequently cause stopwatch timers to catch the wrong tenth.

Scientific counsel on the field would immediately advise judges of the high jump and pole vault to measure heights from the point of take-off instead of from an irrelevant point directly below the bar which should be at the same level but sometimes isn't. Physicists would suggest equipping field judges with surveying instruments for determining after each throw, not only how far the weight traveled but also the relative

[1] Some of the material in this article appeared in a paper by the author in *Scientific American*, April 1937, and is incorporated here by permission of the editors.

Reprinted from *American Journal of Physics* **12**, 7–12 (1944); © American Association of Physics Teachers.

elevation of the landing point and the throwing circle. Certainly it is meaningless if not deceptive to record weight throws to a small fraction of an inch when surface irregularities may be falsifying by inches the true merit of the performance.

In shot-putting, for example, a measured length will be in error by practically the same amount as the discrepancy between initial and final elevations, since the flight of the shot at its terminus is inclined at about 45° to the horizontal. For the discus the effect is some three times as serious because of the flatter trajectory employed with this missile, while broad jumpers under usual conditions must be prepared to give or take as much as 0.5 ft, according to the luck of the pit. Meanwhile, the achievements in these events go down in the books with the last eighth or even the last sixteenth of an inch recorded.

At the 1932 Olympic Games an effective device was used to grade the broad-jumping pit to the level of the take-off board before each leap, but the practice has not become general. Athletic regulations, indeed, recognize the desirability of proper leveling in nearly all the field events, but in actual usage not enough is done about it. Since sprinters are not credited with records achieved when blown along before the wind, there is no obvious reason why weight hurlers should be permitted to throw things down hill.

The rule books make no specification as to the hardness of the surface upon which weights shall be thrown, but this property has a significant effect upon the measured ranges of the shot and hammer, since it is prescribed that measurement shall be made to the near side of the impression produced by the landing weight. In a soft surface this impression may be enlarged in the backward direction enough to diminish the throw by several times the ostensible precision of the measurement.

A physicist would never check the identity of three or four iron balls as to mass by the aid of grocers' scales or the equivalent and then pretend that there was any significance in the fact that one of them was thrown a quarter of an inch farther than the others. In measuring the length of a javelin throw, no physicist who wanted to be right to $\frac{1}{8}$ in. would be content to establish his perpendicular from the point of fall to the scratchline by a process of guesswork, but this is the way it is always done by field judges, even in the best competition.

Among the numerous errors afflicting measurements in the field sports, there is none which is more systematically committed, or which could be more easily rectified, than that pertaining to the variation of the force of gravity. The range of a projectile dispatched at any particular angle of elevation and with a given initial speed is a simple function of g. Only in case the end of the trajectory is at the same level as its beginning does this function become an inverse proportionality; but in any case the relationship is readily expressed, and no physicist will doubt that a given heave of the shot will yield a longer put in equatorial latitudes than it would in zones where the gravitational force is stronger. Before saying that the 55-ft put achieved by A in Mexico City is a better performance than one of 54 ft, 11 in. which B accomplished in Boston, we should surely inquire about the values of g which the respective athletes were up against, but it is never done. As a matter of record, the value of g in Boston exceeds that in Mexico City by $\frac{1}{4}$ percent, so the shorter put was really the better. To ignore the handicap of a larger value of g is like measuring the throw with a stretched tape. The latter practice would never be countenanced under AAU or Olympic regulations, but the former is standard procedure.

Rendering justice to an athlete who has had to compete against a high value of g involves questions that are not simple. It will be agreed that he is entitled to some compensation and that in comparing two throws made under conditions similar except as to g, the proper procedure would be to compare not the actual ranges achieved, but the ranges which would have been achieved had some "standard" value of g—say 980 cm/sec^2—prevailed in both cases. The calculation of exactly what would have happened is probably impossible to physics. Although it is a simple matter to discuss the behavior of the implement after it leaves the thrower's hand and to state how this behavior depends upon g, the dependence of the initial velocity of projection upon g depends upon the thrower's form and upon characteristics of body mechanics to which but little attention has so far been devoted.

The work done by the thrower bestows upon the projectile both potential and kinetic energy. In a strong gravitational field, the imparted potential energy is large and one must therefore suppose the kinetic energy to be reduced, since the thrower's propelling energy must be distributed to both. We have no proof, however, that the *total* useful work is constant despite variation of g, nor do we know the manner of its inconstancy, if any. The muscular catapult is not a spring, subject to Hooke's law, but a far more complicated system with many unknown characteristics. The maximum external work which one may do in a single energetic shove by arms, legs or both obviously depends partly upon the resisting force encountered. Only a little outside work can be done in putting a ping-pong ball because the maximum possible acceleration, limited by the masses and other characteristics of the bodily mechanism itself, is too slight to call out substantial inertial forces in so small a mass. The resisting force encountered when a massive body is pushed in a direction that has an upward component, as in shot-putting, does of course depend upon g; and until we know from experiment how external work in such an effort varies with resisting force, we shall not be able to treat the interior ballistics of the shot-putter with anything approaching rigor.

Several alternative assumptions may be considered. If we suppose that the *velocity* of delivery, or "muzzle velocity," v, of the missile is unaffected by variations of g, we have only the external effect to deal with. Adopting the approximate range formula $R=v^2/g$ (which neglects the fact that the two ends of the trajectory are at different levels and which assumes the optimum angle of elevation) we find that the increment of range dR resulting from an increment dg is simply $-R dg/g$. On the more plausible assumption that the *total work done on the projectile* is independent of g, this total to include both the potential and kinetic energies imparted, one obtains as a correction formula,

$$dR=-\left(1+\frac{2h}{R}\right)R\frac{dg}{g}, \qquad (1)$$

where h is the vertical lift which the projectile gets while in the hand of the thrower. A third assumption, perhaps the most credible of all, would hold constant and independent of g the *total work done upon the projectile and upon a portion of the mass of the thrower's person.* It is not necessary to decide how much of the thrower's mass goes into this latter term; it drops out and we have again Eq. (1), provided only that the work done on the thrower's body can be taken into account by an addition to the mass of the projectile.

These considerations show that a variation of g affects the range in the same sense before and after delivery, an increase in g reducing the delivery velocity and also pulling the projectile down more forcibly after its flight begins. They indicate also that the latter effect is the more important since, in Eq. (1), $1>2h/R$ by a factor of perhaps five in the shot-put and more in the other weight-throwing events.

One concludes that the *least* which should be done to make amends to a competitor striving against a large value of g is to give him credit for the range which his projectile would have attained, for the same initial velocity, at a location where g is "standard." This is not quite justice, but it is a major step in the right direction. The competitor who has been favored by a small value of g should of course have his achievement treated in the same way.

The corrections so calculated will not be negligible magnitudes, as Fig. 1 shows. They are extremely small percentages of the real ranges, but definitely exceed the ostensible probable errors of measurement. It is not customary to state probable errors explicitly in connection with athletic measurements, but when a throw is recorded as 57 ft, $1\frac{5}{32}$ in., one naturally concludes that the last thirty-second inch, if not completely reliable, must have been regarded as having *some* significance.

ROTATION OF THE EARTH

It is customary to take account of the effects of terrestrial rotation when aiming long-range guns, but athletes and administrators of sport have given little or no attention to such effects in relation to their projectiles. As a matter of fact they should, for at low latitudes the range of a discus or shot thrown in an eastward direction

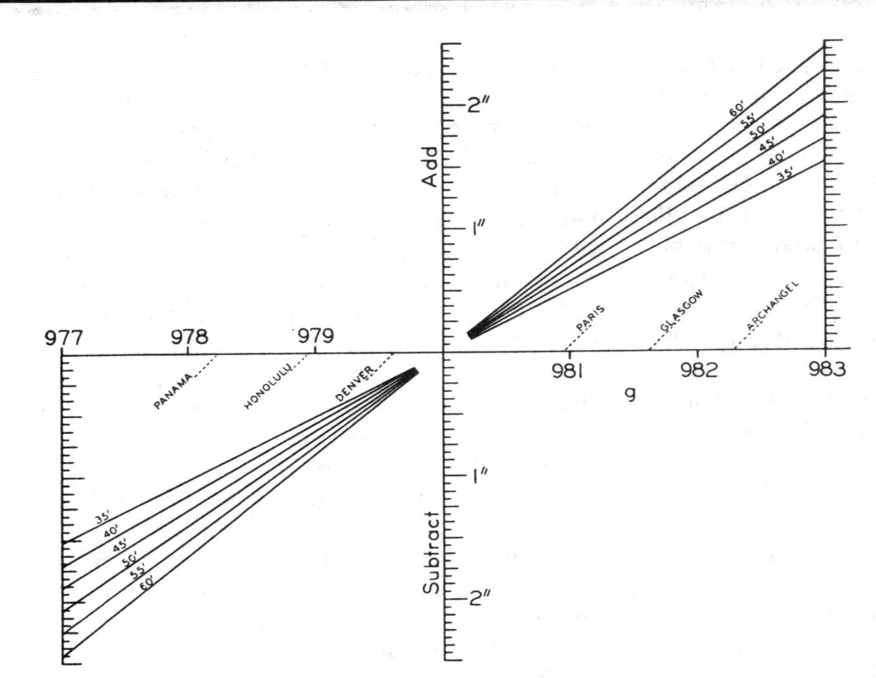

FIG. 1. Graphs for normalizing shot-put ranges to the common value $g=980$ cm/sec². Ranges achieved where $g=980$ cm/sec² are not in need of adjustment, but a range of 50 ft (see inclined line marked 50′) achieved at Glasgow, where $g=981.6$ cm/sec², is entitled to a premium of $1\frac{1}{8}$ in. which should be added before comparing the put with one achieved elsewhere. Distances accomplished where $g<980$ cm/sec² should be subjected to the deductions indicated by graphs in the third quadrant.

exceeds that of a westward throw by more than the ostensible precision of such measurements. The difference between the range of a projectile thrown from the surface of the real earth and the range of one thrown from a nonrotating earth possessing the same local value of g is given by[2]

$$\text{Range}=\frac{V_0^2\sin 2\alpha}{g}+\frac{4\omega V_0^3}{3g^2}\times\sin\alpha[4\cos^2\alpha-1]\cos\lambda\sin\mu, \quad (2)$$

where g is the ordinary acceleration due to weight, V_0 is the initial speed of the projectile, α is the angle of elevation of initial motion (measured upward from the horizontal in the direction of projection), ω(rad/sec) is the angular speed of rotation of the earth, λ is the geographic latitude of the point of departure of the projectile, and μ is the azimuth of the plane of the trajectory, measured clockwise from the north point.

A derivation of this equation (though not the first) is given in reference 2, along with a discussion of its application to real cases. The approximations accepted in the derivation are such as might possibly be criticized where long-range guns are considered, but they introduce no measurable errors into the treatment of athletic projectiles.

The first term of the right-hand member of Eq. (2) is the ordinary elementary range expression, and naturally it expresses almost the whole of the actual range. The second term is a small correction which is of positive sign for eastbound projectiles ($0<\mu<180°$) and negative for westbound. The correction term, being proportional to V_0^3, increases with V_0 at a greater rate than does the range as a whole. Hence the *percentage* increase or decrease of range, because of earth rotation, varies in proportion to V_0 or to the square root of the range itself. Evidently this effect is a maximum at the equator and zero at the poles. Inspection of the role of α shows that the correction term is a maximum for a 30° angle of elevation and that it vanishes when the angle of elevation is 60°.

By the appropriate numerical substitutions in Eq. (2), one may show that a well-thrown discus in tropic latitudes will go an inch farther eastward than westward. This is many times the apparent precision of measurement for this event, and records have changed hands on slimmer margins. Significant effects of the same kind, though of lesser magnitude, appear in the cases

[2] P. Kirkpatrick, Am. J. Phys. 11, 303 (1943).

of the javelin, hammer, shot and even the broad jump, where the east-west differential exceeds the commonly recorded sixteenth of an inch.

Figures 1 and 2 are types of correction charts that might be used to normalize the performances of weight throwers to a uniform value of g and a common direction of projection. Figure 1 has been prepared with the shot-put in mind, but is not restricted to implements of any particular mass. The inclined straight lines of this figure are graphs of $-dR$ *versus* dg from Eq. (1). Values of the parameter R are indicated on the graphs. The uniform value 100 cm has been adopted for h, an arbitrary procedure but a harmless one in view of the insensitivity of dR to h.

Figure 2, particularly applicable to the hammer throw, furnishes means for equalizing the effect of earth spin upon athletes competing with the same implement but directing their throws variously as may be necessitated by the lay-out of their respective fields. An angle of elevation of 45° has been assumed in the construction of these curves, a somewhat restrictive procedure which finds justification in the fact that no hammer thrown at an angle significantly different from 45° is likely to achieve a range worth correcting. These curves are plotted from Eq. (2); their application to particular cases is described in the figure legend.

Upon noticing that some of these corrections are quite small fractions of an inch, the reader may ask whether the trouble is worth while. This is a question that is in great need of clarification and one that may not be answered with positiveness until the concept of the probable error of a measurement shall have become established among the metrologists of sport. Physicists will agree that to every measurement worth conserving for the attention of Record Committees should be attached a statement of its probable error; without such a statement there will always be the danger of proclaiming a new record on the basis of a new performance that is apparently, though not really, better than the old. If the corrections of Fig. 2 exceed the probable error to be claimed for a measurement, then those corrections must be applied.

The aim of the American Athletic Union in these matters is hard to determine. Watches must be "examined," "regulated" and "tested" by a reputable jeweler or watchmaker, but one finds no definition of what constitutes an acceptable job of regulation. Distances must be measured with "a steel tape." The Inspector of

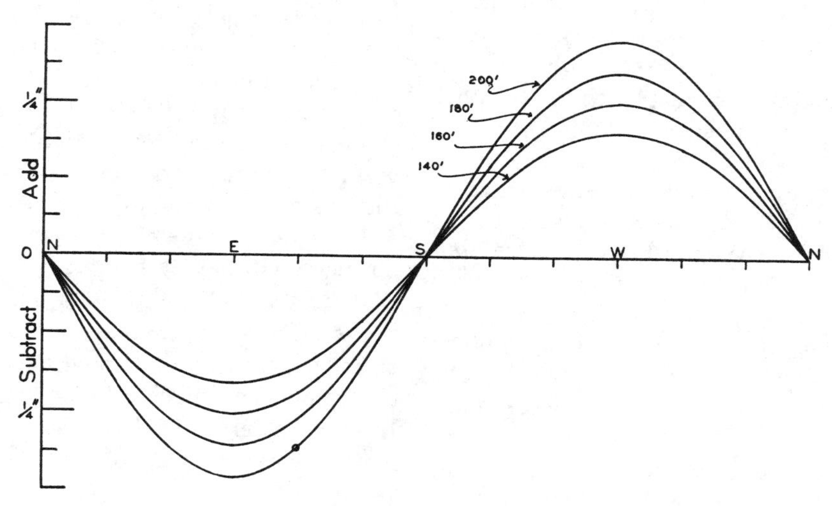

FIG. 2. Curves for rendering throws in various directions comparable. The assumed latitude is 30°, either north or south, and the assumed angle of elevation is 45°. Since the range has a maximum for about this angle of elevation, the curves also apply well to angles several degrees on either side. The curves show, for example (circled point), that a missile thrown 200 ft in a direction 30° south of east should have $\frac{5}{16}$ in. subtracted from its range in order to bring it into fair comparison with unadjusted northward or southward throws or with throws in any other direction which have been adjusted by reference to curves of this type appropriately constructed for their respective latitudes.

Implements must find the weights of the implements "correct." Such ideals of perfection are not realistic, and the only alternative is to recognize the existence of error and state its magnitude. The minimum permissible weight for each implement is prescribed both in pounds and in kilograms by AAU rules, but in no instance are the prescriptions exactly equivalent. A discus thrower whose implement just satisfies the metric specification will use a discus 4 gm, or $\frac{1}{5}$ percent, lighter than that of a competitor whose discus just passes as judged by an inspector using perfect scales calibrated in British units. Those 4 gm will give the former athlete two or three extra inches of distance, an advantage that might be decisive.

Similar comments could be made about the rules of competition of the ICAAAA, where one reads that the javelin throw is measured from the point at which the point of the javelin first strikes the ground. This is a mark that cannot in general be determined to the often implied $\frac{1}{8}$ in. since it is obliterated by the subsequent penetration of the implement. Any javelin throw as correctly measured by ICAAAA rules will show a greater distance than if measured by AAU rules, but few field judges know this nor could they do much about it if they did. It is probable that the rules do not say what was meant in these cases. It is interesting that whereas the hammer, shot and discus must be thrown upon a level surface, there is no such requirement in the case of the javelin.

Any serious attempt to put the measurements of sport upon a scientific basis would be met with vast inertia if not positive hostility. The training of athletes is still very largely an art, and there is no reason to suppose that those who are at present practicing this art with success will be predisposed to changes involving ways of thought which, however commonplace in other disciplines, are novel in athletic competition. One eminent track and field coach, a producer of national, Olympic and world champions, told the writer that he had no interest in hairsplitting; that leveling the ground accurately would be too much trouble; that common sense is better than a wind gage for estimating the effect of wind conditions on sprinters; that a man can't put the shot by theory—it's all in the feeling; that the exact angle of elevation is unimportant as long as he gets it in the groove.

A few years ago, the writer published some criticisms along the lines of the present article and sent reprints to each of the several hundred National Committeemen of the AAU. One acknowledgment was received, but no reactions to the subject matter. In a sense, this indifference was only just recompense for the writer's habit of ignoring communications from nonphysicists proposing novel theories of the atom, or otherwise instructing the physicist as to the foundations of his science.

There probably exists a general feeling that part of the charm of sport resides in accident and uncertainty. Any discussion of the possibility of replacing the balls-and-strikes umpire in baseball by a robot will bring out the opinion that the fallibilities of the umpire are part of the entertainment for which the public pays. An optical instrument for determining from the sidelines whether or not a football has been advanced to first down was tried out in California a few years ago. It was technically successful, but a popular failure. The crowd was suspicious of a measurement that it did not understand and could not watch; the players begrudged the elimination of the breather which a chain measurement affords; and even the linemen protested the loss of their dramatic moment.

Though entertained by such attitudes, the physicist will hardly be able to dismiss a feeling that in any field of popular importance or interest, it is improper to keep up the appearances of accurate and comparable measurement without doing what might be done to gain the reality. In the matter of athletic records, he and very few others know what to do about it.[3]

---

[3] The author will be pleased to furnish reprints of this article to readers who would find interest in bringing it to the attention of athletic authorities.

Physics of Bows and Arrows

Paul E. Klopsteg
Central Scientific Company, Chicago, Illinois

THERE is presumably no sport that may not in some of its aspects be subjected to physical analysis or experiment, particularly so when certain specialized instruments are used in its pursuit. Golf, baseball, tennis, lawn bowling and boomerang throwing are examples. In each of them a physicist may find plenty of scope for "recreational physics," the application of physics for his own recreation. No sport to my knowledge affords greater opportunity for the exercise of recreational or hobby physics than does archery. For a decade and more it has from time to time provided me with hobby material of several kinds. One of them is the outdoor recreation of target shooting and bow-and-arrow hunting. Another is the collecting of old books and prints. Scientific study and experiment constitute a third. Still another is the application of results of the third to improvement in design, and the exercise of craftsmanship in fashioning better weapons and missiles.

Although much that is interesting could be written on the first two aspects of the hobby, this is not the place to attempt such an undertaking. This is an article for physicists written by one who has experienced great satisfaction in the joining of his vocation with his avocation. In it he hopes to convey some idea of what he has experienced and to present certain material that may prove interesting to them and to their students, whether or not they are archers.

The bow is one of the earliest man-made machines with which it is possible to store energy slowly, and at will transform it into the kinetic energy of a missile that is being projected at a more or less distant object. Muscular effort provides the energy for propulsion; coordination of mind and muscle directs that energy. Although the bow is one of the most ancient weapons, the bow and arrow are as modern as they are ancient. They are with us today in greater numbers than ever and are gaining devotees at a rapidly increasing rate. An explanation of the persistence of interest in archery may be the fascination of seeing the arrow fly or in hearing its "thunk" as it strikes. Possibly love of the bow, like affection for one's dog, or love of an open fire, is an inheritance passed along through generations of ancestors for whom the bow meant subsistence and protection.

It hardly needs saying that the toy archery set procurable in the dime store is not representative of the bows and arrows that we are considering. These are by no means toys. They are implements capable of storing and converting large amounts of energy. Even the smaller sizes, for childrens' use, can cause injury and damage if carelessly handled, and their use should be permitted only under adult supervision. The larger sizes, as used in hunting, constitute a powerful, lethal weapon. There are many thousands of bow-and-arrow hunters; of these, large numbers have dis-

Reprinted from *American Journal of Physics* **11**, 175–192 (1943);

FIG. 1. A 42-lb yew bow held at full draw while the arrow is being aimed.

carded the rifle for the bow. They have discovered that, although the bag is necessarily limited by the smaller range and accuracy of the bow, there is incomparable compensation in the thrill of a successful stalk—an art not called for in rifle shooting—and in seeing the arrow miss by the narrowest of margins. I have had no greater fun in hunting than to see a broadhead arrow, at 50 yd, overtake a fleeing jackrabbit and pass directly between his ears.

BOW AND ARROW SHOOTING

To provide a basis for an understanding of the discussion to follow, it is desirable to describe briefly the method of shooting an arrow. There are many ways of holding and drawing the bow and "loosing" the arrow. Types of bow and methods of loose have been employed by ethnologists and archaeologists as "tracers" of races and cultures. In this country, as well as in England, from which our archery derived, we employ the three-finger, or Mediterranean, loose in which the string is drawn by hooking the first phalanx of each of the first three fingers upon it, with the arrow between the first and second fingers. The archer stands facing at right angles to the direction in which he intends to shoot, with the target at his left if he is right-handed. As he raises the bow with his left arm, he draws the string with his right, so that the plane of the drawn bow is approximately at right angles to the direction in which he is facing. Without moving his body he turns his head in the direction of the target and aims with his right eye (Fig. 1). His right elbow and forearm are well elevated, and the latter transmits the force from his arm and shoulder muscles through the fingers to the string. The string comes to rest, pressing against his chin. When he is ready to loose the arrow, he continues to exert force on the string with his drawing hand, slowly and deliberately, until the string slides off the ends of the three drawing fingers. In some kinds of field shooting, in hunting and in flight or distance shooting, the draw and loose are a continuous operation without an intervening pause. The practiced archer is careful not to move the bow hand until the arrow is well on its way, to avoid throwing the latter off its intended course.

The bow used in target shooting "weighs" (pulls) anywhere from 20 to 60 lb, depending on the strength of the archer.[1] Its length is usually between 5 and 6 ft. Hunting bows range in "weight" between 50 and 100 lb at full draw, but there are some archers of Ulyssean brawn who claim ability to draw up to 175 lb. Flight bows may draw 80 or 90 lb for regular style shooting, and up to 200 lb or more for use as "foot bows" which the archer straps to his feet and shoots by lying on his back and elevating his feet until the proper initial angle of the arrow for maximum flight is attained.

Bow strings are now almost invariably made of linen thread, although many other fibrous materials have been employed. A sufficient number of threads provides the requisite strength. Strings should be as light as is consistent with strength and durability. Of all available fibers, linen seems best to satisfy this requirement. It stands at the top of the list for its high tensile strength, small stretch and low mass per unit length in a given string.

[1] The "weight" of a bow is the force required to draw it the full length of its arrow.

ARROWS

Accuracy of flight requires straightness in an arrow throughout its 24 to 30 in. of length. Feathers are commonly used as stabilizing vanes to keep the shaft tangent to its trajectory. Because of their natural warp, they cause the arrow to spin, which partly compensates for lack of straightness. The wood of which an arrow is made must, obviously, be strong and tough to withstand impact and other forces without damage. Only a few kinds of wood are suitable because, in addition to strength and toughness, arrows require a high ratio of stiffness to density. In hunting arrows strength is the first requirement and the stiffness-density ratio is less important. Differences in mass produce differences in velocity in arrows shot from the same bow. Hence it is necessary for accurate shooting, when a number of arrows are used in the same set, that they have the same mass. Not only this, but they must have the same dimensions and form, and the same mass distribution. The reasons for these requirements will appear in the subsequent discussion. In recent years good arrows have been produced of steel and aluminum alloy tubing.

Accuracy in the shooting of successive arrows requires not only the aforementioned uniformity of characteristics, but also that the bow have an unvarying force-draw characteristic so that the total energy for propelling the shaft is the same in each shot. The limbs of the bow must work together and without variation in their mode of motion in successive shots. Finally, the 17 things essential to good shooting—which are distributed among the five principal acts of standing, nocking,[2] drawing, holding and loosing—must be done in precisely the same manner each time to achieve a perfect shot. Hot, sultry weather may affect the bow—and the archer—by change of temperature and moisture content. Shooting a bow is a physical experiment, the results of which are most closely reproduced when there are no uncontrolled variables.

The somewhat sketchy information conveyed up to this point may serve to give some appreciation of the fact that the bow and arrow individually, and their use in combination, present a great wealth of problems. We are now prepared to consider those lying in the domain of physics—to indicate what progress has been made in their solution, and how physical research and development have greatly improved the implements of the sport. Applied physics has given archers a weapon of precision, comparable in performance with a target pistol.

THE BOW

Reduced to its simplest terms, what we require of a bow is that it shall shoot an arrow. There are excellent, good and bad bows, and all of them shoot arrows. To differentiate among them it is desirable to establish some criterion of quality, and one element of this criterion we call *cast.* Cast may be defined as the property of a bow that enables it to impart velocity to an arrow, and it may be expressed in terms of the velocity that it imparts to an arrow of stated mass. However, since arrows of different masses acquire different velocities when shot from the same bow, the most complete picture of cast of a given bow is the graph showing the relationship between the mass of an arrow and the velocity given to it by the bow.

The archer is concerned not only with cast but also with the effort needed to produce that cast. To draw a heavy bow requires great muscular exertion. When this becomes excessive it is detrimental to accuracy. The archer therefore should use a bow that will propel an arrow with high speed, but with minimum energy input to produce the speed. The general problem, then, of portraying excellence as applied to bows resolves itself into an understanding of how a bow casts an arrow and what specific things contribute to high efficiency. We know, of course, that all the energy put into a bow in drawing it does not reappear in the arrow as kinetic energy. Some of it remains behind. The *efficiency of a bow* is the ratio of the kinetic energy of the arrow as it leaves the string to the energy imparted to the bow by the archer as he brings it to full draw.

Bows are usually made of wood or of a combination of wood and other materials which have

[2] Nocking means to place an arrow on the string, with its nock, or notch, in the proper position. Roger Ascham, tutor to Elizabeth, later queen of England, wrote in 1544: "Standinge, nockinge, drawinge, holdinge, lowsinge, whereby commeth fayre shootinge . . . ."

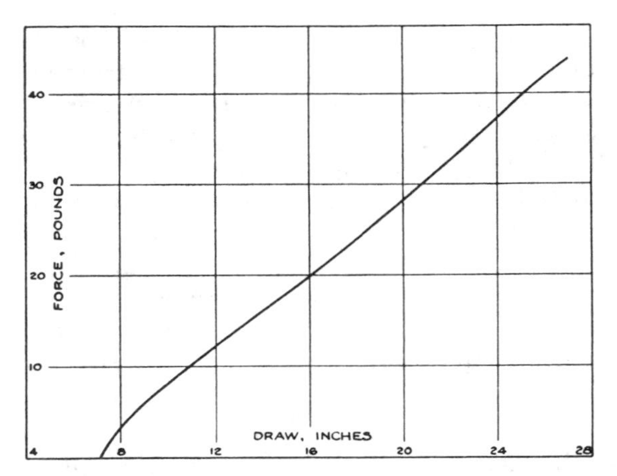

FIG. 2. Force-draw curve for a bow. The area under the curve is a measure of the energy input.

been thought or found suitable. Among them are fibrous materials of high tensile strength which may be glued to the *back*, or convex surface, of the limbs, and materials of high compressive strength which may be glued to the *belly*, or concave surface. Most of the materials of which a bow is made are of organic origin, and their elastic behavior is somewhat indeterminate.

The first cause of reduction in efficiency of a bow is elastic hysteresis. For any bow we may plot a curve with force as ordinates and length of draw as abscissas. This is called the *force-draw curve* for the bow (Fig. 2). Its shape depends mainly on the dimensions and geometry of the bow. A bow with long limbs, straight in their relaxed condition, has a force-draw curve that is nearly linear. If its limbs are straight, but set backward at the handle, the force increases more rapidly at the beginning of the draw and less at the end, than that of the first. If a straight bow has taken a set from much shooting, its draw starts with a reduced rate of increasing force, and shows stiffening towards the end. A straight bow with backwardly curved tips, or "ears," yields a curve similar to the one with the limbs set back. A strongly reflexed bow, of the Turkish type, starts with an exceedingly stiff draw, which eases greatly towards the end. All of these types are illustrated in Fig. 3.

C. N. Hickman[3] has designed a bow in which the limbs are bent backwardly on a circular arc, secured to the rigid handle at a forward angle of about 45°. This bow has a particularly interesting force-draw curve, showing characteristics similar to those of the Turkish bow but accentuated. The first part of the curve rises steeply, then nearly levels off, and again begins rising. The length of draw should be such as to fall within the slowly-rising part of the curve.

Since the energy input, and therefore the energy available for propelling the arrow, are represented by the area under the force-draw curve, it is evident that a study of the curves may teach us things about desirable geometric arrangements of the bow limbs. We must bear in mind, however, that the energy input curve does not tell us how much of that energy gets into the arrow. To be sure, the more energy we can put into the bow, the more will there be for us to

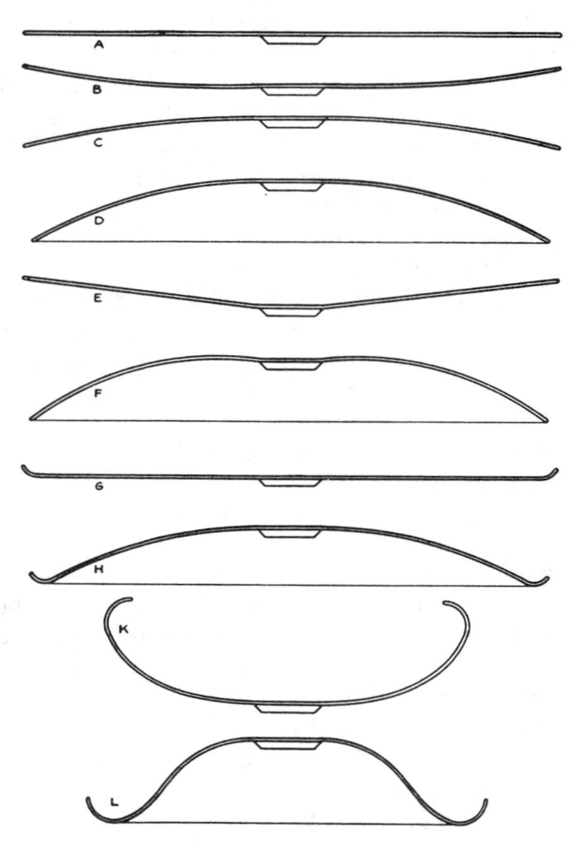

FIG. 3. Types of bows of which all except *K* are in common use. *A*, straight. *B*, reflexed. *C*, type *A* after it has taken a set or "followed the string." *D*, type *A*, *B* or *C* when "braced" or strung; they differ only in their force-draw curves. *E*, straight limbs set backward. *F*, type *E* braced. *G*, straight limbs with "ears." *H*, type *G* braced. *K*, Turkish type, strongly reflexed, with "ears." *L*, type *K* braced.

[3] Hickman, U. S. Patent No. 2,100,317.

"play with" in our endeavor to transfer as much as possible to the arrow.

The force-draw curve shows the value of the long draw. Each inch of draw near the end contributes vastly more to the stored energy than an inch near the beginning, except for special shapes of bows. For the ordinary long bow the energy input is proportional to the square of the draw.

If the bow is kept at full draw for an appreciable length of time, the force corresponding to this amount of distortion ordinarily diminishes. The bow "lets down." Suppose we plot the curve step by step, starting with zero force and extension, increasing the load, holding at full draw for an arbitrary period, say 2.5 sec, and then decreasing the load to zero. Almost invariably we find that the force for a given draw is smaller with decreasing than with increasing load. Although I have made many measurements on many bows, I have found only one in which the hysteresis was not measurable by the method employed. In most bows it lies somewhere between 5 and 20 percent. It is thus apparent that to produce a superior bow we must start with material in which the hysteresis loss is negligible.

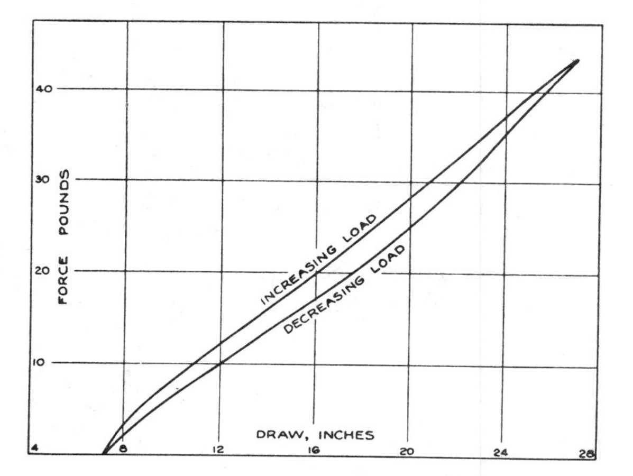

FIG. 4. Hysteresis loop for a bow of laminated bamboo, showing an energy loss of about 12 percent.

The force-draw curves showing hysteresis (Fig. 4) were obtained with the aid of a 16-mm motion picture camera, operating at 16 frames per second. The bow is supported on a dynamometer consisting of a steel or brass ring, within which is mounted a 100-div dial gage on which each division corresponds to 0.0001 in. The ring is calibrated by means of known weights. It is adjusted to read 20 div/kg, and its accuracy is ±1 div or ±0.05 kg.

The device for drawing the string is hooked over the latter in about the manner in which the three drawing fingers would be used in shooting. Attached to this device is a yardstick on which the displacement can be read at the intersection of its scale and the back of the bow. The bow is supported horizontally, and the string is drawn downward by means of a rope passing over a pulley. As the bow is drawn, the camera photographs the entire process from the beginning of the draw through the pause at full draw and back to zero force. The frames of the film provide the time scale, and each frame yields a pair of values for force and draw. This method of obtaining the hysteresis characteristic is much faster than a step-by-step method in which the draw has to be held steady while the dynamometer is being read. Hence the motion picture method more closely approximates actual shooting conditions. In shooting, the arrow is seldom held at full draw as long as 2.5 sec, and upon release the string returns to its zero position in a few hundredths of a second. It is probable that the hysteresis loss measured by the method described is somewhat higher than it is when the bow is normally drawn and loosed. From the hysteresis curve (Fig. 4), the loss is determined by means of a planimeter.

Let W denote the energy required to bring the bow to full draw, and let r represent the fraction of input energy remaining after deduction of the hysteresis loss. The quantity rW is therefore the energy available for propelling the arrow. Since the sound produced and the loss by air resistance to the limbs and string are negligible, it is proper to assume that all the available energy rW is divided between the kinetic energy of the arrow and the kinetic energy of the limbs and string at the instant the arrow leaves the string. Although the acceleration of the arrow is variable, as is also the velocity of each element of mass in the limbs and string, it is the partition of kinetic energy between the bow and the arrow at the instant the latter leaves the string upon which we fix our attention.

If a bow were to have an efficiency of 100 percent, not only would it have to be free of

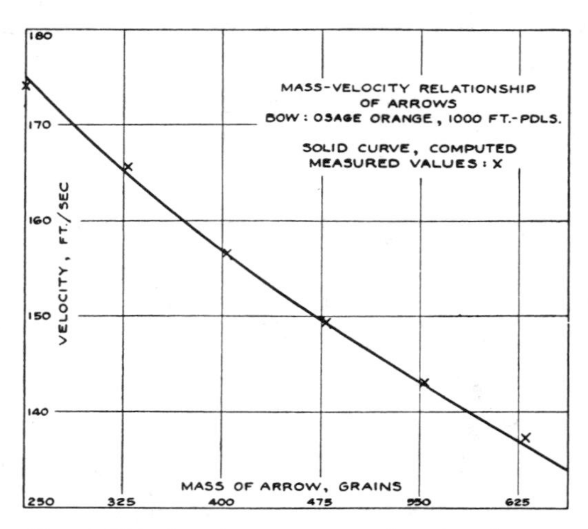

FIG. 5. Velocity of an arrow as a function of mass. $k = 190$ gr.

hysteresis loss, but the kinetic energy of the limbs and string would have to be zero at the instant the arrow leaves the string. Since this is impossible, the best we can do is to strive toward increasing the kinetic energy of the arrow and decreasing that which remains behind after the arrow has taken its departure.

VIRTUAL MASS OF A BOW

In considering the possibility of improving the efficiency of a bow—aside from the obvious necessity of using the best possible materials from the standpoint of hysteresis loss—a most useful concept is that of ***virtual mass***.[4] This I define as a mass which, if it were moving with the speed of the arrow at the instant the latter leaves the string, would have precisely the kinetic energy of the limbs and the string at that instant. Letting K represent the virtual mass of a bow, we may write

$$rW = \tfrac{1}{2}(m+K)v^2, \qquad (1)$$

where m is the mass of the arrow and v is the corresponding velocity. If K is independent of the velocity of the arrow as it leaves the string, then Eq. (1) enables us to determine the curve of cast for a particular bow—namely, the curve of ***v versus m***—provided the available energy rW is known. That the virtual mass K is in fact a constant, has been determined in many measurements with a large number of bows. These measurements consisted of determining by means of a high speed spark chronograph, the velocities imparted by the bow to six arrows ranging in mass from 250 to 700 grains (Fig. 5), in steps of 75 grains.[5] This range includes practically all masses of arrows used with target or hunting bows, as well as most flight arrows. If we measure two velocities v corresponding to two masses m, differing preferably by several hundred grains, we can eliminate K from Eq. (1) and solve for rW, thus obtaining a check on the value of rW, determined as previously described. The values of rW thus found are usually higher than those obtained by either the photographic or the step-by-step method, indicating that the available energy in actual shooting is somewhat higher than that determined by the slower methods.

Having established the constancy of the virtual mass K for a given bow, we can usefully employ this new concept in considering several well-known facts that come out of the archer's experience but the explanation of which has been obscure. When shooting at distant targets, say at 100 yd—a standard distance used in the York Round—it is desirable to use a bow of high cast to avoid such great initial elevation of the arrow as to make aiming difficult. To achieve a flat trajectory, archers have been known to use bows of greatly increased "weight" and have been disappointed when they have found only slight improvement in cast. Other archers who devote their efforts to increasing the maximum distance to which they can shoot an arrow, have found, to their surprise, that doubling the "weight" of a bow by no means doubles the distance. They are doing rather well when doubling the "weight" increases the distance by

[4] This discussion of virtual mass appears here as the first publication on the subject. It developed under the editor's benign prodding for an informative paper on bow-and-arrow physics, which caused me to review experimental data that had lain dormant in my workbooks for more than ten years. I had previously discovered the empirical relationship, which held for most bows I had tested, that the velocity of an arrow is inversely proportional to the cube root of its mass. If Eq. (1) is used to evaluate v and m for several values of rW and K, corresponding to different bows, and the products of v and $m^{\frac{1}{3}}$ are plotted as a function of m, the curves are flat over the ranges of values of m and K employed in practical archery.

[5] Until recently archers expressed masses of arrows in terms of shillings and pence, which were used as balance weights by arrowsmiths in England. In this country the grain is now employed as the unit (15.4 grains = 1 g; 437.5 grains = 1 oz).

20 or 25 percent. These observations from experience are explicable in terms of virtual mass (Fig. 6).

When the virtual mass K of a bow is large, that is to say, comparable to the mass m of the arrow, the efficiency of the bow is about 50 percent. If K equals $\frac{1}{3}m$, the efficiency of the bow is 75 percent, for this represents the division of kinetic energy between the mass of the arrow and the virtual mass of the bow. If, finally, the virtual mass of the bow exceeds that of the arrow, the efficiency of the bow will be less than 50 percent for that particular arrow. As the mass of the arrow is increased, regardless of the virtual mass of the bow, the efficiency for that particular combination is likewise increased.

The recoil, or kick, of a bow is found by experience to be small in bows of good cast, and large in sluggish, heavy bows. This is clearly a matter of efficiency. If the virtual mass is large in relation to arrow mass, the large amount of energy retained in the bow must somehow be dissipated, hence recoil becomes noticeable if not annoying. I have seen archers shooting heavy, steel bows, with sponge rubber padding in the hand to absorb recoil. By contrast, the recoil of a yew target bow of small virtual mass is scarcely perceptible. Doubtless this adds to the enjoyment of shooting.

Another demonstration of recoil is the catastrophic result of loosing a string without an arrow—an accident that sometimes occurs when the arrow falls off the string because of an ill-fitting or broken nock. In this case the bow must dispose of all the energy unaided; almost invariably the bow breaks. If it does not, excessive strength and hence excessive virtual mass are indicated.

The reduction of K to its minimum is obviously desirable, but a lower limit is set by the fact that finite mass in the limbs and string of a bow is indispensable for the storage of the energy with which the arrow is to be propelled. Thus the only possible way of reducing K, other than by reduction of mass in the limbs, is to design the bow so that the velocities of the particles comprising the limbs and string shall be as small as possible at the instant the arrow leaves the string.

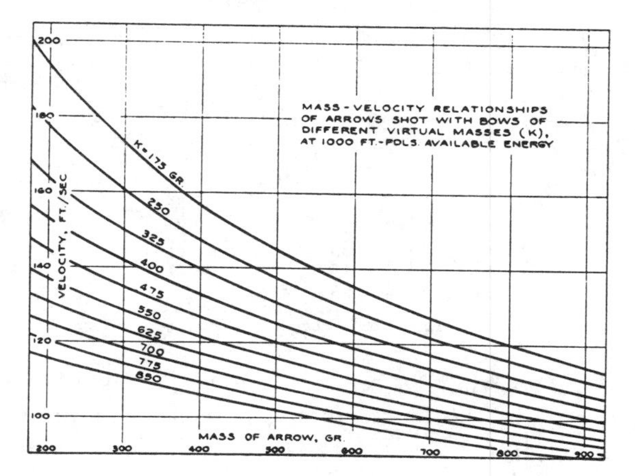

FIG. 6. Family of curves computed from Eq. (1). For values of rW other than 1000 ft pdl, multiply V by $(rW/1000)^{\frac{1}{2}}$.

Much can be and has been done in this direction. Whatever the design, it must provide for keeping the tensile, compressive and shearing stresses at values that are well below the somewhat indefinite elastic limits for such stresses. To produce a bow having maximum cast with a minimum of materials is like skating in the vicinity of a thin spot in the ice. You approach as closely as possible, hoping you won't break through. It is a compromise between using as little material as one *dares*, to produce a small value of K, and using as much as one *must*, to avoid the hazard of breakage. This indicates some of the difficulty in producing a durable bow which, for its "weight" at full draw, has little hysteresis loss and great durability. If one is satisfied with a sluggish bow of low efficiency, the construction is not difficult, but for experienced archers the result would not be satisfying.

FIGURE OF MERIT OF A BOW

Starting again with Eq. (1) and assigning zero value to m, we obtain

$$v = (2rW/K)^{\frac{1}{2}}. \tag{2}$$

This equation establishes a *figure of merit* of a bow from the standpoint of cast. It represents the limiting velocity that the bow could impart to an arrow approaching zero mass. Obviously no arrow shot with this bow could achieve higher velocity than that given by Eq. (2). In the light of what has been said about

TABLE I. Effect of doubling the "weight" and virtual mass of a bow on the velocities of arrows of different masses.

Mass of arrow,	Bow A: 45 lb; K_1 = 0.035 lb; available energy, 1200 ft pdl			Bow B: 90 lb; K_2 = 0.070 lb; available energy, 2400 ft pdl			Difference in velocities,
m (lb)	$m+K_1$ (lb)	v_1^2 (ft²/sec²)	v_1 (ft/sec)	$m+K_2$ (lb)	v_2^2 (ft²/sec²)	v_2 (ft/sec)	v_2-v_1 (percent)
0.050	0.085	28300	168	0.120	40000	200	19
.060	.095	25300	159	.130	37000	192	21
.070	.105	22800	151	.140	34300	185	23
.080	.115	20800	144	.150	32000	179	24
.090	.125	19250	139	.160	30000	173	25

the fact that a certain amount of available energy in the limbs implies a certain minimum value of K, Eq. (2) shows that increasing the "weight" of a bow by increasing the width of its limbs does not appreciably change the figure of merit. We are assuming here that no other dimensions are changed, since, if the bow is properly made, we may neither increase its thickness, nor decrease its length, without risk of breakage. Consequently, doubling the width without changing any other dimensions would also approximately double the value of K, and hence would leave the figure of merit unchanged. This does not mean that there is no advantage in doubling the weight of a bow, as is quickly demonstrated by computing velocities of arrows of several masses shot with two bows, one of which has double the "weight" and double the virtual mass of the other. The result is shown in Table I. This table, which is based on assumed values of m, K and w that are well within the ranges of those used in practice, shows clearly why doubling the weight will not increase the velocity by more than a small fraction of the factor of increase in "weight."

An approach to the problem of the limiting velocity of a given bow with m approaching zero, is to consider the symmetrical limbs of a bow from the standpoint of the free period of vibration of the limbs when the handle of the bow is solidly clamped in a vise. For a given set of dimensions of the bow limbs there is a certain period of vibration. If it be assumed that the dimensions of the limb represent limiting values below which we may not safely go, the only way to increase the stiffness of the limb is to increase its width. Increasing the width, however, does not further decrease the period, a conclusion that bears out the reasoning regarding the figure of merit of two similar bows, one of which has double the "weight" of the other.

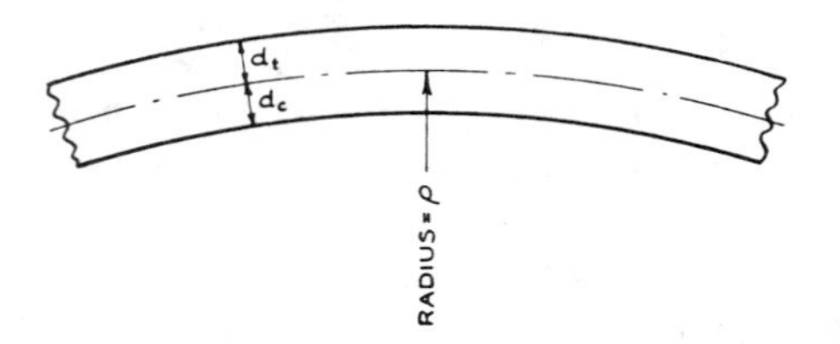

FIG. 7. The maximum fiber stresses in a bow are directly proportional to the distances of the outermost fibers from the neutral axis and to the curvature $1/\rho$.

CONSIDERATIONS IN BOW DESIGN

Among the optimum conditions to be met in a bow so that it may have minimum virtual mass are uniform stress distribution and uniform energy density in the limbs. Consider any given section in a bent limb. Let the distance from the neutral axis of the section to the outermost fiber on the compression (belly) side be d_c and the distance to the outermost fiber of the tension (back) side be d_t (Fig. 7). The stresses in these outermost layers are directly proportional to the aforementioned distances and inversely proportional to the radius of curvature ρ at the point in question. Hence to achieve uniform compressive stress in the belly, the ratio d_c/ρ must be uniform for the entire length of limb; and to achieve uniform tensile stress in the back, the ratio d_t/ρ must likewise be uniform throughout the length of limb.

In wood samples tested at the U. S. Forest Products Laboratory and elsewhere, the strength in tension is between two and three times as great as that in compression. It follows that, for any given radius of curvature, the distance d_t should be less than the distance d_c.

In the traditional English longbow the requirements mentioned are far from being met.

The cross-sectional shape of the English longbow is something approaching semicircular, with the flat part of the section constituting the back, or tension, side of the bow, and the rounded portion constituting the belly, or compression, side. There are variations from the circular shape on the compression side, and there has been much discussion among old-time bowyers and archers whether the correct shape is a semicircle or a parabola or some other curve. None of these, obviously, is correct. If we sketch the cross section and remember that the neutral axis passes through the center of mass of the section, we see that in all these sections the neutral axis is farther from the rounded, or compression, side than it is from the flat, or tension, side. It puts excessive stresses in section areas least able to withstand them. Furthermore, the area of section available to take the compressive stresses at the outermost fiber is greatly reduced by the rounding. Thus the section of the English longbow (Fig. 8) is about as wrong as can be. The consequence of this design is that the wood is usually overworked in compression and underworked in tension. Compression failures are usual and are avoided only by making the limbs so long that the radius of curvature is kept large. Another defect is that, because of the failure to adhere to the principle of uniformly stressing the sections, the radius of curvature of the bent limb at full draw is, in most cases, a variable one with shorter radii and consequent excessive bending in certain portions of the limb, with very little in others. For the reasons mentioned, much of the wood is stressed far below safe limits of strength. This, together with the large length of limb, makes for large virtual mass and a sluggish-acting bow.

Any engineering handbook shows that a cantilever of uniform strength in bending, with the load concentrated at the end, may be designed in one of several ways. One possible design is a beam of constant thickness, tapering in width from maximum at the point of support to zero at the end, its shape an isosceles triangle. Such a beam, when a load is applied at its end to produce a small deflection, bends in a circular arc, provided the material is homogeneous. This fulfils the requirement of uniform stress in tension and compression at any section.

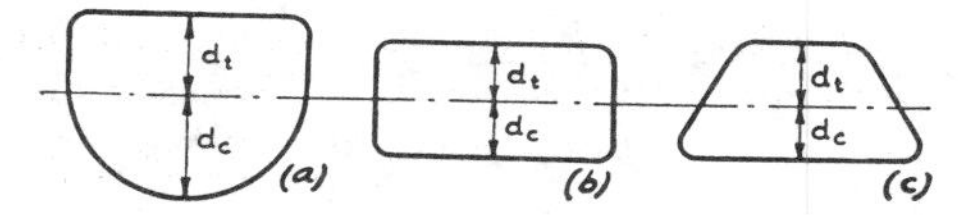

FIG. 8. Section of limbs of bows showing relative distances of outermost fibers from neutral axes. (*a*) Section of the traditional English bow. (*b*) Bow of rectangular section. (*c*) Bow of trapezoidal section, in which the stresses more nearly approximate the ultimate strength characteristics of wood.

A first approximation to scientific bow design would therefore appear to be to start with a rigid section in the middle, supporting the two limbs, each of constant thickness throughout its length, and in the shape of an elongated isosceles triangle tapering from its maximum width at its root, or base, to zero width at its tip.

A limb of this description was found by Hickman to have a free period of vibration, when firmly clamped at its widest point, appreciably shorter than the period of a bar of any other design for which the deflection under a given load is the same. A bow with such limbs, straight from end to end in the completely relaxed condition, is therefore a probable close approach to one with minimum virtual mass.

To make a practicable bow based on the design just mentioned, clearly requires that as the tip of the limb is approached its width be kept greater than if the shape were an elongated isosceles triangle. The width at the tip where the string is attached must be finite, first, so that there may be sufficient material to support the string, and second, to assure stability of the limb, without tendency to twist or distort laterally when the bow is drawn. Another departure from the isosceles triangle is suggested by the fact that the constant radius of curvature holds for small deflections only. Since the condition of maximum stress obtains when the bow is fully drawn, that is, when the load at the end is very large, we must determine the width at any particular section so that the condition of uniform stress is achieved when the radius of curvature is fairly small. This is accomplished in a graphical design worked out by the writer; it is based on Hickman's finding that the tip of a circular-bending, straight limb moves in a path approximately circular, with radius three-fourths the length of the limb and center on the limb at a point one-fourth the length of the limb

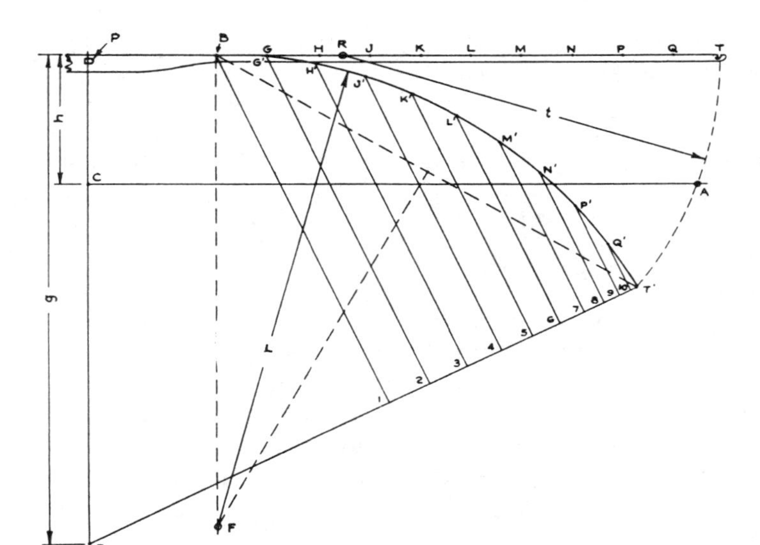

FIG. 9. Scientific design of a bow. BT, the limb, divided into ten equal sections. R, center of path of T, of radius $t[=\frac{3}{4}BT]$. EP, the arrow, of length g. CP, "bracing height" h of the string. CA, $[=ET']$, upper half of the string. Arc BT', the drawn limb; it is the arc of a circle of radius r, with the center on the perpendicular BF to BT and on the perpendicular bisector of chord BT'. The points indicated by primed letters correspond to those with unprimed letters on the straight limb. The lengths $B1$, $G'2$, $H'3$, $\cdots$ give the proper scale of widths of the limb at the corresponding points.

measured from its base. This interesting fact makes it possible to set the specifications of the bow in advance and by methods of geometry to determine the dimensions to meet the specifications. The construction depends also on the fact that in a drawn bow the force in the string is the same throughout, so that the bending moment at a given section is the product of the force in the string and the perpendicular distance from the string to the section in question. For uniform stressing—since the bending moment at any section must be balanced by the elastic torques at the section—the area of the section must be proportional to the bending moment at that section. Since the thickness of limb is constant, it follows that the width at the section in question must be proportional to the distance of its neutral axis from the string. A typical layout following the principles here discussed is given in Fig. 9.

A bow designed according to these principles performs very much as one might expect from theoretical considerations. The limbs may be made shorter than those of the English type longbow by 15 or 20 percent, with less hazard of breakage. The shorter, thinner limbs have greatly reduced virtual mass, and consequently the bow has much higher efficiency than the old style bow. Where, in the older design of bow a rigid section was used barely long enough to provide for a hand grip that gradually merged into the bending limbs, the new design employs a long rigid section which, in fact, may be made of such length that the over-all length of bow is still that of the English longbow. It may be said parenthetically that the force-draw curve for a longbow is nearly linear. The experienced archer likes the "feel" and behavior of such a bow. The long rigid section with the shorter limbs makes it possible to use limbs of equal length and have the arrow "nocked" at the midpoint of the string. In the English longbow design the lower limb was usually made 2 in. shorter than the upper and considerably stiffer, so that neither the undrawn nor the drawn bow manifested symmetry of its limbs with reference to the axis represented by the arrow.

Tests of many bows showed substantially increased cast and efficiency for the new design. An interesting side light is the fact that as late as 1932 practically all bows used in tournaments were of the English longbow type. The design had come down through generations of bowyers from the period in English history in which the archer was the invincible warrior. A tradition of such long standing dies hard. Nevertheless, the following five years, as a result of the publication of a number of articles pointing out the defects of the traditional design and the advantages of the new, witnessed a gradual change at the tournaments to bows in which the rectangular cross section was used. Today the English design of longbow is a rarity at American tournaments.

Limitation of space excludes many details that might prove interesting, particularly about the elastic properties of different kinds of wood with specific reference to their utility in bows. It must suffice to say that in America we have only two or possibly three native kinds of wood which are particularly well suited for the making of bows. They are the western yew, the osage orange and, for the lighter bow, the Tennessee red cedar. Ash, hickory and other kinds of wood are inferior because of large hysteresis and because they take a permanent set much more quickly than the other kinds of wood mentioned. Beginners' bows are usually made of a Cuban wood—*degame*—called lemonwood because of its color. It is the cheapest of the bow woods but yields bows of fair quality, though quite inferior to those of yew and osage orange. We cannot dwell on the merits of the individual species of wood. A stave of sufficient length to make a bow of either yew or osage orange, free from knots, pins, bear scratches, wind shakes and other imperfections and with proper physical characteristics, is a rare and beautiful article. An archer-craftsman will readily pay anywhere from $5.00 to $20.00 for such a stave in the expectation that from it he may be able to fashion a superlative bow. Sometimes he succeeds; more often he is disappointed.

In addition to the so-called *self bows* which are made of a single kind of wood throughout, there are *laminated bows* in which several kinds of wood may be glued together in strips with the purpose of utilizing the tensile and compressive strengths of each to best advantage. Then there are *composite bows* in which fibers of high tensile strength, such as shredded sinew laid in glue, constitute the back, while horn—superb material for compressive strength—is used for the belly. The bows of Ghengis Khan and his hordes were of the composite type. The superior properties of sinew and horn over wood make it possible to use much shorter limbs and to bend them on a much smaller radius than can be done with bows made solely of wood. Turkish bows of four centuries ago still hold the record for distance. In terms of the concept of virtual mass any bow, whatever the shape of its limbs, has very high cast only if its virtual mass is small. In a strongly reflexed bow such as the short Turkish type, the seemingly massive ends of the limbs are moving with small velocity at the instant the arrow leaves the string; notwithstanding the apparent massiveness, the virtual mass is small.

A major contribution to improvement in the straight type of bow is that of Hickman's, who recognized the virtue of the elastic properties and strength of silk, and devised a method of producing backing strips consisting of unspun silk fibers laid in glue. In 1934 the writer pointed out that when sinew or rawhide are used for backing, they should be applied and secured under tension, to assure their best contribution to the performance of the bow. The same procedure is employed in applying silk as backing. The silk replaces a larger mass of wood than its own and stores more energy per unit mass; hence it makes possible a bow of smaller virtual mass. Further reduction of virtual mass may be expected from the use, both for tension and compression, of synthetic materials which have been and are being developed during the war years. A synthetic material having the compressive strength and elastic properties of water buffalo horn—which is difficult to obtain and more difficult to process—would be a great contribution to bow making.

ARROWS AND THE ARCHER'S PARADOX

Earlier in this article certain facts were mentioned regarding the strength and other characteristics of wood which make it especially suitable for arrows. The high acceleration—several hundred times that of gravity—encountered by the arrow as the string urges it forward makes stiffness imperative and strength desirable. Buckling of the "column effect" type should be minimized; but if buckling does take place there should be no breakage. On the other hand, the arrow must not be too stiff, otherwise it will deviate from the line of aim. It will also deviate from this line if it is not stiff enough. These comments regarding stiffness may be summarized by saying that, with respect to stiffness and mass, the arrow must be matched to the bow if it is to fly accurately in the direction in which it is aimed. The reason for these requirements will appear as we consider the

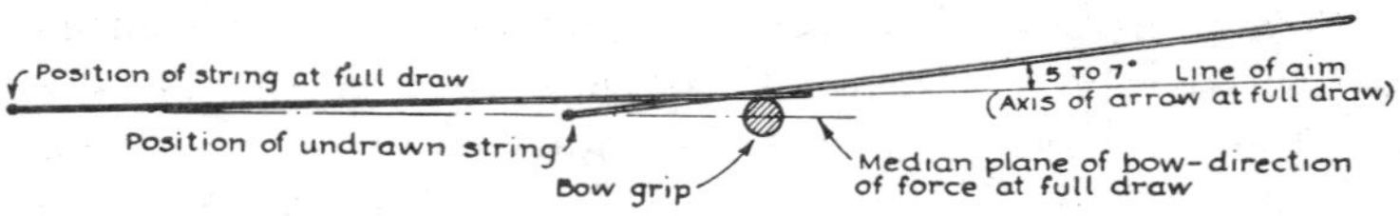

Position of string at full draw
5 to 7° Line of aim
(Axis of arrow at full draw)
Position of undrawn string
Bow grip
Median plane of bow—direction of force at full draw

FIG. 10. Illustrating the archer's paradox.

phenomenon which for a century or more has been known as the *archer's paradox*.

Imagine a bow to be held vertically with an arrow on the string passing the bow on the left side of the latter as the string is drawn back. Suppose further that at full draw the bow and arrow are so oriented in azimuth that the axis of the arrow lies in the vertical plane containing the mark to be hit. Now suppose the bow to be maintained rigidly in that position and the string to be let down gradually by diminishing the force. The string moves forward in the median plane of the bow. Observing the tip of the arrow, we note that as the arrow approaches the undrawn position its tip moves markedly to the left. The axis of the arrow therefore points very appreciably—perhaps 5° to 7°—to the left of where it was aimed. Now, if all conditions are kept as they were except that the arrow is fully drawn and loosed in the regular manner, it will fly accurately to its mark. The archer's paradox is the fact that it does fly to its mark instead of on a line represented by its axis when the string has been gradually let down. This is illustrated in Fig. 10.

The explanation of the archer's paradox was found by means of high speed spark photography which the writer undertook in 1932 to secure direct evidence of what the arrow does as it passes the bow. An electric circuit included a 9-μf condenser, charged to 5600 v with rectified alternating current, and a spark gap of magnesium blocks mounted in a 12-in. reflector. The discharge of the condenser across the gap produced sufficient actinic light to give fully exposed negatives on the fastest photographic plate obtainable at that time. Two cameras were set up, one to photograph the arrow from directly above (Fig. 11), the other to obtain corresponding side views of the bow discharging the arrow. For any particular shot the discharge of the condenser was triggered by a small auxiliary spark gap within the larger one. The arrow penetrated a screen consisting of a sheet of waxed paper having tinfoil on both sides, thus closing a circuit that discharged the condenser through the primary of an air core transformer, to cause a secondary discharge across the trigger gap. Successive photographs of a given arrow shot from a given bow were obtained by placing the circuit-closing screen at increasing distances in successive shots. The pictures thus obtained, one set of which is reproduced in Fig. 12, show that as the string leaves the fingers the arrow is given a lateral impulse at its feathered end. At the same time the tip of the arrow is given a lateral impulse in the same direction by the side of the bow. These lateral forces are sufficient to start an oscillation of the arrow about two nodal points. The arrow oscillates not only while it is passing the bow but for an appreciable number of cycles after it has left the string. An arrow which is properly matched to the bow has a period so related to the time required for it to pass the bow that it apparently "snakes" its way around the handle without once touching

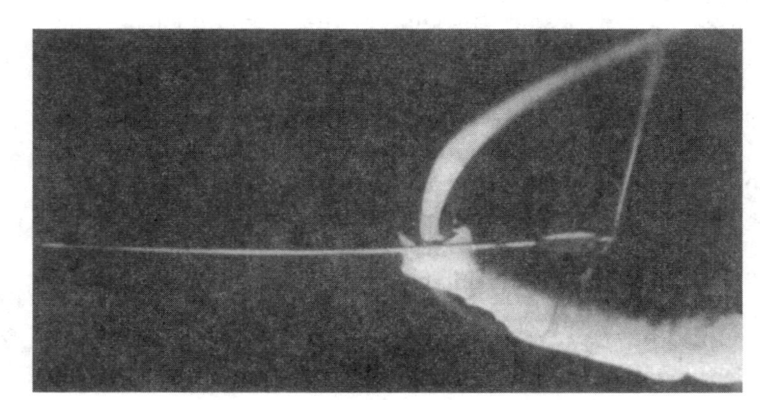

FIG. 11. Speed-flash photograph in a series from which Fig. 12 was compiled. The camera pointed nearly vertically downward, with shutter open, in a darkroom. The arrow released the flash at a predetermined point of its path.

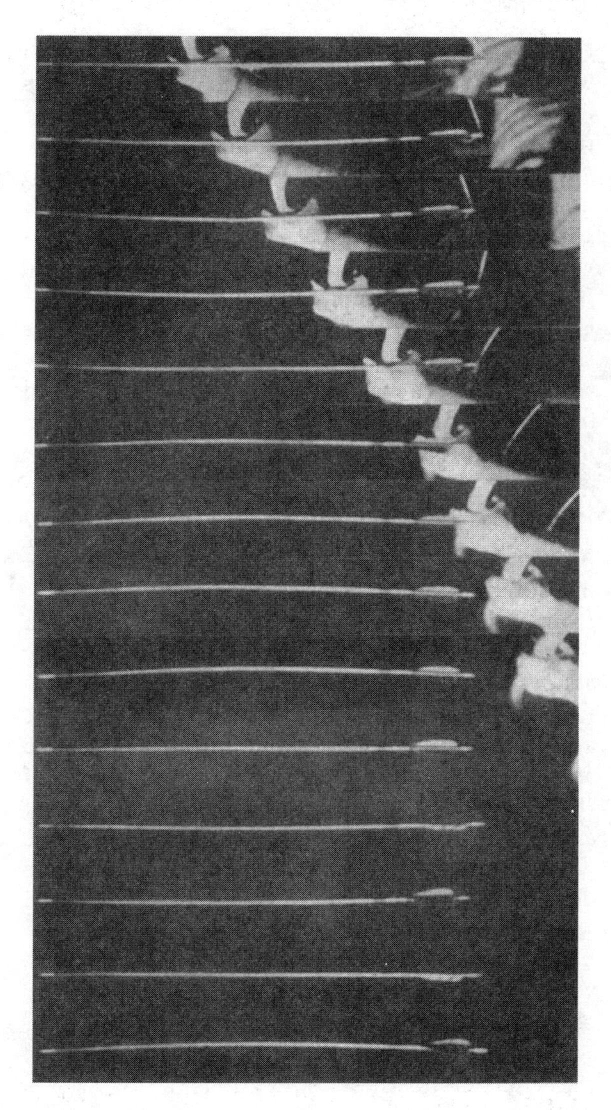

FIG. 12. The archer's paradox explained by speed-flash photography. Note the oscillation of the arrow, so synchronized with its motion past the bow that the arrow bends its way around the bow grip without touching the latter.

the bow after the first lateral impulse at the tip. Figure 13 will help in visualizing how this occurs. The oscillation of the arrow takes place about the line of aim, and since there is no interference to its passage by rubbing or striking the bow, it flies in the direction in which it was aimed. High speed motion pictures by Hickman bear out the findings from spark photographs.

Although conceivably any arrow of appropriate physical characteristics can "snake" its way past any bow handle, the violence of bending at the instant of release may be eased in one of several ways. Thereby the probability of smooth departure and flight is increased, by reduction of amplitude of the oscillation and of interference between the arrow and the bow handle. One help is to narrow the handle at the arrow passage to reduce the "detour" which the arrow must make. Another is to build up the handle on the arrow side of the bow, so that the direction of the force supporting the bow is shifted off the plane of symmetry of the bow and is made to coincide more nearly with the axis of the arrow. The result can also be accomplished by scraping the limbs so that they are slightly unsymmetrical relative to the plane of symmetry of the bow, and are thereby made to exert a force in the direction of the axis of the arrow. However, oscillation cannot be completely eliminated because of the sideward impulse on the nock of the arrow as the string slides off the fingers.

SPINE OF AN ARROW

It can now be readily understood that an arrow which is not properly matched to the bow

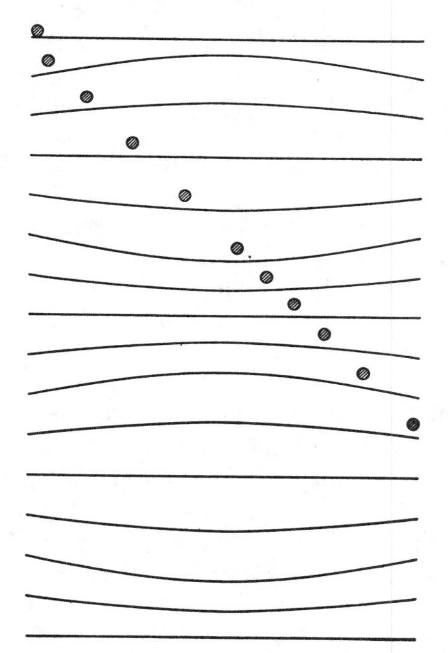

FIG. 13. Schematic representation of the phases of the arrow in its passage by the bow, based on the evidence from speed-flash photography. Figure 12 does not show the first four phases of Fig. 13.

is either too stiff or too limber for that particular bow. If it is too stiff, it oscillates too rapidly, or there is so little amplitude that it does not properly clear the bow, causing deflection to the left of the line of aim (assuming right-handed shooting). If it is not stiff enough, the feathered end fails to move leftward out of the way of the handle as it passes the latter and, consequently, strikes the handle. An error to the right results. For proper matching, then, the arrow should have completed about one full cycle of oscillation by the time it leaves the string. All of this leads us to a consideration of the property of an arrow called *spine*.

When an arrow is shot at long range, the probability of hitting the mark is increased if the velocity is high, because a relatively flat trajectory makes precise aiming less difficult. As we have seen, high velocity for a given bow is achieved when the mass of the arrow is small. Small mass, on the other hand, reduces the energy transferred from the bow to the arrow so that less is available for overcoming air resistance. Furthermore, a very strong bow with which high velocity can be produced in an arrow of larger mass, is for most archers incompatible with accuracy of shooting. All of these facts are cited to show that the optimum conditions for accurate shooting involve a number of compromises among things desirable and things practicable. On account of the improved probability of hitting with an arrow of high velocity, we must compromise on using an arrow of reduced mass, but this in turn requires that, despite lightness, it shall have adequate strength and stiffness. These happen to be the same characteristics which, in combination, give an arrow the quality known to the archer as "good spine."

Spine is somewhat difficult to define because any quantitative expression which would characterize it would not uniquely establish the quality of a particular arrow. This follows because the "take-off" and flight characteristics of an arrow are not uniquely determined by the properties of the arrow itself but in part by the bow with which it is shot. In general, the stiffer and lighter the arrow, the better it is, so long as it is properly matched to the bow with which it is to be used. Since matching to a bow requires that, in a manner of speaking, the arrow should be able to "oscillate its way" past the bow, it is logical that the frequency of the arrow vibrating freely about two nodes with a loop in the middle and two half loops at the ends, should enter into a specification for spine. Furthermore, since the period of the oscillation must be synchronized with the passage of the arrow along the arrow plate of the bow, and since this time of passage is related to the mass of the arrow, this mass should also enter the definition. For a particular arrow, then, we may say that the spine S should be directly proportional to the frequency f and inversely proportional to the mass m of the arrow, or

$$S = kf/m. \tag{3}$$

Although the frequency of the shaft can be measured, stroboscopic or other elaborate apparatus is needed. Equation (3) is therefore a defining formula rather than one also directly adapted to measurement.

It has been stated that for accurate shooting all of the arrows in a set should be as nearly alike as possible. It is not difficult to select shafts of the same mass by accurate weighing. If, out of a large number of such shafts, one selects those with the same stiffness—which can readily be done by measuring deflection under fixed load with the supports placed approximately at the nodal points—and if, further, one selects those with the same balance point, it is reasonably certain that the finished arrows will be matched. Since wood lacks homogeneity, it is possible that, despite all these precautions, a few arrows will not have precisely the same flight pattern as the others. The final test therefore is preferably a shooting test by an expert marksman with the bow for which the arrows are intended. A well-designed shooting machine may also be used to advantage. It can be made to simulate hand shooting, but it obviates the human errors in drawing, aiming, holding and loosing. With such a machine a good bow will shoot a well-made arrow with dispersions within 1 in. at 40 yd. A well-matched set of arrows will group in a 3-in. circle or better at this distance.

The archer, particularly on his "off" days, wonders how much of the variation in his widely scattered hits is caused by differences in mass in his arrows. Should they weigh alike within a grain, or may they differ by more, without appreciable effect on velocity? This question may be answered with the aid of Eq. (1), which yields the expression $v=(2rW/m+K)^{\frac{1}{2}}$. The value of dv/dm obtained from this equation by the assignment of reasonable values to the other quantities, considered to be constant, shows that a change of 1 percent in m produces only about 0.3 percent change in v. Hence precision weighing and adjusting of arrow masses are not required.

From shooting experience one can reasonably well determine the desirable spine characteristics of an arrow. It is thus possible to establish in advance what relative values of stiffness and mass, within some limited range of these values, will be best suited for a particular bow. It is also possible then to examine the characteristics of wood out of which shafts are to be made with reference to the probable spine characteristics of the resulting arrows. This may be done in terms of stiffness of either a cylindrical shaft of arbitrary diameter or a square stick of the wood having arbitrary dimensions, by measuring the density and stiffness. Relative values for spine quality of the wood may then be determined by taking the ratio of the square of the density and the stiffness, expressed in convenient units. Thus different samples and different kinds of wood may be rated in their order of excellence as arrow material.

THE FLIGHT OF AN ARROW

As compared with a high velocity bullet which has a sensibly flat trajectory up to several hundred yards, an arrow is slow-moving indeed, so that even at short ranges there is a large effect of gravitational force on the trajectory. Consideration of the flight of an arrow, particularly of one shot to achieve maximum distance, must take into account this large gravitational effect as well as that of air resistance. Thus far no work has been published, specifically relating to air resistance, that is thoroughly rigorous or entirely valid. The air resistance varies approximately as the square of the arrow velocity. If the proportionality constant for a particular velocity were known, it would be feasible to compute with fair accuracy the trajectory of an arrow having known initial velocity and angle of departure. It could be done by the method of short sections, by which the position, angle and velocity of the arrow at the end of the section are found, assuming constant resistance over the entire short distance. Each section is treated successively in the same way, with the initial conditions established by the computation of the preceding section. Equations for this method have been worked out but have not been applied or checked experimentally. The Didion-Bernoulli method[6] is also applicable. If reliable data on resistance can be obtained—and this is feasible by several methods—we should be able to predict accurately the elements of the trajectory of any kind of arrow shot from a bow of known available energy and virtual mass.

All arrows for target and hunting have stabilizing vanes, usually made of turkey feathers. Since the air resistance to such vanes is quite high, flight arrows during the past several years have had very small stabilizing vanes of Pyralin. Many flight shooters follow the practice of rubbing the arrows to a high gloss and then, before the shot, giving them an extra polishing with graphite. For target shooting this is not necessary since the ranges are relatively short. Good flight shooters today are achieving distances in the neighborhood of 500 yd, and some free-style shooting in tournament competition has exceeded 600 yd.

Hunting arrows usually have fairly large broadhead points, sharpened to a keen edge for penetration. It is difficult to affix these to the shaft so that they are perfectly true and symmetrical, and consequently there may be a slight steering effect by the broadhead. This is usually counteracted by using three and sometimes four large vanes, symmetrically placed on the shaft with great care. In most cases they are attached at a slight angle, to accentuate spinning of the shaft as it flies. To be sure, this dissipates energy at the expense of range; but the hunting bow is much stronger than the target bow and there is energy to spare for the distance at which the hunter may expect to make a successful shot. This distance for a skilful archer may, on the average, be 60 yd or less.

---

[6] Cranz, *Lehrbuch der Ballistik*, vol. 1, pp. 162–166.

PENETRATION OF ARROWS

A hunting arrow with a properly sharpened head and weighing in the neighborhood of 0.07 to 0.10 lb, with a kinetic energy of 50 to 60 ft lb, has very little shocking power but amazing penetration. It will completely penetrate almost any large game animal. Since, for a given resisting force, depth of penetration varies directly as the kinetic energy of the arrow, it is better to use a heavy rather than a light shaft with a given bow. The reason is, as was shown previously, that the heavier arrow takes more of the available energy from the bow than does a light one. Equations (1) and (2) enable one to compute readily the efficiency of a particular bow of known virtual mass with arrows of various masses. From this information relative penetrations can be estimated.

AIMING

Although the subject of aiming comes last in this discussion, it is one of the first things with which the archer must become familiar if he is to become an accomplished marksman. In target shooting, aiming is a most important part of the preparation, and it must be done with precision. Because of the large effect of gravity on the trajectory of the arrow, a system of aiming must be worked out in which the line of aim uniquely establishes initial elevation and azimuth of the arrow to assure its hitting a mark at a particular distance. The arrow is usually drawn so that the index finger is below the chin, pressing upward against the jaw, and with the aiming eye in the vertical plane containing the arrow. One method of sighting is to look over the tip of the arrow at some fixed point, called a *point of aim*. For short distances this is on the ground in front of the target; for long distances it is above the target. This point corresponds to that particular elevation of the arrow which produces an accurate hit at the location of the target. The greater this distance the greater is the elevation, and therefore the greater is the distance or elevation of the point of aim relative to the archer. For a given bow and arrow there are two distances at which the line of aim, as described, intersects the path of the arrow. If the more remote of the intersections occurs at the center of the target, its distance is called the *point-blank range*. For greater distances, the point of aim is above the mark to be hit. For shorter distances the point of aim is below this mark.

Various sights have been devised. These are designed for attaching to the bow. They permit direct aiming at the point to be hit. A bow sight may be as simple as the head of a pin projecting from behind the bow above the arrow, where the pin may be held in a strip of adhesive; or it may be as complicated as a calibrated slide with fine adjustments for the sighting point. However, a mechanical sight is ordinarily not suitable for distances approaching and exceeding the point-blank range. For such cases a sight may be used in which the target is viewed through a prism of small angle, provided with a fixation mark for aiming. Without going into details of construction, it may be said that the prism sight is perhaps the most convenient of the several types of sighting devices, but that many archers still use the point of aim, and indicate it by a marker set in the ground. The latter method obviously requires that the archer stand in a fixed position relative to the marker so that successive shots will fly alike in azimuth as well as in altitude.

One source of error in shooting, causing lateral deviation, comes from tilting the bow by varying amounts from the vertical. Skilful archers have habituated themselves to checking the tilt by sighting the string on some vertical line, like the lines on a building, to keep the bow tilt zero or at a constant, finite angle. Another source of inaccuracy, causing variations in "length," is the failure to hold at full draw before releasing—a malady called "creeping." It comes at the part of the draw where small differences cause large changes in initial velocity. No refinement in sighting means can correct for such faults—and others—in the archer's method.

EXPERIMENTAL APPARATUS

Reference was made in the earlier pages to experimental methods and measurements, in connection with hysteresis data and velocity determinations. A brief description was also given of the method of timing the illumination for photographing arrows in flight.

Among the items of measuring equipment indispensable to the scientific archer are the balance, for weighing arrows, and a "spine tester," which is actually a stiffness tester. The balance should be capable of precision to 0.1 g (about 2 grains). Almost any laboratory balance of the "triple beam" or "trip scale" type is suitable. Stiffness tests are readily made on a set of shafts by the usual laboratory devices for measuring flexure of beams under load. For measuring the small deflections in these tests I have used screw micrometers, an optical lever and an improvised multiplying indicator. The shaft is placed on supports 24 to 26 in. apart, and its deflection measured when a fixed load of a pound or two is applied at the midpoint.

The ballistic pendulum for measuring arrow velocity is readily constructed and used, and it is ideally suited for the purpose. The bob is made of a wooden box with an opening about 10×10 in. and a depth of 12 in. This is filled with sheets of corrugated board, packed tightly side by side edge on, with the corrugations running in the direction of impact of the arrow. The box is supported in the usual way, with two similar bifilar suspensions, spread at the upper ends to promote stability. Somewhat greater accuracy is obtainable with a high speed spark chronograph.

In velocity testing it is desirable to shoot under reproducible conditions. An aid to this end is a shooting machine, of which various experimenters have designed various kinds. All have the common features of providing reproducibility of length of draw and a mechanical release. The clamp for the bow should be adjustable sidewise, to secure flight of the arrow in the direction of aim. As in hand shooting, the same interval should obtain, in successive shots, between completion of the draw and the instant of release.

High speed photography can still contribute materially to an understanding of some of the problems of "internal ballistics" pertaining to both bow and arrow. The modern speed-flash outfit employing Edgerton lamps immediately comes to mind as a superior aid in such studies. High speed motion pictures at 1000 frames per second should be used to supplement the "stop-motion" speed-flash pictures. This rate is easily attained in several designs of motion picture equipment.

CONCLUSION

Much more could be written on the subject of physics as applied to archery. There still remain many interesting problems. A bibliography is appended for those whose interest suggests consulting original articles. What has been written may perhaps justify an opinion held by this writer ever since he became acquainted with archery, that, because of the variety of interests which it offers, and particularly because of the number of problems in physics which it presents, archery constitutes an almost perfect hobby for a physicist. Its problems range from the simple and elementary to those that challenge ingenuity and effort to the utmost. Their solution, aside from the resulting satisfaction, provides a practical guide for correct designs. To the physicist the results are a gratifying demonstration of the efficacy of his science and its methods.

BIBLIOGRAPHY

(Abbreviations: S. A., *Ye Sylvan Archer;* A. R., *Archery Review;* A. B. R., *American Bowman-Review.*)

E. P. Clark, "The cause of 'letting down' of bows in hot, humid weather and a method of preventing it," A. B. R., Apr. 1939.

C. N. Hickman, (1) "Velocity and acceleration of arrows. Weight and efficiency of bows as affected by backing of bow," J. Frank. Inst. **208**, 521 (1929); (2) "A portable spark chronograph for use on either direct or alternating current," J. Frank. Inst. **211**, 59 (1931); (3) "Formulas for strains and stresses in a drawn bow," S. A., Nov. 1930; (4) "Effect of rigid middle section on static strains and stresses," S. A., Feb. 1931; (5) "Effect of bracing height on static strains and stresses," S. A., Mar. 1931; (6) "Effect of string weight on arrow velocity and efficiency in bows," S. A. Apr. 1931; (7) "Effect of permanent set and reflexing on static strains and stresses," S. A., May 1931; (8) "Effect of bow length on static strains and stresses," S. A., Aug. 1931; (9) "Effect of weight and air resistance of bow tips on cast," S. A., Sept. 1931; (10) "Effect of thickness and width of a bow on its form of bending," S. A., Jan. 1932; (11) "The neutral plane of bending of a bow," S. A., Feb. 1932; (12) "Fiber stresses in bows," S. A., Mar. 1932; (13) "A flexible bow handle," A. R., Oct. 1933; (14) "The dynamics of a bow and arrow," J. App. Phys. **8**, 404 (1937).

G. J. Higgins, "The aerodynamics of an arrow," J. Frank. Inst. **210**, 805 (1930).

P. E. Klopsteg, (1) "The measurement of projectile velocity" (with A. L. Loomis), J. A. I. E. E. **39**, 98 (1920); (2) "What is the proper cross section of a bow?" S. A., Dec. 1931; (3) "Constructing a bow with rectangular limb section," S. A. May 1932; (4) "What qualities in bow materials?" A. R., Sept. 1932; (5) "What of the bow?" A. R., Oct. 1932; (6) "Some new light on the paradox of archery," A. R., Dec. 1933 and Jan. 1934; (7) *Turkish archery and the composite bow* (book, 1934); (8) "Photographing the paradox," A. R., Apr. 1933; (9) "The flight of an arrow," A. R., Nov. 1932; (10) "Analyzing the bow sight," S. A., Jan. 1932; (11) "The effect on scores of errors in aiming and holding," S. A., June 1932, A. R., June 1932; (12) "A comparison of the point of aim and sight," A. R., July 1932; (13) "A study of points of aim," A. R., Jan. 1933; (14) "Elements of the prism sight," A. R., Feb. 1933; (15) "Getting the most out of the bowstave," A. R., June 1935; (16) *Science looks at archery* (monograph, 1935); (17) "Mental acrobatics with hits and scores," A. R., Jan. 1936; (18) "A graphic method of designing a bow," S. A., June 1939; (19) "The penetration of arrows," A. B. R., July 1939; (20) "A bend meter and bow weigher using the same indicator," A. B. R., Mar. 1940.

F. Nagler, (1) "Bow design," A. R., June and July 1933; (2) "More about the elliptical bow," A. R., Dec. 1935 and Jan. 1936; (3) "Spine and arrow design" (with W. J. Rheingans), A. B. R., June and July 1937.

H. E. Overacker, "Note on the strength of yew wood," S. A., Apr. 1932.

W. J. Rheingans, (1) "Debunking spine," A. R., May 1933; (2) "Exterior and interior ballistics of bows and arrows," A. R., Mar. and Apr. 1936.

D. Rogers, "The spine-weight coefficient," S. A., Apr. 1933.

W. L. Rodgers, "The bow as a missile weapon," Army Ordnance, May-June 1935.

M. C. Taylor, "Bowstrings," A. B. R., Jan. 1940.

C. Temperley, "Distribution of arrows on the target," A. B. R., Jan. 1941.

Chapter 2
THE NATIONAL PASTIME

CONTENTS

Whoever wants to know the heart and mind of America had better learn baseball.

Jacques Barzun

The above quote is just one of many attempts to describe the powerful attraction which baseball holds for millions of Americans. In view of the special role baseball plays in American life (recall that U.S. presidents often throw out the first ball of the season), it should come as no surprise that the physics of baseball has itself received a good deal of attention. The eight articles in this chapter provide eloquent evidence of that attention and offer key insights into many important aspects of the game.

The first article, "Batting the Ball," contains an excellent analysis by Paul Kirkpatrick on *the* central activity in the game of baseball, the collision (or lack thereof) between a bat and a ball. The difficulty in orchestrating a useful collision, i.e., one that results in a hit, may be inferred by noting that professional athletes who, on the average, *fail* in this endeavor a mere two out of every three tries, remain prime candidates for the Hall of Fame!

In explaining this seemingly generous definition of success, Fitzpatrick points out that the speed and orientation of the bat at the moment of contact with a ball is defined by 13 independent variables, all of which are subject to the batter's discretion. He then notes that:

> In his control of any one of these variables, the batter may err in either the positive or negative sense, so it appears that he is faced at the outset with 26 roads to failure.

In addition, since a timing uncertainty of only ± 0.01 s makes the difference between a hit over second base and a foul near the first or third base lines, timing is clearly of the essence. He concludes then that:

> The high frequency of such fouls shows that even experts do not find it simple to place a complex muscular event in time with a deviation of the order of one-fifth of the athlete's reaction time.

Virtually all aspects of the desired yet elusive collision between bat and ball are discussed in this article, from the stress/strain relation for baseballs (lively balls *do* make a difference), to the relative advantages of light versus heavy bats.

The second and third articles in this chapter deal with the physics of fly balls. "Catching a baseball," by Seville Chapman, contains a fascinating if overly idealized analysis (air resistance being ignored) on the kinematics of a fly ball as seen by the fielder who attempts to catch it. "Looking into Chapman's homer: The physics of judging a fly ball," by Peter Brancazio, remedies the shortcomings of Chapman's article by including the important effects of aerodynamic drag on moving baseballs, and by providing a compelling explanation, based on his improved results, for how an experienced outfielder can quickly and accurately predict the landing point of a fly ball after observing the initial stages of its flight.

The fourth and fifth articles in this chapter document two of the most mysterious and misunderstood pitching phenomena in baseball, the *curve ball* (i.e., the relatively steady lateral deflection of a rapidly spinning baseball) and the *knuckleball* (i.e., the somewhat erratic and unpredictable lateral deflection of a baseball having little or no spin).

The article by Lyman Briggs, "Effect of spin and speed on the lateral deflection (curve) of a baseball; and the Magnus effect for smooth spheres," reports on original wind tunnel experiments in which the lateral deflection of a spinning baseball was measured directly. Briggs, at that time the Director Emeritus of the National Bureau of Standards, had sufficient scientific stature to help dispel, through the publication of his carefully documented results, the myth that a curve ball was an *optical* illusion, a claim familiar from the folklore and popular literature of the time.[1]

In addition, Briggs showed that the magnitude of the lateral deflection of a curve ball varied as the square of the linear speed ($\approx$100 ft/s) and the first power of the rotational speed ($\approx$1800 rpm!). The maximum deflection turned out to be 17 in. which, coincidentally, is the precise width of home plate. The truly insidious nature of the major league curve ball is seen from the fact that a deflection of this magnitude is sufficient to produce not only success but intimidation, viz., a pitch whose curve is sufficient to catch the outside corner of the plate for a called strike even after being launched initially at the batter's head!

In "Aerodynamics of a knuckleball," Robert G. Watts and Eric Sawyer report on wind tunnel experiments in which two possible sources of lateral force imbalance are identified that could account for the well-documented erratic deflection of a knuckleball. (Recall that catchers of knuckleball pitchers routinely employ oversize mitts.)

The first source is seen with a slowly spinning ball and stems from the nonsymmetrical location of the roughness elements (i.e., strings) on the ball's surface. As the ball slowly spins and the orientation of the strings slowly changes, the direction of the deflection force, and hence the direction of the ball itself, also changes.

The second source of lateral force imbalance corresponds to a ball with zero spin and occurs only when the ball is oriented in such a way that a portion of the strings happen to be precisely at a location where boundary layer separation occurs. The results of Watts and Sawyer also show that the measured force on a knuckleball is sufficient to produce a lateral deflection of more than one foot!

"Models of baseball bats," by Howard Brody, is the sixth article in this chapter. It describes experiments in which the vibration frequencies of baseball bats were measured under different circumstances. The results indicate that a baseball bat acts as though it were a free body during the collision between bat and ball, regardless of whether the bat is loosely or tightly gripped by the batter. One major consequence of this fact is the vibration felt in the batter's hands when the ball strikes the end of the bat.

The seventh article in this chapter, "The effect of spin on the flight of batted baseballs," by A. F. Rex, uses the results of Lyman Briggs' article to estimate the probable trajectories of batted balls having different speeds and spin rates. The results indicate that substantial backspin applied to a long fly ball can turn a routine catch at the warning track into a home run, by adding 5 or 6 m to what would otherwise be a horizontal flight of about 100 m. The results also show that backspin is even more important for a low line drive since the direction of the Magnus force, which is always perpendicular to the ball's velocity, is more nearly vertical and produces more lift in that case.[2]

The eighth and final article in this chapter, "The effects of coefficient of restitution variations on long fly balls," by David T. Kagan, calculates the variation in horizontal range due to the allowed variations in the coefficient of restitution of major league baseballs. Interestingly, the magnitude of the maximum variation in range turns out to be 5.1 m, which is the same order of magnitude as the maximum variation in range due to backspin as calculated by Rex. Together these results reveal that, behind every home run, some excellent physics is at work.

In fact, taken as a whole, this chapter reminds us that excellent physics is at work behind virtually every aspect of baseball, a situation that promises no shortage of challenges for physics of sports buffs in the years ahead.

---

[1]See, for example, *Life Magazine*, 83 (15 September 1941).

[2]This effect is even more noticeable in golf where the initial spin lift (Magnus force) on a well-driven golf ball typically exceeds the weight of the ball, resulting in the distinctive concave-upward shape for the first two-thirds of its trajectory. Further discussion on this topic is found in Chap. 3.

Batting the Ball

PAUL KIRKPATRICK
Stanford University, Stanford, California
(Received 10 October 1962; in revised form, 17 April 1963)

The velocity vector of a ball struck by a bat is a stated function of the ball and bat velocities, bat orientation, and certain constants. In the light of the equations of the collision, the operation and the consequences of swinging the bat are analyzed, and the role of the constants is discussed.

IT would be easy to list a dozen popular games in which the central event is a collision between a ball and a bat, club, mallet, racket, paddle, or stick of some sort. Frequently, the player's intention is to transfer to the ball as much momentum as possible, and always he has some problem of directional guidance. We investigate the mechanics of these processes with such generality and realism as may be conveniently introduced. The aim of the study is not to reform batting but to understand it.

The state of the bat at the moment of contact with the ball is defined by 13 independent variables, all of which are subject to the batter's control. These quantities are the 3 positional coordinates of the mass center (or other reference point) of the bat, 3 coordinates of angular orientation, 3 of linear momentum, 3 of angular momentum, and 1 coordinate of time. In his control of any one of these variables, the batter may err in either the positive or the negative sense, so it appears that he is faced at the outset with 26 roads to failure.

It should be said immediately that several of these errors matter so little that, individually, they do not cause failure; and among the others, there is wide variance in the need for precise determination as well as in the difficulties of control. Furthermore, the number which require exact management varies among the several sports. The baseball bat is supposed to be held with the trademark uppermost, but this is one of the simplest of the batter's responsibilities since it is an easy one to fulfill and since other rotational positions of the trademark have almost no effect upon performance unless they are implicated in the breaking of the bat. In tennis, however, the corresponding variable is of much importance since a small angular displacement may make the difference between a perfect placement and a netted ball.

The timing of his stroke makes no demands upon the skill of the golfer since his ball is there to be struck at any time, but the pitched baseball crosses the plate in about 10 msec and the batter must locate his swing in time with approximately this precision. The rotational position of the bat about the vertical axis is a critical matter in all the varieties of tennis, in hockey, and in golf; in baseball, this angle does not require such strict control except in such specialties as place hitting and bunting. There are few cases in which the angular velocities of the bat at the instant of contact are significant in themselves, though they may be practically involved in the production of important linear velocities.

In baseball, the vertical coordinate of the bat at contact is both important and hard to control. Most strike-outs result from its mismanagement, and the 1962 world championship was finally determined by an otherwise perfect swing of a bat which came to the collision 1 mm too high to effect the transfer of title.

STRESSES AND STRAINS

The elasticity of the collisions to be discussed ranges from a relative high in the golf drive, to a low in the return of a handball. Figure 1 shows stress–strain relations for the baseball and golf ball as recorded by an automatic materials testing machine. The lower loop represents the response of a golf ball which was pressed between the flat steel surfaces of the testing machine until its vertical thickness, initially the ball's diameter, was reduced by 0.38 in. during a continuous closure at the constant speed of 2 in./min. At the maximum displacement shown the machine was reversed and the ball allowed to regain its original figure. The narrow loop speaks for the high elasticity of the golf ball, which returned 0.78 of the energy expended in its compression. It should be noted that the work done is represented by the area to the *left* of the curve. High elasticity does not, of course, require Hooke's-law behavior; the type of curvature shown in all four branches is characteristic of the compression of any object which, like the sphere, enlists more and more of its substance to oppose the external force as the distortion proceeds.

The baseball (upper loop) was carried through a cycle like that of the golf ball except that it was pressed between two crossed sections of a regulation bat so that the ball might be cylindrically indented and not merely flattened. This arrangement was employed so that the strains might in some degree resemble those imposed in actual batting, but it must be recognized that while these curves tell us something about the basic properties of the spheres under study they are not directly applicable to the more interesting questions about brief impacts. In the first place, the batted or driven ball is deformed through external contact on *one side only*, and secondly, the internal stresses and yields in the case of the impulsive blow are certainly not duplicated in slow motion in any quasistatic distortion cycle. If a dependable similitude of this kind obtained we should be able to calculate from Fig. 1 the times of contact and coefficients of restitution effective in high power impacts, but it was found that times of contact so computed are too low for agreement with the actual ones, while calculated coefficients of restitution are too high. It appears that the mechanical efficiency of a sudden blow is less than that of a slow one which invests a like amount of work.

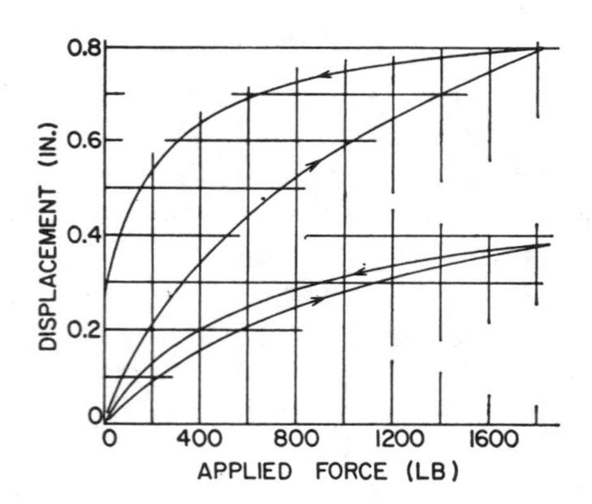

FIG. 1. Stress–strain cycles for the baseball (above) and golf ball.

EFFECT UPON THE BALL

The intention of the batter is to impart a velocity to the ball, often the maximum possible. The elongated body, here called a *bat*, is characterized by a mass M and a moment of inertia $I = Mk^2$ with respect to its mass center. The ball is a smooth sphere of mass m and of unspecified radius. The bat and ball have a mutual coefficient of restitution e which is assumed (somewhat erroneously) to be constant.

For ease of reading, this analysis is confined to motions in a horizontal plane; there is no basic difficulty about a three-dimensional derivation but the additional insight afforded would not be sufficient to justify the extra complication. We choose as x axis a horizontal line passing through the point of contact between ball and bat and disposed normally to the axis of the latter. The x axis is fixed in the observer's space and is positively directed away from the bat. By v and v', we denote, respectively, the velocities of the ball immediately before and after the brief period of contact. As is shown in Fig. 2, the directions of these velocities are given by angles θ and ϕ. We write no equations covering the work done by the muscles of the batter but designate the sig-

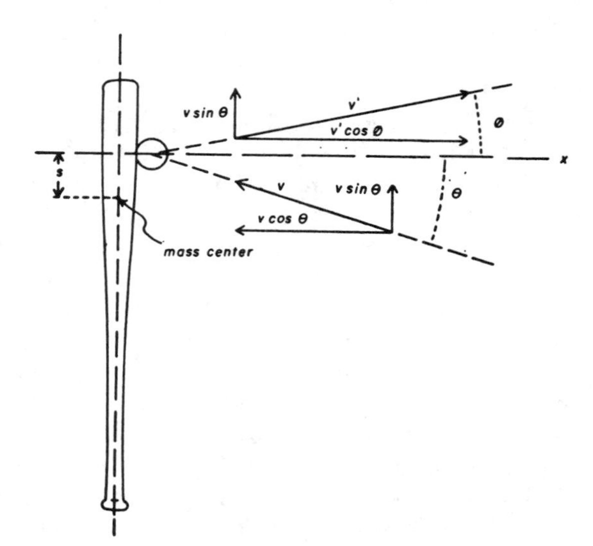

FIG. 2. Illustration of symbols defined and employed in the text.

nificant results of that work by the symbols defined below.

V=x component of the velocity of the mass center of the bat at contact.

ω=angular velocity of the bat at contact with respect to a vertical axis of rotation passing through its mass center.

s=distance of the contact point on the bat from the right section containing the mass center. This quantity is positive when the ball strikes between the mass center and the outer end of the bat.

V'=x component of the velocity of the mass center of the bat as the bat and ball separate.

During the period of contact, the bat expends momentum upon the ball while drawing an additional supply from the continued muscular exertion of the batter, who communicates to it during that period a linear impulse K in the x direction and an angular impulse Ω about the axis of ω. By equating linear impulse to the momentum gained by bat and ball, we obtain

$$K = M(V' - V) + m(v'\cos\phi - v\cos\theta). \quad (1)$$

Treating angular momentum in the same way yields

$$\Omega = ms(v'\cos\phi - v\cos\theta) + I(\omega' - \omega), \quad (2)$$

in which ω' is the angular velocity of the bat as the ball leaves it. We define two further quantities:

u=x-directed velocity of the contact point of the bat when the ball first touches it, and

u'=x-directed velocity of the same point as the ball departs.

By definition of the coefficient of restitution we have

$$e(u - v\cos\theta) = v'\cos\phi - u', \quad (3)$$

and by addition of collinear velocities

$$u = V + \omega s \quad \text{and} \quad u' = V' + \omega' s. \quad (4)$$

These five equations suffice for the determination of the four primed unknowns. In particular, we find for the departing ball

$$v'\cos\phi = \frac{(1+e)(V+\omega s) + v\cos\theta[(m/M) + (ms^2/I) - e] + (K/M) + (s\Omega/I)}{1 + (m/M) + (ms^2/I)}. \quad (5)$$

In applying this equation to a pitched and batted ball, it must be remembered that v is a negative quantity, all others usually being positive.

In many cases, certain quantities may be dropped from Eq. (5) with little effect upon v'. If the collision is of brief duration,[1] as it is with hard and elastic materials, the impulses K and Ω may be neglected. When the collision takes place near to the mass center of the bat, as it very frequently does, all terms containing s may be deleted. As simplified by these approximations Eq. (5) becomes

$$-v'/v = [a(1+e) + (e-r)\cos\theta]/(1+r)\cos\phi \quad (6)$$

in which the new symbols $r = m/M$ and $a = -V/v$ first appear. Eq. (6) contains only dimensionless ratios.

SWINGING THE BAT

Before discussing Eq. (5) we consider the process of endowing the bat with the velocities V and ω. In general, it is achieved by jointed links in a manner approximately illustrated by the carpenter's hammer. In the use of this tool the upper arm, the forearm, and the hammer handle—a system of links hinged together and to a stationary torso—cooperate to impart velocity to the hammer head. Power is fed in at three joints by applied torque at shoulder, elbow, and wrist. This analogy, however, is incomplete since the torso of the batter is far from stationary, and its advance and rotation may even provide the major part of the bat velocity V.

Available scientific analyses of effective batting form are few and qualitative. We cite here only the pamphlet *Science of Baseball*, prepared by Iwanami, Inc. of Tokyo under the supervision of Professor Ichiro Tani and Professor Kyoichi Nitta of Tokyo University. This publication, though nonmathematical, is a serious and well-illustrated study which is of little direct use to the American reader since it is entirely in the Japanese language. It confirms and renders quantitative the general image of the batsman's performance which thoughtful athletes and acute observers already entertain. The surge of output power directed to the activation of the bat, particularly in slugging, starts at ground level and

[1] *Science of Baseball*, cited below, states upon photographic evidence that the duration of contact between the bat and a well-hit ball is about 5 msec.

rises in fluent coordination. First comes the step, carrying the whole body toward the approaching sphere. While the step is in progress, the rotation of the hips begins but, naturally, this process cannot proceed forcibly until the advancing spikes find their grip in the soil. At about that instant, the powerful muscles which control waist rotation take up the duty of bringing the bat (and the whole upper torso) around. Thus far the cocked arms and bat have been merely riders, but now their contribution to angular velocity begins, though hip and torso are still in rotation. The forward arm (left, in the case of a right-handed batter) does the greater share of the work in this phase, pulling the bat around while its outer end swings outward, not so much by propulsion from the wrists as by what is often called centrifugal force. Last, the angle between forearms and bat, already obtuse, is further increased both by wrist flexure and by a push–pull action of the two forearms.

This relay of time-overlapping forces and torques does some 300 *J* of work upon the batter's weapon in a fifth of a second, bringing the trademark across the plate with a speed of some 100 ft/sec. The bat achieves its rendezvous, if all goes well, at a point a few inches in front of the plate where $v = 100$ ft/sec. The ball is crushed to half its regulation diameter, recovers, and takes up its return flight at $v' = 140$ ft/sec possibly.

The performance described above is of course subject to variation. Some batters assume the swinging stance initially and dispense with the step; some space the hands,[2] and some even bat cross-handed. The relative contributions to ultimate bat energy from the component torques naturally vary with physical build and muscular equipment. What we know is derived from laboratory batting: wholehearted swings at fat pitches. So far as known there are no really informative motion pictures of either pitching or batting under combat conditions.

Nice timing of the swing is required for the production of fair balls; from the Iwanami photographs one deduces that a timing uncertainty of ± 0.01 sec makes the difference between a hit over second base and a foul near the first or third base foul lines. The high frequency of such fouls shows that even experts do not find it simple to place a complex muscular event in time with a deviation of the order of $\frac{1}{5}$ of the athlete's reaction time.

It may be and has been argued that the kinetic energy of the bat (or other club) at contact equals the total external work performed at the several bodily joints and should not be affected by the time distribution of power at any of these sites. Why then should the muscular forces be called out sequentially in a particular program rather than activated simultaneously, which would appear to be the woodchopper's time-tried method? Indeed, one may argue that the continuous swing, with all bodily hinges functioning throughout the stroke, *must* produce the greater external kinetic energy since it wastes less energy internally. This seems to follow from the known fact that the maximum external work performed in the contraction of a muscle varies in some inverse manner with the velocity of contraction. High velocities of contraction are, therefore, to be avoided; but this cannot be done by holding the muscle in reserve until the available time for its exercise is largely past.

The refutation of this line of thought is found partly in the fact that the useful work to be done is all against inertia. The work, for instance, put out by the wrists, is proportioned to the resisting force of inertial reaction, and this is called out only by high acceleration, but high acceleration of this kind continued throughout the full duration of the swing would give the bat too much angular displacement, causing it, in the case of a right-handed batter, to point around toward first base. Imparting high velocity to the mass center is only one of the desiderata: this velocity must be attained with the implement in a certain place with a certain orientation. So each of the component torques should be turned on at an instant preceding the instant of contact by just time enough to permit the associated degree of freedom to be exercised throughout its maximum possible extent under maximum muscular exertion.

A more compelling consideration affecting optimum batting form is the fact that the baseball batter must select his procedures under

[2] This form trades off flexible-wrist articulation for more positive bat control. It is a minority choice, particularly in the present era of home-run emphasis, but at least two Hall-of-Fame batters, Wagner and Cobb, elected it.

severe time pressure. The pitched ball, in either the hard or soft form of the game, spends about $\frac{2}{3}$ of a second in its flight to the plate. In this moment the batter must make his observations, complete his forecast of the manner of the ball's arrival, decide upon his consequent plan of action, and get full instructions to his muscles with enough time left to allow them to give the bat suitable values for its thirteen coordinates. Incidentally, these things must be done in hostile surroundings, at appreciable personal risk, under an intense feeling of individual responsibility, and often subject to high-level acoustic annoyances. By comparison, the problems of the tournament golfer at the tee or on the green under conditions of leisure and acoustic sanctity appear simple indeed.[3]

Naturally the batter wishes to postpone his commitment as long as possible so as to have the benefit of the latest and most determinative observations. Every baseball player is so thoroughly familiar with the standard parabolic trajectory that a good look at a few feet of such a flight enables him to extrapolate it to earth, but pitched balls do not always follow such a path and each one is a new ballistic problem to the batter. So he puts off his decision until there is barely enough time left to carry it out. For him the best method of swinging a bat is along the path of minimum time. Being ignorant of the relevant neuromuscular constants of any batter we may not calculate such a path, but we readily see the possible advantage of deferring the wrist break, since an early completion of the work at this joint would increase the moment of inertia of the compound linkage rotating about the shoulder joints or body axis. Batters with strong hands and wrists—and all distinguished sluggers seem to be so equipped—are certainly able to complete the work at the wrist joints in less than the time required to bring the arms about. The baseball batter has another good reason for holding the wrist joints in the cocked position as long as possible: it gives him more time to look over the pitch. If the decision is difficult he may and often does commence the arm swing while continuing his deliberations. If, at the last moment, he decides against a whole-hearted swing no harm has been done him, for the umpire does not regard his effort as a strike unless there has been a break at the wrists, no matter how much bodily rotation and threat of action has been manifested.[4]

COLLISION QUESTIONS

We return to Eqs. (5) and (6) and an inspection of the conditions affecting the velocity of the batted ball. In general, the faster the pitch the faster is the ball's departure, but this is not invariably the case. Evidently the statement should be reversed in the case of a massive and relatively inelastic ball or a ball struck with a light bat at a point far from the mass center. The best technique for hitting an indoor or playground ball which has been softened up by use is different from that which is most effective on a hard ball having a high value of e. In the latter case the duration of contact is short and the impulses K and Ω are, therefore, small. When collision has been initiated there is little more for the batter to do. But with a soft and inelastic ball (low e) the duration of contact may be great enough to make continued application of force to the bat produce important increments of momentum. The softball batter should in no sense throw his bat at the ball but should follow through forcibly.

At what point along the bat should the ball be caused to strike? The answer depends upon whether the batter is interested in giving high velocity to the ball or in protecting his hands from hinge reaction, as many mechanics books seem to assume. Hinge reaction is the lateral force which may be transmitted to the hands from the impact of the ball. There is nothing unpleasant about it, as one may easily demonstrate with the relatively heavy softball. Hinge reaction does not sting the hands. The sting which is sometimes experienced results from the impact of the ball upon an antinode of the bat's transverse vibration. Such impacts should be avoided because they are unpleasant, because they sometimes break bats, and because they divert useful linear kinetic energy into vibrational motion. Of course, Eqs. (5) and (6) do not apply to such collisions.

[3] Opinions differ. One athlete, uniquely qualified by experience in both tournament golf and World Series baseball, found the pressure more severe in the former activity.

[4] This assurance does not hold in the case of an unsuccessful bunt attempt.

The plane motion of the bat just after the collision is given by V' and ω'. We set up a coordinate y to express distances of points along the bat axis from the mass center, choosing the positive direction of y as toward the handle. Then the linear velocity in the x direction for any point on the axis with coordinate y is $V-\omega y$ before the collision, and $V'-\omega' y$ immediately thereafter. By equating these expressions we obtain $y=(V-V')/(\omega-\omega')$, which locates the point where the motion is undisturbed by the collision. By inserting the expressions for V' and ω' which one obtains from Eqs. (1) to (4) we find the undisturbed point to be at $y=k^2/s$. Of course s may be chosen, speaking theoretically, so as to place the undisturbed point at the batter's hands if desired.

It is likely that the batter will prefer to meet the ball at a point yielding a high value of v', but there is little opportunity here for advantageous choice. The terms in Eq. (5) containing s^2 produce an effect which is smoothly symmetrical about the mass center and cannot therefore be of significance for small (positive or negative) values of s. Of the linear s terms, the one involving Ω is inappreciable in hardball situations; the term $s\omega$ does indicate the desirability of meeting the ball at a point outside the mass center, since angular velocity is always present. It would be of interest to observe the values of s utilized by eminent long-ball hitters. Under the dubious assumption that the batter is able to impart to the bat a certain constant amount of kinetic energy (translational plus rotational) it may be shown from Eq. (5) that the velocity of the departing ball has a true maximum when $s=I\omega/MV$.

What are the effects of the mass of the bat, and how should the best mass be selected? If we assume the collision to be essentially instantaneous and hence neglect the impulses K and Ω, then Eq. (6) shows the velocity of the batted ball to be

$$v'=\{V(1+e)+|v|(e-r)\cos\theta\}/(1+r)\cos\phi.$$

Evidently this velocity may be increased by decreasing $r(=m/M)$, that is to say, by using a massive bat, but since r is additively associated with terms of the order of unity the returns diminish after the bat's mass has been increased to a few times that of the ball. Furthermore, as M increases, the velocity V which can be imparted to the bat must decrease, and with it the velocity of the departing ball. At least one may make the qualitative statement that the mass of the bat should be large in comparison to that of the ball yet small in relation to the batter's arms.[5] Such a compromise is possible and is uniformly employed, except that the child batters in Little Leagues often struggle with implements which, for them, are unreasonably massive.

A rough theory of optimum bat mass may be constructed on the assumption that that bat is best which requires the least energy input to impart a given velocity to the ball. This defines and applies a kind of efficiency criterion to bat mass. Rearrangement of Eq. (6) gives the precollision bat velocity as

$$V=[(1+r)v'\cos\phi+(e-r)v\cos\theta]/(1+e). \quad (7)$$

The kinetic energy of the bat, neglecting now the relatively small angular kinetic energy, is $W=MV^2/2$ or, by virtue of Eq. (7),

$$W=[M/2(1+e)^2][(1+r)v'\cos\phi+(e-r)v\cos\theta]^2.$$

It is found that this expression for W has a minimum when

$$r=(v'\cos\phi+ev\cos\theta)/(v'\cos\phi-v\cos\theta)$$
$$\cong(v'+ev)/(v'-v). \quad (8)$$

If the batted ball is to depart with just the reverse of its approach velocity we have $r=(1-e)/2$. This rather typical case is selected for examination because of its arithmetical simplicity. The value of e for baseballs has been measured at the National Bureau of Standards[6]

[5] There is on record (*This Week* magazine, 20 May 1962) a faintly scientific experiment intended to show something about the effect of bat mass upon ball velocity. In this test the distinguished batsman, Roger Maris, batted for distance with 5 different new bats whose weights varied from 33 to 47 oz. These bats, incidentally, were stated by the manufacturer to be reproductions of the favored implements of some of history's greatest home-run hitters. The pitching was by a veteran batting-practice pitcher of the New York Yankees, who attempted to deliver hitable pitches of uniform quality. Maris batted out five long fly balls with each bat and the twenty-five ranges were measured. The correlation coefficient of range with bat mass is found from the published data to be 0.41, indicating a substantial positive relationship. Maris' own favored bat was the lightest of the set, and with it the smallest of the five mean ranges was achieved. This batter, strong enough to wield the heavy, long-range bats, nevertheless prefers a light and maneuverable implement for dealing with unfriendly pitching.

[6] National Bureau of Standards Research Paper RP16 21

where official league balls of the year 1943 were found to have the mean value $e=0.41$. For the case considered this gives $r=0.29$ or $M=3.4\ m$.

It is known that approximations in this theory have had the effect of producing a low value of M, but even the complete elimination of the approximations would not get the theoretical bat mass up to the observed values, which are around $M=7\ m$ in baseball, and $M=5\ m$ in softball. Apparently the principle of least work is not fully binding upon athletes who are much more interested in effectiveness than in efficiency.

The design of bats is presently more traditional than rational, a fact not generally acknowledged or even realized. We read as follows[7]: "Today's bats are scientifically designed and standardized in various sizes, every one shaped and balanced accurately to suit the size, strength, and hitting style of the player."

The credulous present writer followed up this announcement eagerly, asking for references to the researches that had made all this possible and for formulas relating bat constants to measurable human properties. The entire claim was immediately cut down to a statement that a considerable variety of bat sizes and masses is available to the public choice. Bats are not yet sold by prescription.

In the quest for the optimum bat, two opposed desiderata must be compromised. The mass should be zero for maximum wieldiness, yet large (relatively to the ball mass) in order to predominate in the momentum exchange. What one might do is to place the mass where it will do the most good, and the familiar shape of the bat shows some progress in this direction, though not nearly so much as is evident in the golf driver. The mass should be where the collision is to occur. The so-called "bottle bat" once affected by sluggers was a step in just the wrong direction. Of course, batting with a golf driver or other implement with highly concentrated mass would impose upon the already overburdened batter the necessity for precise control of the y coordinate, yet a better concentration than one sees now could surely be achieved by hollow construction of the bat handle and/or use of materials other than the traditional woods.

[7] "Baseball Instructor's Guide," The Athletic Institute, Chicago, Illinois.

Before expending effort upon the improvement of bats the physicist may well ask himself the question, "Why?" Perhaps it is not obvious that the game of baseball would be improved by a scientific disturbance of the existing balance of power between pitcher and batter. There is reason, however, to think that it would. The pitchers' battle is a spectacle more admired than enjoyed. Both spectators and athletes might favor greater general participation, feeling that there is something wrong with a game in which a player may go through an entire contest without once participating in the defensive effort. On the commercial side it is well known that home runs sell tickets.

At least one manufacturer markets bats labeled "Flexible Whip Action." Flexible whip action has aided golfers and pole vaulters and it might help batters. Whip action requires that the acceleration imposed upon the bat by the user's grip on the handle be so great that the massive outer end fails to keep up and bends the implement into a curve, convex on the leading side. The hope is that the energy thus stored is returned at about the instant of collision, but no bats yet introduced are flexible enough to afford much storage. A whip-action bat with which the writer experimented gave indication that its outer end might lag about 2 mm during a high-torque swing.

BATTING SPECIALTIES

The fungo bat, a tool for knocking baseballs about in practice sessions, is a lighter stick than any which would be used on the same ball in competition. Its low mass ($M=4.5\ m$, approximately) is rational in view of the fact that the fungo bat is applied to a ball which has practically zero velocity, a situation in which Eq. (8) prescribes the even lower value $M=m$. The development of a light bat for this specialty was certainly not guided by theory but was perhaps influenced by the fact that a part of the manipulation must be performed with one hand while the user tosses up the ball with the other. Also, this bat is never used in the production of homeruns, where mass has its advantages, but in an activity requiring good control of the ball's velocity vector.

The complementary case, in which the ball has initial velocity but the bat little or none, is bunting. When the bat is presented at rest to receive the ball at the mass center Eq. (6) takes the form $-v'/v=(e-r)\cos\theta/(1+r)\cos\phi$. Inserting the known constants of bats and balls, and assuming $\theta=\phi$, we find that the hard ball loses about 80% of its speed in the bunting collision and the soft ball more than 90%. These estimated loss percentages are on the high side since in the usual execution of a bunt the collision is not completely ballistic. The bat is held in a firm manner which causes it to transfer momentum to the batter and endow the ball with a corresponding additional positive momentum increment. However, this effect is small because of the soft coupling between bat and batter.

The essence of success in bunting is nice placement, which requires better control of v' than is ever thought of in straight-away hitting. If the process described above should seem likely to send the ball too strongly forward the rebound may be reduced in any amount by making the bat velocity negative, a maneuver commonly observed with the baseball but seldom with the softball. For the drop-dead type of bunt we have from Eq. (6), $V=(e-r)v\cos\theta/(1+e)$. Since v is essentially negative, the bat velocity V must also be negative so long as $e>r$, a condition always satisfied except with the lightest of bats and deadest of balls.

DIRECTED BATTING

Contrary to published how-to-bat instructions which might be cited, the ordinary law of reflection does not correctly predict the direction to be taken by the batted ball. This law stands in need of corrections for the effects of spin, friction, imperfect elasticity, yielding of the bat under impact, and bat velocity in the observer's frame of reference. With respect to the last correction, angles of incidence and reflection are equal only in a coordinate system moving with the velocity possessed by the contact spot of the bat at the instant of collision. When these angles are equal in such a moving coordinate system—the bat velocity being positive—the angle of incidence exceeds the angle of reflection if these angles are measured in a frame of reference fixed to the earth. It is for this reason that the distinction between θ and ϕ was retained in the foregoing formulas. Bat and ball velocities are such that this effect alone might cause θ to be twice as large as ϕ if other effects than the motion of the axes did not play compensating roles.

Actually, both the imperfect elasticity of the ball and the transfer of momentum from ball to bat tend to increase the angle of reflection and restore an appearance of validity to the simple reflection law. With all effects operating it is impossible to cover the situation with any simple statement. In bunting, the angle of incidence is the lesser of the two since the reflector is at rest in the observer's space, but in the case of a long drive down a foul line the angle of incidence may be the greater by 25%.

To support these assertions it is economical to write Eq. (6) in the form

$$-v'/v=(D+\cos\theta)/E\cos\phi, \tag{9}$$

in which the meaning of D and E is evident from a glance at Eq. (6). Now ϕ is expressed by $\tan\phi=-v\sin\theta/v'\cos\phi$, as is clear from Fig. 1. For a ball batted in any given direction the sum of θ and ϕ is, of course, known; in the case of a drive down a base line, for eaxmple, we have $\theta+\phi=\pi/4$. The three relations just given suffice for the determination of v'/v and the two angles, on the basis of given or assumed values of r, a, and e. The speed imparted to the ball by a given blow is a function of the angles involved in the collision. By the methods just stated, it may be shown that a drive down a foul line may have over 5% greater velocity than would have been the case with the swing of the bat timed to send the ball over second base. This much added velocity means about 10% more range.

It is unlikely that any batter can be either improved or misled by equations of mechanics except insofar as such equations may lead to revision of the conditions within which he is free to strive. The achievements of the athlete are confined within limits which he has no power to modify. These boundary conditions are established by rules committees, makers of equipment, and the instructors who prescribe athletic form and technique. Scientific descriptions of what the athlete is doing are likely to have for him an interest which is merely academic, in the common, contemptuous sense of that abused word.

Catching a Baseball

SEVILLE CHAPMAN
Cornell Aeronautical Laboratory, Inc., Buffalo, New York 14221
(Received 2 June 1968)

A baseball fielder will arrive at the right place at the right time to catch a fly ball if he runs at the only constant velocity for which the rate of change of tangent of the elevation angle of the ball and the bearing angle of the ball both remain constant. Remarks are made concerning curve balls. Baseball is not a scientific game—although, unknown to the players, many principles of physics can be applied to it.

In his thought-provoking and delightful book, *Science is Not Enough*, Dr. Vannevar Bush[1] devotes an essay to baseball, "When Bat Meets Ball." Baseball, as Bush notes, is not a scientific game.

On p. 216 Bush says that no one knows how an outfielder can take a brief look at a fly ball, run to exactly the right spot, and then look up just in time to catch it...a feat that many good players perform.

It does not seem entirely mysterious. The fielder has played this game many times and has a good memory of fly-ball experiences. He also has two eyes and ears. He knows what a long hit sounds like. His sensory elements are well coupled to his built-in computer, developed by the good Lord over the millenia, which is rather better for the purpose than our present-day machines. The problem of predicting the motion of a target when its laws of motion are known is a standard one for astronomers, ballistic-missile defense engineers, and so on. The question is, what information must the fielder sense in order to know where the ball is going?

We can consider first the more difficult case for the player, that of a ball coming directly at the fielder looking into the sunset. All motion is in one vertical plane. Is it a short pop up to an infielder or a home run over the fence? If the ball lands directly in the fielder's glove without his taking a step, what characterizes the situation?

Let the ball leave the bat (the origin) with an initial speed V at an angle Θ with the ground. As is well known, neglecting air resistance[2] the

[1] V. Bush, *Science is Not Enough* (Wm. Morrow Co., New York, 1967). The book is a series of essays. I recommend it to all scientists, especially now when it is evident that technological problems have social and human ramifications not apparent in engineering equations. It is ably reviewed by R. E. Gibson, Science **159**, 1225–1226 (1968). As for baseball itself, on balance I find the chapter very good, but there are some errors in it ... 100 ft per sec is a more typical speed for a pitched ball than the value given which exceeds the world's record, a typical bat these days weighs in the mid-30-oz range rather than almost twice as much as given. The coefficient of restitution of baseballs is around 0.4–0.45, not 1.0 which Bush grants is an oversimplification, his bunt has the speed of a sizzling grounder, the "reasonable Texas Leaguer" would go over the outfielder's head, ... and several other points. That Bush would publish these statements is interesting in itself, showing that even he is not a computer. In any case, his commentary is stimulating. Of course, his figures were intended to be illustrative rather than representative. All players and the spectators know the difference between a sizzling grounder and a bunt, even if none of them can translate it into feet per second. Why should they?

[2] Aerodynamic forces on a baseball are relatively small and have only a small percentage effect on a trajectory. A small percentage may or may not be important. A normal curve ball deviates from a ballistic trajectory perhaps 1 ft in 60. The CAL team, champion of the Cheektowaga Industrial League, has built a pitching machine used for batting practice that can produce five times this deviation in which case the aerodynamic force is nearly 1 g. Photographs do not show evidence of a "break" (an abrupt change in radius of curvature of the trajectory supposedly only a few feet from the batter). We know of no hard evidence, theoretical or experimental, determining whether the break is a consequence of tension, surprise, perspective, the batter watching the pitcher's followthrough instead of the ball, or whether it physically occurs. Neither we nor other ball players doubt its reality, be this reality physical or psychological. Ball players describe other players as throwing a "light" or "heavy" ball, the effect on catching it being as though the ball were of less than or greater than normal weight. Since the trajectory of a ball with a rise (horizontal spin axis perpendicular to trajectory—ball spinning backwards as from an overhead throw), or a drop (horizontal spin axis perpendicular to trajectory—ball spinning forwards as from a low sidearm throw, or in either case from some added wrist spin applied consciously or unconsciously), differs inappreciably from that of a ballistic trajectory of a faster or slower ball respectively (thrown in normal manner with spin axis nearly parallel to the trajectory—actually downward about 30° and to the right about 30° for a right-handed thrower), the psychological impression is that of a light or heavy ball. This effect is particularly noticeable in the exaggerated drop of the pitching machine.

vertical and horizontal displacements at any time t are:

$$y = (V \sin\Theta)t - 0.5gt^2, \quad (1)$$

$$x = (V \cos\Theta)t, \quad (2)$$

where g is the magnitude of the acceleration of gravity. The time of flight T to return to the initial altitude $y=0$ is

$$T = (2V \sin\Theta)/g, \quad (3)$$

so that the range R is

$$R = (2V^2 \sin\Theta \cos\Theta)/g. \quad (4)$$

Now consider the elevation angle ϕ of the ball as seen by a fielder standing exactly where the ball comes down at a distance R from the origin. It is apparent that

$$\tan\phi = y/(R-x). \quad (5)$$

Modest algebraic manipulation of the preceding equations yields

$$\tan\phi = gt/2V \cos\Theta = (\text{constant})t. \quad (6)$$

Thus, for a fielder in the right place, the *tangent of the elevation angle increases uniformly with time* until the ball is caught. This simple principle applies at any finite range, for any initial speed V, and any initial angle Θ with the ground, for a line drive or for a high fly. At distances other than R the expression is more complicated. (For an external elevator going vertically up a building at a constant rate, $\tan\phi$ also increases uniformly with time.)

In the case of a pop fly to the infield having the *same* initial vertical component of velocity and hence the same initial rate of change of elevation angle as seen by the outfielder, $\tan\phi$ increases at a *decreasing* rate. In the case of the home run, $\tan\phi$ increases at an *increasing* rate.

Some of the relationships are shown in Fig. 1.

How much of the trajectory in the vertical plane needs to be examined in order to characterize the rate of change of elevation angle? Recalling that as a batter the player has perhaps 0.6 sec in which to judge *and hit* a pitched ball, one would expect that as a fielder he would have a good idea of where the batted ball was going in 0.3 sec. Such a response would appear to the spectators to be quite rapid. As the ball gets close no doubt stereoscopic vision aids in the catch.

Now suppose the fielder is not at the correct range for the catch, but rather a distance s from it, and that after a brief time interval τ he runs at a

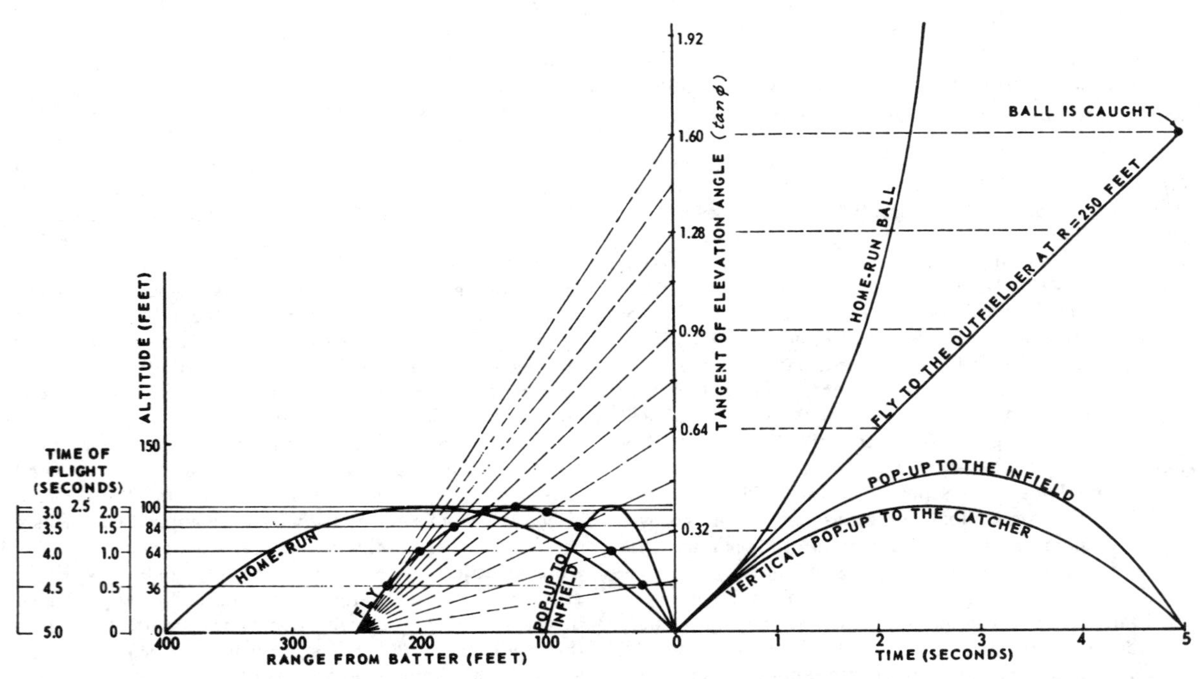

Fig. 1. Trajectories of a home run, a fly ball, and a pop up to the infield (left), and (right) corresponding plots of tan ϕ versus time. As seen from the impact point, tan ϕ = constant × time.

constant speed v in the proper direction (forward or back) so that

$$s=v(T-\tau), \quad (7)$$

which means he reaches the proper spot to catch the ball just as it arrives. What then?

In place of Eq. (5) we now have for the new elevation angle ψ, for the case where the fielder is too close,

$$\tan\psi=y/[R-x-s+v(t-\tau)]. \quad (8)$$

Upon substituting from the preceding equations, and after half a dozen algebraic manipulations we obtain[3]

$$\tan\psi=gt/2[(V\cos\Theta)-v]=(\text{Constant})t. \quad (9)$$

Thus if the player runs at the proper speed so as to maintain a constant rate of increase of $\tan\psi$ he will be in the right place at the right time to catch the ball when it comes down.

The typical outfielder catching a typical fly that stays in the air 5 sec certainly can run a distance of 100 ft in this time. If, however, he is required to move only 50 ft, for instance, he does not run the 50 ft and then stand still to catch the ball. Ordinarily he moves at a constant, in this case, slower rate. Presumably he is adjusting his speed to maintain constancy of the rate of change of $\tan\psi$. Most players prefer to catch the ball at eye level rather than at stomach level, which would seem to offer more latitude for correction in the event of misjudging the ball. It is noted that the condition on $\tan\psi$ for eye level is simpler than for the offset catch at stomach height.

Human factors research[4] by psychologists in perceptual-motor studies of tracking shows that it is easier for an operator to track a target moving at a constant rate than to track one moving with an acceleration. The former task is considered to require a perceptual processing task (i.e., a thinking task) of solving a problem involving a single differential, whereas the latter involves a problem with a double differential. Proficiency comes with practice. Other studies show that an experienced operator can still perform his tracking task well with only an intermittent view of the target's motion. Thus, the normal behavior of a ball player seems to be in conformity with these psychological principles.

Finally, consider the more typical case of a ball hit somewhat to the side (which is much easier to catch). In addition to information about elevation angle and its rate of change, there is now information about azimuth angle and its rate of change. There are now more than enough data (four) to determine the three quantities: time of flight and the two components of displacement information as to where the ball will come down. If the player runs so as to maintain the bearing of the ball constant (i.e., it is always kept, say, due west of him—a type of proportional navigation) and the rate of change of $\tan\psi$ constant, he is in the right place to catch the ball at the right time.

Obviously no ball player ever solves trigonometric equations to catch a ball. What I have tried to show here is that an astonishingly simple amount of information on the constancy of the rate of change of $\tan\psi$ and on the bearing of the ball tell him that he is running at the right speed in the right direction for the catch. Baseball is still a great game, even if the physics[5] in it is unknown to the players.

[3] Some people may regard Eq. (9) as following immediately from Eq. (5) and the principle of Galilean relativity.

[4] The comments in this paragraph are from my colleague Dr. Albert Zavala; R. Chernikoff and F. V. Taylor, "Effects of Course Frequency and Aided Time Constant on Pursuit and Compensatory Tracking," J. Exptl. Psychol. **53**, 285–292 (1957); E. A. Fleishman, "Individual Differences and Motor Learning," in *Learning and Individual Differences*, R. M. Gagné, Ed. (Charles E. Merrill Books, Columbus, Ohio, 1967), Chap. 8, pp. 165–192; J. W. Sanders and A. B. Kristofferson, *The Attentional Demand of Automobile Driving* (Bolt Beranek & Newman Inc., 15 March, 1967), Rept. 1482.

[5] P. Kirkpatrick, "Batting the Ball," Am. J. Phys. **31**, 606–613 (1963); L. J. Briggs, "Effect of Spin and Speed on the Lateral Deflection (Curve) of a Baseball; and the Magnus Effect for Smooth Spheres," Am. J. Phys. **27**, 589–596 (1959); L. J. Briggs, "Methods for Measuring the Coefficient of Restitution and the Spin of a Ball," J. Res. Natl. Bur. Std. Res. Paper 1624 **34**, pp. 1–23 (Jan.–June 1945); J. D. Patterson, "Home Run Hitting," Phys. Teach. **5**, 157 (1967); G. R. Lindsay, "Strategies in Baseball," Operations Res. **11**, 477–501 (1963) with several references; E. Cook, *Percentage Baseball* (Massachusetts Institute of Technology Press, Cambridge, Mass., 1966); D. A. D'Esopo and B. Lefkowitz, "The Distribution of Runs...," Stanford Research Institute, Menlo Park, Calif., presented at Am. Stat. Assoc., 24 August 1960; W. G. Briggs, "Baseball-O-Mation a Simulation Study," presented before the Operations Res. Soc. Am. (1960), revised to appear as a chapter in *Computer Simulation of Human Behavior*, J. M. Dutton and W. H. Starbuck, Eds. (John Wiley & Sons, Inc., New York, 1969); R. E. Trueman, "A Monte Carlo Approach to the Analysis of Baseball Strategy," Operations Res. **7**, Supp. 2, B-98 (1959).

Looking into Chapman's homer: The physics of judging a fly ball

Peter J. Brancazio
Department of Physics, Brooklyn College, City University of New York, Brooklyn, New York 11210

(Received 18 June 1984; accepted for publication 13 September 1984)

How does a baseball player learn to judge a fly ball? An experienced outfielder, observing the initial stages of flight, can predict the landing point rapidly and accurately. In 1968, S. Chapman [Am. J. Phys. **36**, 868 (1968)] proposed that an outfielder unconsciously uses trigonometry to determine the landing point. However, Chapman assumed incorrectly that the effects of aerodynamic drag could be ignored. The trajectory of a baseball is shown to be affected significantly by air resistance, so that the specific trigonometric factor cited by Chapman cannot provide useful cues to the fielder. In evaluating potentially useful cues, we take note not only of aerodynamic drag, but also of the specific neurophysiological processes used for the detection of distance and motion. This study shows that the angular acceleration of the fielder's line of sight to the ball provides the strongest initial cue to the location of the eventual landing point. This suggests that the fielder's vestibular system, responding to the acceleration of the fielder's head as he observes the initial stages of flight, may play a key role in the judgment process.

I. INTRODUCTION

Baseball is a game that calls for the development of some highly specialized and difficult-to-master skills. While pitching and hitting naturally attract the most notice, there are other features of the game that are worthy of attention, particularly from a physicist's point of view. One especially intriguing aspect is the ability to judge a fly ball.

To appreciate this skill, one simply has to observe how an experienced outfielder moves when a fly ball is hit toward him. He reacts almost immediately when the ball is hit, moving smoothly and confidently to the point where the ball will land. If the fielder must run a fairly long distance to make the catch, he typically runs at just the right speed to arrive at the landing point simultaneously with the ball. The best outfielders can even turn their backs to the ball, run to the landing point, and then turn and wait for the ball to arrive. In effect, a fielder observes the motion of the fly ball during the earliest part of its flight—while it is still rising up and away from the batter—and uses this information to compute the range and time of flight of the complete trajectory.

One of the unique aspects of this skill is that there does not seem to be any way to teach someone how to do it—a fact that the author discovered a few years ago when he began coaching Little League baseball. There simply are no coaching techniques or helpful hints that can be used to aid in the learning process. The skill of judging a fly ball, it appears, can only be self-taught on a nonverbal level.

II. CHAPMAN'S THEORY OF TRIGONOMETRIC OUTFIELDING

How does an outfielder know where a fly ball is going to land? What information, taken from his initial observation of the early stages of the trajectory, does the fielder use subconsciously to determine the landing point? In 1968, S. Chapman proposed that an outfielder unknowingly uses trigonometry to direct his movements.[1] Chapman considered the changes that take place in the angle of elevation of the ball (the angle Φ between the horizontal and the fielder's line of sight to the ball). Chapman analyzed specifically the straight-on fly ball—one that is hit directly at the fielder so that he must run either forward or backward (but not sideways) to make the catch. This type of fly ball is usually the most difficult to judge, as the fielder cannot see the arc of the trajectory; the ball appears to him to be rising and falling in a vertical plane. Chapman found that the tangent of the angle of elevation of the ball increases at a steady rate; that is, $d(\tan \Phi)/dt$ is constant if the fielder is standing at the landing point, *or* if the fielder is running at a constant speed toward the landing point such that he will arrive there simultaneously with the ball (see Fig. 1). On the other hand, if the ball is going to drop in front of the fielder, then the angle of elevation appears to increase at a *decreasing* rate, i.e., $d(\tan \Phi)dt$ grows smaller. If the ball is going

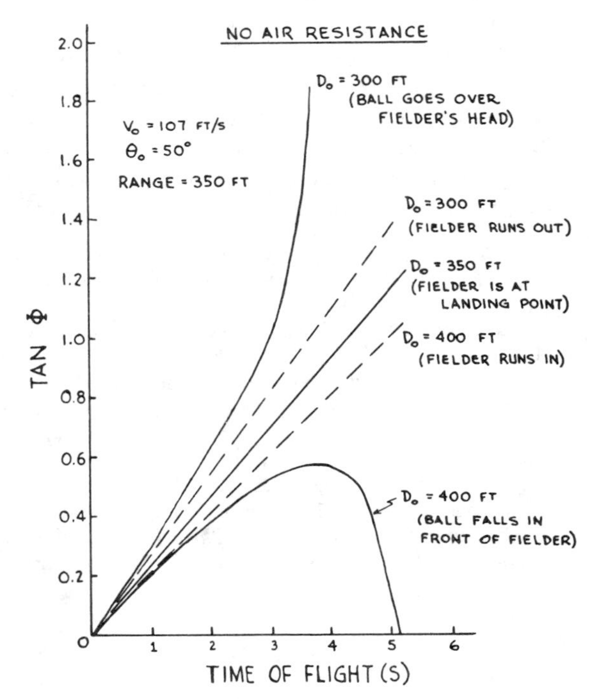

Fig. 1. Variations in tan Φ as seen by fielder for a typical fly ball in the absence of air resistance. D_0 is the initial distance from launching point to fielder. Solid lines: fielder remains stationary; dashed lines: fielder runs at uniform speed to make the catch.

over the fielder's head, then the angle of elevation appears to increase at an *increasing* rate [$d(\tan \Phi)/dt$ increases]. Chapman, therefore, concluded that a fielder subconsciously judges a fly ball by moving (or not moving) so as to maintain a constant rate of increase of the tangent of the angle of elevation.

III. EFFECTS OF AERODYNAMIC DRAG

Unfortunately, there is one serious flaw in Chapman's argument, namely his claim that "aerodynamic forces are relatively small and have only a small percentage effect on a trajectory."[1] As shown in Fig. 2, a baseball in flight normally experiences two main forces: its weight, acting downward vertically; and an aerodynamic drag force acting antiparallel to the velocity vector of the ball. A baseball weighs about $5\frac{1}{4}$ oz and has a diameter of about 2.9 in.[2] It is not unusual for a well-hit baseball to leave the bat at a speed of 80 miles/h or more, and at such speeds the Reynolds number for a baseball in air at 60° F is approximately 2×10^5, a value indicating that the ball is experiencing turbulent flow. Under these conditions, the aerodynamic drag force on a sphere is given by the equation

$$F_D = \tfrac{1}{2}\rho A C_D v^2, \tag{1}$$

where ρ is the air density, A is the cross-sectional area of the sphere, C_D is the drag coefficient, and v is the speed. Frohlich[3] has examined appropriate values for the drag coefficient of a baseball in flight. He has proposed that a baseball may undergo a "drag crisis" (a change from a fully turbulent to a laminar boundary layer) as the ball slows down in flight. However, there is no direct evidence that such a change takes place.[4] Accordingly, we have chosen to use a constant drag coefficient for all calculations. A terminal speed of 140 ft/s for a baseball has been reported[5]; by using the condition that at terminal speed the drag force is equal to the weight of the projectile, a value of $C_D = 0.3$ was computed. It turns out that the drag force on a baseball moving at 80 miles/h is about 2/3 the weight of the ball. Clearly, the effects of air resistance on the flight of a baseball ought to be quite significant.

The equations of motion in two dimensions for a projectile experiencing aerodynamic drag can be written as follows:

$$m\ddot{x} = -F_D \cos\Theta = -F_D(v_x/v),$$
$$m\ddot{y} = -F_D \sin\Theta - mg = -F_D(v_y/v) - mg, \tag{2}$$

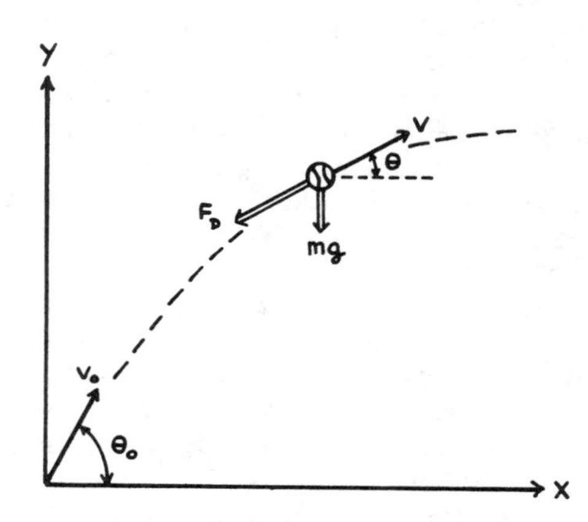

Fig. 2. Forces on a baseball in flight.

where v_x and v_y are the horizontal and vertical components of the instantaneous ball velocity v. To simplify matters, the effects of lateral wind forces and aerodynamic lift have been ignored (in general, the spin rate of a well-hit fly ball is not high enough to generate substantial lift through the Magnus effect). Assuming the drag force given by Eq. (1), Eqs. (2) become

$$\ddot{x} = -Kvv_x, \quad \ddot{y} = -Kvv_y - g, \tag{3}$$

where $K = \rho A C_D/2m = 0.0016\ \text{ft}^{-1}$.

These equations cannot be solved analytically, so trajectories must be obtained by approximation and numerical integration. One method[6] is to integrate the equations directly over small time increments, using the third derivative to improve the approximation, as follows:

$$\Delta x = v_x\Delta t + \tfrac{1}{2}\ddot{x}(\Delta t)^2 + \tfrac{1}{6}\dddot{x}(\Delta t)^3,$$
$$\Delta y = v_y\Delta t + \tfrac{1}{2}\ddot{y}(\Delta t)^2 + \tfrac{1}{6}\dddot{y}(\Delta t)^3, \tag{4}$$

where

$$\dddot{x} = -K(\dot{v}v_x + v\ddot{x}), \quad \dddot{y} = -K(\dot{v}v_y + v\ddot{y}),$$
$$\dot{v} = (v_x\ddot{x} + v_y\ddot{y})/v.$$

New values of v_x $(=\dot{x})$ and v_y $(=\dot{y})$ are computed by using

$$\Delta v_x = \ddot{x}\Delta t + \tfrac{1}{2}\dddot{x}(\Delta t)^2, \quad \Delta v_y = \ddot{y}\Delta t + \tfrac{1}{2}\dddot{y}(\Delta t)^2.$$

An alternate method[7,8] reduces the solution to the numerical evaluation of definite integrals:

$$x = -\int_{\Theta_0}^{\Theta} \frac{v^2}{g}\, d\Theta, \quad y = -\int_{\Theta_0}^{\Theta} \frac{v^2}{g} \tan\Theta\, d\Theta,$$
$$t = -\int_{\Theta_0}^{\Theta} \frac{v\, d\Theta}{g\cos\Theta},$$

where

$$v^2 = \sec^2\Theta \{(1/v_{x0}^2) - (K/g)[F(\Theta) - F(\Theta_0)]\}^{-1},$$
$$F(\Theta) = \tan\Theta\sec\Theta + \ln(\tan\Theta + \sec\Theta). \tag{5}$$

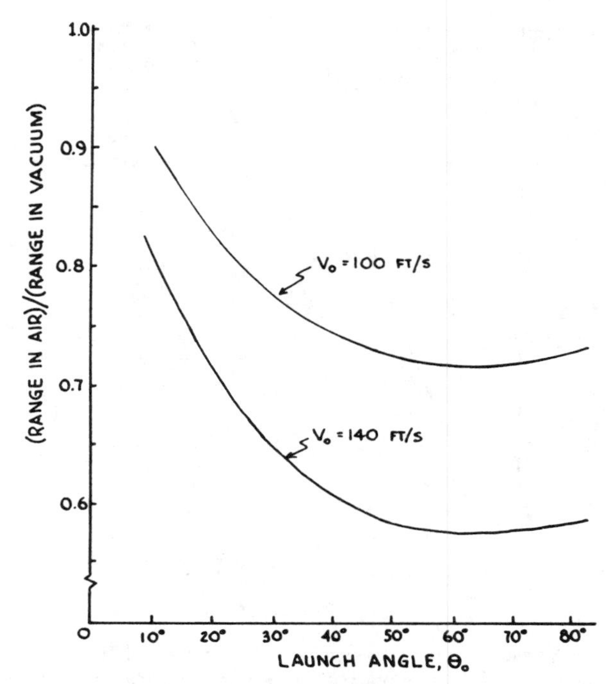

Fig. 3. Effect of air resistance on the horizontal range of a baseball trajectory.

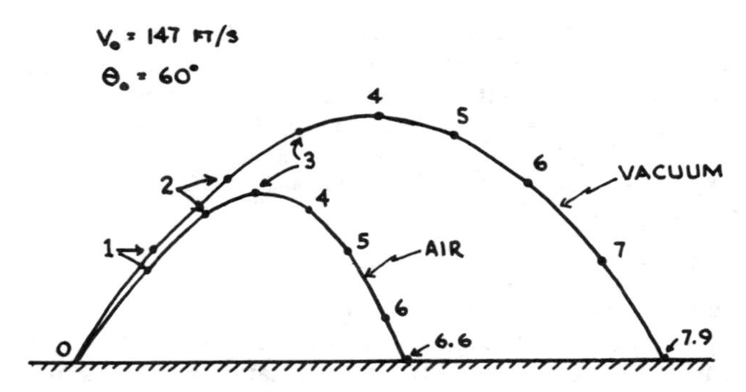

Fig. 4. Effect of air resistance on a baseball trajectory. The numbers indicate the elapsed time in seconds. Range: air, 323 ft; vacuum, 581 ft. Maximum height: air, 174 ft; vacuum, 252 ft.

Computer programs written in BASIC were devised to generate trajectories by each of the two methods outlined above, and solutions were obtained rather readily using a TRS-80 model III microcomputer. The numerical results produced by the two methods agreed to within 0.5% when 1°-angle increments (for the integral method) and 0.1-s time increments (for the stepwise integration method) were used.

A. Results

The calculations show, as predicted, that air resistance significantly affects the flight of a baseball. The extent to which the horizontal range of a fly ball is reduced by aerodynamic drag is shown in Fig. 3, where the ratio (range in air)/(range in vacuum) is shown for various combinations

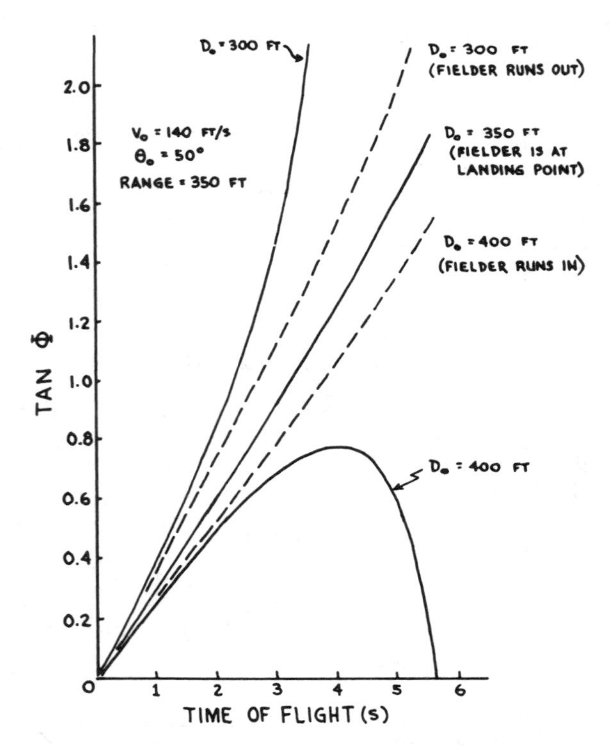

Fig. 5. Variations in tan Φ as seen by fielder for a typical fly ball affected by air resistance. D_0 is the initial distance from launching point to fielder. Solid lines: fielder remains stationary; dashed lines: fielder runs at uniform speed to make catch.

of launching speed and launching angle. Figure 4 demonstrates the extent to which the range, height, and time of flight are reduced for a typical set of launching conditions.

Aerodynamic drag causes the trajectory to lose its parabolic shape. The descent of the ball from the peak of its trajectory is steeper and slower than the rise. Accordingly, the tangent of the angle of elevation does not vary in the way described by Chapman (who derived his results based on parabolic trajectories). Figure 5 shows the actual changes in tan Φ for a typical fly ball in air (compare with Fig. 1). It is evident that a fielder who has judged a fly ball correctly does *not* see tan Φ increasing at a uniform rate; instead, tan Φ tends to increase at an increasing rate as the ball descends. The rate of change of the tangent of the angle of elevation is therefore not a useful cue for judging a fly ball.

IV. THE PERCEPTION OF MOTION IN DEPTH

It seems evident that the question of how a fielder judges a fly ball cannot be answered without some consideration of the process of visual perception. How does the human visual system track the position of an object moving rapidly? How does the brain make use of this visual information to program and coordinate the muscular responses that bring the body to meet a moving object at a common point in space and time?

At the present time, psychologists and physiologists do not fully understand these processes. It seems certain, however, that the visual system does not react as a camera or video recorder does, i.e., by storing sequences of static visual images. Rather, the system receives information through several different channels and creates a mental image of the scene. For example, we know that the visual system is able to take two slightly different views, one from each eye, and create a single, integrated three-dimensional picture. The phenomenon of depth perception derives from binocular vision in two ways: through parallax (each eye sees the object against a different point of the background) and through vergence (the eye muscles must align the eyes in slightly different directions so that their lines of sight converge on the object). Yet, binocular vision is not a necessary condition for depth perception. A one-eyed outfielder can learn to judge a fly ball! An individual with poor or no vision in one eye can still sense depth by way of a different visual cue, namely, the size of the image of the object on the retina, which governs the decrease in apparent size with distance.

Neurophysiological and psychophysical studies show that the detection of *motion* by the visual system proceeds via different mechanisms and neural pathways than those used to detect distance alone.[9] The rapid changes in light flux on the cells of the retina as an object passes across the field of view provide the data used by the brain to create the awareness of motion. The rate of change in the size of the image on each retina serves specifically as an indicator of motion in depth.

There is a source of *nonvisual* sensory information for the detection of motion, and it turns out to be surprisingly powerful. The source is the vestibular system, located in the inner ear, which provides the information used to maintain balance, tell up from down, and detect both linear and angular acceleration. One of its functions is to coordinate eye and head movements, enabling an observer to keep his gaze fixed on some object even though his head is in motion. If, for example, you fix your eyes on some stationary object and turn your head rapidly from side to side, the object will remain sharply in focus. To maintain fixed lines of sight, your eyeballs must rotate at the same rate as your head is turning, but in the opposite direction. Yet if your head is held motionless and the object is moved rapidly from side to side, it appears as a blur; your eyeballs cannot follow its motion rapidly enough to keep it in focus.

This apparent violation of the principle of relative motion occurs because the feedback signal that triggers the eye muscles into action is generated by different pathways according to the source of motion.[10] With the head in motion (and the object at rest), the movement of the eyeballs is directed by signals from the vestibular system, generated by the motions of the head. With the object in motion and the head stationary, there is no signal from the inner ear; the motion of the eyes is then triggered by purely visual stimulation. It turns out that vestibular signals are processed relatively rapidly by the lower brain (the cerebellum), whereas visual signals are processed, more slowly, by the cerebral cortex. The fact that vestibular feedback is faster than visual feedback explains why the object can be kept in sharp focus when the head is moving, but not when the head is stationary.

V. THE MATHEMATICS OF JUDGING A FLY BALL

The preceding survey of the perceptual cues that are used to detect distance and motion serves to sharpen our focus in considering how a fielder judges a fly ball. For example, there does not seem to be any reason why a fielder ought to react to the rate of change of the tangent of the angle of elevation, and thus there is no perceptual basis for Chapman's hypothesis. However, several other geometric or trigonometric features of the trajectory of a fly ball now seem worthy of consideration. These include the following: (1) the player–ball distance D, which is inversely proportional to the apparent size of the ball as seen by the fielder; (2) its rate of change D; (3) the component of the instantaneous ball velocity perpendicular to the fielder's line of sight v_p (which represents the speed with which the ball appears to move across the sky); (4) its rate of change $\dot{v}_p$; (5) the angular velocity ω; and (6) angular acceleration α of the fielder's line of sight to the ball.

Mathematical expressions for each of these quantities can be written without difficulty. As shown in Fig. 6, x and y are the horizontal and vertical coordinates of the ball measured from the launching point, u is the instantaneous horizontal distance from the fielder to the ball, R is the horizontal range of the fly ball, and x_0 is the distance from the fielder's initial position to the eventual landing point (x_0 is considered negative if the ball is destined to go over the fielder's head). It is assumed that the fielder, after an initial reaction time delay T_D, runs at a constant speed v_F to arrive at the landing point simultaneously with the ball (v_F is positive as the fielder runs inward). Thus if T is the time of flight of the ball from launch to landing,

$$v_F = x_0/(T - T_D). \tag{6}$$

The horizontal distance at time t after the ball is launched is

$$u = \begin{cases} R + x_0 - x, & t \leqslant T_D, \\ R + x_0 - x - v_F(t - T_D), & t > T_D. \end{cases} \tag{7}$$

The time derivatives of u and y are:

$$\dot{u} = \begin{cases} -\dot{x} = -v_x, & t \leqslant T_D, \\ -(\dot{x} + v_F) = -(v_x + v_F), & t > T_D, \end{cases}$$

$$\dot{y} = v_y, \quad \ddot{u} = -\ddot{x} = Kvv_x, \quad \ddot{y} = -Kvv_y - g, \tag{8}$$

where v_x and v_y are the horizontal and vertical components of the instantaneous ball velocity v. The second derivatives, $\ddot{x}$ and $\ddot{y}$, were previously defined by Eq. (3).

The ball velocity relative to the fielder may be resolved into components v_p and v_R perpendicular to and along the fielder's line of sight to the ball:

$$v_P = (v_x + v_F)\sin\Phi + v_y \cos\Phi,$$
$$v_R = (v_x + v_F)\cos\Phi - v_y \sin\Phi. \tag{9}$$

The relations $\sin\Phi = y/D$, $\cos\Phi = u/D$, and $D = (u^2 + y^2)^{1/2}$ lead to

$$v_P = [uv_y + (v_x + v_F)y]/D = (u\dot{y} - y\dot{u})/D,$$
$$v_R = [u(v_x + v_F) - yv_y]/D = -(u\dot{u} + y\dot{y})/D. \tag{10}$$

As expected, $v_R = -\dot{D}$. Moreover, since $\Phi = \tan^{-1}(y/u)$, differentiation with respect to time yields

$$\omega = \dot{\Phi} = (u\dot{y} - y\dot{u})/D^2 = v_P/D, \tag{11}$$
$$\alpha = \ddot{\Phi} = (u\ddot{y} - y\ddot{u} - 2D\dot{D}\dot{\Phi})/D^2. \tag{12}$$

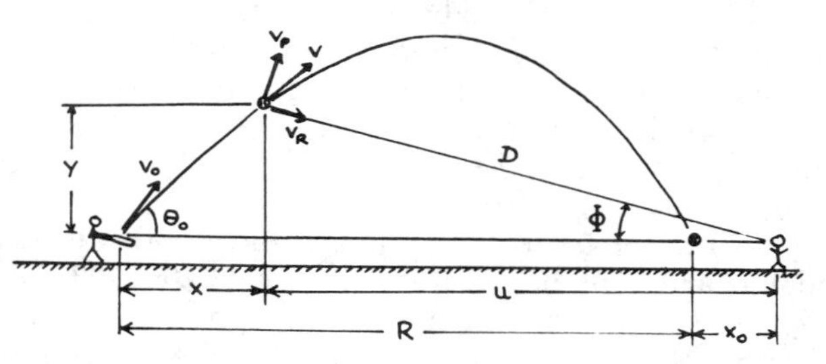

Fig. 6. Coordinates and symbols used to describe trajectory parameters.

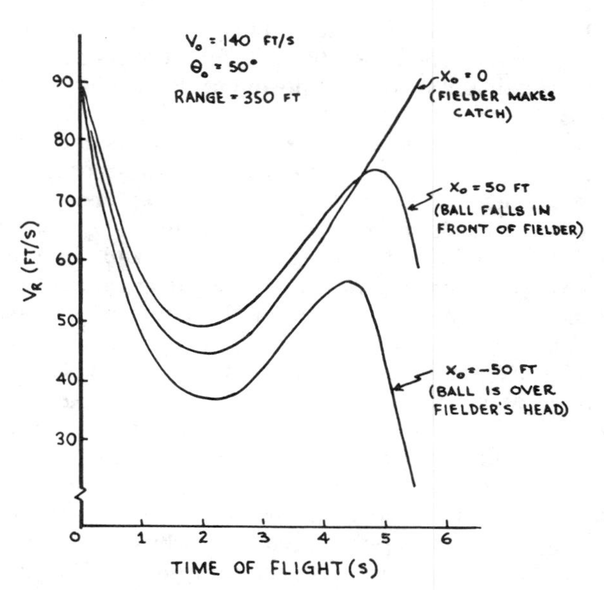

Fig. 7. Variations in v_R (the velocity component along the line of sight) as seen by the fielder at different distances (x_0) from the landing point. Note that $-v_R = \dot{D}$, the rate of change of the ball–fielder distance.

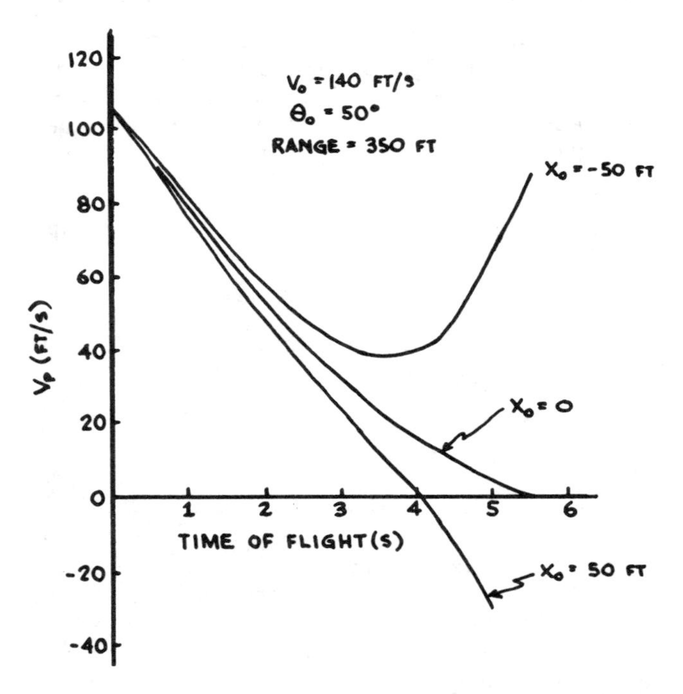

Fig. 8. Variations in v_P (the velocity component perpendicular to the line of sight) as seen by the fielder at different distances (x_0) from the landing point. Note that v_P goes to zero when the fielder is positioned properly to make the catch.

Finally, since $v_P = \omega D$ from (11),

$$\dot{v}_P = \alpha D + \omega \dot{D}. \tag{13}$$

Using the computer-generated values of x and y and their derivatives, any of the above quantities may be calculated and plotted for various initial launching conditions (v_0 and Θ_0) and initial fielder distances from the landing point (x_0). Do any of these quantities show characteristic variations that could be used subconsciously by a fielder, *particularly in the early stages of the flight*, to judge the location of the landing point?

Figure 7 shows the variations in v_R ($= -\dot{D}$), the rate of change of the ball–fielder distance, for a typical fly ball as they would appear to a nonmoving fielder ($v_F = 0$) who is standing either at the landing point, 50 ft in front of this point, or 50 ft beyond it. There are no distinctive differences between the three curves until several seconds have elapsed—long after the fielder must make his decision whether to run in or out to make the catch. This is strong evidence that the rate of change of distance, and along with it the rate of change of the apparent size (angular diameter) of the ball, play little or no role in the judgment process.

The corresponding curves for v_P (Fig. 8) are also quite similar to one another in the first few seconds and begin to diverge dramatically only toward the end of the flight. Note, however, that v_P goes to zero for a fly ball that is being caught. This same result occurs when the fielder runs at constant speed (v_F) toward the landing point to make the catch. We are led to conclude that while the perpendicular velocity component—the speed with which the ball appears to move across the sky—offers no *initial* information to the fielder, it later provides him with a strategy for making the catch, namely, *the fielder moves in such a way as to reduce the perpendicular velocity to zero.*

A. Initial values of trajectory parameters

The experienced outfielder normally knows within the first second of flight where to go to make the catch. Thus it remains for us to identify the key perceptual cues presented at the beginning of the ball's trajectory.

We may approach this question analytically by determining the initial values of the relevant quantities. At $t = 0$, we have the following:

$$u_0 = R + x_0, \quad \dot{u}_0 = -v_{x0} = -v_0 \cos\Theta_0,$$
$$\ddot{u}_0 = -Kv_0^2 \cos\Theta_0, \quad y_0 = 0,$$
$$\dot{y}_0 = v_{y0} = v_0 \sin\Theta_0, \quad \ddot{y}_0 = -(Kv_0^2 \sin\Theta_0 + g).$$

Thus by substitution into Eqs. (10)–(13),

$$v_{P0} = v_0 \sin\Theta_0, \tag{14}$$
$$\dot{D}_0 = -v_0 \cos\Theta_0, \tag{15}$$
$$\omega_0 = (v_0 \sin\Theta_0)/(R + x_0), \tag{16}$$
$$\alpha_0 = [v_0/(R + x_0)]^2 [\sin 2\Theta_0 - (\mathrm{K} \sin\Theta_0 + g/v_0^2)(R + x_0)], \tag{17}$$
$$\dot{v}_{P0} = [v_0^2/(R + x_0)] [\tfrac{1}{2} \sin 2\Theta_0 - (K \sin\Theta_0 + g/v_0^2)(R + x_0)]. \tag{18}$$

Equations (14) and (15) show that v_{P0} and $\dot{D}_0$ are independent of the fielder's distance, and thus are not informative. However, the other three quantities do depend on the fielder's position, and so they may potentially be used by the fielder to determine the direction in which he should start running. When the values of ω_0, α_0, and $\dot{v}_{P0}$ as viewed at various distances from the landing point are compared with their respective values viewed *at* the landing point (where $x_0 = 0$), a rather striking result is revealed: *The differences are considerably larger for α_0 than for ω_0 or $\dot{v}_{P0}$*. By way of demonstration, Table I contains some numerical comparisons for several typical fly balls, as seen by an outfielder who is 300 ft from the batter. For each of the sample fly balls, the values ω_0, α_0, and $\dot{v}_{P0}$ have been computed (1) as measured at a point 300 ft from the batter and (2) as measured at the eventual landing point. In Table I, the ratios of these two numbers has been tabulated. In each case, we see that this ratio is largest, by a considerable factor, for α_0. On this basis, we may conclude that *the angular acceleration provides the most valuable initial cue as to the location of the landing point and the direction in which the fielder must move.*

B. The angular acceleration of the line of sight

A survey of the properties of the initial angular acceleration α_0, computed for an extensive range of fly balls and fielder positions, leads to the following general conclusions.

(1) If a fielder is located beyond the landing point of a fly ball, so that he must run *in* to make the catch ($x_0 > 0$), the angular acceleration of his line of sight to the ball will exhibit a relatively large, *negative* value. If the fielder must run *outward* to make the catch ($x_0 < 0$), then he will observe a large *positive* α_0.

(2) When the fielder is standing at the eventual landing point of a fly ball, the initial angular acceleration of his line of sight is observed to be quite small, typically 0.5°/s² or less, except for the steepest launching angles. If the initial launch angle Θ_0 is less than the angle for maximum hori-

Table I. Initial values of trajectory parameters.

Characteristics of sample fly balls ($C_D = 0.3$; $K = 0.0016\ \text{ft}^{-1}$).

Trajectory number	Launch speed v_0(ft/s)	Launch angle Θ_0	Horizontal range R (ft)	Maximum height H (ft)	Time of flight T (s)	Comments
1	165	55°	401	185	6.76	home run
2	140	50°	351	129	5.66	catchable fly ball
3	120	60°	250	128	5.64	catchable fly ball
4	100	30°	209	34	2.90	line drive base hit
5	80	60°	137	65	4.02	infield pop fly

Parameter values seen by fielder 300 ft from batter and at landing point. ω_0 = initial angular velocity of line of sight, α_0 = initial angular acceleration of line of sight, and $\dot{v}_{PO}$ = initial rate of change of velocity perpendicular to line of sight.

	ω_0(°/s)		α_0(°/s²)		$\dot{v}_{PO}$ (ft/s²)	
Trajectory number	Fielder at landing pt.	Fielder at 300 ft	Fielder at landing pt.	Fielder at 300 ft	Fielder at landing pt.	Fielder at 300 ft
1	19.3	25.8	− 0.59	3.32	− 36.0	− 25.2
2	17.5	20.5	− 0.20	1.55	− 28.7	− 24.1
3	23.8	19.8	− 0.51	− 2.02	− 27.2	− 31.4
4	13.7	9.5	0.34	− 2.16	− 19.5	− 25.8
5	28.9	13.2	− 0.28	− 4.31	− 20.9	− 31.8

Parameter ratios: R = (value at 300 ft)/(value at landing point).

Trajectory number	$R(\omega_0)$	$R(\alpha_0)$	$R(\dot{v}_{PO})$
1	1.34	− 5.62	0.70
2	1.17	− 7.73	0.84
3	0.83	3.95	1.15
4	0.70	− 6.32	1.32
5	0.46	15.58	1.53

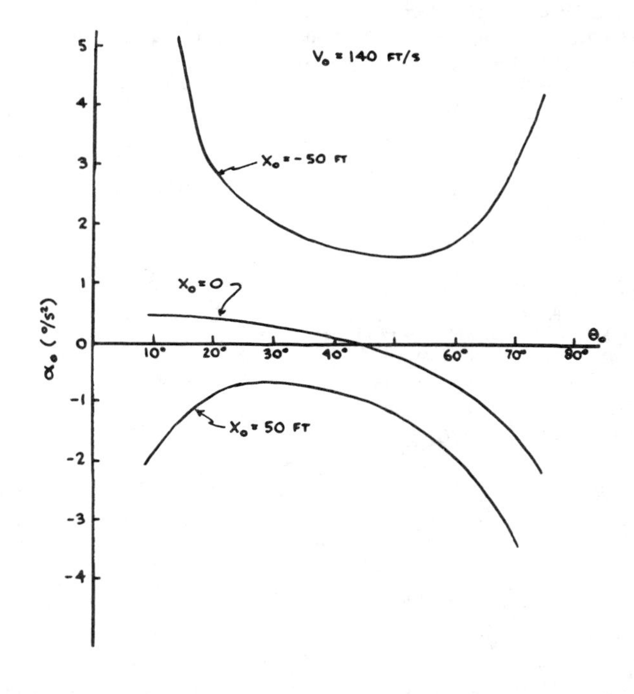

Fig. 9. Dependence of initial angular acceleration α_0 on launching angle (Θ_0) and fielder distance (x_0).

zontal range Θ_{0m} for the given launch speed, then α_0 will have a small positive value. If Θ_0 is greater than Θ_{0m}, α_0 will have a small negative value. (In a vacuum, $\Theta_{0m} = 45°$, but aerodynamic drag on a baseball typically reduces Θ_{0m} by 2° to 5°.)

These results are depicted graphically in Fig. 9, which shows the dependence of α_0 on the launching angle and fielder distance for a given launching speed. These results were, as noted, obtained from computer simulations of the effects of aerodynamic drag. It is of some interest to note that in a vacuum, where the horizontal range $R = v_0^2 \sin 2\Theta_0/g$, Eq. (17) reduces to

$$\alpha_0 = -gx_0/(R + x_0)^2.$$

Thus for an outfielder located at the landing point ($x_0 = 0$) of *any* fly ball, α_0 is identically zero in the absence of air resistance. Figure 9 shows the extent to which aerodynamic drag causes α_0 to deviate from zero.

VI. DOES AN OUTFIELDER JUDGE FLY BALLS BY EAR?

The mathematical analysis of a fly ball reveals two visual cues that may be used by a fielder to judge the landing point. The angular acceleration of the fielder's line of sight to the ball provides the strongest cue at the very beginning of the flight. A large *negative* angular acceleration informs the fielder that he must run in to make the catch, whereas a large *positive* acceleration sends the fielder outward.. A near-zero acceleration tells the fielder that he is close to or at the landing point. In the later stages of the flight, the fielder uses the perpendicular component of the ball velocity v_P as a guide; he runs at a speed that brings v_P to zero at the instant that the catch is made.

In support of this hypothesis, we can show that the detection of both α_0 and v_P are related to standard perceptual processes. As noted earlier, v_P is the speed with which the ball appears to move across the outfielder's field of vision, a motion which is detected directly by the visual system. Thus an outfielder can easily establish a feedback loop by which his own motion tends to reduce the ball's motion across his visual field. But the detection of angular acceleration proceeds via an entirely different pathway. If the fielder fixes his gaze on the ball as it moves, then the angular acceleration of the ball in flight must be matched by an equal angular acceleration of his head. As noted earlier, this motion is detected by the vestibular system of the inner ear.

It seems reasonable, therefore, to propose the following hypothesis: *The rapid reaction of an experienced outfielder to the initial motion of a fly ball is triggered by vestibular feedback*. When a fly ball is hit, the fielder's first response is to tilt his head backward, that is, to give it an angular acceleration, to follow the rising flight of the ball. The response of the vestibular system to the rate of acceleration triggers the fielder's body into motion. Thus through a process of trial-and-error learning, the nervous system of the fielder becomes programmed to respond to the vestibular signals generated by his first head movements. These signals then become coordinated with the muscular movements needed to direct the fielder's body to the eventual landing point.

In suggesting that a fielder judges a fly ball "by ear," we are, of course, referring to the ear as the site of the vestibular system and not to its auditory functions. Baseball fans sometimes notice how an outfielder seems to "take off at the crack of the bat." However, the sound of the bat–ball collision is not always audible (for example, a softball does not produce a loud sound when hit), and even if it is audible, there is about a $\frac{1}{3}$-s travel time for the sound to reach the fielder. Moreover, a fielder can judge a *thrown* fly ball just as easily as a batted ball. It should be evident that auditory signals play no part in the judgment process.

Whether or not the vestibular system, acting in response to the fielder's initial head motions, aids in the process of judging a fly ball remains to be documented by experiment. In any case, the result would have no practical value, since it is hard to imagine how it could be applied to the teaching of the skill or the improvement of technique. Baseball players must continue to teach themselves how to judge a fly ball through the time-honored method of trial and error, involving many hours, if not years, of practice. If anything, this inquiry has served to deepen my respect for this unique and underappreciated skill, and for the remarkable complex of processes that take place in the brain of an outfielder whenever he catches a "routine" fly ball.

ACKNOWLEDGMENTS

I wish to thank Professors Matthew Kleinman, Donald Landolphi, and Theodore Raphan of Brooklyn College for some highly instructive conversations. David Brancazio (Bronx High School of Science) and Lawrence Brancazio (Hunter College High School) graciously allotted me generous amounts of computer time and provided valuable programming assistance.

[1]S. Chapman, Am. J. Phys. **36**, 868 (1968).

[2]Baseball fans, scientists included, as a rule are much more familiar with baseball-related data measured in US customary units rather than in metric. Accordingly, we have used US customary units throughout.

[3]C. Frohlich, Am. J. Phys. **52**, 325 (1984).

[4]W. F. Allman, Science 83 **4**, 92 (1983).

[5]L. J. Briggs, Am. J. Phys. **27**, 589 (1959).

[6]C. Frohlich, Am. J. Phys. **49**, 1125 (1981).

[7]G. W. Parker, Am. J. Phys. **45**, 606 (1977).

[8]J. L. Synge and B. A. Griffith, *Principles of Mechanics, 2nd Edition* (McGraw-Hill, New York, 1949), pp. 154–157.

[9]D. Regan, K. Beverley, and M. Cynader, Sci. Am. **241** (1), 136 (1979).

[10]E. Bizzi, Sci. Am. **231** (4), 100 (1974).

Effect of Spin and Speed on the Lateral Deflection (Curve) of a Baseball; and the Magnus Effect for Smooth Spheres

LYMAN J. BRIGGS
National Bureau of Standards, Washington, D. C.
(Received March 26, 1959)

The effect of spin and speed on the lateral deflection (curve) of a baseball has been measured by dropping the ball while spinning about a vertical axis through the horizontal wind stream of a 6-ft tunnel. For speeds up to 150 ft/sec and spins up to 1800 rpm, the lateral deflection was found to be proportional to the spin and to the square of the wind speed. When applied to a pitched ball in play, the *maximum* expected curvature ranges from 10 to 17 in., depending on the spin. The deflections of rough baseballs accord in direction with that predicted by the Magnus effect. But with *smooth* balls at low speeds the deflection is in the *opposite* direction. This is studied with an apparatus specially designed to measure the pressure at any point in the equatorial plane of the rotating ball.

INTRODUCTION

EVERYONE who has played golf or baseball or tennis knows that when a ball is thrown or struck so as to make it spin, it usually "curves" or moves laterally out of its initial vertical plane.

How is this lateral deflection related to the spin and speed of the ball? An experimental answer to this question was sought for baseballs.

The first explanation of the lateral deflection of a spinning ball is credited by Lord Rayleigh[1] to Magnus,[2] from whom the phenomenon derives its name, the "Magnus effect."

The commonly accepted explanation is that a spinning object creates a sort of whirlpool of rotating air about itself. On the side where the motion of the whirlpool is in the same direction as that of the wind stream to which the object is exposed, the velocity will be enhanced. On the opposite side, where the motions are opposed, the velocity will be decreased. According to Bernoulli's principle, the pressure is lower on the side where the velocity is greater, and consequently there is an unbalanced force at right angles to the wind. This is the Magnus force.

In the case of a cylinder or a sphere, the so-called whirlpool, or more accurately the circulation, does not consist of air set into rotation by friction with the spinning object. Actually an object such as a cylinder or a sphere can impart a spinning motion to only a very small amount of air, namely to that in a thin layer next to the surface. It turns out, however, that the motion imparted to this layer affects the manner in which the flow separates from the surface in the rear, and this in turn affects the general flow field about the body and consequently the pressure in accordance with the Bernoulli relationship. The Magnus effect arises when the flow follows farther around the curved surface on the side traveling with the wind than on the side traveling against the wind. This phenomenon is influenced by the conditions in the thin layer next to the body, known as the boundary layer, and there may arise certain anomalies in the force if the spin of the body introduces anomalies in the layer, such as making the flow turbulent on one side and not on the other. As we shall see, a reverse Magnus effect may occur for smooth spheres. Rough balls, such as baseballs and tennis balls, do not show this anomalous effect.

The ingenious experiments which led Magnus to the discovery of the effect were made chiefly with a small cylinder rotating about a vertical axis in a horizontal wind and so mounted that it was free to move laterally across the wind, but not downstream. The pull of a cord wrapped around the axis served to give the cylinder its initial spin. Magnus makes no comment about the smoothness of the surface of the cylinder. The boundary-layer concept, which was introduced by Prandtl in 1904, was of course not available to Magnus.

---

[1] Lord Rayleigh, "*On the irregular flight of a tennis ball*," Scientific Papers 1, 344 (1869–81).

[2] G. Magnus, "*On the deviation of projectiles; and on a remarkable phenomenon of rotating bodies*." Memoirs of the Royal Academy, Berlin (1852). English translation in Scientific Memoirs, London (1853), p. 210. Edited by John Tyndall and William Francis.

Reprinted from *American Journal of Physics* **27**, 589–596 (1959);

The beautiful photograph made by Professor F. N. M. Brown of the University of Notre Dame in his low-turbulence wind tunnel illustrates what has just been said (see Fig. 1). Here the wind with its smoke filaments is coming from the right at 60 feet per second. The ball is stationary in the tunnel but spinning counterclockwise at 1000 revolutions per minute about a horizontal axis at right angles to the wind. The crowding together of the smoke filaments over the *top* of the ball shows an increased velocity in this region and a corresponding decrease in pressure, which according to the Bernoulli principle, would tend to deflect the ball *upward* across the wind stream.

It will also be noted that the *wake* of the ball has been deflected *downward*. According to the principle of the conservation of momentum, this must likewise be accompanied by a corresponding *upward* thrust on the ball.

Put in other words, if the wind speed is from east to west and the ball is spinning counterclockwise about a *vertical* axis, then the Magnus force on the ball is directed towards the north.

For a further discussion of the flow past rotating cylinders, including many photographs, see Prandtl[3] and Goldstein.[4]

PART I. EXPERIMENTS WITH BASEBALLS

Air-Gun Experiments

My first measurements were made with an air-gun which had earlier been constructed at the National Bureau of Standards to measure the coefficient of restitution of baseballs.[5] The ball was mounted on a spinning tee located in front of the muzzle. The wooden projectile from the air-gun drove the spinning ball a distance of 60 ft (the distance from the pitcher's rubber to the home plate) where it made an imprint on a vertical target.

The spin of the ball before impact was measured with a Strobotac. The speed could (in theory) be computed by measuring the drop of the ball, i.e., the vertical distance of the projected horizontal axis of the gun above the point of impact.

The direction of the spin about the vertical axis could be reversed at will and measurements were made with the ball first spinning clockwise (looking down) and then counterclockwise. One-half of the horizontal distance between the two target imprints gave the lateral displacement sought.

This setup was simple, and at first sight appeared usable. The observed deflections were in the expected direction, and shifted to the other side of the target when the spin was reversed. But the results were erratic. A stroboscopic camera was then installed 30 ft above the floor looking down on the last half of the flight path, and the position of the ball was photographed at exact 0.05-sec intervals against the scale on the floor.

These photographic measurements gave the trajectory of the ball, its speed and the drag; but they also indicated that the spin of the ball was greatly reduced when it was distorted by the impact of the projectile; and that the reaction between the spinning ball and the projectile gave rise to a small component of velocity normal to the flight path, which contributed to the observed lateral deflections. The trajectory was in fact that of a batted ball instead of a pitched ball. This line of attack was consequently abandoned in favor of wind tunnel measurements.

Wind Tunnel Measurements

In these measurements, the spinning ball was dropped from the upper side of the NBS 6-ft

FIG. 1. Showing airflow past spinning ball in wind tunnel. Wind coming from right, 60 ft/sec. Spin 1000 rpm, counterclockwise, about a horizontal axis at right angles to wind. Magnus force, upward. Courtesy of Professor F. N. M. Brown, University of Notre Dame.

[3] Ludwig Prandtl, *Essentials of Fluid Dynamics* (Hafner Publishing Company, Inc., New York, 1952).
[4] S. Goldstein, *Modern Developments in Fluid Dynamics* (Clarendon Press, Oxford, England, 1938), Vols. 1 and 2.
[5] Lyman J. Briggs, J. Research Natl. Bur. Standards **34**, 1 (1945).

octagonal wind tunnel across a horizontal wind of known velocity. By coating the bottom of the ball lightly with a lubricant containing lamp-black, its point of impact was recorded on a sheet of cardboard fastened to the tunnel floor. The lateral deflection, which is of immediate interest, was taken as one-half of the measured spread of the two points of impact, with the ball spinning first clockwise and then reversed.

The spinning mechanism was mounted outside on the top of the tunnel with its hollow shaft projecting vertically downward one-half inch through the tunnel wall. A concentric suction cup to support the ball was mounted on this shaft. The spinning ball was released by a quick-acting valve which cut off the suction and opened the line to the atmosphere.

The spinner was belt-driven by a small dc motor with its armature current supplied from a potentiometer circuit to secure the desired speed range. The angular speed (rpm) was measured with a calibrated Strobotac which illuminated a rotating target on the spinner mechanism. The ball was shielded by a thin-walled cylinder (4 in. o.d., 4 in. long) mounted on the inner wall of the tunnel, concentric with the spinner shaft. While this introduced some additional turbulence over that created by the bare ball on its spinner, it gave more consistent

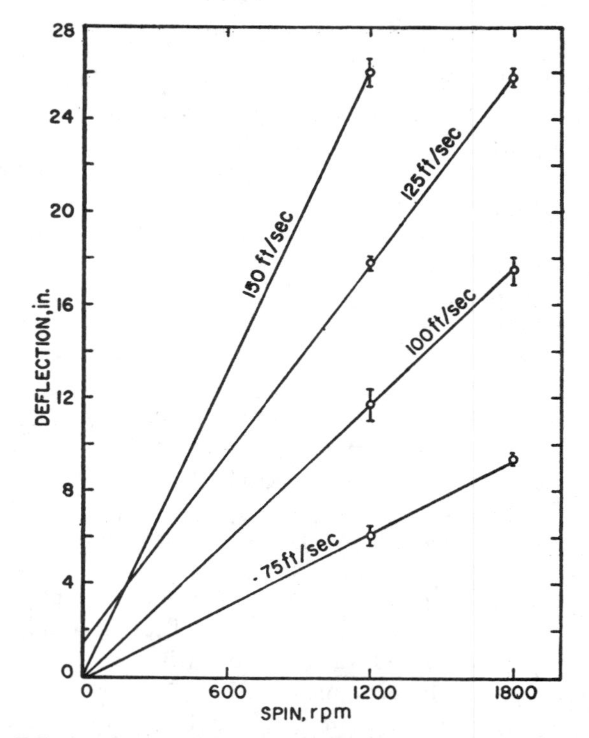

FIG. 2. Lateral deflection of a baseball, spinning about a vertical axis, when dropped across a horizontal windstream. These values are all for the same time interval, 0.6 sec, the time required for the ball to cross the stream.

results, and prevented irregularities caused by the ball being torn off its support by the wind stream during release. Official American League balls were used throughout the measurements.

TABLE I. Lateral deflection of a spinning baseball in a 6-ft drop across the tunnel windstream at various spins and speeds.

Spin rpm	Speed ft/sec	Deflection, inches	Ratio of deflections	Ratio of (speeds)²
1200	125 100	17.8 11.7	1.52	1.56
1200	150 125	26.0 17.8	1.46	1.44
1200	150 100	26.0 11.7	2.22	2.25
1200	100 75	11.7 6.1	1.92	1.77
1200	125 75	17.8 6.1	2.91	2.79
1200	150 75	26.0 6.1	4.25	4.0
1800	125 100	25.8 17.5	1.47	1.56
1800	125 75	25.8 9.4	2.98	2.79
1800	100 75	175 94	1.81	1.77

Owing to the method of construction, the center of gravity of a baseball often does not coincide exactly with its geometrical center. As a result of this asymmetry, the ball rotating on its spinner is subject to a lateral centrifugal force. This causes the ball to depart from a truly vertical fall when there is no wind. The departure may be upstream, laterally, or downstream, depending on the angular position of the heavy side of the ball when released, and results in a scatter of impacts for the same spin and windspeed.

To minimize this effect, the ball was turned in different positions while being placed in the suction cup of the spinner, until a position in which the center of gravity appeared to fall on the spin axis was found by trial.

At least three measurements were made for each spin and wind speed, as well as for the spin reversed. The mean values are given in Table I

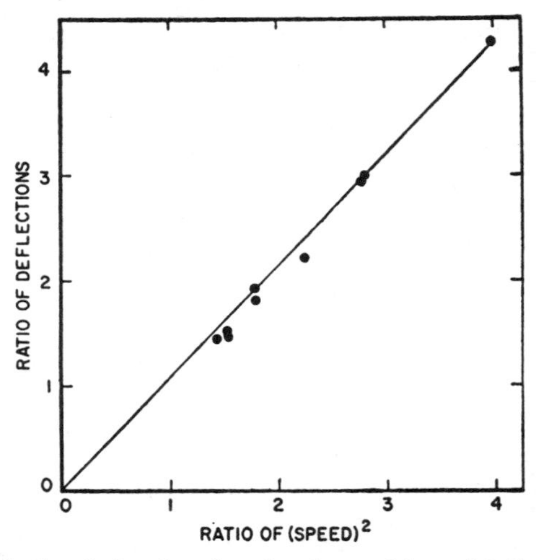

FIG. 3. Graph showing that the observed lateral deflections are proportional to the *square* of the wind speed.

and shown graphically in Fig. 2 together with the standard deviations. The lateral deflection in all instances accorded in direction with that expected from the Magnus effect.

It will be noted from Fig. 2 that straight lines drawn through the observed deflections at spins of 1200 and 1800 rpm for different wind speeds pass nearly through the origin. In other words, within experimental limits, the lateral deflection is directly proportional to the spin.

The effect of wind speed on the lateral deflection is shown in Table I. The fourth column of the table gives the ratio of the observed deflections at known wind speeds. For comparison, the last column gives the ratio of the corresponding wind speeds squared. These results are plotted in Fig. 3. Subject to experimental errors, the square relationship is seen to hold.

We conclude then that for speeds up to 150 ft/sec and spins up to 1800 rpm, the lateral deflection of a baseball spinning about a vertical axis is directly proportional to spin and to the square of the wind speed.

Maximum Curve Expected for Pitched Baseballs in Play

All the lateral deflections shown in Fig. 2 refer to what took place in 0.6 sec, the time required for the ball to fall across the wind stream of the 6-ft tunnel. We have now to convert these measurements into what deflection would be expected if the ball were traveling the 60 ft from the pitcher's rubber to the plate at various speeds and spins.

The results are summarized in Table II. A ball thrown at a speed of 100 ft/sec would travel the 60 ft from rubber to plate in 0.6 sec, which is the time required for the ball to fall across the wind stream in the tunnel measurements. Consequently in this case the observed lateral deflection in the tunnel would be equal to that of the ball in play. In other cases a correction is necessary.

The lateral deflecting force of the tunnel stream is practically constant for a given spin and speed, so that the lateral acceleration of the ball is constant and the distance traveled is proportional to the square of the elapsed time. At 125 ft/sec, the thrown ball travels 60 ft in 0.48 sec, while the dropped ball takes 0.6 sec to cross the tunnel. Hence the deflection in 0.48 sec would be

$$17.8 \text{ in.} \times (0.48/0.60)^2 = 11.4 \text{ in.}$$

It will be noted from Table II that the amount the ball curves in 60 ft is proportional to the spin, but is practically independent of the speed, namely, about 11 in. at 1200 rpm and 17 in. at 1800 rpm. This result may seem surprising until it is recalled that at the higher speeds, the ball is in the 60-ft zone for a shorter time and that the lateral displacement is proportional to the time squared.

These measurements were all made with the ball spinning about a vertical axis, which gave the maximum lateral deflection. Usually in play the spin axis is inclined, which reduces the effect. If the spin axis were horizontal and normal to the flight path, no lateral deflection would take

TABLE II. Curving of baseballs in play.

Speed ft/sec	Spin rev/sec	Turns in 60 ft	Curve (lat. def.) in., in 60 ft
75	20	16	10.8
75	30	24	16.7
100	20	12	11.7
100	30	18	17.5
125	20	9.6	11.4
125	30	14.4	16.5
150	20	8	11.6

place. With clockwise spin (seen from right) the pitch would be a drop.

Notes Pertaining to Baseballs in Play

These measurements were designed to cover simply the range of conditions encountered in play. The following records are of interest in this connection. Bob Feller, former Cleveland pitcher, in 1947 threw a baseball across the plate at a speed of 98.6 mph (144 ft/sec), as measured with electronic instruments. J. G. Taylor Spink, Editor of the *Sporting News*, states that this is the accepted world record for the fastest pitch.[6] The next fastest pitch of record is 94.7 mph (138 ft/sec) by Atley Donald, former New York Yankee pitcher, in 1939.

Dr. H. L. Dryden kindly arranged for the measurement of the "terminal velocity" of a baseball, that is its maximum speed after falling from a great height to the ground.[7] This measurement was carried out in a vertical wind-tunnel at the National Advisory Committee for Aeronautics, the wind speed being adjusted until the ball just floated in the windstream. The terminal velocity was about 140 ft/sec. The celebrated catch by Charles Street, of the Washington Ball Club, of a ball dropped from a window of the Washington Monument gave a computed velocity *in vacuo* of 179 ft/sec. The NACA measurements show that, owing to the resistance of the air, the actual speed could not have exceeded 140 ft/sec. However, home runs *batted* into the stands must have an *initial* velocity considerably higher than this.

With the cooperation of the pitchers of the Washington Ball Club, the spin of a pitched ball was measured. This was done by fastening one end of a long tape to the ball and then laying the rest loosely (but untwisted) on the ground between the rubber and the plate, the free end being pegged down. After the ball was caught, the number of turns was counted. The highest spins measured were 15.5–16 turns in 60 ft and the lowest 7–8. Assuming the speed of the pitch to be 100 ft/sec, the maximum spin measured would be about 1600 rpm. These spins are covered by the wind tunnel observations.

[6] J. G. Taylor Spink (personal communication).
[7] H. L. Dryden (personal communication, with permission).

PART 2. SMOOTH BALLS AND THE MAGNUS EFFECT

The experiments with *rough* baseballs (Part 1) all showed lateral deflections in conformance with the Magnus effect. But with *smooth* balls, especially at low wind speeds, an anomaly is encountered. The deflections are usually in the *opposite* direction.

Maccoll, using a 6-in. diam wooden sphere, rotating on a wind tunnel balance, was apparently the first to demonstrate the existence of a small negative lift on a sphere at low wind speeds.[8] His C_L curve, negative at first, crosses the *no lift* axis when the equatorial speed/wind speed $= U/V = 12.3/24.6 = 0.5$.

Davies, in his experiments with smooth and dimpled golf balls, found that for the smooth ball the lift was negative at all rotational speeds below 5000 rpm (equatorial speed, 2100 ft/min).[9] Above this, the lift was positive, but was less than for the standard ball.

In Davies' measurements, the ball was dropped across the horizontal wind stream of an open tunnel operating at 105 ft/sec. The axis of spin was horizontal and normal to the wind stream. A quick-acting device released the spinning ball. The vertical drop was 0.67 to 1.3 ft. Spins up to 8000 rpm could be obtained.

It will be noted that in Davies' experiments the point of impact on the tunnel floor represents the combined effect of spin and drag, both being directed downstream. To get the lift, the point of impact had to be reduced by the drag with no spin; whereas in my baseball measurements the lateral deflection, being at right angles to the drag, could be measured directly.

Brown has recently measured the lift coefficient of a sphere of 3.36-in. diam mounted on a balance in a low-turbulence tunnel, and rotating at fixed speeds ranging from 700 to 4500 rpm.[10] His results all show a negative lift at low wind speeds, the graphs crossing the no-lift axis at values of V/U ranging from 0.1 to 0.5, where U is the wind speed and V is the peripheral speed.

[8] Maccoll, J. Roy. Aeronaut. Soc. **32**, 777 (1928).
[9] J. M. Davies, J. Appl. Phys. **20**, 821 (1949). This paper contains references to articles not here recorded.
[10] F. N. M. Brown, University of Notre Dame, Notre Dame, Indiana (personal communication, 1958; unpublished data, with permission).

TABLE III. Lateral deflection (inches) of smooth Bakelite ball, 6-ft drop.

Wind speed, ft/sec	Spin, rpm	Deflection, in.	
75	1200	1.1	Pro-Magnus
100	1200	0.5	Pro-Magnus
125	1200	0.6	Anti-Magnus
150	1200	2.2	Anti-Magnus
75	1800	1.4	Pro-Magnus
100	1800	0.5	Pro-Magnus
125	1800	0.8	Anti-Magnus
150	1800	7.3	Anti-Magnus

Smooth Rubber Ball

The writer has measured the lateral deflection of a *smooth* rubber ball, using the same setup employed with baseballs. The ball, which had been cast in an accurately spherical mold, was 2.88 in. in diam, practically that of a baseball, but was heavier (wt 188 g; baseball about 145 g). Its center of gravity agreed closely with its geometrical center and it had a good bounce.

This smooth ball was deflected laterally *opposite* in direction to that of the baseballs. The deflection was small (partly owing to the increased weight) but increased steadily from 3.6 in. at 75 ft/sec (1800 rpm) to 8.8 in. at 150 ft/sec (all negative Magnus). At 1800 rpm the rotational velocity of a point on the equator was 1350 ft/min. Davies states that his smooth golf ball began to develop a positive lift at 5000 rpm (equatorial speed, 2100 ft/min).

Bakelite Sphere

Similar tests were made with a smooth Bakelite sphere (good ground finish) of 3-in. diam, sphericity 0.005 in., wt 312 g. The Reynolds number at 150 ft/sec was 2.4×10^5. At this speed the ball was presumably passing out of the critical Reynolds-number range for the drag coefficient for spheres.[11] See Goldstein.[4]

The lateral deflections, which are very small, are given in Table III. The unexpected result was that at the lower wind speeds the ball had a Magnus deflection, crossing to the anti-Magnus regime between 100 and 125 ft/sec. Here we seem to have a new effect for *smooth* balls, which has been masked in the measurements reported above.

[11] The Reynolds number is the product of the speed of the wind and the diameter of the sphere divided by the kinematic viscosity of the air.

It seems now to be well established by the results of four different laboratories that a spinning *smooth* ball at low wind-tunnel speeds usually does not conform with Magnus effect but exerts a "negative" lift. It appears likely that this occurs in a certain Reynolds-number range where it is possible for the boundary layer on the side moving with the wind to remain laminar while that on the opposite side becomes turbulent. Since a turbulent layer will in general follow farther around the surface before separating than a laminar layer, the crosswind force is reversed, if the rate of spin is not too high. We would expect to pass from this condition into the Magnus regime when the spin is sufficiently increased or when the flow on both sides becomes turbulent, as it will when the Reynolds number is sufficiently increased.

Evidently we may also pass into the Magnus regime at low Reynolds numbers by arranging conditions so that the flow is laminar on both the approaching and receding sides, the requirements now being that disturbances, such as vibration of the sphere and turbulence in the wind stream, are sufficiently reduced. This is believed to be the explanation of the pro-Magnus results of Table III.

In general, then, we find that the sign and magnitude of the effect of spin are dependent on dynamic conditions as well as on the "smooth-

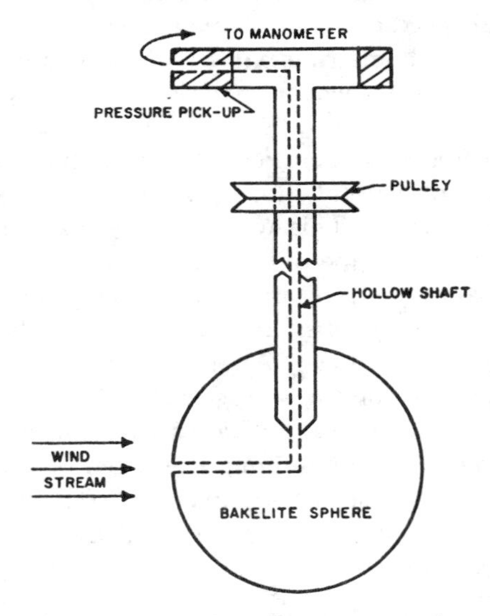

FIG. 4. Schematic drawing of apparatus used to measure the pressure at various points on a ball spinning in a wind stream.

ness" of the ball. The foregoing explanations have been offered as the most reasonable ones on the basis of the information available. The author knows of no detailed investigations of the flow in these cases.

PART 3. PRESSURE DISTRIBUTION OVER A SPINNING SPHERE IN ITS EQUATORIAL PLANE

To learn more about the forces acting on the surface of a smooth rotating sphere, the following apparatus, shown schematically in Fig. 4, was constructed.

The smooth Bakelite sphere (3-in. diam) used earlier in the free-fall measurements was mounted on a hollow vertical shaft ($\frac{3}{8}$-in. diam), the center of the unshielded sphere projecting downward ($2\frac{1}{2}$ in.) into the horizontal wind stream of the tunnel. A $\frac{1}{8}$-in. hole drilled along an equatorial radius connected the surface of the sphere with the hollow shaft.

The shaft extended upward through the top wall of the 6-ft octagonal tunnel, where it was expanded into a cylindrical head ($1\frac{1}{8}$-in. diam), which was drilled radially to lie directly above the hole in the sphere. The head was surrounded by a closely fitted pickup sleeve, which through a drilled radial hole ($\frac{1}{16}$-in. diam) connected the ball with the manometer. The pickup sleeve could be rotated and held independently in any angular position about its vertical axis, so that the pressure on the ball at any point in its horizontal equatorial plane could be measured.

The oppoiste leg of the manometer was connected to a static pressure orifice in the wall of the tunnel. The wind speeds (75 and 125 ft/sec) were based on impact pressure measurements in the empty tunnel. The hollow shaft was belt-driven by a dc motor on a potentiometer circuit. The spin, measured by a stroboscope, was roughly 1800 rpm.

As far as the writer knows, this is the first time the pressure at the equatorial surface of a spinning sphere has been measured directly. It provides a point-to-point determination of the pressure around the entire periphery of the sphere and eliminates the disturbance set up by the introduction of an external probe. By drilling the entrance hole in the sphere at higher latitudes, the pressure distribution over the entire

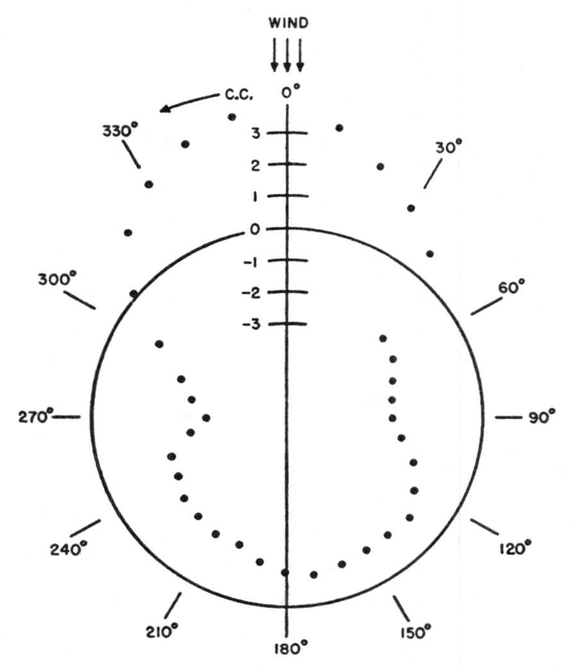

FIG. 5. Polar diagram of pressure distribution around smooth Bakelite ball (looking down along vertical spin axis). Ball, 3-in. diam; counter-clockwise spin, about 1800 rpm; wind speed, 125 ft/sec; pressure, inches of water.

surface of the sphere could be found. Only the equatorial pressures have been measured in the present study.

Two examples are given of the pressure distribution around the spinning sphere. In Fig. 5, the observed pressures at 125 ft/sec and 1800 rpm are shown in a polar diagram. It will be seen that the pressure is above atmospheric for 40°–50° on either side of the impinging windstream and that for this region the difference in pressure of symmetrical points would tend to force the ball to the right in Fig. 5, an anti-Magnus effect. The cosine projection of these above-atmospheric pressures on a horizontal diameter normal to the windstream is, however, small. For the remainder of the diagram, the pressures in the left half are consistently lower than corresponding points in the right half, tending to force the ball to the left, thus giving a positive Magnus effect.

If we sum up the pressure differences of corresponding points in Fig. 5 (after a cosine projection on a transverse equatorial diameter as illustrated in Fig. 6), we come out with a value of -5.78 pressure units, the minus sign indicating that the resultant transverse force is in the

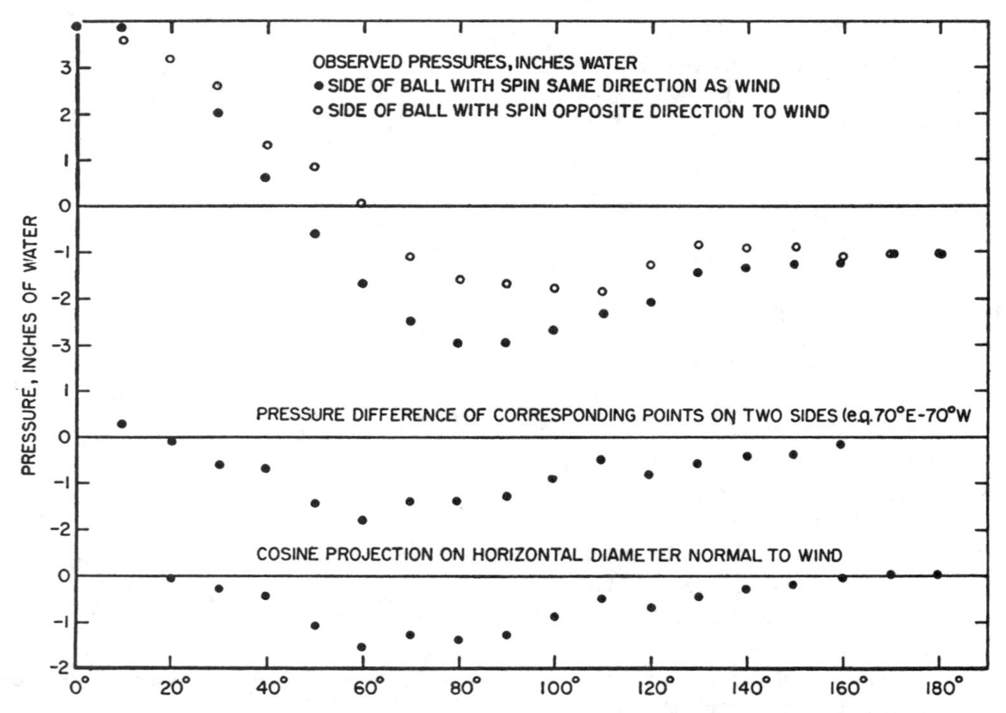

FIG. 6. Rough Bakelite ball, 3-in. diam; wind speed, 125 ft/sec; spin, about 1800 rpm. Meridional rubber bands stretched over ball to simulate roughness.

Magnus direction at 125 ft/sec and 1800 rpm. The corresponding measurement at 75 ft/sec is +1.77 pressure units, an anti-Magnus effect. These results are opposite in sign from those obtained when the Bakelite ball was dropped across the windstream. Here, however, we are dealing only with the pressures in a narrow equatorial belt on the ball ($\frac{1}{8}$-in. wide) and not with its entire surface.

Rough Bakelite Ball

Finally, the surface of the same ball used in the preceding measurements was roughened by attaching rubber bands along meridional lines. Figure 6 shows: (a) the observed pressures at 10° intervals measured from the direction of the wind; (b) the pressure difference of corresponding points; and finally (c) the projection of these differences on a horizontal diameter normal to the wind. The resultant pressures are consistently in accord with the Magnus effect, but it will be noted that this effect is substantially reduced by what takes place on the opposite side of the ball. Their sum, −10.8, taken in conjunction with appropriate units of area, gives a measure of the resultant force on the ball at the equator. The corresponding figure for the smooth Bakelite ball at the same speed is −5.78.

ACKNOWLEDGMENT

I would like to express my indebtedness to G. B. Schubauer, B. L. Wilson, R. H. Heald, R. J. Hall, and G. H. Adams for valued assistance in various ways.

Aerodynamics of a knuckleball

Robert G. Watts
Eric Sawyer
Department of Mechanical Engineering
Tulane University
New Orleans, Louisiana 70118
(Received 30 July 1974; revised 2 December 1974)

The nature of the forces causing the erratic motion of a knuckleball has been investigated by measuring the lateral force on a baseball in a wind tunnel. We have identified two possible sources of a lateral force imbalance that can give rise to the observed erratic changes in the direction of a knuckleball as it moves through the air. One of these results because the nonsymmetrical location of the roughness elements (strings) gives rise to a nonsymmetrical lift (lateral) force. A very slowly spinning knuckleball will have imposed upon it a lateral force that changes as the positions of the strings change. A knuckleball whose spin is identically zero has a constant lateral force unless a portion of the strings is precisely at a location where boundary layer separation occurs. If this happens, the point of boundary layer separation switches alternatively from the front to the rear of the strings, shifting the wake from one position to another, and thereby giving rise to a second possible alternating force imbalance. A two-dimensional analysis of the trajectory of the baseball indicates that the measured force can cause a deflection of the baseball's trajectory of more than a foot. An effective knuckleball should be thrown so that it rotates substantially less than once on its path to home plate.

INTRODUCTION

The knuckleball is the name given to a type of pitch in the game of baseball. The ball is held with the first knuckles or the fingernails (hence, the name knuckleball). As it is thrown, the fingers are extended in such a fashion as to inhibit the backspin normally possessed by a thrown ball. Indeed, it is commonly believed that a properly thrown knuckleball should have no spin at all. As a knuckleball approaches home plate, it changes directions erratically in an apparently random manner.

The lateral deflection of a spinning ball (curve ball) has been studied by Briggs[1] and others and appears to be well understood. It results when the spin of the baseball causes boundary layer separation to occur further upstream on the portion of the ball's surface that has relative motion against the flow of air past the ball. The wake then shifts towards that side of the baseball and the ball is deflected towards the opposite side.

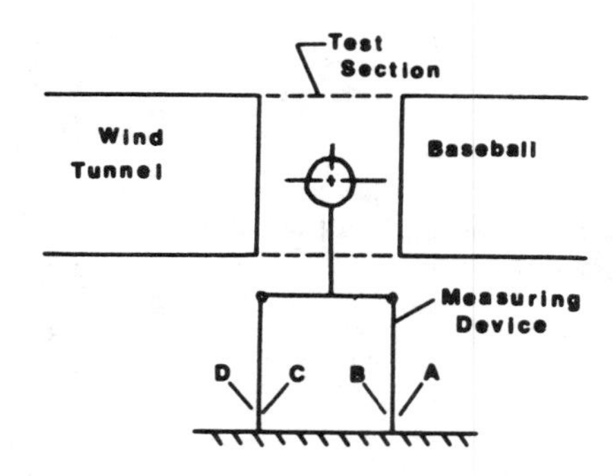

Fig. 1. Schematic of apparatus. The measuring device is in the position used to measure drag.

In an effort to determine precisely the nature of the forces that cause the erratic motion of the knuckleball, we sought to model the situation as nearly as we could in the fluid mechanics laboratory. The results of these experiments are described herein.

APPARATUS AND EXPERIMENTS

The experimental arrangement that we used consisted of a subsonic wind tunnel, a device for measuring lift and drag, and a strip chart recorder for measuring and recording the lift and drag forces.

A drag and lift measuring device was used to measure the forces on the baseball in the wind tunnel. The device (Fig. 1) consisted of two beams rigidly attached at the base and pinned at the top. Foil strain gauges located at A, B, C, and D were placed on each side of the beams and were connected to a Baldwin–Lima–Hamilton micro strain indicator in such a fashion that the total strain output was four times the strain measured by one of the gauges.

A system of known weights was attached by a pulley to the test stand in order to calibrate the measuring device. The resulting curve confirmed that the strain was directly proportional to the force component.

A baseball mounted on this device and placed in a wind tunnel will experience the same flow history as a knuckleball moving through still air at the velocity of the

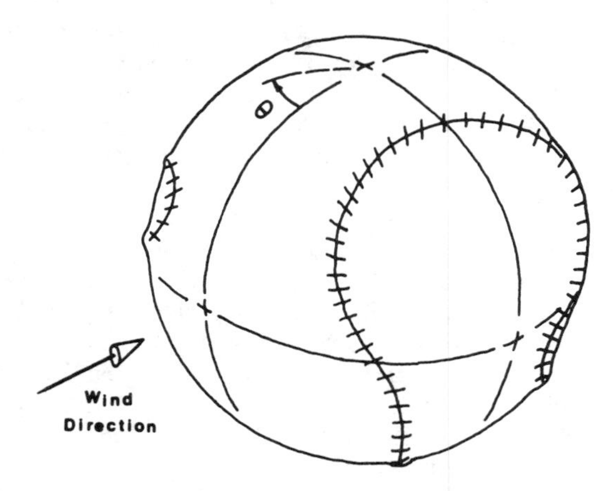

Fig. 2. Position of baseball at $\theta = 0°$. The ball can be rotated in the direction θ to a new position.

Reprinted from *American Journal of Physics* **43**, 960–963 (1975);

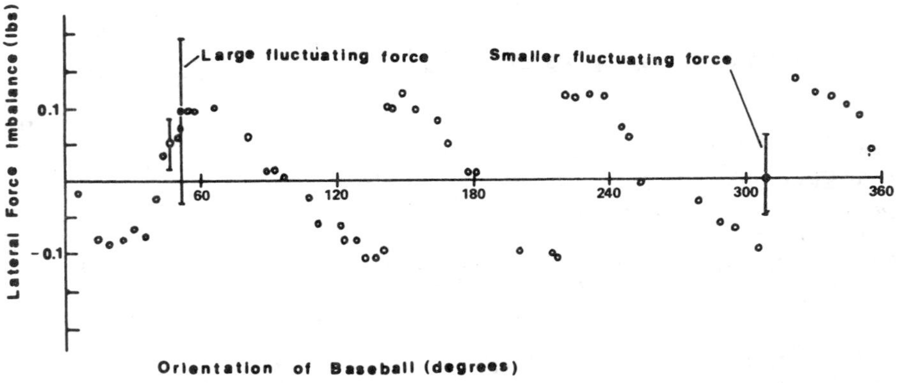

Fig. 3. The variation of the lateral force imbalance with orientation of the baseball. (See Fig. 2 for definition of θ.)

air in the wind tunnel. The velocity profile across the duct of the wind tunnel was measured by using a static Pitot tube. The velocity at a distance 3 in. from the center line of the test section was within 1% of the center line velocity. The velocity across the test section was therefore assumed to be independent of radial position.

With the ball in place and the tunnel velocity at about 50 ft/sec, the chart recorder indicated a lateral force with high-frequency noise superimposed. The frequency of the noise was at least 10 cycles/sec. The noise was probably associated with the turbulence in the wind tunnel. By inserting grids of various sizes in the wind tunnel upstream of the baseball, we were able to decrease the magnitude of the noise by a factor of 2 or 3. In any case, the frequency of the noise was so high that it could cause no measurable deflection of the baseball. We shall give an estimate of this in a later section of this paper. The constant lateral force that we measured was of sufficient magnitude (~ 0.05–0.1 lb, the same order of magnitude as the drag) to cause a significant deflection of the baseball in the distance of 60.5 ft from the pitcher's mound to home plate. However, the force was *constant*. If the ball did not spin, it could give rise to a laterally curved trajectory, but not to the erratic motion associated with the knuckleball.

The lateral force that we measured resulted from the fact that the strings on the surface of the ball form a roughness pattern that is nonsymmetric. The flow pattern about the ball (including the wake) will therefore be nonsymmetric. This will naturally cause a nonsymmetric lateral force distribution and result in a net force in one direction or another. This explanation seems to be justified intuitively. Furthermore, by changing the orientation of the baseball, we were able to change the measured drag force by as much as 50% and to cause the lateral force to change sign.

While we were investigating the effect of the location of the strings on the drag and lateral forces, we discovered that when the ball was oriented in a certain way the strip chart began to record a randomly changing lateral force with a frequency low enough (< 1 cycle/sec) to cause a significant lateral deflection of the baseball. This only occurred for certain special orientations, and we concluded rather quickly that these orientations place a portion of the strings of the baseball at approximately the position where boundary layer separation occurs, at an angular position about 105°–115° from the front stagnation point.

We suspected that the oscillatory force resulted from the point of boundary layer separation changing from the front to the back of the stitches, thereby causing an oscillating wake. In order to check this theory, fine wool threads were individually glued to the rear portion of the ball (in the wake). The ball was then reinserted in the wind tunnel. The location of the wake could be detected as that portion of the rear of the ball where the wool threads indicated backflow.

As we expected, one could clearly detect the fluctuation of the edge of the wake as it moved from the front edge of the stitches to the rear edge and back again.

LATERAL FORCE

It now remained to obtain quantitative measurements to be sure that the forces we observed could cause the ball to deflect as much as a foot or two as they are observed to do for real baseballs. We also wanted to find out how the magnitude and frequency of the fluctuating force varied with velocity, and how the constant lateral force varied with velocity and with the orientation of the baseball.

A standard 2.88-in.-diam baseball was inserted in the wind tunnel in the position shown in Fig. 2. An isometric sketch is shown in order to establish the initial position of the strings. With the ball oriented in this way the lateral force was zero. The ball was then turned about the vertical axis. The lateral force was recorded for various values of the angle θ between 0 and 2π rad. The velocity was

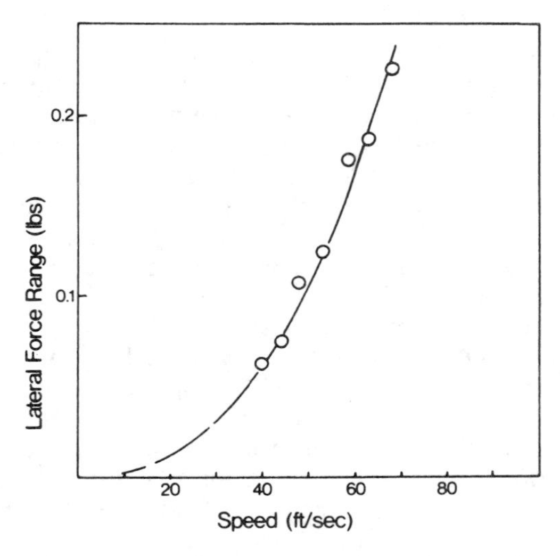

Fig. 4. Variation of difference between the maximum and minimum lateral forces.

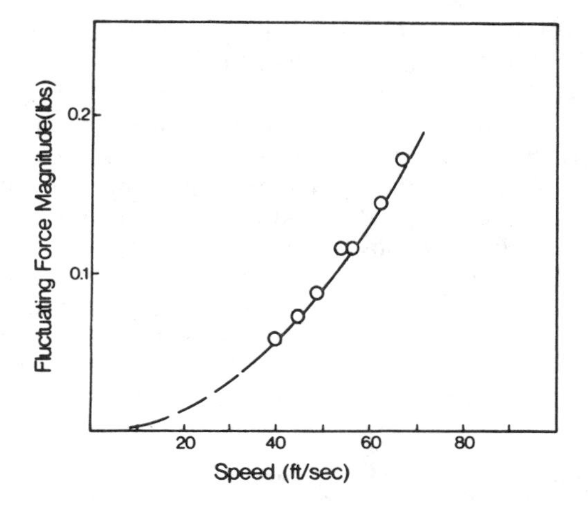

Fig. 5. Variation of the magnitude of the fluctuating force.

68 ft/sec. The results are shown in Fig. 3.

Figure 3 reveals several interesting features, some of which we did not anticipate. The lateral force varied between about 0.1 lb towards the left and 0.1 lb towards the right. At angles of approximately $\theta = 140°$ and $220°$ there was a nearly discontinuous jump in the lateral force from about 0.08 lb in one direction to a similar value in the opposite direction. Also, at approximately $\theta = 52°$ and $310°$ there was an instability that caused the lateral force to alternate from left to right with an amplitude of about 0.18 lb and a frequency of 0.5–1 cycle/sec. This alternating force occurred when a portion of the strings was located just at the point where boundary layer separation takes place. As we pointed out above, it resulted when the point of separation changed from one side of the strings to the other. *It did not always occur* when the strings lay at the edge of the wake. For example, at $\theta = 140°$ and $220°$, there occurred a practically discontinuous change in the lateral force, indicating that the separation point had moved from the front to the rear of the strings (or vice versa) and remained there. The data near all four of the positions $\theta = 52°$, $140°$, $220°$, and $310°$ were quite repeatable.

Figures 4 and 5 give an indication of how the various forces change with velocity. In the Reynolds number range in question ($\mathrm{Re} \approx 10^5$) the drag *coefficient* for a sphere is approximately constant.[2] The drag *force* therefore increases as the square of the velocity. It is therefore not too surprising that the magnitudes of the various lateral forces increased as approximately the square of the velocity. Figure 4 shows such a variation in the difference between the maximum and minimum lateral forces on the baseball for a given velocity. Figure 5 shows how the magnitude of the *fluctuating* force near $\theta = 52°$ and $310°$ varied with the velocity. The frequency of this fluctuating force did not appear to vary appreciably with velocity, nor did the value of θ for which it occurred.

THE TRAJECTORY

We now wish to compute possible deflections in the trajectories of thrown balls in response to the forces measured, and to draw some conclusions about the possible origin of the erratic motion of knuckleballs. We begin with a simple force balance on the baseball in the direction mutually perpendicular to the original direction of level flight of the ball and the gravitation vector. Although the lateral force is actually perpendicular to the instantaneous direction of flight of the ball, it serves our present purpose best to use the approximation, accurate for small deflections,

$$F = m\frac{d^2x}{dt^2}, \qquad (1)$$

where F is the lateral force, m is the mass of the baseball, and d^2x/dt^2 is the lateral acceleration of baseball. The steady-state solution that results when F is a periodic force $F_0 \sin(\omega t)$ is

$$x = \frac{F_0}{m\omega^2}\sin(\omega t + \pi). \qquad (2)$$

The magnitude of the displacement from the undisturbed trajectory of the baseball therefore varies inversely with the square of the frequency of the imposed force. Taking the mass of the baseball to be 10^{-2} slugs and the force to have a magnitude of 0.05 lb, we find that a deflection of 1 ft or more can be obtained only if the frequency is less than about 0.2 cycle/sec. In particular, the high-frequency force that we attributed to wind tunnel noise could cause a deflection of only about 5×10^{-4} ft if we assume its frequency is 10 cycles/sec and its amplitude is 0.02 lb.

If a lateral force $F(t)$ is suddenly applied to a ball whose initial lateral velocity and displacement are zero, the lateral displacement after a time t will be approximately

$$x(t) = m^{-1}\int_0^t \int_0^\tau F(\lambda)\,d\lambda\,d\tau. \qquad (3)$$

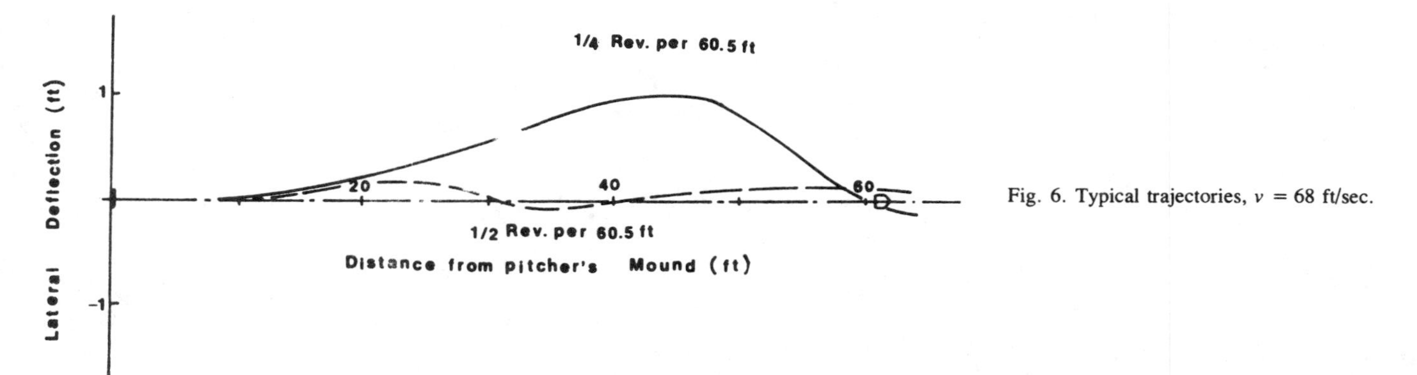

Fig. 6. Typical trajectories, v = 68 ft/sec.

In particular, if $F(t)$ is a constant, F_0,

$$X(t) = \tfrac{1}{2}(F_0/m)t^2. \quad (4)$$

If the forward speed of the ball is constant, the time to reach home plate will be D/V, where D is the distance traversed by the ball and V is its speed. Hence, the distance the ball will curve by the time it reaches home plate is

$$X = \tfrac{1}{2}(F_0/m)(D/V)^2. \quad (5)$$

When we recall that the lateral force F_0 is (roughly speaking) proportional to the square of the velocity, we obtain the somewhat surprising result that the lateral deflection does not vary with speed.

If the knuckleball is thrown in such a fashion that it has no spin at all, it can only curve laterally in one direction. The maximum deflection will be dependent only on the initial orientation of the baseball (i.e., the initial locations of the strings). The single exception can occur in the remote possibility that the strings are initially positioned so that they disturb the point of boundary layer separation and bring about the oscillating wake phenomena described above. This could cause the erratically fluctuating lateral motion that is actually observed to occur with real knuckleballs.

A much more plausible argument is that the ball spins very slowly as it approaches the plate, thereby exposing the ball to a varying lateral force depending on the orientation of the ball at any given instant. If the ball spins too fast, we can see from Eq. (2) that the deflection will be small. For example, if we approximate the curve in Fig. 3 by a sine wave of amplitude 0.08 lb and we assume the baseball is thrown at 60 ft/sec and undergoes two complete revolutions before reaching home plate, the amplitude of the deflection is only 0.048 ft. The spin of the ball should therefore be substantially slower than this.

An important feature of Fig. 3 is the nearly discontinuous change in the lateral force at $\theta = 140°$ and $220°$. If the ball should move through this orientation at some time during its flight towards home plate, the induced slow curvature caused by the slowly varying lateral force will suddenly become a very sharp "break," that is, a sudden change in curvature.

We have numerically integrated Eq. (3) using our data for the cases where the ball is initially oriented such that $\theta = 90°$ and the spin is such that the ball undergoes a quarter- and a half-revolution, respectively, during the time required to reach home plate. The results are shown in Fig. 6. Clearly, the first is the better pitch.

We wish to make one additional remark regarding the physical nature of curving baseballs. It is frequently stated by those who actually play the game that a curving baseball "breaks" rather than simply deflecting with constant curvature. If a lateral force $F(t)$ is applied to a baseball so that it is always perpendicular to the instantaneous direction of flight, the baseball will curve in such a fashion that $F(t)$ is equal to the instantaneous centripetal acceleration $mv^2\kappa(t)$, where $\kappa(t)$ is the instantaneous curvature of the path of the baseball. Thus, a constant lateral force would give rise to a constant curvature

$$\kappa = F_0/mv^2. \quad (6)$$

No "break" occurs. From the batter's vantage point, however, things appear quite different. As the approximate solution given as Eq. (4) shows, what the batter sees is a projectile whose lateral deflection is *changing at an accelerating rate*. How much worse if the deflecting force is erratic!

SUMMARY AND CONCLUSIONS

We can summarize our most important results as follows:

(i) There are two possible mechanisms for the erratic lateral force that causes the fluttering flight of the knuckleball. A fluctuating lateral force can result from a portion of the strings being located just at the point where boundary layer separation occurs. A far more likely situation is that the ball spins very slowly, changing the location of the roughness elements (strings), and thereby causing a nonsymmetric velocity distribution and a shifting of the wake.

(ii) An effective knuckleball should have a slight spin. Too much spin could prove disastrous, however, since the inertia of the ball would not allow a significant deflection.

(iii) The magnitude of the lateral force increases approximately as the square of the velocity. This results in a total lateral deflection that is independent of the speed of the pitch.

[1]L. J. Briggs, Am. J. Phys. **27,** 589 (1959).
[2]L. Prandtl, *Essentials of Fluid Dynamics* (Hafner, New York, 1952), p. 191.

Models of baseball bats

Howard Brody
Physics Department, University of Pennsylvania, Philadelphia, Pennsylvania 19104

(Received 14 July 1989; accepted for publication 30 October 1989)

By observing the vibrations of a hand-held baseball bat, it is possible to show that the bat behaves as if it were a free body at the impact of the bat and the ball. The hand-held bat shows none of the behavior of a bat with one end firmly clamped in a vise.

It has been conjectured, but never verified, in several articles in this Journal[1,2] that a hand-held baseball bat behaves as if it were a free body when it strikes the ball. Recently, Weyrich *et al.*[3] attempted to determine the effect of bat grip firmness on the post-impact velocity of a baseball. This was done using both a freely suspended bat and a bat with its handle firmly clamped in a massive vise. What was not done was to determine which of these two limiting cases more closely resembles the actual case of a hand-held bat, or whether a loosely held bat corresponds to a freely suspended bat and a tightly gripped bat corresponds to a bat in a vise. Such an investigation has been carried out for tennis rackets,[4] and rackets have been found to act as if they are free bodies when the ball impacts on them. However, a tennis racket is very flexible and quite light compared to a baseball bat, and in addition, it is usually held by just one hand. Therefore, it is not clear whether both implements will behave in the same way during the very short time that the ball is in contact with the bat or racket.[5]

This investigation shows that the freely suspended bat (essentially a free body) corresponds to both the loosely held and very tightly gripped bat, and a bat firmly clamped in a vise does not behave the way a hand-held bat does. It also shows how well the hands tend to damp out the vibrations of the bat that occur when the bat is struck at a location that is not near the node.

One of the basic differences in behavior between a free bat and a bat with one end firmly clamped is the allowable normal modes of oscillation of the bat. Figure 1 shows the lowest modes of oscillation for these two cases. For the clamped bat, the fundamental mode of oscillation (often called the "diving board" mode) has a node only at the clamp. The next higher mode of oscillation of the clamped bat has a node well beyond the center of mass of the bat in addition to the one at the clamp. For the free bat, the mode of oscillation corresponding to the fundamental mode of the clamped bat is not allowed. The lowest frequency of oscillation has two nodes, a period not greatly different from the period of the first harmonic of the clamped bat, and an outer node that is located near the position of the node of the first harmonic of the clamped bat. If a clamped bat is struck at an antinode of the first harmonic (such as the tip), both the low-frequency (diving board) mode and the higher-frequency first-harmonic mode should be present. If a free bat is struck at the same location (near the tip), because the low-frequency (single-node) oscillation is not excited, the lowest observed frequency should be the one corresponding to the two-node mode. Therefore, if a hand-held bat is struck near the tip and the low-frequency (single-node) oscillation is not observed, that would be good evidence that the hand-held bat is acting as if it were a free body. If the low-frequency mode of oscillation is observed in a hand-held bat, that would be good evidence that the bat was acting as if one end were clamped.

This experiment consisted of measuring the natural frequencies of oscillation of baseball bats that were free, clamped, and hand held with various degrees of firmness. It was then possible to determine which situation (free or clamped) more accurately described the hand-held bat, and how grip firmness affected this result. To accomplish this, a small Kynar thin-film piezoelectric vibration sensor[6] was taped to the handle of the bat, approximately 0.35 m from the butt end. The output from the Kynar was fed directly into an oscillosocpe and the resultant traces photographed. Two bats were used, an aluminum softball bat and a wooden softball bat. In each case the bat was struck by a ball at the desired location. Since the Kynar sensor produces a measurable voltage for a very small amplitude of vibration, it was possible to carry out the experiment by hitting the bat with a ball that was hand held rather than having to swing the bat at the ball. This allowed a certain degree of reproducibility in the results as well as introducing a modicum of safety into an experiment that you might want students to do.

The results obtained using the aluminum bat and the wooden bat were essentially the same, and so no distinction will be made between the two bats in describing the experiment and the conclusions drawn, even though there were substantial differences in the frequencies of vibrations of the two bats. Note that there is a major controversy in baseball concerning whether aluminum bats (which are not allowed in the Major Leagues) are better than wooden bats, but that problem will not be addressed in this article.

Figure 2 shows the output of the Kynar vibration sensor when a bat, with its handle firmly clamped in a vise, is struck with a ball near the node on the first harmonic and then struck near the tip. In both cases a low-frequency oscillation (27 Hz) is shown and, in addition, there is a higher-frequency (317 Hz) signal present in the off-node hit

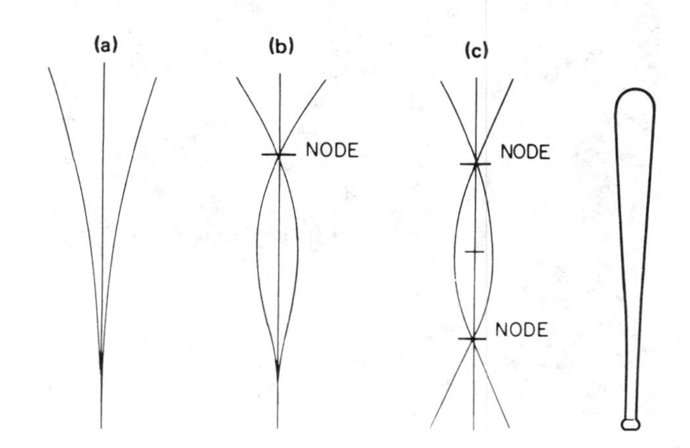

Fig. 1. Oscillations of a baseball bat (a) and (b) with one end clamped and (c) with both ends free.

Reprinted from *American Journal of Physics* **58**, 756–758 (1990); © American Association of Physics Teachers.

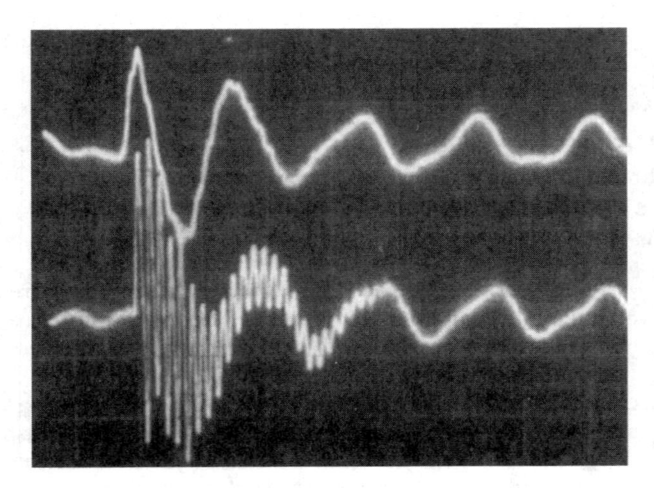

Fig. 2. Oscilloscope traces showing the output of a Kynar vibration sensor fastened to a bat with its handle clamped in a vise. The bat was struck near the node (upper trace) and near the tip (lower trace). The sweep speed was 20 ms/div (200 ms full scale), and the vertical gain of the scope was 0.2 V/div.

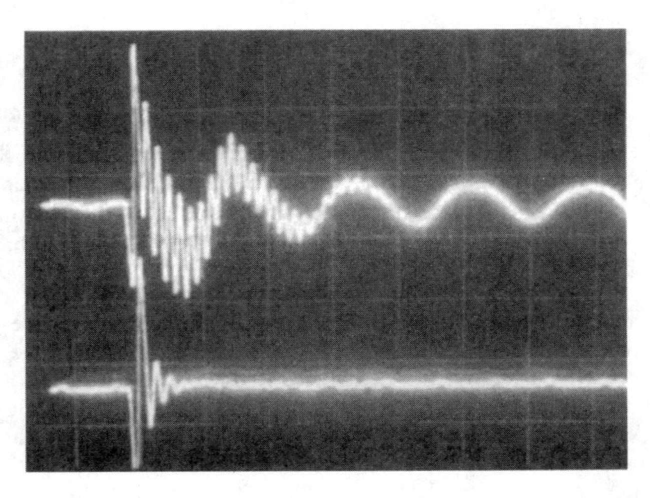

Fig. 4. Oscilloscope traces showing the output of Kynar vibration sensor fastened to a bat with its handle clamped in a vise (upper trace) and hand held (lower trace). In both cases the bat was hit near the tip, and the sweep speed was 20 ms/div with a vertical gain of 0.2 V/div.

that is much larger in amplitude than the higher-frequency oscillation that is present when the bat is struck near the node. The lower-frequency oscillation corresponds to the mode of vibration shown in Fig. 1(a), while the higher-frequency oscillation corresponds to the vibration shown in Fig. 1(b).

If the bat is freely suspended and hit with a ball near the tip, the oscillations produce the signal shown in the upper trace of Fig. 3. The frequency of this oscillation is about 20% lower than the high-frequency oscillation shown in Fig. 2 (clamped handle), and there is no visible sign of the low-frequency oscillation that is very obvious in Fig. 2. These free bat oscillations correspond to the vibrations shown in Fig. 1(c). The lower trace in Fig. 3 shows the same bat held in a vise and struck at the same location. It is also clear from these scope traces that the amplitude of the higher-frequency vibrations damp out quicker than the low-frequency (fundamental) oscillations when the bat has its handle clamped in a vise.

The oscillations of a hand-gripped bat are shown in the lower trace of Fig. 4. There is no sign of the low-frequency oscillation that is evident when the bat handle is in a vise, as is shown in the upper trace of Fig. 4. This means that a bat that is firmly held by hands acts as if it were a free bat, as far as the allowable modes of vibration are concerned. It is clear, by the rate at which the vibrations damp out, that the hands do affect the subsequent vibration of the bat. This can be seen quite easily in Fig. 5 where the oscillations of a bat that is loosely held and tightly gripped are shown. A number of additional pictures were taken at various sweep speeds of the oscilloscope and hitting the bats at various locations while gripping the handles very tightly. None of these pictures showed the low-frequency oscillation that was present when the bat handle was clamped in a vise.

The frequencies of oscillation, as determined from the scope traces, as well as other measured parameters of the

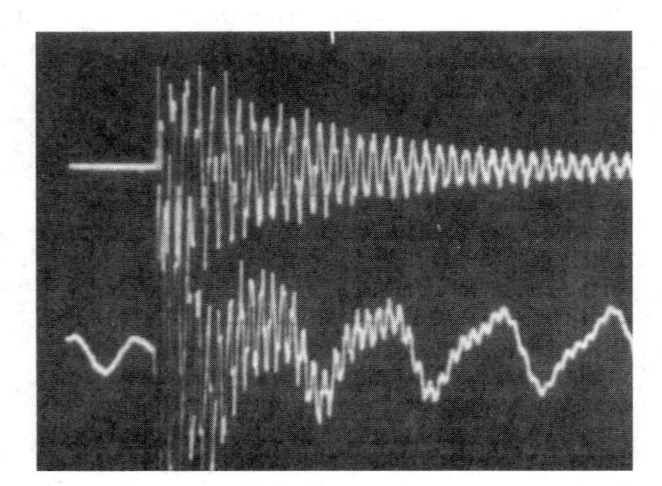

Fig. 3. Oscilloscope traces showing the output of Kynar vibration sensor fastened to a bat with both ends free (upper trace) and with its handle clamped in a vise (lower trace). The bat was hit near the tip in both cases, and the sweep speed was 20 ms/div with a vertical gain setting on the oscilloscope of 0.2 V/div.

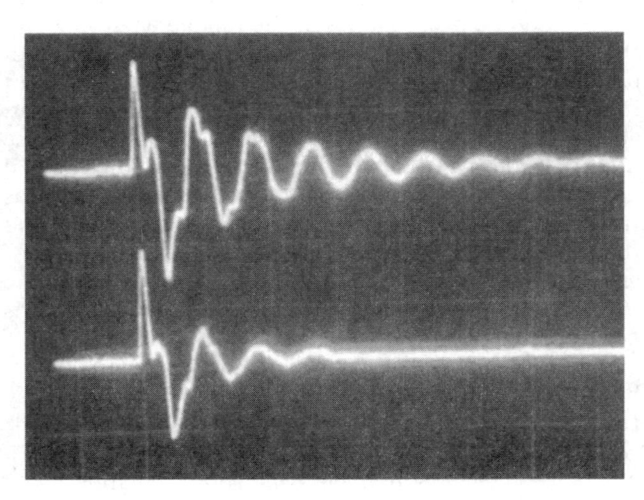

Fig. 5. Oscilloscope traces showing the output of a Kynar vibration sensor fastened to a bat that was hand held loosely (upper trace) and tightly gripped (lower trace). For these data, the sweep speed was 5 ms/div (50 ms full scale), and the vertical gain was 0.2 V/div.

Table I. Bat parameters.

	Aluminum bat	Wooden bat
Mass (kg)	0.825	0.846
Length (m)	0.81	0.84
Moment of inertia about cm (kg m^2)	0.046	0.047
Distance butt to cm (m)	0.46	0.51
Fundamental frequency of oscillation clamped handle (Hz)	27.2	18.2
First harmonic, clamped handle (Hz)	317	209
Lowest frequency free bat (Hz)	242	163
Deformation under 98-N load (m)	1.5×10^{-4}	4.2×10^{-4}

bat that may be of interest, are given in Table I. Since the measured length and inertial properties of the two bats are rather similar and their frequencies of vibration differ by 50%, the stiffness of the two bats must be quite different. Because one expects the natural frequency to be proportional to the square root of the stiffness, the aluminum bat should be more than at least twice as stiff as the wooden bat. It was noted that it required higher gain on the oscilloscope when the oscillations of the aluminum bat were being observed than when the oscillations of the wooden bat were being observed (0.2 vs 1 V/div), and this is a good indication that the aluminum bat is stiffer. To obtain a quantitative comparison, each bat was subjected to a downward force near its center with the ends of the bat blocked up off a table. Under comparable loads (98 N), a point on the wooden bat deformed approximately two and a half times as much as the corresponding point on the aluminum bat.

These tests were conducted with softball bats struck by softballs. When the same bats were struck with a hardball, the results led to the same conclusions—the hand-held bat behaves as if it were a free body. This is to be expected, since the contact time of a hardball on a bat has been determined to be about 1.5 ms (Ref. 5) compared to the 3.5-ms contact time of a softball on a bat.

These data lead to some interesting conclusions. Grip firmness at impact time should not influence the post-impact velocity of the ball. Grip firmness probably does have a significant influence on the bat velocity and bat position control up to the impact. However, if the batter were to release the bat completely at the time of impact, the subsequent trajectory of the ball should be the same as if the bat were firmly gripped throughout the swing. The subsequent trajectory of the bat would be quite different in these two cases, much to the dismay of the pitcher. In addition, the concept of adding some fraction of the hand and arm mass to the bat mass to get an effective mass or striking mass[7,8] seems to be contradicted by the results presented in this article.

[1]P. Kirkpatrick, "Batting the ball," Am. J. Phys. **31**, 606–613 (1963).

[2]H. Brody, "The sweet spot of a baseball bat," Am. J. Phys. **54**, 640–643 (1986).

[3]A. Weyrich, P. Messier, B. Ruhman, and M. Berry, "Effects of bat composition, grip firmness, and impact location on postimpact ball velocity," Med. Sci. Sports Exercise **21**, 199–205 (1989).

[4]H. Brody, "Models of tennis racket impacts," Int. J. Sports Biomech. **3**, 293–296 (1987).

[5]S. Plagenhoef, *Patterns of Human Motion* (Prentice-Hall, Englewood Cliffs, NJ, 1971), p. 71. Based on 4000-frame/s photography, the contact time of a softball on a bat was found to be 3.5 ms and the contact time of a tennis ball on the strings was 4–5 ms. A similar result for tennis balls was found using a laser and an oscillosocpe by H. Brody, Am. J. Phys. **47**, 482–487 (1979).

[6]Kynar piezoelectric film is available form Pennwalt Corp., P. O. Box C, King of Prussia, PA 19406.

David Griffing, *The Dynamics of Sports* (Dalog, Oxford, OH, 1987), 3rd ed., pp. 101–103.

[8]S. Plagenhoef, *Patterns of Human Motion* (Prentice-Hall, Englewood Cliffs, NJ, 1971), pp. 61–63.

The effect of spin on the flight of batted baseballs

A. F. Rex
Physics Department, University of Puget Sound, Tacoma, Washington 98416

(Received 30 July 1984; accepted for publication 12 November 1984)

Distances traveled by batted baseballs with various spin types and rates have been calculated using a computer. It is found that balls hit with backspin tend to travel farther than balls hit with little spin or topspin, particularly for lower trajectories. Two types of hits are considered specifically: the long fly ball of potential home run distance and the low line drive to the outfield.

I. INTRODUCTION

Recently, Frohlich[1] discussed an important factor affecting the flight of baseballs: the roughness of the surface of the ball. Frohlich's calculations suggest that relatively minor changes of the surface may have a dramatic effect on the trajectory of a batted ball. While great care has been taken to ensure uniformity of the size, weight, and coefficient of restitution of baseballs, such care has not always been taken with the surface material. This factor alone could result in a great variation in the number of home runs hit in different years as changes are made to the surface of the ball.

The purpose of this paper will be to study another major factor affecting the trajectory of batted baseballs: spin. The effect of spin on pitched balls has been studied most thoroughly by Selin,[2] who used high speed photography, and by Briggs,[3] who used wind tunnel measurements to calculate the lateral deflection of pitched balls with various speeds and spin rates. In this paper the results of Briggs are used to estimate the lateral force on a ball with given speeds and spin rates. The force obtained in that manner is then used to calculate the probable trajectory of batted balls of different realistic speeds and spin rates.

II. METHODS AND CALCULATION

For a pitched ball suffering lateral deflection d, it may be assumed that

$$d \cong at^2/2 \,,$$

where a is the lateral acceleration and t is the time it takes

Reprinted from *American Journal of Physics* **53**, 1073–1075 (1985);

for the pitched ball to travel to home plate. This approximation is good as long as $d \ll x$ (the distance from the mound to home plate) and the acceleration is constant. Let the lateral force be F ($= ma$). Now note that $t \cong x/v$, where v is the speed of the pitch, again with the assumption that $d \ll x$. Thus the lateral "Magnus" force is

$$F \cong 2mv^2 d/x^2 . \tag{1}$$

Briggs showed that d is proportional to the rate of spin but independent of v. Therefore, the force is proportional to the spin rate and the square of the velocity, and we may write

$$F = kv^2 \nu , \tag{2}$$

where ν is the spin rate in hertz and k is a constant to be determined from the calculations of Briggs. Combining Eqs. (1) and (2), the constant k is found to be

$$k = 2md/x^2 \nu . \tag{3}$$

The mass of the ball is 0.145 kg. The value for x used by Briggs was 60 ft, or about 18.3 m. For all values of d and ν given by Briggs, the best value of k is 1.1×10^{-5} N·s^3/m^2.

The direction of the Magnus force on a batted ball depends on the direction of the spin. For simplicity it will be assumed that the spin is either topspin or backspin, i.e., the axis of rotation is perpendicular to the plane in which the ball flies. This means that the ball will not leave the plane in which it moves initially, so that the problem will remain two-dimensional. It should be noted that sidespin is sometimes substantial when a ball is hit near the end of the bat or when the batter swings much too early or much too late. For a well-struck ball, however, backspin or topspin usually dominates.

The direction of the Magnus force changes as the ball flies through the air since under the conditions stated the force vector will remain perpendicular to the velocity. Using Eq. (2) the components of the Magnus force may be written as

$$F_x = -k\nu v\dot{y}, \quad F_y = k\nu v\dot{x} . \tag{4}$$

In Eqs. (4) it is assumed that the positive x axis is directed outward horizontally from home plate and the positive y axis is directed vertically upward. Also, the (arbitrary) designation that $\nu > 0$ corresponds to backspin and $\nu < 0$ to topspin has been adopted.

In order to give realistic results, it is necessary to include the drag force. Since this paper is concerned with the effects of spin, a constant drag coefficient $C_d = 0.5$ will be assumed. The components of the drag force are[1]

$$F_x = -\rho A C_d v\dot{x}/2, \quad F_y = -\rho A C_d v\dot{y}/2 , \tag{5}$$

where A is the cross-sectional area of the ball (approximately 42 cm^2) and ρ is the fluid density of air (1.29 kg/m^3). Combining Eqs. (4) and (5) yields the equations in motion:

$$m\ddot{x} = F_x = -k\nu v\dot{y} - \rho A C_d v\dot{x}/2 ,$$
$$m\ddot{y} = F_y = k\nu v\dot{x} - \rho A C_d v\dot{y}/2 - mg . \tag{6}$$

Since these equations cannot be solved in closed form for $x(t)$ and $y(t)$, the trajectories were calculated numerically using a BASIC program on an Apple IIe computer. The program used Taylor series expansions with the method outlined by Frohlich.[1,4] Letting $B = k\nu/m$ and $C = \rho C_d A/2m$,

$$\ddot{x} = -Bv\dot{y} - Cv\dot{x}, \quad \ddot{y} = Bv\dot{x} - Cv\dot{y} - g .$$

Then

$$\dddot{x} = -B(\dot{v}\dot{y} + v\ddot{y}) - C(\dot{v}\dot{x} + v\ddot{x}) ,$$
$$\dddot{y} = B(\dot{v}\dot{x} + v\ddot{x}) - C(\dot{v}\dot{y} + v\ddot{y}) ,$$

where $\dot{v} = (\dot{x}\ddot{x} + \dot{y}\ddot{y})/v$.

Finally, the Taylor expansions

$$\Delta x = \dot{x}\Delta t + \tfrac{1}{2}\ddot{x}\Delta t^2 + \tfrac{1}{6}\dddot{x}\Delta t^3 ,$$
$$\Delta \dot{x} = \ddot{x}\Delta t + \tfrac{1}{2}\dddot{x}\Delta t^2$$

were used to calculate the horizontal increments in position and velocity, with similar formulas used for the vertical variables. It was found that time increments of 0.005 s were sufficient to yield final horizontal distances accurate to within several centimeters of those computed using much smaller intervals. Therefore, $\Delta t = 0.005$ s was used throughout.

III. RESULTS AND DISCUSSION

As one would expect, balls hit with backspin were generally found to travel farther than balls hit with little spin or topspin. The calculated distances of balls hit with reasonable speeds (35 and 40 m/s) are shown in Tables I and II as a function of spin rate and initial angle above the horizontal. A range of initial ("take off") angles of 30°–60° was chosen

Table I. Distance traveled (meters) by batted balls as a function of spin rate and initial angle above the horizontal for balls with an initial speed of 35 m/s.

	Spin Rate (Hz)				
Angle (degrees)	30	20	0	− 20	− 30
60	95.4	95.5	95.7	95.8	95.9
54	105.6	105.5	105.4	105.1	104.8
48	111.7	111.3	110.8	110.2	109.7
45	113.0	112.5	111.8	110.7	110.5
44	113.2	112.7	112.0	110.8	110.6
43	113.3	112.8	111.8	110.7	110.2
42	113.0	112.8	111.8	110.7	110.2
36	111.0	109.1	108.1	106.5	106.2
30	102.6	101.7	99.8	98.4	98.0
15	64.6	64.0	62.8	61.6	61.0
10	48.0	47.4	46.1	45.4	44.8

Table II. Distance traveled (meters) by batted balls as a function of spin rate and initial angle above the horizontal for balls with an initial speed of 40 m/s.

	Spin Rate (Hz)				
Angle (degrees)	30	20	0	− 20	− 30
60	119.7	119.9	120.5	120.8	120.9
54	132.8	132.7	132.5	132.3	132.1
48	140.7	140.4	139.3	138.6	137.8
45	142.7	141.8	140.9	139.5	139.0
44	142.8	142.3	140.9	139.4	138.9
43	143.3	142.3	140.8	139.2	138.7
42	143.2	142.1	140.6	139.0	138.3
36	139.4	138.7	136.6	134.5	133.2
30	130.4	128.9	126.9	124.4	122 8
15	83.0	81.7	79.7	77.7	76.3
10	61.2	60.4	58.3	56.8	56.1

so that the flight of long flv balls (potential home runs) could be examined. It was assumed that balls were struck from a height of 1.0 m above the ground and that the spin rate remained constant throughout the flight. Although no data on rotation rates of batted balls is available, it is not unreasonable to believe that they should be comparable with pitched balls. Briggs measured rotation rates in the range 20–30 Hz for pitched balls. The "zero-spin" case is presented for reference. Actually a ball with zero spin would be aerodynamically unstable.[5] The zero-spin trajectories calculated will be very close to those of balls hit with just enough spin to be stable.

Clearly the effect of spin is greater as the take-off angle is decreased. This can be understood by considering the direction of the Magnus force. When a ball with backspin follows a lower trajectory, the Magnus force is more nearly vertical, thus producing greater lift. For higher trajectories the vertical component of the force is smaller. When a ball is hit with topspin it becomes advantageous to use a higher trajectory, for in this case the vertical component of the force (which is now negative) should be minimized. Therefore, it is not surprising that the take-off angle which produces the greatest distance for a given initial speed is slightly less for balls with backspin (42°–43°) than for balls with topspin (44°–45°). It should be noted, however, that it would be very difficult to hit a ball with a high trajectory and topspin (try it!). The higher topspin values become more realistic at lower take-off angles. At first glance it may appear that the added distance of up to 5 or 6 m gained by healthy backspin is insignificant for a flight of over 100 m. Such distances are crucial in baseball, however (a "game of inches"). Five to 6 m is just the difference between home run distances of a "hitter's park" (e.g., Wrigley Field, Chicago) and a "pitcher's park" (e.g., Busch Memorial Stadium, St. Louis).

Since the effects of spin become more pronounced for lower trajectories, additional calculations were performed for 10° and 15° take-off angles. Initial speeds of 35 and 40 m/s again produced reasonable distances. In a game situation these cases correspond to hard hit " line drives" which strike the ground in the outfield. Such balls hit with topspin produce the well-known "sinking line drive" effect. From calculations presented in Tables I and II it is clear that this effect is substantial. These results demonstrate how difficult it may be for an outfielder to judge a line drive properly and how crucial it is for him to determine the nature of the spin of the batted ball. Since the ball is initially too far away for the fielder to see the spin clearly, he apparently must use the flight of the ball to make that determination.

Finally, it should be noted that it would be interesting to perform similar calculations for softballs, should wind tunnel data become available. Frohlich[1] has suggested that aerodynamic drag is probably stronger for softballs than for baseballs, due to the larger area-to-mass ratio. If that is so, then it is likely that spin will also be more significant for the softball. At least one experienced softball power hitter[6] has estimated that the proper backspin may improve the distance traveled by a softball by at least 10 m for a given initial angle and speed.

[1]C. Frohlich, Am. J. Phys. **52**, 325 (1984).
[2]C. Selin, Res. Q. **30**, 232 (1959).
[3]L. J. Briggs, Am. J. Phys. **27**, 589 (1959).
[4]C. Frohlich, Am. J. Phys. **49**, 1125 (1981).
[5]R. G. Watts and E. Sawyer, Am. J. Phys. **43**, 960 (1975).
[6]C. Grobmyer, private communication (1984).

The effects of coefficient of restitution variations on long fly balls

David T. Kagan
Physics Department, California State University, Chico, Chico, California 95929-0202

(Received 2 December 1988; accepted for publication 14 March 1989)

The coefficient of restitution of major league baseballs is required to be 0.546 ± 0.032. These allowed variations affect the launch velocity and ultimately the range of fly balls. The variations in the range for well-hit balls are calculated here to be on the order of 15 ft. These calculations provide an interesting collection of mechanics problems that might be of interest for undergraduate students.

I. INTRODUCTION

The coefficient of restitution for a collision is the ratio of the final relative velocity to the initial relative velocity of the colliding objects. The coefficient of restitution of a major league baseball is required to be 0.546 ± 0.032.[1,2] The question raised here is how will the allowed variations in the coefficient of restitution affect the distance traveled by a well-hit ball?

The first step toward the answer is to find the effect of coefficient of restitution variations on the velocity of the ball as it leaves the bat, referred to here as the launch velocity. Next, the range for a baseball as a function of the launch velocity must be studied. To keep some resemblance to reality, air resistance must be included. Finally, the two results can be put together to find the variations in the range due to fluctuations in the coefficient of restitution. Since the interest here is just the variations in the range, as opposed to the values of the range, and because it is often easier and frequently more accurate to calculate variations, as opposed to the actual values themselves, only the variations will be found.

II. THE VARIATION OF THE LAUNCH VELOCITY WITH THE COEFFICIENT OF RESTITUTION

The collision viewed from a coordinate system moving with the center of mass of the bat at the moment just before impact is shown in Fig. 1. According to the standard assumptions, the bat can be treated as a free body.[3] The equations needed to describe the ball–bat collision come from the conservation of linear momentum, the conservation of angular momentum, and the definition of the coefficient of restitution. These equations are

$$mv_0 = MV - mv, \tag{1}$$

$$I\omega_0 - mv_0 b = I\omega + mvb, \tag{2}$$

$$e = (v + V - \omega b)/(v_0 + \omega b), \tag{3}$$

where m is the mass of the ball, M is the mass of the bat, I is the moment of inertia of the bat about the center of mass, v_0 and v are the initial and final velocities of the ball, V is the final velocity of the center of mass of the bat, ω_0 and ω are the initial and final angular velocities of the bat, b is the

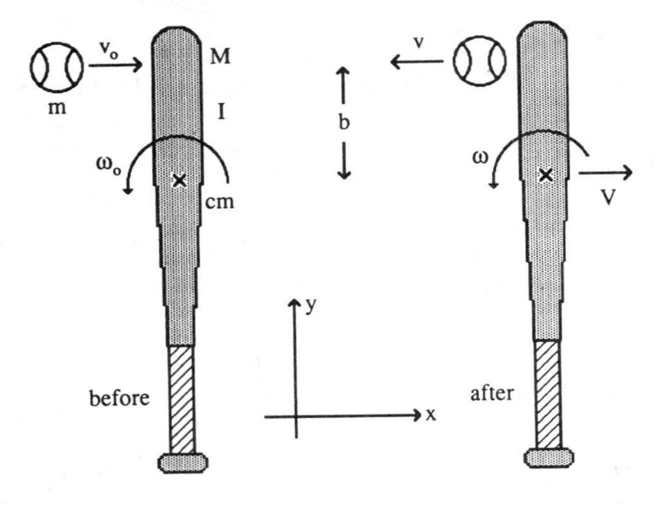

Fig. 1. The ball–bat collision viewed from a coordinate system moving with the center of mass of the bat at the moment just before impact.

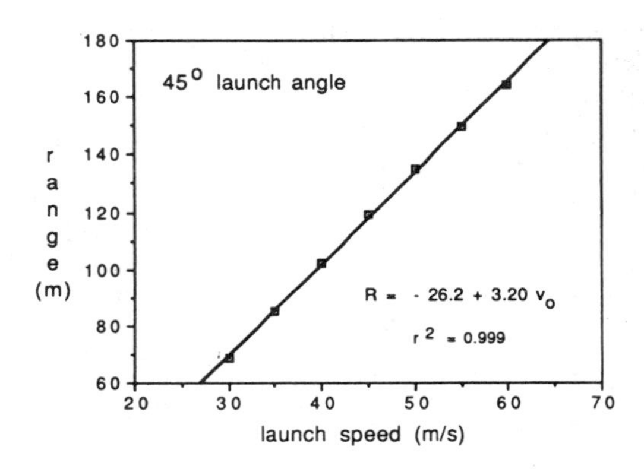

Fig. 2. Range versus launch speed for 45° including the least-squares-fit line.

distance between the center of mass of the bat and the point where the ball collides, and, finally, e is the coefficient of restitution. Equations (1), (2), and (3) assume that all the motion is happening in a plane, which is only approximately correct.

Eliminating V and ω and solving for the final velocity of the ball gives

$$v = [I\omega_0 b(1+e) - mb^2 v_0 - Iv_0(m/M - e)] \times [mb^2 + I(1+m/M)]^{-1}. \quad (4)$$

This is the launch velocity of the ball in the c.m. frame. Equation (4) can be differentiated to find the variation of the launch velocity of the ball with the coefficient of restitution,

$$\frac{dv}{de} = \frac{\omega_0 b + v_0}{mb^2/I + (1+m/M)}. \quad (5)$$

Using some typical values such as $m \approx 0.15$ kg, $M \approx 1.0$ kg, $b < 0.10$ m, $v_0 \approx 50$ m/s, $\omega_0 \approx 10$ rad/s, and $I \approx 0.05$ kg m^2, the terms in Eq. (5) are found to be on the order of

$$\omega_0 b \approx 1 \text{ m/s},$$
$$v_0 \approx 50 \text{ m/s},$$
$$mb^2/I \approx 0.03,$$
$$1 + m/M \approx 1.2.$$

Table I. The range in meters for a baseball calculated for various angles and launch speeds in meters per second. The calculations were done using $\mathbf{a} = -\mathbf{g} - cv\mathbf{v}$, where $c = 0.0050\ \text{m}^{-1}$.

v	30°	35°	40°	45°	50°	55°	60°
30	63.248	67.018	69.327	68.712	66.800	63.535	57.895
35	79.982	83.443	86.893	85.379	83.716	79.050	72.788
40	97.342	102.185	103.210	102.367	99.582	94.779	86.862
45	112.931	119.295	121.277	119.388	115.409	109.254	99.814
50	130.664	134.463	137.533	134.708	131.006	123.475	112.517
55	146.189	151.331	151.832	149.722	144.918	136.167	124.870
60	161.502	166.018	167.476	164.353	158.428	148.463	135.827

The point is that the first term in the numerator and the first term in the denominator of Eq. (5) can both be neglected, leaving the result

$$\frac{dv}{de} = \frac{v_0}{1+m/M}. \quad (6)$$

It is interesting to note that this is the same result as for the collision of two billiard balls. This is due to the fact that the collisions where the ball is well hit occur near the center of mass of the bat.

It is easiest to transform out of the c.m. frame and into the "home plate" frame at this point. The variation of the launch velocity with the coefficient of restitution is now given by

$$\frac{dv}{de} = \frac{v_0 + v_b}{1+m/M}, \quad (7)$$

where v is now the launch velocity, v_b is the speed of the center of mass of the bat, and v_0 is now the speed of the pitch. It will be assumed that Eq. (7) is valid even when the ball is launched at an angle.

III. THE RANGE AS A FUNCTION OF THE LAUNCH VELOCITY

The range of a projectile in the absence of air resistance is given by

$$R = (v^2 \sin 2\theta)/g.$$

However, air resistance cannot be neglected in the flight of baseballs. Assuming that the air resistance is proportional to the velocity squared, the components of the acceleration of the ball can be written as

$$a_x = -cvv_x,$$
$$a_y = -g - cvv_y,$$

where g is the acceleration due to gravity, v is the speed of the ball, v_x and v_y are the horizontal and vertical components of the velocity of the ball, and c is the drag factor.

Using the value $c = 0.0050$/m given by Brancazio,[4] the range of the ball for various launch speeds and firing angles was calculated numerically (e.g., see Swartz).[5] The results are summarized in Table I. Notice that the maximum range does not occur at the 45° from the no-air resistance case, but is closer to 40°. This is consistent with the idea that the air resistance causes a projectile to "run out of gas" and "drop like a rock." That is why outfiedlers are taught to get "under the ball."

Figure 2 is a typical plot of the range versus the launch speed for a given launch angle. The slope of the curve generated by these points is the variation of the range with the launch speed. The calculated points are remarkably close to a straight line. While least-squares fitting is usually reserved for fitting data, as opposed to calculated points, it provides a convenient method for finding the slope. Note that the points do not form a straight line exactly. They are above the line in the middle and below at the extremes.

Table II contains the least-squares fit values of the slopes as well as the correlation coefficients for the fits. The slopes vary from 2.60 to 3.32 s with the angle while the correlation coefficients are surprisingly high considering that the range of launch speeds covers a factor of 2.

IV. THE VARIATIONS IN RANGE DUE TO FLUCTUATIONS IN THE COEFFICIENT OF RESTITUTION

The variations in the range due to the fluctuations in the coefficient of restitution can now be found in a straightforward manner,

$$\frac{dR}{de} = \left(\frac{dR}{dv}\right) \cdot \left(\frac{dv}{de}\right). \tag{8}$$

While the calculation indicated by Eq. (8) can be repeated for many combinations of pitch speed, bat speed, and launch angle, it is more instructive just to use some typical values to get an idea of the size of the effect. Using an average value of 3 s for the variation of the range with the launch speed and the variation of the launch speed due to the coefficient of restitution from Eq. (7) yields

$$\frac{dR}{de} \approx (3\ \text{s}) \cdot \frac{v_0 + v_b}{1 + m/M}. \tag{9}$$

Typical values of $v_0 + v_b \sim 60$ m/s, $m \approx 0.15$ kg, and $M \approx 1.0$ kg give a value of

$$\frac{dR}{de} \approx 160\ \text{m}.$$

The value allowed by major league baseball of $de = 0.032$ results in range variations of

$$dR \approx 5.1\ \text{m} \approx 17\ \text{ft!}$$

This is about equal to the width of the warning track.

Table II. The least-squares-fit slopes for the range versus launch speed plots at various angles.

Launch angle	Slope $= \frac{dR}{dv}$	Correlation coefficient r^2
30°	3.29 s	1.000
35°	3.32 s	0.999
40°	3.28 s	0.999
45°	3.20 s	0.999
50°	3.06 s	0.999
55°	2.84 s	0.998
60°	2.60 s	0.998

V. CONCLUSIONS

These results bring up some interesting (if meaningless) questions. Is the difference between a warning track out and a home run really influenced by slight variations in the baseball? Is there a way that a pitcher can tell before he throws a pitch if he has a live or a dead ball? Could Dennis Eckersley have thrown Kirk Gibson, in the vernacular of baseball, "a good ball to hit"?

In all seriousness, the physics problem described here is a good fraction of an upper division undergraduate mechanics course. Its solution involves rotational collisions, the application of the conservation laws for angular and linear momentum, the concept of the coefficient of restitution, transformations to and from center-of-mass coordinates, projectile motion including air resistance, numerical analysis of motion, least-squares fitting, and, perhaps most importantly, it produces interesting results!

ACKNOWLEDGMENTS

The author wishes to thank his colleagues, Fred Boos and Eric Dietz, for their useful insights and critiques.

APPENDIX

It is appropriate to make some comments on the limitations of the model presented here. The value of the coefficient of restitution specified by the major leagues is prescribed to be measured by firing a ball at 85 ft/s (≈ 56 mph ≈ 26 m/s) at a wall of ash, the material of which the bats are constructed. The value of the coefficient of restitution is then the ratio of the rebound to incident velocities.[2] This value may be different from the value in a ball–bat collision due to several factors. First, the bat may absorb energy differently than the wall of ash (for example, by vibrating). Second, the actual relative velocities during the collisions discussed here are much higher than 26 m/s. The coefficient of restitution of a baseball is known to vary with impact velocities.[6] In general, the higher the impact velocity the lower the coefficient of restitution. Third, the tolerance in the coefficient of restitution of ± 0.032 is not necessarily a manufacturing tolerance. That is, the variations from ball to ball may actually be much smaller than this.

Approximations have also been made in describing the flight of the ball. The drag factor for a baseball is not well known. In addition, and perhaps more importantly, it has been suggested that the drag factor varies dramatically during the flight as the air passing the ball changes from partially turbulent flow to fully turbulent flow and back.[7,8] Also, the spin imparted to the ball during the collision has a substantial effect on the range.[9]

In summary, the answers to the equations raised in the conclusion are "probably not," "almost certainly not," and "definitely not, the baby cleared the fence by 25 ft!"

[1]Jeffrey Kluger, "What's behind the home run boom?" Discover, 78–79 (April 1988).

[2]Peter J. Brancazio, *Sportscience* (Simon & Schuster, New York, 1984), p. 219.

[3]Howard Brody, "The sweet spot of a baseball bat," Am. J. Phys. **54**, 640–643 (1986).

[4]Peter J. Brancazio, "Trajectory of a fly ball," Phys. Teach. **23**(1), 20–23

(1985).

[5]Clifford E. Swartz, *Used Math* (Prentice-Hall, Englewood Cliffs, NJ, 1973), p. 202.

[6]Stanley Plagenhoef, *Patterns of Human Motion* (Prentice-Hall, Englewood Cliffs, NJ, 1971), p. 83.

[7]Reference 2, p. 328.

[8]Cliff Frohlich, "Aerodynamic drag crisis and its possible effect on the flight of baseballs," Am. J. Phys. **52**, 325–334 (1984).

[9]Robert G. Watts and Steven Baroni, "Baseball–bat collisions and the resulting trajectories of spinning balls," Am. J. Phys. **57**, 40–45 (1989).

Chapter 3
THE GOLF BALL AND OTHER SPHERICAL OBJECTS

CONTENTS

This chapter contains six articles in all, one each on golf, bowling, "super ball," and shot put, and two on somewhat different aspects of basketball.

Golf

The first article in this chapter is "Maximum projectile range with drag and lift, with particular application to golf," by Herman Erlichson. In it, the author computes trajectories for driven golf balls[1] moving under the combined influence of gravity, drag (air resistance), and spin lift (Magnus force). He also calculates an optimum projection angle for maximum range and compares it with the loft angle built into typical golf driver clubs.

Unlike a baseball (Chap. 2) for which the air resistance is known to be *quadratic* in speed over the entire range of speeds encountered in play, the precise dependence of air resistance on speed for a golf ball has been the subject of some debate.[2] Erlichson's trajectories, for example, are calculated on the assumption that spin lift and drag are both *linear* in the velocity. His results agree fairly well with available empirical studies with one notable exception: The optimum angle of launch according to his calculations is 16° while golf driver clubs, which impart launch angles approximately equal to the loft angle of the club face, are typically designed with loft angles closer to 11°.

One reason for the difference between theory and practice may lie in the distinction between range in air, or "carry" as it is known in golf, and the total distance, or "drive," which includes both carry and "run" (the added distance achieved by a bounding ball after striking the ground). Clearly, while a higher launch angle may produce a longer carry, it might not necessarily produce a longer drive.[3]

It has also been recently suggested that the discrepancy stems from Erlichson's assumption of linear rather than quadratic velocity dependence for spin lift and drag.[4] Although some discussion continues on the precise character of drag and spin lift on a golf ball for speeds typically encountered in play, it appears that a complete and accurate picture of golf ball trajectories is currently emerging.

Bowling

"Bowling frames: Paths of a bowling ball," by D. C. Hopkins and J. D. Patterson, is the second article in this chapter. The authors calculate the path of a bowling ball launched horizontally with arbitrary speed and spin on a surface whose coefficient of kinetic friction is assumed constant. The calculations show that the ball moves initially in a parabolic path, and then a straight line as it begins to roll without slipping. This would seem to describe a fairly gentle curve as seen from the vantage point of the bowler, especially when one considers that a standard bowling lane is quite long and narrow (63 ft × 3.5 ft).

As cited in the article, however, the folklore of bowling includes a characterization of a "hook" ball as one "which starts rolling straight but which *breaks sharply* before reaching the pins" (emphasis added).

Reconciliation of these apparently different descriptions may not be all that difficult. First of all, because a bowling ball tends to be very massive, the spin-induced lateral force of friction on the ball requires a fairly long time to overcome the ball's inertia. As a result, the ball, although moving in a parabolic path, begins at a point where the curvature is almost totally imperceptible, thereby giving the initial impression that the ball is moving "straight." Parabolic motion also means that the lateral deflection is quadratic in both time and distance and, as a result, the greatest deflection tends to occur both later in time and farther down the lane, i.e., closer to the pins.

Also, as Hopkins and Patterson point out, some possible variability in the coefficient of kinetic friction, together with some initial bouncing of the ball, might result in more curvature later in the path, thereby creating or adding to the impression of a "hook."

Super Ball

The third article in this chapter, "Kinematics of an ultraelastic rough ball," by Richard L. Garwin, calculates some of the familiar, if bizarre, characteristics of the "Super Ball." It is also shown that these characteristics are largely predictable on the basis of few assumptions, the principal one being conservation of kinetic energy (i.e., perfectly elastic collisions). Garwin provides a key physical insight into the workings of the Super Ball when he states in his concluding remarks that:

> the striking behavior of the Super Ball is due not so much to an extremely low internal dissipation for elastic waves as it is to a near equality between periods for a bounce of the ball from the wall and for a torsional oscillation of the ball.

Basketball

"Physics of basketball," by P. J. Brancazio, and "Kinematics of the free throw in basketball," by A. Tan and G. Miller, are the fourth and fifth articles in this chapter. Brancazio determines that there is an optimum shooting angle in basketball which employs the lowest launch speed, requires the smallest shooting force, and provides the greatest margin for error. He also determines that the most significant effect of air resistance on a basketball is to "aid the shooter by increasing the margins for error!"

Tan and Miller focus on the kinematics of the two basic styles of free throw in basketball, the familiar and almost universally used overhand push shot, and the very unusual, rarely used underhand loop shot made famous, although for different reasons, by Wilt Chamberlain and Rick Barry.[5] According to Tan and Miller, kinematic analysis favors the overhand push shot despite the fact that several empirical studies cited in their article indicate that free throw percentages tend to be higher with the underhand loop shot.

Shot Put

The final article in this chapter, "Maximizing the range of the shot put," by D. B. Lichtenberg and J. G. Wills, addresses the question of the optimum angle at which a shot should be launched to achieve maximum horizontal range. While their initial claim may appear startling (viz., that the optimum launch angle is 42° rather than 45°), the reason for the discrepancy is shown to be the difference in the *vertical* location of the shot at the beginning and end of its motion. Stated another way, if the shot were launched from ground level (rather than six or seven feet above it), the optimum angle of launch would be 45°. Interestingly enough, air resistance *does* shorten the range of the shot (by about 1/2 ft for a 70-ft put) but, otherwise, has very little effect on the optimum angle of launch.[6]

---

[1]The trajectory characteristics of a well-driven golf ball are quite unique, beginning with a very low angle of launch (typically 10°–12° above the horizontal), a straight to slightly concave-upward shape for the first two-thirds of its flight, a very steep convex-downward shape for the last third of its flight, a fairly low maximum vertical height, plus an inordinately long time of flight (around 7 s!) P.G. Tait ["Long driving," *Badminton Mag.* (March, 1896)] noted that the combination of unusual shape, measured carry ($\approx$200 yards), and time of flight ($\approx$6.5 s) for a well-driven golf ball *could not be reconciled* by any ideal (parabolic) trajectory, and he was apparently the first person to correctly identify air resistance and spin lift (along with gravity) as the forces responsible for such a trajectory.

[2]William M. MacDonald, "The physics of the drive in golf," *Am. J. Phys.* **59**, 213 (1991), and references cited therein.

[3]A higher launch angle means a steeper angle of descent, with less run and possibly less drive a consequence.

[4]See Ref. 1. MacDonald, citing experiments performed more recently than those on which Erlichson based his work, calculates golf ball trajectories using a quadratic model of lift and drag. He also compares his results to those of Erlichson, and obtains results largely in agreement with empirical evidence on golf ball carry cited by D. Williams in "Drag force on a golf ball in flight and its practical significance," *Q. J. Mech. Appl. Math.* **XII** (3), 387 (1959).

[5]As discussed in the Introduction, Chamberlain resorted to the underhand shot in an unsuccessful attempt to raise an embarrassingly poor foul shooting percentage (51%). Barry's use of the underhand shot is memorable because his career foul shooting percentage (90%) has yet to be improved upon!

[6]Lichtenberg and Wills report that the presence of air resistance changes the optimum angle of launch for a shot put by about 0.13° which, in turn, changes the horizontal range by a rather insignificant 0.01 in.

Maximum projectile range with drag and lift, with particular application to golf

Herman Erlichson
Department of Applied Sciences, The College of Staten Island, Staten Island, New York 10301

(Received 23 October 1981; accepted for publication 16 June 1982)

This paper explores the interesting problem of projectile motion without the vacuum idealization. Particular attention is paid to golf ball trajectories with and without lift. No lift trajectories with linear and quadratic drag are considered first. Then, trajectories with lift and linear drag are investigated. Projection angles for maximum range are determined for all these cases. Computer solutions are used throughout, with a Runge–Kutta routine used for all cases except for the well-known closed solution for the no lift, linear drag projectile.

I. INTRODUCTION

One of the interesting questions which is usually left unanswered in introductory or intermediate mechanics courses[1] is "What angle do you need for maximum projectile range if you are not in vacuum?" Virtually everybody has had the experience of hitting a golf ball, or a baseball, or throwing a football, and there is a natural curiosity about how to project a ball in air in order to achieve maximum range.

Three previous papers in this Journal have addressed the problem of the projection angle for maximum range for the shot put,[2] the discus,[3] and for round pebbles or stones.[4] In the case of the shot put the air resistance has little effect on the range, whereas in our analyses the drag is always a significant factor. In the case of the discus analysis, the author assumes quadratic drag and lift forces, whereas our analysis of the golf ball problem uses linear drag and lift forces. The discus paper also has the effect of wind as a major factor, in our paper we assume no wind. It is interesting to note that the optimum projection angles for the quadratic drag golf ball with no lift found in this paper (38° and 35°) are similar to the optimum discus projection angles found by Frohlich[3] (33° to 39°) for no wind conditions. The paper on throwing pebbles assumes a quadratic drag force and no lift.

The author's interest in these matters was all the greater because he has been an avid golfer for a long time. The large discrepancy between the approximately 11 deg of loft for the golf driver club and the 45 deg maximum range angle for a vacuum was the motivation to begin a study of the question of maximum projectile range in the presence of air resistance, with particular application to the flight of a golf ball.

Of course, the first question which should be addressed is the nature of the resisting, or drag force. There was evidence in the literature[5] that the drag force on a golf ball is linear. This case is well known and can be solved exactly if

Reprinted from *American Journal of Physics* **51**, 357–362 (1983);

there is no lift. However, the aerodynamic lift on a golf ball due to the backspin of the ball is a major force on the ball. Thus to do anything with the golf ball problem requires an analysis with three forces: gravity, drag, and lift. In this paper we start with the treatment of no lift projectiles, and then turn to the problem of trajectories with lift.

In 1949 Davies[6] investigated the lift and drag forces on spinning golf balls. He used the B. F. Goodrich wind tunnel with a wind stream velocity of 105 ft/s, and dropped rotating golf balls into this wind stream. The rotational speeds used were up to 8000 rpm. Davies found that the drag "increased nearly linearly from about 0.06 lb for no spin to about 0.1 lb at 8000 rpm" and that "the lift varied with the rotational speed as $L = 0.064[1 - \exp(-0.000\,26N)]$" ($L$ in lb and N in rpm).

In 1959 Williams[5] carried out an analysis which utilized data on golf ball carry as a function of projection velocity. The Reynolds number for Davies's work was 88 000; Williams's work was for the higher Reynolds numbers actually achieved in golf play. He used initial velocities in the range from 150 to 225 ft/s (Reynolds numbers from about 1.25×10^5 to 1.9×10^5). Williams showed that in this range of Reynolds numbers the drag force varied linearly with the velocity. The drag force is customarily written as

$$D = C_D A\,(1/2)\,\rho v^2\,, \tag{1}$$

where C_D is the drag coefficient, A is the cross-sectional area of the ball, ρ is the air density, and v is the ball speed. Williams showed that $C_D = (46/v)$ where v is in ft/s. Thus the drag coefficient drops significantly for higher speeds, giving the hard hitter an advantage.[7] A similar drop in the drag coefficient for smooth spheres occurs at a Reynolds number just above 2×10^5.[8] Since the drag coefficient for the golf ball goes inversely with v, Williams found that the drag force varied linearly with the speed, specifically that the drag force D was

$$D = 0.000\,783v \text{ lb}\,. \tag{2}$$

Williams's calculations were for the British ball whose diameter is 1.62 in., as compared to a diameter of 1.68 in. for the American ball.

II. LINEAR DRAG

As a first step, we consider the case of a nonspinning golf ball with linear air resistance. The component dynamical equations for this case are

$$-cv_x = m\frac{dv_x}{dt}\,, \tag{3}$$

$$-mg - cv_y = m\frac{dv_y}{dt} \tag{4}$$

with the well-known solutions

$$x = \frac{m}{c}v_{x_0}(1 - e^{-(c/m)t})\,, \tag{5}$$

$$y = \frac{m}{c}\left(\frac{mg}{c} + v_{y_0}\right)(1 - e^{-(c/m)t}) - \frac{mgt}{c}\,. \tag{6}$$

In this solution the golf ball starts at $x = y = 0$ with initial velocity components v_{x_0} and v_{y_0} at $t = 0$.

The weight of a golf ball is 1.62 oz and the projection speed of a good drive is about 200 ft/s.[9] The value of c, as given by Eq. (2) above, is $c = 0.000\,783$ lb/(ft/s). Hence,

$$\frac{c}{m} = \frac{0.000\,783}{(1.62/16)/32} = 0.25\ \text{s}^{-1}\,. \tag{7}$$

The terminal velocity of a freely falling golf ball is, therefore,

$$v_t = \frac{mg}{c} = \frac{32}{0.25} = 128\ \text{ft/s} = 87\ \text{mph}\,. \tag{8}$$

If we substitute the value for c/m given by Eq. (7) in Eqs. (5) and (6), and we use $v_0 = 200$ ft/s, we get

$$x = 800(\cos\theta_0)(1 - e^{-t/4})\,, \tag{9}$$

$$y = 4(128 + 200\sin\theta_0)(1 - e^{-t/4}) - 128t\,. \tag{10}$$

Figure 1 shows a plot of these coordinates for projection angle θ_0 in 5-deg intervals from 10° to 50°. Maximum range is achieved for an angle between 30° and 35°. Additional computer runs show that the maximum range of 503.9 ft occurs for $\theta_0 = 32°$.

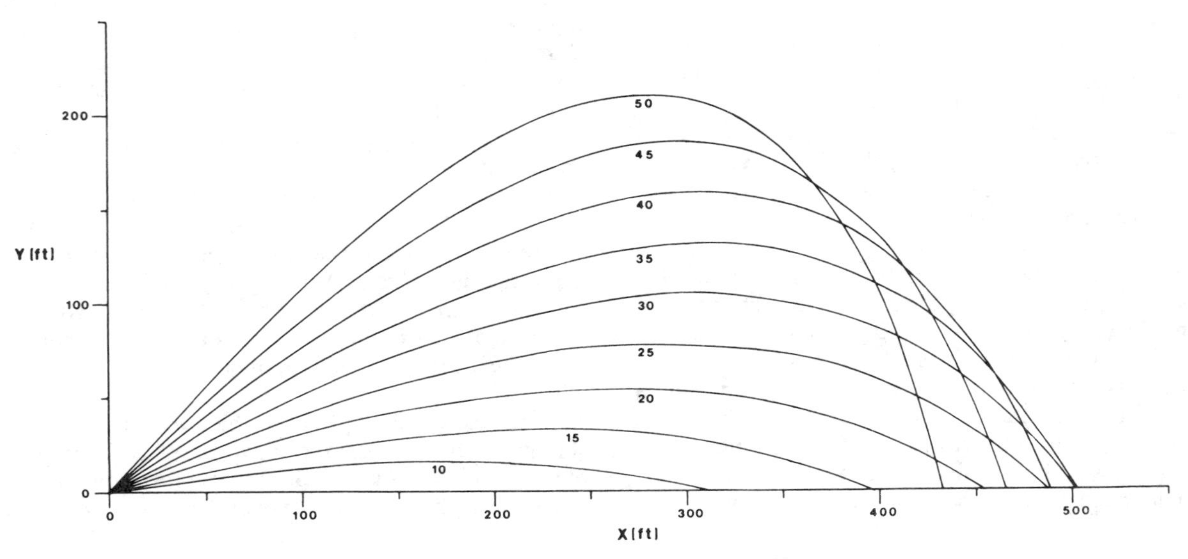

Fig. 1. Trajectories for various projection angles, linear drag, $c/m = 0.25$.

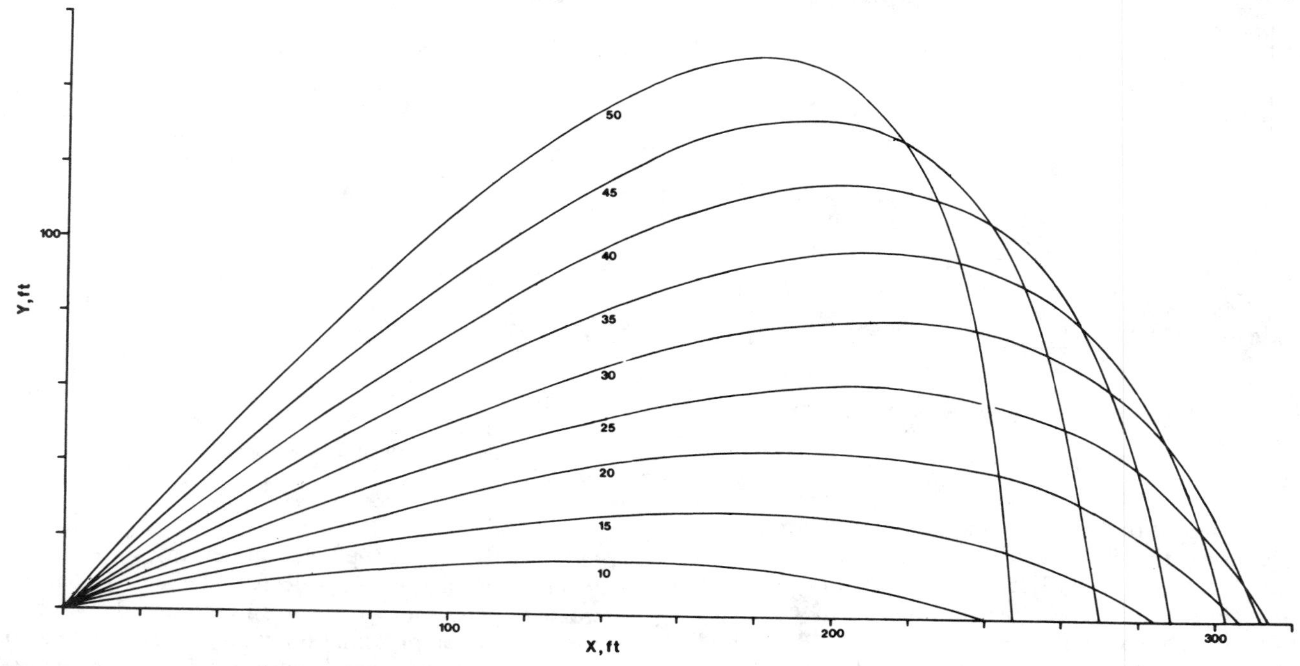

Fig. 2. Trajectories for various projection angles, linear drag, $c/m = 0.5$.

It is of interest to examine the effect of varying the drag coefficient c. Suppose, for example, that the drag is doubled so that $c/m = 0.5$. Under these circumstances we have

$$x = 400 \cos \theta_0 (1 - e^{-t/2}) , \tag{11}$$

$$y = 2(64 + 200 \cos \theta_0)(1 - e^{-t/2}) - 64t . \tag{12}$$

Figure 2 shows a plot of Eqs. (11) and (12) for θ_0 in 5-deg intervals from 10° to 50°. The maximum range is now 314.2 ft, and it occurs for a projection angle of 26°.

III. QUADRATIC DRAG

For low speeds in air the drag force goes as the square of the speed. Although the well-driven golf ball faces linear drag, the terminal velocity of 128 ft/s found in Eq. (8) can be used to get a hypothetical quadratic drag coefficient for a slowly moving golf ball as follows:

$$c_2 v_t^2 = mg ,$$

$$\frac{c_2}{m} = \frac{g}{v_t^2} = \frac{32}{(128)^2} = \frac{1}{512} \text{ ft}^{-1} . \tag{13}$$

[10]

The equations of motion for the case of quadratic drag are

$$- c_2 v^2 \cos \theta = m \frac{dv_x}{dt} , \tag{14}$$

$$- mg - c_2 v^2 \sin \theta = m \frac{dv_y}{dt} \tag{15}$$

or

$$\frac{dv_x}{dt} = - \frac{c}{m} \sqrt{v_x^2 + v_y^2}\, v_x , \tag{16}$$

$$\frac{dv_y}{dt} = - g - \frac{c}{m} \sqrt{v_x^2 + v_y^2}\, v_y . \tag{17}$$

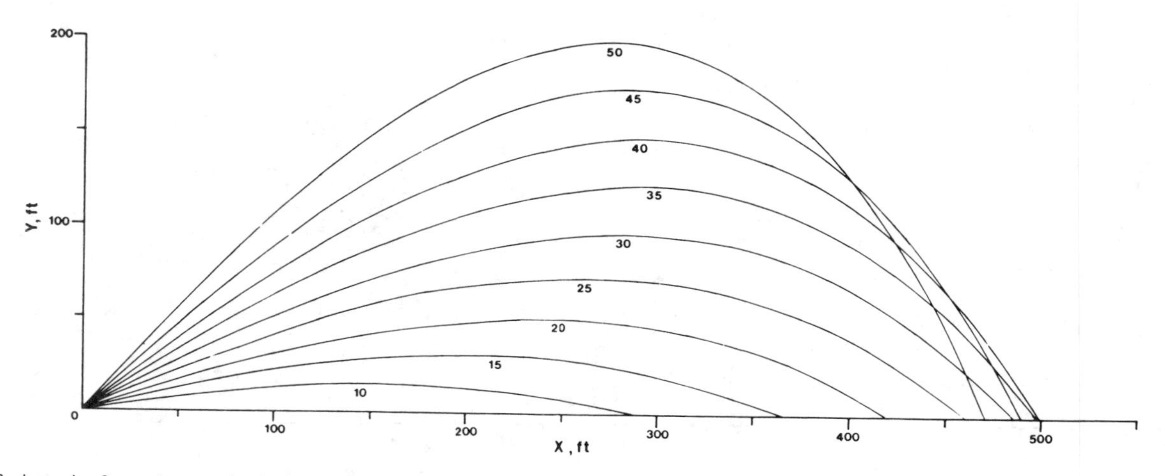

Fig. 3. Trajectories for various projection angles, quadratic drag, $c/m = 1/512$.

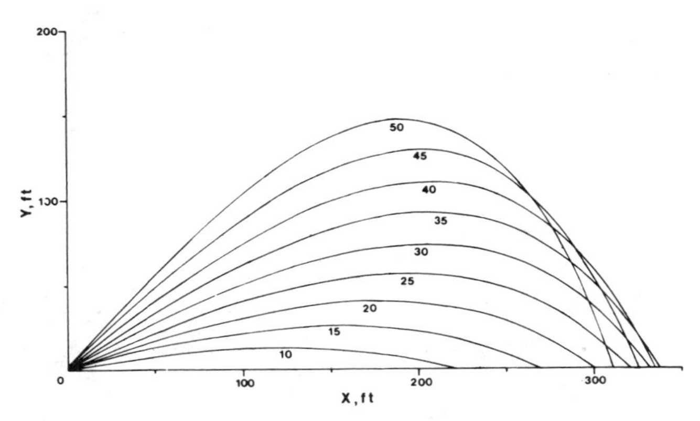

Fig. 4. Trajectories for various projection angles, quadratic drag, $c/m = 1/256$.

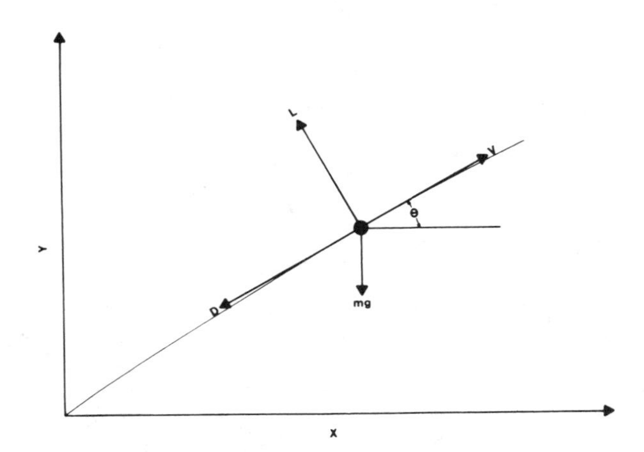

Fig. 5. Forces on a golf ball hit with backspin.

These coupled differential equations were solved numerically using the Runge–Kutta method. The subroutine RKGS from the IBM System/360 Scientific Subroutine Package, Version III, was used for this purpose.

Figure 3 shows the Runge–Kutta solution. A maximum range of 499.1 ft was achieved at an initial projection angle of 38 deg. The drag coefficient was then doubled so that

$$\frac{c_2}{m} = \frac{1}{256}\ \text{ft}^{-1}. \tag{18}$$

Figure 4 shows the Runge–Kutta solution for this case. Now the maximum range was 336.6 ft for a projection angle of 35 deg.

IV. CHECK OF RUNGE–KUTTA ROUTINE

The Runge–Kutta routine RKGS was submitted to an accuracy check by comparing its solution for the linear drag case with the exact solution. The Runge–Kutta solution agreed to at least four places with the exact solution.

V. GOLF BALL WITH LIFT AND LINEAR DRAG

The forces on a golf ball hit with backspin are shown in Fig. 5. The lift force can be written as

$$L = C_L A \tfrac{1}{2} \rho v^2 . \tag{19}$$

This lift force is due to a pressure difference caused by the spinning ball. The backspin of the ball increases the air speed over the top of the ball and decreases it below the ball. Thus the air pressure is decreased above the ball and increased below the ball, providing an aerodynamic lift force. As a first approximation, it seems reasonable to assume that this varies linearly with ω, the backspin angular speed, and v, the speed of the ball. The linear dependence on ω is essentially in agreement with the experimental work of Davies quoted earlier. We also assume that C_L varies inversely with v, just as Williams found for C_D. Since ω does not change appreciably during the flight of the ball,[13] we will take it to be constant. This gives us a lift force which varies linearly with the speed v,

$$L = kv . \tag{20}$$

Newton's equations in the x and y directions are

$$-cv_x - kv \sin\theta = m\frac{dv_x}{dt},$$

$$-\frac{c}{m}v_x - \frac{k}{m}v_y = \frac{dv_x}{dt}, \tag{21}$$

$$-cv_y + kv\cos\theta - mg = m\frac{dv_y}{dt},$$

$$-\frac{c}{m}v_y + \frac{k}{m}v_x - g = \frac{dv_y}{dt}. \tag{22}$$

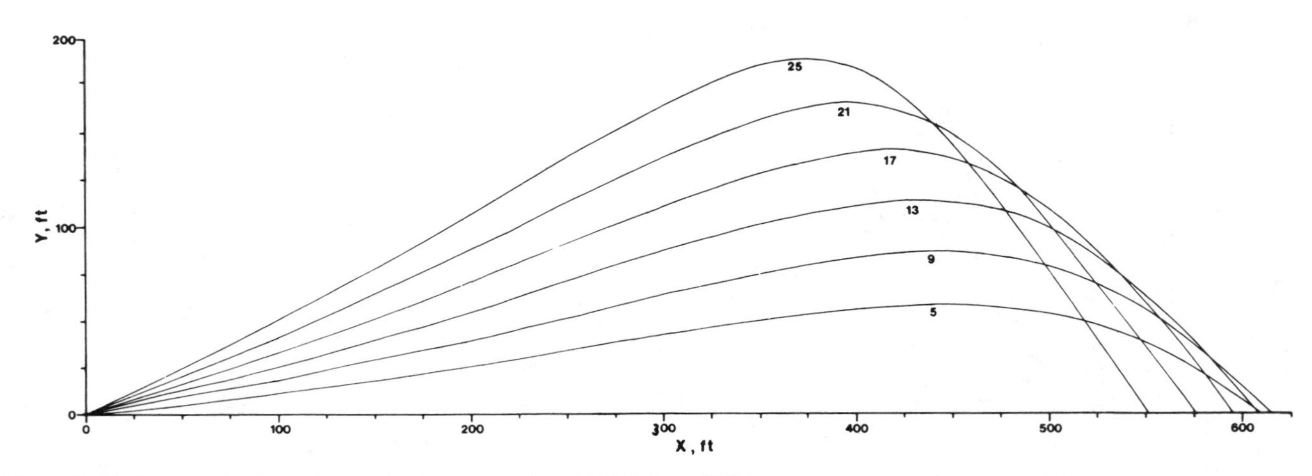

Fig. 6. Golf ball trajectories for various projection angles, $c/m = 0.25$, $k/m = 0.247$.

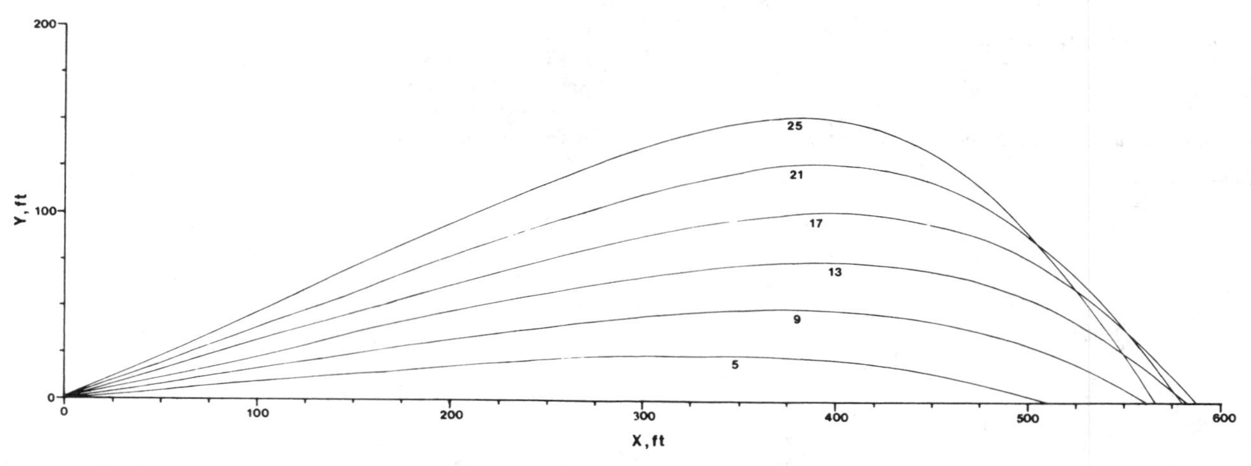

Fig. 7. Golf ball trajectories for various projection angles, $c/m = 0.25$, $k/m = 0.179$.

The initial lift on the golf ball is about 2.5 oz.[13] Hence, k/m can be calculated as follows:

$$kv_0 = 2.5/16 \text{ lb},$$

$$\frac{kv_0}{m} = \left(\frac{2.5}{16}\right) \left[\left(\frac{1.62}{16}\right)/32\right],$$

$$\frac{k}{m} = \frac{2.5(32)}{1.62(200)} = 0.247 \text{ s}^{-1}. \tag{23}$$

We once again take $c/m = 0.25$, as in Eq. (7). The Runge–Kutta solution of Eqs. (21) and (22) with these values for k/m and c/m is shown in Fig. 6. The maximum range of 614.6 ft is obtained for $\theta_0 = 9°$.

The values used above for lift and drag coefficients are very nearly the same. On the other hand, Cochran and Stobbs say: "For instance, as the ball flies away from the tee, drag may be at three to four ounces and lift at about two and a half ounces."[14] If we take the ratio of lift to drag coefficients to be 2.5/3.5 we obtain

$$\frac{k}{m} = \left(\frac{2.5}{3.5}\right)(0.25) = 0.179 \text{ s}^{-1}. \tag{24}$$

The computer solution of Eqs. (21) and (22) with $k/m = 0.179$ and $c/m = 0.25$ is shown in Fig. 7. The maximum range of 587.6 ft is now achieved for $\theta_0 = 16°$. This compares with Cochran and Stobb's investigation which found maximum carry at $\theta_0 = 20°$.[15] These last results seem to call for club manufacturers to fashion drivers at more than the approximate 10° that is common. However, Cochran pointed out that the use of a more lofted driver would reduce ball projection velocity and increase backspin, with an overall reduction in range.[16] Moreover, although the normal driver swung to catch the ball at a higher projection angle will maximize carry, it may not maximize total length since the ball will now hit the ground at a steeper angle. Cochran and Stobbs conclude:

> In most combinations of conditions, the longest drive will often be the one that goes off at 12° or 13°—a little higher than the starting angle of most people's drives, most of the time, but not very much so.
>
> In general, in fact, comparatively little may be gained in total distance off the tee even by the most strenuous and drastic attempts to alter the trajectory. For the vast majority of golfers, hitting the ball true and in good timing will prove much more important to good driving than "flying this one a bit higher."[17]

ACKNOWLEDGMENTS

The author acknowledges the assistance of Alan Benimoff of the College of Staten Island in the drawing of the figures. The kind help of Ed Van Dyke, Director, R & D and Quality Control of the Ben Hogan Company, in providing reference material is also acknowledged.

[1]A typical case is K. R. Symon, *Mechanics* (Addison-Wesley, Reading, MA, 1971), which treats the case of linear drag but does not consider the problem of maximum range, except for an unexplained Fig. 3.29 (p. 115), which shows "trajectories for maximum range for projectiles with various muzzle velocities."

[2]D. B. Lichtenberg and J. G. Wills, Am. J. Phys. **46**, 546 (1978).

[3]C. Frohlich, Am. J. Phys. **49**, 1125 (1981).

[4]J. A. Zufiria and J. R. Sanmartín, Am. J. Phys. **50**, 59 (1982).

[5](a) D. Williams, Quart. J. Mech. Appl. Math. **XII** (3), 387–392 (1959). However, (b) A. Cochran and J. Stobbs in *The Search for the Perfect Swing* (Lippincott, Philadelphia and New York, 1968), p. 160 indicate that the drag increases a little faster than the speed. They say "as a ball's speed doubles from, say 100 to 200 feet per second, the drag generated goes up about two and a half times." This would give a drag force which goes as $v^{1.3}$. *The Search for the Perfect Swing* presents the results of a team of researchers in Great Britain who conducted a scientific study of various aspects of golf. This investigation was sponsored by the Golf Society of Great Britain. The book was written for a popular audience and therefore does not contain many of the research details which would be presented in journal papers. A further volume written for scientists is promised (p. 227) but has not yet appeared to this author's knowledge.

[6]J. M. Davies, J. Appl. Phys. **20**, 821 (1949).

[7]This advantage comes from the horizontal distance covered by the hard hitter's golf ball during the time its velocity is falling to the lower one of the average hitter. If the drag coefficient were to remain constant instead of dropping with v, Williams calculated that 43 yards would be covered by the ball in going from 225 to 190 ft/s. The experimental difference in carry for these two projection velocities was found to be 52 yards. Thus the hard hitter has "gained" $52 - 43 = 9$ yards, due to the drop in the drag coefficient. Williams concluded "It can be said therefore that only some 17 per cent of the long hitter's advantage in distance can be attributed to the drop in drag coefficient. The rest apparently he gains by honest effort" (p. 392).

[8]See, for example, V. L. Streeter and E. B. Wylie, *Fluid Mechanics*, 6th ed. (McGraw-Hill, New York, 1975), Fig. 5.21 on p. 279.

[9]Reference 5(b), p. 163.

[10]The drag coefficient C_D corresponding to this value of c_2/m for the

British ball can be readily determined:

$$\frac{c_2}{m} = \left(\frac{C_D}{m}\right) A \tfrac{1}{2}\rho, \quad A = 0.0143 \text{ ft}^2 ,$$

$$\rho = 0.002\,37 \text{ slugs/ft}^3, \quad m = (1.62/16)/32 \text{ slugs} ,$$

$$C_D = 0.36 .$$

[11]Reference 5(b), p. 161.
[12]Reference 5(b), p. 160.
[13]Reference 5(b), p. 164.
[14]Reference 5(b), pp. 164–165.
[15]Reference 5(b), p. 165.

Bowling frames: Paths of a bowling ball

D. C. Hopkins and J. D. Patterson
Physics Department, South Dakota School of Mines and Technology, Rapid City, South Dakota 57701
(Received 26 April 1976; revised 14 October 1976)

We consider a bowling ball which starts with arbitrary linear and angular velocity and for which the coefficient of kinetic friction is assumed to be constant. We solve the resulting equations in closed form as well as suggest a geometrical construction of the path which may be qualitatively useful. We also numerically evaluate several parameters of the path for typical initial conditions. We find that the description of the problem is greatly simplified by using a coordinate system which is rotated with respect to the bowling alley by an angle which depends on the initial conditions. The ball describes a parabolic path up to a certain characteristic time (which we calculate), and then it rolls without slipping in a straight line. We contrast this solution with some of the folklore of bowling as it relates to "hooks."

I. INTRODUCTION

One of the problems always faced by teachers is how to interest the students in the subject matter. We have found that beginning or intermediate level mechanics can often be applied to certain games and such application often enhances the desire of the students to learn the material. When discussing rotational motion, for example, it is natural to relate the material to games involving balls. We discuss here the use of mechanics in finding the path taken by a spinning bowling ball. Much of our discussion can be qualitatively tested by any student the next time he goes bowling.

Many physics textbooks consider the problem of the motion of a spinning ball when it is set down on a flat surface or the problem of a ball originally given a translational velocity without rotation on a flat surface.[1] The bowling ball problem can be regarded as a generalization of these types of problems. In the bowling ball problem we give the ball an arbitrary (planar) initial vector linear velocity and an arbitrary initial vector angular velocity. This generalized problem is still readily solvable, and it seems to us to be an interesting example of motion with kinetic friction. It also provides a worthwhile example of the simplification which may result from using the most appropriate reference frame. We prove that the initial conditions determine a constant direction of frictional force on the sliding ball. Thus, the description of the problem is simplified by using a coordinate system in which this frictional force lies along one of the axes. We find that the bowling ball describes a parabolic path as long as it is slipping, and after that it simply rolls in a straight line. As a further point of interest, we compare this solution with some of the folklore of "hooks" in bowling[2] and notice that there are some variances. Finally, we discuss some of the idealizations that we make and comment about what deviations we might expect between the actual motion of a bowling ball and our calculations.

II. ANALYTICAL DESCRIPTION OF PATH

As shown in Fig. 1, a bowling alley is long and narrow. Crudely put, the object is to roll a ball, without stepping over the foul line, so as to knock over as many pins as possible. In this paper we will confine ourselves to describing how the initial conditions determine the path of the ball.

Figure 2 sets the geometry for our problem. The y axis can be regarded as being along the bowling lane. Throughout the paper v_x, v_y will denote linear velocity and ω_x, ω_y will denote angular velocity. The initial values of these quantities at time $t = 0$ will be denoted by v_{0x}, v_{0y}, ω_{0x}, ω_{0y}. For convenience we will assume the ball starts at $x = y = 0$ when $t = 0$. Any angular velocity ω_z that the ball might have has no effect upon its path. Of course, an $\omega_z \neq 0$ may affect the way the pins are scattered when the ball hits them.

If r is the radius of the bowling ball, the velocity relative to the floor of the point on the ball in contact with the floor has the components

$$v_x^b = v_x - r\omega_y, \tag{1a}$$

$$v_y^b = v_y + r\omega_x. \tag{1b}$$

When the ball rolls without slipping, $\mathbf{v}^b = 0$.

If m is the mass of the ball, g the acceleration due to gravity, and μ the coefficient of kinetic friction, then the frictional force has magnitude μmg and direction $-\mathbf{v}^b$. Since the total horizontal force arises from friction, we have by Newton's second law

$$\dot{v}_x = F_x/m = -\mu g v_x^b/v^b, \tag{2a}$$

$$\dot{v}_y = F_y/m = -\mu g v_y^b/v^b, \tag{2b}$$

where

$$v^b = [(v_x^b)^2 + (v_y^b)^2]^{1/2}, \tag{3}$$

and the dot means differentiation with respect to time. If the bowling ball is assumed to be a uniform sphere, then the moment of inertia (I) for any axis passing through the center of the sphere is equal to $(2/5)mr^2$. A well-known theorem[3] states that the external torque about the center of mass of an object equals the time derivative of the angular momentum ($\mathbf{L}_c$) about the center of mass of the object. Since for a sphere $\mathbf{L}_c = I\boldsymbol{\omega}$, where $\boldsymbol{\omega}$ is the vector velocity about an axis through the center of mass, we obtain the angular equations

$$r\dot{\omega}_x = r(rF_y)/[(2/5)mr^2] = (5/2)\dot{v}_y, \tag{4a}$$

$$r\dot{\omega}_y = -r(rF_x)/[(2/5)mr^2] = -(5/2)\dot{v}_x. \tag{4b}$$

Differentiating Eqs. (1) with respect to time and using Eqs. (2) and (4), we find

$$\dot{v}_x^b = -(7/2)\mu g v_x^b/v^b, \tag{5a}$$

$$\dot{v}_y^b = -(7/2)\mu g v_y^b/v^b. \tag{5b}$$

Reprinted from *American Journal of Physics* **45**, 263–266 (1977); © American Association of Physics Teachers.

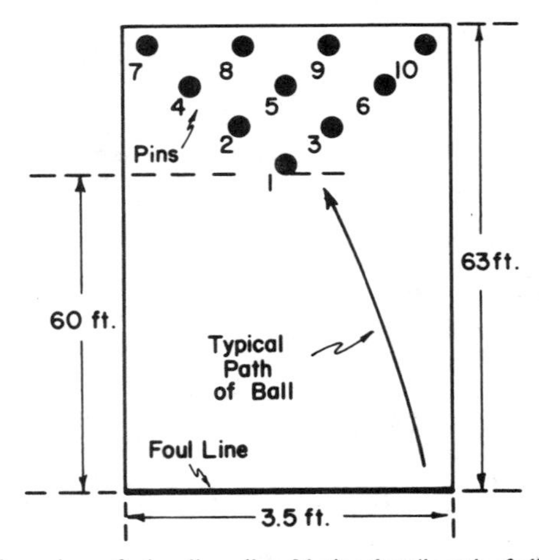

Fig. 1. Dimensions of a bowling alley. Notice that (length of alley)/(width of alley) $\gg 1$, and so the drawing has not been made to scale. The pit is directly in back and there are gutters on either side.

From Eqs. (5) we see

$$\dot{v}_x^b/v_x^b = \dot{v}_y^b/v_y^b, \tag{6}$$

which implies

$$v_x^b = cv_y^b. \tag{7}$$

The constant c can be evaluated from initial conditions, and we find

$$c = v_{0x}^b/v_{0y}^b = (v_{0x} - r\omega_{0y})/(v_{0y} + r\omega_{0x}). \tag{8}$$

Equation (7) gives us the important result that $\mathbf{v}^b$ is constant in direction, and since we have assumed a constant coefficient of friction, the frictional force is constant in both direction and magnitude.

Since the frictional force is constant in direction, we can simplify the equation of motion if we choose one of the axes to lie along the direction of this force. As shown in Fig. 3, if we rotate the coordinate system by an angle defined by (clockwise angles are defined to be negative)

$$\tan\theta = -v_x^b/v_y^b = -c, \tag{9}$$

then $\mathbf{v}_b$ is along the y' axis and the frictional force is along the $-y'$ axis. Since typically $v_{0y}^b \gg v_{0x}^b$, θ will be a rather small angle of the order of a few degrees (see Tables I and II). In the rotated coordinate system the acceleration is given by $a_x' = 0$ and $a_y' = -\mu g$ (since we know that the

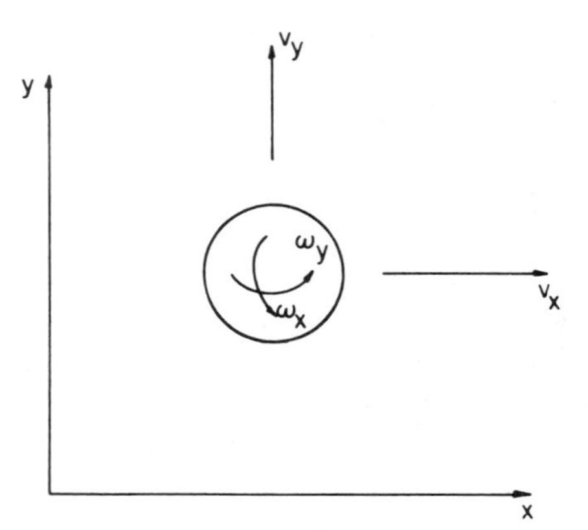

Fig. 2. Coordinate system for bowling ball paths.

Table I. Fixed quantities and initial conditions.[a]

Case	v_{0x} (ft/sec)	v_{0y} (ft/sec)	ω_{0x} (rad/sec)	ω_{0y} (rad/sec)
1	0	18	0	4
2	0	18	4	4
3	0.5	18	4	4
4	0.5	24	4	4

[a] For all cases $g = 32.17$ ft/sec², $\mu = 0.09$, and $r = 0.354$ ft were used.

magnitude of the force is μmg). Thus in the rotated system, the velocities are

$$v_x' = v_{0x}', \tag{10a}$$

$$v_y' = v_{0y}' - \mu g t, \tag{10b}$$

and the coordinates are

$$x' = v_{0x}'t, \tag{11a}$$

$$y' = v_{0y}'t - (\mu g/2)t^2. \tag{11b}$$

Equations (11) constitute an exact solution for the path of the ball as long as it is doing some sliding relative to the floor. However, they are not necessarily in the most convenient form for application. In particular, we would like to obtain the coordinates of the ball in the nonrotated coordinate system. Any vector $\mathbf{A}$ has components in the rotated and nonrotated coordinate systems that are related by

$$A_x = A_x' \cos\theta - A_y' \sin\theta = (A_x' + cA_y')/(1 + c^2)^{1/2}, \tag{12a}$$

$$A_y = A_x' \sin\theta + A_y' \cos\theta = (-cA_x' + A_y')/(1 + c^2)^{1/2}, \tag{12b}$$

so we can relate coordinates and velocities in the two coordinate systems as needed.

Eliminating t from Eqs. (11), the equations of the path in the rotated coordinate system take the following simple parabolic form:

$$(y' - Y_v') = -[Y_v'/(X_v')^2](x' - X_v')^2, \tag{13}$$

where

$$X_v' = v_{0x}'v_{0y}'/\mu g, \tag{14}$$

$$Y_v' = (v_{0y}')^2/2\mu g, \tag{15}$$

and where by Eqs. (12) $v_{0x}' = (v_{0x} - cv_{0y})/(1 + c^2)^{1/2}$, $v_{0y}' = (cv_{0x} + v_{0y})/(1 + c^2)^{1/2}$. The physical meaning of X_v' and Y_v' should be clear from Figs. 4(a) and 4(b). It should be noted that these figures are not drawn to scale, as typically X_v' will be considerably smaller than Y_v' (see, e.g., Table II). Figure 4(a) would be appropriate for a left-handed bowler who started the ball on the left part of the lane, whereas Fig. 4(b) would be appropriate to a right-handed bowler who started the ball on the right side of the lane.

Equations (10), (11), and (13) are only valid as long as the ball slides relative to the floor. The ball stops all sliding and begins rolling in a straight line where $\mathbf{v}^b = 0$. We can easily calculate v_x^b and v_y^b. By Eqs. (3), (5), and (7) we can write

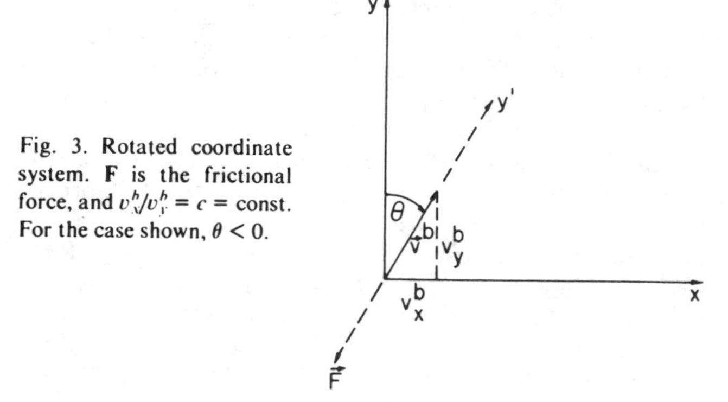

Fig. 3. Rotated coordinate system. **F** is the frictional force, and $\dot{v}_x^b/\dot{v}_y^b = c =$ const. For the case shown, $\theta < 0$.

$$\dot{v}_x^b = -(7/2)\mu g c/(1+c^2)^{1/2}, \qquad (16a)$$

$$\dot{v}_y^b = -(7/2)\mu g/(1+c^2)^{1/2}. \qquad (16b)$$

Thus

$$v_x^b = v_{0x}^b - (7/2)\mu g c t/(1+c^2)^{1/2}, \qquad (17a)$$

$$v_y^b = v_{0y}^b - (7/2)\mu g t/(1+c^2)^{1/2}. \qquad (17b)$$

It is then elementary to show that v_x^b and v_y^b both become zero at the same time T_s. T_s thus denotes the time during which the ball slides relative to the floor, and Eqs. (17) imply

$$T_s = 2v_0^b/7\mu g, \qquad (18)$$

where

$$v_0^b = [(v_{0x}^b)^2 + (v_{0y}^b)^2]^{1/2}. \qquad (19)$$

It is also interesting to compute the location of the ball and the slope of its path at the time that it stops sliding. The location (X_s', Y_s') in the rotated coordinate system is clearly given by Eqs. (11) with $t = T_s$, and the angle θ' of the slope is easily obtained from Eqs. (10) via

$$\theta' = \tan^{-1}(v_y'/v_x') \quad \text{at} \quad t = T_s. \qquad (20)$$

Since θ is a small angle and since the bowling alley is long and narrow, one would expect $|\theta'|$ to be fairly close to 90°. The calculations given in Table II for some fairly typical initial conditions verify this belief. Figures 4(a) and 4(b) also show θ and θ'. Finally, for the purpose of geometrical construction, it is useful to give a rough estimate for X_s'/X_v'. From Eqs. (11a),

$$X_s'/X_v' = T_s/T_v, \qquad (21)$$

where T_v is the time it would take the ball to reach the vertex of the parabola if it did some sliding all the way. From Eqs. (11a) and (14), T_v is clearly

$$T_v = X_v'/v_{0x}' = v_{0y}'/\mu g. \qquad (22)$$

By Eqs. (18), (21), and (22) we thus have

$$X_s' = (2/7)\, X_v'\, (v_0^b/v_{0y}'). \qquad (23)$$

In descending powers of v_{0y} the ratio v_0^b/v_{0y}' is given in terms of the initial conditions by

$$\frac{v_0^b}{v_{0y}'} = 1 + \frac{r\omega_{0x}}{v_{0y}} + \frac{r^2\omega_{0y}{}^2 - rv_{0x}\omega_{0y}}{v_{0y}{}^2} + \cdots. \qquad (24)$$

The initial speed of the ball toward the pins, v_{0y}, will be much larger than the other speeds involved for any reasonable bowler. Therefore, as a good rule of thumb, $X_s' \cong (2/7)X_v'$. X_s' and Y_s' are also shown in Figs. 4(a) and 4(b).

III. QUALITATIVE DISCUSSION AND EXACT NUMERICAL RESULTS

With the above as a background, one can readily construct the paths of the ball as shown in Fig. 4. Although θ and θ' can be either both positive or both negative depending typically on whether the bowler is left- or right-handed, $|\theta| + |\theta'|$ will always be near 90° for typical bowlers. To construct the path we need only calculate $\theta, X_v', Y_v', X_s', Y_s'$, and then construct the parabola and tangent shown. Changes in the path due to changes in initial conditions can be deduced qualitatively from this geometrical construction. For example, if $v_{0x} = 0$ (and $v_{0y}, \omega_{0y} > 0$ with ω_{0x} small), then $c < 0$ and Fig. 4(a) applies. Let us suppose Fig. 4(a) is a plot for $\omega_{0x} = 0$ (no back spin). If we now add a small back spin, c is reduced slightly in magnitude. Hence, by Eqs. (12) and (15), there is no change in Y_v' to first order in c. However, to first order in c, X_v' decreases and so does θ. So the effect of back spin is to rotate the parabola clockwise (the direction of decreasing θ) and to narrow it. As indicated numerically below, this may result in giving added "hook" to the ball. Similar comments can be made about Fig. 4(b). For this figure $c > 0$, so with $v_{0x} = 0$ we require $\omega_{0y} < 0$ and $v_{0y} + r\omega_{0x} > 0$. Again it needs to be emphasized that neither of these figures is drawn to scale. The bowling alley is

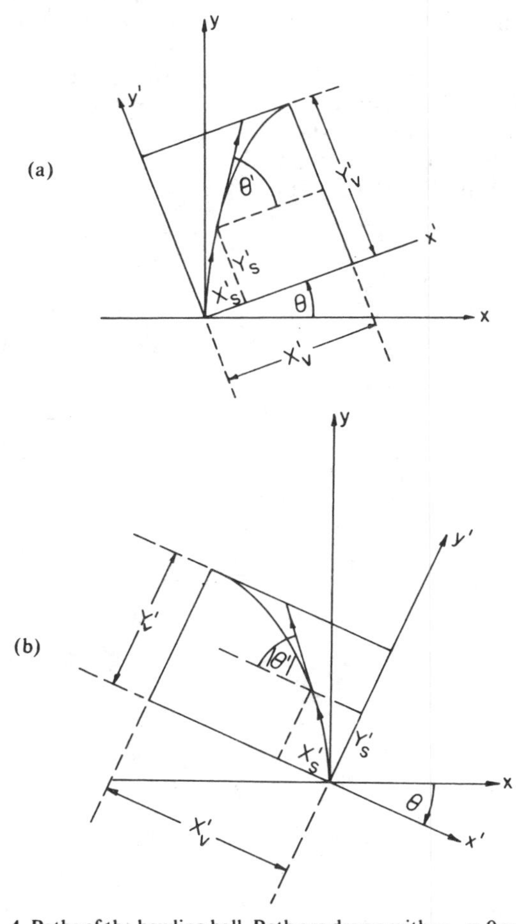

Fig. 4. Paths of the bowling ball. Both are drawn with $v_{0x} = 0$ so the paths start off tangent to the y axis. (a) $c < 0$, so $\theta > 0$ and $X_v' > 0$. (b) $c > 0$, so $\theta < 0$ and $X_v' < 0$. After the ball stops slipping, it rolls along the tangent shown. Note $X_s' \cong (2/7)\,X_v'$, for all typical initial values.

Table II. Computed quantities for paths.

Case	v^b_{0x} (ft/sec)	v^b_{0y} (ft/sec)	v^b_0 (ft/sec)	c	v_{0x}' (ft/sec)	v_{0y}' (ft/sec)
1	−1.416	18.000	18.056	−0.0787	1.412	17.944
2	−1.416	19.416	19.468	−0.0729	1.309	17.952
3	−0.916	19.416	19.438	−0.0472	1.348	17.957
4	−0.916	25.416	25.432	−0.0360	1.363	23.966

Case	T_s (sec)	T_v (sec)	θ (deg)	θ' (deg)	X_v' (ft)	Y_v' (ft)
1	1.782	6.198	4.500	83.698	8.752	55.611
2	1.921	6.201	4.169	83.969	8.117	55.661
3	1.918	6.203	2.701	83.798	8.361	55.692
4	2.510	8.278	2.062	85.334	11.283	99.200

Case	X_s' (ft)	Y_s' (ft)	$\theta + \theta'$ (deg)	X_s (ft)	Y_s (ft)
1	2.516	27.378	88.198	0.360	27.491
2	2.515	29.149	88.138	0.389	29.255
3	2.586	29.123	86.499	1.210	29.213
4	3.421	51.035	87.396	1.583	51.125

very long and narrow, and consequently the path of the ball must also be long and narrow.

Perhaps a feeling for the path can be obtained most clearly by examining some numerical results for some typical initial conditions. These are summarized in Tables I and II. The coefficient of sliding friction, μ, is poorly known and possibly changes from place to place, but we have assumed it to be constant. Our value is based on some rough measurements. Also, we have assumed that the ball does not bounce. As the tables show, one does not expect a path with very much overall curvature. In Table II, X_s and Y_s are the coordinates in the unrotated coordinate system at which the ball starts to roll without slipping. Notice that $\theta + \theta'$ is the angle with respect to the x axis at which the ball strikes the pins; as expected, it comes out to be nearly 90°. Comparing case 1 and case 2, we see for these typical examples that $\theta + \theta'$ is decreased slightly by adding back spin ($\omega_{0x} > 0$) to the ball. Thus, in this case, the effect of back spin is to give the ball "more hook."

Finally, it is worthwhile to make a few comments about how our solution compares to some of the folklore of bowling as it relates to hooks. A fairly recent book on bowling[2] characterizes a hook ball as one which starts rolling straight but which breaks sharply before reaching the pins. This may well be the bowler's perception of what the ball does, but it should be clear from our analysis that, if the ball is going to break at all, it must start out in a parabola and hence "break" all the way to the point (X_s, Y_s). The same book attempts to make a distinction between a curve ball (which has a curvature throughout the path) and a hook ball. This seems to us to be an artificial distinction. One point of our analysis is perhaps a little weak. This is our assumption that the coefficient of sliding friction is a constant. It might well happen that the wax interface between the ball and the floor is affected by v_b. Perhaps when v_b approaches zero the coefficient of sliding friction changes somewhat (increases perhaps), and this could affect the amount of curvature near the end of the sliding part of the path. Also, the ball may bounce slightly in the beginning. A combination of these two effects could act to give less curvature in the beginning and more curvature later in the path. Thus the true nature of a hook may possibly be intermediate between the folklore[2] and our mathematical solution.

[1]See, e.g., R. A. Becker, *Introduction to Theoretical Mechanics* (McGraw-Hill, New York, 1954), pp. 209 and 219.

[2]K. Kidwell and P. Smith, Jr., *Bowling Analyzed* (Brown, Dubuque, IA, 1960), p. 55ff.

[3]P. A. Tipler, *Physics* (Worth, New York, 1976), p. 340.

Kinematics of an Ultraelastic Rough Ball

RICHARD L. GARWIN
IBM Watson Laboratory, Columbia University, New York, New York 10025

(Received 6 June 1968; revision received 11 September 1968)

A rough ball which conserves kinetic energy exhibits unexpected behavior after a single bounce and bizarre behavior after three bounces against parallel surfaces. The Wham-O *Super-Ball®* (registered by Wham-O Corporation, 835 E. El Monte St., San Gabriel, Calif. 91776), appears to approximate this behavior and provides an inexpensive and readily available model of kinematics quite different from that of a point mass or smooth ball. The analysis is most strikingly illustrated by the fact that the ball returns to the hand after three collisions with the floor, the underside of a table, and the floor. Some questions are raised concerning the dynamics of the collision.

INTRODUCTION

"Angle of incidence equals angle of reflection" is a commonly quoted but, of course, not universal result, and it is useful to demonstrate to a freshman or high school physics class that real objects exhibit behavior equally predictable but in some cases quite different. In particular, a perfectly rough ball which conserves kinetic energy behaves in such an unexpected way that it is difficult to pick up after it has bounced twice upon the floor, and, more bizarre, it returns to the hand on being thrown to the floor in such a way that it bounces from the underside of a table as in Fig. 1.[1] It turns out that the two assumptions of conservation of kinetic energy and no slip at the contacting surfaces predict qualitatively the observed behavior of the *Super Ball*. It is not at first clear, however, how even a microscopically perfectly elastic body can conserve kinetic energy in such collisions, and this is discussed later.

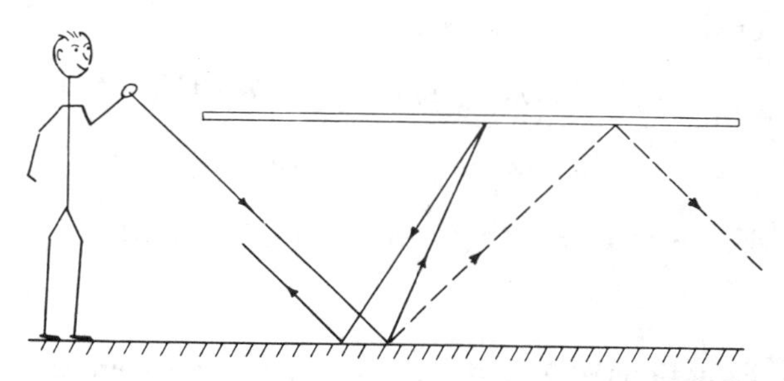

FIG. 1. A *Super Ball* seems to return to the hand after bouncing against the underside of a table, while the expectation is for it to continue bouncing between the floor and the table as shown by the dotted line.

[1] This was first demonstrated to me by L. W. Alvarez with a Wham-O *Super Ball*.

Two primary assumptions suffice to determine the trajectories of the *Super Ball*: (a) Kinetic energy is conserved during a collision (the rotational plus the translational energy of the ball is the same after collision as before). (b) There is no slip at the point of contact (the ball is "perfectly rough"). Two further conditions follow from the laws of mechanics: (c) Angular momentum L about the point of contact is conserved during the collision. (d) The normal component of velocity is reversed by a collision. (c) follows from the approximation that the contact occurs at a point and that all forces act through that point, so their moment about that point is zero. (d) follows from the assumption of linear equations of motion and the observation that, in the special case of normal incidence without spin, all kinetic energy after the collision must again be in the normal velocity, which is therefore preserved with change of sign, as a consequence of (a). Since the spin, the normal velocity and the tangential velocity after collision are all assumed to be linear functions of the velocities before the collision, the cross coupling between normal velocity and the other two degrees of freedom is shown by this special case to be zero under *all* initial conditions.

Figure 2 shows the elementary collision against a hard surface oriented in the xz plane. Before collision, the body has a velocity V_b along x and an angular velocity ω_b along z. After the bounce, these variables become V_a and ω_a, respectively.

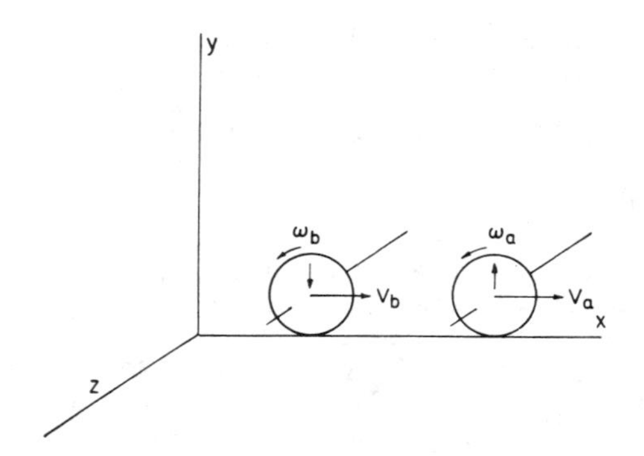

FIG. 2. The *Super Ball* has velocity V_b and spin ω_b before bouncing from the xz plane. After the bounce the velocity and spin are V_a and ω_a respectively. The normal velocity along y is simply reversed in the collision.

The subscriptions b and a refer to conditions *before* or *after* a given collision.

I. COLLISION KINETICS

Let the moment of inertia about the z axis be $I \equiv \alpha MR^2$, with $\alpha = \frac{2}{5}$ for a uniform sphere. M is the mass, and R the radius of the ball. Then the kinetic energy, including only the spin ω and the x component of the velocity V, is

$$K = \tfrac{1}{2}MV^2 + \tfrac{1}{2}I\omega^2 = M/2(V^2 + \alpha\omega^2R^2)$$
$$= M/2(V^2 + \alpha C^2), \qquad (1)$$

where $C = \omega R$ has been introduced to simplify the equations but in fact represents the (signed) peripheral velocity of the ball's surface adjacent to the wall, relative to the center.

The angular momentum about the point of contact is then (for the ball close to the wall)

$$L = I\omega - MRV = (\alpha MR^2\omega - MRV)$$
$$= MR(\alpha C - V), \qquad (2)$$

in which L is expressed as the angular momentum about the center of the ball plus that of the center of mass about the point of contact.

Thus, equating the kinetic energy and the angular momentum before and after a bounce, we have

$$K_b = M/2(V_b^2 + \alpha C_b^2) = M/2(V_a^2 + \alpha C_a^2) = K_a, \qquad (3)$$

and

$$L_b = MR(\alpha C_b - V_b) = MR(\alpha C_a - V_a) = L_a, \qquad (4)$$

in which the velocities *before* are V_b, C_b and those *after* are V_a, C_a. Equations (3) and (4) become

$$V_b^2 + \alpha C_b^2 = V_a^2 + \alpha C_a^2, \qquad (5)$$

and

$$V_b - \alpha C_b = V_a - \alpha C_a; \qquad (6)$$

and simplification gives

$$\alpha(C_b - C_a)(C_b + C_a) = (V_a - V_b)(V_a + V_b), \qquad (5')$$

$$\alpha(C_b - C_a) = -(V_a - V_b). \qquad (6')$$

Dividing, we obtain an interesting result,

$$C_b + C_a = -(V_a + V_b),$$

or

$$V_b + C_b = -(V_a + C_a), \qquad (7)$$

showing that the parallel velocity of the contacting portion of the ball, $S \equiv V + C$, is precisely *reversed* by a bounce.

Subtracting Eq. (7) from Eq. (6) we have

$$(\alpha - 1)C_b - (\alpha + 1)C_a = 2V_b,$$

or

$$C_a = [(\alpha - 1)/(\alpha + 1)]C_b - [2/(\alpha + 1)]V_b; \qquad (8)$$

and resubstituting in Eq. (7) we find

$$V_a = -[2\alpha/(\alpha + 1)]C_b - [(\alpha - 1)/(\alpha + 1)]V_b. \qquad (9)$$

Equations (8) and (9) thus give the spin and velocity after collision.

For example, if a spherical ball approaches the wall with $\omega = 0$, $V_b = V_\perp$, i.e. with no spin and at an angle $\theta_b = 45°$ with respect to the normal, then $C_b = 0$ and

$$C_a = (-10/7)V_\perp, \qquad V_a = \tfrac{3}{7}V_\perp,$$

so that the ball on bouncing makes an angle $\theta_a = \tan^{-1}(3/7) = 23.2°$ with the outward normal from the wall. [See Fig. 3(a).] Similarly, a ball

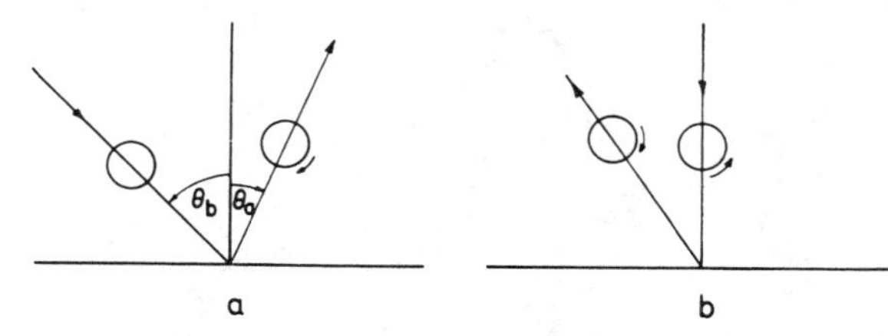

FIG. 3. (a) A *Super Ball* with zero spin bounces from a wall with $\tan\theta_a = 3/7 \tan\theta_b$ and with spin velocity 10/7 of its initial horizontal velocity. (b) A ball with initial spin has its spin velocity reduced to 3/7 and reversed, and acquires transverse velocity as shown.

with initial spin velocity C_b but with no translational velocity V_b leaves the wall with

$$C_a = -\tfrac{3}{7}C_b, \qquad V_a = -\tfrac{4}{7}C_b$$

as indicated in Fig. 3b.

II. TRAJECTORIES

Equations (8) and (9), and assumption (d) suffice to allow the path of a *Super Ball* to be traced through any number of collisions with rigid walls. One simply uses the geometry to determine V_b and $V_\perp$ before any new collision, from the V_a and $V_\perp$ after the previous one. One then applies Eqs. (8) and (9), and assumption (d) and reprojects the velocity to obtain the initial velocities for the next collision.

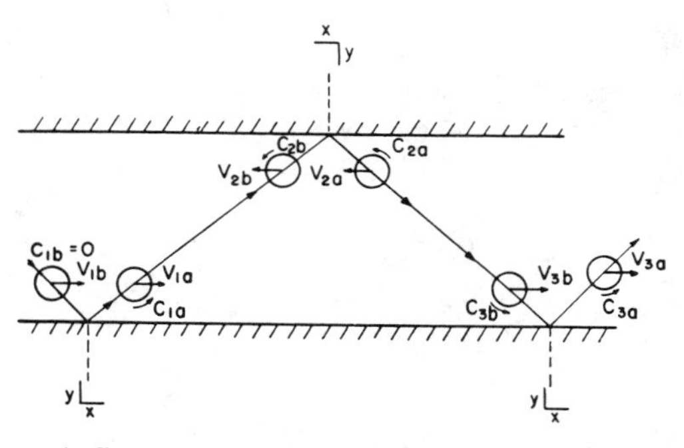

FIG. 4. A *Super Ball* makes three successive bounces between two parallel walls. The positive sense of velocity and spin is shown before and after each of the three collisions.

If we consider throwing the ball to the floor and allowing it to bounce against the underside of a table before bouncing again from the floor, we have the situation of Fig. 4, with the coordinate system of Fig. 1, except that for bounce 2, the coordinates used are shown in the inset, in order to employ the standard Eqs. (8) and (9). For a ball without initial spin ($C_{1b} = 0$), repeated application of these equations gives V_{1a}, C_{1a}; V_{2b}, C_{2b}; V_{2a}, C_{2a}; V_{3b}, C_{3b}; and finally V_{3a}, C_{3a}.

$$V_{1a} = -[(\alpha-1)/(\alpha+1)]V,$$

$$C_{1a} = [-2/(\alpha+1)]V,$$

$$V_{2b} = [(\alpha-1)/(\alpha+1)]V,$$

$$V_{2a} = -\left(\frac{2\alpha}{\alpha+1}\right)\left(\frac{-2}{\alpha+1}\right)V - \left(\frac{\alpha-1}{\alpha+1}\right)^2 V$$

$$= [V/(\alpha+1)^2][4\alpha-(\alpha-1)^2],$$

$$C_{2a} = \left(\frac{\alpha-1}{\alpha+1}\right)\left(\frac{-2}{\alpha+1}\right)V - \left(\frac{2}{\alpha+1}\right)\left(\frac{\alpha-1}{\alpha+1}\right)V$$

$$= [V/(\alpha+1)^2][-2(\alpha-1)-2(\alpha-1)],$$

$$V_{3b} = [V/(\alpha+1)^2][(\alpha-1)^2-4\alpha],$$

$$C_{3b} = C_{2a}.$$

$$V_{3a} = [V/(\alpha+1)^3][-\alpha^3+15\alpha^2-15\alpha+1]. \quad (10a)$$

$$C_{3a} = [V/(\alpha+1)^3][-6\alpha^2+20\alpha-6]. \quad (10b)$$

For a sphere $\alpha = \tfrac{2}{5}$, and

$$V_{3a} = -\frac{333}{343}V, \qquad C_{3a} = \frac{130}{343}V.$$

Thus a sphere *returns* after three bounces with but 3% lower velocity than it had when it started. An accurate reproduction of its trajectory is presented in Fig. 5. It is worth noting[2] that a

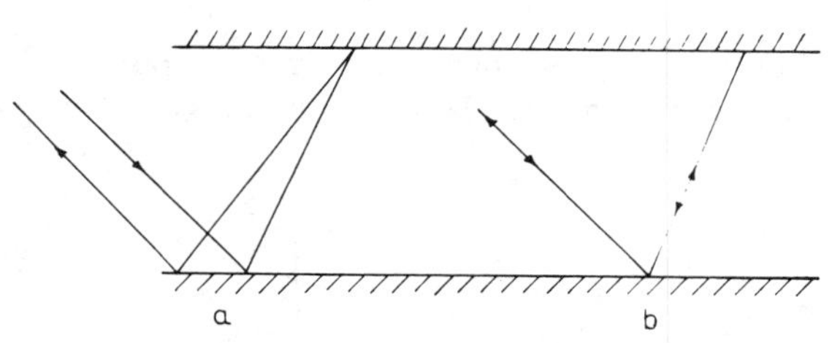

FIG. 5. A *Super Ball* thrown without spin will follow the path indicated in 5(a) in bouncing from the floor to the underside of a table and back to the floor. The tangent of the angle of bounce is 3% greater than that of the angle of incidence. For comparison, the trajectory of a body with $I = \tfrac{1}{3}MR^2$ is shown in 5(b)—it returns precisely along its initial path.

[2] R. Friedberg, private communication, December 1967.

body with $\alpha=\frac{1}{3}$ (e.g., a ball containing a central high-density mass $\frac{1}{5}$ that of the original ball) returns without spin and parallel to the original direction. Although gravity has been neglected in this calculation, its inclusion changes nothing so long as the ball does in fact strike the underside of the table.

It is difficult to make all the required substitutions accurately, and advantage may be taken of the linear form of Eqs. (8) and (9) to define a collision matrix M such that the velocity after a collision $\mathbf{V}_a$ is related to the velocity before, $\mathbf{V}_b$, by

$$\mathbf{V}_a=\mathbf{M}\cdot\mathbf{V}_b, \tag{11}$$

in which $\mathbf{V}_a$ and $\mathbf{V}_b$ are row vectors with three elements representing the parallel velocity V, the peripheral velocity C, and the normal velocity V_n, respectively. The collision matrix is then

$$\mathbf{M}=\begin{bmatrix}(1-\alpha)/(\alpha+1) & -2\alpha/(\alpha+1) & 0\\ -2/(\alpha+1) & (\alpha-1)/(\alpha+1) & 0\\ 0 & 0 & -1\end{bmatrix}$$

$$=-1/(\alpha+1)\begin{bmatrix}\alpha-1 & 2\alpha & 0\\ 2 & 1-\alpha & 0\\ 0 & 0 & \alpha+1\end{bmatrix}. \tag{12}$$

Equation (12) has been used in a computer program to provide the data for Fig. 6, which shows the trajectory of a *Super Ball* and of a smooth ball within a unit square.[3] In computing the trajectories of Fig. 6, before applying the collision matrix, the computer after each collision determines the point of impact of the next collision and transforms the components of the velocity vector to a new wall-oriented frame of reference. This velocity transformation is done by matrix multiplication by matrices $\mathbf{D}_1$, $\mathbf{D}_2$, $\mathbf{D}_3$, according as the next side struck is at angle $\pi/2$, π, or $3\pi/2$ with respect to the previous side of the square.

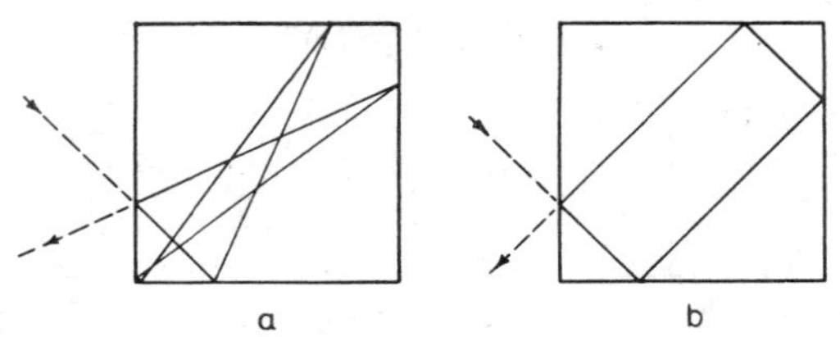

FIG. 6. (a) Shows the path of an ultraelastic rough ball. (b) Shows the path of a smooth one inside a rigid square.

Following Romer[4] we could have obtained Eqs. (10) by repeated matrix multiplication: $\mathbf{V}_{1a}=\mathbf{M}\cdot\mathbf{V}_{1b}$; $\mathbf{V}_{2b}=\mathbf{D}_2\cdot\mathbf{V}_{1a}$; $\mathbf{V}_{2a}=\mathbf{M}\cdot\mathbf{V}_{2b}$; $\mathbf{V}_{3b}=\mathbf{D}_2\cdot\mathbf{V}_{2a}$; $\mathbf{V}_{3a}=\mathbf{M}\cdot\mathbf{V}_{3b}$; or $\mathbf{V}_{3a}=\mathbf{M}\mathbf{D}_2\mathbf{M}\mathbf{D}_2\mathbf{M}\mathbf{V}_{1b}$. Here

$$\mathbf{D}_2=\begin{bmatrix}-1 & 0 & 0\\ 0 & 1 & 0\\ 0 & 0 & -1\end{bmatrix}.$$

$\mathbf{D}_2$ does not commute with $\mathbf{M}$, the product of the five matrices, giving

$$\mathbf{V}_{3a}=1/(\alpha+1)^3=\begin{bmatrix}-\alpha^3+15\alpha^2-15\alpha+1 & -6\alpha^3+20\alpha^2-6\alpha & 0\\ -6\alpha^2+20\alpha-6 & \alpha^3-15\alpha^2+15\alpha-1 & 0\\ 0 & 0 & -(\alpha+1)^3\end{bmatrix},$$

which reduce to Eqs. (10) for the special case $C_{1b}=0$.

[3] The program was written in APL\360 with the help of T. M. Garwin. A copy of the program and of the output for the cases in Fig. 6 is available to anyone inquiring.

[4] R. H. Romer, Amer. J. Phys. **35**, 862 (1967). I am indebted to the referee, who called to my attention this paper. My purpose is to explain the behavior of the *Super Ball*, but Romer's elegant matrix treatment of binary one-dimensional collisions could be applied directly to the exchange of energy between the two internal degrees of freedom (velocity and spin) by a simple transformation. I record the algebra leading to Eqs. (10) to motivate the use of the matrices **M** and **D**.

III. DISCUSSION

In going from Eqs. (4) and (6) to Eq. (7), we have tacitly excluded one solution, i.e. $C_a = C_b$, $V_a = V_b$. In addition, we have nowhere overtly used assumption (b) that there is no slip at the point of contact. In fact, conservation of energy [assumption (a)] as well as conditions (c) and (d) are consistent with this excluded solution, which would indeed be predicted if there were a frictionless slip between ball and wall. The question is more complicated, however, as is apparent from the result of Eq. (6), namely that the parallel velocity of the point of contact is just *reversed* by the chosen rough energy-conserving collision. Clearly, if the rough ball were again brought in contact with the wall immediately after such an energy-conserving collision and were held there for a similar time, the parallel surface velocity would once again reverse in this second collision, and the ball would rebound just as if it had been perfectly smooth and slipped freely.[5] Equally clearly, if the period of contact were insufficient for the shear wave to penetrate through the ball, the ball would be left in a state of internal excitation (torsional vibration) which would contradict assumption (a), which assumes that no energy is present in modes other than those of uniform translation and of spin. Thus, it seems likely that the striking behavior of the *Super Ball* is due not so much to an extremely low internal dissipation for elastic waves as it is to a near equality between periods for a bounce of the ball from the wall and for a torsional oscillation of the ball. It is this further requirement on the relative speeds of compression and shear waves which leads me to use the term "ultraelastic" in this paper. Even for zero internal friction, the bounce can never really be perfectly elastic because the higher modes of vibration are not harmonically related to the fundamental, and so it retains some excitation after the bounce.

The conceptual model of a collision thus has the normal velocity of the ball, upon contact, undergoing a half-period of oscillation and thus reversing. At the same time, the surface of the ball in contact with the wall comes abruptly to rest and a shear wave propagates into the ball, and, reflecting from the free surface, reverses the initial velocity of the surface near the wall. The process is easy to imagine for a rough elastic slab colliding with a wall, or for a spinning sphere or disk, mounted on a fixed axis, the surface of which is suddenly brought to rest. It is simpler to think about a ring rather than a sphere or other solid body. For a ring, $\alpha = 1$, and Eq. (12) shows that such a body transforms *all* its energy of translation into spin (and vice versa) at each collision.

Although the problem is too complex to be treated here, it would be of some interest to consider the dynamics of a collision in more quantitative detail, probably taking into account the complete modal spectrum, in order to understand ultimate limits on elasticity of a collision.

[5] Indeed, $\mathbf{M}^2 = \mathbf{I}$, so that two bounces on the floor should restore the initial parallel velocity and spin. I do not know to what precision a real *Super Ball* recovers its initial spin on the second bounce if, e.g., the initial parallel velocity is zero.

ACKNOWLEDGMENTS

I am indebted particularly to R. Friedberg, and N. M. Kroll for their suggestions, and to T. M. Garwin for his help in programming.

Physics of basketball

Peter J. Brancazio
Department of Physics, Brooklyn College, City University of New York, Brooklyn, New York 11210
(Received 29 November 1979; accepted 11 August 1980)

Does a knowledge of physics help to improve one's basketball skills? Several applications of physical principles to the game of basketball are examined. The kinematics of a basketball shot is studied, and criteria are established for determining the best shooting angle at any given distance from the basket. It is found that there is an optimum shooting angle which requires the smallest launching force and provides the greatest margin for error. Some simple classroom illustrations of Newtonian mechanics based on basketball are also suggested.

I. INTRODUCTION

Anyone who has followed sports closely over the past few decades cannot help but notice the evergrowing influence of science and technology. Athletes have benefited greatly from advances in sports medicine, training techniques, and equipment design. In a number of sports, detailed technical and statistical analyses of patterns of play have led to more sophisticated game strategies. What has yet to be fully developed, however, is an understanding and appreciation of the physical principles underlying various sports.

The scientific literature of the physics of sports is curiously sparse. In the past this Journal has contained articles on baseball,[1-4] tennis,[5] bowling,[6] golf,[7] springboard diving,[8] and (miraculously) only one brief article on jogging.[9] However, there seems to be no published research on the physics of basketball. Even within the field of physical education, remarkably little has been done on this subject; the *Research Quarterly* (of the American Alliance for Health, Physical Education, and Recreation) has published only one article on the topic, and that in 1951.[10] The following is an attempt to fill this gap.

The author freely admits to a lifelong obsession with basketball as a spectator and particularly as a still-active participant in playground pickup games. His fellow players have frequently reminded him (with considerable ironic intent) that his professional knowledge of physics ought to give him an unique advantage. In truth, the major purpose of this research was to find some means to compensate for the author's stature (5′10″ in sneakers), inability to leap more than 8 in. off the floor, and advancing age. With this impetus, the author has been able to demonstrate with some success that one's performance on a basketball court can be improved through the study and application of kinematics and Newtonian mechanics. Using the equations of projectile motion, a mathematical analysis of basketball shooting has been developed that can be used to determine the optimum angle and speed with which to shoot a basketball from any point on the court. But first let us consider some simple qualitative applications of physical principles to basketball play. These examples serve an additional purpose in that they can also be used to enliven the introductory physics classroom as illustrations of basic Newtonian physics.

II. WARM-UPS: PHYSICS OF SOME BASKETBALL FUNDAMENTALS

Given the interest in basketball among students at many colleges (not to mention the growth of women's basketball in recent years), illustrations of physical principles based on basketball can be used effectively in the physics lecture room as illuminating and entertaining demonstrations of how physics can be applied to seemingly unrelated situations. What follows may also be of particular interest to junior high school and high school science teachers in the inner city schools (where interest in basketball is particularly intense) as a way to motivate and interest their students.

A. Physics of the layup shot

One aspect of the principle of inertia is that if an object is initially fixed with respect to a moving reference frame and is suddenly released, the object will continue to move with the same velocity as the reference frame at the instant it is released. An early illustration of this principle was provided by Galileo in in his *Dialogues Concerning the Two Chief World Systems,* in which he pointed out that a ball released from the top of the mast of a moving ship is seen to land at the foot of the mast, and therefore must be moving horizontally as it falls with the same horizontal speed as the ship.[11]

In a basketball game, this phenomenon is illustrated by any shot taken on the run. For example, suppose that an offensive player breaks past the defense and dribbles the ball directly toward the basket as fast as he can for a layup. A few feet from the basket, he shoots the ball while still running. The inexperienced player will tend to push his shot *toward* the basket, and the result is that the ball will overshoot the rim and/or slam off the backboard with too great a speed. The player has failed to take inertia into account; by pushing his shot toward the basket he has unwittingly added to the velocity that the ball already possessed by virtue of the player's running motion. The proper way for him to take this shot is to *shoot the ball vertically upward with respect to his own body* (i.e., so the ball has no horizontal component of motion in the moving reference frame) when he is about 2–3 feet from the basket. If this is done properly, the ball should arch nicely into the basket. A

Reprinted from *American Journal of Physics* **49**, 356–365 (1981);

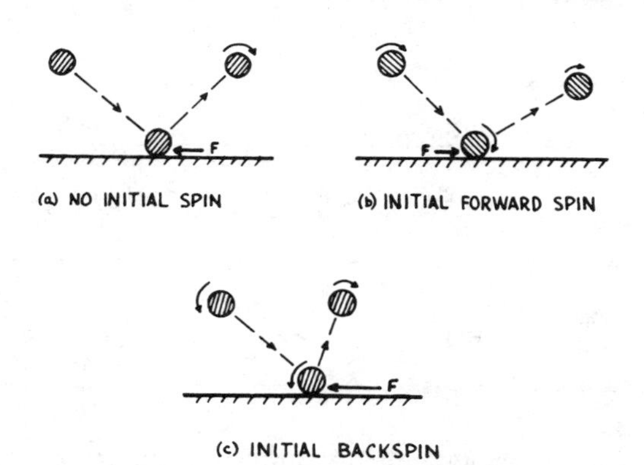

Fig. 1. Changes in translational and rotational motion of a ball when bounced on a rigid horizontal surface for various initial spin conditions.

similar situation arises when a player is moving cross court, parallel to the baseline, and takes a running shot (usually a hook shot) as he crosses the free-throw lane. If the player is moving, say, from left to right and aims his shot directly at the center of the basket, the ball will continue its crosswise motion and will hit to the right of the center of the basket. To compensate for inertia, the player must aim the shot at the left-hand side of the rim.

B. Value of backspin

An essential principle of basketball shooting is that the ball should be pushed by the fingertips rather than by the palm of one's hand. First of all, shooting with the fingertips gives the shooter much better control over the path of the shot. Secondly, a ball shot from the fingertips and released with a slight flick of the wrist will automatically have a backspin.

Why is backspin so important? According to one of the greatest basketball coaches, Arnold "Red" Auerbach,

> The fingertips ... help to impart backspin, which makes the shot softer and helps the shot to be "lucky".... A ball that strikes the rim and then stops has good backspin. Some say this is luck. But why is it that the great shooters always seem to make more of these so-called lucky shots?[12]

The value of backspin in basketball shooting is the result of good physics rather than good luck. To understand why, let us make a simple analysis of what happens when a ball bounces. We consider three cases, in which a moving ball strikes a horizontal surface with (a) no initial spin; (b) a forward spin; and (c) a backspin, as shown in Fig. 1. In each case the ball has an initial translational velocity at some angle to the surface. In what follows we shall consider just the horizontal component (parallel to the surface) of this velocity.

When the ball strikes the surface, a friction force arises at the point of contact which alters both the translational and rotational motions of the ball. In general, there will be a transfer of energy between the translational and rotational modes, with an overall net loss in the total kinetic energy. At the simplest level, it is easy to determine in each case the direction of the friction force and its effect on the motion of the ball. If the ball has no initial rotation [Fig. 1(a)] then the friction force opposes the forward translational motion. This force also creates a torque about the center of gravity of the ball, so that the ball gains angular momentum. Hence there is a loss of translational energy and an increase in rotational energy. Thus we expect a ball that is bounced forward with no initial spin to rebound with a forward spin and a slower translational speed.

Now suppose that the ball is projected with an initial forward spin [Fig. 1(b)]. At the point of contact the rotational velocity ($v = R\omega$) will be opposite to the horizontal translational velocity component (v_x) of the ball. If the spin rate is large enough ($R\omega > v_x$) the friction force on contact will act in the forward direction. In this case, there will be an increase in the horizontal velocity and a decrease in the angular momentum; rotational energy will be converted to translational energy. The ball will appear to gain speed on the bounce, and will skip forward at a lower angle to the ground.

If the ball is thrown with a backspin [Fig. 1(c)] then the friction force at the point of contact opposes both the translational and rotational motions of the ball, and it will be larger than in the previous two cases. The result is a relatively large decrease (and possible reversal of direction) of both the translational and rotational motions. Consequently, a ball projected forward with backspin will lose considerable speed on the bounce and may even bounce backwards.

The above discussion can be presented at the most elementary classroom level as an illustration of Newton's laws of motion. Most students have played tennis, ping pong, or various wall sports (handball, paddleball, etc) at one time or another and are familiar with the way a ball behaves when it is hit with a slice (giving it a backspin) or given a topspin (forward spin). They are usually quite surprised and pleased to find that they can readily understand the physics behind these phenomena.

A more detailed quantitative analysis can be performed at the introductory college (trigonometry or calculus based) physics level. For simplicity, we assume that the bounce is elastic in the vertical direction and consider only the motion parallel to the surface. As shown in Fig. 2, the ball has an initial horizontal speed v_0 and angular velocity ω_0 (clockwise rotation is considered positive). After the bounce, the ball has a final horizontal speed and angular velocity of v_f and ω_f, respectively. It is further assumed that the ball does not skid: the point of contact comes to an instantaneous stop, so that $v_f = R\omega_f$. The equations for linear and angular momentum then yield

$$F\Delta t = \Delta(mv_x) = m(v_f - v_0),$$

$$FR\Delta t = \Delta(I\omega) = 2mR^2(\omega_0 - \omega_f)/5,$$

where Δt is the time during which the ball (radius R) is in contact with the surface. The two equations are then combined to eliminate $F\Delta t$:

$$v_f = v_0 + 2R(\omega_0 - \omega_f)/5.$$

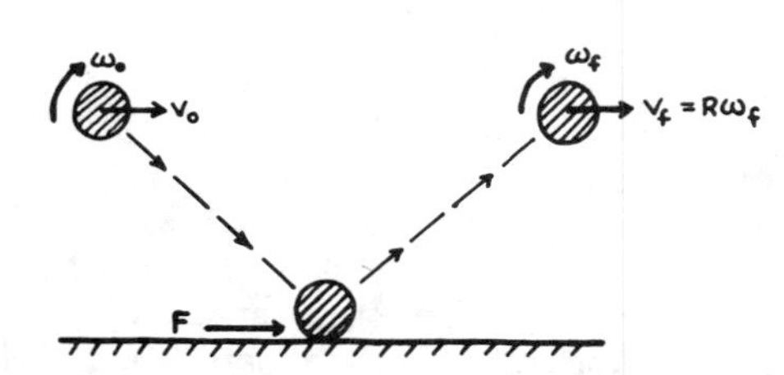

Fig. 2. Notation used in analysis of energy changes for a ball bounced on a rigid horizontal surface. It is assumed that the bounce is elastic in the vertical direction and that the ball bounces without skidding.

Table I. Energy changes for a bouncing ball (no skidding).

	$\Delta KE_{\text{translation}}$	$\Delta KE_{\text{rotation}}$	ΔKE_{total}
No initial spin	$-\frac{12m}{49}v_0^2$	$\frac{5m}{49}v_0^2$	$-\frac{m}{7}v_0^2$
Initial forward spin	$-\frac{2m}{49}(6v_0^2 - 5R\omega_0 v_0 - R^2\omega_0^2)$	$\frac{m}{49}(5v_0^2 + 4R\omega_0 v_0 - 9R^2\omega_0^2)$	$\frac{m}{7}(v_0 - R\omega_0)^2$
Initial backspin	$-\frac{2m}{49}(6v_0^2 + 5R\omega_0 v_0 - R^2\omega_0^2)$	$\frac{m}{49}(5v_0^2 - 4R\omega_0 v_0 - 9R^2\omega_0^2)$	$\frac{m}{7}(v_0 + R\omega_0)^2$

The no-skidding condition $v_f = R\omega_f$ can then be substituted to eliminate either v_f or ω_f from the above equation. We are particularly interested in the energy changes that occur; i.e., the changes in the translational, rotational, and total kinetic energies of the ball. The results are as follows:

$$\Delta KE_{\text{trans}} = (1/2)mv_f^2 - (1/2)mv_0^2 = -2m(6v_0 + R\omega_0)(v_0 - R\omega_0)/49,$$

$$\Delta KE_{\text{rot}} = (1/2)I\omega_f^2 - (1/2)I\omega_0^2 = m(5v_0 + 9R\omega_0)(v_0 - R\omega_0)/49,$$

$$\Delta KE_{\text{tot}} = \Delta KE_{\text{trans}} + \Delta KE_{\text{rot}} = -m(v_0 - R\omega_0)^2/7.$$

Table I summarizes the energy changes in the three cases of (a) no initial spin; (b) initial forward spin; and (c) initial backspin. A comparison of the results clearly reveals that *a backspinning ball always experiences a greater decrease in translational energy and in total energy than a forward-spinning ball.* In addition, the energy losses suffered by a forward-spinning ball are always less than those experienced by a ball with no spin.

With these results in mind, we can now understand why a basketball launched with backspin is more likely to go in the basket. When the ball hits the rim or backboard it experiences a change in speed, spin, and energy. We have seen that a backspinning basketball always has a greater decrease in speed and energy than a forward-spinning ball does under the same conditions. This is what makes the shot seem to be "softer" and more likely to drop in the basket after it hits the rim.

A ball can also be thrown with a sideways spin. When the ball hits a surface it will veer sharply to the left or right, depending on the spin direction. Good shooters learn to make effective use of spin. One example is the reverse layup, in which a player dribbles along the baseline and under the basket. As he comes out from under the basket, he spins the ball upward against the backboard. On contact, the ball veers sharply backward into the basket.

III. BASKETBALL SHOOTING: WHAT IS THE BEST LAUNCHING ANGLE?

In this section we will use the equations of projectile motion to describe the trajectory of a basketball. The purpose of this analysis will be to determine how to take the "best" shot from any given location on the court. That is, at a given distance from the basket there are an infinite number of trajectories, each with a specific initial speed and launching angle, that connect the shooter's hand with the center of the basket. We shall attempt to identify and establish criteria to determine the "best" trajectory; i.e., the one that is most likely to result in a basket. We shall then see whether or not the theory conforms with the real world. Since most basketball players develop their shooting ability by trial and error and constant practice, it will be interesting to find out if the "best" trajectories developed experimentally by basketball players conform to those developed theoretically.

A. Projectile motion equations

We begin by presenting the equations of motion for a body projected in a uniform gravitational field with initial speed v_0 at an angle with the horizontal of θ_0, as shown in Fig. 3. The equations for the x and y coordinates after a time t has elapsed are

$$x = v_0 t \cos\theta_0,$$

$$y = v_0 t \sin\theta_0 - (1/2)gt^2,$$

where g, the acceleration of gravity, is taken to be 32.2 ft/sec^2. The velocity components of the projectile after a time t are

$$v_x = v_0 \cos\theta_0,$$

$$v_y = v_0 \sin\theta_0 - gt.$$

The angle θ that the velocity vector of the projectile makes with the horizontal at any point on its path is given by the equation

$$\tan\theta = v_y/v_x = \tan\theta_0 - gt/v_0\cos\theta_0.$$

Through some simple manipulations of the above equations, we can eliminate the time t and arrive at the following useful relationships:

$$\tan\theta = 2y/x - \tan\theta_0, \quad (1)$$

$$v_0^2 = \frac{gx}{2\cos^2\theta_0(\tan\theta_0 - y/x)}. \quad (2)$$

The highest point of the trajectory ($y = y_{\max}$) occurs when $v_y = 0$ and $\tan\theta = 0$. It can then be shown that

$$y_{\max} = v_0^2 \sin^2\theta_0/2g. \quad (3)$$

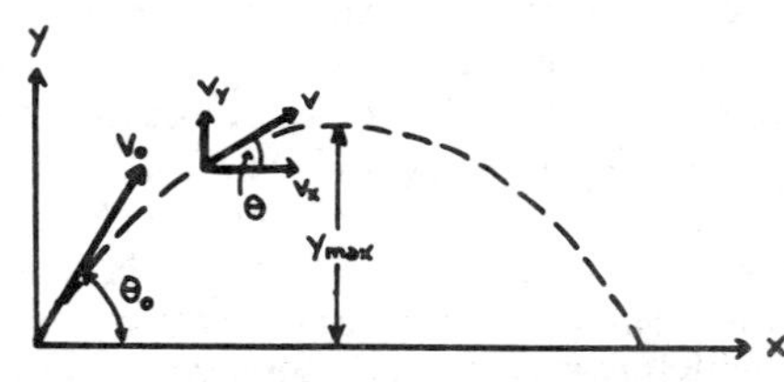

Fig. 3. Coordinate system and symbols used for description of projectile motion.

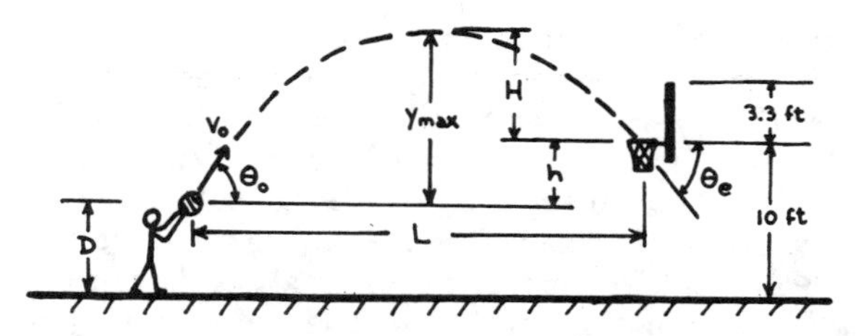

Fig. 4. Diagram and symbols used for description of trajectory of a basketball shot.

These equations describe the parabolic motion of a projectile in the absence of air resistance or other retarding forces. A typical basketball shot has a low air speed (20–30 ft/sec) and a short time of flight (~1 sec). On the other hand, a basketball is rather light in weight and has a relatively large surface area for a projectile. As a result, the effect of air resistance is small, but not negligible. Detailed calculations taking air resistance into account reveal that basketball trajectories do deviate somewhat (of the order of 5–10%) from the parabolic. Nevertheless, the assumption of parabolic motion (ignoring air resistance) does not affect the qualitative findings. We will therefore use the above equations to describe the trajectory of a basketball, and will take note afterwards of the quantitative changes produced by air resistance.

Figure 4 illustrates the path of a successful shot in the absence of air resistance. It has a trajectory which leaves the shooter's hand with an initial speed v_0 and launching angle θ_0, and passes through the point ($x = L, y = h$), where L is the horizontal distance from the point of release to the center of the basket and h is the vertical distance between the rim of the basket and the point of release. Equation (2) becomes

$$v_0^2 = \frac{gL}{2\cos^2\theta_0(\tan\theta_0 - h/L)}. \tag{4}$$

In this equation, v_0 and θ_0 are the variables; for any given launching angle θ_0 within limits to be specified there is an unique positive value of v_0 that will give the desired trajectory. Equation (4) thus describes a family of parabolas that connect the point of release with the center of the basket.

Figure 5 is a graph of the relationship between v_0 and θ_0 for a typical set of values of h and L. Note that there is a minimum in the curve; in fact it can be demonstrated that there is an unique minimum initial speed for every (h,L) combination. The existence of this minimum speed v_{0m} can be shown by the standard technique of setting the first derivative $dv_0/d\theta_0 = 0$. The result obtained is that the minimum occurs at the angle θ_{0m} given by

$$\theta_{0m} = 45° + (1/2)\arctan(h/L) \tag{5a}$$

or

$$\tan\theta_{0m} = h/L + (1 + h^2/L^2)^{1/2}. \tag{5b}$$

Substitution of Eq. (5b) into Eq. (4) then yields

$$v_{0m}^2 = gL\tan\theta_{0m} = g[h + (h^2 + L^2)^{1/2}]. \tag{6}$$

Thus of all the possible trajectories linking the point of release to the center of the basket, there is one launching angle θ_{0m} which gives the minimum-speed trajectory. The curve shown in Fig. 5 is symmetrical about a vertical axis through θ_{0m}. Note also that in order for v_0 to be finite and real, θ_0 must be in the range $90° > \theta_0 > \arctan h/L$.

A further constraint on the allowed values of θ_0 is established by the fact that the ball must *drop* into the basket; that is, the ball must be on the descending part of its parabolic trajectory when it reaches the basket. To state this criterion mathematically, let us define the *angle of entry* θ_e as the angle between the horizontal and the tangent to the trajectory as the ball crosses the plane of the rim, as shown in Fig. 4. We can use Eq. (1) to obtain an expression for θ_e. Since θ_e is measured positively *below* the horizontal and θ is measured positively *above* the the horizontal, it follows from Eq. (1) that

$$2h/L - \tan\theta_0 = \tan\theta = \tan(-\theta_e) = -\tan\theta_e$$

or

$$\tan\theta_e = \tan\theta_0 - 2h/L. \tag{7}$$

Hence the condition $\tan\theta_0 > 2h/L$ must be satisfied for the ball to be on the descent when it reaches the basket. However, there is a further restriction on θ_0, as we shall now demonstrate.

B. Margin for error

According to official basketball rules, the basket must be 18 in. (1.5 ft) in diameter, and the basketball must be between $29\frac{1}{2}$ and 30 in. in circumference. Hence the diameter of a basketball is about 9.5 in. (0.79 ft), or slightly more than half the diameter of the basket. As a result, the center of the ball does not have to pass through the exact center of the basket for a score. However, it must pass through in such a way that it clears both the front and back rims. Figure 6 shows a side view of the rim; D_r is the diameter of the rim (=1.5 ft). $C'C$ is the path of the center of a ball that passes through the center of the basket at an angle of entry θ_e. $A'A$ is the parallel path of a ball whose center falls short of the basket center by a distance ΔL but whose edge just clears the front rim. $B'B$ is the path of a ball whose center overshoots the basket center by a horizontal distance ΔL but just misses the back rim.[13] The lines FG and AE are drawn perpendicular to $A'A$, $B'B$, and $C'C$. The ball has

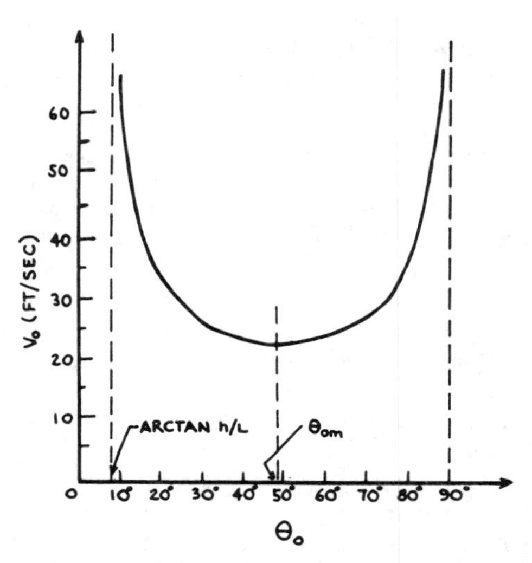

Fig. 5. Graph of relationship between v_0 and θ_0 [Eq. (4)] for a trajectory with h = 2 ft, L = 13.5 ft.

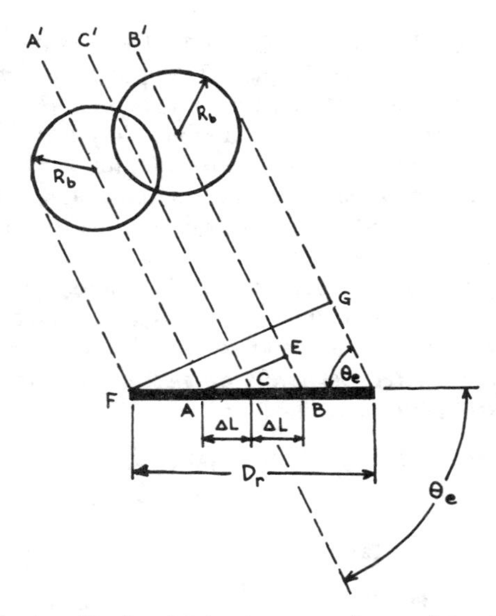

Fig. 6. Side view of basket rim showing range of successful paths of a ball passing through the basket at a given entry angle θ_e.

a diameter D_b (=0.79 ft) and radius R_b. From the geometry, we see that

$$FG = D_r \sin\theta_e,$$

$$AE = FG - 2R_b = D_r \sin\theta_e - D_b,$$

and

$$AB = 2\Delta L = AE/\sin\theta_e.$$

Combining these results we obtain

$$\Delta L = (1/2)(D_r - D_b/\sin\theta_e). \qquad (8)$$

Here ΔL represents a "margin for error" in shooting; the horizontal distance L from the shooter to the basket can be changed by a maximum of $\pm\Delta L$ and the ball will still go in the basket. We note, however, that the margin for error disappears ($\Delta L = 0$) if $\sin\theta_e = D_b/D_r = (0.79)/(1.5) = 0.53$, which corresponds to $\theta_e = 32°$.

It follows that the angle of entry must be at least 32°; if θ_e is less than 32° the ball cannot clear both rims. It is possible, of course, that the ball may still bounce off the rim and/or backboard and rebound into the basket, but a shot cannot go *cleanly* through if the entry angle is less than 32°. From Eq. (7) the *minimum launching angle* θ_{0L} is therefore given by

$$\tan\theta_{0L} = \tan 32° + 2h/L = 0.62 + 2h/L. \qquad (9)$$

The launching angle θ_0 must lie in the range $\theta_{0L} \le \theta_0 < 90°$ for a successful shot. For every angle θ_0 within this range there exists a specific launching speed v_0 [given by Eq. (4)] that will make the ball pass through the center of basket. However, we have also shown that the shot will still be successful if the horizontal distance is short or long by an amount ΔL. We will now relate this allowable deviation in distance to the corresponding margins for error in speed and angle.

First, let us assume that the shooter launches the ball at the correct launching angle for a center-of-basket trajectory for the given height and distance. By what amount can his launching speed differ from the correct launching speed (v_0) such that the ball will still go in the basket? Thus for a fixed θ_0 there is a range of values $v_0 \pm \Delta v$ that leads to a successful shot, as shown in Fig. 7. We define Δv as the *margin for error in speed.* It can be calculated directly from Eq. (4) as follows: First, v_0 is calculated for the given values of h, L, and θ_0. We replace L by $L + \Delta L$ to calculate $v_+ = v_0 + \Delta v_+$; then we replace L by $L - \Delta L$ to calculate $v_- = v_0 - \Delta v_-$. A detailed mathematical analysis shows that to a good approximation, $\Delta v_+ = \Delta v_-$. Therefore, we define $\Delta v = \Delta v_+ = \Delta v_-$.

Now let us assume that the shooter launches the ball at the correct launching speed v_0 for a center-of-basket trajectory. By what amount can the launching angle differ from θ_0 for a successful shot? There is a *margin for error in angle* $\Delta\theta_\pm$ such that if the launching angle lies between $\theta_0 - \Delta\theta_-$ and $\theta_0 + \Delta\theta_+$ the horizontal distance will fall between $L - \Delta L$ and $L + \Delta L$. Once again, Eq. (4) is used to calculate $\Delta\theta_-$ and $\Delta\theta_+$ by consecutive substitution of $L - \Delta L$ and $L + \Delta L$ for fixed values of v_0, h, and L. The behavior of $\Delta\theta_\pm$ is more complicated than that of Δv. In general, $\Delta\theta_+$ and $\Delta\theta_-$ will *not* be equal. Moreover, it can be shown that if θ_0 is less than θ_{0m} (the minimum-speed angle) then an increase in the angle causes the shot to travel a greater distance. However, if θ_0 is more than θ_{0m}, an increase in the angle makes the shot fall short, as shown in Fig. 8. Also if $\theta_0 = \theta_{0m}$, both an increase and a decrease in the launching angle will cause the shot to fall short [Fig. 8(c)].

These two margins for error, Δv and $\Delta\theta_\pm$, will serve as criteria for selecting the best trajectory. The larger the margins for error for a given shooting angle, the more freedom the shooter has to deviate from the precise values of v_0 and θ_0 needed for a center-of-basket trajectory. The aim is to find the launching angle and speed combinations that maximize the margins for error.

Thus far we have considered only the two-dimensional motion of the basketball. For a successful shot, the trajectory cannot deviate too far to the left or right of the center of the basket, or the ball will fall outside or hit the side of the rim. Consequently there is a "lateral margin for error" to be considered. This can be derived rather easily. In order for the ball to just clear the side of the rim, the center of the ball cannot deviate laterally from the center of the basket by more than $R_r - R_b$, where R_r and R_b are the radii of the rim and ball, respectively (see Fig. 9). Using the numbers given previously, we find $R_r - R_b = 0.36$ ft. This lateral displacement is over a straight-line distance from the point of release to the center of the basket of $(h^2 + L^2)^{1/2}$. Thus the *margin for error in lateral angle* $\Delta\psi$ is given by

$$\tan\Delta\psi = 0.36/(h^2 + L^2)^{1/2}. \qquad (10)$$

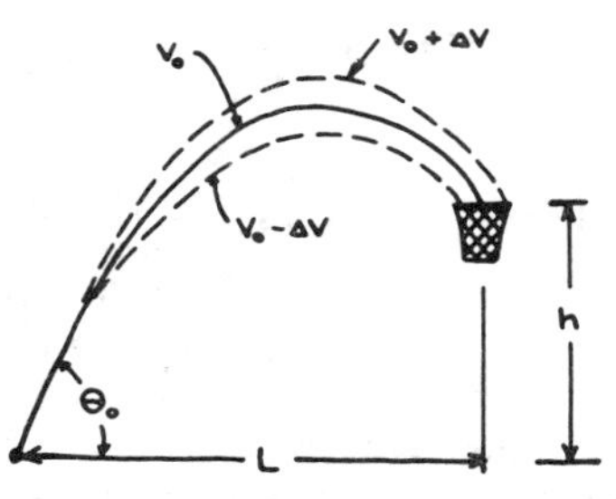

Fig. 7. Margin for error in speed: for a given launching angle θ_0, there is a range of launching speeds $v_0 \pm \Delta v$ that will result in a successful shot.

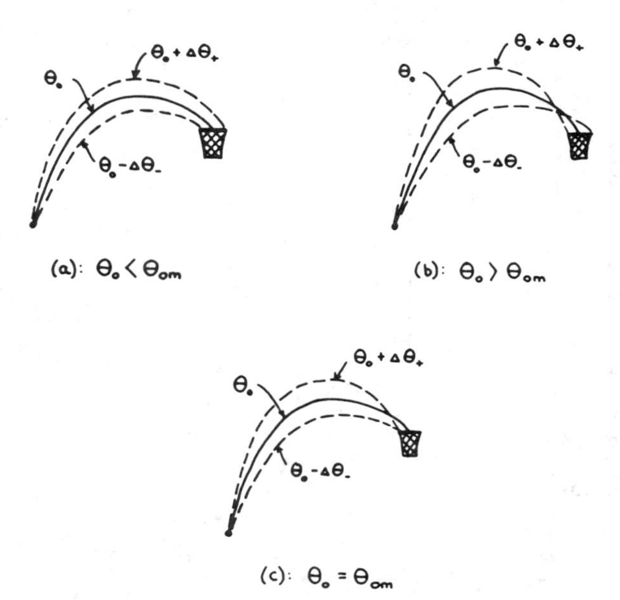

Fig. 8. Margin for error in angle: for a given launching speed v_0, any launching angle within the range $\theta_0 - \Delta\theta_-$ to $\theta_0 + \Delta\theta_+$ will result in a successful shot.

It is evident from Fig. 9 that this represents an upper limit to the margin for error on the assumption that the shot does not fall long or short of the center of the basket. The behavior of $\Delta\psi$ is rather simple because $\Delta\psi$ depends only on the launching height and distance and is independent of the launching speed and angle. Thus the closer the shooter is to the basket and the higher the launching point, the larger the margin for error and the greater the likelihood of success.

We complete the mathematical preliminaries by deriving equations for two useful quantities. The first is an estimate of the *average launching force F*. The basketball is accelerated from rest to a speed at release of v_0, so the work done on the ball is $W = (1/2)mv_0^2$, where m is the mass of the basketball (the change in gravitational potential energy during the launching of the ball can be neglected). The work done can in turn be equated to Fd, where F is the average force exerted and d is the distance the ball is moved by the shooter's hand from rest to the point of release. The distance d depends upon the player's particular shooting style. Some shooters release the ball with a sharp flick of the wrist while pivoting the forearm slightly about the elbow. Others may start the ball at the shoulder and release it when the arm is fully extended. Hence d may be between 6 in. and 3 ft. We will adopt a "typical" value of $d = 2$ ft. Also, the official weight of a basketball is 20 to 22 oz, corresponding to an average mass of 0.041 slugs. Letting $W = Fd = (1/2)mv_0^2$ and substituting numerical values, we obtain

$$F = mv_0^2/2d \approx 0.01v_0^2. \tag{11}$$

It follows from this result that a shot launched from a given point with the minimum speed v_{0m} will require the least amount of force.

Finally, we define H, *the maximum height above the rim*:

$$H = y_{\max} - h = (v_0^2 \sin^2\theta_0/2g) - h. \tag{12}$$

This quantity will be used to determine the launching angle. When examining the trajectory of any shot, either on sight or on film, one finds that it is rather difficult to estimate or measure the launching angle directly. However, H can be measured with little difficulty by lining up the peak of the trajectory (viewed from the side) with the plane of the rim, using the backboard as a scale of reference (see Fig. 4). For a standard rectangular (6 ft × 4 ft) backboard, the top of the backboard is 40 in. (3.3 ft) above the plane of the rim. Substituting for v_0 from Eq. (4) into Eq. (12) and solving, we obtain the following equation for θ_0 in terms of h, L, and H:

$$\tan\theta_0 = \frac{2}{L}(H + h)\left[1 + \left(\frac{H}{H + h}\right)^{1/2}\right]. \tag{13}$$

C. Numerical results

We will now proceed to use the equations derived in Sec. III B to give numerical values for the relevant quantities. First, let us establish a reasonable range of values for the two parameters h and L. We have defined h as the vertical distance between the point of release and the rim of the basket, as shown in Fig. 4. The rim is 10 ft above the floor, and the point of release is at a height D above the floor. The value of D depends upon the height of the shooter and the type of shot taken. Generally, the ball is released at a point about 1–2 ft above the shooter's head. If the shooter is taking a jump shot, the point of release may be raised by an additional foot or so. The tallest players on occasion are actually shooting down at the basket! However, for the vast majority of players of all ages, D ranges from 6 to 9 ft. Hence h will be generally between 1 and 4 ft.

Figure 10 shows a plan of a basketball court. Most jump shots or set shots are taken at distances of 10–20 ft from the basket. A player who is closer than 10 ft to the basket is more likely to bank the shot off the backboard rather than to use a parabolic trajectory. Shots are rarely taken at distances greater than 25 ft; these are invariably desperation shots taken as time is running out. For the 1979 season, the National Basketball Association instituted a "three-point basket" for shots taken beyond a line which averages 23 ft from the basket. Only a few of these are attempted in a typical game. For our purposes, then, we will let L range from 10 to 25 ft. For the chosen ranges of h and L, the minimum-speed angle θ_{0m} lies between 45° and 55°. As the

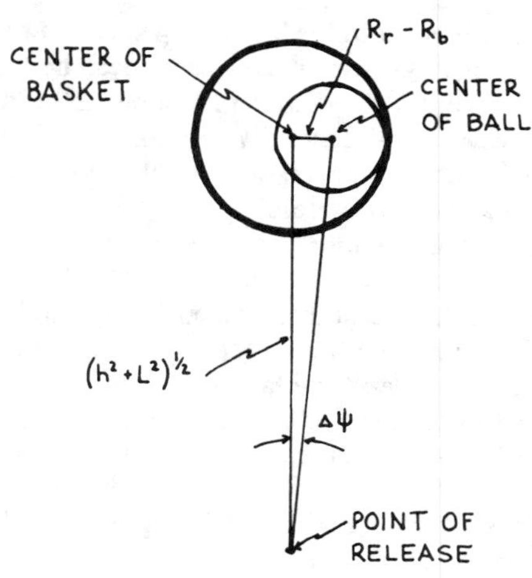

Fig. 9. Diagram for determination of margin for error in lateral angle $\Delta\psi$ (top view of basket).

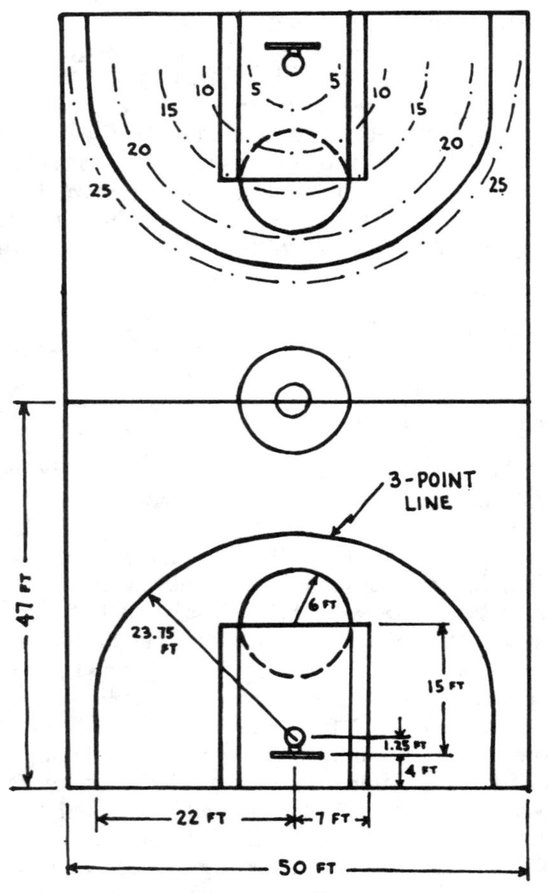

Fig. 10. Dimensions of a basketball court (professional rules). Dot-dash lines show distance from the basket in 5-ft intervals.

shooting distance and launching height increase, θ_{0m} approaches 45°.

Now let us examine the behavior of the margins for error in speed and in angle. Figure 11 shows how these quantities vary with launching angle for the specific distances $h = 2$ ft, $L = 13.5$ ft—equivalent to a shot taken from the free-throw line by a six-foot-tall player. For comparison purposes, we have plotted the *fractional* errors ($\Delta v/v_0$, $\Delta\theta_+/\theta_0$, and $\Delta\theta_-/\theta_0$). We see first of all that the margin for error in speed increases slowly as θ_0 increases and is very small, exceeding 0.01 only above 60°. This means that the shooter has very little leeway in his launching speed. That is, for a given launching angle the difference in speed between a shot that passes through the center of the basket and one that just clears either rim is generally less than 1%. To maximize the margin for error in speed, one should shoot with as high a launching angle as possible.

By comparison, the margins for error in angle tend to be much larger. Note that $\Delta\theta_+/\theta_0$ and $\Delta\theta_-/\theta_0$ each show a very sharp "peak" (actually a discontinuity) and have their largest values within a few degrees of θ_{0m}, the minimum-speed angle. To understand how these curves should be interpreted, let us take the following example. At a launching angle of 48° for the given h and L there is a specific launching speed ($v_0 = 22.46$ ft/sec) that will result in a center-of-basket trajectory. At this angle, $\Delta\theta_+/\theta_0 = 0.119$ and $\Delta\theta_-/\theta_0 = 0.038$. This means that for the same launching speed, the launching angle can be increased by 11.9% (up to 53.7°) or decreased by 3.8% (down to 46.2°)

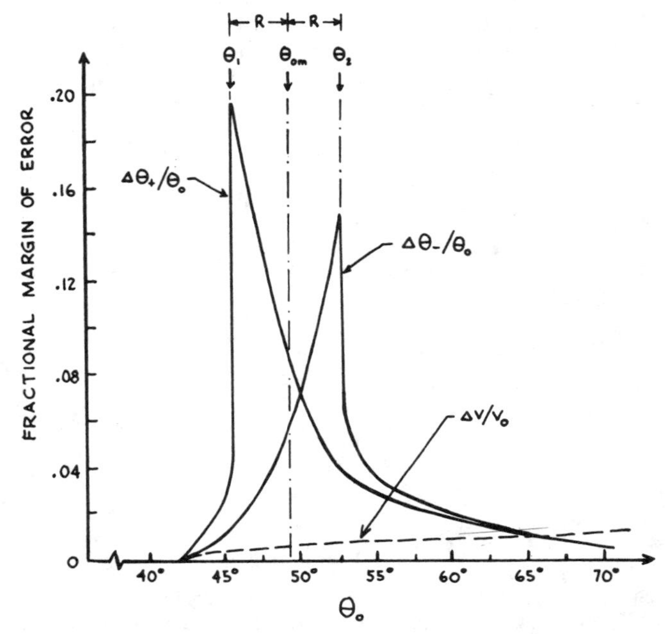

Fig. 11. Margins for error in speed and angle for trajectories with $h = 2$ ft, $L = 13.5$ ft.

and the ball will still go through the basket. As can be seen in Fig. 11, the margin for error in angle will be large if the launching angle is between θ_1 (=45.8°) and θ_2 (=52.6°), or specifically in the range 49.2 ± 3.4°, where 49.2° is the minimum-speed angle.

The behavior of the margin for error is angle near θ_{0m} can be understood by referring to Fig. 12. This shows a more detailed version of the bottom part of the $v_0 - \theta_0$ curve originally shown in Fig. 5. The solid lines contain the region of all successful shots, while the dashed line represents only the center-of-basket trajectories. It can be seen that the margin for error in angle at the particular launching speed v_1 is much larger than for any speed greater than v_1. Note that this speed corresponds to the center-of-basket trajec-

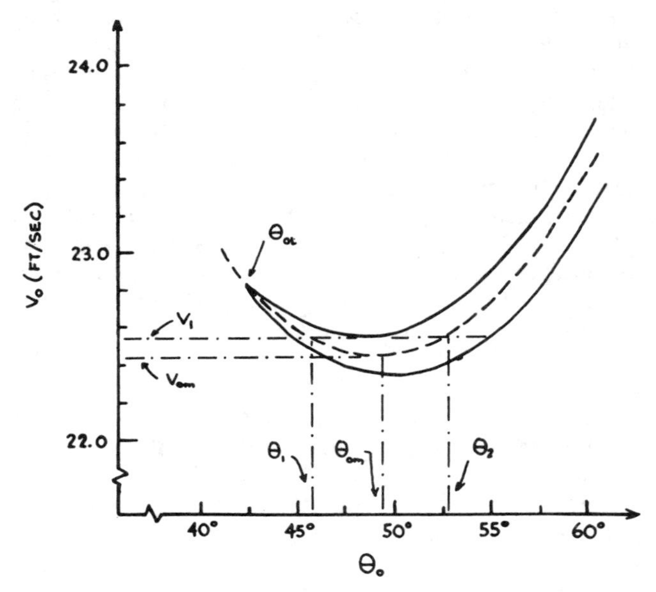

Fig. 12. Graph of v_0 vs θ_0 for $h = 2$ ft, $L = 13.5$ ft. The solid lines enclosed all points corresponding to a successful shot. The dashed line represents those points corresponding to a center-of-basket trajectory.

tory at the launching angles θ_1 and θ_2. Figure 12 demonstrates graphically why the margin for error in angle is largest in the range between θ_1 and θ_2 center on θ_{0m}. Let us define the *angular range R* as

$$R = \theta_{0m} - \theta_1 = \theta_2 - \theta_{0m}.$$

To maximize the margin for error in angle, the launching angle should therefore lie in the range $\theta_{0m} \pm R$. On the other hand, the margin for error in speed is maximized by making the launching angle as large as possible. However, $\Delta v/v_0$ varies very slowly with launching angle, while $\Delta\theta_\pm/\theta_0$ peaks very sharply near θ_{0m}. Moreover, it was noted previously that a shot launched with the minimum speed requires the smallest launching force. Taking everything into account, it seems rather clear that *the minimum-speed angle is the best launching angle*.

Several interesting aspects of basketball shooting emerge when we compare the relevant quantities for shots of different distances and heights. In Table II(a) we compare shots taken at the minimum-speed angle at a distance of 15 ft and for different values of h. The results show the advantage of a higher point of release (smaller h). As h decreases, the launching speed and force become smaller; all margins for error as well as the angular range become larger. Thus the higher the point of release, the more likely it is that the shot will be successful. This shows that given equal shooting ability, a taller player has an advantage over a shorter one *at any distance from the basket* in that the taller player has a greater margin for error. Fortunately for the shorter players, most taller players tend to develop their rebounding and under-the-basket play at the expense of their longer-distance shooting skills.

Table II(b) compares shots taken at different distances from the basket for the same launching height. As expected, the margins for error generally get smaller as the distance to the basket increases.

D. Effects of air resistance

The numerical results presented in Sec. III C were based on the assumption that air resistance is negligible and the trajectory is parabolic. As noted previously, air resistance *does* have a measurable effect on the trajectory of a basketball. To assess the effects of air resistance, the equations of motion for a projectile subject to a small aerodynamic drag force were calculated using the approximation methods outlined by Parker.[14] Since these calculations are too detailed to present here, we shall simply summarize the results.

The drag force on a sphere moving through air at the speeds of interest is given by the equation $F_D = (1/2)C_D \rho A v^2$, where C_D is the drag coefficient, ρ the density of air, and A the corss-sectional area of the sphere. For a basketball (weight 21 oz, diameter 0.79 ft) at speeds of 20–30 ft/sec, $C_D = 1/2$. Using $\rho = 1.37 \times 10^{-3}$ slugs/ft^3 for air, we obtain $F_D = 0.000\,29 v^2$ lb. For an air speed of 25 ft/sec, the deceleration due to drag is about 4 ft/sec^2 tangent to the path of the basketball.

A basketball launched with the speed and angle determined by the drag-free equations will typically fall short of the center of the basket by a foot or more because of the drag force. To compensate for this, the launching speed must be increased on the order of 5% for a given launching angle. The more time the ball spends in flight, the greater the effect of drag, so that more compensation is required as the launching angle and/or distance from the basket increase. In general, the minimum-speed angle will be about 1° lower than the drag-free value for a given h and L. The drag force changes the trajectory shape so as to make the descent slightly steeper than the ascent. As a result the angle of entry θ_e is increased by 2°–3°. This in turn lowers the minimum launching angle θ_{0L} by 1°–2°, and increases the margin for error in distance ΔL. The margins for error in speed and angle and the angular range R are corre-

Table II. Effect of launching height and distance on minimum speed trajectories.

(a) Same distance to basket (L = 15 ft)										
h	θ_{0L}	θ_{0m}	R	v_{0m}	F	H	$\Delta\psi$	$\Delta v/v_{0m}$	$\Delta\theta_+/\theta_{0m}$	$\Delta\theta_-/\theta_{0m}$
1	37.0	46.9	3.9	22.72	5.16	3.28	1.37°	0.005	0.099	0.073
2	41.6	48.8	3.4	23.49	5.52	2.85	1.36°	0.004	0.085	0.058
3	45.6	50.7	2.8	24.27	5.89	2.47	1.35°	0.003	0.073	0.045
4	49.1	52.5	2.3	25.07	6.29	2.14	1.33°	0.003	0.062	0.033

(b) Same launching height (h = 2 ft)										
L	θ_{0L}	θ_{0m}	R	v_{0m}	F	H	$\Delta\psi$	$\Delta v/v_{0m}$	$\Delta\theta_+/\theta_{0m}$	$\Delta\theta_-/\theta_{0m}$
10	45.6	50.7	3.5	19.82	3.93	1.65	2.02°	0.005	0.094	0.052
15	41.6	48.8	3.4	23.49	5.52	2.85	1.36°	0.004	0.085	0.058
20	39.4	47.9	3.1	26.68	7.12	4.08	1.03°	0.004	0.078	0.058
25	38.0	47.3	2.9	29.53	8.72	5.31	0.82°	0.003	0.072	0.056

h = vertical distance; point of release to rim of basket (ft).
L = horizontal distance; point of release to center of basket (ft).
θ_{0L} = minimum launching angle.
θ_{0m} = minimum speed angle.
R = angular range.
v_{0m} = minimum launching speed (ft/sec).
F = average launching force (lb).
H = maximum height above rim (ft).
Δv = margin for error in speed (ft/sec).
$\Delta\psi$ = margin for error in lateral angle.
$\Delta\theta_\pm$ = margin for error in angle.

spondingly increased. Thus the most significant effect of air resistance is that it tends to *aid* the shooter by increasing the margins for error!

IV. BASKETBALL SHOOTING: THEORY AND PRACTICE

Our development of the theory of basketball shooting has been based on the assumption that a shooter cannot always launch his shot at the precise speed and angle needed for a center-of-basket trajectory. The launching parameters can deviate somewhat from the precise values, and the shot will still be successful because of the simple fact that the ball is smaller than the basket. Our goal has been to find the conditions that maximize the margins for error, and on this basis we have concluded that the minimum-speed angle is the best launching angle. Now we must face the question as to whether this theory bears any relationship to actual basketball play.

Most basketball players are not formally taught how to shoot. Instead they learn by imitation, trial-and-error, and constant practice until they develop a "kinesthetic memory" of what they consider to be the proper launching speed and angle to be used for any given distance. The best shooters develop this ability to an extraordinary degree. We saw previously that the margin for error in speed is very small—almost always less than 1%. Thus a player who finds that his shots are falling short is able to adjust his shot by increasing the launching speed by no more than a fraction of a percent. This demonstrates the exceptional skill involved in having a good "shooting touch." The average professional player hits about 50% of his shots under game conditions; for open (unguarded) shots or in practice the average increases to 70% or more. When shooting a free throw (foul shot) the player is allowed time to relax and concentrate on his shot. Even allowing for shots that hit the rim or backboard and rebound into the basket, the margins for error are still rather small. Yet the best professionals hit this shot 90% of the time![15] It is rather remarkable that the human body can be trained to reproduce the required movements so precisely.

What launching angles do most basketball players prefer to use? The author's observations on this matter are based on countless hours of playing, watching, and coaching basketball, as well as on some experimentation with his own shooting. First, let us consider the "high-arch" shot—one that is launched at an angle well above the minimum-speed angle. The high-arch shot in theory has the advantage of a larger margin for error in speed. It is also more difficult to block (it can be launched over the outstretched hand of a taller defender). However, in practice it is a rather difficult shot to launch and to aim. For example, a shot taken by a six-footer (h = 2 ft) from the free-throw line and launched at a 65° angle would have a maximum height H above the rim of 5.77 ft, or 2.44 ft above the top of the backboard. In fact, one rarely ever sees a shot from this distance go higher than the top of the backboard (H = 3.33 ft). Using Eq. (13) we find that this corresponds to a launching angle of 50° to 60°, depending on the height of the shooter. As a general rule, in actual play no shot is ever launched at an angle greater than 60° in the 10–25-ft shooting range.

At the other extreme is the "flat" or low-arch shot, which is aimed almost on a line at the basket, and is launched at an angle well below the minimum-speed angle. (It may even be launched at an angle below θ_{OL}, so that the entry angle is less than 32° and a good bounce is needed for the shot to be successful.) This type of trajectory is commonly seen in playground basketball. It is used especially by youngsters who, because of a lack of height and strength, have developed a habit of aiming the ball *at* the rim instead of trying to arch the shot *over* the rim. The theory shows, however, that a flat shot has smaller margins for error and actually requires more launching force in comparison to a shot taken with a higher arch (closer to the minimum-speed angle).

The style of play in many ways determines the type of shot to use. In the early days of basketball, team play was slower and more deliberate. Players tended to run in fixed patterns and the ball was passed frequently until one player managed to get free for an open shot. The most common type of shot was the two-handed set shot, a relatively high-arch shot launched from the chest or over the head with both feet on the floor. Since the late 1950s the style of play has changed noticeably. There is now a greater emphasis on speed and individual "one-on-one" play. The most common shot is the jump shot, which is launched (usually one-handed) while in midair. Often, a player will dribble to a specific spot on the floor, come to a sudden stop, leap vertically or slightly backwards, and release his shot. A jump shot must be launched quickly with a minimum of effort. Obviously, the minimum-force feature of the minimum-speed angle is a big advantage here. The jump shot has become so popular that many players use it even when they are unguarded, in preference to a set shot taken with both feet on the floor (presumably giving the shooter better balance and control). We have seen that there is a theoretical advantage to this, because the margins for error increase as the launching height increases [see Table II(a)]. However, there is a tendency to launch a jump shot on a flat trajectory. This is a disadvantage to the shooter, in that a slightly higher-arch shot can be launched with less force and better accuracy.[16] There is also a tendency to launch all shots at the same angle regardless of distance. However, the minimum-speed angle increases as the distance decreases [see Table II(b)]. Therefore, one should learn to shoot at a higher angle as one gets closer to the basket.

As a simple test of the theory, a brief study was made to determine the launching angles used by good shooters. A number of Brooklyn College students (all of them considered to be good playground ballplayers) were filmed using a super-8 movie camera as they took set and jump shots at various distances from the basket. The developed film was then run at slow speed through a film editor–viewer, and the trajectory of each shot was traced out on a transparent plastic sheet placed over the viewer screen. The launching angle could then be calculated with reasonable accuracy from the shape and maximum height of the trajectory. Nearly 80 shots were analyzed in this fashion. The results showed that the successful shots were being launched at angles very close to the minimum-speed angle and within the angular range for the given height and distance. These very limited and relatively crude measurements seem to indicate that the better shooters do learn to shoot at or near the minimum-speed angle.

A more detailed study of preferred shooting angles might be of particular interest to basketball coaches as well as to physical education faculty in the areas of motor learning, kinesiology, and biomechanics. Results could be used to improve the teaching of basketball shooting to youngsters. One technique suggested by Mortimer[10] is to suspend strings or other markers at the appropriate height and distance of the trajectory peak to be used as aiming targets.

V. SUMMARY AND CONCLUSIONS

The major purpose of this work has been to show how a knowledge of physics can be used to understand certain aspects of basketball play and to improve one's performance as a player. We have seen how an awareness of the principle of inertia and an understanding of the effects of spin can help to improve specific shooting techniques. Our analysis of the trajectory of a basketball has demonstrated the existence of a best shooting angle—namely, the minimum-speed angle. There are clear advantages to shooting at this angle in comparison to the more commonly used flat trajectory.

Thus a knowledge of physics *can* make one a better basketball player. The results obtained here should be useful in teaching youngsters how to play the game. An additional benefit is that many of the examples used in this work can be presented in the physics classroom as interesting illustrations of basic Newtonian principles.

[1]L. J. Briggs, Am. J. Phys. **27**, 589 (1959).
[2]P. Kirkpatrick, Am. J. Phys. **31**, 606 (1963).
[3]S. Chapman, Am. J. Phys. **36**, 868 (1968).
[4]R. G. Watts and E. Sawyer, Am. J. Phys. **43**, 960 (1975).
[5]H. Brody, Am. J. Phys. **47**, 482 (1979).
[6]D. C. Hopkins and J. D. Patterson, Am. J. Phys. **45**, 263 (1977).
[7]T. Jorgenson, Am. J. Phys. **38**, 644 (1970).
[8]C. Frohlich, Am. J. Phys. **47**, 583 (1979).
[9]J. D. Memory, Am. J. Phys. **41**, 1205 (1973).
[10]E. M. Mortimer, Res. Q. **22**, 234 (1951).
[11]G. Galilei, *Dialogues Concerning the Two Chief World Systems*, translated by Stillman Drake, 2nd ed. (University of California, Berkeley and Los Angeles, 1967), p. 126 and 144.
[12]A. Auerbach, *Basketball for the Player, the Fan, and the Coach* (Pocket, New York, 1976), p. 80.
[13]A less stringent criterion would be that the center of the ball (rather than the edge) must pass inside the edge of the rim, since the ball would then hit the rim and probably drop into the basket. This would allow a somewhat larger margin for error. However, we shall adopt the purist approach accepted by the best shooters, who like to make the ball "swish" through the basket without touching the rim.
[14]G. W. Parker, Am. J. Phys. **45**, 606 (1977).
[15]According to the *Guinness Book of World Records* (Bantam, New York, 1980), the record for consecutive free throws is 2036, set by Ted St. Martin in 1977. In 1978, Fred L. Newman made 88 consecutive free throws while blindfolded!
[16]The author's own experience serves to verify this conclusion. For many years he used a flat shot, with consistently mediocre results. A shift in shooting technique to a higher launching angle led, after only a few weeks of practice, to a significant and very satisfying improvement in shooting accuracy.

Kinematics of the free throw in basketball

A. Tan[a)] and G. Miller
Department of Physics, University of Florida, Gainesville, Florida 32611
(Received 16 June 1980; accepted 12 September 1980)

The kinematics of the two basic styles of free throw in basketball are discussed. It is shown that from a purely kinematic and trajectory point of view, the overhand push shot is preferable to the underhand loop shot. The advantages of the underhand shot lie in the actual execution of the shot.

I. INTRODUCTION

The free throw is an important and oftentimes decisive element in a basketball game. According to one study,[1] approximately one-fourth of all victories were decided by the free throws. Of the various styles used in the free throw, two basic styles are recognized. In the *overhand push shot*, the ball is moved in a straight line from near the shoulder to the release point (Fig. 1). In its most common form, the overhand shot is mainly a one-handed shot, the other hand going to a supportive and directive effort. The second category of the free throw is the *underhand loop shot*, in which the ball is swung in an arc from above the knees to the release point (Fig. 1). This is a two-handed shot where the ball is held symmetrically between the two hands during the entire motion. Data collected by Hobson,[2] Burke,[3] and Beetz[4] all show that the percentages of the free throws made were the highest for the underhand shot. Although the underhand shot has fallen into disuse, at one time it was the only shot recommended from the foul line. It is interesting to note that in the professional NBA league today, only one player (Rick Barry) uses the underhand shot and yet he makes the highest percentage of the free throws.[5] In this paper, we investigate the kinematics of the two styles of free throws and their relative advantages and disadvantages.

II. METHOD OF STUDY

In the basketball free throw, the player stands behind the foul line located at a horizontal distance of 13.8 ft from the center of the hoop and is allowed to shoot a 30-in.-circumference ball into an 18-in.-diameter hoop. The hoop is situated at a height of 10 ft above the floor level. For a *clean* shot (that is, the ball does not hit the rim), the smallest angle of entry of the ball is $\sin^{-1} 30/18\pi \simeq 32°$. With a flatter trajectory, the ball might still go through the hoop after hitting the rim once or more, but its chances decrease because the effective cross section of the hoop gets smaller.

The problem is one of a projectile in an inclined plane from the release point to the basket. Referred to the release point as the origin and neglecting air resistance,[6] the horizontal and vertical components of the velocity of the ball are given by

$$v_x = dx/dt = v_0 \cos\alpha \quad (1)$$

and

$$v_y = dy/dt = v_0 \sin\alpha - gt, \quad (2)$$

where v_0 is the velocity of projection, α the angle of projection, and g the acceleration of gravity. If θ is the angle of the incline, the range on the incline is found to be[7]

$$r = 2v_0^2 \cos\alpha \sin(\alpha - \theta)/(g \cos^2\theta). \quad (3)$$

The optimum projection velocity and projection angle for a projectile on a plane have been found by Durbin.[8] Differentiating (3) with respect to α and v_0, we get

$$dr/d\alpha = 2v_0^2 \cos(2\alpha - \theta)/(g \cos^2\theta), \quad (4)$$

$$dr/dv_0 = 4v_0 \cos\alpha \sin(\alpha - \theta)/(g \cos^2\theta). \quad (5)$$

For a given v_0, the maximum range R is obtained when $dr/d\alpha = 0$ and $d^2r/d\alpha^2 < 0$. These conditions give

$$\alpha = \pi/4 + \theta/2 \quad (6)$$

and

$$R = (v_0^2/g)/(1 + \sin\theta). \quad (7)$$

Conversely, for a given range R, the minimum projection velocity is obtained for the projection angle (6). Durbin[8] has discussed why the condition of maximum range is also the condition for optimum range in a basketball shot. First, a minimum projection velocity would mean a minimum muscular effort and a steadier hand. A minimum velocity would also mean a minimum air resistance. Secondly, Eq. (4) tells us that the change in range due to an error in projection velocity is the smallest when the projection velocity is the slowest. All three considerations imply that Eq. (6) is the desired projection angle.

With this condition, the coordinates of the center of the hoop are given by

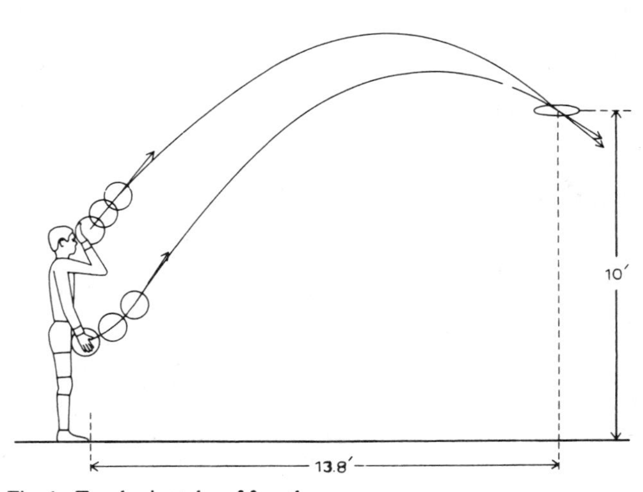

Fig. 1. Two basic styles of free throw.

Reprinted from *American Journal of Physics* **49**, 542–544 (1981); © American Association of Physics Teachers.

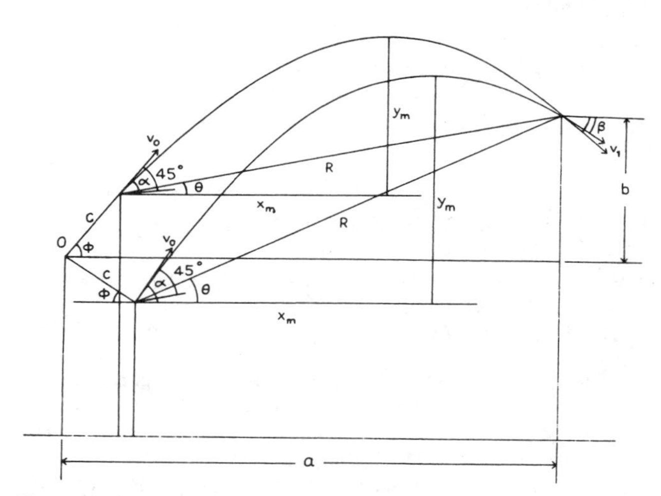

Fig. 2. Trajectories of the ball from the free-throw line.

$$x_1 = R\cos\theta = (v_0^2/g)(1 - \sin\theta)/\cos\theta, \tag{8}$$

$$y_1 = R\sin\theta = (v_0^2/g)\sin\theta/(1 + \cos\theta). \tag{9}$$

The time of flight is given by

$$t_1 = x_1/v_x = v_0(1 - \sin\theta)/(g\cos\theta\cos\alpha). \tag{10}$$

The angle of entry of the ball into the basket can be found from Eqs. (1), (2), (6), and (10):

$$\beta = \theta/2 - \pi/4. \tag{11}$$

The coordinates of the apex of the path are obtained from the condition $dy/dx = 0$. Equations (1), (2), and (4) give

$$x_m = v_0^2\cos\theta/(2g), \tag{12}$$

$$y_m = v_0^2\sin^2\alpha/(2g). \tag{13}$$

The velocity of entry into the basket can be calculated from Eqs. (1), (2), (4), and (11):

$$v_1 = v_0\cot\alpha. \tag{14}$$

Finally, the length of the trajectory is given by

$$s = \int_{t=0}^{t_1} (v_x^2 + v_y^2)^{1/2}\, dt.$$

Substituting from Eqs. (1) and (2), integrating and putting limits, we arrive at

$$s = \frac{v_0^2\sin\alpha}{2g} + \left(\frac{t_1}{2} - \frac{v_0\sin\alpha}{2g}\right) \times (g^2t_1^2 - 2gt_1v_0\sin\alpha + v_0^2)^{1/2} + \frac{v_0^2\cos^2\alpha}{2g} \times \ln\left(\frac{(g^2t_1^2 - 2gt_1v_0\sin\alpha + v_0^2)^{1/2} + gt_1 - v_0\sin\alpha}{v_0(1 - \sin\alpha)}\right). \tag{15}$$

The angle θ and range R depend upon the release point and are therefore different for different styles of free throw. In Fig. 2, O is the fulcrum of the arm of the player and c is the length of the arm from the fulcrum to the center of the palm. b is the difference in the vertical heights of O and the hoop, while a is the horizontal distance between O and the vertical from the center of the hoop. The angle made by the extended arm at the moment of projection is denoted by ϕ. For the two styles of free throw, $\phi = \pi/4 + \theta/2$ and $\phi = \pi/4 - \theta/2$, respectively.

For the overhand shot, we have the relations

$$c\cos\phi + R\cos\theta = a, \tag{16}$$

$$c\sin\phi + R\sin\theta = b. \tag{17}$$

Eliminating R between Eqs. (16) and (17), we get

$$c\sin(\theta/2 - \pi/4) = a\sin\theta - b\cos\theta. \tag{18}$$

Equation (18) can be solved either graphically or numerically to give the value of θ. The range R can then be determined from either of the Eqs. (16) or (17).

For the underhand shot, Eq. (16) is still valid but Eq. (17) is replaced by

$$R\sin\theta - c\sin\phi = b. \tag{19}$$

In this case, we get, instead of Eq. (18),

$$c\sin(\theta/2 + \pi/4) = a\sin\theta - b\cos\theta. \tag{20}$$

III. RESULTS AND DISCUSSION

The average professional player today stands 6 ft 6 in. tall and wears size 13 shoes. That means, the perpendicular from O (which falls on the heels) is about 9.5 in. away from the foul line and $a = 14.6$ ft. The fulcrum O is about 5.5 ft above floor level (the feet are not allowed to leave the floor) and therefore $b = 4.5$ ft. Also, for the average player, $c = 2.5$ ft. On substitution of these values, we can solve for θ from Eqs. (18) and (20). We get, $\theta = 12°$ for the overhand shot and $\theta = 25°$ for the underhand shot. The quantities α, β, R, s, y_m, x_m, v_0, v_1, and t_1 are readily calculated from Eqs. (4)–(15). The results are listed in Table I.

A comparison of the flight parameters enables us to determine the relative advantages and disadvantages of the two styles of free throw.

(1) The overhand shot requires a lower projection velocity than the underhand shot and this is a distinct advantage for the former, as discussed earlier.

(2) The entry velocity, on the other hand, is slower for the underhand shot. This means that there is less chance of bouncing off the rim when a clean shot is missed.[8] Here the advantage goes to the underhand shot.

(3) The entry angle is larger for the overhand shot and therefore the projected cross section of the hoop is larger. This is another advantage of the overhand shot. For the underhand shot, the entry angle is just larger than the minimum angle for a clean shot. For this reason, a slightly

Table I. Trajectory parameters.

Quantity	Definition	Overhand push shot	Underhand loop shot
θ	Angle of incline	12°	25°
α	Angle of projection	51°	57.5°
β	Angle of entry into basket	39°	32.5°
R	Inclined range	13.5 ft	14.0 ft
s	Length of trajectory	15.5 ft	15.9 ft
y_m	Maximum altitude from level of projection	5.0 ft	7.1 ft
x_m	Position of maximum from point of projection	8.0 ft	9.0 ft
	Maximum altitude from floor level	12.4 ft	11.2 ft
	Position of maximum from the foul line	8.8 ft	10.2 ft
v_0	Velocity of projection	22.9 ft/sec	25.3 ft/sec
v_1	Velocity of entry into basket	18.5 ft/sec	16.1 ft/sec
t_1	Time of flight	0.916 sec	0.933 sec

higher angle of projection is recommended for the underhand shot. However, this in turn requires a higher projection velocity and results in a higher entry velocity as well.

(4) The range, length of trajectory, and time of flight are smaller for the overhand shot. These factors all go indirectly in favor of the latter style of throw. A smaller range would require a slower projection velocity. A shorter trajectory and time of flight all mean less encounter with air resistance.

(5) The location of the apex of the trajectory also indirectly affects the accuracy of the shot. For the underhand shot, the apex is situated closer to the basket, which in turn accounts for a smaller velocity of entry.

On the whole, the overhand shot is preferable to the underhand shot from a purely kinematic point of view. However, the strength of the underhand shot lies in other areas that may more than compensate for its inferior trajectory. First, the ball is evenly balanced between the two hands and the muscles are more relaxed. The accuracy in the forward direction is much greater than that of the overhand shot. The error in the projection angle is also smaller since the entire length of the arms go into the swing. These advantages are lacking in the overhand shot, where the extensor muscles used are more difficult to control. The mainly one-handed nature of the shot makes it prone to greater error in the forward direction. The arm is moved in a more complicated manner, where the hand is straightened while the ball is pushed along a straight line. In conclusion, if the underhand loop shot is superior as evidence indicates,[2–5] it is not because of its trajectory, but in spite of it.

[a]Present address: Dept. of Mathematics and Physics, Newberry College, Newberry, SC 29108.

[1]J. W. Bunn, *Basketball Technique and Team Play* (Prentice-Hall, Englewood Cliffs, NJ, 1964), p. 89.

[2]H. Hobson, *Scientific Basketball* (Prentice-Hall, Englewood Cliffs, NJ, 1964), p. 124.

[3]J. W. Burke, *Reports of 125 Prominent Coaches* (Springfield College, Springfield, MA, 1950).

[4]R. Beetz, *A Study of Two Methods of Free-Throwing*, Master's thesis, Springfield College, 1953 (unpublished).

[5]N. McWhirter, *Guinness Book of World Records* (Sterling, New York, 1980), p. 526.

[6]The more realistic problem of including air resistance can only be solved by numerical integration [see, for example, G. W. Parker, Am. J. Phys. **45**, 606–610 (1977)]. The effects of the air resistance are to reduce the range and increase the angle of entry. However, for slow speeds, these effects would not be appreciable to affect the conclusions of this paper.

[7]R. E. Becker, *Introduction to Theoretical Mechanics* (McGraw-Hill, New York, 1964), p. 126.

[8]E. J. Durbin, Sports Illus. **47**, 25 (1977); **47**, 124 (1977).

Maximizing the range of the shot put

D. B. Lichtenberg and J. G. Wills
Physics Department, Indiana University, Bloomington, Indiana 47401
(Received 12 July 1976)

We address the problem of the optimum angle at which a putter should release the shot in order to achieve maximum distance.

I. OPTIMAL ANGLE NEGLECTING AIR RESISTANCE

In most elementary physics textbooks, it is stated that the maximum range of a projectile is attained when the angle of firing is 45°. However, film studies by Cureton[1] of shot putters in action have shown that the best putters release the shot at an angle of between 40° and 42° from the horizontal. We have spoken to several people, both in physics and in coaching, and they seemed genuinely puzzled by the discrepancy. (A common reaction among physicists was to attribute the difference to the effect of air resistance, but as we shall show in Sec. II, air resistance does not appreciably affect the optimum angle.)

The most recent film study of shot putters was carried out by Dessureault.[2] In filming 13 shot putters, from novices to world record holders, he found that the angle of release of the best put was 43° and that the mean angle was 38°. Dessureault gave no explanation for these angles, stating that "biomechanics is still in the measurement period," and quoting Booysen[3] to the effect that "only by measuring of thousands of the world's best in every track and field event can the sought-for definite standard, an objective possibility of comparison, be found."

On the contrary, it is easy to show by using elementary mechanics that the best angle of release of the shot put is around 42°. The reason that the angle is not 45° is that the shot is released at a finite (nonzero) height above the ground. We assume that a putter releases the shot with initial speed v at a height h above the ground making an angle θ with the horizontal. We treat h, v, and θ as independent variables, although in actual practice, the putter may have to compromise in achieving the highest possible values of h and v, and optimum value of θ. The horizontal distance or range the shot goes before striking the ground we call R. The quantities h, v, θ, and R are shown in the diagram of Fig. 1.

For simplicity, we first treat the problem neglecting air resistance. The horizontal distance x the shot travels during a time t is

$$x = vt \cos\theta. \tag{1}$$

The shot is subject to the acceleration of gravity g. Its vertical position z at any instant is

$$z = h + vt \sin\theta - (1/2)gt^2. \tag{2}$$

The shot hits the ground at a time T, when $z = 0$. At the time T, x is equal to the range R. Then, from Eqs. (1) and (2), we get

$$R = vT \cos\theta \tag{3}$$

$$0 = h + vT \sin\theta - (1/2)gT^2. \tag{4}$$

We solve Eq. (3) for T and substitute into Eq. (4) to eliminate T. We obtain

$$0 = h + R \tan\theta - (1/2)gR^2 \sec^2\theta/v^2. \tag{5}$$

Solving for R, we get

$$R = v^2 \cos\theta[\sin\theta + (\sin^2\theta + 2gh/v^2)^{1/2}]/g. \tag{6}$$

By inspection of Eq. (6), we see that increasing h and v will increase R. Thus, as expected, the putter should strive to maximize both the speed and height of release.

To find the maximum range as a function of θ, we must take the derivative of R with respect to θ and set it equal to zero. We obtain

$$\sin^2\theta_m = (2 + 2gh/v^2)^{-1}, \tag{7}$$

where θ_m is the angle which gives the maximum range R_m. From Eq. (7), if we know the height at which the shot leaves the hand and the speed of release, we can calculate the angle which will give the maximum range. We can use that angle in Eq. (6) to find the range. Usually, however, the initial speed v is not known and may be inconvenient to measure. We can eliminate v by solving Eq. (7) for v and substituting into Eq. (6) with $R = R_m$, $\theta = \theta_m$. We obtain the surprisingly simple result

$$R_m = h \tan 2\theta_m. \tag{8}$$

Equation (8) is a convenient one to obtain the best angle. As an example, suppose the shot leaves the putter's hand at a height of 7 ft above the ground. (This is about the average height of release.) Also suppose that the putter wants to set a world record of about 73 ft. Substituting these numbers into Eq. (8), we find

$$\theta_m = 42.3°.$$

It can be easily seen from Eq. (8) that for a weaker shot putter, the angle should be somewhat less. However, the best angle does not change very much. For example, for a put of 50 ft with a height of release of 7 ft, the best angle is 41°; for a put of 35 ft, the optimum angle is 39°.

We can use Eqs. (6) and (7) to obtain the maximum range in terms of the initial speed and height. We get

$$R_m = v^2(1 + 2gh/v^2)^{1/2}/g. \tag{9}$$

We note that if $h = 0$, Eqs. (8) and (9) reduce to the elementary textbook formulas $\theta_m = 45°$, $R_m = v^2/g$, as they should.

We can obtain another useful equation by eliminating h between Eqs. (8) and (9). We get

$$R_m = v^2 \cot\theta_m/g. \tag{10}$$

Although we are primarily interested in the effect of θ on R, we briefly comment on the effects of h and v. Because the optimum angle is near 45° and because h is small compared to R, an increase in height Δh increases R by

Reprinted from *American Journal of Physics* **46**, 546–549 (1978);

about the same amount. Similarly, R goes approximately as v^2, so that a 1% increase in the speed of release increases R by almost 2%. Both of these results can be easily seen by expanding Eq. (9), assuming $v^2 \gg gh$, which is equivalent to assuming $R_m \gg h$. We get

$$R_m \cong v^2/g + h. \tag{11}$$

It is interesting to write R_m in terms of the kinetic energy $K = (1/2)Mv^2$ of the shot at release and the potential energy $V = Mgh$, where M is the mass of the shot. Then Eq. (11) becomes

$$R_m \cong (2K + V)/(Mg). \tag{12}$$

Thus, if the putter can give the shot a certain amount of energy, he is twice as well off by giving the shot kinetic rather than potential energy. But, of course, the putter cannot freely trade off between these two types of energy. As an extreme example, if the putter released the shot at ground level, his motion would be so awkward that he would be able to give the shot only very little kinetic energy.

To increase either v or h, the putter must impart additional energy to the shot. This, in general, costs him additional energy. On the other hand, by releasing the shot closer to the optimum angle, the putter increases the range without delivering additional energy to the shot. To a first approximation, this latter option should not cost him additional energy.

Because the range is stationary at the optimum angle, small errors in θ do not appreciably affect R. On a put of about 70 ft, an error of 1° shortens R by about 1/2 in.,[4] and an error of 3° shortens R by about 4 in. As the angle deviates more and more from the optimum, the effect on R gets progressively worse. For example, a 10° error in θ shortens R by about 3 1/2 ft.

We next treat the problem including air resistance. We find that although the range of a 70-ft put is reduced by about 1/2 ft in still air, the optimum angle is changed by about 0.13°, which is a negligible amount. (A 0.13° error in θ changes R by about 0.01 in.)

II. EFFECTS OF AIR RESISTANCE

In this section, we shall show that air resistance shortens the range of a 70-ft put by about 1/2 ft, but has a negligible effect on the optimal angle. If we include air resistance, additional variables enter the problem: namely, the angular velocity of the shot and the axis of rotation. However, we shall not consider these variables, assuming that for a 16-lb shot, any slight rotation imparted by the putter has a negligible effect on the range.

The equation of motion of the shot is

$$M\,d\mathbf{u}/dt = -M\mathbf{g} - D(\mathbf{u} - \mathbf{w})/|\mathbf{u} - \mathbf{w}|, \tag{13}$$

where D is the drag arising from the air, $\mathbf{u}$ is the instantaneous velocity of the shot, and $\mathbf{w}$ is the wind velocity. The drag of an object can be written

$$D = (1/2)cA\rho(\mathbf{u} - \mathbf{w})^2, \tag{14}$$

where A is the cross-sectional area of the object in the direction of motion, ρ is the density of the resistive medium, and c is a dimensionless quantity called the drag coefficient. The drag coefficient is a complicated function of the shape of the object and of the Reynolds number, which in turn, is proportional to the velocity. However, for a sphere the size

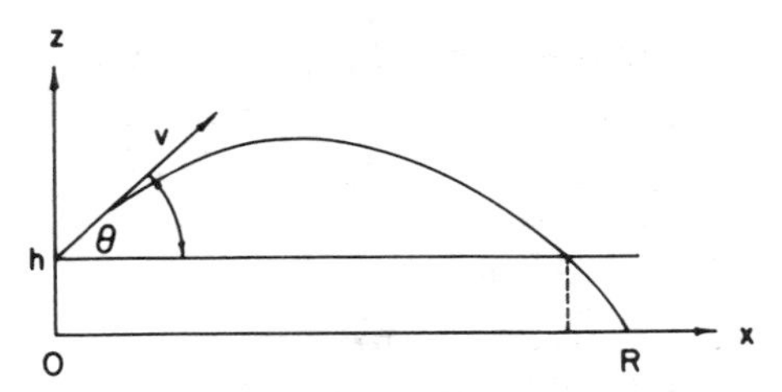

Fig. 1. Kinematic variables involved in the shot put.

of the shot traveling at the speeds that putters normally achieve, c is approximately a constant[5] with the value $c \approx 1/2$.

Substituting Eq. (14) into Eq. (13) and writing the equation in component form we obtain

$$du_x/dt = -(1/2)cA\rho|\mathbf{u} - \mathbf{w}|(u_x - w_x)/M, \tag{15}$$

$$du_y/dt = -(1/2)cA\rho|\mathbf{u} - \mathbf{w}|(u_y - w_y)/M, \tag{16}$$

$$du_z/dt = -g - (1/2)cA\rho|\mathbf{u} - \mathbf{w}|(u_z - w_z)/M, \tag{17}$$

where the y direction is perpendicular to the initial velocity $\mathbf{v}$. Equations (15)–(17) are coupled differential equations which cannot be integrated analytically in terms of elementary functions.[6] We have solved Eqs. (15)–(17) numerically in the case of still air. However, rather than present the details of our numerical solution, we find it more instructive to give an approximate analytic solution, since the effects of the drag are small in any case.

We begin our approximate treatment by writing

$$|\mathbf{u} - \mathbf{w}| \simeq [(u_x - w_x)^2 + w_y^2 + u_z^2 + w_z^2]^{1/2}. \tag{18}$$

We have dropped the term u_y because it is very small (being perpendicular to the direction of v), and have dropped the term $-2u_zw_z$ because u_z is in the same direction as w_z for about half the trajectory and opposite to w_z for the other half. We are, of course, assuming that $\mathbf{w}$ remains constant while the shot is in the air. We next replace our approximate expression for $|\mathbf{u} - \mathbf{w}|$ by some appropriate constant value γ. We can get upper and lower limits on the effect of the drag. We get an upper limit on γ by replacing the components of $\mathbf{u}$ by the components of $\mathbf{v}$ in Eq. (18), as u exceeds v only near the end of the trajectory. We get a lower limit by taking only u_x different from 0 and replacing u_x by $v_x = v\cos\theta$, as $v\cos\theta$ is approximately the minimum speed of the shot. Putting in these limits, we obtain

$$v^2\cos^2\theta + w^2 - 2vw_x\cos\theta < \gamma^2 < v^2 + w^2 - 2vw_x\cos\theta. \tag{19}$$

Thus we will be able to bracket the shortening of the range, and the average of our upper and lower limits will be pretty close to the actual value.

With these approximations, Eqs. (15)–(17) become

$$du_x/dt = -\alpha(u_x - w_x), \tag{20}$$

$$du_y/dt = -\alpha(u_y - w_y), \tag{21}$$

$$du_z/dt = -g - \alpha(u_z - w_z), \tag{22}$$

where

$$\alpha = (1/2)cA\rho\gamma/M. \tag{23}$$

Equations (20)–(22) may be readily integrated in terms of elementary functions. Putting in the initial conditions, we obtain

$$x = (v \cos\theta - w_x)(1 - e^{-\alpha t})/\alpha + w_x t, \quad (24)$$

$$y = -w_y(1 - e^{-\alpha t})/\alpha + w_y t, \quad (25)$$

$$z = h + (v \sin\theta + g/\alpha - w_z)(1 - e^{-\alpha t})/\alpha + (w_z - g/\alpha)t. \quad (26)$$

As before, it is possible to set $x = R$ and $z = 0$ when $t = T$, and to substitute the expression for T from Eq. (24) into Eq. (26). However, the resulting expression is rather unwieldy. It is better to remember that α is small and to expand Eqs. (24) and (26) to first order in α.

We get

$$x = vt \cos\theta + (1/2)\alpha t^2(w_x - v \cos\theta), \quad (27)$$

$$z = h + vt \sin\theta - (1/2)gt^2 + (1/2)\alpha t^2[w_z - v \sin\theta + (1/3)gt]. \quad (28)$$

We now set $x = R$, $z = 0$ when $t = T$. Then Eq. (27) becomes

$$T = R \sec\theta/v + (1/2)\alpha T^2(1 - w_x \sec\theta/v). \quad (29)$$

Since we are interested in T only to order α, we can substitute $T = R \sec\theta/v$ in the second term on the right-hand side of Eq. (29). We substitute the resulting expression into Eq. (28) with $z = 0$, and $t = T$. We obtain to order α

$$0 = h + R \tan\theta - (1/2)gR^2 \sec^2\theta/v^2 + \alpha R^2 \sec^2\theta[(1/2)gw_x R \sec^2\theta/v^2 - (1/2)w_x \tan\theta + (1/2)w_z - (1/3)gR \sec\theta/v]/v^2. \quad (30)$$

We are interested in the change in range when α is small. We get this quantity by taking the derivative of R with respect to α and evaluating when $\alpha = 0$. We obtain the following rather complicated expression

$$dR/d\alpha = R^2(2gRv \cos\theta + 3v^2 w_x \sin\theta \cos\theta - 3gw_x R - 3v^2 w_z \cos^2\theta)/6v^2 \cos^2\theta(v^2 \cos\theta \sin\theta - gR). \quad (31)$$

We are interested in change in R at the optimum angle. In this case, R is given approximately by Eq. (10). Substituting Eq. (10) into Eq. (31), and letting $d\alpha = \alpha$, we get the simpler expression

$$dR_m = \frac{-\alpha v^3}{g^2 \sin\theta_m}\left(\frac{1}{3 \sin\theta_m \cos\theta_m} - \frac{w_x}{2v \sin\theta_m} - \frac{w_z}{2v \cos\theta_m}\right), \quad (32)$$

where α is given by Eq. (23).

Let us evaluate Eq. (32) in the case of still air. If we take

$$v = 46 \text{ ft/s}, \quad \theta_m = 42°, \quad A = 0.12 \text{ ft}^2$$

$$c = 0.5, \quad \rho = 0.071 \text{ lb/ft}^3, \quad M = 16 \text{ lb} \quad (33)$$

with $v \cos\theta_m < \gamma < v$, we find

$$-7 < dR_m < -5 \text{ in.} \quad (34)$$

for a put of about 73 ft. Thus air resistance shortens the range by about 1/2 ft. This value agrees very well with the results of our numerical calculation which gives $dR_m = -5.9$ in. For shorter puts, the effect of air resistance is considerably less, since as we can see from Eq. (32), dR_m goes approximately like v^4 (remember that α goes approximately like v) or like R_m^2.

Because the correct value for γ in the case of still air is so close to halfway between the limits, let us approximate γ^2 by taking the mean of the upper and lower limits of γ^2. This gives

$$\gamma^2 = (1/2)v^2(1 + \cos^2\theta) + w^2 - 2vw_x \cos\theta. \quad (35)$$

Using this expression for γ, we evaluate the change in range as an example if there is a wind of speed half the speed of the shot or about 23 ft/s. We keep the conditions of Eq. (33). We find that the range is shortened a little less than 2 in. for a tail wind, about 14 in. for a head wind, and about 7 in. for a cross wind. These values of course should be compared with a shortening of 6 in. in still air. It is amusing that a cross wind shortens the range compared to the range in still air.

We next consider the change in the optimum angle arising from air resistance. It suffices to consider the case of still air. We set $\mathbf{w} = 0$ in Eq. (30), take the derivative of R with respect to θ, and set it equal to zero. This gives us the condition

$$R_m = v^2 \cot\theta_m(1 - \alpha v \csc\theta_m/g)/g. \quad (36)$$

Substituting Eq. (36) back into Eq. (30) to eliminate R_m, we obtain an expression for θ_m to first order in α. Taking the derivative of θ_m with respect to α in this expression, evaluating it at $\alpha = 0$, and again setting $d\alpha = \alpha$, we get

$$d\theta_m = -\alpha v(\cos\theta_m - \sec\theta_m/3)/g, \quad (37)$$

where $d\theta_m$ is expressed in radians. Putting in the numbers of Eq. (33) and changing to degrees, we find

$$d\theta_m = -0.13°. \quad (38)$$

As we have remarked, this change in the optimum angle has a negligible effect on the range. This can be verified directly by evaluating Eq. (6) with the optimum angle and with the angle changed by 0.13°. It would be wrong to use Eq. (8) or (10) to find the change in R arising from a small change in θ, because these equations are correct only at $\theta = \theta_m$, and, unlike Eq. (6), are not stationary with respect to small changes from the optimum.

According to the rules governing the shot put, the diameter of the shot may not be less than 4.331 in. nor more than 5.118 in. There is a 40% difference in area between the two extremes. We see from Eq. (32) that the loss in range is directly proportional to α, which, in turn, is directly proportional to the cross sectional area A. In computing a loss of range $dR_m = -1/2$ ft, we used a value of A about midway between the two extremes. The actual value of dR_m can vary by ±20% just because of the size of the shot. Thus, a world-class putter who uses the smallest permissible shot will have a little more than a 2-in. advantage over a putter who uses the largest allowed shot, all other things being equal.

In all our computations, we used a nominal value of $g = 32$ ft/s^2. Actually, g varies from place to place on the earth's surface. Similarly, the density of air depends on weather and altitude. All these variables together can change the range of a put by several inches, but our conclusions about the optimum angle remain valid.

In all our considerations we have been concerned with the path of the shot from the moment of release until the time of landing. We have not considered how, in fact, the putter achieves the maximum v and h and the optimum θ. The principles of mechanics can help here too, but the problem is much more difficult. At present, a large part of our understanding of this part of the problem must come, as Dessureault and Booysen believe, from observing the best shot putters in action.

ACKNOWLEDGMENTS

We should like to thank Professor John M. Cooper and Professor Emil Konopinski for valuable discussions. Also Martin G. Olsson has kindly pointed out that there is an approximate relation between the static strength of the putter and the velocity of the shot. He writes: "If the putter can lift a dumbbell of weight W from his shoulder he will be able to impart an acceleration

$$a = W/m = (W/mg)g$$

to the shot (neglecting the weight of the shot). If his arm has length l the final velocity will be

$$v = [2gl(W/mg)]^{1/2}$$

using $W = 150$ lb, $l = 3$ ft, $mg = 16$ lb, we find

$$v \simeq 42 \text{ ft/s}.$$

Since the putter imparts additional velocity from his initial movement in the ring this would probably account for the required 47 ft/s. Although this estimate is rough it shows directly the basic strength required to set a world record."

[1]T. K. Cureton, "Elementary Principles and Techniques of Cinematographic Analysis," Res. Quarterly **10,** 3 (1939).
[2]Jacques Dessureault, "Selected Kinetic and Kinematic Factors Involved in Shot Putting," Ph.D. thesis (Indiana University, 1976).
[3]Hannes Booysen, "Thoughts on Shot Put Technique," Track Technique **43,** 1365 (1971).
[4]Distances in the shot put are measured to the nearest ¼ inch.
[5]L. Prandtl and O. G. Tietjens, *Applied Hydro- and Aerodynamics,* (McGraw-Hill, New York, 1934), p. 100.
[6]J. M. J. Kooy and J. W. H. Uytenbogaart, *Ballistics of the Future* (Technical Publishing Co. H. Stam, Haarlem, Holland, 1946), p. 123.

Chapter 4
RUNNING AND JUMPING

CONTENTS

The first article in this chapter, "Physics of sprinting," by Igor Alexandrov and Philip Lucht, proposes a simple two-parameter model of running for race distances of 290 m or less. The two parameters in question are f, the propulsive force per unit mass of the athlete, and σ, the resistive force per unit momentum of the athlete, presumed constant, or nearly so, for a given runner.

Values of f and σ were deduced for a given athlete by measuring the times for two straight-line races run at different distances. As a test of the model, these values were then used to predict, rather successfully, the time for that athlete to run a sprint of still *different* length on a *curved* track, where centripetal forces significantly affect the outcome.

The second article in this chapter, "A theory of competitive running," by Joseph B. Keller, proposes a model of running which is based on four physiological quantities: (1) the maximum force a runner can exert; (2) the resistive force opposing the runner; (3) the rate at which energy is supplied by the oxygen metabolism (i.e., respiration and circulation); and (4) the initial amount of energy stored in the athlete's body at the start of a race. These four quantities were treated as parameters to be determined from a least-squares comparison between theory and world record data for races at 22 different lengths ranging from the 50-yard dash to the 10 000-m run.

One object of the theory was to predict the "endurance curve" for running—the empirical relationship between average velocity ($V = D/T$) and race distance D, which is characterized by an initial increase and subsequent decrease in average running velocity with increasing race distance (see Fig. 3, p. 45 of Keller's article). The theoretical curve is found to agree closely with average velocities calculated from the 22 world records, with the maximum difference for longer races being 3.1% and the maximum difference for sprints being only 2.1%.

The theory also predicts the existence of a critical race distance, $D_c = 291$ m, which divides sprints, in which the successful strategy is to employ maximum acceleration for the entire duration of the race, from runs, in which the successful strategy is to employ maximum acceleration for only the first one or two seconds, followed by a constant speed throughout the race and then a slight slowing down in the last one or two seconds.

The existence of the critical distance D_c and the need for a different strategy for races longer than this distance both stem from limitations imposed on even the best-conditioned athletes by human physiology. In effect, sprints are anaerobic races, i.e., activities that are sufficiently short to be run primarily on the oxygen present in the body at the start of the race. Runs, on the other hand tend to be aerobic in character, relying on timely conversion of chemical potential energy to kinetic energy through the oxidation of the body's metabolic fuels.[1]

The third article in this chapter, "Physics of long-distance running," by J. Strnad, proposes a two-parameter model of running which, while somewhat more accurate than the four-parameter Keller model, is limited to races of 2000 m or more. Strnad notes that there is a strong empirical correlation between average velocity V and race duration T for long distance races in a variety of sports, including ice skating, running, walking, and free-style swimming.

In particular, Strnad notes that empirical plots of $\ln V$ versus $\ln T$ (where $V = D/T$, and D is race length) may be least-squares fitted by straight lines of *identical* slope regardless of the sport in question! These results may also be summarized by the equation $V = V_1/T^s$, where V_1 and s are constants; $s = 0.069$ for all cases considered, and the value of the parameter V_1 varies from 18.17 m/s for men's ice skating, to 10.13 m/s for men's running, to 2.63 m/s for men's free-style swimming.

According to Strnad, the constancy of s both within a given sport at different distances, as well as across different sports confirms two things:

a. that "extremely trained athletes achieving records show approximately equal performance,"

and

b. that "power output for a given time T is approximately independent of the type of locomotion."

The fourth article in this chapter, "Power demand in walking and pace optimization," by A. Bellemans, presents a simple model of walking which, while not completely accurate, does offer agreement with certain experimental studies involving subjects on a treadmill. Bellemans apportions the total power expenditure to two different aspects of walking: the motion (i.e., acceleration and deceleration) of the legs, and the lifting and lowering of the body's center of mass (i.e., work involving gravity). Bellemans predicts an optimal pace length for walking which (a) increases linearly with walking speed, (b) corresponds to minimal power demand, and (c) approximates empirical studies of pace length versus walking speed.

"Effect of wind and altitude on record performance in foot races, pole vault, and long jump," by Cliff Frohlich, is the fifth article in this chapter. In it, the author first develops a simple model of sprinting which takes both wind speed and altitude into account. This is done essentially by incorporating the aerodynamic drag force F_D into the model, a force that depends on both the relative speed between the sprinter and the air (and hence the wind speed relative to the ground), and the density of air (and hence the altitude above sea level, with which it varies readily).

Since the energy needed for successful pole vaulting and long jumping comes at the expense of the sprinter's kinetic energy, Frohlich's models for pole vaulting and long jumping follow directly from his model for sprinting. These three models are then used to predict the effects of various wind speeds and air densities on a 6.10-m pole vault effort, an 8.83-m long jump effort, and a 200-m dash run in 20.0 s.

Frohlich shows that athletic performance in moderate winds of 2.0 m/s or at altitudes comparable to Mexico City differ by several percent from performances at sea level or in still air. As a consequence, his calculations suggest that Carl Lewis' 1983 long jump of 8.80 m at sea level was actually superior to Bob Beamon's 8.90-m jump[2] at the 1968 Olympic Games in Mexico City!

"Could an athlete run a 3-m radius 'loop-the-loop?'," by Angelo Armenti, Jr., is the sixth article in this chapter. By adapting the sprinting model of Alexandrov and Lucht (the first article in this chapter) to a vertical track of circular cross section, Armenti concludes that the answer to the question in the article's title is "Probably (and almost certainly) not." The results suggest basically two reasons why such an attempt would fail: (1) before reaching the top of the loop, the normal force between the track and the athlete's shoes would fall below the minimum needed to maintain traction, and (2) the legs of a running athlete could not sustain the approximately 4 g's of acceleration required near the bottom of the loop.

The seventh and final article in this chapter, "Fast running tracks," by Thomas A. McMahon and Peter R. Greene, describes the design and construction of a "tuned" indoor (220 yard) track at Harvard whose compliance (i.e., "springiness") was adjusted to match the mechanical properties of the human runner. The authors first develop a mechanical model of a human runner based on a rack and pinion element connected in series with a damped spring, and then cite model results predicting that a properly tuned track should enhance running speeds by between 2% and 3%, results that were soon empirically verified. In their conclusion, the authors predict that if the technology used in their relatively short indoor track were extended to a standard-length outdoor track of optimum mechanical design, the world record for the mile could be improved by as much as seven seconds!

---

[1]The existence of $D = 291$ m as a dividing line between sprints and runs has been discussed widely in the literature and is closely connected to the different ways in which the human body funds energy consumption for muscular activity of ever-increasing duration. During the first 15 to 20 seconds of vigorous exercise, the primary source of motive power comes from the splitting of creatine phosphate (CP). After about 30 s (approximately the time needed to sprint 291 m), CP splitting is rapidly replaced as the dominant energy source by the anaerobic breakdown (glycolysis) of glycogen to lactic acid, the rapid buildup of which forces the sudden reduction in muscle activity familiar to weekend athletes. Oxidation (i.e., aerobic breakdown) of glycogen is the primary source of energy in 5000- and 10 000-m runs (13 to 30 min in duration), and the oxidation of fat is the primary energy source for the 42 195-m marathon (somewhat more that two hours in duration). For more on muscle metabolism and related topics, see the *Interface Series on Energy and Exercise* and, in particular, Henry A. Bent, "VI. Reactions for every occasion," *J. Chem. Educ.* **55**, 796 (1978).

[2]Bob Beamon's legendary long jump performance, a record unrivaled for 23 years, was finally broken by Mike Powell on August 30, 1991 with a new world record jump of 8.95 m.

Physics of sprinting

Igor Alexandrov[a] and Philip Lucht
Department of Physics, University of Utah, Salt Lake City, Utah 84112
(Received 11 February 1980; acepted 18 June 1980)

Sprinting is described by a simple physical model. The model is used to predict the differences between the recorded times for races on a straight track and on a curve. It is shown that the choice of the running lane makes a non-negligible difference.

I. INTRODUCTION

In the past few years there has been a growing interest in the application of physical models to problems of animal locomotion in general[1,2] and to human athletic performance in particular.[3-5]

A problem of special interest is to devise a model that represents with reasonable accuracy the propulsive and resistive forces acting on a human runner.

One common model assumes that the sum total of resistive forces F_r acting on a runner can be represented by

$$F_r = -\sigma M v, \tag{1}$$

where M is the mass of the runner, v is the speed, and σ is a parameter of the model, presumed constant or nearly constant for a given runner.[6] Although such a form is suggested by analogy with viscous forces, it is far from clear that it can adequately represent human runners where the forces are rapidly varying and, in general, complex. Nonetheless such a form was used by Keller[3] to model competitive running and draw certain conclusions regarding optimal running strategy and maximal future performances. One significant conclusion reached by Keller is that a well-conditioned athlete is able to sustain a maximum and nearly constant muscular effort for races over distances of 290 m or less. We shall define all such races as sprints.

A basic question is: What are typical values of the parameter σ? It is to be expected that σ will depend on the runner, his speed, the type race, and possibly on external conditions such as the track surface, altitude, etc. Keller[3] uses a value of $\sigma = 0.44\ \text{sec}^{-1}$ while Whitt and Wilson[5] quote similar values. The latter, however, were measured at walking speeds. To our knowledge there exists no measurement of σ at sprinting speeds.

We propose that for a given sprinter a measurement of σ and the propulsive force parameter f (defined below) can be performed in principle by measuring that sprinter's running times for two straight line races of different distances. Our model will then allow us to use the values of these parameters to predict that runner's performance for a sprint run on the curve.

II. MODEL

We assume as does Keller[3] that a sprinter of mass M is subject to two horizontal forces. One is the resistive force expressed by Eq. (1) and the other is a constant propulsive force F_p which we write as

$$F_p = fM, \tag{2}$$

where f is the force per unit mass of the athlete.[6] The equation of motion for the sprinter is then

$$\frac{d^2x(t)}{dt^2} = \frac{dv(t)}{dt} = f - \sigma v(t), \tag{3}$$

where $x(t)$ is the distance measured from the start of the race and $v(t)$ is the sprinter's instantaneous velocity. The solution to this equation subject to the initial conditions $x(0) = 0$, $v(0) = v_0$ is

$$v(t) = (f/\sigma)(1 - e^{-\sigma t}) + v_0 e^{-\sigma t}, \tag{4}$$

$$x(t) = (f/\sigma)t + (f/\sigma^2 - v_0/\sigma)(e^{-\sigma t} - 1). \tag{5}$$

These well-known solutions describe the motion of a particle sedimenting under the action of a constant force of "gravity" Mf and a viscous force $-\sigma M v$; the particle (sprinter) approaches the terminal velocity $v_t = f/\sigma$ asymptotically.

Let us suppose that the times t_1 and t_2 for an athlete's performances over two different straight-line sprinting distances x_1 and x_2 are known. Then Eq. (5) can be used (with $v_0 = 0$) to solve for f and σ (see Appendix).

We give in Table I the best running times over two different distances for four world-class sprinters from the 1967–69 period.[7,8] (We note that two of them are still holders or coholders of three separate world records.) The computed values of σ and f for each runner are shown in columns 6 and 7, respectively.

The last entry in Table I is Tommie Smith who has the distinction of having a recorded time for the straight-track 100-m race and also for the rarely run straight-track 220-yard race for which he holds the world's record. In addition he is the holder of the world's record for the 200 m on the curve which he established in Mexico City in 1968.[7]

The computed values of the parameters f and σ differ widely from runner to runner. We have no way of knowing how reliable the data are but if we assume that the computed values of f and σ for Tommie Smith are accurate, then we can extend our theory to predict his performance in a 200-m race on the curve. The closeness of our prediction to the actual recorded time will be a test of the validity of the basic assumptions of the theory embodied in Eqs. (1) and (2).

III. RUNNING THE CURVE

In a typical 200-m race the starts are staggered so that each runner runs the first 100 m of the race on the curve. However, the radius of curvature of the curved portion of each lane is different. The lanes are 1.22 m wide so that the inner radius of the nth lane is

$$R(n) = 100/\pi + (n-1)(1.22)\ \text{m}.$$

Let the runner be running with speed v on a curve of

Reprinted from *American Journal of Physics* **49**, 254–257 (1981); © American Association of Physics Teachers.

Table I. Computed values of the parameters f and σ for various runners (*indicates a current world record holder). The times are the runners best times over the given distances.[7,8]

Runner	x_1	t_1(sec)	x_2	t_2(sec)	σ(sec^{-1})	f(N/kg)	v_t(m/sec)	$\lvert\delta\sigma/\sigma\rvert$	$\lvert\delta f/f\rvert$	$\lvert\delta v_t/v_t\rvert$
John Carlos	60 yards	6.0	100 yards	9.0	0.667	8.13	12.19	13%	10%	3%
Bill Gaines	60 yards	5.9	100 yards	9.3	1.250	13.45	10.76	25%	23%	3%
Jim Hines*	100 yards	9.2	100 m	9.9	0.581	7.10	12.22	65%	51%	14%
Tommie Smith*	100 m	10.1	220 yards (straight)	19.51	1.252	13.46	10.75	19%	18%	1%

constant radius R and subject to the forces (1) and (2). Now, however, the propulsive force must supply a component that gives rise to the centripetal acceleration of magnitude v^2/R. The athlete's equation of motion then becomes

$$dv/dt + \sigma v = (f^2 - v^4/R^2)^{1/2}. \quad (6)$$

Equation (6) is to be solved with given parameters f, σ, and R subject to the initial conditions $v(0) = 0$, $x(0) = 0$.

An approximate analytical solution can be obtained by noticing that for typical situations the second term under the square root is small, $v^4/R^2 \ll f^2$, and by expanding the square root we obtain

$$dv/dt + \sigma v \cong f - (1/2)v^4/fR^2. \quad (7)$$

As it stands, Eq. (7) is still not subject to a simple analytical solution. We can, however, solve it by treating the term in v^4 as a small perturbation (see Appendix). The solution so obtained subject to the initial conditions $v_c(0) = 0$, $x_c(0) = 0$ gives the speed $v_c(t)$ and the distance travelled along the curve $x_c(t)$ as:

$$v_c(t) = (f/\sigma)(1 - Z^{-1}) - 1/2(f/\sigma)(fR^{-1}\sigma^{-2})^2F_1(Z), \quad (8)$$

$$x_c(t) = (f/\sigma^2)(\ln Z - 1 + Z^{-1}) - (1/2)(f/\sigma^2)(fR^{-1}\sigma^{-2})^2F_2(Z), \quad (9)$$

where $Z = e^{\sigma t}$ and

$$F_1(Z) = 1 + (10/3 - 4\ln Z)Z^{-1} - 6Z^{-2} + 2Z^{-3} - 1/3\, Z^{-4},$$

$$F_2(Z) = -37/12 + \ln Z + (2/3 + 4\ln Z)Z^{-1} + 3Z^{-2} - 2/3\, Z^{-3} + 1/12\, Z^{-4}.$$

The accuracy of our perturbation procedure depends on the smallness of the dimensionless parameter $\epsilon = (fR^{-1}\sigma^{-2})^2/2$ which for the case of interest to us has a value of 0.031 39. For running times on the order of 10 sec, terms in Z^{-1} and higher powers of Z^{-1} are negligible and can be dropped from the expressions for F_1 and F_2.

With these approximations we can write expressions for $x_c(t)$ and $v_c(\infty)$, the latter being the terminal speed of the runner on the curve. We note that with our typical value of σ the sprinter attains 98% of his terminal speed within approximately 3 sec:

$$v_c(\infty) \cong (1 - \epsilon)v_t, \quad (10)$$

$$x_c(t) \cong v_t(1 - \epsilon)t - (v_t/\sigma)(1 - 37/12\,\epsilon). \quad (11)$$

Equations (10) and (11) are useful in that they allow us to bring out the effect of curvature of the track represented by the parameter ϵ. Physically, 2ϵ is the square of the ratio of the centripetal acceleration v_t^2/R calculated at terminal speed to the propulsive force per unit mass f. With our parameters the difference between v_t and $v_c(\infty)$ is of the order of 0.3 m sec^{-1}. Thus the runner accelerates slightly upon entering the straightaway.[10]

Equations (10) and (11) allow us to compute the total time t_T for the race in a closed (approximate) form

$$\begin{aligned} t_T &= t_{100} + 100/v_t + \epsilon/\sigma \\ &= 2(100)/v_t + 1/\sigma + \epsilon[(100/v_t) - 13/(12\sigma)] \\ &\quad + \epsilon^2[(100/v_t) - 25/(12\sigma)] + O(\epsilon^3). \quad (12) \end{aligned}$$

Equation (12) then can be used to predict the running time, for example, for Tommie Smith running a 200-m race on the curve. We use the parameters of Table I and assign him to lane 3, the lane in which he ran for his world record. Then $R = 34.27$ m, $\epsilon = 0.0314$, $v_t = 10.75$. We obtain

$t_{100} = 10.35$ sec,
$v_{100} = 10.41$ m/sec,
time to run the last 100 m = 9.33 sec,
total predicted time for the race = 19.68 sec.

In the Mexico City Olympics on 16 October 1968 in the men's 200-m final, Tommie Smith running in lane 3 set the still-current world record of 19.83 sec.

In order to test the accuracy of our calculation we numerically integrated the exact Eq. (6) and computed the time necessary to cover the first 100 m of the course on the curve. The comparison of this calculation with the results of our approximate analytical solution is shown in Table II. The two procedures give results that differ at most by 0.3%. This surprisingly close agreement lends support to the validity of the approximations made in the calculations. The close agreement between our "predicted" and the measured value of Tommie Smith's performance in the 200-m race on the curve comes as somewhat of a surprise considering the meagerness of the data and the crudeness of the model.

Finally, we apply our model to the following problem. Since the value of the radius of curvature R enters Eqs. (8)–(12) we would expect that athletes running in the outside lane (lane 8) would enjoy an advantage over those in the inside lane.

In fact, it is well known that sprinters dislike running in

Table II. Comparison of numerical solutions with approximate analytical solutions. Times are calculated for the first 100 m of the race on the curve; $f = 13.46$ N/kg, $\sigma = 1.252$ sec^{-1}.

Lane	R(m)	t_{100}(sec) numerical solution	t_{100}(sec) approximate solution	difference (%)
1	31.83	10.361	10.389	0.27%
3	34.27	10.327	10.348	0.20%
8	40.37	10.266	10.277	0.11%
straight track	∞	10.100	10.100	

the inside lane complaining that the turn is "too tight." Our calculation bears this out. We find that, with the parameters we used for our test case, the total recorded time for a runner in lane 1 would have been 19.72 sec whereas for lane 8 it would have been 19.60 sec. At finishing speeds of about 11 m/sec that difference in time translates into a distance difference of over a meter. Since many sprint races are lost or won by mere centimeters the above effect is quite significant.

Curiously enough sprinters do not consider lane 8 to be the most advantageous one. Because of the staggered start runners in lane 8 run for half the race ahead of the other competitors. This, they claim, puts them at the psychological disadvantage of not being able to see what their competition is doing.

IV. ERRORS

We wish to make a few brief comments about the "experimental" errors. Since neither the records books nor the sport press include error bars with their reported results we will limit ourselves to a few crude estimates. We assume that most of the error is the result of inaccurate time measurements. Many of the times given in Table I were probably the averages of several hand-recorded times. Other times might have been recorded electronically and then rounded off to the nearest tenth of a second.

For the purposes of this estimate, we assume that the accuracy in the recorded times is ±0.05 sec. We notice that the resulting variation in the calculated values of f and σ are quite large. This is due in a large part to the fact that the calculation involves a denominator which is a difference between two numbers numerically close to one another. One consequence of this effect is that data for races over similar distances such as 50 and 60 yards or 100 yards and 100 m are practically useless.

However, as is apparent from Eq. (12), the total time t_T depends primarily on v_t, so to some degree the errors in f and σ tend to cancel one another. The uncertainties included in Table I were calculated according to the assumption $\Delta t_1 = -\Delta t_2 = \pm 0.05$ sec. Assuming this error in the measurement of time the resulting error in the predicted t_T for T. Smith is

$$\Delta t_T = \pm 0.054 \text{ sec.}$$

A less conservative error estimate of $\Delta t_1 = -\Delta t_2 = \pm 0.11$ sec brings the predicted and observed t_T within the range of "experimental" error.

V. CONCLUSIONS

In summary we have found that the forces acting on a sprinter at typical sprinting speeds can be reasonably well modeled by Eq. (3). The use of that equation over two different distances can determine the sprinter's parameters f and σ. The values of the parameters that we calculate for a few sample cases are in some disagreement with those commonly found in the literature. Our model, extended to include centripetal effects, then allows us to predict an athlete's time for a race run on the curve. This in turn can be used to compute the difference in the times for two comparable athletes running in different lanes. This difference is solely due to the different radii of curvature of their respective lanes. We find its magnitude to be non-negligible.

Clearly our results suggest a line of further investigation for those interested in the biomechanics of sports. To the extent that a sprinting human can be characterized by parameters f and σ it might be of interest to measure these parameters under carefully controlled conditions and correlate them with the physical characteristics of the sprinters, such as height, weight, type of build, reaction time, strength of leg muscle, etc. It is conceivable that these parameters f and σ might also depend on external conditions such as track surface, altitude, and other factors.

Finally, we wish to comment that it has been our experience that there is a fairly large audience among today's college students interested in learning about the biomechanics of sports from the point of view of fairly sophisticated physics. Such a trend is certainly indicated by the amount of research done in many countries in the area of sports medicine and technique.

APPENDIX

In order to determine the parameters f and σ for a given sprinter, let us assume that the sprinter has recorded times t_1 and t_2 over two straightaway distances of lengths x_1 and x_2. Using Eq. (5) twice with $v_0 = 0$ and neglecting the exponential term we solve for f and σ to obtain

$$\sigma = (1/t_1)[(x_2/x_1) - 1]/[(x_2/x_1) - (t_2/t_1)], \quad (13)$$

$$f = \sigma^2 x_2/(\sigma t_2 - 1). \quad (14)$$

In solving Eq. (7) we procede as follows. First, we solve it with the v^4 term set equal to zero. This solution which we call w is the same as Eq. (4) with $v_0 = 0$,

$$w(t) = (f/\sigma)(1 - e^{-\sigma t}).$$

Next, we write the solution to Eq. (7) as $v = w + u$ where we treat u as a small perturbation. u then satisfies

$$du/dt + \sigma u = -1/2\, v^4/(fR^2). \quad (15)$$

Replacing v in Eq. (15) by its approximation w, we integrate twice to obtain the solutions (8) and (9).

For our test case (Tommie Smith) the parameters f and σ are directly determined from Eqs. (13) and (14) using the data given in Table I.

Equations (8) and (9) can be further simplified if we notice that for our test case $\sigma \simeq 1.25\ \text{sec}^{-1}$ and for running times of the order of 10 sec, $Z^{-1} \sim 10^{-6}$ so that F_1 and F_2 can be approximated by

$$F_1 \simeq 1, F_2 \simeq t\sigma - 37/12.$$

This gives Eqs. (10)–(12). These equations then predict the time to complete the curve portion of the race t_{100}, the speed at the end of 100 m, and the total time.

For comparison we find that, using the same values of f and σ, a 200-m race run entirely on the straightaway would have taken 19.40 sec as compared with our lane 3 calculated time 19.68 sec. Even for a runner in lane 8 the straightaway time is faster by 0.20 sec.

[a]On leave of absence from Department of Physics–Astronomy, California State University at Long Beach, Long Beach, CA 90840.

[1]K. Schmidt-Nielsen, Science **177**, 222 (1972).

[2] *Mechanics and Energetics of Animal Locomotion*, edited by R. M. Alexander and G. Goldspink (Halsted, New York, 1977).
[3] J. B. Keller, Phys. Today **26**(9), 42 (1973).
[4] H. Lin, Am. J. Phys. **46**(1), (1978).
[5] F. Whitt and D. Wilson, *Bicycling Science* (MIT, Cambridge, MA, 1974).
[6] We use mks units throughout so that M = kg, x = m, v = m/sec, F = N, f = N/kg, and σ = sec^{-1}.
[7] N. McWhirter, *Guinness Book of World Records* (Sterling, New York, 1979).
[8] *Sports Illustrated* (New York, NY) and *Track and Field News* (Los Altos, CA) miscellaneous issues between 1967 and 1969.
[9] Personal observation (IA).
[10] We wish to thank an anonymous referee for pointing this out to us.

A theory of competitive running

Using simple dynamics one can correlate the physiological attributes of runners with world track records and determine the optimal race strategy.

Joseph B. Keller

World records for running provide data of physiological significance. In this article, I shall provide a theory of running that is simple enough to be analyzed and yet allows one to determine certain physiological parameters from the records. The theory, which is based on Newton's second law and the calculus of variations, also provides an optimum strategy for running a race.

The records give the shortest time T in which a given distance D has been run. Our theory determines this function theoretically in terms of the following physiological quantities: the maximum force a runner can exert, the resistive force opposing the runner, the rate at which energy is supplied by the oxygen metabolism, and the initial amount of energy stored in the runner's body at the start of the race. By fitting the theoretical curve to four observed records, these quantities can be determined and the other records can be predicted. Alternatively, by measuring these quantities by independent physiological studies, one can use the theory to predict all the track records to which it applies.

The theory accounts for the main features of the records at distances from 50 meters to 10 000 meters. However, it does not account for the records at larger distances. These range up to 59 days for 5560 miles, the distance from Istanbul to Calcutta and back, set by M. Ernst (1799–1846).[1] Other physiological factors must be included to account for these records.

Optimal running strategy

A runner's speed varies during a race; we assume that the speed, $v(t)$, is chosen in the way that minimizes the time T required to run the distance D, subject to physical and physiological limitations. We shall determine this optimal speed variation $v(t)$ by formulating and solving a mathematical problem in optimal control theory.

The theory predicts that the runner should run at maximum acceleration for all races at distances less than a critical distance D_c = 291 meters. Thus the races at distances less than 291 meters should be classified together as "short sprints" or "dashes." For D greater than 291 meters the theory predicts maximum acceleration for one or two seconds, then constant speed throughout the race until the final one or two seconds and finally a slight slowing down. This result confirms the accepted view that a runner should maintain constant speed to achieve the shortest time, and refines that view by fitting the constant speed to appropriate variable speeds during the initial and final seconds.

The author is professor of mathematics at the Courant Institute of Mathematical Sciences, New York University.

The mathematical solution for the optimal velocity

For $D \le D_c$ the critical distance we have $f(t) = F$, and equations 2 and 3 yield

$$v(t) = F\tau(1 - e^{-t/\tau}) \qquad \text{(A1)}$$

Then equation 1 becomes

$$D = F\tau^2\left(\frac{T}{\tau} + e^{-T/\tau} - 1\right) \qquad \text{(A2)}$$

This gives the relation between D and T for $D \le D_c$. To find $E(t)$ we use equation Al for $v(t)$ in equations 5 and 6 to obtain

$$E(t) = E_0 + \sigma t - F^2\tau^2\left(\frac{t}{\tau} + e^{-t/\tau} - 1\right) \qquad \text{(A3)}$$

By setting $E(T_c) = 0$ we get the largest value of t for which the assumption $f = F$ is consistent with equation 7. Then D_c is the value of D given by equation A2 with $T = T_c$.

For $D \ge D_c$, $v(t)$ is given by A2 for $0 \le t \le t_1$, and by $v(t) = v(t_1)$ = constant for $t_1 \le t \le t_2$. In the interval $t_2 \le t \le T$, $v(t)$ is obtained by setting $E(t) = 0$ in equation 5, which yields $f = \sigma/v$. With this value of F, equation 2 can be solved with the result

$$v^2(t) = \sigma\tau + [v^2(t_1) - \sigma\tau]e^{-2(t_2 - t)/\tau} \qquad \text{(A4)}$$

The times t_1 and t_2 can be found by computing D from equation 1, using the three expressions for $v(t)$ just given, and then maximizing D with respect to t_1 and t_2. The fact that $v(t)$ equals a constant in the middle interval can be proved by the methods of the calculus of variations.

Reprinted from *Physics Today* **26(9)**, 42–47 (1973); © American Institute of Physics.

UNITED

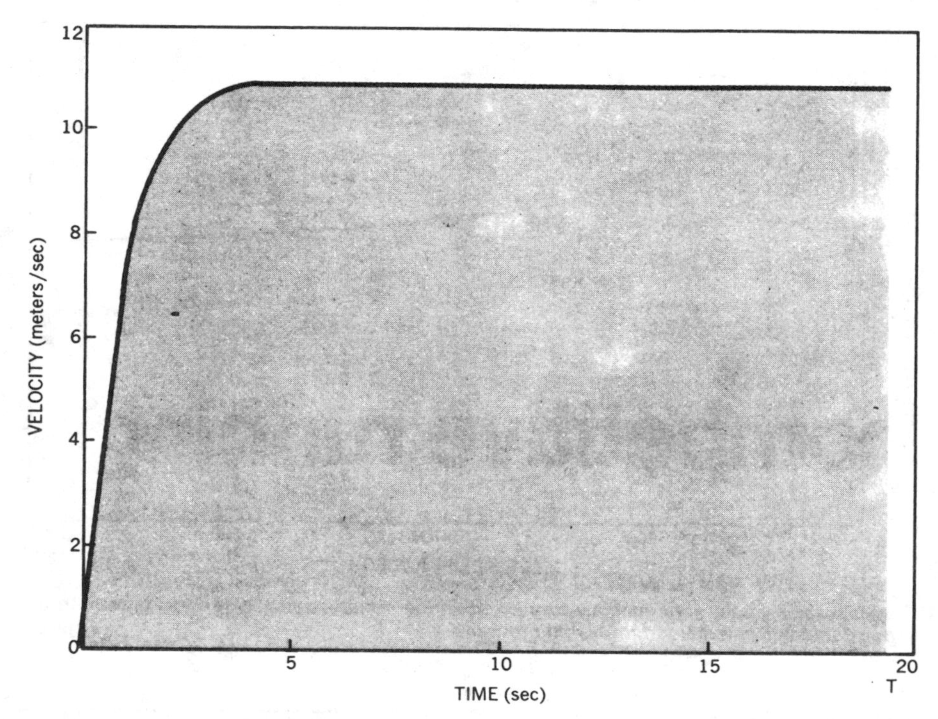

The 220-yard dash. The optimal velocity $v(t)$ is plotted versus t. The propulsive force $f(t) = F$ is maximum throughout the race.
Figure 1

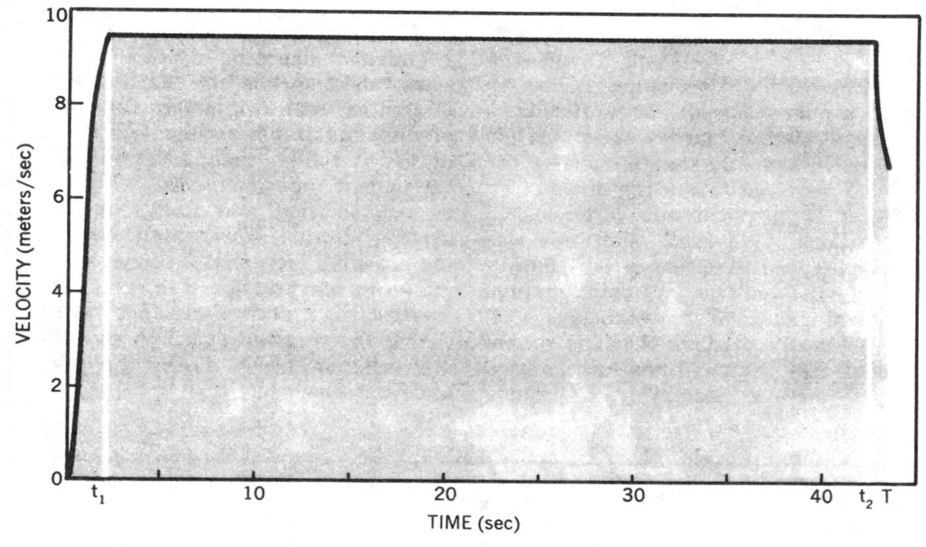

The 400-meter run. The optimal velocity $v(t)$ is plotted versus t. The propulsive force $f(t) = F$ is a maximum during the initial 1.78 seconds. After this initial acceleration, v remains constant until 0.86 seconds before the end of the race when the oxygen supply E becomes zero. Finally $E(t)$ remains zero during the last 0.86 seconds. This can be understood by thinking about a car with a limited amount of gas traveling over a distance D in the shortest time; the fuel should be used up shortly before the end.
Figure 2

Runners at distances greater than 291 meters often finish with a kick rather than with the negative kick of the optimal solution. This discrepancy indicates either that they are not doing as well as they could or that the theory is inadequate. Presumably their goal is to beat competitors rather than to achieve the shortest time, and that goal influences their strategy. But if they ran at the optimal speed determined by the theory, they might do even better at beating competitors. Some trials in which runners attempt to follow the optimal strategy might determine which is the correct explanation, and whether the optimal solution is better than the usual strategy.

In our theory we assume that the resistance to running at speed v is proportional to v. Another assumption, suggested by measurements of R. Margaria and his collaborators,[2,3] is that the resistance is a constant independent of the velocity. We have worked out the theory in this second case for comparison and found that it leads to a quite unsatisfactory prediction, which indicates that it must be rejected if the other assumptions of the theory are correct.

Formulation of the theory

The length D of a race is related to the time T required to run it by the equation

$$D = \int_0^T v(t)dt \qquad (1)$$

The velocity $v(t)$ is determined by the equation of motion, which we assume to be

$$\frac{dv}{dt} + \frac{v}{\tau} = f(t) \qquad (2)$$

In this equation $f(t)$ is the total propulsive force per unit mass exerted by the runner, part of which is used to overcome the internal and external resistive force v/τ per unit mass. It is an assumption that the resistance is a linear function of v and that the damping coefficient τ is a constant. Initially the runner is at rest; so

$$v(0) = 0 \qquad (3)$$

The force $f(t)$ is under the control of the runner; so we may think of it as the control variable. The runner must adjust it so that T, determined by equation 1, is as small as possible when $v(t)$ is the solution of equations 2 and 3. There are two restrictions on $f(t)$. First, there is a constant maximum force per unit mass F that the runner can exert; so f must satisfy the inequality

$$f(t) \leq F \qquad (4)$$

Second, the rate fv of doing work per unit mass must equal the rate at which the body supplies energy. This rate is limited by the availability of oxygen for the energy-releasing reactions, which we shall now consider.

Initially there is a certain quantity of available oxygen in the muscles, and more oxygen is provided by the respiratory and circulatory systems. It is convenient to measure the quantity of available oxygen in units of the energy it could release upon reacting. Thus we denote by $E(t)$ the energy equivalent of the available oxygen per unit mass at time t, by E_0 the initial amount, and by the constant σ the energy equivalent of the rate at which oxygen is supplied per unit mass in excess of the non-running metabolism. Then the equations of energy or oxygen balance can be written in the form

$$\frac{dE}{dt} = \sigma - fv \qquad (5)$$

In addition E satisfies the initial condition

$$E(0) = E_0 \qquad (6)$$

Because the energy equivalent of the available oxygen can never be negative, E also must satisfy the inequality

$$E(t) \geq 0 \qquad (7)$$

This is, indirectly, the second restriction of $f(t)$.

Now the runner's problem and ours is to find $v(t)$, $f(t)$ and $E(t)$ satisfying equations 2 through 7 so that T, defined by equation 1, is minimized. The four physiological constants τ, F, σ and E_0 are given, and so is the length of the race D. In other words, the problem is to find the rate of consumption of the initial oxygen supply in order to run the distance D in the shortest time.

My solution to this problem gives the result that for D not greater than D_c, where D_c is the critical distance mentioned above, $f(t) = F$, and v increases monotonically. For D greater than D_c, $v(t)$ increases for t less than t_1, v is constant for t between t_1 and t_2, and v decreases for t greater than t_2 until the end of the race, T. Figure 1 shows the optimal velocity $v(t)$ as a function of t for the 220-yard dash, which is a short sprint. In figure 2 the optimal $v(t)$ is shown for the 440-meter run, (D greater than D_c).

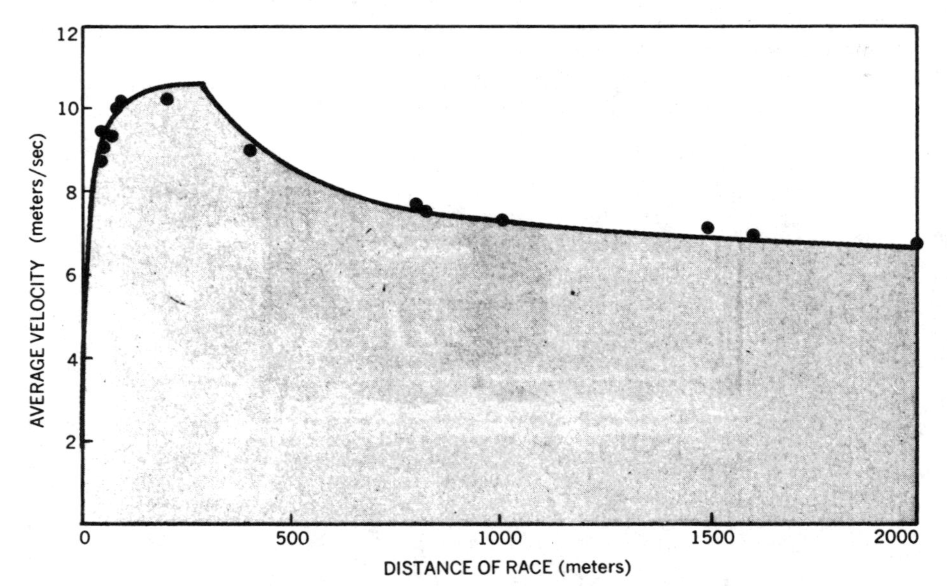

The average velocity for running a race. The theoretical curve is seen to agree with average velocities calculated from world records. Figure 3

Comparison of theory and observation

A. V. Hill first pointed out the physiological significance of track records.[3] Since then, many investigators have tried to extract physiological information from these records,[1] but they were hampered by a lack of any theory of running that correlates the data. Twenty-two world records for distances from 50 yards to 10 000 meters are shown in Table 1. The first four are taken from data published by B. B. Lloyd,[4] and the others are from the *Reader's Digest Almanac 1972*, page 980. We have determined the two constants τ and F to yield a least-squares fit of the times given by the theory to the record times for the first eight races, which we assume to be short sprints. Then we determined σ and E_0 to give a least-squares fit of the times given by the theory for the remaining 14 races. In both cases we minimized the sum of the squares of the relative errors. The values of the four physiological constants obtained in this way, appear in Table 2. Also shown there is the value of D_c computed from the theory with these constants. We see that D_c = 291 meters is between 220 yards and 400 meters; so the first eight races are short sprints and the other 14 are not. The ratio of the initial oxygen supply, E_0, to the rate of oxygen supply σ is E_0/σ = 58 seconds. Thus the initial supply is equivalent to the oxygen that would be supplied by respiration and by circulation is 58 seconds.

Table 1. Track Records

Distance D	Time T (record) min:sec	Time T (theory) min:sec	Error (per cent)	Average velocity D/T (theory) m/sec	t_1 sec	T-t_2 sec
50 yd	5.1	5.09	−0.2	8.99		
50 m	5.5	5.48	−0.4	9.12		
60 yd	5.9	5.93	0.5	9.26		
60 m	6.5	6.40	−1.5	9.38		
100 yd	9.1	9.29	2.1	9.85		
100 m	9.9	10.07	1.7	9.93		
200 m	19.5	19.25	−1.3	10.39		
220 yd	19.5	19.36	−0.7	10.39		
400 m	44.5	43.27	−2.8	9.24	1.78	.86
440 yd	44.9	43.62	−2.9	9.22	1.77	.86
800 m	1:44.3	1:45.95	1.6	7.55	1.07	1.08
880 yd	1:44.9	1:46.69	1.7	7.54	1.06	1.08
1000 m	2:16.2	2:18.16	1.4	7.24	0.98	1.16
1500 m	3:33.1	3:39.44	3.0	6.84	0.88	1.31
1 mile	3:51.1	3:57.28	2.7	6.78	0.87	1.34
2000 m	4:56.2	5:01.14	1.7	6.64	0.84	1.43
3000 m	7:39.6	7:44.96	1.2	6.45	.80	1.60
2 miles	8:19.8	8:20.82	0.2	6.43	.80	1.63
3 miles	12:50.4	12:44.89	−0.7	6.31	.77	1.80
5000 m	13:16.6	13:13.11	−0.4	6.30	.77	1.82
6 miles	26:47.0	25:57.62	−3.1	6.20	.75	2.10
10000 m	27:39.4	26:54.10	−2.7	6.20	.75	2.12

By using the constants in Table 2, we have computed the time given by the theory for each race. The results are shown in Table 1, together with the average velocity D/T given by the theory and the values of t_1 and $T - t_2$ for the races with D greater than D_c. The error in time between the theoretical value and the record is also shown in Table 1 as a percentage. We see that for the short sprints the error is at most 2.1%. However, for the longer races, it reaches 3.1% for the six-mile race. The average velocity given by the theory is plotted against distance in figure 3, which also shows the average velocities computed from the record times. Note that the initial increase and ultimate decrease of the average velocity is predicted by the theory quite satisfactorily. In comparing the theory with the actual records, it must be borne in mind that the theory

Table 2. Physiological Constants

τ = 0.892 sec
F = 12.2 m/sec^2
σ = 9.93 calories/kg sec
E_0 = 575 calories/kg
D_c = 291 m

uses a single set of physiological constants at all distances, while the record holders at these distances undoubtedly had somewhat different constants from one another. The theory I have presented is too simple, because it omits various important mechanical and physiological effects. It ignores the up-and-down motion of the limbs; it fails to distinguish between internal and external resistance; it does not take into account the depletion of the fuel that uses the least oxygen and the transfer to the use of less efficient fuels; it ignores the accumulation of waste products and the mechanisms of removing them, and it probably ignores some other effects as well. A better theory incorporating some of these effects might be able to account for the records at longer distances, as well as those considered here. In support of such a new theory, measurements of the resistive force and the other physiological parameters are needed.

Nevertheless, this theory yields some definite results, and it would be of interest to adapt it to other types of races. Ice skating, swimming and bicycle races, for example, could be studied, taking into account the special features of each type of race. Another interesting problem would be to determine the influence of hills and valleys on the optimal velocity in longer races; this would require only a modification of the present theory.

* * *

This work was supported in part by the National Science Foundation. I thank Cesar Levy and Clyde Kruskal for performing the calculations, and Thomas J. Osler, who won the A.A.U. 50-km race in 1967 and who is now professor of mathematics at Glassboro State College, New Jersey, for his comments on the manuscript.

References

1. B. B. Lloyd, Advancement of Science **22,** 515 (1966).
2. R. Margaria, P. Cerretelli, P. Aghemo, G. Sassi, J. Appl. Physiol. **18,** 367 (1963).
3. R. Margaria, P. Cerretelli, F. Mangilli, P. E. Di Pramero, "The energy cost of sprint running," ler Congres Europêen de Medicine Sportive, Prague, (1963).
4. A. V. Hill, *Muscular Activity: Herter Lectures, 1924,* Williams and Wilkins, Baltimore (1926). □

Physics of long-distance running

J. Strnad
Department of Physics and J. Stefan Institute, University of Ljubljana, 61111 Ljubljana, Yugoslavia

(Received 2 December 1983; accepted for publication 20 March 1984)

The physics of sprinting over distances less than 290 m has been discussed in this journal at some length.[1,2] Here we tackle the problem of running from the other end at distances of 2000 m and more. A very simple phenomenological model fits the current track world records for men surprisingly well. Track records for women and records for ice skating, walking, and swimming are considered for comparison. The model offers an easy overview over various types of locomotion and may be of interest as a starting point for further consideration.

Suppose that a runner's center-of-mass is moving in a horizontal direction uniformly and rectilinearly with average velocity V. Then the external forces acting on the runner, averaged over one step, are in equilibrium. In the direction of motion the force of the track on the runner's feet is acting and in the opposed direction, the component of weight, against which the center-of-mass is lifted, and the air resistance. Actually, the center-of-mass is in accelerated motion in the horizontal as well as vertical direction. We, however, consider the effective horizontal force in the opposed direction of running averaged over one step, F.

To a good approximation this force can be taken as independent of velocity. The air resistance in long-distance running represents a small fraction of F only and its dependence on V can be neglected owing to the small velocity interval in question, from 5.5 to 6.9 m/s. This is in accord with detailed measurements undertaken in a few cases. R. Margaria *et al.* found directly that the effective average force is independent of velocity.[3] T. Fukunaga *et al.*, by measuring the mechanical power output in the forward direction, obtained a quadratic dependence on velocity.[4] However, their individual measurements are scattered appreciably and a power proportional to V, i.e., a constant force, fits the data on the stated interval even better.

The average velocity is given as the quotient of duration and length:

$$V = L/T.$$

For a constant force F the excess average locomotive power of the runner is proportional to velocity: $P = FV$. The corresponding excess work is proportional to the length:

$$W = FL.$$

In steady exercise energy is supplied by aerobic processes in a runner's muscles.[5] The oxygen stored in the muscles can be neglected. The increase of the duration from T to $T + dT$ is accompanied by an increase of excess work from W to $W + dW$. In developing a simple phenomenological model it is natural to suppose that not all this increase is utilized for locomotion, a relative part s of it being consumed for maintenance of the engine, i.e., mainly for preparation of metabolites and the transport of the metabolites as well as oxygen into muscles and of water and carbon dioxide out of them, for internal friction, etc. In this case the equation $dW = (1 - s)P\,dT$ or

$$dW/W = (1 - s)dT/T \qquad (1)$$

is valid. Its solution reads

$$W \propto L \propto T/T^s.$$

The velocity $V = L/T$ depends then on T as

$$V = V_1/T^s, \quad T \text{ in seconds.} \qquad (2)$$

From equation $dP/P = -s\,dT/T$ the same dependence follows for power, $P \propto T^{-s}$.

In the corresponding equation for velocity, $dV/V = -s\,dT/T$, the velocity V in the denominator can be approximated by $V_0 = \text{const}$. The resulting equation $V = -sV_0 \ln(T/T_0)$ reproduces track records for men only slightly worse than Eq. (2). This is due to the short interval of velocities.

If s would be set equal to zero, velocity and power would be independent of the duration or length, which for long distances would be an acceptable zeroth approximation.

A power law like (2) was introduced by A. E. Kennelly as early as 1906 to fit measurements with racing animals.[6] He put the equation in the form $V = V_1'/L^{1/8}$, which in our terms and assumptions reads $V = V_1/T^{1/9}$ with $V_1 = 7.5$ m/s.

J. G. Purdy in a review quoted 19 equations devised to fit experimental data.[7] Some of them are quite complicated containing five terms with ten parameters. Besides exponentials polynomials are met as well. The majority of equations gives the "running curve" $V(L)$ and only one, not applicable for long distances, connects V and T directly.

Instead of a "horizontal analysis," i.e., the study of world records at a given instant, the development of world records in time is often studied in a "vertical analysis." In this case power laws or polynomials are used also.[8]

Table I. Track world records for men as of autumn 1983, the average velocity $V = L/T$, the velocity V_c calculated with Eq. (1) ($s = 0.069$ and $V_1 = 10.35$ m/s), and the relative error $(V_c - V)/V$. These and other data in this article were compiled by the reporters of RTV Ljubljana, M. Rožman, M. Trefalt, and S. Urek, to whom the author is indebted for help.

L km	T h	min	s	V m/s	V_c m/s	Relative error percent
2		4	51.40	6.86	6.84	−0.2
3		7	32.10	6.64	6.65	0.1
5		13	00.41	6.41	6.40	−0.1
10		27	22.40	6.09	6.08	−0.2
20		57	24.20	5.81	5.78	−0.6
25	1	13	55.80	5.64	5.68	0.7
30	1	29	18.80	5.60	5.60	0.05
42.195	2	08	13.00	5.49	5.46	−0.4

Reprinted from *American Journal of Physics* **53**, 371–373 (1985);

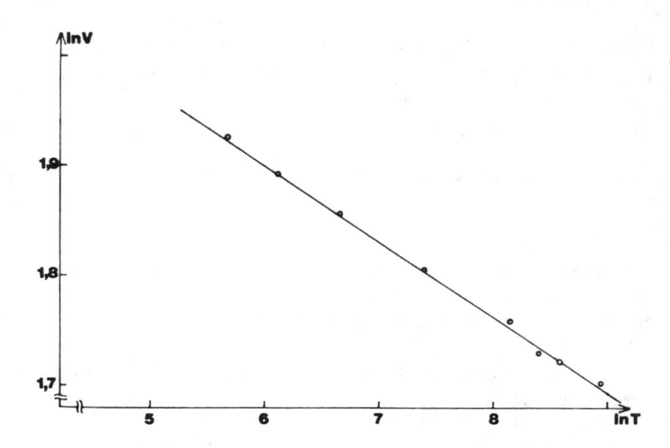

Fig. 1. The dependence of the average velocity on the duration for track world records for men as of autumn 1983 (see Table I). The straight line corresponds to Eq. (1) with $V_1 = 10.31$ m/s and $s = 0.069$.

As to my knowledge Eq. (1) was not considered in the literature. B. B. Lloyd came close to it, but he complicated his calculations by a polynomial dependence of power on velocity.[9] One cannot help observing that simple models, applicable on a restricted domain, are consequently avoided. An extensive model was produced by J. B. Keller.[10] This model, which includes the equation of motion, is not appropriate for very long distances.

M. H. Lietzke, following Kennelly in studying $L \propto T^k$ and $V \propto L^{k'}$, considered $k' = (k-1)/k$ to be the exhaustion constant "giving a quantitative measure of the rate of fatigue."[11] Our considerations show, however, that it would be preferable to call $s = k - 1 = k'/(1-k')$ the maintenance constant.

In spite of the simplifying assumptions in its derivation, Eq. (2) with $s = 0.069$ and $V_1 = 10.31$ m/s fits track records for men surprisingly well (Table I, Fig. 1). On the domain from less than 5 min for 2 km to more than 2 h for the Marathon race (42.195 km) the maximal relative error amounts to 0.7 percent. This accuracy is appreciably better than for a much more complicated model.[10]

The lower limit for the validity of our model is dictated by anaerobic processes becoming important at short and intermediate distances. Due to a restricted supply of energy there is an upper limit as well which, however, cannot be as easily recognized. Races over much longer distances do not take place on level ground and/or are interrupted by periods of rest. For $L = 100$ km the equation $T = (L/V_1)^{1/(1-s)}$ gives a time of 5 h 30 min and a velocity of 5 m/s. In 1983 the winner of the Biel 100-km race achieved a time of 6 h and 47 min and a velocity of 4.10 m/s. On level ground a shorter time would be expected but it is doubtful whether our model would be valid in this case.

Similar results are obtained for track records for women using the same coefficient $s = 0.069$ and fitting only constant $V_1 = 9.0$ m/s. In passing we notice the development of records in time: the current value for women, $V_1 = 9.0$ m/s, coincides with the one for men around 1912.

For comparison let us add data for a few other types of locomotion: ice skating, walking, and swimming. In these cases fewer data exist and the agreement is, in general, worse, the error amounting to a few per cent, at most. Approximations are obtained with the same value of s, fitting only the constant V_1 (Table II, Fig. 2). This implies that extremely trained athletes achieving records show approximately equal performance.

Table II. The constant V_1 (extrapolated velocity at $T = 1$ s) for various types of locomotion.

	V_1
ice skating—men	18.17 m/s
running—men	10.13
—women	9.0
walking—men	7.35
free-style	
swimming—men	2.63
—women	2.46

We can assume that the power output for a given time T is approximately independent of the type of locomotion. Then for equal efficiency a greater velocity is associated with a smaller force: $F = P/V$. Hence the force F is greater for running and walking than for ice skating. At ice skating the steps are larger and the lifting of the center-of-mass smaller. In free-style swimming, however, there is no lifting of the center-of-mass at all and the comparatively great force is due to water resistance only.

A universal coefficient s is only a first approximation. Lietzke even stated explicitly that in a log–log plot running and swimming show distinct slopes.[11] Certainly, in a more sophisticated model different values of s for various types of locomotion should be introduced and interpreted. To that end our simple model is an appropriate starting point. On the other hand, models not containing, at least asymptotically, a corresponding term (1), are scarcely applicable for long-distance running.[10,12]

In concluding, a pedagogical remark, describing the way to the model in this rather marginal subject, may be to the point. In an attempt to make physics more attractive, even to pupils of primary school, in the spirit of R. Sexl's *Physics in Sports*,[13] the velocity in running and swimming world record events was considered.[14] Plotting a diagram, the

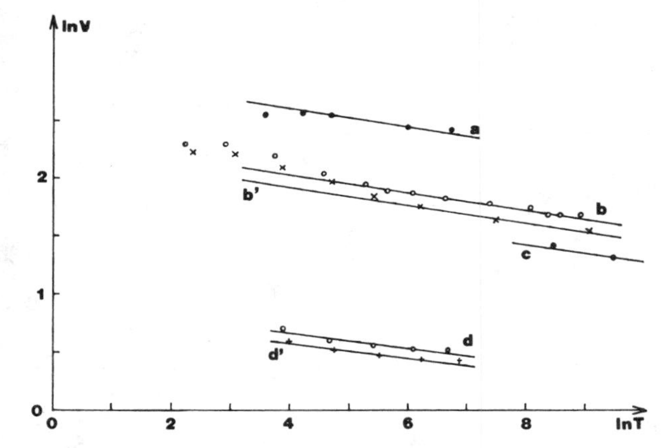

Fig. 2. The dependence of the average velocity on the duration for world records: ice skating—men (a, 500, 1000, 1500, 5000, and 10 000 m); running—men (b, 100, 200, 400, 800, 1500, 2000, 3000, 5000, 10 000, 20 000, 25 000, 30 000, and 42 195 m) and women (b', 100, 200, 400, 800, 1500, 3000, 10 000, and 42 195 m); walking—men (d, 20 and 50 km); and free-style swimming—men (d, 100, 200, 400, 800, and 1500 m) and women (d', 100, 200, 400, 800, and 1500 m). The constant s is taken equal to 0.069; for the constant V_1 see Table II.

time interval was found to be unpleasantly wide, so ln T was used instead of T. With some surprise the straight line on the side of great time, i.e., long distance, was observed. A model was devised to relate ln T to V. After some further deliberation ln V was considered instead of V. It was found that $\ln V \propto -\ln T + \text{const}$ reproduces data somewhat better than $V \propto -\ln T + \text{const}$. The former was accepted and Eq. (1) set up, the decisive proof of the model being the agreement with experimental data. The results were found to be interesting enough to be written up for publication at the level of *Physics in Sports*. Further references were brought to the attention of the author by a referee. Only then did some merits of the simple phenomenological model as a research tool became apparent.

The way to the results is characteristic for the method of physics as the following two quotations show. L. Boltzmann wrote: "Allegedly Gauss should have answered a friend's question concerning some urgent work: 'All formulas and results are ready, I have only to find a way that will lead me to them.' I do not believe that Gauss said this, he was not open-minded enough, but he has certainly often thought so... ."[15] G. Holton put it this way:"... has thus been characterized by the American nuclear physicist, H. D. Smyth, 'We have a paradox in the method of science. The research man may often think and work like an artist, but he has to talk like a bookkeeper, in terms of facts, figures, and logical sequences of thought.' For this reason we must not take very seriously the chronology or method outlined in the original publication of scientific results... . It is a part of the game to make the results in retrospect appear neatly derived from clear fundamentals... ."[16]

ACKNOWLEDGMENT

Thanks are due to A. Kodre for enlightening discussions.

[1] I. Alexandrov and P. Lucht, Am. J. Phys. **49**, 254 (1981).

[2] H. Lin, Am. J. Phys. **46**, 15 (1978).

[3] R. Margaria, P. Cerretelli, P. Aghemo, and G. Sassi, J. Appl. Physiol. **18**, 367 (1963).

[4] T. Fukunaga, A. Matsuo, K. Yuasa, H. Fujimatsu, and K. Asahina, Ergonomics **23**, 123 (1980).

[5] R. Margaria, Sci. Am. **226**, 88 (1972).

[6] A. E. Kennelly, Proc. Am. Acad. Arts Sci. **42**, 275 (1906).

[7] J. G. Purdy, Res. Quarterly **45**, 224 (1974).

[8] P. Mognoni, C. Lafortuna, G. Russo, and A. Minetti, Eur. J. Appl. Physiol. **49**, 287 (1982).

[9] B. B. Lloyd, Circ. Res. Suppl. I **XX**, 218 (1967).

[10] J. B. Keller, Phys. Today **26**, 42 (1973).

[11] M. H. Lietzke, Science **119**, 333 (1954).

[12] See also P. S. Riegel, Am. Sci. **69**, 285 (1981) and D. F. Griffing, *The Dynamics of Sports* (Mohican, Loudonville, OH, 1982).

[13] R. U. Sexl, A. Wehrl, and A. Pflug, *Physik im Sport* (University of Vienna, preprint, 1982).

[14] J. Strnad, Presek **11**, 136 (1984).

[15] L. Boltzmann, *Vorlesungen über Maxwells Theorie der Elektricität und des Lichtes* (A. Barth, Leipzig, 1891), p. III.

[16] G. Holton and S. G. Brush, *Concepts and Theories in Physical Science* (Addison-Wesley, Reading, MA, 1973), p. 188.

Power demand in walking and pace optimization

A. Bellemans
Department of Physics, Université Libre de Bruxelles, CP 223, 1050 Brussels, Belgium
(Received 20 February 1980; accepted 22 May 1980)

The power expenditure in walking is presented as the sum of the work required for accelerating and decelerating the legs, plus the gravitational work associated to the lift of the trunk at each step. These two contributions are, respectively, inversely and directly proportional to pace length. As a consequence this model predicts an optimal pace, for a given speed, which corresponds to minimal power demand. This is in agreement with experiments on subjects walking on a treadmill. The model is primarily of pedagogical interest and its quantitative predictions not very accurate.

I. INTRODUCTION

The aim of this paper is to present an elementary formulation of the work expenditure corresponding to the most common physical exercise, i.e., *walking*. Many factors should be taken into account; the most obvious ones are speed, pace length, body mass, and slope; but the nature of the floor, the morphology of the subject, its clothing, and the ambiant conditions play a role as well. Reasonably good experimental data are actually available for standardized conditions, i.e., people in gym clothes walking on a treadmill driven at controlled speed, the power demand being deduced from oxygen consumption.[1] In what follows we shall express that power as the sum of (a) the work of acceleration of the lower limbs and (b) the gravitational work performed at each step. A consequence of that model is the existence of an optimal step length, corresponding to minimal power for a given walking speed. The quantitative predictions and the weaknesses of that model will also be discussed.

II. ELEMENTARY FORMULATION OF POWER DEMAND IN WALKING

Consider a subject walking on the flat at constant speed v. At first sight (if one neglects air resistance) it would seem that no work at all is needed to maintain its motion. However, each time a foot strikes the ground, it stops momentarily while the rest of the body continues to move forwards. Hence when that foot leaves the ground, the whole leg must be accelerated forwards in order to overtake the body. Next it will be decelerated until the foot comes again to rest on the ground. During that process, two sets of antagonistic muscles of the leg act successively, each of them performing work equal to the corresponding change of kinetic energy of the limb. We shall not attempt to formulate this work precisely at present but it must clearly be proportional to v^2, times some effective mass of the leg, since at each step, the foot velocity varies back and forth between 0 and some value of the order of v (Ref. 2). Given the step frequency v/a, where a is the length of a single step, it then follows that the power input to both legs, when walking, may be expressed as

$$P_1 = \alpha M v^2(v/a) = \alpha M v^3/a, \qquad (1)$$

where M is the total body mass and α is a numerical factor smaller than one.

Besides the work associated to leg motion, gravitational work is also performed while walking, as the trunk moves up and down during each step. Assuming both legs to be fully extended in the astride position, the lift H of the trunk is determined by a and the leg length L [see Fig. 1(a)] i.e., $H = L(1 - \cos\gamma)$ or, to a sufficient approximation for the present purpose

$$H = L\gamma^2/2 = a^2/8L. \qquad (2)$$

The lift work should be of the order of MgH, hence the corresponding power

$$P_2 = (\beta Mga^2/L)(v/a) = \beta Mgva/L, \qquad (3)$$

where the constant β should be close to 0.1 according to (2).

The total mechanical power required for walking is then the sum of P_1 and P_2, i.e.,

$$P(v,a) = \alpha M v^3/a + (\beta Mgv/L)a. \qquad (4)$$

Enlarging the pace at constant v lowers the first term and increases the second one, so that formula (4) predicts the existence of an optimal pace $a^*(v)$, minimizing $P(v,a)$. Deriving (4) with respect to a and equating to 0, one finds

$$a^*(v) = (\alpha/\beta)^{1/2}(L/g)^{1/2}v \qquad (5)$$

and

$$P^*(v) = 2(\alpha\beta)^{1/2}(g/L)^{1/2}Mv^2 \qquad (6)$$

as the corresponding minimal power.

One would therefore expect that, when walking *naturally* on the flat at a fixed velocity, a subject will adjust its pace automatically to the optimal value corresponding to the minimal work expenditure. This has indeed been verified experimentally.[1,3] For example, Fig. 2 shows power consumption versus step length for a subject walking on a

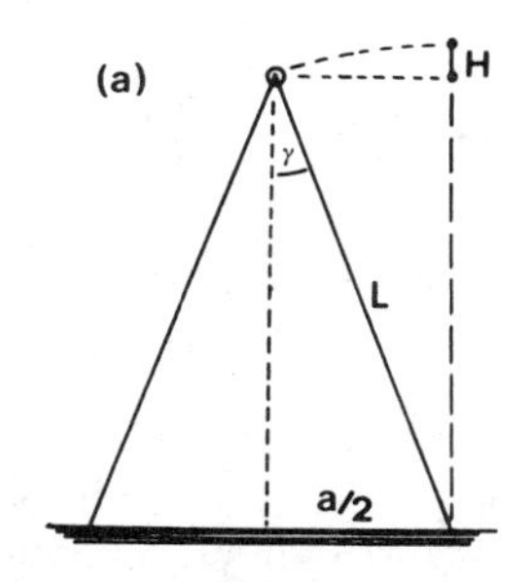

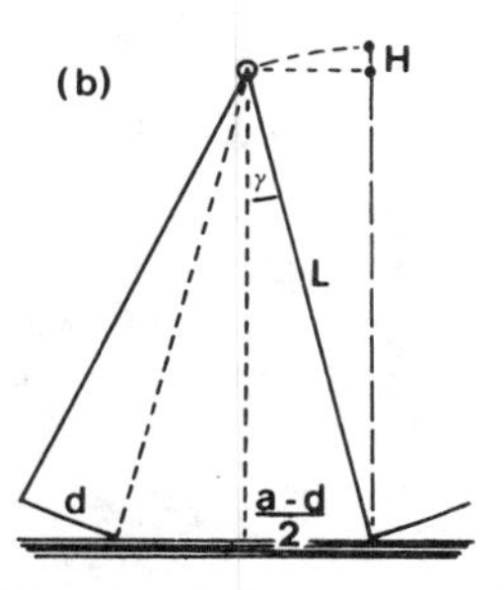

Fig. 1. Vertical lift of the trunk in walking, in relation to pace and leg length: (a) crude and (b) refined models.

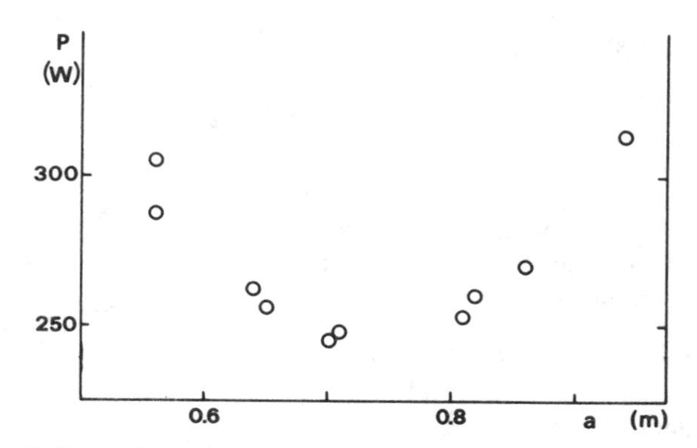

Fig. 2. Power demand versus pace length, for a subject walking on a treadmill at 4 km/h (data extracted from Fig. 8(a) of Ref. 1).

treadmill at approximately 4 km/h; when walking *naturally* at that same speed, the pace length and power consumption of that subject were close to 250 W and 0.70, m, respectively.

Note also that expression (6) is qualitatively correct (on the average) in predicting an increase of the power demand with increasing body mass of the subjects[1]; it also indicates a slight lowering of that power (at constant body mass) with increasing leg length. (This seems reasonable but difficult to investigate experimentally.)

III. QUANTITATIVE COMPARISON WITH EXPERIMENT

In the Appendix we show that a reasonable value for α is 0.1. On the other hand, Eq. (2) implies $\beta = 1/8$ but this value is too high, first, because the body does not rise as a whole at each step and, second, because some energy, stored in the form of elastic strain in muscles and tendons as a lower limb strikes the ground, is recovered in the subsequent recoil.[4,5] We therefore plausibly adopt $\beta = 0.05$. Further, as muscle efficiency usually amounts to 20% only, Eq. (6) must be multiplied by 5 in order to be compared to the actual power expenditure, calculated from the oxygen consumption (1 l/min of O_2 being equivalent to ~340 W).

We limit the comparison to the data relative to one particular subject (No. 2), among the eleven ones studied by Cotes and Meade,[1] because of the abundance of data relative to him. Its mass and leg length are 68 kg and 0.92 m, respectively, and its power expenditure at rest (which must of course be added to the power required for walking) amounts to 110 W. We then obtain the following numerical expressions from (5) and (6)

$$a^*(v) \simeq 0.43\, v, \tag{7}$$

$$P^*(v) \simeq 110 + 157\, v^2 \text{ (in watts)}. \tag{8}$$

These expressions are plotted on Figs. 3 and 4, respectively (full lines), together with the experimental data. Of course nobody would seriously expect quantitative agreement but there are also qualitative differences. In both figures the experimental points approximately follow straight lines which, when extended to the vertical axis, intercept it, respectively, in $a \simeq 0.33$ m and $P \simeq 167$ W (a notably higher value than the resting power). Actually the experimental observations of the lift H per step, by Cotes and Meade, do not agree with the simple picture of Fig. 1(a); a better one is that of Fig. 1(b) where the foot length d is taken into account.[1] Equation (2) should then be replaced (to the same degree of accuracy) by

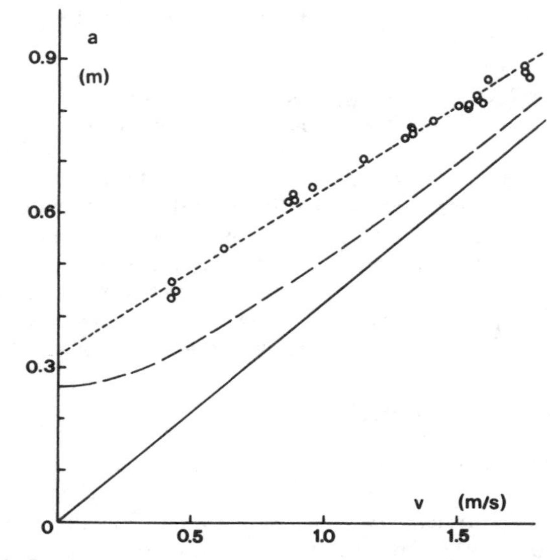

Fig. 3. Pace length versus speed. The experimental points are deduced from Table 3 of Ref. 1 (subject No. 2): plain line—Eq. (7); broken line—Eq. (11); dotted line—least-squares fit of data.

$$H = (a - d)^2/8L, \tag{9}$$

which in turn leads to

$$P(v,a) = \alpha M v^3/a + (\beta M g v/L)(a - d)^2/a \tag{10}$$

and

$$a^*(v) = [d^2 + (\alpha/\beta)(L/g)v^2]^{1/2}, \tag{11}$$

$$P^*(v) = 2(\beta M g v/L) \times \{[d^2 + (\alpha/\beta)(L/g)v^2]^{1/2} - d\}. \tag{12}$$

For the particular subject studied, d is nearly equal to 0.27 m[1] and one obtains the broken lines of Figs. 3 and 4 (again allowing for a 20% muscle efficiency and for a resting power of 110 W).

The improvement is slight but noticeable. The tricky point remains that the resting power is seemingly not ap-

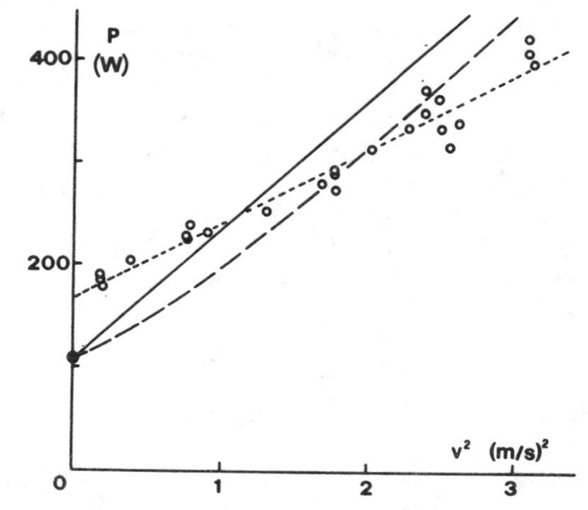

Fig. 4. Power expenditure versus v^2. Same origin of data as for Fig. 3: plain line—Eq. (8); broken line—Eq. (12); dotted line—least-squares fit of data.

proached at small speeds.[1] One possibility is that some additional power, almost independent of v, is required to mobilize muscles insuring the correct balance of the body when walking. Another possibility is that power optimization is only achieved at normal speeds, i.e., 4–6 km/h (and probably also at greater ones where the power demand becomes high) but not at low speeds, like 2 km/h, to which the subject is not accustomed and which are not very demanding.

On the other hand, one should not worry too much about the slope discrepancy in Fig. 4; somewhat different values of α and β could improve the situation but it is our opinion that the crudeness of the model does not justify such an exercise. Any quantitative theory probably requires a precise biomechanical analysis of walking and, in this respect, it is worth to mention that, during the last few years, much care has been taken in analyzing human gait and modelling it on computers.[6]

FINAL REMARKS

The model presented here is mainly of pedagogical interest and is included in our physics course to freshmen in Physical Education and Physical Therapy, the goal being to show how simple physical ideas help to clarify complicated phenomena.

Note that we did not consider inclined floors. For a small slope of angle θ, it is sufficient to include an additional power $Mgv \sin\theta$, but for important slopes, the simple pictures of Fig. 1 are inadequate.[7] This is also true for very high walking speeds (10 km/h or more).

ACKNOWLEDGMENT

We are indebted to J. Van Fraechem, Institute of Physical Education and Kinesitheray, University of Brussels, for introducing us to the subject and for helpful discussions.

APPENDIX: ESTIMATION OF α

The relative motion of a lower limb, with respect to the trunk, will be approximated here as the rotation of a rod of length L and moment of inertia I around the hip joint. The muscular work for (i) accelerating one leg to its maximum angular speed ω_m, (ii) decelerating it again to zero, is then equal to $2\,(I\omega_m^2/2)$ and we shall tentatively write $\alpha Mv^2 \simeq I\omega_m^2$. In order to overtake the trunk, the rear foot must move at the average speed $2\,v$ and its maximum speed, relative to the trunk, could possibly be of the order $1.5\,v$. This implies in turn $\omega_m \simeq 1.5\,v/L$. Finally, the mass m of a lower limb (on the average) is equal to 16% of M and its radius of gyration $(I/m)^{1/2}$ is close to $0.45\,L$.[8] Hence

$$\alpha \simeq I\omega_m^2/Mv^2 \simeq (I/ML^2)(\omega_m L/v)^2 \simeq 0.16(0.45)^2(1.5)^2,$$

i.e., $\alpha \simeq 0.07$. This is of course a very crude estimate; we shall actually adopt the slightly higher value $\alpha = 0.1$, allowing for some contribution of the upper limbs, the motion of which help to control the balance of the body.

[1] J. E. Cotes and F. Meade, Ergonomics **3**, 97 (1960); further references may be found in this paper and in P. O. Astrand and K. Rodahl, *Textbook of Work Physiology* (McGraw-Hill, New York, 1970), Chap. 16.

[2] A. H. Cromer, *Physics for the Life Sciences* (McGraw-Hill, New York, 1974), p. 88 and 115.

[3] H. Magne, J. Physiol. Path. Gen. **18**, 1154 (1920).

[4] G. A. Cavagna, F. P. Saibene, and R. Margaria, J. Appl. Physiol. **19**, 249 (1964).

[5] R. McN. Alexander and H. C. Bennet-Clark, Nature **265**, 114 (1977).

[6] Most of the relevant literature may be found in the *Journal of Biomechanics*, e.g., D. A. Winter, A. O. Quanbury, and G. D. Reiner, J. Biomech. **9**, 253 (1976); M. Y. Zarrugh and C. W. Radcliffe, *ibid.* **12**, 99 (1979).

[7] M. A. Townsend and T. C. Tsai, J. Biomech. **9**, 227 (1976).

[8] M. Williams and I. Lissner, *Biomechanics of Human Motion* (Saunders, Philadelphia, 1977), Appendix A, Tables A-1, A-2, A-5, and A-7.

Effect of wind and altitude on record performance in foot races, pole vault, and long jump

Cliff Frohlich
Institute for Nongeophysics, University of Texas, Austin, Texas 78712

(Received 13 January 1984; accepted for publication 23 August 1984)

Using only elementary physics, one can estimate the effect of wind and altitude on performance in several track and field events. Experiments have shown that the power lost to aerodynamic drag forces is about a tenth of the total power expended in running at sprint speeds. From this observation one can calculate the effect of wind or of air density changes on sprinting speed. In pole vaulting, the sprinter converts his kinetic energy into potential energy to clear the bar. In long jumping, he is a projectile, but he is prevented from reaching his optimum distance expected for his initial velocity by the height which he can attain during his jump. For each of these events, performance in moderate winds of 2.0 m/s or at altitudes comparable to Mexico City differ by several percent from performances at sea level or in still air. In longer running races and in bicycle races, aerodynamic forces play an important role in racing strategy. However, since the athletes perform in groups it is difficult to calculate the effect on individual performances.

I. INTRODUCTION

Aerodynamic forces significantly affect strategy and record performance in numerous sporting events. Indeed, articles in this journal alone have discussed the shot put,[1] for which the effects are minor, and also baseball,[2–4] golf,[5] and discus throw,[6] where aerodynamic forces are enormously important. For baseball, golf, and the discus, the interesting physics occurs because the athlete can utilize variable aerodynamic lift forces or take advantage of sudden changes in aerodynamic drag which occur at certain air speeds. Since these complications significantly affect quantitative analysis of these events, iterative numerical calculations are required to evaluate the effect on performance. For this reason, students in introductory physics courses may not appreciate the analysis.

Aerodynamic drag also affects performance in sporting events such as running and jumping. The primary purpose of this article is to present simple models of performance in selected track and field events in order to evaluate the effects of wind and altitude on record performances. There is no question that aerodynamic forces affect record performance. For example, the 1968 Olympic games were held in Mexico City at an altitude of 2340 m where the air density is about 75% of that at sea level. At these games, men set world records at distances of 100, 200, 400, and 800 m as well as in the pole vault, long jump, and triple jump. Of these records, those at 100 m, 400 m, and the long jump were still the world records in January 1983, and the listed records at 200 m and the triple jump were performances in Mexico City in subsequent meets.

The physics necessary to quantitatively evaluate these effects is at the level usually presented in a first-year university course. For example, during most of a long jumper's leap, he is a projectile acted upon only by the constant, downward force of gravity and a much smaller drag force. Thus we can analyze his performance reasonably well using formulas from ballistics. Since running consists of repetitive motions, the speed a runner can maintain over a set distance is limited primarily by the power his body can produce over that distance. Of course, performance in both running and jumping also depends on numerous other factors, including some which are definitely not part of introductory physics, such as biochemistry, neurophysiology, and psychology. However, in spite of their simplicity, the models presented here do predict record performance surprisingly well, and provide a reasonably accurate estimate of the effects of wind and altitude on record performance.

The only necssary formula which is not usually presented in introductory courses is that for the aerodynamic drag force F_D affecting a moving object or person:

$$F_D = \tfrac{1}{2} C_D \rho A V^2. \quad (1)$$

Here ρ is the density of air (1.225 kg/m^3 at sea level and 16 °C), V is the velocity relative to the air, and C_D is the drag coefficient, a dimensionless, empirically determined quantity. A is the projected area, or the area of the shadow if the object or person is illuminated along the direction of air flow.

Although Eq. (1) may not be familiar to beginning students, they can understand its basic form. Suppose air consists of mutually noninteracting molecules, and air drag comes about as an object collides with these molecules, imparting momentum to them. The drag force F_D is just the momentum imparted per unit time. The volume per unit time swept out by the object is AV, and in this unit time the object collides with particles of mass ρAV. If each collision is elastic, it will change the momentum (on the average) by twice V times the particle mass. This argument finds the drag to be

$$F_D = 2\rho A V^2.$$

Streamlining and other shape related factors reduce the drag force, making it useful to introduce a fudge factor, the drag coefficient C_D. The factor one-half is associated with C_D for historical reasons. Unless indicated otherwise in this paper we will use the values presented in Table I for A, C_D, and other relevant quantities.

II. SPRINTING

Experimental studies of human beings running on treadmills[7] have shown that the power P expended by a runner is proportional to his velocity V. Although not all the power expended goes into mechanical work, that which does is proportional to the runner's mass. For example, consider

Reprinted from *American Journal of Physics* **53**, 726–730 (1985); © American Association of Physics Teachers.

Table I. Parameters describing a typical world-class athlete, and used for all calculations in this study unless otherwise indicated.

M	mass	75 kg
A	projected area	0.5 m^2
h_{cm}	height cm above ground	1.0 m
C_d	drag coefficient	0.9
h_{jump}	height cm raised in vertical jump	1.0 m
Δh_{LJ}	decrease in cm height during long jump	0.5 m
P_s	spring power capacity	3.0 kW
P_m	marathon power capacity	1.5 kW
R	power constant [Eq. (2)]	3.634 W/kg-m/s
V_s	sprint velocity, no wind [derived from Eq. (2)]	10.0 m/s
V_m	marathon velocity, no wind [derived from Eq. (2)]	5.35 m/s

energy to accelerate and decelerate the legs, energy to raise the runner's center of mass on each stride, and momentum lost as the forward foot strikes the ground. Thus a simple model to describe the power expended is

$$P = MRV,$$

where R is a constant. Experimentally, the power expended is determined by having the runner breathe into a special apparatus, measuring the amount of oxygen he consumes at different speeds, and comparing this to the oxygen he consumes when at rest. Note that the above equation is equivalent to saying the energy expended (power × time) is proportional to distance traveled, regardless of speed.

When a human runs on a track, he must overcome the drag resistance of air, which requires an additional power output of VF_D. If there is a wind of velocity V_w the relative velocity is $V - V_w$ and thus

$$P = MRV + \tfrac{1}{2}\rho AC_D(V - V_w)^2 V. \tag{2}$$

This equation governs all our calculations concerning the effect of wind and altitude on running. The actual speed for a particular race is determined by the power output that the athlete can maintain over the race distance.

There have been several empirical studies[7–11] to determine the values of the parameters in Eq. (2). These show generally that for an athlete sprinting and exerting power P_s, the power lost to air resistance is about one-tenth of the first term. This suggests that athletes could run about 10% faster with a wind at their back which exactly matched their running speed, i.e.,

$$\text{no wind:} \quad P_s = MRV_0 + \tfrac{1}{2}\rho AC_D V_0^3,$$

$$\text{matching wind:} \quad P_s = MRV_1 + \tfrac{1}{2}\rho AC_D(V_1 - V_1)^2 V_1.$$

If $V_1 = V_0 + \Delta V$, one finds

$$\frac{\Delta V}{V_0} = \frac{\tfrac{1}{2}\rho AC_D V_0^2}{MR} = 0.1.$$

To determine the effects of smaller winds, one could differentiate Eq. (2), remembering that the sprinter's total power output P_s should be unaffected by wind. However, the wind velocities of interest are often a substantial fraction of V_0. For this reason it is preferable to solve Eq. (2) analytically, which is possible since it is a cubic equation in V, to obtain the results in Table II.

Table II. Effect of wind and air density on performance in three events: the 200-m dash, the pole vault, and the long jump. These results are calculated by solving Eq. (2) to find the change in velocity under the stated conditions, and then calculating the effect on performance using Eqs. (4) and (5). In addition, the calculation for the long jump includes the correction for the drag during the jumper's trajectory. The calculations assume that at sea level in no wind, the athlete's running speed is 10 m/s, allowing him to run 200 m in 20.0 s, pole vault 6.10 m, and long jump 8.83 m. The table gives the difference between the athlete's performance and these marks under the condition stated.

Wind velocity in running direction (m/s)	200-m dash (seconds)	Pole vault (meters)	Long jump (meters)
+8	−1.675	+0.975	+0.979
+6	−1.415	+0.807	+0.817
+4	−1.045	+0.578	+0.593
+2	−0.571	+0.304	+0.317
0	0	0	0
−2	+0.666	−0.323	−0.351
−4	+1.424	−0.656	−0.728
−6	+2.276	−0.989	−1.126
−8	+3.221	−1.317	−1.539
Air density relative to sea level = 1.00			
0.70	−0.485	+0.257	+0.326
0.75	−0.401	+0.211	+0.269
0.80	−0.318	+0.166	+0.213
0.85	−0.237	+0.123	+0.158
0.90	−0.157	+0.081	+0.104
0.95	−0.078	+0.040	+0.052
1.00	0	0	0
1.05	+0.077	−0.043	−0.051
1.10	+0.153	−0.077	−0.100

Because the effect of changes in air density in Eq. (2) is linear and relatively small, differential methods produce a more reasonable result:

$$P_s = MR(V_0 + \Delta V) + \tfrac{1}{2}(\rho - \Delta\rho)AC_D(V_0 + \Delta V)^3.$$

After expanding, subtracting the terms from Eq. (2), and dropping the higher order terms, one obtains

$$\frac{\Delta V}{V} = \frac{\Delta\rho}{\rho}\left(\frac{1}{MR/\tfrac{1}{2}\rho AC_D V_0^2 + 3}\right) \approx 0.087\frac{\Delta\rho}{\rho}. \tag{3}$$

At Mexico City, where air density is about 25% lower than that at sea level, this predicts that sprint velocity will be improved by 2.2%. An exact solution of Eq. (2) finds that $\Delta V/V$ is 0.0200.

III. POLE VAULT

What height h_{PV} could a jumping human attain if he could convert his kinetic energy developed from sprinting into gravitational potential energy? The change in potential energy $\tfrac{1}{2}MV_0^2$ due to the change in height $\Delta h_{c.m.}$ of his center of mass will be $Mg\Delta h_{c.m.}$, and so

$$h_{PV} = h_{c.m.} + \Delta h_{c.m.} = h_{c.m.} + \tfrac{1}{2}V_0^2/g. \tag{4}$$

Since a man's standing center of mass $h_{c.m.}$ is approximately 1.0 m above the ground, this suggests that with a velocity of 10 m/s he could high jump about 6.10 m, or 20 ft. In practice men can only high jump about 2.3 m (7 ft, 6½ in.), showing that Eq. (4) does not explain the physics of high jumping. This is also clear from observing high jumpers, since they exert no special effort to develop great speed before they jump.

However, Eq. (4) provides a remarkably accurate prediction for performance in pole vaulting. Current vaulting records stand at about 5.8 m, corresponding to a velocity in Eq. (4) of 9.8 m/s. This is pretty good sprinting considering that the vaulter must run while carrying the vaulting pole. Whereas a casual observer might think that pole vaulters use their strength to lift themselves over the bar, in fact the pole is a tool to help them convert kinetic energy to potential energy. A good vaulter needs excellent sprint speed together with the gymnastic ability to effectively utilize the pole. At the top of the vault, some kinetic energy must remain so that the vaulter does not fall into the bar. However, if his lateral velocity at the top is 1.0 m/s, this will only reduce the height predicted from Eq. (4) by about 5.0 cm.

Wind and altitude affect pole vaulting performance because they affect the vaulter's approach speed. Applying differential methods to Eq. (4), we find that if the vaulter can make the necessary gymnastic adjustments to utilize his changed speed,

$$\Delta h_{PV} = V_0 \Delta V/g = 2(h_{PV} - h_{c.m.})\Delta V/V_0.$$

This shows that pole vaulters should be able to vault about 0.21 m (8 in.) higher at Mexico City than at sea level, simply because they can run 2% faster. Similarly, wind also significantly affects optimum vault performance (Table II).

IV. LONG JUMP

A projectile with velocity V_0 will travel the maximum distance on level ground if its initial takeoff angle is $\pi/4$, and this distance is

$$d_{LJ} = V_0^2/g.$$

This suggests that a sprinter running at 10 m/s could jump 10.2 m (33 ft, 5½ in.). This is somewhat longer than the actual world record of 8.9 m, set by Bob Beamon at the 1968 Olympics at Mexico City.

Although this is an attractive model, there are serious problems. At the peak of his jump the athlete's center of mass would reach a height of 2.55 m above its initial position. This suggests that a long jumper could high jump to a height of about 3.5 m (11 ft, 8 in.) by extending his body and diving over the high-jump bar. As stated previously, this is definitely not the strategy used by high jumpers.

Indeed, humans are such inefficient jumpers that they are unable to convert even half of their kinetic energy into potential energy unless they use tools such as a vaulting pole.[12] In fact, the height h_{jump} by which a human can raise his center of mass seems to be an important factor limiting horizontal jumping distance. Suppose we model long jumping (Fig. 1) as a straightforward ballistics problem where the jumper has initial velocity V_0, jumps to raise his center of mass an amount h_{jump} at the apex of his trajectory, and lands after his center of mass has fallen a distance Δh_{LJ} below its initial value. With these assumptions, the distance traveled is

$$d_{LJ} = (V_0^2 - 2h_{jump}g)^{1/2}\{(2h_{jump}/g)^{1/2} + [2(h_{jump} + \Delta h_{LJ})/g]^{1/2}\}. \quad (5)$$

Note that the left-hand part of this expression is the jumper's horizontal speed and the right-hand term is just his time in the air. For h_{jump} equalling 1.3 m and Δh_{LJ} of 0.5 m, the distance d_{LJ} is 9.68 m. However, it is unlikely that a man at full sprint can achieve a h_{jump} equal to that necessary to equal a world record performance high jump. If h_{jump} is 1.0 m, the distance is 9.01 m or 29 ft, 7 in., quite close to actual record performance. In this case the jumper is in the air almost exactly 1.0 s.

At sea level, what is the influence of aerodynamic drag on this performance? Clearly drag affects performance in two ways, by slowing the jumper down after he is airborne and by affecting his initial takeoff speed. After he is airborne, the drag force F_D is given by Eq. (1). For a man at sea level with mass M of 75 kg, this would provide a deceleration of 0.348 m/s^2. Over 1.0 s this reduces the distance jumped by $aT^2/2$, which is 0.174 m or 6.8 in.

Of course, this is only an estimate, since the above calculation assumes that his velocity throughout his path equals his initial, highest velocity, and that the drag force acts only along the horizontal direction. In practice the trajectory includes up and down motion as well. A more exact numerical calculation gives nearly the same result, or 0.184 m. To perform the numerical calculation the author utilized a program developed to study the trajectory of baseballs,[4] but incorporated the same physical parameters used in the approximate calculation above.

Since the reduction of distance ($aT^2/2$) is proportional to the drag force, we can evaluate the effect of changes in wind

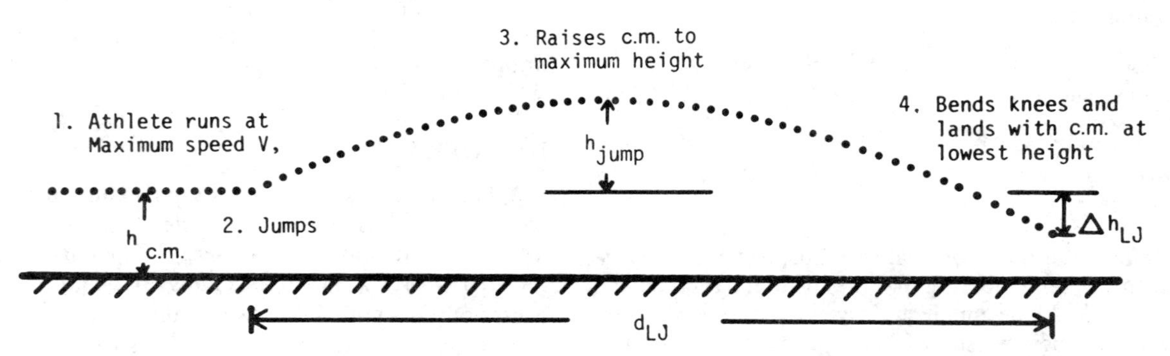

Fig. 1. Principal parameters affecting distance in long jump are the athlete's maximum running speed V, the amount h_{jump} by which he can raise his center of mass by jumping, and the amount Δh_{LJ} that his center of mass falls below the initial value upon landing. In the present model the distance jumped is given by Eq. (5) in the absence of air resistance. In practice air affects both the athlete's initial speed V and his trajectory after takeoff.

velocity on air density. For example, at Mexico City the air density and thus the distance reduction will be less by 25% or 4.5 cm. In effect, an athlete in Mexico City will jump 4.5 cm further than an athlete with identical mass and takeoff speed jumping at sea level.

However, as noted previously, wind and air density also affect a long jumper's approach velocity. If the approach speed is changed by ΔV, this will change the distance Δd_{LJ} of the jump:

$$\Delta d_{LJ} = \frac{\partial d_{LJ}}{\partial V_0}\Delta V = \frac{V_0 \Delta V}{(V_0^2 - 2gh_{jump})} \times \left[\left(\frac{2h_{jump}}{g}\right)^{1/2} + \left(\frac{2(h_{jump} + \Delta h_{LJ})}{g}\right)^{1/2}\right] = \frac{V_0 \Delta V d_{LJ}}{V_0^2 - 2h_{jump} g}.$$

Thus

$$\frac{\Delta d_{LJ}}{d_{LJ}} = \frac{\Delta V}{V_0} \frac{V_0^2}{V_0^2 - 2h_{jump} g}.$$

For an increased speed of 2.0% as calculated from Eq. (2) for Mexico City, there is an increase in distance of 2.49%, or about 0.222 m for Bob Beamon's 8.90-m jump. Together, the contributions from increased takeoff velocity and decreased in-flight drag are about 0.267 m ($10\frac{1}{2}$ in.) suggesting that Bob Beamon's jump of 8.90 m corresponds to a sea-level jump of 8.63 m. A somewhat more exact calculation, proceeding directly from Eq. (5), rather than the differential form in Eq. (6), yields a decrease of 0.269 m (Table II). In any case, these calculations suggest that Carl Lewis' 1983 jump of 8.80 m at sea level is a superior performance to Beamon's 8.90-m jump at Mexico City.

V. LONGER RACES AND CYCLING

As the length of a race becomes longer, the power level that a runner can maintain decreases. Marathon runners can produce power P_m at about half of the level of power produced in sprinting. However, because the power dissipated to air drag in the absence of wind goes as the cube of the running velocity [Eq. (2)], a smaller fraction of their energy is spent in overcoming air resistance. For example, if there is no wind and the runner exerts power $P_m = 1.5$ kW, the velocity is 5.35 m/s and the second term in Eq. (2) is about 3% as large as the first term.

Furthermore, for various reasons neither wind nor increased altitude is likely to improve the runner's time. In long races the runners usually run around a track or loop course, and the effect of running against the wind over a portion of the loop is always larger than the benefit gained by running with the wind over another portion (see Table II). The effect of altitude is more complicated. Although the runner does face decreased air resistance, he must breathe air at lower pressure. For physiological reasons[13] this reduces the power he can exert, resulting in slower performances at higher altitudes for all races longer than about 800 m.

Several studies[7,14] have shown that running directly behind another runner reduces the aerodynamic drag force to about one-fifth to one-half of the force experienced by the front runner. Thus if runners are of approximately equal ability, the runner who follows throughout most of the race will win because he has saved more of his energy for the final sprint. This becomes especially important if there is a stiff wind.

Although the energy saved by following is only a few per cent of the total energy expended in running, this apparently becomes a significant factor affecting strategy in middle distance races such as the 1500-m race. Races between world-class athletes are generally slow, tactical races because none of the runners wishes to lead, or if one is forced to lead he will run slowly to conserve energy for the final sprint. Record-setting races are nearly always run when there is only one world-class runner and when he employs a "rabbit" to lead and set the pace for about three-quarters of the race. The energy thus saved allows the following runner to run the final lap a few seconds faster than otherwise would have been possible. Although scientists are in universal agreement about the value of shielding, top runners often attribute the value of the "rabbit" to reducing the psychological effort necessary to set a record-level pace.

However, there is total agreement about the value of shielding in competition bicycling.[14–16] Since cycling speeds are more than twice running speeds over the same distance, the drag force from Eq. (1) on a leading cyclist is approximately eight times higher than on a runner, while a following cyclist reduces his drag by at least one-third.[14] Cyclists further reduce drag by leaning forward to reduce their frontal area. Experimental bicycles not allowed in regular competition also utilize fairings shaping the air flow and sharply reducing the drag coefficient C_D (see Ref. 16). However, in competition the drag forces are still so large that it becomes nearly impossible for a lone cyclist to outrun two or more cyclists who work together as a group. For this reason, virtually all of the strategy in competition cycling concerns avoiding air drag by "drafting" or following other cyclists, and trying to "break away" from the main pack with a small group of elite cyclists who together have the ability to stay well in front of the pack.

VI. DISCUSSION

There exists an enormous literature on the biomechanics of track and field events (e.g., see Hay[17]), however, most of this material is concerned with the details of technique rather than the basic physics underlying performance. In the present analysis the models presented ignore technique and concentrate on describing optimum performance in terms of simple observable quantities such as the athlete's sprint speed, power expenditure, vertical jump, etc. In spite of their simplicity, the models do agree quantitatively with the observed performance of world class male athletes. They also give us an appreciation for world class performances in the pole vault and long jump, where athletes are achieving marks very close to the optimum predicted from their sprint speed.

Notwithstanding the success of the models presented here, in some ways they are oversimplifications. For example, Hay[18] summarizes results of film analyses of world class long jumpers. The films show that long jumpers use a lower, flatter trajectory than assumed in the present paper, achieving values of h_{jump} of only about 0.5 to 0.7 m. In addition, their measured jump length exceeds the distance traveled by their center of mass during flight because they have a slight forward lean on takeoff and a pronounced backward lean upon landing. Furthermore, although Eq. (2) describing running speed is based on sound experimental observations of oxygen consumption, its descriptive nature gives little insight into the physical processes controlling running. Our model for pole vaulting ignored the possibility that energy could be lost to the pole, or that a man might use arm strength to push upward off the pole near the top of his trajectory. Indeed, it is even possible for a man's body to clear a vaulting bar while his center of mass passes beneath the bar. However, although the models are perhaps too simple, the simplifications do not significantly affect the calculations concerning the effect of wind and altitude on performance.

As much as possible the model parameters used in the calculations (Table I) were selected to be simple, whole numbers. One justification for this was because whole numbers are more convenient for classroom examples. However, another reason is that published measurements of many of the parameters such as A, C_D, P_S, and P_M cover a range of values. Although the author has made some analytical approximations in the calculations, these approximations cause smaller systematic errors than the uncertainties in the parameters themselves. Students may find it of interest to vary the parameters to model performance of other types of athletes, such as female athletes or "average" high school athletes. Students could also modify Eq. (2) to predict the performance of runners racing up hills. They may also wish to extend the models to make them more realistic using information from descriptive sources such as Hay.[18]

The results of the present study suggest that one can use elementary physics to evaluate the effect of environmental conditions on record performances in some track and field events. Strangely enough, the rules are not consistent concerning whether performances aided by wind are acceptable for records. In the sprints and long jump, records cannot be accepted if the performance occurred with a following wind exceeding 2.0 m/s. However, performances made at any altitude are acceptable for records. For the pole vault, a performance in any wind or at any altitude is acceptable.

However, the results of Table II suggest that it is appropriate that the rules ignore the influence of altitude on records. Even at the altitude of Mexico City, the improvement in performance over that at sea level is slightly less than the improvement caused by a 2.0-m/s following wind. Variations in air temperature and humidity also affect air density. Equation (2) can be used to calculate how this influences performance as well. However, for all realistic situations this influence will be significantly smaller than that caused by a 2.0-m/s wind.

[1]D. B. Lichtenberg and J. G. Wills, Am. J. Phys. **46**, 546 (1978).
[2]L. J. Briggs, Am. J. Phys. **27**, 589 (1959).
[3]R. G. Watts and E. Sawyer, Am. J. Phys. **43**, 960 (1975).
[4]C. Frohlich, Am. J. Phys. **52**, 325 (1984).
[5]H. Erlichson, Am. J. Phys. **51**, 357 (1983).
[6]C. Frohlich, Am. J. Phys. **49**, 1125 (1981).
[7]L. G. C. E. Pugh, J. Physiol. **213**, 255 (1971).
[8]A. V. Hill, Proc. R. Soc. London B **102**, 380 (1928).
[9]L. G. C. E. Pugh, J. Physiol. **207**, 823 (1970).
[10]J. R. Shanebrook and R. D. Jaszczak, Med. Sci. Sports **8**, 43 (1976).
[11]C. T. M. Davies, J. Appl. Physiol. **48**, 702 (1980).
[12]P. Kirkpatrick, Am. J. Phys. **25**, 614 (1957).
[13]L. G. C. E. Pugh, J. Physiol. **192**, 619 (1967).
[14]C. R. Kyle, Ergonomics **22**, 387 (1979).
[15]C. T. M. Davies, Eur. J. Appl. Physiol. **45**, 245 (1980).
[16]F. R. Whitt and D. G. Wilson, *Bicycling Science* (MIT, Cambridge, MA, 1982).
[17]J. G. Hay, *A Bibliography of Biomechanics Literature* (published privately, Muscatine, IA, 1980), 4th ed.
[18]J. G. Hay, *The Biomechanics of Sports Techniques* (Prentice-Hall, Englewood Cliffs, NJ, 1978), 2nd ed.

Could an athlete run a 3-m radius "loop-the-loop?"

Question:

Everyone has observed carnival rides, cyclists, skateboarders, etc., do a "loop-the-loop." The theoretical minimum velocity is given from the equality of the gravitational and centripetal forces as $v = \sqrt{gR}$. For a loop of radius $R = 3$ m this is approximately 5.5 m/s. Could an athlete run such a 3-m radius "loop-the-loop?" Good sprinters can run well above 5.5 m/s, perhaps 9 m/s.

This interesting question was asked by **Kevin Braun**, *a physics student of* **Fr. Earl W. Meyer** *at Thomas More Prep, Hays, Kansas 67601. For our answer we turned to* **Angelo Armenti, Jr.** *of Villanova University. Professor Armenti teaches a course in the physics of sports, which has been described in this journal [12, 349 (1974)].*

Answer:

On a qualitative level, it is clear that the runner must slow down as he runs uphill since a portion of his essentially constant power output would have to be expended doing work against gravity and hence would not be available to overcome the higher internal resistance forces associated with higher running speeds. Consequently, the runner's speed will suffer on the loop, especially near the top, possibly to the point where contact with the track would be lost. Even if this critical situation is not reached, there is an additional problem associated with the friction between runner and track needed for propulsion. As the runner's speed decreases, so too will the normal force between track and runner. This in turn will reduce the frictional force between the runner and the track with slipping the likely result. On the qualitative level then, it would appear doubtful that an athlete could successfully run the loop.

On a quantitative level, a model is needed to describe the physics of sprinting. A very useful model is that of Alexandrov and Lucht.[1] These authors assume a constant propulsive force on the sprinter F_p along with a variable resistive force F_r where these forces are given by

$$F_p = fM, \qquad \text{and} \qquad F_r = -\sigma M v$$

M is the mass of the sprinter, v the instantaneous speed of running, and the two parameters f and σ characterize the athlete in question. Clearly, F_r is related to the internal mechanics of the human body rather than air resistance or any other external resistive force. Typical values for world class sprinters are

$$f = 10 \text{ N/kg} \qquad \text{and} \qquad \sigma = 0.9 \text{ s}^{-1}$$

We will use these values to make predictions. Note that, according to the model, the *maximum* running speed of the athlete is

$$v_t = f/\sigma \simeq 11 \text{ m/s}$$

for the parameters we have chosen; the average running speed would be somewhat smaller.

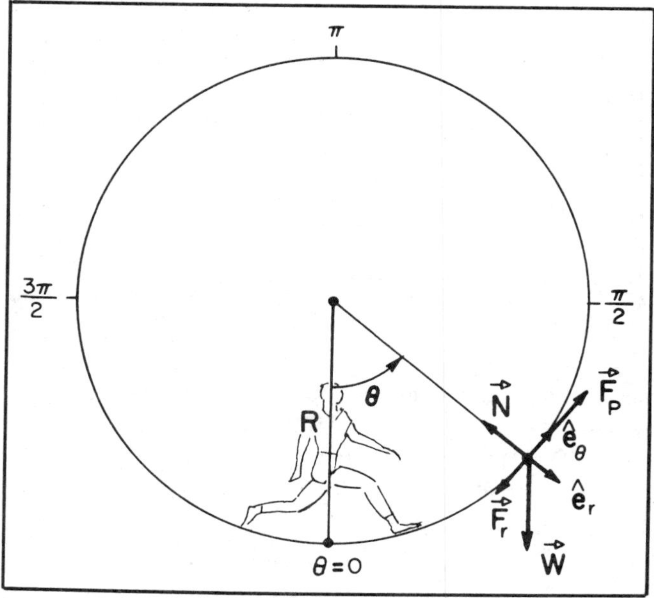

Fig. 1. Force diagram for a runner on a "loop-the-loop." The forces on the runner are $\vec{F}_p$, the propulsive force, $\vec{F}_r$, the internal resistance force, $\vec{N}$, the normal force exerted by the track, and $\vec{W}$, the weight of the runner. $\hat{e}_\theta$ and $\hat{e}_r$ are, respectively, the tangential and radial unit vectors.

In Fig. 1, we show the force diagram for the runner. $\hat{e}_\theta$ is the unit vector tangent to the loop in the counterclockwise sense while $\hat{e}_r$ is the unit vector normal to the loop in the radially outward direction. The net force on the sprinter in the $\hat{e}_\theta$ direction is

$$f_\theta \equiv |F_p| - |F_r| - Mg \sin\theta \qquad (1)$$

while the net force on the sprinter in the $\hat{e}_r$ direction is

$$f_r \equiv Mg \cos\theta - N \qquad (2)$$

Now the radius vector from the center of the loop to the runner is given by

$$\vec{r} = R\,\hat{e}_r$$

The velocity of the runner is given by

$$\vec{v} = \frac{d\vec{r}}{dt} = R\,\dot{\theta}\,\hat{e}_\theta \qquad (3)$$

while the acceleration of the runner is given by

$$\vec{a} = \frac{d\vec{v}}{dt} = R\,\ddot{\theta}\,\hat{e}_\theta - R\,\dot{\theta}^2\,\hat{e}_r \qquad (4)$$

where

$$a_\theta \equiv R\,\ddot{\theta}$$

is the tangential component of the acceleration and

$$a_r \equiv -R\,\dot{\theta}^2$$

is the radial component of acceleration (i.e., centripetal

Reprinted from *The Physics Teacher* **19**, 624–625 (1981); © American Association of Physics Teachers.

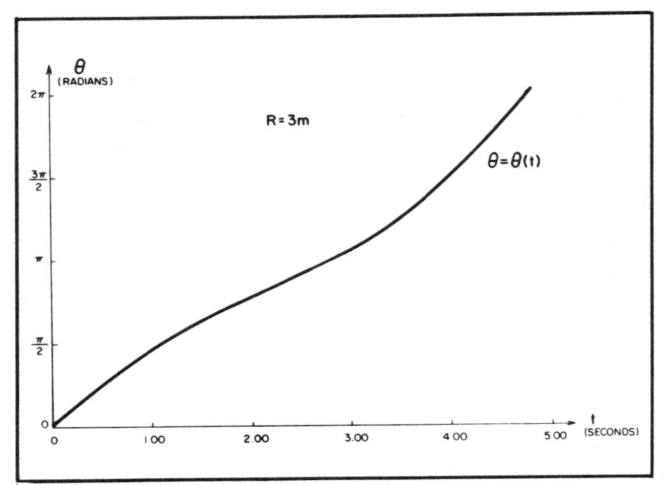

Fig. 2. Numerical solution $\theta = \theta(t)$ for Eq. (7). The initial data used were v_o and $R\dot{\theta}_o = 9$ m/s and $\theta_o = 0$ at $t = 0$, with $R = 3$ m.

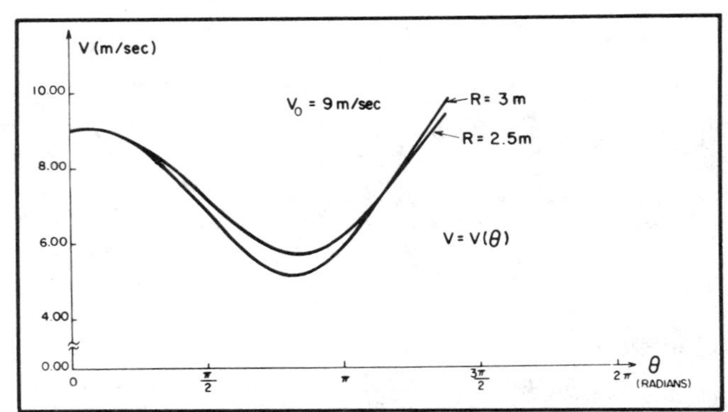

Fig. 3. Speed versus position on the loop for $R = 3$ m and $R = 2.5$ m. For the $R = 3$ m case, the speed falls below the critical value given by $v = \sqrt{gR} \simeq 5.5$ m/s. For the $R = 2.5$ m case, the runner's speed exceeds the critical value ($v \simeq 4.9$ m/s) at all times.

acceleration) and $\dot{\theta}$ and $\ddot{\theta}$ are, respectively, the first and second derivatives of θ with respect to time.

From Newton's second law we have that

$$f_\theta = Ma_\theta \quad \text{and} \quad f_r = Ma_r$$

Hence,

$$fM - \sigma MR\dot{\theta} - Mg \sin\theta = MR\ddot{\theta} \tag{5}$$

and

$$Mg \cos\theta - N = -MR\dot{\theta}^2 \tag{6}$$

These two equations may be rearranged to read

$$\ddot{\theta} = \frac{1}{R}(f - \sigma R\dot{\theta} - g \sin\theta) \tag{7}$$

and

$$\eta \equiv \frac{N}{Mg} = \cos\theta + \frac{R\dot{\theta}^2}{g} \tag{8}$$

Eq. (7) may be integrated numerically. (We used the Initial Half-Step Method[2] on a Hewlett-Packard 9830 Programmable Calculator/Plotter.) The numerical solution to this nonlinear differential equation is given in Fig. 2.

Figure 3 shows the speed profile of the runner around the loop. For the 3-m loop, the speed drops below the critical value $v = \sqrt{gR}$ before the top of the loop is reached. For a 2.5-m loop, the speed stays above the critical value. However, there are other problems as can be seen in Fig. 4.

In Fig. 4 we show the normal force (in units of Mg) as a function of θ. Note that for the 3-m loop, the normal force goes to zero before θ reaches the value $\theta = \pi$. For the 2.5 m loop, η stays positive but falls below unity. This almost certainly would mean a loss of the necessary frictional force required for propulsion.

Clearly, the propulsive force on the runner is related to the frictional force between the runner and the track. In fact, one can argue that

$$F_p \leq \mu N$$

where μ is the coefficient of static friction between the track and the runner's shoes. At the point where slipping begins,

$$F_p = fM = \mu N$$

Hence,

$$\eta = N/Mg = f/\mu g$$

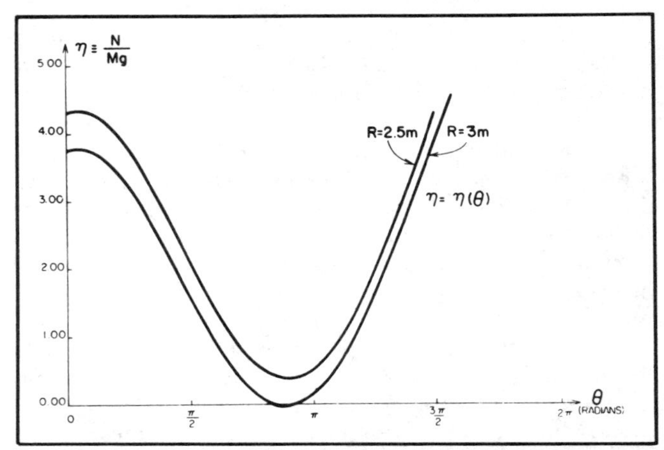

Fig. 4. Normal force (in units of Mg) versus position on the loop for $R = 3$ m and $R = 2.5$ m. For the $R = 3$ m case, η falls to zero (runner loses contact with track). For the $R = 2.5$ m case, η stays positive but falls below unity (runner loses traction and begins to slip).

Since f is on the order of 10 N/kg and $g = 9.8$ m/s^2, the critical value of normal force is on the order of

$$\eta_c \simeq 1/\mu$$

If the traction is excellent, $\mu \cong 1$ and the minimum acceptable value of η is

$$\eta_c \simeq 1$$

The point of all this is that, even if the necessary speed is maintained, the normal force can fall below that needed to maintain traction — and the runner fails. Notice also that the "g-forces" involved are considerable.

So, our answer to the question "Could an athlete run a 3-m 'loop-the-loop' " is "*Probably (and almost certainly) Not*".

References

1. Igor Alexandrov and Philip Lucht, "Physics of Sprinting," Am. J. Phys. **49**, 254 (1981).
2. John R. Merrill, *Using Computers in Physics* (Houghton Mifflin, Boston, 1976), Ch. 1.

Fast Running Tracks

On a springy new indoor track at Harvard University runners can run faster than they can on standard tracks. The design of the track was arrived at through a close analysis of the mechanics of human running

by Thomas A. McMahon and Peter R. Greene

In October of last year Harvard University opened a new indoor sports facility housing a six-lane 220-yard running track. The track has a supporting substructure consisting primarily of wood and a synthetic surface covering. Most people find running on the track a pleasant and unusual experience. It feels particularly springy and therefore is comfortable to run on, and members of the Harvard track team have sustained considerably fewer injuries since they began practicing on the track. What is most remarkable, however, is that an analysis of the 1977–78 records of Harvard runners and their competitors shows that the track substantially enhances a runner's performance, reducing the time of, say, a well-run mile by several seconds. By all estimations the new track is fast.

One of the most important parameters of track design is compliance, and it is mainly in this respect that the new track differs from other indoor tracks. Just how compliant—or, alternatively, how stiff—should a track be? In 1976, when the new track was being designed, the Harvard track coach, Bill McCurdy, and the Harvard Planning Office sought our advice on this question. At that time there was no information available concerning the effect of a given track surface on running speed and comfort. It is intuitively clear, however, that running on a compliant surface such as grass is less conducive to injury than running on a hard surface such as asphalt. Indeed, if comfort were the only consideration, all tracks would be extremely springy, perhaps as springy as a diving board. Intuition also suggests, however, that on a springy track the time spent rebounding from the surface is increased, so that a runner is slowed down. Hence if a running track is to be fast, its compliance must be limited.

One might therefore suppose the hardest track surfaces are the fastest, but that is not the case. A compliant surface acts as a spring, and we found that if the stiffness of the spring is closely tuned to the mechanical properties of the human runner, the runner's speed can be increased. In other words, there is a specific intermediate track compliance at which running speed is optimized. At that optimum compliance the runner stepping down onto the track stores elastic energy in its surface, much as a pole vaulter stores elastic energy in bending a fiberglass pole, which is recovered as the pole propels him upward. A track whose compliance is in the theoretically optimum range can be called a tuned track.

A tuned track is special only in that its properties are particularly compatible with those of the runner. Hence our first step in designing such a track was to develop a model of the mechanics of running, one that was complex enough to be realistic but simple enough to be mathematically manageable. Running speed is determined by several parameters. With the physiological simplifications introduced in the model it was possible to derive mathematical expressions for those parameters as functions of track stiffness. In this way we were able to predict the effect various track compliances would have on running speed. Our final step was to test the model predictions with runners and experimental tracks. The ultimate test was provided by the new Harvard track, whose wood substructure is designed to achieve a compliance in the range we found to be optimum. The results of these tests, which are discussed below, are in good agreement with the theoretical predictions.

Running is essentially a series of bounds. As the runner's foot strikes the running surface the antigravity muscles that support his skeleton contract and ultimately reverse the downward velocity of his body. To understand how our model was devised and why the tuned track works, it is necessary to understand how the physiological properties of these muscles and the reflexes that control them determine the functioning of the runner as a mechanical system.

Consider first an isolated muscle, removed from an animal and placed in an apparatus that measures its force at a constant length. With its blood supply cut off, a muscle from a mammal such as a rat will quickly fatigue and die, but in a suitably oxygenated bath a muscle from a small amphibian such as a frog will remain alive for hours or even days. A muscle that is not being stimulated to contract is said to be resting. When the force developed in the resting muscle is plotted against the length of the muscle, the force-length relation curves upward, that is, the force increases with length. The slope dF/dx of the force-length curve is the stiffness of the muscle, and the curve shows that it too increases with length. (Resting muscle is stiffer at greater lengths for the same reason a nylon stocking feels stiffer as it is stretched. As the muscle is lengthened more of the collagen fibers in the connective tissue associated with it become taut and begin bearing load.)

If a resting muscle is given a small electric shock, either directly or through the nerve that supplies it, the force in the muscle rises above the resting level and then quickly drops. This momentary rise in force, called a twitch, is the basic event in muscle contraction. In a sustained activity such as running, a train of twitches can merge at an elevated level of force; in this state the muscle is said to be tetanized. The force-length curve for tetanized muscle shows that, as in resting muscle, force increases with length. The same property is exhibited by a mechanical system: a coiled spring. It was this fact that suggested to us a model of running in which the muscles would be represented by springs.

The difference in force between tetanized muscle and resting muscle is termed developed tension. An important phenomenon having to do with developed tension is observed when tetanized muscle is placed in another type of apparatus: one in which the muscle is allowed to shorten and pick up a weight so that its shortening velocity can be determined. This experiment reveals that the shortening velocity does not increase throughout the shortening period; rather, after a brief starting period the mus-

Reprinted from *Scientific American* **239(6)**, 148–163 (1978); © Scientific American, Inc.

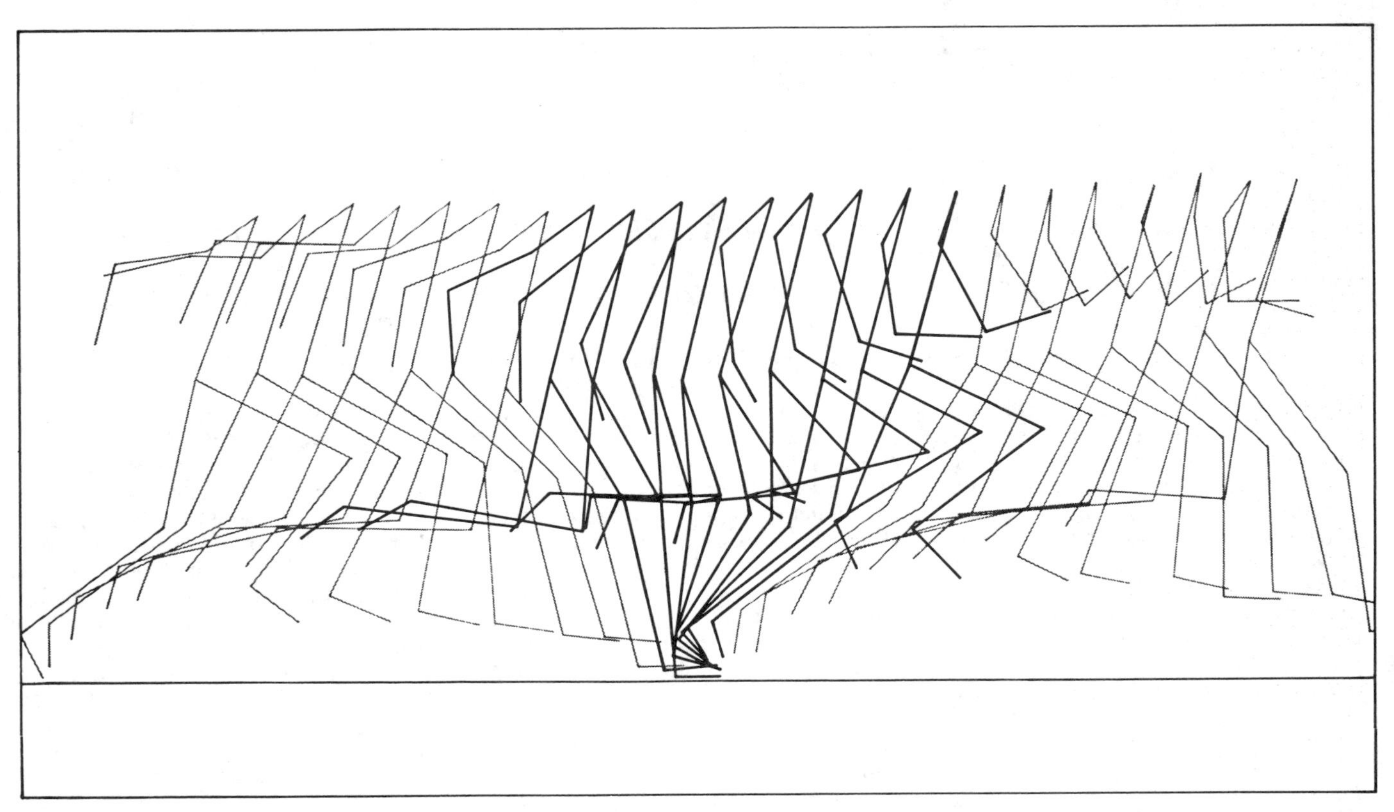

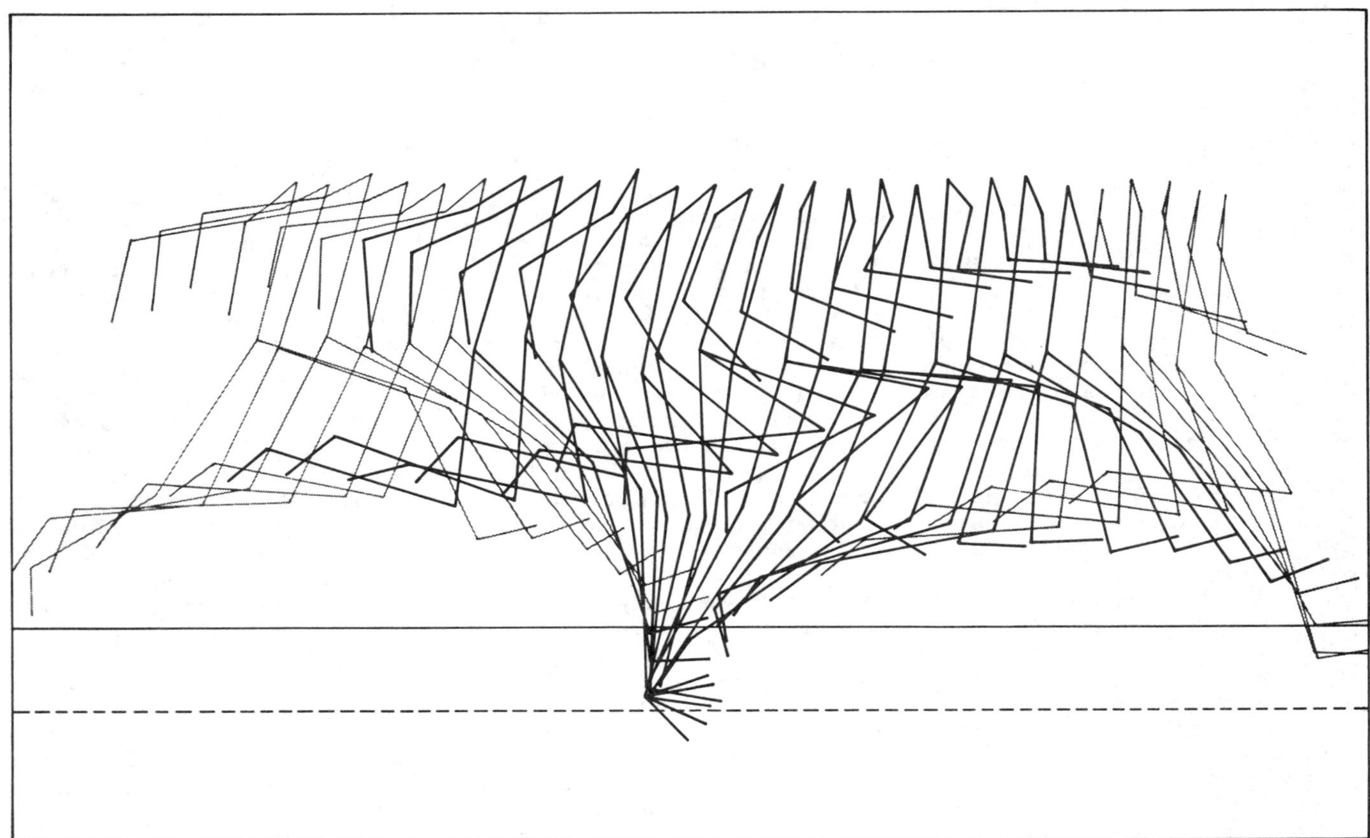

COMPUTER SEQUENCES showing running positions are obtained by digitizing the points marking a runner's major limb joints in a motion picture of running and then connecting those points on a digital plotter. In the sequences shown here the points mark the runner's right hip, shoulder, ear, elbow, wrist, knees, ankles and big toes. The same runner is shown on a hard, or stiff, surface (*top*) and on a soft, or springy, one (*bottom*). Running speed can be viewed as the ratio of the runner's step length (the distance the body travels while one foot is on the ground) to his ground-contact time (the time the foot is on the ground). The sequences show that a runner's step length and ground-contact time are both increased on a soft surface, with the net result that his running speed is decreased. A very hard surface, however, is not the fastest. The authors found that there is a range of intermediate track stiffnesses at which running speed is enhanced. The computer analysis for this illustration was performed in the Gait Analysis Laboratory of the Children's Hospital Medical Center in Boston.

cle shortens at a constant speed. When the shortening velocity is measured with different weights, it becomes apparent that muscle shortens more rapidly when it is working against lighter weights. It is as if a damping element—a dashpot, or shock absorber—were subtracting some of the developed tension. When the damping phenomenon was first discovered some 50 years ago, it was suggested that something like a mechanical dashpot was actually present in muscle, but it is now clear that the damping effect is a property of the force-generating mechanism itself, specifically of the interaction of actin and myosin, the contractile proteins of muscle.

Hence an isolated muscle shortening against a constant load behaves like a damped spring, or, more precisely, like a stretched spring acting in parallel with a dashpot. In the body, however, the functioning of the muscles is controlled by the nervous system. It is conceivable that under such conditions the spring-and-dashpot system might not apply. What we found is that with certain elaborations the model derived from the inherent properties of isolated muscles also describes muscles and reflexes as they function in the feedback control system of the body. (It should be noted that a mathematically linear spring, which has constant stiffness at any length, is utilized in the model, even though an isolated muscle acts as a mathematically nonlinear spring, which has variable stiffness. Recent experiments concerning reflexes, which we shall discuss, serve to justify this simplification.)

Muscle fibers receive their commands, or electrochemical stimuli, directly from the specialized nerve cells in the spinal cord called alpha motoneurons. The traffic of impulses along the axon, or long fiber, of an alpha motoneuron is efferent: it moves outward from the spinal cord and the rest of the central nervous system. Information comes back to the central nervous system along afferent pathways whose nerve-cell bodies are found in the dorsal-root ganglion: a swelling of the peripheral nerve just outside the spinal cord. The fibers of these nerve cells terminate in receptors that are specialized to respond most effectively to certain types of stimuli. Receptors in the loco-

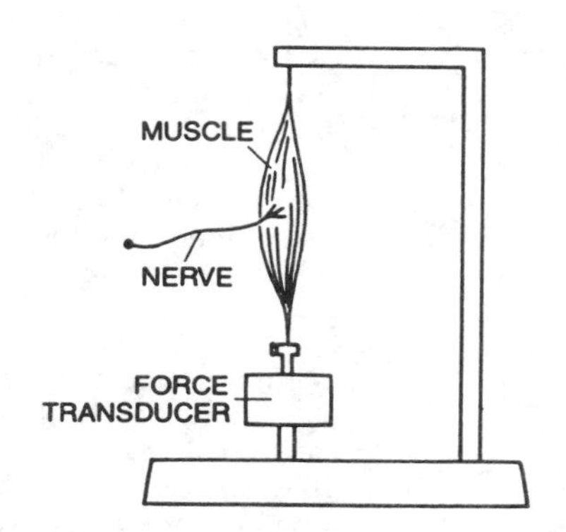

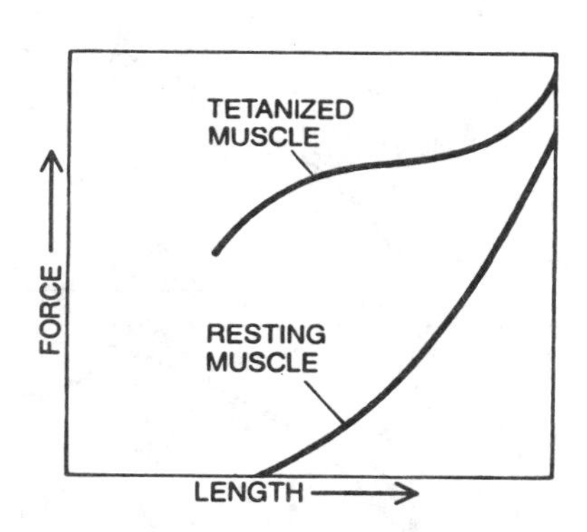

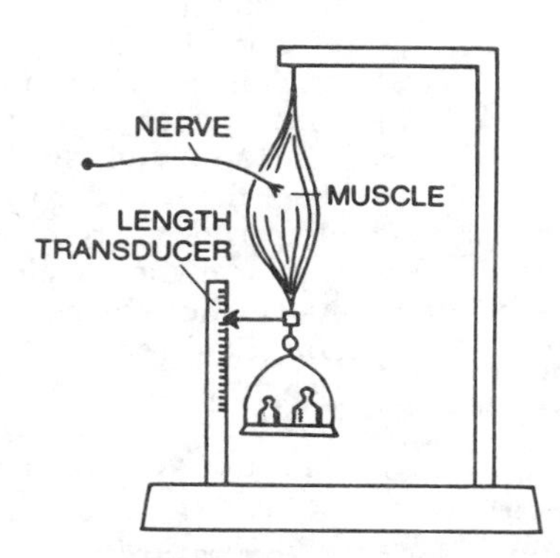

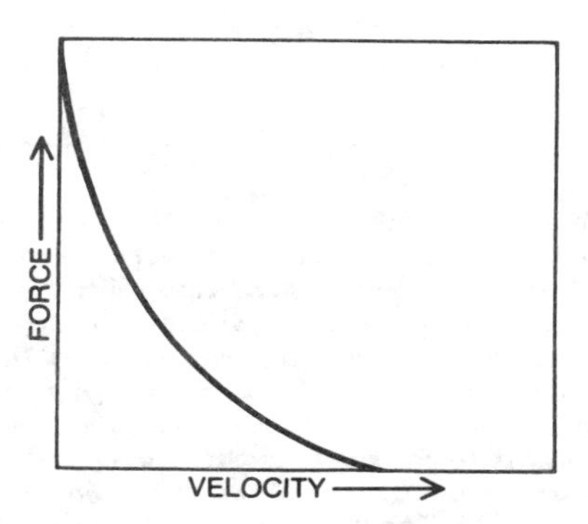

MUSCLE REMOVED FROM THE BODY of a frog and held at co[illegible] length in an apparatus that measures force (*top*) develops more force at greater length[illegible]her it is resting (engaged in minimal electrochemical activity) or tetanized (engaged in sustained electrochemical activity). A mechanical device that develops more force at greater lengths is a coiled spring. When the muscle is allowed to shorten and pick up a weight so that its shortening velocity can be measured (*bottom*), it is discovered that if the amount of force developed in the muscle (the weight lifted) is decreased, the shortening velocity is increased. The mechanical device that creates such a damping effect is a dashpot, or shock absorber. In the authors' mechanical model they represented muscles employed in running as spring connected in parallel with dashpot.

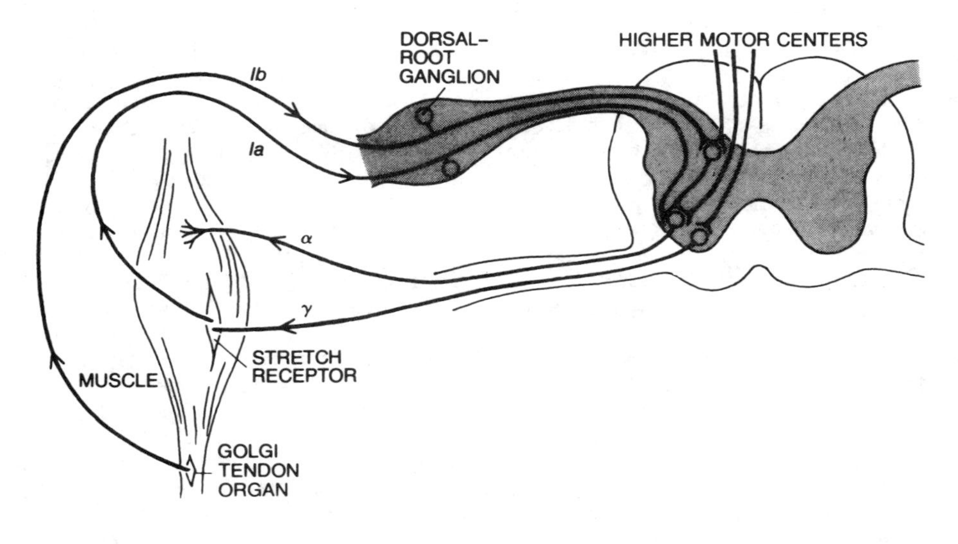

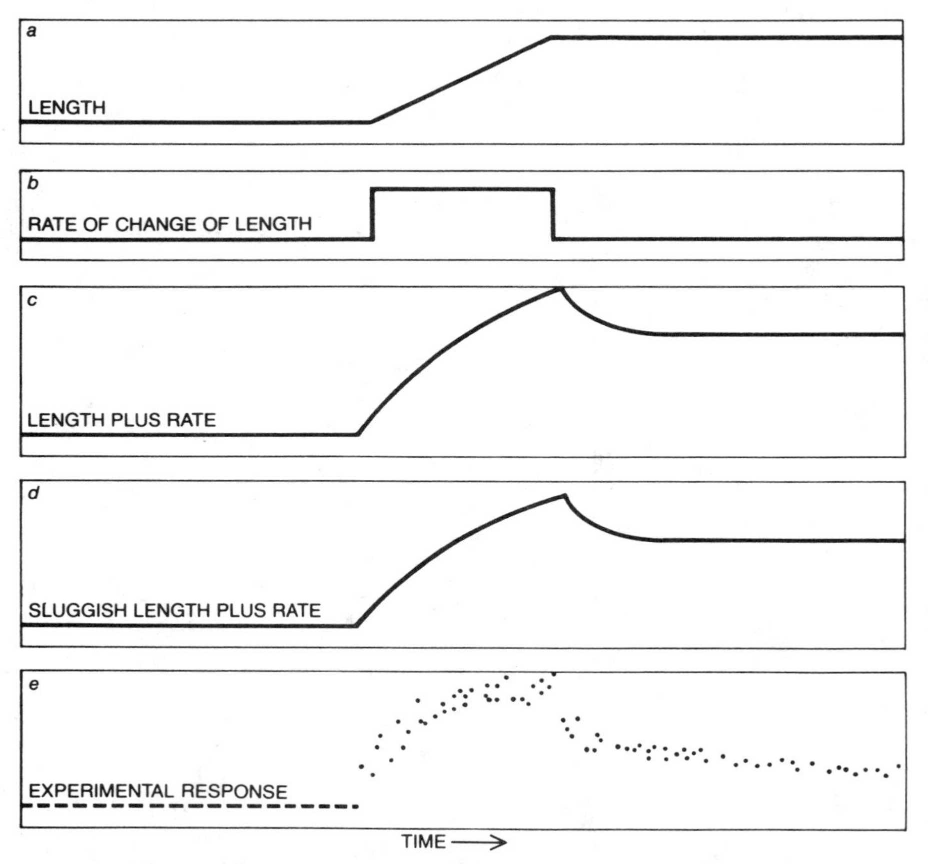

MUSCLES IN THE BODY are regulated by the feedback-control system of the reflexes: commands in the form of electrochemical impulses are sent to the muscles along the axon, or long fiber, of the specialized nerve cells in the spinal cord called alpha motoneurons; information about muscle activity is sent back to the central nervous system by stimulus-sensitive receptors. Impulses from the receptors travel along the fibers of nerve cells found in the part of the peripheral nervous system known as the dorsal-root ganglion. These pathways can be divided into two types: the *Ia* pathways innervate stretch receptors (*shown greatly enlarged here*), which are primarily sensitive to changes in muscle length, and the *Ib* pathways innervate the receptors called Golgi tendon organs (*also shown greatly enlarged*), which are primarily sensitive to changes in muscle force. The stretch receptors also act as muscles, innervated by the nerve cells called gamma motoneurons. In addition the stretch receptors are rate sensitive, that is, they are responsive to the rates of change in their stimuli. The rate sensitivity of the reflexes is another way the muscle system is damped. The graphs in the lower part of the illustration show how this phenomenon affects the stretch receptors called spindle organs. When a spindle organ is steadily lengthened starting from a resting position, the magnitude of its response is proportional not to the sum (*c*) of the "ramp" function representing the change of length (*a*) and the step function representing the change of rate (*b*) but to a damped, or sluggish, version of the sum (*d, e*). These findings suggest that dashpot introduced into the model of running to represent the force-velocity relation of muscle also serves to represent the rate sensitivity of reflexes. Graphs are adapted from work of Tristan D. M. Roberts of University of Glasgow.

motor system provide the central nervous system with the information that makes smooth movement possible. In control engineering the route from a receptor to the central nervous system over afferent pathways and from the central nervous system to an effector organ (such as a muscle) over efferent pathways is called a feedback loop; in physiology it is called a reflex arc.

For example, when a physician strikes the large tendon below a patient's kneecap with a hammer, he is testing the muscle-control system known as the stretch reflex. When the hammer strikes the tendon, the quadriceps group of muscles is stretched momentarily. As the muscles lengthen, stretch receptors in them (spindles) also lengthen and transmit a burst of impulses. This information travels along an afferent pathway to the quadriceps alpha motoneurons, eliciting the kick that tells the physician the patient's neuromuscular wiring is in good condition.

Two groups of afferent pathways designated *Ia* and *Ib* are the most important ones in the muscle-control system. The *Ia* pathways innervate the stretch receptors. Since stretch receptors are connected in parallel with muscles, they are primarily sensitive to muscle length. For example, in the stretch-reflex test an increase in the length of the quadriceps muscles sends information along an afferent *Ia* pathway. The *Ib* pathways innervate the receptors known as Golgi tendon organs, which are mainly sensitive to muscle force.

Stretch receptors respond to dynamic stimuli as well as static ones: they are sensitive to the rates of change in the strength of stimuli as well as to the changes themselves. In mechanical terms the stretch receptors encode and send back to the spinal cord not only position information but also velocity information.

Consider a spindle-organ receptor that is being steadily lengthened. The rate of change in length begins at zero, rises to a constant level and then falls back to zero when the lengthening stops. Experimental results show that the magnitude of the spindle response is proportional not to the sum of the "ramp" function representing the change in length and the step function representing the rate of change in length but to a sluggish version of the sum. The sluggishness is due partly to a property of all muscles, including spindles, known as series elasticity, and partly to the dashpot property of the receptors. In other words the sluggishness is an intrinsic property of the rate sensitivity of the reflex.

The spindle organ is itself composed of muscle cells and is separately innervated by the nerve cells called gamma motoneurons. Hence the length at which

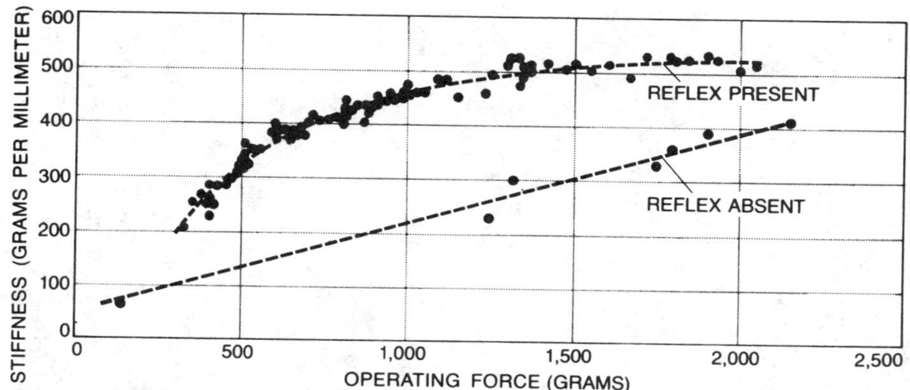

STIFFNESS OF AN ISOLATED MUSCLE (the change in the force developed by the muscle when it is stretched, divided by its change in length) increases as the muscle is stretched. Recent results indicate that in the body the reflexes maintain the stiffness of an actively contracting muscle at an almost constant level. J. A. Hoffer and S. Andreassen of the University of Alberta measured the increase in the stiffness of the cat soleus muscle (a plantar-flexor muscle in the ankle) as a function of force. When the pathways leading from the receptors back to the central nervous system were severed, the stiffness rose in proportion to the force exerted by the muscle (*broken black line*). With those pathways intact, however, the muscle stiffness was higher. (A closed-loop negative-feedback control system is always stiffer than an open-loop one.) After rising with low forces the stiffness reached a plateau at moderate and high forces (*broken color curve*). These findings suggest that spring, which has constant stiffness, is a suitable representation of the muscles involved in running, particularly in moderate-to-large range of forces.

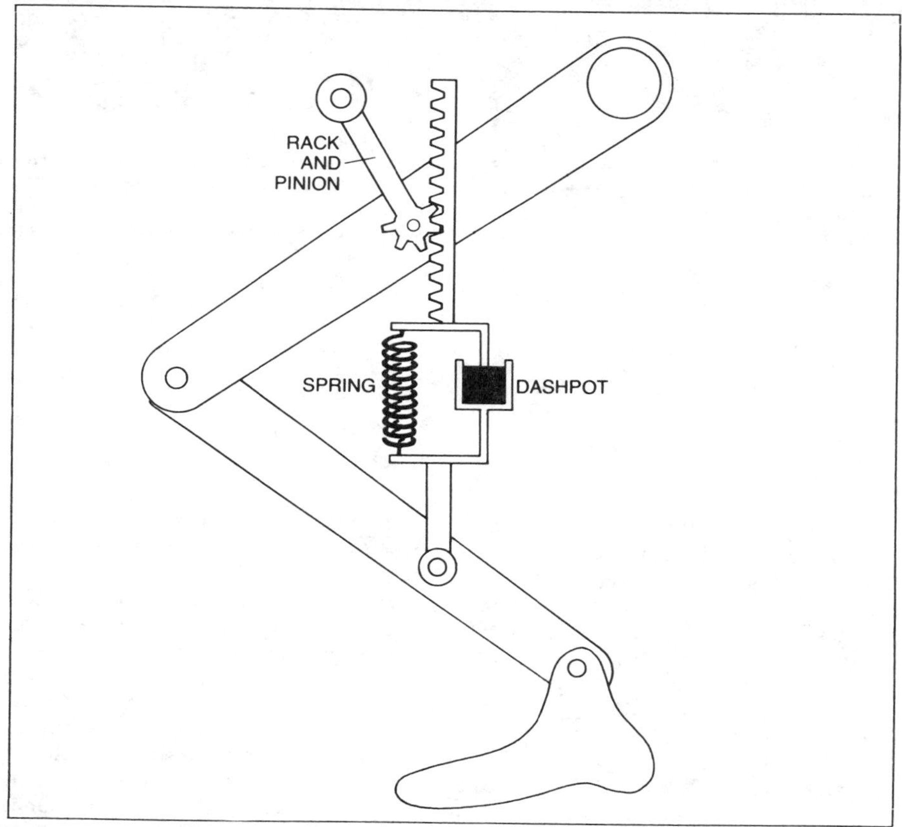

MODEL OF THE MUSCLES AND REFLEXES involved in running consists of a rack-and-pinion element connected in series with a damped spring. The dashpot damping effect is created by a piston moving in a cylinder containing a viscous fluid such as oil (*color*). In the model the rack and pinion represents the role of movement commands from higher motor centers, the spring represents the muscle stiffness (assumed to be constant) and the dashpot represents both the force-velocity relation of muscle and the rate sensitivity of the reflexes. In other words, the setting up of leg positions is modeled by the motion of the rack and pinion, but the reaction to external force disturbances, such as the impact of the track on the contact leg, is modeled by the deflection of the damped spring. Hence the leg and hip positions at the beginning and end of the ground-contact phase, which determine step length, depend only on the movement of the rack and pinion, whereas the body motion during that phase, which determines ground-contact time, depends only on the deflection of the damped spring. With these simplifications, once runner's inherent spring stiffness and damping constants have been determined experimentally, effect of track stiffness on ground-contact time and step length can be calculated.

it sends a given intermediate frequency of impulses to the spinal cord can be changed by input from the higher motor centers. That makes the spindle organ a specialized type of strain gauge: a length-reporting device that converts mechanical stretch into nerve impulses, which in this case can be recalibrated to accommodate large-scale changes such as a change in the position of the body.

Consider again the stretch-reflex arc, where the final response to an imposed lengthening of muscle was the patient's kick reaction. Because the imposed muscle lengthening ultimately produced this muscle shortening, the reflex arc is a negative-feedback control system. Both the length and the rate of change of the length of the quadriceps muscles are reported along the *Ia* pathway as negative-feedback signals. Considering the mass of the leg, one can correctly predict that the normal overall response to the hammer blow will be a heavily damped, decaying oscillation of the knee angle, with the velocity feedback from the muscle giving rise to the damping effect. If the stretch receptor were sensitive only to muscle length, there would be no feedback damping, and in the absence of friction the oscillation would continue indefinitely.

What is the control function of the Golgi tendon organs, the group of receptors sensitive to muscle force? For many years it was believed their only role was to turn off the alpha-motoneuron activity and cause a muscle to relax whenever a critical muscle force, which might be capable of injuring the muscle, was exceeded. Newer evidence shows, however, that feedback from the Golgi tendon organs may also help to modify the stretch reflex in such a way that muscle stiffness, rather than muscle force or length, is the controlled property. J. A. Hoffer and S. Andreassen of the University of Alberta measured the increase in the overall stiffness of the plantar-flexor muscles in the ankle of the cat as a function of muscle force. In their first set of experiments the afferent pathways carrying information from the muscle spindles and Golgi tendon organs were intact. At low forces stiffness rose with force, but at moderate and large forces it reached a plateau. In their second set of experiments the afferent pathways were severed: the stiffness of the same muscles was always lower (because a closed-loop negative-feedback control system is always stiffer than an uncontrolled system), but it continued to increase no matter how much force was applied. In other words, through the regulation of the muscle spindles and Golgi tendon organs, muscle stiffness at all forces except very low ones was moderately independent of muscle force.

We were able to get similar results

INDOOR TRACK AND TENNIS FACILITY at Harvard University opened in October of last year. Its six-lane 220-yard track features banked turns, a polyurethane top surface and a substructure made primarily of wood. The substructure is designed to convey a stiffness between two and three times the stiffness of an average athlete. Stiffness is in the low end of the range authors found to be optimum with respect to running speed. Facility was designed and built by T.A.C., Inc., and the Turner Construction Company of Boston.

with the reflexes controlling the antigravity muscles of the human leg. In our experiments weight lifters bearing loads of up to twice their body weight on their shoulders executed small bouncing motions on a long plank. The subjects kept their knees bent to the extent they would in running. Knowing the total mass of the subject and the weight (and therefore the force exerted on the muscles), the effective mass of the plank and the natural frequency of vibration, we could use a simple mathematical procedure to calculate the stiffness of the subjects. We found that at knee-flexion angles in the range appropriate for running, the controlled, or reflex, stiffness is virtually independent of load and consequently of muscle force.

The observation that the reflexes maintain muscle stiffness at an almost constant level turns out to have considerable practical value. Remember that the problem with using a mathematically linear spring to represent the muscles involved in running was that the spring has constant stiffness and the muscles do not. The results of the experiments imply that the change in muscle stiffness with muscle force observed in experiments with isolated muscles can be ignored. Actually during the first quarter or so of the ground-contact period in a running step the reflexes have not had time to act, and so the intrinsic properties of the muscles alone determine the stiffness and damping of the leg. In the second half of the contact period, however, the stiffness and damping are almost certainly under reflex control and therefore are reasonably well represented by a damped linear spring.

We are now able to complete our model of the mechanics of running. We shall view the muscles and reflexes of the leg as a system consisting of a rack-and-pinion element acting in series with a damped spring [*see bottom illustration on page 154*]. The rack-and-pinion element is included to emphasize the role of movement commands descending from higher motor centers; the spring represents the assumed constant stiffness of the muscles and reflexes, and the dashpot damping mechanism represents the intrinsic force-velocity property of the muscles and the rate sensitivity of the reflexes. In other words, the setting up of leg positions is modeled by the motion of the rack and pinion, and the effect of external force disturbances such as the impact of the track on the runner's leg is modeled by the deflection of the damped spring. As we shall now show, the simplifications introduced by the model make it possible to calculate a runner's speed on a track of arbitrary compliance with only a few items of information: the runner's mass, the length of his leg, his spring stiffness and his step length and top running speed on a hard surface.

What value of track compliance—if any—will optimize running speed? For our purposes the most useful way to measure running speed is as the ratio of step length (the distance the body moves forward while one foot is on the ground) to ground-contact time (the time the foot is on the ground). On a tuned track the ground-contact time should be minimized and the step length should be maximized. In the model it is primarily the runner's mass and the motion of the damped spring that determine the body motion during the ground-contact phase and therefore the ground-contact time; it is the motion of the rack-and-pinion element that determines the leg and hip positions at the beginning and end of the ground-contact phase and therefore the step length.

Consider first the relation of track stiffness to ground-contact time. To derive an expression for ground-contact time as a function of track stiffness it is necessary to assign to a runner a spring stiffness and also to determine the extent to which the combined effects of the force-velocity relation and the velocity feedback from the reflexes act to damp his spring. We observed that once a runner's muscles have guided his foot into contact with the ground their most important role is to contract as forcefully and as soon as they can in order to reverse the downward velocity of the body. For this reason we chose to make the somewhat arbitrary assumption that the rack and pinion cannot be disturbed by external forces while the foot is on the ground, so that the ground-contact time on a hard surface depends entirely on the runner's mass and the properties of the damped spring. Hence the numerical values denoting the runner's spring stiffness and damping can be derived from the measurement of his ground-contact time when he is running on a hard surface. (The matter is complicated by the fact that a runner's spring stiffness is affected by his effort. Muscle strength varies greatly among runners, and the stronger a runner is, the more muscle fibers he can recruit in parallel and the stiffer his spring is. For this reason we chose to determine a runner's spring stiffness by measuring his ground-contact time as he was running at top speed on a hard surface.)

Once a runner's stiffness and damping constants have been determined it is a straightforward matter to calculate how his ground-contact time is changed on tracks of various stiffnesses. This result was the most important in predicting the effect of track stiffness on running speed. It can be assumed the runner is tuned to the track so that he rebounds from it in half a period of the lowest natural mode of vibration. As intuition suggests, the calculations show that the ground-contact time on soft surfaces is long. If there were no damping element in the spring of human muscle, the ground-contact time would decrease continuously from that value as track compliance decreases, reaching a minimum at the compliance of infinitely hard surfaces. In that case a running surface such as concrete would be the fastest. Our findings show, however, that the ground-contact time does not decrease over the entire range of compliances. Since the runner is a damped system, his ground-contact time reaches a minimum at an intermediate compliance; at a track stiffness between two and four times greater than the runner's the ground-contact time is shorter than it is on harder surfaces. Therefore on tracks of precisely the right stiffness enhanced speeds should be possible.

To check these results we built several test tracks, ranging from wood tracks with an adjustable compliance to pillows or large foam-rubber blocks laid end to end. Test subjects were not given any information concerning our expectations. They were simply instructed to run at top speed, first on an experimental track and then on the concrete

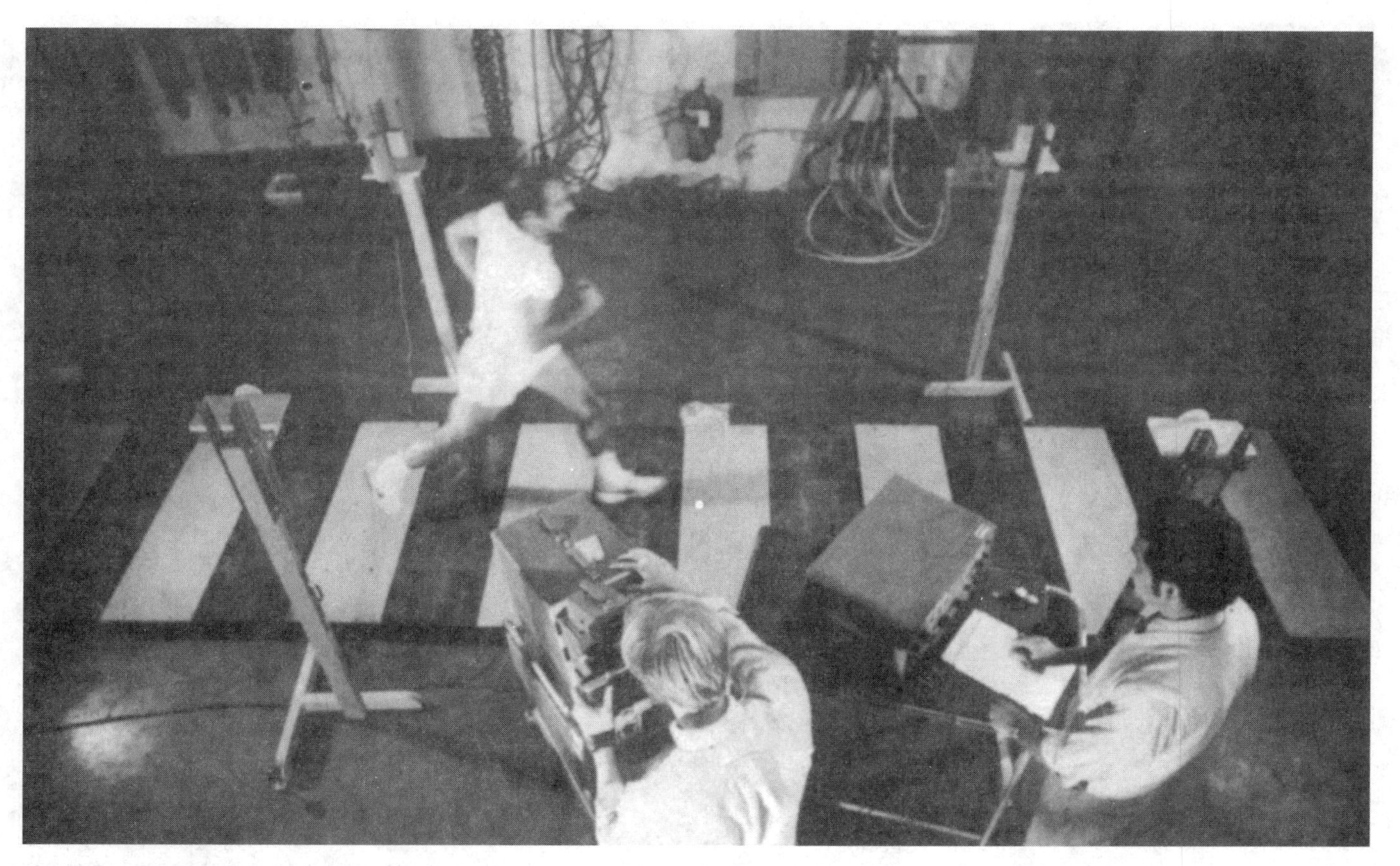

MODEL RESULTS WERE TESTED with runners on tracks of various degrees of stiffness. In the experimental setup shown here a force platform under one of the panels of the track (not visible) records the beginning and end of a ground-contact phase. The results are plotted on the oscilloscope shown at the bottom of the photograph. Running speed is recorded with the electronic timer at the lower right and the two pairs of timing lights arranged on each side of the track. (The lights are placed at the level of the runner's neck to avoid the triggering of false timing readings by the runner's arms and legs.) The running trials were also recorded on high-speed motion-picture film.

RUNNING FILMS show that whether a runner is on a fairly stiff wood track (*top*) or on a soft pillow track (*bottom*) the trajectory of his head is nearly level. Therefore for the purposes of the model it can be assumed that the hip follows a level path no matter what surface runner is on. Assumption simplifies calculation of effect of track stiffness on the step length (*see bottom illustration on opposite page*).

floor beside it. The trials were recorded with motion-picture cameras in order to measure the ground-contact time. Photoelectric timers and a precision force platform were also used. On all the tracks the measurements agreed well with the model predictions.

Running speed is not entirely determined by ground-contact time, however; step length is also an important factor. Remember that step length is the distance traveled while the foot is on the ground, as opposed to the distance between footprints, which is stride length. We noticed that as our subjects ran on the more compliant experimental tracks, such as the one consisting of foam-rubber blocks, their step length was greater. The explanation of this phenomenon has to do with the position of the contact leg at the beginning and the end of the contact phase. At those points the leg is fully extended, and it can be assumed there are no loads acting on the spring, that is, the geometry of the leg as it strikes the ground depends solely on the rack-and-pinion element.

For our investigations of the rack-and-pinion motion we made another assumption based on a feature we observed in the running films. No matter how hard or soft the running surface was, the trajectory of each runner's head was nearly level, as if his postural controllers were trying to give the acceleration-sensing mechanism of the inner ear a smooth ride. In any case it is certain that at the frequency in question, with the modest value of the acceleration due to gravity and the short time between footfalls, the vertical motion of the body's center of mass cannot be great. Hence we chose to assume that the hip follows a level path whether the runner is on a hard surface or a soft one.

The step length on a track of arbitrary compliance can be calculated as follows. In each running step one of the runner's legs makes contact with the ground as the other swings over the running surface. As the swing leg moves forward it stays above the undeflected surface of the track, whether the track is hard or soft. On a hard track the contact leg also stays above the surface. On a soft track, however, the contact leg sinks below the surface. Therefore it is necessary to visualize the runner moving across a hypothetical surface a distance δ below the undeflected one, where δ is the average track deflection over a complete stride cycle. (The stride cycle includes not only the ground-contact phase but also the aerial phase; if the runner were standing motionless on the compliant track, he would be standing in a well of depth δ.)

At the beginning (and end) of the contact phase the runner's contact leg is fully extended, and we shall call the full hip-to-heel length l. At mid-stance,

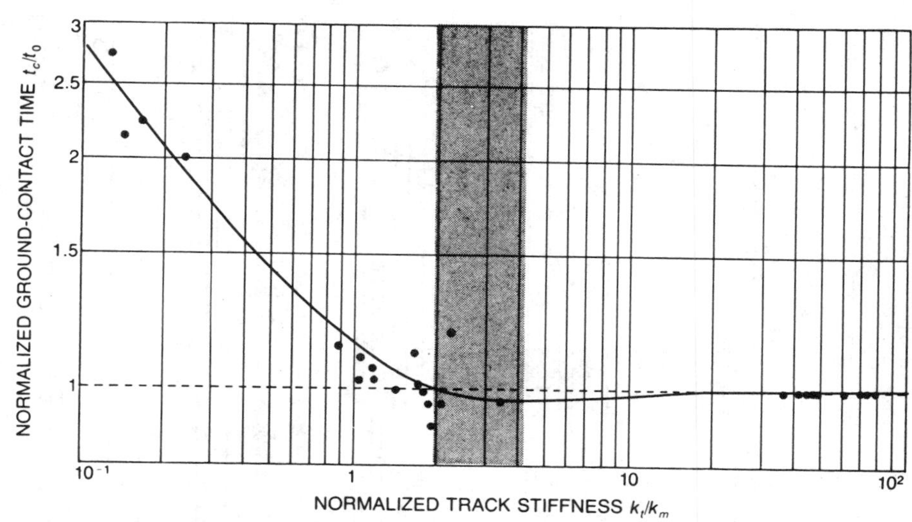

GROUND-CONTACT TIME can be derived from the model as a function of track stiffness. In this graph of the function, normalized, or dimensionless, values of each parameter have been used so that theoretical results (*curve*) can be compared with experimental findings for many runners (*black dots*). Ground-contact time has been normalized by dividing the time on a compliant surface t_c by the time on a hard surface t_0; track stiffness has been normalized by dividing the track stiffness k_t by the runner's stiffness k_m. Because of the damping of human muscle ground-contact time does not decrease continuously as track stiffness increases but reaches minimum at stiffnesses between two and four times that of average athelete (*color*).

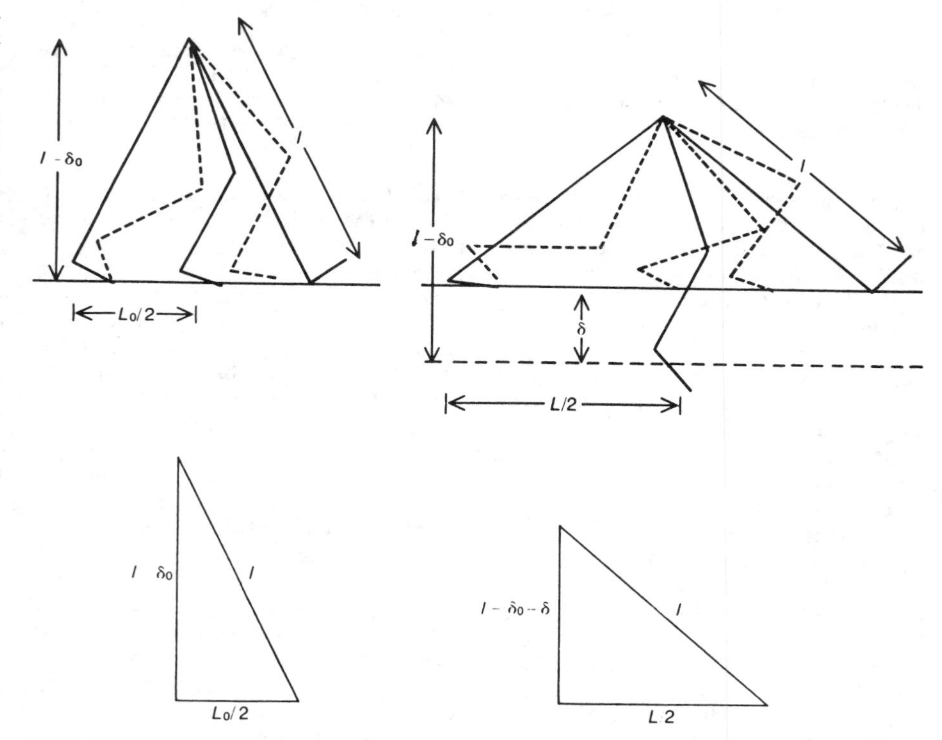

STEP LENGTH IS LONGER on a compliant surface (*right*) than on a hard one (*left*) because the bending angle of the hip joint is greater, as is shown by these schematic representations of a runner on the two surfaces. At the beginning and end of either running step the contact leg (*solid black line*) is fully extended, with length l. After heel contact (*leg position farthest to the right*) the contact leg moves back (to the left) as the swing leg moves forward. At mid-stance, when the body has traveled a distance of half a step length, the contact leg is flexed, that is, shortened by a distance of δ_0, which is taken to be independent of the track compliance. Hence on a hard surface, which has step length L_0, the distance from the hip to the track surface is $l - \delta_0$. On a compliant surface, however, the flexed contact leg sinks below the undeflected track surface (*solid color line*). It is as if the runner were moving across a hypothetical surface (*broken color line*) a distance δ below the undeflected surface, where δ depends on the compliance of the track and the mass of the runner. Therefore the distance from the hip to the undeflected surface on a compliant track is $l - \delta_0 - \delta$. The right triangles below each figure in the illustration show how these factors result in a greater step length L on a compliant surface. Since the distance δ is equal to the runner's weight divided by the track stiffness, and since L_0 and l can be measured directly, the step length on a track of arbitrary stiffness can be calculated by applying Pythagorean theorem to the two triangles (*see illustration on page 162*).

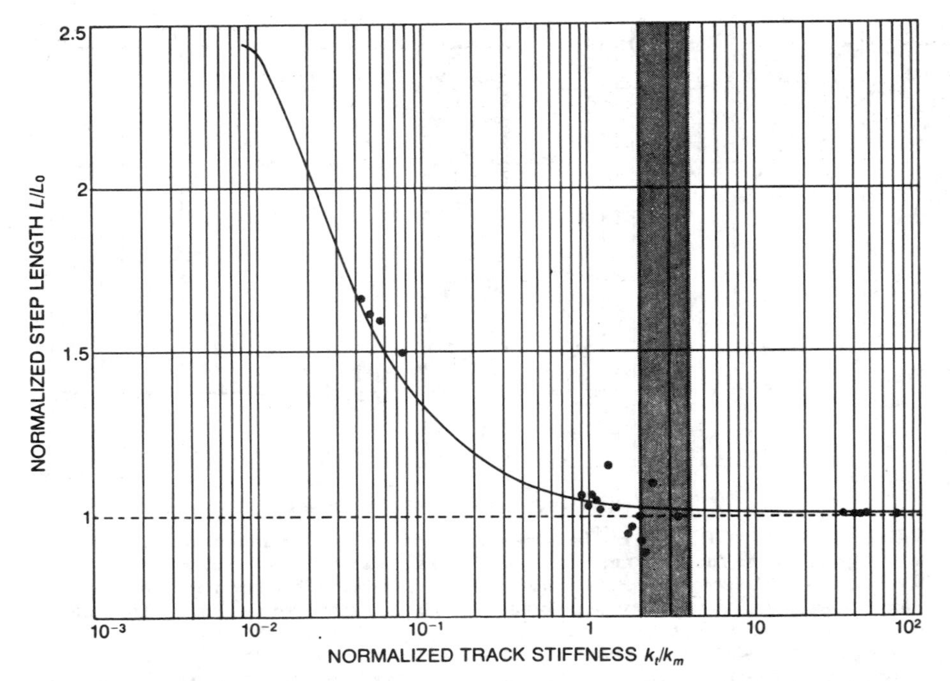

MODEL PREDICTIONS of the effect of track stiffness on step length (*curve*) are in close agreement with experimental findings (*black dots*). Step length is greatest on softer surfaces, where ground-contact time is also greatest. In the range of stiffnesses where ground-contact time is minimized and therefore optimum with respect to running speed (*color*) step length is still slightly increased over its hard-surface value. Hence in that range, and particularly in the low end of the range, running speed is enhanced. (In the graph step length has been normalized by dividing step length L on a surface of arbitrary stiffness by step length L_0 on a hard surface; track stiffness has been normalized by dividing track stiffness k_t by runner's stiffness k_m.)

when a distance of half a step has been traveled, the contact leg is flexed, shortened by, say, a length of δ_0 on a hard track, so that the distance between the hip and the hard-track surface is $l - \delta_0$. Assuming that the hip follows a level path, this situation is described by a right triangle with hypotenuse l and sides $l - \delta_0$ and $L_0/2$, where L_0 is the step length on a hard surface. On a compliant track, however, the comparable triangle has sides l, $L/2$ and $l - \delta_0 - \delta$, where L is the step length and $l - \delta_0 - \delta$ is the distance from the hip to the undeflected track surface. The average deflection δ is determined by two measurable quantities: the runner's body weight and the track stiffness. The leg length l and the step length on a hard track L_0 can be measured directly. Hence by applying the Pythagorean theorem to the two triangles it is a simple matter to calculate what a runner's step length will be on a track of arbitrary stiffness.

We found that the model predictions agreed well with our experimental measurements of step length on compliant surfaces. Just as we had calculated, the foam-rubber blocks increased a subject's step length by approximately 50 percent, partly compensating for the more than 100 percent increase in ground-contact time with the same surface. The net effect was that the runners were slowed to only about 70 percent of their hard-surface speed; if their step length had stayed constant on the foam rubber, they would have been slowed to about 50 percent of that speed.

Thus step length is most increased (a desirable effect) in the range of track compliances where ground-contact time is increased (an undesirable effect). It turns out, however, that in the range of compliances that are optimum with respect to ground-contact time, step length is still slightly increased over its hard-surface value. Therefore in this range of compliances—at a stiffness slightly more than twice the stiffness of the runner—the two factors determining running speed come together in an optimum way. Our model results showed that on tracks in this range of compliances the combined effect of decreased ground-contact time and increased step length should lead to a speed enhancement of between 2 and 3 percent.

At the end of the 1977–78 track season, the first season the track was in service, we compared the overall performance of members of the Harvard track team on the new track with their performance at the Intercollegiate Amateur Athletic Association of America championships held at Princeton University and the Heptagonal championships held at Cornell University. The Princeton and Cornell tracks are almost identical with the new Harvard track in size, but both of those tracks have a hard substructure with a synthetic top surface. (The Harvard track is also the only one of the three with banked turns. There is disagreement over the value of this feature of track design: banked turns increase the comfort and safety of running at high speeds, but they work to the disadvantage of runners moving at speeds lower than those for which the banking was designed: a fast quarter-mile.)

The analysis of the track records leaves little doubt about the effects of the tuned track. Although the two meets away from Harvard were championships, where runners are challenged to make their best showing, every Harvard runner except one ran his best time of the year on the new Harvard track. The runners' average speed advantage on the new track was 2.91 percent, which is in close agreement with the predictions derived from our model. In the preceding year the team ran an average .26 percent slower at home (on a cinder track) than they did away, and so it seems unlikely that these statistics reflect a home advantage. Moreover, similar statistics on 11 of the top runners from four other institutions show that they experienced the same kind of improvement, running an average of 2.12 percent faster on the new Harvard track than they did at the championships.

In the same season an individual Harvard runner's best time on the new track was an average of 3.87 percent faster than his best time off it and an average 2.04 percent faster than his best time of the preceding season. In terms of competition, however, the real advantage of a properly tuned track is probably not its speed enhancement; after all, the opposing teams run faster on it too. It is the comfort and safety of a tuned track that gives the home team an edge, enabling them to train much harder than they could on an ordinary track. A coach who can train ambitiously without fear of injuring his runners during practice can build a better team. Far fewer runners have been injured on the new track, although the Harvard team trained very hard last year.

In view of the discussion so far it may come as a surprise that no one expects a world record to be set on the Harvard track. That is because the International Amateur Athletic Federation does not recognize records for indoor running. Indoor tracks must usually be fitted into existing structures and so there is little standardization of design; for example, the length of such tracks and the radii of their turns vary greatly. Hence official world records can be set only on outdoor tracks.

No outdoor track of the optimum mechanical design has yet been built. If such a track is built, we predict that the world record for the mile could be improved by as much as seven seconds. The opportunity stands as a challenge.

EVENT	RUNNER	1977–78 SEASON: AVERAGE TIME AT HOME/ NUMBER OF MEETS	1977–78 SEASON: AVERAGE TIME AWAY/ NUMBER OF MEETS	1977–78 SEASON: SPEED ADVANTAGE AT HOME (PERCENT)	1976–77 SEASON: AVERAGE TIME AT HOME/ NUMBER OF MEETS	1976–77 SEASON: AVERAGE TIME AWAY/ NUMBER OF MEETS	1976–77 SEASON: SPEED ADVANTAGE AT HOME (PERCENT)
TWO MILE	EICHNER	9:05.56/4	9:32.5/1	4.9	9:29.0/3	9:28.0/2	−0.2
	MEYER	9:18.7/6	9:36.5/2	3.2			
	FITZSIMMONS				9:02.5/2	9:12.5/4	1.8
	DUNN				9:20.7/1	9:28.3/3	1.4
ONE MILE	SHEEHAN				9:25.2/1	9:33.4/1	1.4
	EICHNER	4:13.4/3			4:21.35/2	4:17.93/3	−1.3
	McNULTY	4:11.56/7	4:08.74/2	−1.1			
	CAMPBELL				4:15.0/2	4:12.5/3	−1.0
	FINN				4:26.4/1	4:24.8/3	−0.6
1,000 YARDS	CHAFEE	2:15.95/2			2:15.3/1	2:15.4/2	
	DOLSON	2:14.58/5	2:18.1/1	2.6	2:16.0/1	2:15.5/1	
	CAMPBELL				2:15.3/1	2:13.78/4	−1.1
880 YARDS	CHAFEE	1:54.92/6	1:56.43/2	1.3			
	DOLSON	1:55.14/2	1:57.33/1	1.9			
	SELLERS	1:57.94/6	2:01.1/1	2.7			
600 YARDS	SCHMIDT				1:13.8/2	1:13.4/4	−0.6
	SELLERS				1:19.3/2	1:17.4/1	−2.5
440 YARDS	LAMPPA	50.46/7	51.7/1	2.5			
	NICODEMUS	50.36/7	53.3/1	5.8			
	SCHMIDT	48.93/2	51.5/1	5.3			
				AVERAGE 2.91%			AVERAGE −0.26%

HARVARD TEAM'S TRACK RECORDS for the first season on the new track (*color*) show that the track substantially enhances running performance. Each Harvard runner's performance on the new track is compared with his performance at the Intercollegiate Athletic Association of America and the Heptagonal meets, which were run on standard tracks. The new track conveyed an average speed advantage of nearly 3 percent. Further analysis of these statistics shows the probability is about 93 percent that any given individual will run faster on the new track. Similar statistics are given for preceding year (*white*), when Harvard team's home meets were run on cinder track.

Chapter 5
TENNIS

CONTENTS

Five of the six articles in this chapter were authored by Howard Brody, a recognized expert on the physics of tennis.[1] The first article, "Physics of the tennis racket," explores some of the physical characteristics of a tennis racket in an effort to shed light on such questions as a racket's *optimal* size, weight, and shape. In particular, by studying the location of the center of percussion, the contact time between ball and strings, the period of oscillation of a vibrating racket, and the coefficient of restitution (COR) of a tennis ball, Brody suggests ways in which a tennis racket with improved playing characteristics might be designed.

One of the major conclusions of the article centers on the merits of the Prince oversize racket, whose initial design was pioneered by Howard Head.[2] Brody notes that:

> The tests and calculations show that a racket with a larger head seems to have definitive advantages. It is not clear whether the Prince racket is the optimum size, since an even larger racket may eventually prove to be better. However, if a racket of conventional length, weight, and balance is desired, the Prince is an improvement over standard rackets.

The second article in this chapter, "Physics of the tennis racket II: The 'sweet spot,'" observes that there are several definitions of the term "sweet spot," and that opinions differ as to whether that spot is a point or a region on the face of the racket. In addition, while almost everyone agrees that the sweet spot is a place on the racket where it "feels good" to hit the ball, this article develops definitions, based on physical principles, that help clarify the origin of those good feelings.

In particular, Brody focuses on three major definitions of sweet spot, with each definition being based on a different physical principle. The first definition limits the sweet spot to that portion of the racket face where the COR exceeds a predetermined threshold value. As Brody points out, a better description of this area might be "power region," since this is the area from which, for a given incoming ball velocity and precollision racket motion, the greatest ball return velocity would be generated.

The second definition relates the sweet spot to the so-called *center of percussion,* a point on the racket face at which ball collisions of any speed produce just the right combination of translational and rotational racket motion so that the amount of collision force transmitted to the hand of the person gripping the racket is precisely zero.

The third definition connects the sweet spot to a point on the racket face at which the *vibrations* resulting from ball collsions at that point lead to a lack of vibration (node) on the racket handle at the location on the player's hand.

Clearly, whether due to the satisfaction of hitting a return shot with great power, or of experiencing the minimal force or vibration on the racket hand, each of the three sweet spots defined above would qualify as places where it "feels good" to hit the ball.

In addition, by showing that the restoring force of a tennis string is *directly* proportional to string *tension* and *inversely* proportional to string *length,* Brody provides an elegant explanation for why oversize rackets, with their substantially longer strings, must be strung so much more tightly than standard rackets in order to "feel" the same.

Also, as many recreational tennis players have discovered through random empirical study (i.e., by accident on the tennis court), the sweet spot for most rackets falls not at the center of the racket face but at a location much closer to the handle! The reason for this surprising result stems from the fact that the racket is secured by the players hand *not* at the location of the racket's frame, but near the end of the handle, some distance away. In fact, were a racket to be secured at the location of the frame, the sweet spot would occur precisely at the center of the racket face.

In the third article in this chapter, "The moment of inertia of a tennis racket," Howard Brody focuses on a racket's moment of inertia about various axes in an effort to relate this physical parameter to how the racket plays and feels.

The moment of inertia of a tennis racket about its polar (i.e., long) axis is one of the most interesting and important ones for the game of tennis. This moment determines the racket's resistance to twisting when the ball strikes the racket off center, a fairly frequent event for all but the most accomplished players. The larger this moment of inertia, the greater the rotational inertia and, hence, the greater its resistance to twisting. In practice, racket manufacturers enhance this particular moment in one of two ways: (1) by adding weights to the edges of standard size rackets (e.g., the Wilson Perimeter Weight System racket), or (2) by making the racket head wider, as Howard Head did with the Prince racket. Since the moment of inertia is proportional to the first power of the mass and the second power of the mass distance from the axis, this second method is clearly very effective, i.e., small increases in racket width are capable of producing large increases in polar moment of inertia.

This article also explores the moment of inertia of a tennis racket about two axes at right angles to the polar axis. The first is the moment about an axis *parallel* to the racket face, a moment which determines the racket's resistance to being swung in a normal fashion. This moment is sometimes referred to as the "swing weight" of the racket since it is related to how "heavy" the racket feels when swung.

The second moment is the one about an axis *perpendicular* to the racket face, a moment which defines the racket's resistance to being moved in a "chopping" mode, a motion often employed by servers in an effort to apply side spin to the ball.

One of the most important yet least regulated aspects of the game of tennis is the interaction between the tennis ball and the court surface on which it is played. The fourth article in this chapter, "That's how the ball bounces," by Howard Brody, attempts both to understand and to quantify this interaction which, historically, has been characterized by such rather vague terms as "fast" and "slow." Brody's analysis is based on two key parameters, the coefficient of kinetic friction and the coefficient of restitution between tennis ball and court surface.

In the course of quantifying the speed of tennis courts, primarily in terms of the coefficient of kinetic friction, Brody uncovers a number of fascinating insights into the collision between ball and court, two of which include the critical role played by spin, and the large accelerations (150 g's or more) present even during "gentle" bounces.

The fifth article in this chapter, "The aerodynamics of tennis balls: The topspin lob," by Antonin Štěpánek, combines wind-tunnel measurements with theoretical trajectories in an effort to model one of the most difficult and interesting strokes in the game of tennis, viz., the topspin lob. The theoretical analysis assumes motion under the action of three forces—gravity, plus aerodynamic drag (air resistance), and lift (Magnus force). One of the major contributions of this article is an empirical determination of the dependence of C_D and C_L, the drag and lift coefficients, on the linear and spin velocities of the tennis ball. Another interesting experimental observation established that, when hit by professional tennis players, topspin lobs can achieve spin rates of 3500 rpm.

The sixth and final article in this chapter, "The tennis-ball bounce test," by Howard Brody, analyzes the kinematics of a bouncing tennis ball. In order to be approved for tournament play, tennis balls dropped from a standard height of 100 in. onto a concrete floor must rebound not less than 53 nor more than 58 in. Brody describes and compares two different methods for measuring the bounce characteristics of tennis balls. One method involves a direct photographic measurement of the rebound height using a high-speed videotape camera. The other method involves measuring the elapsed time between the arrival of acoustic signals corresponding to the first and second bounces of a tennis ball dropped from the standard height. The rebound height could then be inferred from the time measurements and compared with the results of the first method. The measured distribution of rebound heights for the balls tested fell within the limits specified by the tennis sanctioning bodies.

---

[1]See Chap. 10 for a partial listing of the many contributions by Howard Brody to the physics of sports. See, in particular, *Tennis Science for Tennis Players* (University of Pennsylvania, Philadelphia, 1987).

[2]Howard Head became a pioneer in the physics of sports, not so much by writing books or articles on the subject, but by inventing sports implements based on physical principles. In addition to developing the oversize Prince tennis racket in the early 70s, Head was also known for his invention of the first aluminum snow ski in 1950.

Physics of the tennis racket

H. Brody
Physics Department, University of Pennsylvania, Philadelphia, Pennsylvania 19104
(Received 23 October 1978; accepted 8 February 1979)

Several parameters concerning the performance of tennis rackets are examined both theoretically and experimentally. Information is obtained about the location of the center of percussion, the time a ball spends in contact with the strings, the period of oscillation of a tennis racket, and the coefficient of restitution of a tennis ball. From these data it may be possible to design a racket with improved playing characteristics.

The physics of the tennis racket is a subject that very few people paid attention to until Head produced his oversized Prince racket. In fact, in reading various tennis magazines, looking at advertisements for tennis rackets, and talking to tennis players, tennis professionals, or sports equipment sales people one gets such conflicting statements that one realizes that no one seems to understand the physics of the tennis racket.

When a tennis racket is examined, one notes that the basic shape and size have not changed in over half a century. It is probable that these parameters were not determined solely by playing characteristics but also by structural considerations imposed by the strength of wood. When metal rackets became popular a few years ago (and composite fiberglass, boron, or carbon filament recently) the manufacturers initially copied the general shape and size of the wooden rackets, since they probably assumed that they were optimum—or that any radical change might not sell. The same might be said for the strings—where gut has successfully withstood the challenge of all the modern synthetic materials for tournament play.

There seems to be no information in the published physics or tennis literature about the optimization of size, shape, weight, etc., of a tennis racket. Since everyone learned to play with essentially the same type of racket, any radical change would feel wrong and require the player to relearn to some degree. Under these conditions the design of rackets might evolve and improve slowly—but there is no way to know if a racket that was radically different might not prove better if a player originally learned with it.

One obvious change was the introduction of the Prince racket with an oversized head. Head, its inventor, was very careful to produce a racket of the same overall length, weight, and balance as a conventional racket so that it would "feel" the same as a normal racket. To prove this, any average player can pick one up and play with it immediately. However to play with it in an optimum manner does take some retraining since it was designed to give its best performance with the ball striking it 5 or 6 cm closer to the handle than on a normal racket. To understand the advantages of the Prince racket (without playing with it) some simple kinematics must be investigated.

Consider a tennis racket of mass M suspended freely in space. If a ball strikes it and imparts a momentum $+\Delta p$ to the racket, the center of mass of the racket will move with a velocity $V = +\Delta p/M$.

If the ball hits the racket at the racket center of mass, the racket will translate but not rotate. If the ball hits the racket at a distance b from the center of mass, the racket will rotate around the center of mass as well as translate. From conservation of angular momentum, the angular velocity of the racket about the c.m. will be $\omega = \Delta pb/I_{\text{c.m.}}$ where $I_{\text{c.m.}}$ is the moment of inertia of the racket about its center of mass. Assume the racket has been struck by a ball, the racket (c.m.) is translating to the right with velocity V and the racket is rotating clockwise about the c.m. with angular velocity ω [Fig. 1(a)]. There then will be one point in the racket which is instantaneously at rest, if the racket handle is long enough. If this point is a distance a from the center of mass then the condition for that point to have no velocity is $V = \omega a$. In other words, the motion due to the rotation exactly cancels the overall translation (of the racket) at that one point. Then

$$a = \frac{V}{\omega} = \frac{\Delta p/M}{\Delta pb/I_{\text{c.m.}}} = \frac{I_{\text{c.m.}}}{b\,M}.$$

The value of a is independent of how hard the ball hits the racket (Δp) as it should be. Then $ab = I_{\text{c.m.}}/M = k^2$ (radius of gyration squared).

If a ball hits a racket at a distance b from the c.m. (at point B) and the racket is held a distance a from the c.m. (at point A), no force or impulse from the hand need be imparted to the racket, since, in a frame of reference initially moving with the racket, the point A remains at rest. In the frame of reference with the racket handle being swung with velocity V', the point A continues to move with velocity V' with no external force applied to it.

If the racket is held at point A then point B is called the center of percussion and it is of some interest to determine the distance $a + b$. This distance $a + b$ can be obtained with a simple experiment that uses the racket as a physical pendulum. If the racket is allowed to swing freely about the point B, the frequency of the oscillation (for small amplitude) can be calculated and also measured.

In Fig. 1(b), the restoring torque is $-Mg\,b \sin\theta \sim -Mg\,b\,\theta$ (for small θ). Then $\tau = I_B\,(d^2\theta/dt^2)$, where I_B is the moment of inertia about B. Using the parallel axis theorem, $I_B = I_{\text{cm}} + Mb^2$ and substituting this into the $\tau = I\alpha$ equation

$$-Mg\,b\,\theta = (I_{\text{cm}} + Mb^2)\,\frac{d^2\theta}{dt^2}.$$

The solution to this equation is a simple harmonic motion with angular frequency $\omega = [Mg\,b/(I_{\text{cm}} + Mb^2)]^{1/2}$ but $I_{\text{c.m.}} = Mk^2$, so $\omega = [gb/(k^2 + b^2)]^{1/2}$. However, since $k^2 = ab$, $\omega = [\text{gb}/(\text{ab} + \text{b}^2)]^{1/2} = [\text{g}/(\text{a} + \text{b})]^{1/2}$. Conse-

Reprinted from *American Journal of Physics* **47**, 482–487 (1979); © American Association of Physics Teachers.

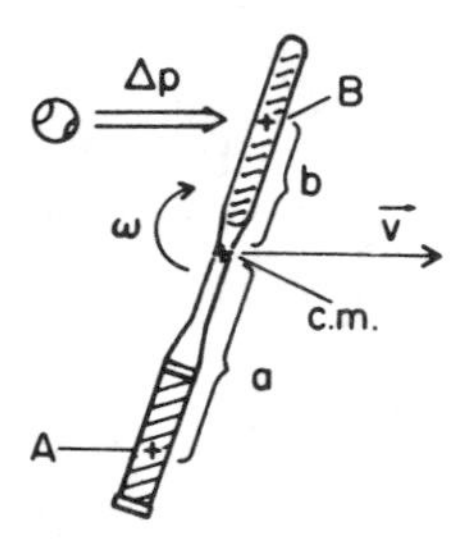

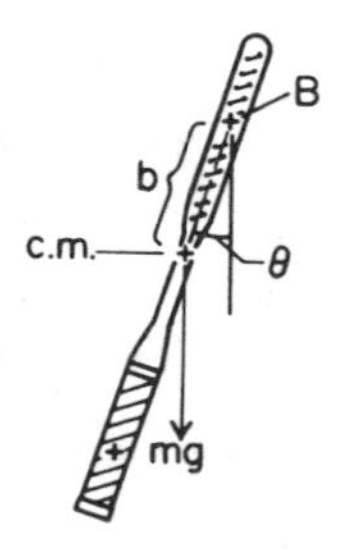

Fig. 1. (a) Tennis racket suspended freely in space; (b) Tennis racket as a physical pendulum with pivot at center of head.

quently, from a measurement of the angular frequency of the racket used as a pendulum and a knowledge of g, $a + b = g/\omega^2$ can be found.

In actually doing the measurement, care must be exercised to make the plane of swing perpendicular and not parallel to the face of the racket since the moments of inertia are different. For a normal racket held at the handle, the center of percussion is 5 or 6 cm below the center of the head of the racket. (Since these equations are symmetric in a and b, $a + b$ could also be found by allowing the racket to pivot about the point where the racket is held.) Three rackets (Spalding Smasher, Davis TAD, and Prince) were suspended with a pivot point at the center of their head and allowed to oscillate as a pendulum. Their period was measured and the distance ($a + b = g/\omega^2$) calculated. The distance of the conjugate center of percussion point from the butt end of the racket was then obtained. These data are shown in Table I. If a racket was held at the conjugate point, then the center of percussion would be at the center of the head of the racket. If you desire to hit a ball at the center of percussion of a normal racket you should either "choke up" by moving your hand up the handle 5 or 6 cm and hit the ball at the center of the head or keep your hand in its normal position and hit the ball close to the throat of the racket. Neither of these is recommended.

A third choice is to build a different racket so that when the racket is held in the normal place, the center of percussion is at the center of the stringed region of the head. This can be done by making the tip of the racket very massive or by extending the head of the racket considerably in the direction of the handle. It is this latter technique that Head employed in his Prince racket.

There is, of course, the question as to whether the center of percussion is the optimum location to hit a ball. If the ball spends an appreciable amount of time in contact with the strings of the racket (a long dwell time) then it might be possible for a player to add momentum and energy to the racket, hence the ball, during the dwell time. However, because of the conjugate nature of the two points, the motion of the center of percussion will not be affected by a simple force applied at the handle, hence it may not be possible to affect the ball during its dwell. A torque about the center of the handle will, however, affect the ball. Since the dwell time of a tennis ball on the strings of a racket is about 5 msec, the idea of hitting the ball at the center of percussion of the racket bears further examination but not in this paper.

If a ball is hit at the center of percussion there will be no net force on the hand. However, in the hand frame of reference the racket will attempt to rotate about an axis perpendicular to the handle at the point where it is held. If the ball is hit further away from the handle than the center of percussion, as it is in most rackets, then the racket (in the hand frame of reference) will attempt to rotate about an axis passing through the handle at a point closer to the head of the racket. On forehand there will then be a net force on the hand (due to the ball interaction) that will tend to pull the racket forward out of the player's hand (tending to force open the fingers and thumb).

If the ball is hit closer to the handle than the center of percussion (which is almost impossible on a normal racket) then the racket will attempt to rotate about a point near the butt end of the handle. There will again be a net force or impulse on the hand, only now (for forehand) it will be directed against the palm and base of the fingers. From this analysis it would seem advantageous to hit the ball closer to the handle rather than farther away, and this is reinforced by the following analysis. If the racket is treated as a free body (not held in a hand) it is clear from the two body kinematics that the closer to the center of mass of the racket that the ball strikes, the higher will be the ball's rebound velocity. This is because less energy goes into the rotation of the racket when the ball hits closer to the c.m. Consequently, enlarging the head by extending it toward the handle has definite advantages. This would lead to a very elongated shape unless the head were made wider, and there is a good argument for doing this as well.

Many players, particularly beginners, have a problem that the racket twists or rotates in their hand. This is because the ball has hit off center (not along the polar axis) and a net angular impulse is imparted to the racket. The further from the axis the ball hits, the more the racket will tend to twist. This twisting can be reduced by increasing the polar moment of inertia of the racket. This can be accomplished by increasing the mass along the sides of the head or by keeping the mass fixed and increasing the distance from axis to the side of the head (in other words, making the head larger).

The Wilson company has chosen the former by adding tungsten weights in a perimeter weighted racket. Head has opted for the latter in his Prince racket. Since the moment of inertia goes as mR^2, an increase in size of 40% results in

Table I. Racket parameters.

Racket	Davis TAD	Spalding Smasher	Prince
Length (cm)	68.9	67.0	68.6
Period of oscillation about center of head (sec)	1.26	1.28	1.30
$a + b = g/\omega^2$ (cm) distance between pivot and center of percussion	39.4	40.7	42.0
Distance between center of head and butt end (cm)	54.6	54.0	51.1
Distance between center of percussion conjugate point and butt end (cm)	15.2	13.3	9.1

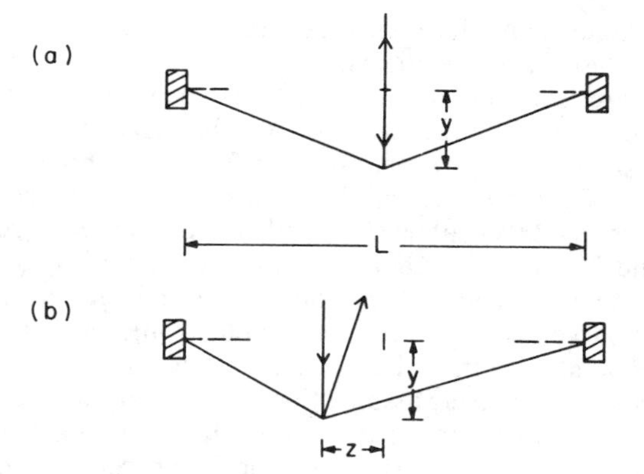

Fig. 2. String deflection when ball hits (a) at center of strings and (b) closer to one side of racket.

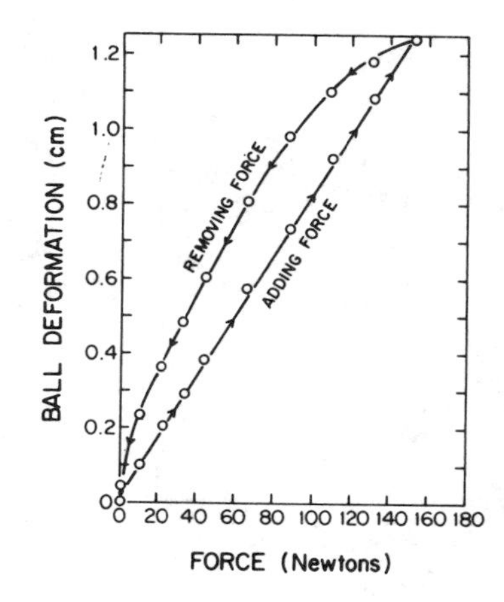

Fig. 4. Deformation of a tennis ball as a function of the applied force. The ball was placed in a rigid hemispherical cup so that only one side deformed.

a doubling of the moment of inertia. It is also possible to increase the diameter of the racket handle so that the torque tending to prevent twisting is increased without increasing the forces applied by hand.

Even if the ball is hit off center and the racket twists a little, there is another effect that will tend to compensate for it. If the ball hits off center, the deflection of the strings is asymmetric. Neglecting spin of the ball, gravity, etc., in the rest frame of the racket, it is expected that relative to the plane of the strings the angle of the ball's rebound will be equal to the angle of its incidence. This will be true only if the strings deflect symmetrically [Fig. 2(a)] which occurs for hits in the center. For off center hits, the asymmetrical string deflection tends to deflect the ball toward the center [Fig. 2(b)]. Head was able to demonstrate this using a high-speed motion picture camera with the racket held in a vise. The data he obtained allowed him to determine the angular error of the rebounding ball as a function of position and map out contours of angular error regions with error less than 10°, between 10° and 20° and greater than 20°.[1]

This angular error depends upon several factors. If a string of length L deflects (perpendicular to the plane of the strings) by a distance y when hit by the ball and the ball misses the center by a distance z, the angular error will be proportional to $(z/L)\ (y/L)$. There is not much a physicist can say about reducing z, hence reducing the error, but y and L are subject to analysis. The maximum value of the string deflection can be obtained if the effective spring constant of the strings is known (the slope in Fig. 5), the ball momentum specified and the assumption that the motion is simple harmonic is made. Then $\int F\,dt$ over $\frac{1}{2}$ cycle equals Δmv. Since $F = -ky$ and $y = A \sin \omega t$, the amplitude A of the deflection can be obtained: $A = (\Delta v/2)(m/k)^{1/2}$, where m is the mass of the ball. (The mass of the strings is neglected.) The value of k is a linear function of T/L, the string tension divided by the string length and also depends upon the effective number of strings that deflect. When all of this is put together, the angular error $\Delta\theta = (Cz\Delta v/L)\ (m/LT)^{1/2}$, where C is a constant.

It is then quite clear that increasing the size of the racket head (increasing L) reduces this error, hence increases the effective area of the racket within which this error is tolerable. It is also clear that an increase in string tension will reduce this error somewhat.

Another interesting result is that this error is proportional to Δv, the change in the ball velocity. Consequently, when the ball is hit hard, unless it is hit dead center on the racket, it may not go exactly where it is aimed. This compounds the difficulty of hard hits, which, due to their kinematics, already have very little margin for error.

The string tension also influences how long a ball spends in contact with the strings (dwell time) and the velocity at which the ball leaves the racket.[2] To optimize these parameters various measurements have to be made, including a determination of some of the properties of tennis balls.

The most surprising thing is that it appears that stringing the racket looser rather than tighter will actually lead to slightly higher ball rebound velocities (more "power"). This is due to the fact that tennis balls have a rather low coefficient of restitution and a dwell time on the strings which is short when compared to half of the natural period of vi-

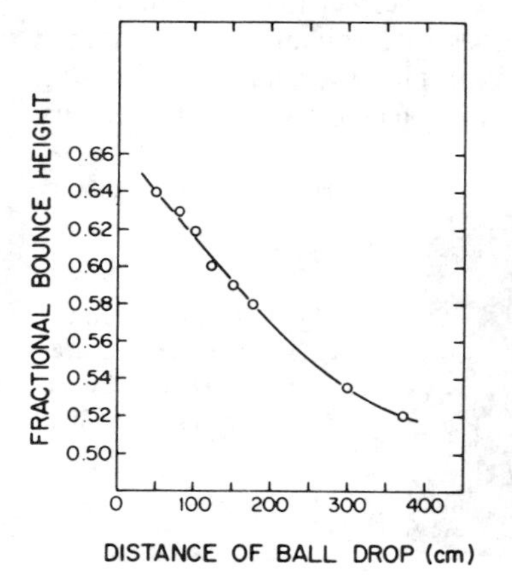

Fig. 3. Ratio of rebound height to drop distance as a function of drop distance. Ball was dropped onto a hard surface.

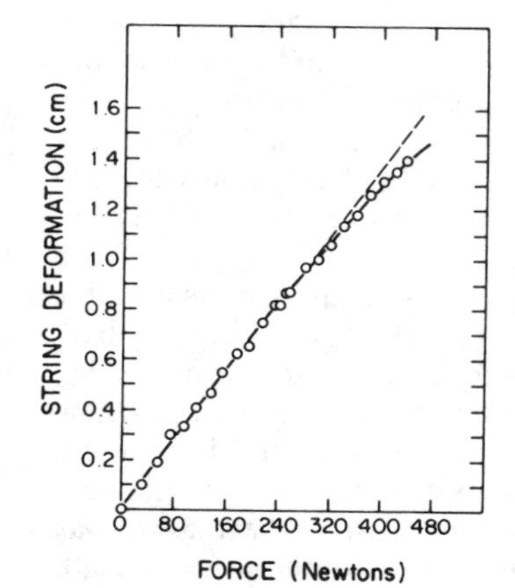

Fig. 5. String deformation as a function of applied force. The racket head was braced and the force was applied over a circular area of 12 cm².

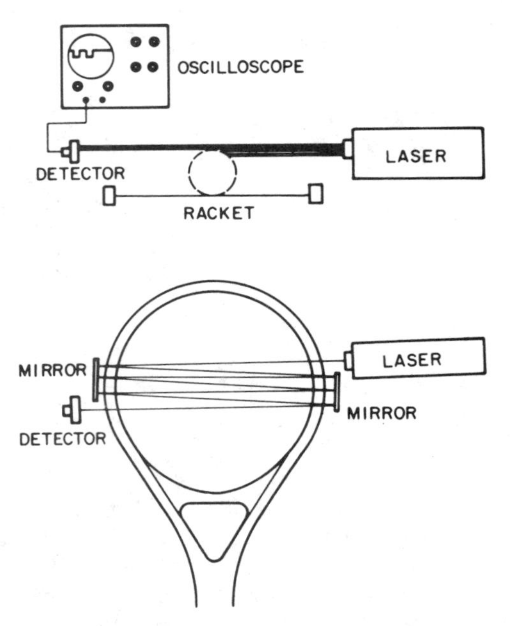

Fig. 6. Method for measuring the dwell time of a tennis ball on the strings of a racket. The head of the racket is firmly clamped in place and the laser beam is parallel to and one ball diameter above the plane of the strings.

bration of many tennis rackets. To see how this all fits together, let us examine some simple tests on balls and rackets that were made.

When a tennis ball is dropped from a specified height onto a hard surface, the height of the ball's rebound is determined by its coefficient of restitution (COR). The ratio of final height to initial height is $(COR)^2$ if air resistance is neglected. The COR of several tennis balls was measured by this method and also measured directly by determining the ball velocity using electronic techniques. It was discovered that the higher the ball velocity upon impact, the lower the COR for that collision, and that about half of the kinetic energy of a ball is lost when it hits a hard surface. A measurement of ball deformation versus applied force shows a large hysteresis loop, which is indicative of a large loss of energy. The strings (and the racket frame) have relatively little loss of energy when they are deformed. If the strings can take up a larger fraction of the energy, the ball will deform less and lose less energy. Consequently the ball will rebound with more energy (fed to it by the strings) and will have a higher velocity. The less the tension in the strings, the more they deform the larger the fraction of energy they store (and subsequently give back), the less the ball deforms, the less kinetic energy the ball dissipates. There is a limit to how loose the strings should be strung, since a butterfly net is clearly not any good. Once the tension is reduced to a point where the strings being to move in a direction parallel to the plane of the racket head, the energy that goes into that mode is probably not recovered and the strings will probably wear out because they are rubbing against each other.

The above analysis will only hold if the strings are able to feed the energy they absorb (or store) back into the ball. This means that the ball cannot come off of the strings before the strings come back to their undeformed position. This could only happen if the time required for the deformed ball to revert back to its original spherical shape is short compared to the time required for the strings to snap back. Electronic measurement of the time the ball spends in contact with a hard surface and the strings on a racket show that this is not the case.

There is a much easier way to experimentally demonstrate the above concept. If a tennis ball is dropped onto a hard surface from a height of 100 cm, it rebounds to a height of about 62–65 cm. If the same ball is dropped from a height of 100 cm above the head of a tennis racket (resting on the floor) it will rebound to a height of slightly over 80 cm. Measurement of the ratio of rebound velocity to the incident velocity, obtained using electronic means are in agreement with the rebound height measurements.

One would naively assume that a similar argument would hold with respect to a stiff versus a flexible racket frame. By the above arguments, a more flexible racket would lead to more energy being returned to the ball. However, due to its mass, the natural half period of vibration of a racket can be long compared to the dwell time of the ball on the strings. Consequently, most of the energy that goes into deflecting or deforming the racket is *not* fed back into the ball. The ball is gone by the time the racket snaps back. Therefore a stiffer racket would provide more power since less energy would go into the racket frame deformation and also the natural period of oscillation is reduced. If the racket oscillation time can be reduced so that half a period is comparable to the ball dwell time, the energy from the racket deformation will be fed back into the ball's kinetic energy.

When the tension in the strings is reduced, the dwell time of the ball is increased, and this can also improve the match between dwell time and racket oscillation period.

The great loss of energy to the racket deformation can be seen by bouncing a ball off of the racket with the handle in a vise measuring the COR as a function of the position on the head where the ball hits. The COR varies from 0.6 very close to the throat of the racket to 0.2 near the tip of the racket with values between 0.4 and 0.5 at the center of the head. This is because the stiffness of the racket changes with position, with little flex close to the throat and much deflection near the tip. Consequently more energy goes into racket deformation when the ball is struck near the tip.

If the COR tests are done with the head of the racket clamped (instead of the handle) the COR is much higher and it maximizes at the center of the head. This is because that is where the strings are effectively least tightly strung and they deflect most. Moving away from center, the strings deflect less, the ball deforms more and more energy is lost.

In actual play, several other factors enter that do not show up in a bench test. As was stated before, if the ball struck closer to the neck, then it is closer to the racket center of mass and less energy goes into racket rotation, hence more energy goes into the ball. However, a racket is swung in an arc with a point in the body as the pivot, so the velocity

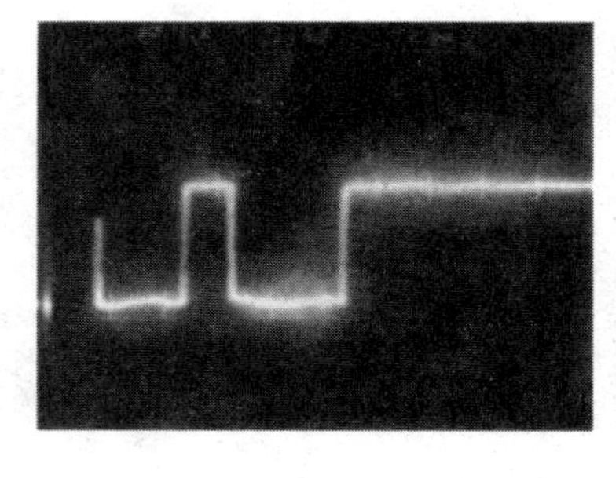

Fig. 7. Picture of an oscilloscope trace measurement of dwell time. Sweep speed is 5 msec per division.

of a particular point on the racket is a function of its distance from that point. Moving the impact point several inches toward the body should reduce the velocity of the racket at the impact point by less than 10%, so this effect can probably be ignored. (It may be somewhat more significant in the serve when the wrist is snapped—which makes the pivot to impact distance considerably shorter.)

The length of time that a ball spends in contact with the strings of the racket can be determined by direct measurement or it can be approximately calculated from static measurements made on the ball and strings. If the string-ball system is treated as a simple harmonic oscillator, then the dwell time of the ball on the strings will be half the natural oscillation period of the system. The effect "k" (spring constant) of the strings is a function of the stringing tension, gauge and type of string, size of the racket head, etc. This k was determined for the strings of several different rackets by measuring the delection of the strings as a function of applied force. Typical measured values ranged from 2 to 3.5×10^4 N/m (the slope of Fig. 5). A tennis ball has a mass of 0.060 kg, leading to a natural frequency of oscillation of $100h$ [$f = \omega/2\pi = (1/2\pi)(k/M)^{1/2}$] and a dwell time of 4.5×10^{-3} sec.

The direct measurement of dwell time can be accomplished in a number of ways, i.e., by using high-speed motion picture photography (greater than 1000 frames/sec), by taking a series of single flash pictures of a repetitive event—each picture at a slightly different electronic delay or by a method using a laser, photodetector, and an oscilloscope that I have developed.

The apparatus for the laser method is shown in Fig. 6. The ball is placed on the racket strings and the laser is adjusted in position so that its beam is parallel to the rebounding surface and is exactly one ball diameter above it (part of the beam is intercepted by the ball at rest). A photodetector (silicon solar cell) is positioned in the beam beyond the ball and the output of the detector fed to an oscilloscope on internal trigger. The ball is then removed.

The ball is then dropped onto or propelled at the rebounding surface. The ball passing through the laser beam blocks the light from the detector and triggers the scope sweep. The light will again hit the detector when the ball has passed completely through the beam. However, due to the position of the laser and rebounding surface, this will only occur while the ball is in contact with the surface (strings). As the ball leaves the surface, it again intercepts the laser beam and cuts off the light to the detector. The resulting scope trace shows a period of time with no light (ball moving toward surface), a period of time with light (ball in contact) and a second period of no light (ball rebounding). This method not only gives the dwell time, but, since the ball diameter is known, it gives the velocity of incidence and velocity after rebounding—hence the coefficient of restitution. If the ball does not cut through the laser beam at a diameter (the center of the ball misses the beam by a distance Δx) then there will be an error in the timing due to the fact that the beam will be cut by a chord rather than a diameter. The chord, being shorter than the diameter by a distance $4(\Delta x)^2/D$ will cutoff the light for less time and will also increase the apparent dwell time correspondingly. This timing error will be $4(\Delta x)^2/vD$ where v is the velocity of the ball and D its diameter. The fractional error in the time the light is cut off is then $4(\Delta x)^2/D$ and the fractional error in the dwell time (t) will be $4(\Delta x)^2/Dvt$.

Table II. Coefficients of restitution.[a]

Surface	Ratio of rebound height to drop height	COR	COR from velocity measurements
Lead brick	0.520	0.721	0.767
Prince racket (70-lb tension)	0.716	0.846	0.887
Prince racket (50-lb tension)	0.730	0.854	0.901
Spalding Smasher (unknown tension)	0.726	0.852	0.895

[a]These data were taken with the ball dropped from a height of 3.7 m above the rebounding surface. It is the average of data taken with several types of tennis balls (pressureless Tretorn, Penn ball, and Spalding Australian ball). The COR values obtained by direct velocity measurements (laser method) are consistently higher than the rebound measurements by about 0.05. This is due to the air resistance which reduces the kinetic energy of the ball during both its fall and rise, and therefore give lower values for the COR. For these data the racket head was clamped.

For Δx of order 1 cm, $D = 13.2$ cm, $v = 800$ cm/sec and $t = 5 \times 10^{-3}$ sec the time errors are 2% and 7.6%, respectively. To reduce this error, a pair of plane mirrors was set up and the laser beam multiply reflected before hitting the detector. The spacing between adjacent beams was of order 1 cm which leads to a maximum spatial error $\Delta x = 0.5$ cm and a maximum error in dwell time of less than 2%.

A second error in dwell time is present if the laser beam is not exactly one ball diameter above the rebounding surface. If the placement error is Δy, then the dwell time will be in error by $2\,\Delta y/v$. An error Δy of 2 mm then leads to an error in timing of 0.5×10^{-3} sec, so an effort was made to have the beam within 1 mm of the ball top. The results of these measurements with several different rebounding surfaces are shown in Table III.

The predicted dwell time of 5×10^{-3} sec seems to be confirmed by these measurements. The tennis rackets used were those that were readily available. It is clear that these measurements should be repeated using a number of identical rackets, each strung with different material and a variety of tensions.

The tennis racket frame also has a natural period of vibration which is determined by the mass and mass distribution of the frame, the elasticity or stiffness and the damping of the oscillations. The parameter of interest is

Table III. Typical dwell times (msec).[a]

Rebounding surface	Penn	Type of ball: Spalding Australian	Tretorn (pressureless)
Lead brick	3.9	4.5	4.5
Prince (70 lb)	6.1	6.3	6.3
Prince (50 lb)	6.5	6.7	6.4
Spalding Smasher	6.4	6.8	6.6

[a]All balls were dropped from a height of 3.7 m. Their measured velocity at impact (8.3 m/sec) was slightly lower than the theoretical value (8.5 m/sec) because of air resistance. One Prince racket was strung with a synthetic at 70-lb tension, the other Price with gut at 50-lb tension and the Spalding Smasher with synthetic at an unknown tension. The racket head was clamped for these data.

Table IV. Period of oscillation of tennis rackets.[a]

Type of racket	Frequency of oscillation (h)	Time for half-period (msec)
Spalding Smasher	32.8	15.3
Wilson T-3000	25.6	19.6
Prince	31.3	16

[a]These data were taken with the handle clamped firmly to a massive table and the racket struck at the center of the strung area by a ball.

actually a half period ($T/2$) which is the time it takes for the racket to go from its undeflected configuration to fully deflected by the ball and then return to its undeflected or equilibrium position.

The period of oscillation of several tennis rackets was measured by placing the racket handle in a vise with the head free to vibrate. A small mirror (mass $\ll$ 1 g) was taped to the tip of the head. A laser was positioned so that its beam reflected off of the mirror and onto a photodetector with the racket at rest. The output leads from the cell were connected to a scope, internally triggered and set for single sweep at either 10 or 20 msec/division. The racket was then deflected, the trace photographed, and the period of oscillation measured.

The half period of the several rackets measured came out to be approximately twice the dwell time of the ball on the strings. If this is the mode of oscillation that a racket undergoes when it strikes a ball, then all the energy that goes into deforming the racket is lost. This is because the ball has left the strings before the racket can begin to snap back. This energy loss can be eliminated by making a much stiffer racket—one with a natural period of oscillation half as long as a normal racket, stringing the strings at much lower tension, using thinner or more elastic strings, or some combination of all of these. The result will be a racket that will propel the ball with a higher velocity for the same swing by the player.

If a hand held racket tends to oscillate with a node at the center of mass and not at the handle, then the oscillation period is much shorter and the energy of deformation might be fed back into the ball. Unfortunately my measurements do not distinguish between these two cases since they were done with the racket firmly clamped and not hand held.

The tests and calculations show that a racket with a larger head seems to have definitive advantages. It is not clear whether the Prince racket is the optimum size, since an even larger racket may eventually prove to be better. However, if a racket of conventional length, weight and balance is desired, the Prince is an improvement over standard rackets. One possible modification would have been to produce a racket with an oversized head where the center of the strings coincided with the center of percussion and they were located at the same distance from the butt end as the center of the head is in a conventional racket. This would lead to a racket with somewhat different weight and balance, but retain the hitting distance that players have learned to play with.

ACKNOWLEDGMENTS

I would like to thank Jim Hobbs and Michael Procario for their help in taking some of the data presented in this paper.

APPENDIX

The calculation of dwell time assumes a simple harmonic motion with the restoring force of the strings on the ball proportional to their deflection out of the plane of the racket. This is a good approximation for small deflections (Fig. 5), but is the deflection small when a ball is actually hit? If there is an appreciable deflection, then the cubic term in the force-deflection relationship becomes important and the strings tend to act as if they are stiffer (the effective k increases). The maximum deflection of the strings can be calculated assuming a linear relationship, but knowing that the actual displacement from equilibrium will be less:

$$\Delta mv = \int F dt = \int -kx dt,$$

where x is the displacement from equilibrium. If $x = A \sin \omega t$ where $\omega = (k/m)^{1/2}$, the integral becomes $\Delta mv = \int -kA \sin\omega t dt = -kA/\omega \cos \omega t$. Evaluating this over a full half cycle $\Delta mv = 2kA/\omega = -2kA\ (m/k)^{1/2}$. Then $A = (\Delta v/2)\ (m/k)^{1/2}$. Putting in the measured values for the k of the strings (from Fig. 5), $k = 2 \times 10^4$ N/m, the measured mass of a tennis ball, $m = 0.060$ kg and a $\Delta v = 40$ m/sec yields a maximum deflection $A = 3.5$ cm. The $\Delta v = 40$ m/sec corresponds to a well hit ground stroke with the ball approaching at 40 miles/h and leaving at 50 miles/h.

As is evident from Fig. 5 a deflection of 3.5 cm is well into the region where the cubic term becomes important. This means that the ball will spend less time in contact with the strings than the linear calculation would predict. The agreement between the calculation and the dwell time measurements that were made is because the measurements were made with ball velocities that corresponded to 1–1½ cm deflections of the strings where the linear approximation is good.

In the limit of infinite tension in the strings, the dwell time is determined by the ball only. Measurements made under these conditions (bouncing the ball off a brick) gave dwell times of about 4 msec.

[1]Howard Head, U.S. Patent No. 3999756, Dec. 28, 1976.

[2]It is interesting to note that string tension is the one parameter that the average player can easily vary a great deal since a racket can be strung with anywhere from 35 to 75 lb of tension. The other parameters (racket weight, balance, handle circumference, head size, flex, or whip) have a much narrower range of variations—among the readily available commercial rackets. Yet the string tension is probably selected by many players in a completely irrational manner—"Gee, that sounds good."

Physics of the tennis racket II: The "sweet spot"

H. Brody
Physics Department, University of Pennsylvania, Philadelphia, Pennsylvania 19104
(Received 22 September 1980; accepted 17 November 1980)

The term sweet spot is used in describing that point or region of a tennis racket where the ball should be hit for optimum results. There are several definitions of this term, each one based on different physical phenomenon. In this paper the different definitions are discussed and methods are described to locate the points corresponding to each one.

One of the problems in trying to analyze the "sweet spot" of a tennis racket is that, not only are there several different definitions of sweet spot, but some people consider it a point and others claim that it is a region on the face of the racket. To add to the confusion, most people who use the term, particularly in advertisements, never bother to define it at all.

Howard Head, in his ads for the Prince racket and in his patent,[1] defines the sweet spot to be that region on the racket face where the coefficient of resitution (COR) is above some arbitrary value when the racket handle is firmly clamped. To reduce confusion, this area should be called the power region and the place where the COR maximizes should be called the power point.

A second definition claims that the sweet spot is the center of percussion. This is a well-defined point (or pair of conjugate points) for a rigid body and the location of this point can be found by using the racket as a physical pendulum and measuring its period. Since the center of percussion was covered in the previous paper,[2] it will not be discussed further here.

A third definition has the sweet spot as the point where, when the ball hits, the resulting higher frequency vibrations or oscillations are a minimum. This point is the node and the region around the node where the vibration is below some arbitrary value will be designated as the nodal region.

Many people use the term sweet spot to designate that point or area on the racket where it feels good when you hit the ball. This is quite subjective and hard to define, but it is probably closely related to some combination of the previous definitions.

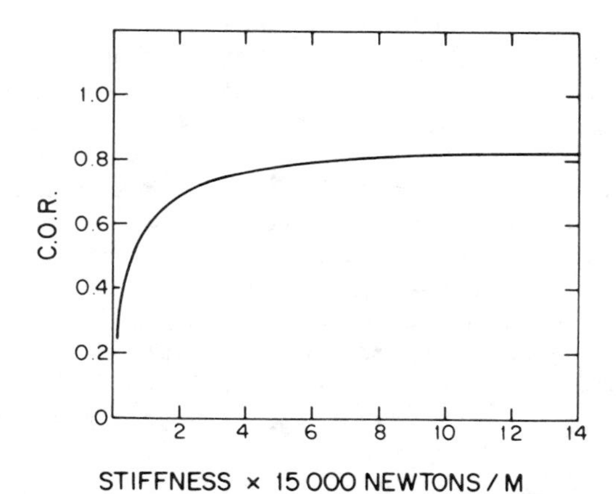

Fig. 1. Coefficient of restitution verses racket stiffness for a ball hitting at the center of the string area when the racket handle is firmly clamped. A typical racket will deflect 1 cm for a force of 100 N, hence will have a stiffness of 10 000 N/m.

POWER REGION

There is usually one area on the stringed portion of the face of a racket where the coefficient of restitution (COR) is a maximum. Contours of equal COR can be determined experimentally by having a tennis ball strike the racket at various places and measure the ratio of rebound velocity to incident velocity with the racket handle firmly clamped. Another way to accomplish this is to drop a ball on the face of the racket and measure the ratio of rebound height h_r to drop height h_d The COR is $\sqrt{h_r/h_d}$. It is found that the COR tends to maximize along the longitudinal axis of the racket and peaks fairly close to the throat—not near the center of the head. If the ball, string, and racket system is analyzed, it is quite clear why this is the case. It was shown in paper I that any energy that goes into the racket's deformation is lost since the ball has left the strings by the time the racket springs back. It was also shown that the higher the string tension the more the ball deforms upon impact hence the more energy the ball dissipates. Therefore, to increase the COR, more energy should be absorbed by the strings (lower tension) and less energy go into racket deformation (stiffer frame). Using this simple model, the data for Fig. 1 (COR versus racket stiffness) was generated.

The reason that the maximum COR is along the longitudinal axis and displaced several centimeters up from the throat end of the face is due to the strings being "softest" at the center of the head and the frame being stiffest near the clamped handle. For the strings only (racket head clamped) the power spot would be in the center of the head. This last fact can be demonstrated experimentally by applying a known force to the strings at various locations and measuring the deflection or by calculating the deflection y as a function of positions for a given external force F. Referring to Fig. 2, with a single string of length L and tension T, the restoring force $F = T_1 \sin\theta_1 + T_2 \sin\theta_2$. Setting $T_1 = T_2 = T$ and $\sin\theta = \tan\theta$ gives

$$F = \frac{Ty}{L/2 - z} + \frac{Ty}{L/2 + z} = Ty\left(\frac{1}{L/2 - z} + \frac{1}{L/2 + z}\right) = \frac{TyL}{L^2/4 - z^2}.$$

If $z \ll L$, this can be expanded giving

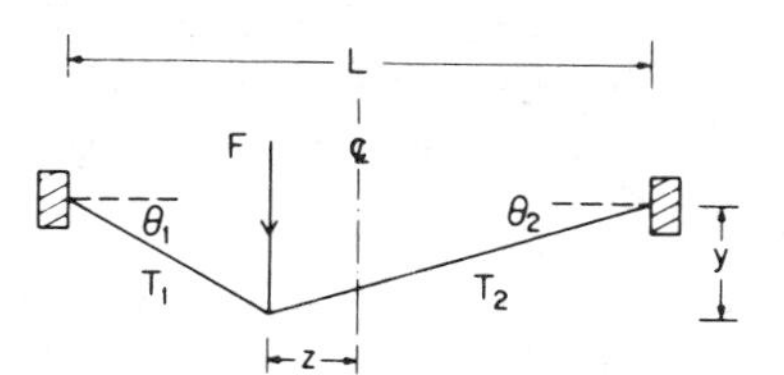

Fig. 2. Deflection y of a string of length L due to a force F applied at a distance Z from its center.

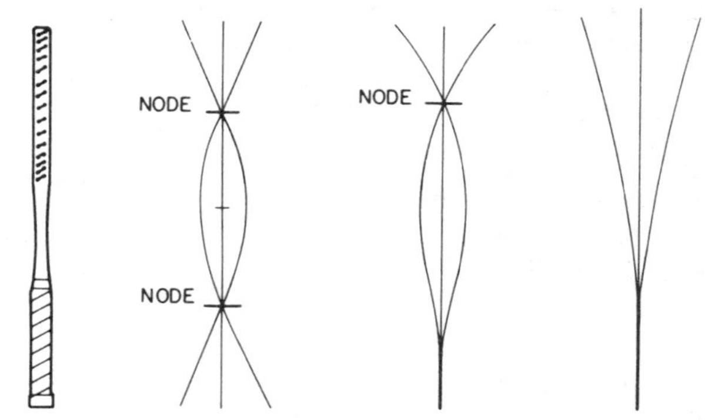

Fig. 3. Vibrations of a tennis racket for the free-free and the clamped-free modes.

$$F = (4Ty/L)\,(1 + 4z^2/L^2 + \cdots).$$

For a given deflection y, the force will be a minimum at the head center for two separate reasons. The strings that meet at the center of the head are the longest ones in the racket and this reduced the $4Ty/L$ term. (It is assumed that T, the tension, is the same for all strings.) At the center of the head $z = 0$, so the $4z^2/L^2$ term is zero. This would produce a maximum of the COR in the center of the head if the frame were absolutely rigid (or the head clamped). Since it is the handle that is clamped in the COR tests, the frame flexibility must be taken into account. The racket frame deflection of several rackets was measured for a given force as a function of longitudinal position x from the clamp. It varied for different rackets, but the deflection was of the form kx^m where m was of order 2. The racket becomes much stiffer closer to the clamped end. This frame flexibility term dominates over the previously determined string term, shifting maximum COR close to the throat of the racket.

There are a number of ways to increase the COR of a racket. When the head of a racket is enlarged by extending it toward the handle (without increasing the overall racket length) a region of higher COR is produced compared to a racket of standard head size with the same flexibility. To make use of this region of higher COR in actual play, the ball must be hit closer to the handle of the racket. This will lead to a higher ball velocity if the racket motion is translational and all points on the head have about the same velocity. If the racket is rotating (such as in a serve where it is whipped or snapped), the higher COR region, being closer to the effective axis of rotation, is moving with a lower velocity than the region near the racket tip. On this type of swing, maximum ball velocity is not obtained by hitting the ball at the point in the racket where the COR maximizes, but at a point where the racket velocity is high if the racket is not too flexible. It is interesting to note that some people, upon examining their racket, will find two separate areas of wear on the strings. One area is above the center of the head and indicates where they have been hitting the serve. The second area of wear is below the center and is the result of ground strokes.

A second method to increase COR is to make the frame itself stiffer by increasing its cross-sectional area or by going to a new material. When wood was the only material used in racket construction, a stiffer racket meant a heavier racket. With the availability of composite materials, it is now possible to have a racket that is both light and quite stiff while being capable of standing up to the pounding that a racket is subject to.

A third way to increase a racket's COR is to use strings at reduced tension, however, that is not the only important parameter. In examining the formula for string deflection $F = 4T/L\,y\,(1 + 4z^2/L^2 + \cdots)$ it is clear that what is important is the string tension divided by string length and not the tension alone. It is then obvious why, to get a similar "feel," a larger headed racket is strung at a higher tension.

NODE

If the handle end of the racket is firmly clamped, and the racket is struck, it will oscillate. The fundamental mode of vibration has no nodes and a frequency of 25–40 Hz depending upon the mass, mass distribution, and stiffness of the racket (Fig. 3). The next higher mode of oscillation has a single node and, if the racket were a uniform beam of length L, the frequency would be about six times higher than the fundamental and it would have a node a distance $L/5$ from the tip. If the racket were to be considered a free uniform beam of length L (no clamping) its fundamental frequency would also be about six times the clamped fundamental and it would have two nodes, $L/5$ from each end. Many people consider the excitations of this higher mode of oscillation to be undesirable—leading to loss of control, fatigue by the player, and a generally unsatisfactory feeling when the ball is hit. The node is a clearly defined point (or line) while the size of the nodal region is subjective since it is determined by how much amplitude of the higher mode of oscillation one is willing to accept. The next higher modes above these for both the free-free and free-clamped beam are sufficiently high in frequency and damp out so quickly that they can be ignored.

Since a racket is not a uniform beam, but a relatively complicated structure, the frequency and amplitude of the oscillations for the free-free racket and free-clamped racket must be determined experimentally. This has been done in the case of the free-free racket by Lacoste[3] and Hedrick *et al.*[4] using similar techniques. Hedrick freely suspended a racket, mounted an accelerometer on the handle, and hit the strings at various locations. The accelerometer recorded an impulse plus an oscillation. The impulse is not present when the ball hits the center of percussion (COP) and the oscillation is not present when the ball hits the position of the node of the fundamental free-free mode of oscillation. This latter point is then called the sweet spot by those authors and they have determined its position for a large number of rackets. They did not report the distance between the COP and the node.

To measure the oscillations of a racket with the handle clamped, a method was developed to determine the position of the tip of the racket as a function of time. This method is illustrated in Fig. 4 and a series of photographs was obtained corresponding to a ball hitting at different locations along the polar (longitudinal) axis of a racket. When the ball struck at the location of the node of the higher mode of oscillation, the tip executed a pure simple harmonic motion at the (clamped-free) fundamental frequency [Fig. 5(b)]. As the location of the ball impact moved away from the node, an increase in amplitude of the higher frequency becomes apparent [Fig. 5(a) and (c)]. In addition, it is quite

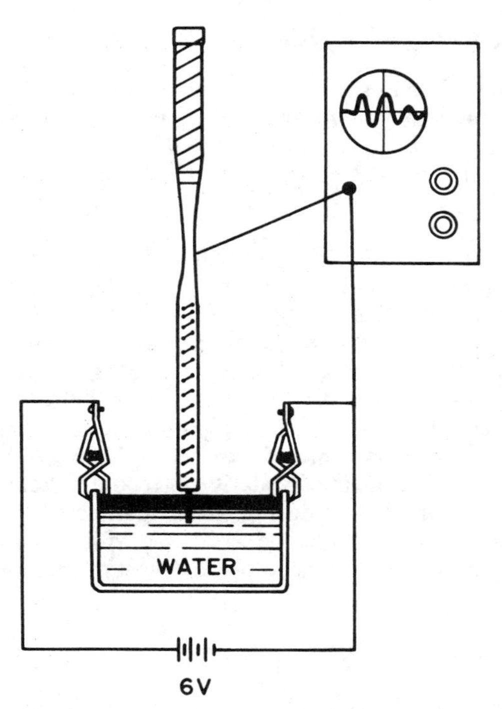

Fig. 4. Method used to determine the position of the racket head as a function of time. The racket handle was clamped (not shown) and the face of the racket hit by a ball. A light-photodetector system (not shown) triggered the scope.

clear that the phase of the higher-frequency oscillation relative to the fundamental frequency of oscillation changes when going from one side of the node to the other.

The apparatus that was used consisted of a glass beaker filled to the top with tap water, a pinch of salt, two clip leads, a battery, and a fine, stiff wire that was taped to the tip of the racket and the other end connected to an oscilloscope. Less than a centimeter of wire protruded into the water. The scope was triggered by a light source—photodetector system that was interrupted by the ball just before it hit the strings. The ball was suspended on a thread (as a pendulum) and released each time from the same height to give reproducible results.

The free-free and free-clamped experiments should give roughly the same answer for frequency and the node location except that when the racket is clamped, its length is now shorter than the length of the free racket by about 10 cm. This means that these two methods will give slightly different results for the location of the node, but the discrepancy should be no more than 2 or 3 cm. However, since in actual use a racket is neither free nor clamped but hand held, to get a true answer Enoch Durbin of Princeton has mounted an accelerometer on the handle of a hand-held racket and determined the node under those conditions.[5] One would expect the values he obtains to be between the free-free and clamped-free values, but detailed comparisons on similar rackets have not been performed.

Since the hand is a sensitive detector of vibrations, it is possible to get a rough location of the node and the size of the nodal area without any instrumentation. Hold the racket in your hand, face up, and drop a ball onto the strings at various locations. With a ball drop of 20 cm or so you should be able to excite a large enough oscillation to be easily detected.

A series of measurements was made on a number of rackets to determine the location of the node, (with the handle clamped) and the frequencies of oscillation. The data for a few selected rackets are given in Table I. The actual position of the node for a given racket is a function of the mass distribution of the frame and the relative flexibility of various parts of the frame. (Some rackets have stiff shafts and a flexible head, some are the opposite, and some are uniform.) In addition, the flex of the head is influenced by the string tension and pattern, so this can move the node position. To obtain the size of the nodal region, the amplitudes of the oscillation would have to be determined. To do

Table I. Racket oscillation parameters.

Racket	Frequency of fundamental mode (Hz)	Frequency of high mode (Hz)	Distance of node from racket tip (cm)
Prince Pro	36.2	168	13.5
Prince Classic	31.0	127	13.5
Wilson T-2000	26.0	132	13
Head Master	32.5	168	13
Durbin	30.6	138	13
Durbin (with 10 g on tip)	29.2	133	11.5

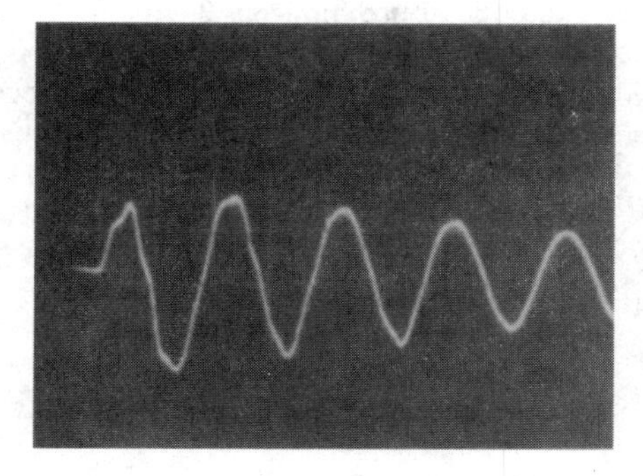

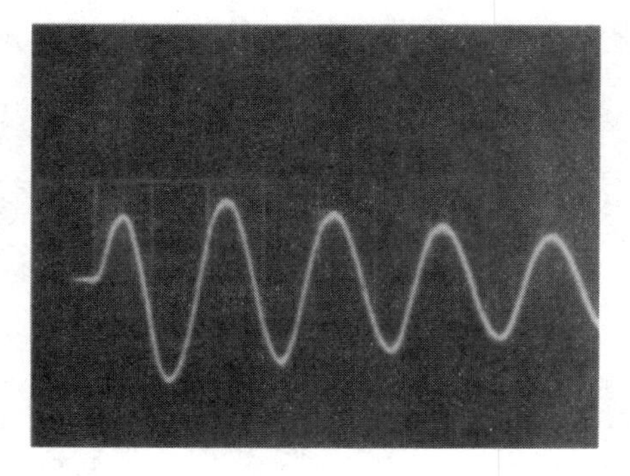

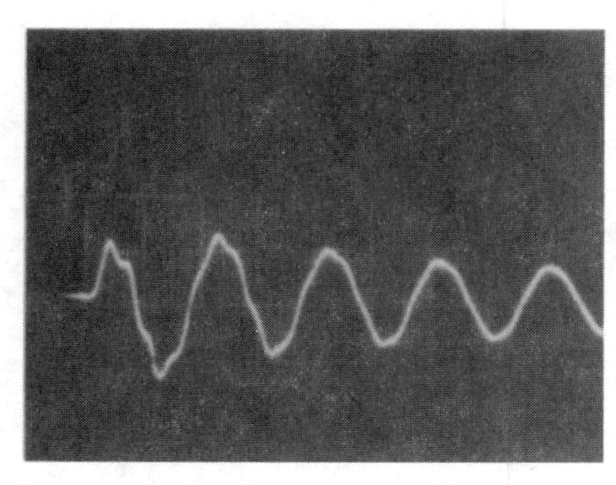

Fig. 5. Typical oscilloscope traces showing racket head position as a function of time. Sweep speed is 20 msec/div.

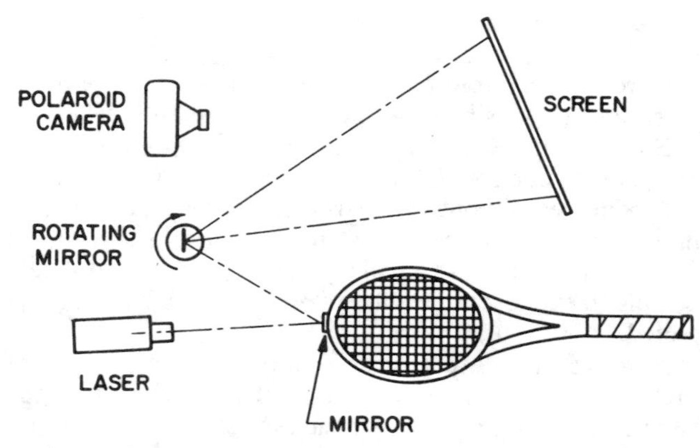

Fig. 6. Alternate method to determine position of racket head as a function of time.

this the resultant output from the liquid potentiometer should be frequency analyzed to get the ratio of the amplitude of the higher mode to the amplitude of the fundamental mode.

The shift in the position of the node was also determined for a single racket when a small mass was taped to the tip. 10 g moved the node approximately 2 cm.

The three definitions of sweet spot all have merit and, in general, the points corresponding to them are located in different places. It is assumed that if the ideal tennis racket could be designed, it would have all three of these points located at the center of the stringed area and have a power region and nodal region covering most of the face of the racket.

ACKNOWLEDGMENTS

I would like to thank Enoch Durbin for several helpful discussions and Jim Hobbs and Oleh Hnatiuk for their help in taking some of the data presented in this paper.

APPENDIX

A second method for observing both the fundamental and higher modes of oscillation with the handle clamped was used and it has the virtue of not requiring an oscilloscope, scope camera, or explanation of why the water beaker system works. (The liquid potentiometer is a pedagogical gem in this author's opinion.) This second method requires a light source (laser), mirror chip glued to the racket tip, rotating mirror, screen, and a camera (Fig. 6). It does require a certain amount of dexterity and coordination to have the ball hit the racket during the time interval when the beam of light is traversing the screen, but with some practice it was possible to get a good picture most of the time.

[1]H. Head, U.S. Patent 3999756, 28 December 1976.
[2]H. Brody, Am. J. Phys. **47**(6), 482 (1979).
[3]F. R. Lacoste, U.S. Patent 3941380, 2 March 1976.
[4]K. Hedrick, R. Ramnath, and B. Mikic, World Tennis **27**(4), 78 (1979).
[5]E. Durbin, patent pending and private communication.

The moment of inertia of a tennis racket

HOWARD BRODY

The moment of inertia of a tennis racket about its principal axes should be of interest to tennis players in that it influences how the racket plays and how the racket feels when it is swung. To the best of my knowledge, this information is not available from any of the conventional sources (racket retailers, racket manufacturers, tennis professionals, tennis books, or the magazines that test and evaluate rackets). The measurement of these moments is a very straightforward exercise that can be done easily in an introductory-level physics laboratory. If several different shape or size rackets are available to be measured, the results can be compared and the variation of the moments with the obvious physical dimensions of the rackets can be observed.

The basic measurements are accomplished by suspending a racket on a wire (Fig. 1), measuring the period of the torsional oscillations, and then converting these results into the values of the moments of inertia of the rackets. The equipment needed is a wire, a clock or stopwatch, a method of securing the rackets to the wire, and an object of a mass comparable to a racket, but of known moment of inertia to calibrate the torsional constant of the wire. This latter object can be a rectangular block disk or other object the value of whose moment can be calculated or found in one of the standard introductory textbooks. Having the mass of the calibration block comparable to the mass of the racket insures that the torsional constant of the wire is the same for both the calibration and data taking (most tennis rackets have a mass between 0.350 and 0.40 kg).

The period, T, of a torsional oscillator is

$$2\pi\sqrt{I/k}$$

where I is the moment of inertia of the object and k is the torsional constant of the wire. Solving for I gives

$$I = k\,T^2/4\pi^2 \tag{1}$$

The value of k is determined using an object of known moment in the oscillator (I used a 0.5986-m uniform rod of mass 0.3743 kg and of moment $(1/12)\ \mathrm{m}\ L^2 = I_o = 0.01118$ kg-m^2) which had a measured period of oscillation $T_o = 29.58$ s. Substituting this into Eq. (1) gives the value of the unknown moment, I_x

$$I_x = I_o T_x^2/T_o^2 = 1.278 \times 10^{-5}\ \mathrm{kg\ m^2}\ T_x^2$$

The moments are measured about three perpendicular axes passing through the center of mass of the racket and then the parallel axis theorem is used to determine the moment about other axes, such as one passing through the handle. It is also instructive to see if the sum of the moments about the two axes in the plane of the racket is equal to the moment about the third axis, which is a theorem that holds for a two-dimensional object (see Appendix 1).

The moment along the long axis (Polar or roll moment of inertia)

This moment determines the racket's resistance to twisting when a ball is hit off center. The larger the moment, the more stable the racket will be when the ball is not hit along the long (polar) axis of the racket. This moment can be increased by adding discrete masses to the outside edge of the head (as Wilson has done in their Perimeter Weight System rackets), or by increasing the width

Howard Brody *received his S.B. from MIT and his PhD from Cal-Tech in physics. He is a professor of physics at the University of Pennsylvania where he does research on elementary particles. He has written several articles on the physics of tennis, lectured extensively on the subject, and is seen occasionally on the tennis courts. (Department of Physics, University of Pennsylvania, Philadelphia, Pennsylvania 19104)*

Reprinted from *The Physics Teacher* **23**, 213–216 (1985); © American Association of Physics Teachers.

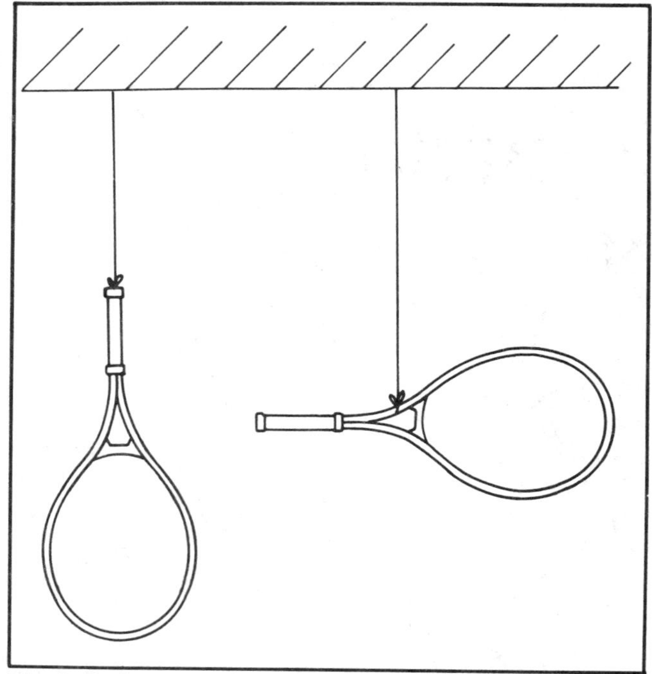

Fig. 1. Racket suspension to measure the polar moment of inertia (a) and the moment about the other in-plane axis (b).

Table I

Polar moment of inertia for several tennis rackets

Racket	Mass (kg)	Maximum head width (m)	Polar moment of inertia ($kg\ m^2$)
Head Master	0.365	0.238	0.00117
Wilson T-2000	0.375	0.242	0.00121
durbin aluminum	0.399	0.249	0.00136
Prince Graphite 90	0.357	0.265	0.00138
Prince Pro 110	0.357	0.275	0.00155
Prince Classic 110	0.372	0.278	0.00170
Prince Graphite 110	0.388 0.380	0.277 0.283	0.00175 0.00179
Prince Graphite 125	0.350	0.301	0.00191
Weed aluminum	0.354	0.316	0.00224

of the head (as Howard Head has done in the Prince racket or Tad Weed has done in the Weed racket). Because the moment goes as the mass times radius squared, this second method of increasing the polar moment is quite effective as can be seen by the data in Table I.

Since it is unlikely that you will be able to do this experiment when you are in a sports shop trying to buy a racket, the measured moments of several rackets are plotted in Fig. 2 versus the mass times the maximum width of the head squared. As you can see, all the rackets tested fall along a single line, and therefore it is possible to get a good idea of the polar moment of a racket by simply measuring the head diameter and using Fig. 2. (The Wilson PWS rackets may not fall on this line because they should have a larger polar moment than a conventional racket for a given maximum head diameter due to the extra discrete masses added at the edges of the frame.) The conclusion is clear. The wider the head of the racket, the less likely it is that the racket will twist in your hand on an off-axis hit.

The other moments of inertia

The moment of inertia around an axis perpendicular to the handle can be determined by two methods. The torsional oscillator method described above can be used to get the moment about the axis through the c.m., perpendicular to the previous axis, and then the parallel axis theorem ($I_d = I_{cm} + m\,d^2$) can be used to determine the value about an axis passing through the handle (Fig. 3). (It is very difficult to use the torsional method about an axis not going through the c.m.) A second method for determining the moment about an axis through the handle is to turn the racket into a physical pendulum with the desired axis as the pivot axis. The period of this pendulum is given by the formula

$$T = 2\pi\sqrt{I/mgd}$$

where m is the racket mass, $g = 9.8\ m/s^2$ and d is the distance between the pivot point and the c.m. Solving this for I yields

$$I = T^2 mgd/4\pi^2$$

These two methods will work whether the axis selected is in the plane of the tennis racket or perpendicular to the plane of the racket.

The moment of inertia about the x-axis through the handle and in the plane of the racket is known colloquially as the "swing weight," because it is a measure of how "heavy" the racket feels when you swing it and what the racket's "hitting mass" is when the ball is struck. There are reasons for wanting to use a large swing weight and there are reasons for wanting to have a small swing weight, depending on your own athletic ability, strength, and the style of tennis that you play. A person who hits mainly ground strokes from the baseline and plays on "slow" courts may profit by using a racket which has a larger swing weight, or is head-heavy. A player who likes to serve and volley or plays on "fast" courts, should use a smaller moment, lower swing-weight racket to optimize performance. Table II gives the measurements made in determining this moment for several rackets about an axis 0.076 m (3 in.) from the butt end of the handle. This axis could be at the butt end of the racket or through a point where your hand grips the racket (0.076 m from the butt). This second location is selected here so that the center of percussion

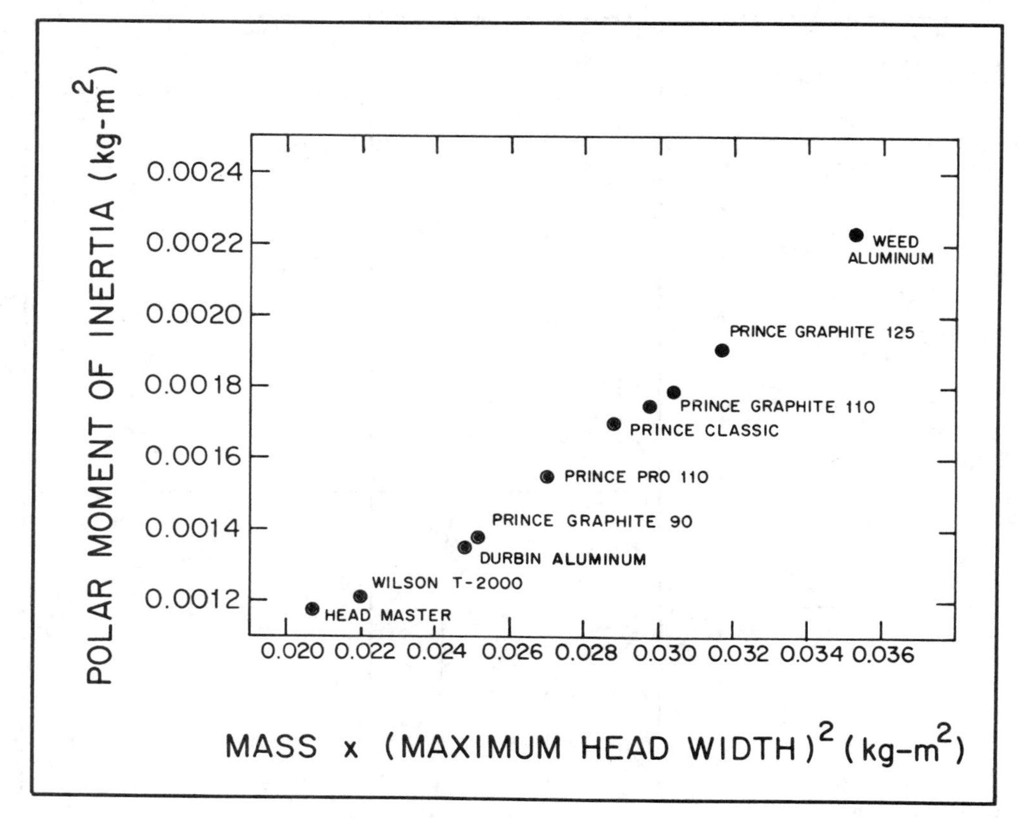

Fig. 2. Measured polar moment of inertia for several rackets versus the mass of the racket times its head diameter squared.

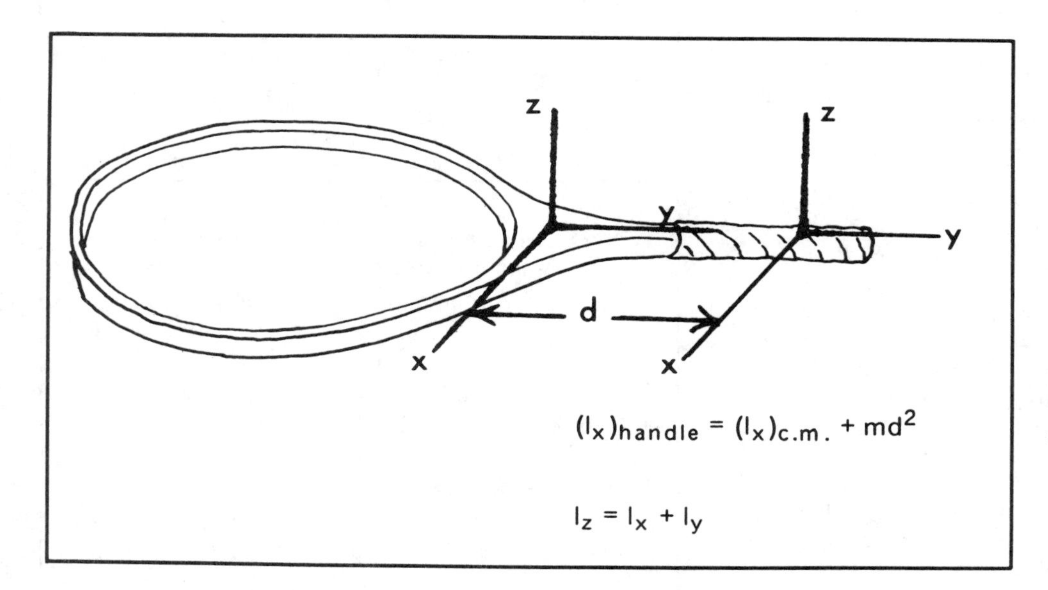

Fig. 3. The axes of a tennis racket through the c.m. and the axes that a tennis player tends to rotate the racket around.

point as defined by one tennis manufacturer[1] can be determined from this data.[2]

The moment about the z-axis perpendicular to the plane of the racket is of concern only when spin is being applied to the ball with a shot such as in a "slice" or "chop" or when serving. Measurements on the moments around the other two axes through the handle but in the plane of the racket, show that the moment about the x-axis is about 20 times the moment about the y-axis. Therefore, the moment about the axis through the handle, but perpendicular to the racket plane, being the sum of the other two moments, will only be about 5% greater than the "swing weight." For this reason it was only measured for a few rackets.

Table II

The "swing weight" of several tennis rackets

Racket	Moment about axis through c.m. (kg m^2)	m d^2 (kg m^2)	"swing weight" (kg m^2)
Wilson T-2000	0.0164	0.0260	0.0424
Prince Graphite 90	0.0143	0.0218	0.0361
durbin	0.0185	0.0228	0.0413
Prince Classic	0.0158	0.0236	0.0394
Prince Graphite 110	0.0175	0.0232	0.0407
Prince Graphite 125	0.0160	0.0213	0.0373
Weed	0.0154	0.0209	0.0394

When the parallel-axis theorem is used to calculate the value of the moment about the axis through the handle, the md^2 term is slightly larger than the term for the moment about the axis through the c.m. Therefore, when you are buying a racket, if you know the mass and the distance from the axis to the c.m. (a relatively simple measurement to perform in a store), you can get some estimate of the swing weight from these static measurements.

Appendix

For a rigid body that is two dimensional (lies in a plane), the sum of the moments of inertia about two perpendicular axes that lie in the plane is equal to the moment of inertia about the axis that is perpendicular to the plane at the point where the other two axes intersect. Figure 4 shows a schematic racket made up of many mass points, m_i. The moment about the x axis, I_x is $\Sigma m_i y_i^2$ and about the y axis, I_y is $\Sigma m_i x_i^2$. The moment about the z axis, I_z is $\Sigma m_i r_i^2$. As can be seen in Fig. 4, $r_i^2 = x_i^2 + y_i^2$, so $I_z = I_x + I_y$.

This was verified by determining the moment about the three axes through the center of mass of several rackets (Table III). (The theorem holds for axes through any point, so the c.m. was used because of the ease of doing the measurement at that point.) Since the same torsion wire is used to measure all three moments, it is not even necessary to calibrate the wire, and showing that $T_z^2 = T_x^2 + T_y^2$ (where T is the period of oscillation about the ith axis) is sufficient to show that $I_z = I_x + I_y$.

As an example, for the Wilson T-2000, the sum of the periods squared about the in-plane axes is 1376.7 ± 3.6 s^2, while the period squared about the axis perpendicular to the racket plane is 1367.5 ± 1.9 s^2. Even though these agree to within one percent, they differ by more than the experimental uncertainty. For this theorem to hold, the object must be two dimensional and if it has a thickness, you would expect the sum of the moments of inertia about the in-plane axes to be greater than the moment about the third axis. The maximum thickness of a racket (the handle) is about 0.04 m, and the maximum width and length of a racket are 0.27 m and 0.68 m, respectively. The square of the ratio of the thickness to the other dimensions is small, but obviously cannot be neglected at the level of a few tenths of a percent.

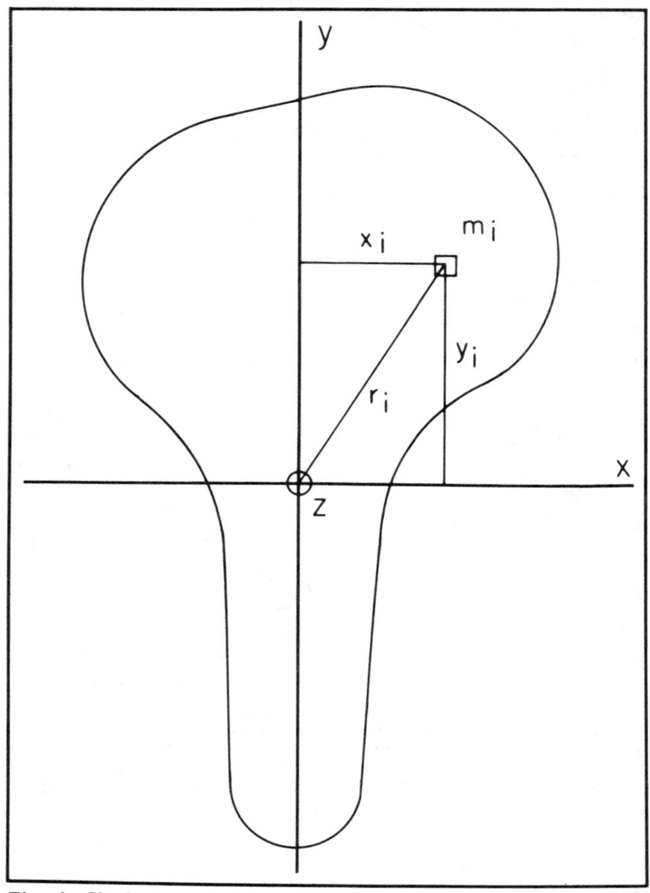

Fig. 4. Single mass point (m_i) at a distance y_i from the x axis, x_i from the y axis, and r_i from the z axis.

Table III

The period squared of oscillation about the three principal axes of several tennis rackets

Racket	Period^2 x axis (s^2)	Period^2 y axis (s^2)	Sum (s^2)	Period^2 z axis (s^2)
Prince Graphite 110	1370.7 ± 3.7	136.7 ± 0.4	1507.2 ± 3.7	1487.6 ± 3.9
Prince Classic	1238.3 ± 1.8	133.4 ± 0.6	1371.7 ± 1.9	1369.0 ± 1.9
Wilson T-2000	1281.6 ± 3.6	94.5 ± 0.5	1376.1 ± 3.6	1367.5 ± 1.9

References

1. H. Head, U.S. Patent No. 3999756, Dec. 28, 1976.
2. H. Brody, Am. J. Phys. 47, 482 (1979).

That's how the ball bounces

HOWARD BRODY

A tennis court's dimensions are specified exactly to the centimeter by the rules of the game, but nowhere do those rules mention the composition of the surface material that the game is to be played on. This leads to a variety of types of bounce because tennis is played on clay, wood, synthetic and rubber carpets and tiles, concrete, and even grass. Some of these surfaces are characterized as being "fast," some are called "slow" – but this is quite subjective. This article is an attempt to quantify the interaction of a tennis ball and the court surface, and show how, by making a few simple measurements, one court surface can be compared to another court surface.

The result of the interaction of a ball and a surface can be analysed to a first-order approximation if two physical characteristics of the ball and surface are known – the coefficient of kinetic (sliding) friction and the coefficient of restitution. The values of both these coefficients have been obtained for several types of surfaces and a description of these measurements is given in the appendix.

When a tennis ball, initially having no spin, strikes the ground, it will acquire a forward spin due to the frictional force between the ball and the surface of the court. If this frictional force is large enough and the angle at which the ball strikes the surface is also large, the ball will begin to roll before losing contact with the ground. If the frictional force is small or the ball hits the court at a small (glancing) angle, the ball will slide or skid and not go into the rolling mode before losing contact with the ground. It is possible to calculate the rebound angle (ϕ) and the reduction in the ball's forward velocity in terms of the incident angle (θ), the coefficient of restitution (e), and the coefficient of friction (μ), for both these cases. (The fractional reduction in the magnitude of the vertical velocity component upon bouncing is given by the coefficient of restitution.) Note that unlike the custom with light rays, the incident angle here is measured with respect to the surface.

Howard Brody *received his S.B. from MIT and his PhD from CalTech in physics. He is a professor of physics at the University of Pennsylvania where he does research on elementary particles. He has written several articles on the physics of tennis, lectured extensively on the subject, and is seen occasionally on the tennis courts. (Department of Physics, University of Pennsylvania, Philadelphia, Pennsylvania 19104)*

Analysis

When a ball strikes the gound there are three forces acting on it, gravity (mg), a normal force (N), and friction (μN). As shown in Fig. 1, the normal force and the gravitational force are perpendicular to the surface while the friction force is parallel to the surface and opposing the direction of the ball's motion. The normal force, N, is equal to the gravitational force plus the mass of the ball times its acceleration in the vertical direction as it comes to rest and then shoots up again.

$$N = mg + m\, dv_y/dt \qquad (v_y \text{ is the vertical component of the ball velocity.})$$

For a tennis ball in normal play, the mg term can be neglected with respect to the $m\, dv_y/dt$ term. In play, a tennis ball must cross the net that has a height, at its lowest point, of three feet (0.91 m). Assuming the ball trajectory has 1 m as its highest point, the vertical velocity component of the ball (v_y) will be 4.4 m/s when it hits the ground. A tennis ball dropped from a height of 1 m has a coefficient of restitution (e) of 0.78 and a contact time, t, with the ground of 5×10^{-3} s.[1] With this data we can calculate the average acceleration, dv_y/dt, while the ball is in contact with the ground. The vertical velocity of the ball when it leaves the ground after bouncing, v_{fy} is minus e times the ball's vertical velocity v_{iy} just before hitting the ground. Therefore the difference in vertical velocity is v_{iy} $(1 + e)$, and the acceleration will be v_{iy} $(1 + e)/t = 1600$ m/s^2. Even for this "gentle" bounce, the $m\, dv_y/dt$ term is 150 times as large as the mg term. Consequently, N can be set equal to $m\, dv_y/dt$ with essentially no

Reprinted from *The Physics Teacher* **22**, 494–497 (1984); © American Association of Physics Teachers.

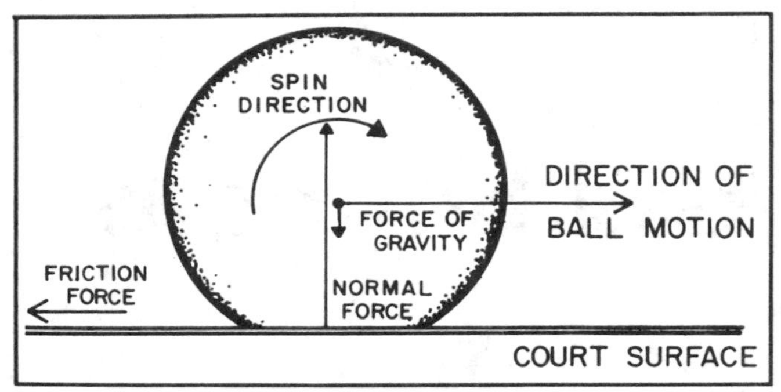

Fig. 1. Forces on a ball as it hits the ground.

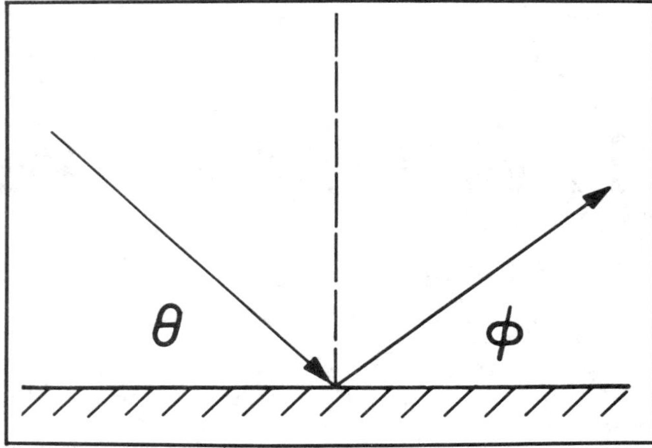

Fig. 2. Incident angle θ and rebound angle ϕ.

error. If the highest point in the trajectory is greater than 1 m or if the ball is hit with an initial downward velocity component (which does not happen often), this approximation becomes better.

The approximation will also be made that the ball remains essentially spherical in shape (does not deform) during the time that it is in contact with the ground. To check this assumption, let us calculate the average normal force of the court on the ball (N) during contact for a 0.057 kg tennis ball.

$$N = m\, dv_y/dt = 0.057 \text{ kg} \times 1600 \text{ m/s}^2$$
$$= 90 \text{ N}$$

Rule 3 of the USTA[2] states that a ball shall be between 6.35 and 6.67 cm in diameter and under a load of 8.165 kg (an 80-N force), shall deform between 0.56 and 0.72 cm. Therefore, the average ball deformation is of order ten percent of its diameter, so calling the ball spherical is not too bad an approximation.[3] If the highest point in the trajectory is greater than 1 m, this approximation becomes worse.

Case 1. Friction small — The ball does not acquire enough angular velocity to roll.

The change in the x (horizontal) momentum component of the ball is equal to the impulse due to friction, where v_{iy} and v_{ix} are the initial vertical and horizontal components of the velocity just before striking the court, and v_{fy} and v_{fx} are the vertical and horizontal components of the velocity just after striking the court.

$$m\,(v_{ix} - v_{fx}) = \int F_x\, dt = \int \mu N\, dt$$

Since N is equal to $m\, dv_y/dt$, this may be substituted into this equation, giving

$$m\,(v_{ix} - v_{fx}) = \int \mu m\,(dv_y/dt)\, dt = \mu m \int dv_y$$
$$= \mu m\,(v_{fy} - v_{iy}) = \mu m\, v_{iy}\,(1 + e)$$

Note that this expression is independent of the time that the ball spends in contact with the court surface.

Dividing through this equation by m yields

$$v_{ix} - v_{fx} = v_{iy}\,(1 + e)\mu \qquad (1)$$

Dividing this equation through by v_{iy}, referring to Fig. 2, and noting that

$$\tan\theta = v_{iy}/v_{ix},\ \tan\phi = v_{fy}/v_{fx},\ \text{and } e = v_{fy}/v_{iy}$$

gives

$$\cot\theta - (e)\cot\phi = (1 + e)\mu$$

Solving this equation for $\tan\phi$ yields

$$\tan\phi = e/\left[\cot\theta - (1 + e)\,\mu\right] \qquad (2)$$

Dividing Eq. (1) by v_{ix} yields the fractional change in the horizontal velocity

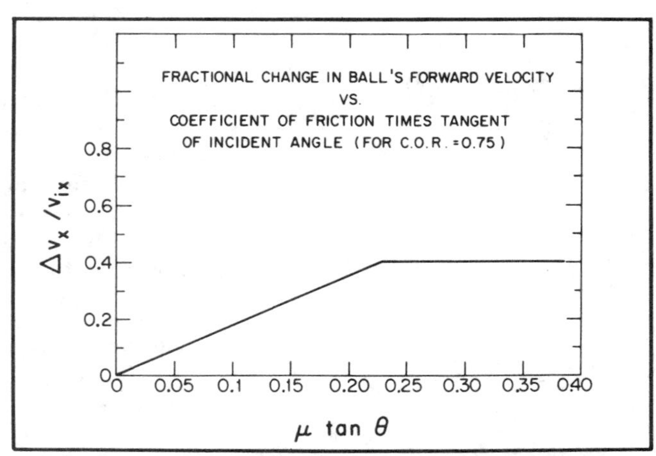

Fig. 3. Fractional change in a ball's forward velocity versus the product of the coefficient of friction and the tangent of the incident angle θ for a coefficient of restitution, $e = 0.75$.

$$(v_{ix} - v_{fx})/v_{ix} = (1 + e)\,\mu\, v_{iy}/v_{ix} = (1 + e)\,\mu\tan\theta \qquad (3)$$

and this is shown in Fig. 3 for $e = 0.75$.

The fractional change in the horizontal velocity is then proportional to the coefficient of friction and decreases as the angle of incidence decreases. This means that a ball that has a large initial horizontal velocity will have a smaller fractional velocity change than a softly hit ball. All of this will change, however, if the ball goes into the rolling mode as we will see later.

Since the coefficient of restitution enters the above relationship as $(1 + e)$, the observed variation in e from court to court will only change the velocity term by about three percent. The variations in μ, the coefficient of friction, observed for different surfaces will lead to changes that are an order of magnitude larger.

Using a typical value of the e (0.75), Eq. (1, 2, and 3) now become

$$v_{ix} - v_{fx} = 1.75\,\mu\, v_{iy} = 1.75\,\mu\sqrt{2gh}$$
$$\tan\phi = 0.75/(\cot\theta - 1.75\,\mu)$$
$$(v_{ix} - v_{fx})/v_{ix} = 1.75\,\mu\tan\theta$$

where h is the height of the ball (in meters) at the peak of its trajectory.

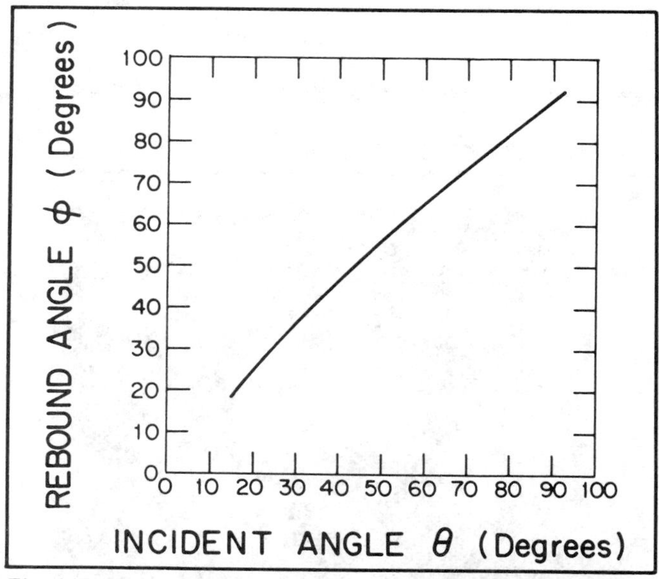

Fig. 4. Rebound angle of the ball ϕ versus the incident angle θ for a ball that has begun to roll before leaving the court surface when coefficient of restitution $e = 0.75$.

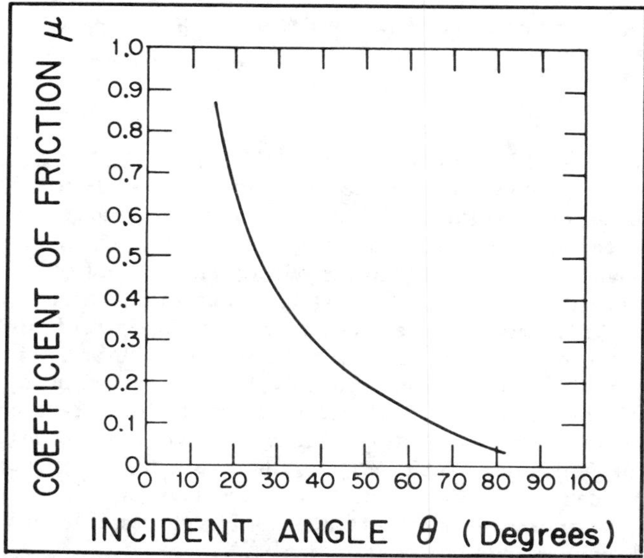

Fig. 5. Minimum coefficient of friction required to produce rolling versus the incident angle θ for coefficient of restitution, $e = 0.75$.

Case 2. Ball begins to roll before losing contact with the ground.

As in Case 1, the change in the horizontal momentum is equal to the impulse $\int F_x\,dt$ which is $\int \mu N\,dt$. However, since the ball changes from the sliding, skidding mode to rolling while in contact with the ground, the limits on the integral are not the full time of the ball contact with the ground. The $\int F_x\,dt$ term must be evaluated by making use of the fact that once the ball begins to roll, its angular velocity ω is v_{fx}/R, where R is the ball's radius and v_{fx} is the horizontal component of the ball's center of mass velocity after the sliding or skidding has ceased and rolling has begun. It is assumed that once rolling has begun, the ball will not slow down any more. The angular velocity ω is related to $\int F_x\,dt$ since $\omega = \int \alpha\,dt$ over the same limits of integration, where α is the ball's angular acceleration and it is caused by the torque produced by the friction between the ball and the ground.

$$\alpha = \tau/I = F_x R/I$$

where I is the ball's moment of inertia about its center. If the ball is considered a hollow sphere (which is a good description of a tennis ball), substituting $2\,m\,R^2/3$ for I yields

$$\alpha = 3\,F_x\,R/2\,m\,R^2 = 3\,F_x/2\,m\,R$$

$$\omega = \int \alpha\,dt = \int (3\,F_x/2\,m\,R)\,dt = (3/2\,m\,R)\int F_x\,dt$$

But since $\omega = v_{fx}/R$, this allows the evaluation of

$$\int F_x\,dt = 2\,m\,v_{fx}/3 = m\,(v_{ix} - v_{fx})$$

Dividing through by m yields

$$v_{ix} - v_{fx} = 2\,v_{fx}/3$$

$$v_{fx} = 0.6\,v_{ix} \tag{4}$$

The angle of the rebound ϕ can be obtained since $v_{fy} = (e)\,v_{iy}$

$$\tan\phi = v_{fy}/v_{fx} = e\,v_{iy}/(0.6\,v_{ix}) = e/0.6\cot\theta \tag{5}$$

Using the value 0.75 for e gives

$$\tan\phi = 1.25\tan\theta$$

and this is illustrated in Fig. 4.

The limiting case

The value of μ_m, the minimum coefficient of friction required to just bring the ball to rolling is then a function of the angle of incidence, θ, and the coefficient of restitution, e. This can be seen by setting the result of Case 1 (Eq. (2)] equal to the result of Case 2 [Eq. (5)], eliminating ϕ, and solving for μ.

$$\tan\phi = e/0.6\cot\theta = e/(\cot\theta - (1+e)\,\mu)$$

$$\mu_m = 0.4\cot\theta/(1+e) \quad \text{and if } e = 0.75, \quad \mu_m = 0.23\,cot$$

If you examine Fig. 3, you will notice that this corresponds to the point where the break in slope in the fractional change in forward velocity occurs ($\mu\tan\theta = 0.23$).

Figure 5 shows the minimum coefficient of friction μ_m versus the incident angle θ, for rolling to occur when $e = 0.75$.

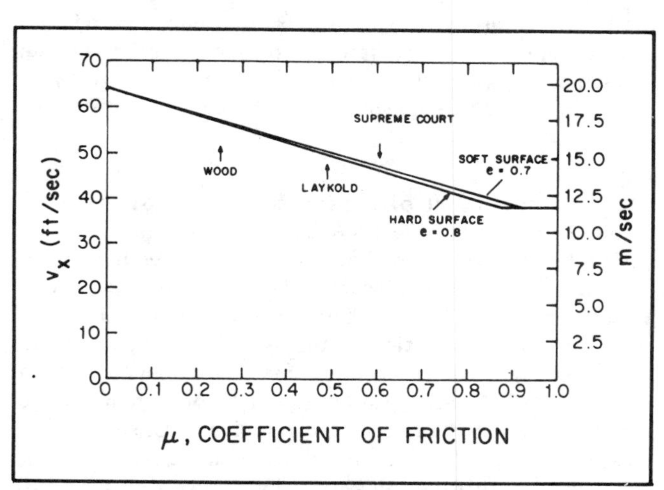

Fig. 6. Horizontal ball velocity after the bounce versus the coefficient of friction. The incident velocity is 19.5 m/s in the horizontal direction and 4.9 m/s in the vertical direction immediately before the bounce. The two lines are for courts with coefficients of restitution, e = 0.70 and 0.80. The three arrows indicate the measured coefficient of friction for a tennis ball on polished wood, Laykold and Supreme Court.

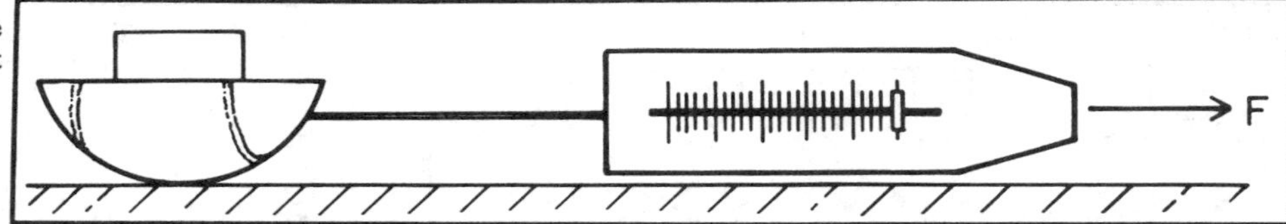

Fig. 7. Device used to measure the force needed to keep a weighted ball sliding at constant velocity.

The speed of a tennis court

Some tennis courts are considered "fast" while others are said to be "slow." This description refers to the change in the horizontal component of the ball velocity at the bounce. A large change in the velocity means a slow court, while if there is a small change in the horizontal velocity, it is a fast court. The analysis done in the first part of this paper then shows that for a tennis court surface, this "speed" is determined primarily by the coefficient of friction. A court is clearly "slow" when the ball goes into the rolling mode. Equation (4) shows that when this occurs, the horizontal velocity after the bounce is 0.6 times the horizontal velocity before the bounce. However, this does not happen for most of the ball trajectories that one encounters in tennis.

A well hit shot has a (downward) vertical velocity of about 5 m/s and a horizontal velocity of about 25 m/s (60 miles/hour) when it lands. This gives a cot $\theta = 5.0$. In order to get into the rolling mode, μ would have to be greater than 0.23 cot θ, or greater than unity. Working backwards and selecting a rather large value of μ (0.6, which is typical of a "slow" court), the ball would have to have a cot $\theta < 2.6$ ($\theta > 21$ degrees) to go into the rolling mode. Even for a ball with a moderate horizontal velocity of 20 m/s, this would correspond to a vertical velocity of 8 m/s at ground level. A vertical velocity of 8 m/s would require a peak in the trajectory of 3.2 m, which is a soft lob type of shot. Clearly it is not a typical drive or groundstroke. If μ were less than 0.6, the ball would have to have an even higher maximum in its trajectory in order to roll before leaving contact with the ground. Therefore, since most shots do not clear the net by more than a meter, when they bounce, the ball will remain in the sliding, skidding mode and the resulting loss of horizontal velocity will be proportional to the coefficient of friction.

On high lobs (shots that have a very high maximum in their trajectory) the ball will probably go into the rolling mode for almost any value of friction encountered in a normal tennis court. Therefore, on such shots, there is no difference between the "speed" of a "fast" court and a "slow" court. The ball will lose 40 percent of its forward velocity in both cases.

Appendix

The coefficient of friction is the ratio of the force required to keep an object moving at constant velocity on a horizontal surface to the force normal (perpendicular) to the surface – which is usually the weight of the object. Tests were conducted by slicing a new tennis ball in half and then making additional cuts in the rubber and felt so that it could be flattened out somewhat by masses placed within the half ball. The horizontal force required to maintain uniform motion of the ball on a court surface was measured (Fig. 7) and the coefficient of friction was determined by dividing this force by the weight of the ball plus added mass. Since there is the possibility that the usual approximation made in friction calculations (the coefficient of friction is a constant) may not be reliable in this case because the area of the ball in contact with the court surface changes as the ball flattens somewhat, measurements were made with masses of from 1 to 5.6 kg sitting on the ball. The extra weight tended to flatten the ball more, but it did not seem to change the coefficient of friction.

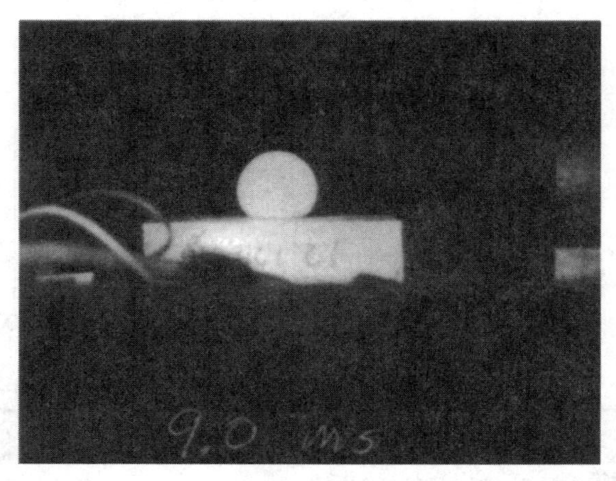

Fig. 8. Photograph of a ball at its maximum compression after being dropped from a height of about a meter onto a lead brick.

Surface	Coefficient of friction (μ)
Wood (Basketball court)	0.25
Laykold (Hard court, acrylic coated)	
near baseline	0.49
in alley, near net	0.53
Supreme court (Synthetic Carpet)	0.61

The coefficient of restitution (e) is the ratio of the vertical component of ball velocity immediately after the bounce to the vertical velocity immediately before hitting the surface. It varies with surface and with incident ball velocity and it is normally measured by dropping a ball from a set height (h_d) and measuring its rebound height (h_r). The coefficient of restitution is the square root of h_r/h_d. For a ball to meet USTA standards,[2] it must have an e between 0.73 and 0.76 when dropped from a height (h_d) of 254 cm onto a hard surface. The USTA specifies the temperature, barometric pressure, and humidity for the ball coefficient of restitution test, and I had always assumed this was because air resistance might affect the results. When I did a calculation for a 254-cm ball drop with and without air resistance, there was only a 4 percent change in the apparent coefficient of restitution. Therefore, the atmospheric specifications are in the tennis rules because of the effect of temperature, pressure and humidity on the properties of the ball itself, and not the air resistance.

When a ball is dropped onto a hard surface from a height which is similar to the heights encountered in a normal game, the measured e will be somewhat higher (about 0.78). Soft surfaces may have a lower e – for example a Supreme Court had a e of 0.70 with a drop height of 127 cm.

References

1. H. Brody, Am. J. Phys. 47 (6), 482 (1979).
2. USTA Yearbook.
3. These results seem contrary to what is expected because high speed photographs of tennis balls being squashed flat come to mind. Those pictures were taken of balls undergoing changes in velocity an order of magnitude greater than the change that occurs when a ball hits the ground at 4 m/s (Fig. 8).

The aerodynamics of tennis balls—The topspin lob

Antonín Štěpánek
ČVUT Faculty of Mechanical Engineering, Department of Physics, Suchbatarova 4, 166 07 Prague, Czechoslovakia

(Received 1 May 1986; accepted for publication 30 April 1987)

A general description is presented of the calculation of the ballistic trajectory of a flying spinning ball acted on, in addition to the forces of gravity and drag, by the so-called Magnus foree. By applying the regression analysis to results of wind-tunnel measurement of the drag and lift coefficients of a spinning ball, a calculation of the nonlinear differential equation of the hodograph was carried out by means of the Runge–Kutta method. The theoretical results that can be used to calculate the ballistic trajectories for any ball game were applied to one of the most difficult and most interesting tennis strokes, i.e., to the topspin lob. Practical results obtained for various distances are presented in a table as well as in graphical form.

LIST OF SYMBOLS

v	ball flight velocity
w	equatorial velocity of the spinning ball
C_D	drag coefficient
C_L	lift coefficient
R	ballistic trajectory radius of curvature
m	ball mass
G	ball weight
g	gravity acceleration
τ	angle between the velocity vector and a horizontal plane
α	initial stroke angle
s	length of the ballistic trajectory
ρ	air density
D	drag force
M	Magnus force
D^*, M^*	dimensionless drag and Magnus force
x,y	Cartesian coordinate of the ballistic trajectory
y_0	initial ball height
Re	Reynolds number
n	ball revolutions

I. INTRODUCTION AND HISTORICAL BACKGROUND

If the Coriolis force is neglected, the shape of the ballistic trajectory of a flying rotating ball is essentially affected by three forces, i.e., the force of gravity G, the drag force D, and the so-called Magnus force M, which was explained by Magnus as early as 1853.[1] This force always acts perpendicular to the vector of the flying ball velocity and its axis of rotation. Its magnitude and, in particular, the direction of the ball rotation substantially affects the shape of the ball's trajectory in many games such as tennis, golf, ping-pong, baseball, etc.

A considerable amount of literature dealing with the force effects on a rotating cylinder has been published over the years; experimental studies of rotating spheres, on the other hand, have been available in a limited number only, e.g., Maccoll[2] and Höerner[3] measured the force effects on a rotating smooth sphere and Davis[4] those acting on golf balls with various surfaces. A feature common to these studies was the determination of C_D and C_L, the drag and lift coefficient characteristics of a rotating sphere.

In contrast to these studies and their quantitative results, one finds in numerous publications only qualitative results and an analysis of the reason of curving of a spinning ball trajectory made on the basis of simple considerations founded on the Bernoulli equation; see, for example, Refs. 5 and 6.

II. EQUATION GOVERNING THE TRAJECTORY OF FLIGHT—THEORETICAL INTRODUCTION

Consider the case of a rotating ball projected with an initial velocity v_0 at an angle α, the axis of rotation of which is parallel to the horizontal plane. In the case of a lifted or topspin stroke (lob), the direction of the rotation is such that the vector of the angular velocity $\boldsymbol{\Omega}$, when following the flying ball, lies along an axis parallel to the horizontal plane and aims from right to left. The Magnus force acts towards the center of curvature of the ballistic trajectory thus increasing the trajectory curving and shortening the range compared with the trajectory of the nonrotating ball.

The calculation of the flight trajectory starts from the equilibrium of forces into the normal (Fig. 1),

$$mv^2/R = mg\cos\tau + M. \tag{1}$$

Noting that τ decreases as the arc length s increases we have $R = -ds/d\tau$, recalling that $dt = ds/v$, $dx = ds\cos\tau$, and $dy = ds\sin\tau$, one obtains after elimination of R from Eq. (1) and integration, the parametric equation of the ballistic trajectory in the form

$$x = -\frac{1}{g}\int_{\alpha}^{\tau}\frac{v^2\cos\tau}{\cos\tau + M^*}\,d\tau, \tag{2}$$

$$y = y_0 - \frac{1}{g}\int_{\alpha}^{\tau}\frac{v^2\sin\tau}{\cos\tau + M^*}\,d\tau. \tag{3}$$

The time from instant of striking necessary to reach these coordinates is obtained by the expression

$$t = -\frac{1}{g}\int_{\alpha}^{\tau}\frac{v}{\cos\tau + M^*}\,d\tau, \tag{4}$$

where $M^* = M/mg$ is dimensionless Magnus force referred to the unit weight of the ball. To be in position to carry out the integration in Eqs. (2)–(4) one must first determine the dependence of the ball velocity on angle τ, i.e., write the equation of the hodograph $v = v(\tau)$. One proceeds from the equation of motion that, for the tangen-

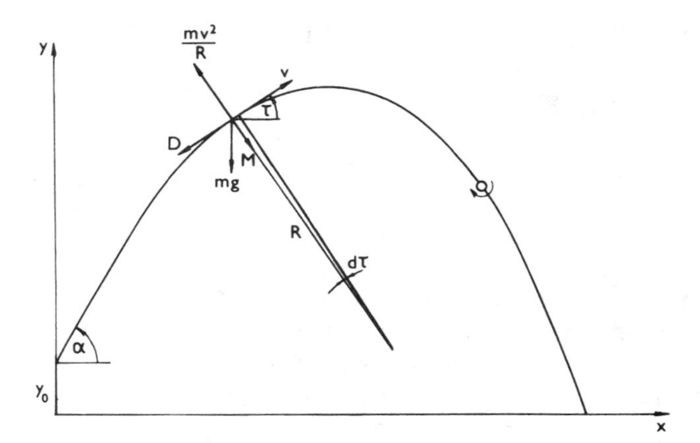

Fig. 1. Ballistic trajectory and forces acting on a flying and rotating ball.

tial direction at a point of the ballistic trajectory (Fig. 1), has the form

$$m\frac{dv}{dt} = -D - mg\sin\tau. \qquad (5)$$

Substituting for dt in Eq. (5) and some manipulation leads to the differential equation of the hodograph in the form

$$\frac{dv}{d\tau} = \frac{\sin\tau + D^*}{\cos\tau + M^*}v, \qquad (6)$$

where similarly $D^* = D/mg$ is the dimensionless drag force referred to the unit weight of the ball. For the case $M^* = 0$, Eq. (6) corresponds to the usually used hodograph; see, e.g., Ref. 7.

A dimensional analysis, e.g., Ref. 8, in the case of a flying ball leads to the conclusion that the dimensionless drag and Magnus force turn out to be

$$D^* = C_D(\pi d^2/8mg)\rho v^2 \qquad (7)$$

$$M^* = C_L(\pi d^2/8mg)\rho v^2, \qquad (8)$$

where d is the ball diameter and ρ is the air density. The drag and lift coefficients C_D and C_L for a spinning ball can be considered $C_D = f(w/v,\mathrm{Re})$, $C_L = f(w/v,\mathrm{Re})$, where w is the equatorial velocity of a flying ball and Re is the Reynolds number. The effect of the Mach number is practically negligible up to $\mathrm{Ma} < 0.3$.

III. MEASUREMENT OF THE LIFT AND DRAG COEFFICIENTS

A device for ejecting spinning balls in the aerodynamic tunnel is shown in Fig. 2. In the experiments, balls of the Tretorn trademark were used. These balls, manufactured pressureless, were fastened into a special fixture, after which coaxial holes were drilled through the ball. Into the holes, miniature brass bearings were glued in such a way as to enable the ball to be fixed between steel pointed centers. In comparison with other, mostly pressurized, balls, its properties remain unchanged.

Ball spinning was initiated through a small electric motor with a foam rubber conical follower that was pressed against the ball surface. After releasing the pin that was pressing the electric drive into engagement with the ball, a trigger was synchronously released, which in turn immediately detracted, through wires, the steel centers allowing the ball to fall freely into the air stream of the tunnel.

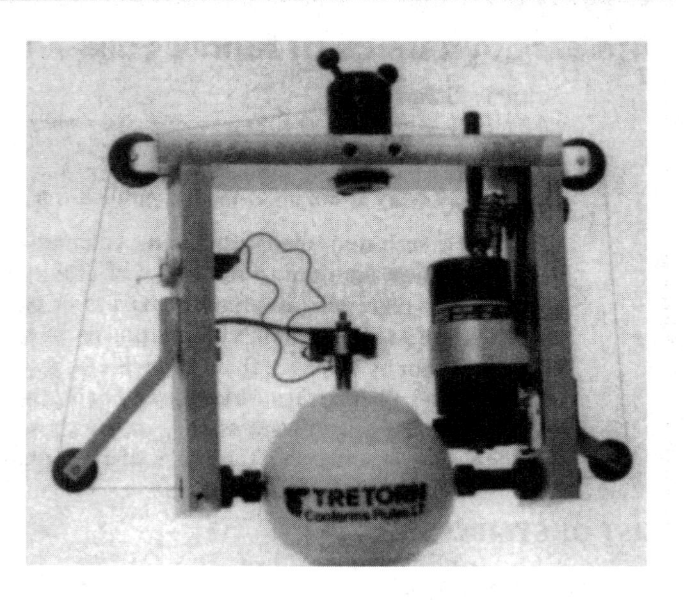

Fig. 2. Device for rotating and releasing the ball.

Spinning speed of the ball was measured by an induction sensor located closely above the ball, see Fig. 2, picking up signals from a small ferrite magnet, glued into a hole drilled on the ball periphery. Simultaneously, the spinning speed was measured by a stroboscope. Since a complete agreement between both values was found to exist in the whole

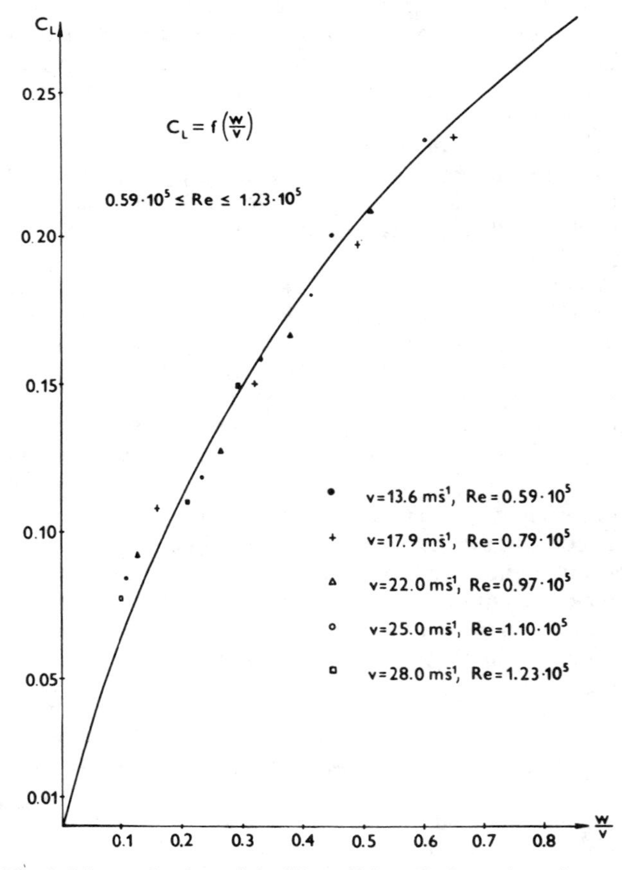

Fig. 3. Measured values of the lift coefficient C_L for various air stream velocities. The resulting regression curve applies for all data.

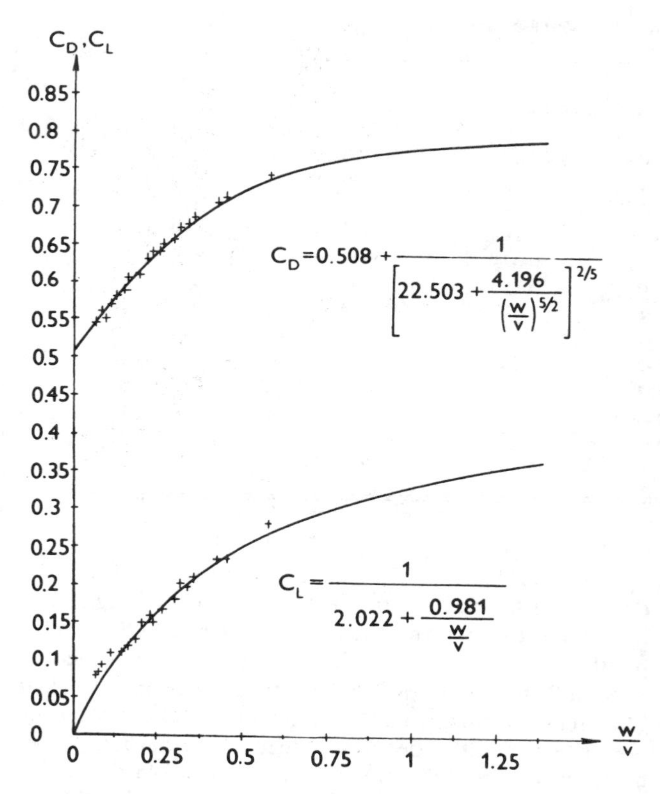

Fig. 4. Values of the drag and lift coefficient as a function of the w/v ratio.

range of the spinning speed used in the experiments, the induction sensor was removed, thus ensuring a more realistic flow pattern around the ball, especially in the initial stages of each experimental run.

Measurements were carried out in the open-type aerodynamics tunnel of the Research Institute for Aeronautics in Prague, which has a diameter of 1.8 m. The drag and lift coefficients C_D and C_L can be determined using a simple procedure described, e.g., in Ref. 4. Measurements were made at air velocities $13.6 \leqslant v \leqslant 28$ m s^{-1} and ball revolution $800 \leqslant n \leqslant 3250$ rpm. Measured values of the coefficients were plotted in the form of $C_D = C_D(w/v)$ and $C_L = C_L(w/v)$ relations in which the Reynolds number, in accordance with dimensional considerations, is a parameter. However, results of measurements in the aerodynamics tunnel for the above-mentioned range of air velocities and revolution were so scattered, see Fig. 3 (similar scatter was obtained for the C_D value) that a distinct dependence upon the air velocity, and thus upon Re, could not be obtained. Therefore we can conclude that, within the limits of measurement precision, the dependence of C_D and C_L upon Re may be neglected, which greatly simplifies the forthcoming calculation.

Therefore, in order to smooth the measured values, it was possible to make use of a one-dimensional regression function whose general form was chosen to be $y = (a_0 + a_1 x^{k_1} + a_2 x^{k_2})^{k_3}$, see Ref. 9. Using a nonlinear regression, employing the method of least squares, 50 curves were obtained in which the standard deviation $s(y_i)$ was chosen as the criterion for a best fit. Using this procedure, the following equations for the C_D and C_L coefficients were obtained, see Fig. 4:

$$C_D = 0.508 + \frac{1}{[22.503 + 4.196(w/v)^{-5/2}]^{2/5}}, \tag{9}$$

$$C_L = \frac{1}{2.202 + 0.981(w/v)^{-1}}, \tag{10}$$

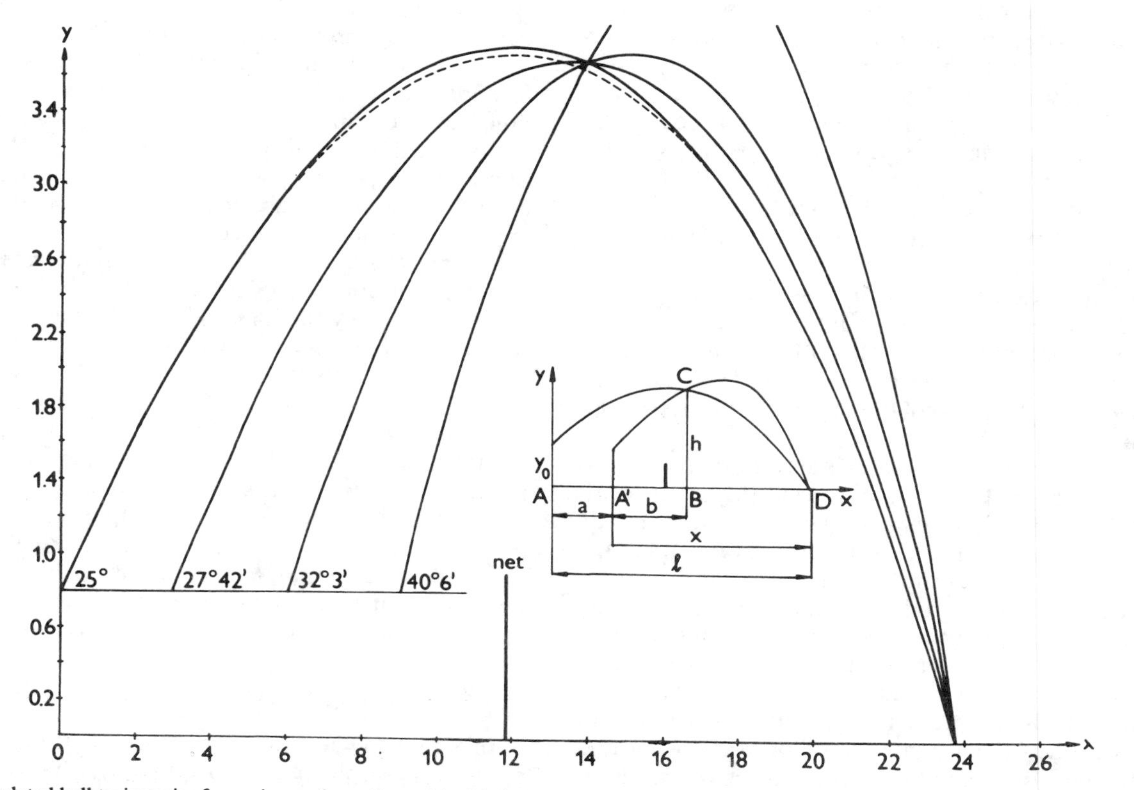

Fig. 5. Calculated ball trajectories for various values of a and b with $l = 23.77$ m, $y_0 = 0.8$ m, and $H = 3.7$ m.

Table I. Numerical results of the mathematical model for $l = 23.77$ m, initial height $y_0 = 0.8$ m, and obstacle height $H = 3.7$ m.

a	b	α	n	t_C	t_{max}	v_0	w/v	Re$\times 10^{-5}$
			0	0.873	1.672	19.35	0	0.87–0.59
0	13.885	25°	2500	0.751	1.480	23.00	0.37–0.63	1.04–0.62
			3500	0.729	1.440	23.75	0.51–0.87	1.07–0.63
			0	0.776	1.668	17.19	0	0.78–0.54
3	10.885	27°42″	2500	0.675	1.505	20.14	0.43–0.70	0.91–0.56
			3500	0.661	1.474	20.59	0.59–0.96	0.93–0.57
			0	0.654	1.680	15.21	0	0.69–0.47
6	7.885	32°3′	2500	0.581	1.551	17.31	0.50–0.79	0.78–0.49
			3500	0.572	1.531	17.61	0.69–1.09	0.80–0.50
			0	0.511	1.745	13.26	0	0.60–0.38
			2500	0.462	1.654	14.71	0.59–0.96	0.66–0.40
9	4.885	41°6′	3500	0.458	1.642	14.81	0.82–1.34	0.67–0.41
			4500	0.457	1.636	14.86	1.05–1.71	0.67–0.41
			6000	0.455	1.633	14.91	1.39–2.28	0.68–0.41

with standard deviations of $s(C_D) = 0.086$ and $s(C_L) = 0.11$. Equations (9) and (10) are very similar to the course of relationships obtained by Davies[4] for dimple and mesh golf balls. These relations also have their limiting values $(C_D)_{\lim} = 0.796$ and $(C_L)_{\lim} = 0.494$.

IV. NUMERICAL CALCULATION

As a first step, the differential equation of the hodograph (6) has been solved, using the Runge–Kutta method and the initial conditions $\tau = \alpha$, $v = v_0$. The starting value

$$v_0 = \sqrt{\frac{(g/2)x_{max}^2 (1 + \tan^2 \alpha)}{y_0 + x_{max} \tan \alpha}} \tag{11}$$

corresponds to a velocity at which the ball would travel, from an initial height y_0 and an initial projection angle α, a distance x_{max} in a vacuum. On the basis of discrete velocity value calculated from Eq. (6), it is not difficult to determine, through numerical integration, coordinates of the ballistic trajectory from Eqs. (2) and (3) and the corresponding time necessary for traveling to that point from Eq. (4). It is clear that using Eq. (6), respecting variable air resistance and trajectory curvature due to the Magnus force, the ball cannot reach the distance x_{max}. Therefore, in subsequent calculations, the initial velocity v_0 is increased by an increment $\Delta v = 1$ m s^{-1} till the required value x_{max} was exceeded. After that, velocity was diminished by one step and a finer value $\Delta v/10$ introduced into the algorithm. The whole procedure was repeated till the prescribed x_{max} was reached with a preset accuracy. In Table I, corresponding to Fig. 5, results of the above procedure are summarized as an illustration. Here, the calculations were made for one of the most difficult and interesting tennis strokes, i.e., for the topspin lob. If the length of the tennis court is 23.77 m and, while playing doubles, the opposing netman stands 2 m behind the net (see point B in Fig. 5), it is possible to determine from Table I, for a given distance of the striking A player, the initial angle α and initial velocity v_0 for a given spin, which for an obstacle of the height h (e.g., $h = 3.7$ m) would result in a most rapid stroke falling just upon the opposer's baseline, point D.

In the computer program, the input values are distances a and b, court length l, height h, revolutions of a ball n, and the initial height y_0. For these input values the program determines the angle α, the velocity v_0, the total flight time t_{max}, and the important value of time necessary to reach the point C, t_C. Moreover, Table I presents ranges of the w/v ratio and the Reynolds number Re as a check. It is clear that on increasing the ball spin, the time necessary to reach the critical point C decreases steadily so that the player must react more and more rapidly. If, e.g., a player intends to lob the opposing netman from the baseline, $b = 13.885$ m, the time without rotation is $t_C = 0.873$ s. However, playing a lifted stroke with $n = 3500$ rpm, t_C decreases to 0.729 s. This means that the opposer's reaction must be approximately 0.14 s more rapid. The effect of the topspin lob becomes even more pronounced, if played from somewhere within the court, e.g., from point A′. In this case the ball must be played under a greater angle. If, e.g., $a = 9$ m and $b = 4.88$ m, one can see from Table I that for $n = 3500$ rpm we must react at the latest in 0.46 s, regardless of the fact that the ball ascends steeply with the angle $\alpha = 40°6'$ so that it appears visually to fall well behind the baseline. This fact, together with the need for an extraordinarily fast reaction, influences in the initial stage of the ball flight, the decision making of even experienced players. It happens thus that the opposing player does not react properly to the ball's flight in spite of its reaching a lower height than in C, at which it could be returned.

It is interesting to note that for a constant distance b, the total flight distance being the same, all ballistic trajectories are approximately the same. They differ insignificantly in the vicinity of their maxima where all trajectories for rotating balls are below the ballistic curve without rotation, i.e., for $n = 0$.

In Fig. 5, for $a = 0$, $b = 13.88$ m, the dashed line corresponds to a ball rotating with $n = 3500$ rpm. The maximum of this trajectory lies only 4.1 cm below the curve for $n = 0$. The difference diminishes, attaining for $a = 9$ m only 1.4 cm.

For other values of l, h, a, b, n, and y_0, that might be of interest, a FORTRAN IV program is available and may be obtained from the author on request.

V. COURT—EXPERIMENT

In order to obtain real values of the tennis-ball rotational speed for a topspin lob played optimally, a high-speed camera STALEX with a maximum frequency of 3000 frames per second was employed. In shooting the movie, Czechoslovak Davis Cup player and doubles specialist Pavel Složil played this difficult stroke repeatedly with an effort to achieve a maximum possible spin.

After evaluating all the film material it became clear that the highest rotation obtained was around 3500 rpm. Although this is probably not the final limit of human possibilities these days, it is sufficient to play a fast and effective lob stroke. Therefore, the limiting value of $n = 3500$ rpm also closes the fourth column in Table I. Only for $a = 9$ m, as an illustrative example, the calculation was made for higher spin values of $n = 4500$ rpm and $n = 6000$ rpm. It may be seen that increasing the spin further above 3500 rpm results in accelerating the ball into point C by 0.001 s or by 0.003 s at 6000 rpm. From both viewpoints, i.e., what is practical and possible, it is clear that this insignificant acceleration does not produce any appreciable time gain for the attacking (or defending) player. On the other hand, it can only play a significant role after contacting the playground where it causes the ball to bounce off fast and high with a higher spin requiring more skill returning the ball.

ACKNOWLEDGMENTS

The author is indebted to Pavel Složil and to Jiří Valta, Director of the Tennis Center in Prague, for their generous help in shooting the movie mentioned in the experimental part of the article. I am also very grateful to my son Tony for continuous assistance and to Z. Kober from the Research Institute for Agriculture for excellent and professional cameraman's work.

Further, my thanks are due to E. Bornhorst and Dr. M. Jirsák, from the Research Institute for Aeronautics in Prague for their help in making experiments in the aerodynamic tunnel. Last but not least I should like to thank C. Suk, chairman of the Czechoslovak Tennis Association, Dr. Vl. Šafařík, Head of the Tennis Department on the Institute of Sports in Prague, and the professional photographer Pavel Štecha for the documentation and the photography.

[1]H. G. Magnus, Poggendorf's Ann. Phys. Chem. **88**, 1 (1853).
[2]J. W. Maccoll, J. R. Aeronaut. Soc. **32**, 777 (1928).
[3]S. F. Hoerner, *Fluid-dynamic Drag*, published by the Author, 1965.
[4]J. M. Davies, J. Appl. Phys. **20**, 821 (1949).
[5]D. Maier, Modellflugzeug, München, 1978.
[6]W. Abe, *Tennis, Forum der Experten* (Verband Deutscher Tennislehrer, Hannover, West Germany, 1983), Vol. 4.
[7]J. L. Synge and B. A. Griffith, *Principles of Mechanics* (Toronto, 1949).
[8]H. Schlichting, *Boundary-Layer Theory* (McGraw-Hill, Karlsruhe, 1968).
[9]K. Květoň, *Set-Up of Empirical Models Using Nonlin and Atlas Programes*, Research Report (Faculty of Mech. Eng., Prague, 1986).

The Tennis-Ball Bounce Test

Howard Brody

University of Pennsylvania, Philadelphia, PA 19104

In order for a particular brand of tennis ball to be approved for tournament play by the ITF (International Tennis Federation) and the USTA (United States Tennis Association), a representative sample of those balls must pass a number of tests, one of which is the rebound test.[1,2] To pass this test, a tennis ball, when dropped from a height of 100 in (254 cm) onto a concrete floor, must rebound not less than 53 in (135 cm) nor more than 58 in (147 cm) with all measurements taken from the bottom of the ball.[3] The USTA has a committee (The Tennis Ball Testing and Equipment Committee) that meets regularly to test tennis balls submitted to them by manufacturers desiring approval of their product. The approval of the USTA and the ITF, which is printed on the tennis-ball container, is a player's assurance that the balls meet the set standards.[2] The USTA has constructed a rather complicated electro-optical device for automatically doing these measurements in order to reduce the chance for human error and to speed up the procedure. It is also possible to determine the rebound height of a ball by measuring the time between the first and second bounce, which is much easier to do experimentally.

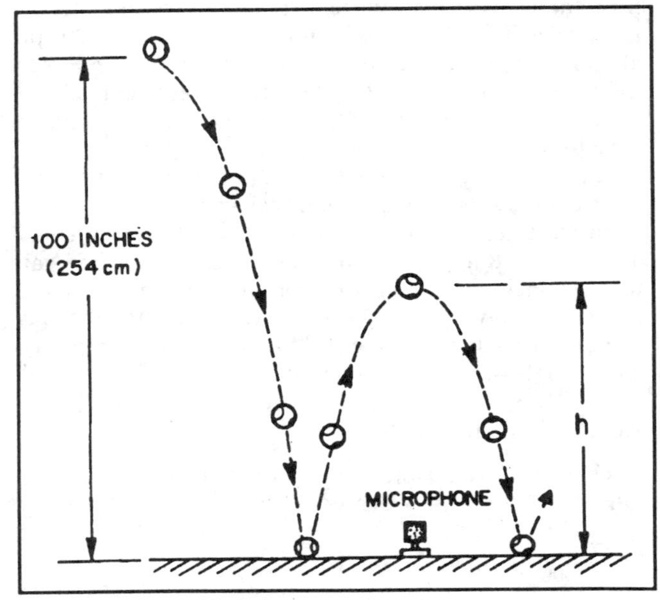

Fig. 1. Diagram of ball drop experiment.

Procedure

In our experiment, a dozen Nassau brand non-pressurized tennis balls[4] were dropped one at a time from a height of 100 in onto a concrete floor covered with vinyl tile[5] and allowed to bounce at least twice, as is shown in Fig. 1. A microphone was placed on the floor approximately equidistant from the two impact points. If the microphone is not equidistant, a systematic error in timing will be present, since the velocity of sound in air is only 330 m/s. The resultant signal (the noise of the ball impacting on the floor) was amplified and then displayed on a scope. The sweep gate outputs of the scope were used to turn on and then turn off a commercial timer[6] running at 100 kHz. The time between the bounces was recorded for each event, and at the same time a Camcorder[7] with a high-speed electronic shutter (1/2000 s) was aimed at the location of the peak of the ball's rebound. A 2-m stick was mounted vertically in the field of view to allow the height of the bounce to be measured. The entire sequence of ball releases was carried out several times in order to get enough data to see if the timing method of determining height of the rebound agreed with the visual (video) observations.

The videotape obtained in the experiment was played back, advancing the tape a single field (1/60 s) at a time,[8] and the maximum height of the bottom of the ball was recorded for each event. Since the Camcorder has audio input, it was used to make sure that the event being observed corresponded to the event recorded by the timer. This correspondence could also have been done by placing a different printed number in the field of view for each event, but the audio method was much more convenient since the experimenter was on a ladder dropping the balls. (The electronic counter reset itself to zero after a preset time, so all data could be recorded from the ladder position.)

Analysis

The data taken with the video system are shown in Fig. 2; it is clear that the balls meet the USTA specifications. The predicted bounce height, h, was calculated for each time difference recorded in this experiment. If air resistance is neglected and the time between bounces is t, then it is possible to relate the bounce height to time t. Between bounces, the ball will spend half of the time ($t/2$) rising and half of the time ($t/2$) falling. The maximum height the ball will rise to, h, will be equal to the distance a ball will fall in time $t/2$, and this is given by $h = \frac{1}{2} g (t/2)^2 = gt^2/8$, where $g = 9.80$ m/s^2. These results and the results obtained with the video system were entered in a spreadsheet program and compared. Both the average of the difference and the average of the absolute value of the difference between the video data and the heights obtained by timing

Reprinted from *The Physics Teacher* **28**, 407–409 (1990); © American Association of Physics Teachers.

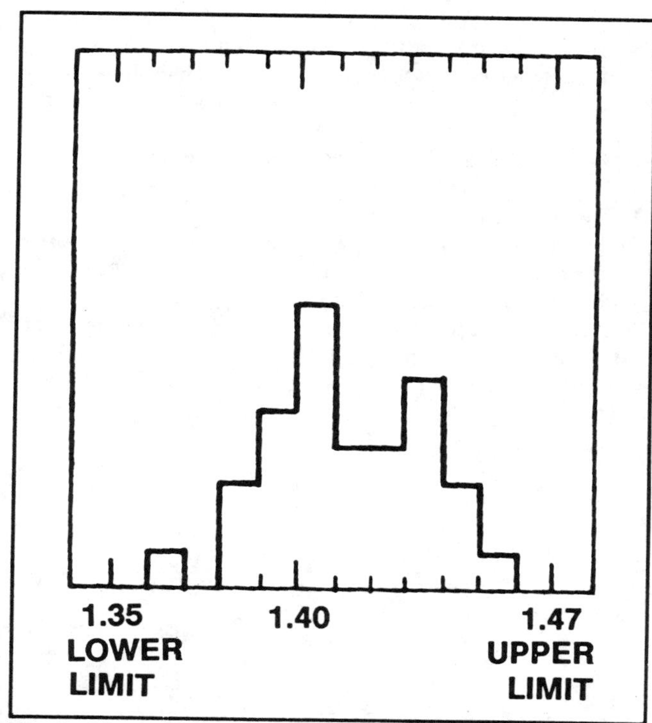

Fig. 2. Measured distribution of rebound heights for a tennis ball dropped from 2.54 m above a rebounding surface.

were 0.014 m. Examination of the data revealed that in every event measured, the value obtained by timing was larger than the value obtained from video measurements. This indicated that there was a small systematic error in the experiment, and a number of potential sources of error were investigated to discover the problem.

In obtaining the data, the optical axis of the video camera was set parallel to the ground approximately 1.41 m above the rebounding surface. Because the 2-m stick was behind the plane determined by the bouncing ball, each height measurement had to be corrected for parallax in the optical system. This correction was almost 0.01 m at the extremes of the bounce heights observed, but it reduced the average difference between the two sets of data only from 0.0140 to 0.0139 m. Since the average rebound height (1.409 m) was approximately the same as the height of the camera axis, parallax was not a source of systematic error, and by including this correction the average values changed by at most a millimeter.

Because air resistance was not included in this calculation, a numerical computer program was run for a few sample ball drops to see if this would be an important effect. At the velocities encountered in this experiment, air resistance makes a significant difference in the value of the coefficient of restitution obtained from the bounce height, compared with the actual coefficient of restitution (0.75 vs 0.78, for example), but a difference of less than a millimeter in the predicted maximum height for a given time difference between bounces. This first fact is probably one of the reasons why the USTA bounce test uses rebound height and not the coefficient of restitution in its specifications.

The cause of the systematic shift between the data sets was the fact that the measured times were from ball impact to ball impact, while calculation assumed they were from the ball leaving the surface to the ball impacting the surface. To correct for this, the "dwell time" of the ball in contact with the floor had to be subtracted from each timing difference. The approximate value of this dwell time for a nonpressurized ball dropped onto a hard surface is known to be about 5 ms.[9] Using the spreadsheet program, the value of the ball dwell time was adjusted to minimize the average difference between the two methods of determining the bounce height. The results, shown in Table I, are in agreement with the previous data. If this experiment were done with "normal" pressurized tennis balls, the value of the dwell time would have been one or two milliseconds shorter, but still significant.

Table I. Calculation of the difference in height between the sound (timing) analysis method and the visual measurements for various values of dwell time (34 events).

dwell time (seconds)	average difference (meters)	average absolute difference (meters)
0.0	−0.0139	0.0139
0.003	−0.0060	0.0063
0.004	−0.0033	0.0044
0.005	−0.0007	0.0034
0.0053	+0.0001	0.0033
0.006	+0.0019	0.0036
0.007	+0.0046	0.0052

It is clear from the analysis of the data that timing of the impact sound can be used to determine the rebound height of a tennis ball with sufficient accuracy to make it useful as a test for tennis ball certification. It is also an easy way to determine ball rebound properties as a function of ball wear, temperature, age, etc., or an interesting laboratory experiment at the introductory level.[10] ♦

Acknowledgment

Some of the television equipment used in this experiment was provided by a grant from the University of Pennsylvania Research Foundation. The balls were provided by Nassau through the United States Professional Tennis Registry.

References

1. *Official USTA Tennis Yearbook 1988* (H.O. Zimman, Inc., Lynn, MA 01901).
2. Some of the specifications included in USTA Rule 3 (The Ball) are: The ball shall have a bound of more than 53 in (135 cm) and less than 58 in (147 cm) when dropped 100 inches (254 cm) upon a concrete base. Unless otherwise specified, all tests shall be made at a temperature of approximately 68°F (20°C) and a relative humidity of approximately 60 percent. All balls should be removed from their container and kept at the

recognized temperature and humidity for 24 hours prior to testing. Unless otherwise specified, the limits are for a test conducted in an atmospheric pressure resulting in a barometric reading of approximately 30 in (76 cm). Before any ball is tested it shall be steadily compressed by approximately 1 in (2.54 cm) in each of three diameters at right angles to one another in succession; this process to be carried out three times (nine compressions in all). All tests to be completed within two hours of precompression. Measurements are to be taken from the concrete base to the bottom of the ball.

3. Note that the metric requirements of the rebound test are more stringent than the requirements given in inches.
4. Pressurized balls are the balls normally used in tournaments. Nonpressurized balls were used in this experiment to allow the experiment to be repeated at a later time. Pressurized balls gradually lose their bounce over a few weeks due to the porosity of the rubber, which allows the gas inside to slowly seep out. Pressurized balls only remain "fresh" as long as they are kept under pressure comparable to their internal pressure.
5. It was suggested that the author not rip out the tiles for the purpose of performing this experiment.
6. Hewlett-Packard 5512A Electronic Counter.
7. RCA CPR 300 Camcorder.
8. Panasonic PV 4760 VCR.
9. H. Brody, *Am. J. Phys.* **47**, 485 (1979); H. Brody, *Tennis Science for Tennis Players* (University of Pennsylvania Press, Philadelphia, 1987), pp. 11–13.
10. I would like to thank Peter Brancazio for pointing out to me that the measurement of the COR of balls using sound has been previously reported by Alan D. Bernstein, *Am. J. Phys.* **45**, 41 (1977).

Chapter 6
THE BICYCLE

CONTENTS

The first article in this chapter, "The stability of the bicycle," by David E. H. Jones, explores the role of *self-stability* in the age old question of what makes a bicycle ridable. As the author notes, one well-known theory[1] postulates that a leaning bicycle is righted by the centrifugal force generated when the rider, sensing a fall, intervenes by steering into the direction of the fall, thereby forcing the bicycle into a curved (saving) path.

Jones points out, however, that although this theory does account for three familiar attributes—namely, a bicycle cannot be ridden if the front forks are not free to swivel; the faster a bicycle moves the easier it is to ride; and a stationary bicycle cannot be balanced—it does not account for the fact that even a riderless bicycle is capable of remarkable self-stability, as evidenced by its ability to move in a curved path for 20 s or more before collapsing. The author also explores, both analytically and experimentally, the contribution to bicycle stability that results from the gyroscopic action of the front wheel.[2]

The most significant result of the article, however, may be the author's development of a model in which stability is characterized and predicted on the basis of just two parameters, the front projection (castor) of the front wheel and the fork angle (degree of tilt) of the steering axis.

The second article in this chapter, "The stability of bicycles," by J. Lowell and H. D. McKell, expands on the model developed by David Jones by showing that a bicycle is almost, though not quite, self-stable,[3] and confirming that the most important parameter governing its stability is the "castor" of the front wheel. The article also demonstrates that bicycle stability consists of two different aspects—resistance to sideways deflections, and resistance to the buildup of oscillations—the first of which is *helped* by gyroscopic effects while the second is *exacerbated* by them. In the authors' words:

> Although the bicycle is stable in the sense that it does not fall over when pushed sideways, it is unstable against oscillation in the front wheel direction (α) and the resulting side-to-side oscillation of the machine as a whole. Gyroscopic effects make these oscillations grow much more rapidly than they would otherwise do, and Coulomb friction does not damp them out. It appears that the magnitude of "castor" required is governed by the need to keep the growth rate of this oscillatory instability manageably small. But for this, reasonable stability could be achieved with a much smaller castor.

The authors conclude with the following observation:

> The fairly large castor of real bicycles ensures a slow growth of the oscillatory instability, so that the rider can be quite leisurely in his intervention.

An important corollary to all of this is that bicycles with small castor are very difficult to ride.

"More bicycle physics," by Thomas B. Greenslade, Jr., is the third article in this chapter and describes three bicycle problems suitable for inclusion in the mechanics portion of an introductory physics course. The first problem involves a bicycle-riding broom seller who negotiates turns while balancing a load of brooms on his head! Using typical data and a number of reasonable assumptions, Greenslade develops a self-consistent model for the broom seller's achievement.

The second problem deals with the legendary tendency of the Penny-farthing bicycle to toss its rider over the handlebars at the slightest obstacle encountered by the large front wheel. Greenslade notes that this type of bicycle is always on the verge of instability due to the fact that the center of mass of rider plus machine needs to be raised only a small distance to reach a state of unstable equilibrium.

The third problem proposes a model of bicycle dynamics in which the effects of rolling friction, air resistance, and gravity are included. Greenslade uses this model to predict maximum speeds for coasting down a hill as well as the maximum speed for pedaling up a hill of given incline.[4]

"Bicycles in the physics lab," by Robert G. Hunt, is the fourth and final article in this chapter. In it, the author describes a number of laboratory exercises, at varying levels of difficulty, that students with an interest in bicycles and bicycle physics would definitely enjoy.

In one experiment, students use spring balances to compare the force applied at the pedal to the force generated at the rear wheel. For a constant applied force of 100 N at the pedal, the force at the rear tire typically varied from 14–35 N, depending on the gear employed.

In another experiment, the ratio of output force to input force was calculated by comparing the measured radii of the various gears employed. The article also pursues such questions as the maximum upward incline on which a bicycle can be negotiated at constant velocity. Typical results for a ten-speed bicycle: 17° in low gear and 6° in high gear.

---

[1]See, for example, S. Timoshenko and D. H. Young, *Advanced Dynamics* (McGraw-Hill, New York, 1948).

[2]As a bicycle tilts, the angular momentum vector of the front wheel also tilts, causing the front wheel to precess about the steering axis (much like a gyroscope) and to steer the bicycle into a curve that counteracts the initial tilt.

[3]As Lowell and McKell point out, it is the *near* self-stability of conventional bicycles that makes them so easy to ride.

[4]This portion of the article contains a small error that was later corrected. See "More on bicycle physics," by Brynjolf Dokken, and "The author replies," by Thomas B. Greenslade, *Phys. Teach.* **22**, 138 (1984). Copies of these brief comments are reproduced on page 187.

THE STABILITY OF THE BICYCLE

Tired of quantum electrodynamics, Brillouin zones, Regge poles? Try this old, unsolved problem in dynamics—how does a bike work?

David E. H. Jones

ALMOST EVERYONE can ride a bicycle, yet apparently no one knows how they do it. I believe that the apparent simplicity and ease of the trick conceals much unrecognized subtlety, and I have spent some time and effort trying to discover the reasons for the bicycle's stability. Published theory on the topic is sketchy and presented mainly without experimental verification. In my investigations I hoped to identify the stabilizing features of normal bicycles by constructing abnormal ones lacking selected features (see figure 1). The failure of early unridable bicycles led me to a careful consideration of steering geometry, from which—with the aid of computer calculations—I designed and constructed an inherently unstable bicycle.

The nature of the problem

Most mechanics textbooks or treatises on bicycles either ignore the matter of their stability, or treat it as fairly trivial. The bicycle is assumed to be balanced by the action of its rider who, if he feels the vehicle falling, steers into the direction of fall and so traverses a curved trajectory of such a radius as to generate enough centrifugal force to correct the fall. This

David E. H. Jones took bachelor's and doctor's degrees in chemistry at Imperial College, London, and has since alternated between the industrial and academic life. Currently he is a spectroscopist with ICI in England.

Reprinted from *Physics Today* **23(4)**, 34–40 (1970); © American Institute of Physics.

"UNRIDABLE" BICYCLES. David Jones is seen here with three of his experimental machines, two of which turned out to be ridable after all. At top of this page is URB I, with its extra counter-rotating front wheel that tests the gyroscopic theories of bicycle stability. At left is URB III, whose reversed front forks give it great stability when pushed and released riderless. URB IV (immediately above) has its front wheel mounted ahead of the usual position and comes nearest to being "unridable." —FIG. 1

theory is well formalised mathematically by S. Timoshenko and D. H. Young,[1] who derive the equation of motion of an idealized bicycle, neglecting rotational moments, and demonstrate that a falling bicycle can be saved by proper steering of the front wheel. The theory explains, for example, that the ridability of a bicycle depends crucially on the freedom of the front forks to swivel (if they are locked, even dead ahead, the bicycle can not be ridden), that the faster a bicycle moves the easier it is to ride (because a smaller steering adjustment is needed to create the centrifugal correction) and that it can not be balanced when stationary.

Nevertheless this theory can not be true, or at least it can not be the whole truth. You experience a powerful sense, when riding a bicycle fast, that it is inherently stable and could not fall over even if you wanted it to. Also a bicycle pushed and released riderless will stay up on its own, traveling in a long curve and finally collapsing after about 20 seconds, compared to the 2 sec it would take if static. Clearly the machine has a large measure of self-stability.

The next level of sophistication in current bicycle-stability theory invokes the gyroscopic action of the front wheel. If the bike tilts, the front wheel precesses about the steering axis and steers it in a curve that, as before, counteracts the tilt. The appeal of this theory is that its action is perfectly exemplified by a rolling hoop, which indeed can run stably for just this reason. A bicycle is thus assumed to be merely a hoop with a trailer.

The lightness of the front wheel distresses some theorists, who feel that the precession forces are inadequate to stabilize a heavily laden bicycle.[2,3] K. I. T. Richardson[4] allows both theories and suggests that the rider himself twists the front wheel to generate precession, hence staying upright. A theory of the hoop and bicycle on gyroscopic principles is given by R. H. Pearsall[5] who includes many rotational moments and derives a complex fourth-order differential equation of motion. This is not rigorously solved but demonstrates on general grounds the possibility of self-righting in a gyroscopically stable bicycle.

A non-gyroscopic bicycle

It was with vague knowledge of these simple bicycle theories that I began my series of experiments on bicycle stability. It occurred to me that it would be fun to make an unridable bicycle, which by canceling the forces of stability would baffle the most experienced rider. I therefore modified a standard bicycle by mounting on the front fork a second wheel, clear of the ground, arranged so that I could spin

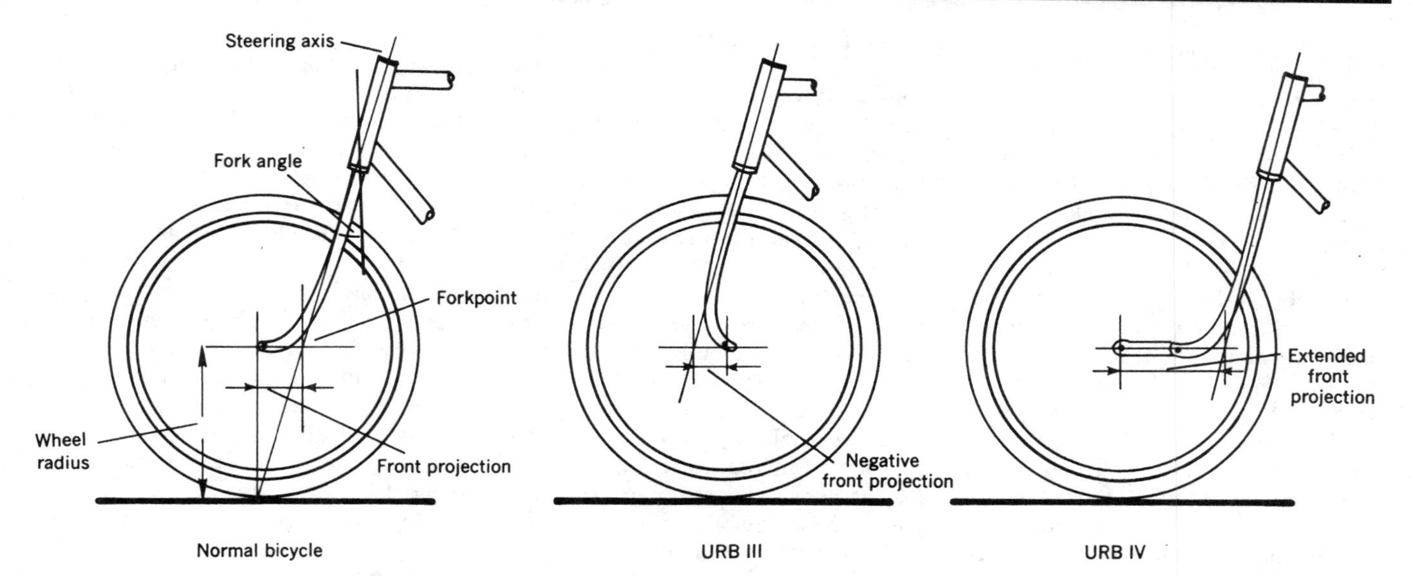

FRONT-FORK GEOMETRY. On left is a normal bicycle. Center shows URB III with reversed forks giving a negative front projection, and on right is URB IV with extended front projection. —FIG. 2

it against the real front wheel and so oppose the gyroscopic effect. This creation, "Unridable Bicycle MK I" (URB I), unaccountably failed; it could be easily ridden, both with the extra wheel spinning at high speed in *either* direction and with it stationary. Its "feel" was a bit strange, a fact I attributed to the increased moment of inertia about the front forks, but it did not tax my (average) riding skill even at low speeds. The result resolves the ambiguity admitted by Richardson: the gyroscopic action plays very little part in the riding of a bicycle at normally low speeds.

This unexpected result puzzled me. If the bicycle, as seemed likely, is a hoop with a trailer but is not gyroscopic, perhaps the hoop is not gyroscopic either? I repeated the experiment of URB I on a hoop by constructing one with an inner counter-rotating member, and this collapsed gratifyingly when I tried to roll it. The hoop is a bona fide gyroscope.

Then I tried to run URB I without a rider, and its behavior was quite unambiguous. With the extra wheel spinning against the road wheels, it collapsed as ineptly as my nongyroscopic hoop; with it spinning the same way it showed a dramatic slow-speed stability, running uncannily in a slow, sedate circle before bowing to the inevitable collapse.

These results almost satisfied me. The light, riderless bicycle is stabilized by gyroscopic action, whereas the heavier ridden model is not—it requires constant rider effort to maintain its stability. A combination of the simple theories accounts neatly for all the facts. But the problem of why a ridden bicycle feels so stable, if in fact it is not, remains. There was one more crucial test: Could URB I be ridden in its disrotatory mode "hands off"? For about the only sensible theory for riding with "no hands" supposes that the rider tilts the frame by angular body movements and thus steers by the resulting front-wheel precession.[3]

Gingerly, and with great trepidation, I tried the experiment—downhill, to avoid complicating the effort with pedalling. URB I is not an easy bicycle to ride "hands off" even with the front wheel static; it somehow lacks balance and responsiveness. In the disrotatory mode it was almost impossible and invited continual disaster, but it could, just, be done. I was thus led to suspect the existence of another force at work in the moving bicycle.

More theories

In the preliminary stages of this investigation, I had pestered all my acquaintances to suggest a theory of the bicycle. Apart from the two popular theories that I have mentioned already, I obtained four others, which I shall call theories 3, 4, 5 and 6:

3. The bicycle is kept upright by the thickness of its tires (that is, it is a thin steamroller).
4. When the bicycle leans, the point of contact of the front tire moves to one side of the plane of the wheel, creating a frictional torque twisting the wheel into the lean and stabilizing the bicycle, as before, by centrifugal action.
5. The contact point of the bicycle's front tire is ahead of the steering axis. Turning the front wheel therefore moves the contact point with the turn, and the rider uses this effect, when he finds himself leaning, to move his baseline back underneath his center of gravity.
6. The contact point of the bicycle's tire is behind the steering axis. As a result, when the bicycle leans a torque is developed that turns the front wheel.

I suspect that theory 3 is not really serious. Theories 5 and 6 raise the question of steering geometry, which I was later to look at in this work—note that the gyro theory is silent on why all front forks are angled and all front forks project forward from them. To test this matter I made URB II.

URB II had a thin front wheel, only one inch in diameter (an adapted furniture castor) mounted dead in line with the steering axis, to test any steering-geometry theory. It looked a ludicrous contraption. URB II was indeed hard to ride, and collapsed readily when released, but this was at least in part because it could negotiate no bump more than half an inch high. The little front wheel also got nearly red hot when traveling fast.

I abandoned URB II as inconclusive, but preferred theory 6 to theory 5, because in all actual bicycles the front wheel's contact point is behind the intersection of the steering axis with the ground. Theory 6 is also advocated by the only author who supports his hypothesis with actual measurements.[6] But I could not see why this force should vanish, as it has to, once the bicycle is traveling in its equilibrium curve. I had grave suspicions of theory 4, for surely this torque acting across less than half the width of the tire would have a very small moment, and would depend crucially on the degree of inflation of the tire? Besides, I did not want nasty variable frictional forces intruding into the pure, austere Newtonian bicycle theory towards which I was groping.

Steering geometry

The real importance of steering geometry was brought home to me very dramatically. I had just completed a distressing series of experiments involving loading URB I, with or without its extra gyro wheel, with some 30 pounds of concrete slabs and sending it hurtling about an empty parking lot (there are some tests one can not responsibly carry out on public roads). The idea was to see if the extra weights—projecting from the front of the frame to have the maximal effect on the front wheel—would prevent the gyro effect from stabilizing the bicycle, as anticipated from the difference between ridden and riderless bikes. It appeared that the weights made the bicycle a little less stable, and the counter-rotating wheel still threw it over almost immediately. But the brutal effects on the hapless machine as it repeatedly crashed to earth with its burden had me straightening bent members and removing broken spokes after almost every run.

It occurred to me to remove the handlebars to reduce the moment of inertia about the steering axis; this meant removing the concrete slabs and the brake assembly, which incidentally enabled the front wheel to be turned through 180 deg on the steering axis, reversing the front-fork geometry (see figure 2). I had tried this experiment once before, calling the result URB III; that machine had been strangely awkward to wheel or ride, and I had noted this result as showing that steering geometry was somehow significant. Idly I reversed the forks of the bike and pushed it away, expecting it to collapse quickly. Incedibly, it ran on for yards before falling over! Further tests showed that this new riderless bicycle was amazingly stable. It did not merely run in a curve in response to an imposed lean, but actively righted itself—a thing no hoop or gyro could do. The bumps and jolts of its progress did not imperil it, but only as it slowly lost speed did it become unstable. Then it often weaved from side to side, leaning first one way and then the other before it finally fell over. This experiment convinced me that the forces of stability were "hunting"—overcorrecting the lean at each weave and ultimately causing collapse. Once or twice the riderless disrotatory URB I had shown momentary signs of the same behavior in its brief doomed career.

Why does steering geometry matter? One obvious effect is seen by wheeling a bicycle along, holding it only by the saddle. It is easy to steer the machine by tilting the frame, when the front wheel automatically steers into the lean. This is not a gyroscopic effect, because it occurs even if the bike is stationary. A little study shows that it occurs because the center of gravity of a tilted bicycle can fall if the wheel twists out of line. So here was a new theory of bicycle stability—the steering is so angled that as the bike leans, the front wheel steers into the lean to minimise the machine's gravitational potential energy. To check this theory I had to examine the implications of steering geometry very seriously indeed.

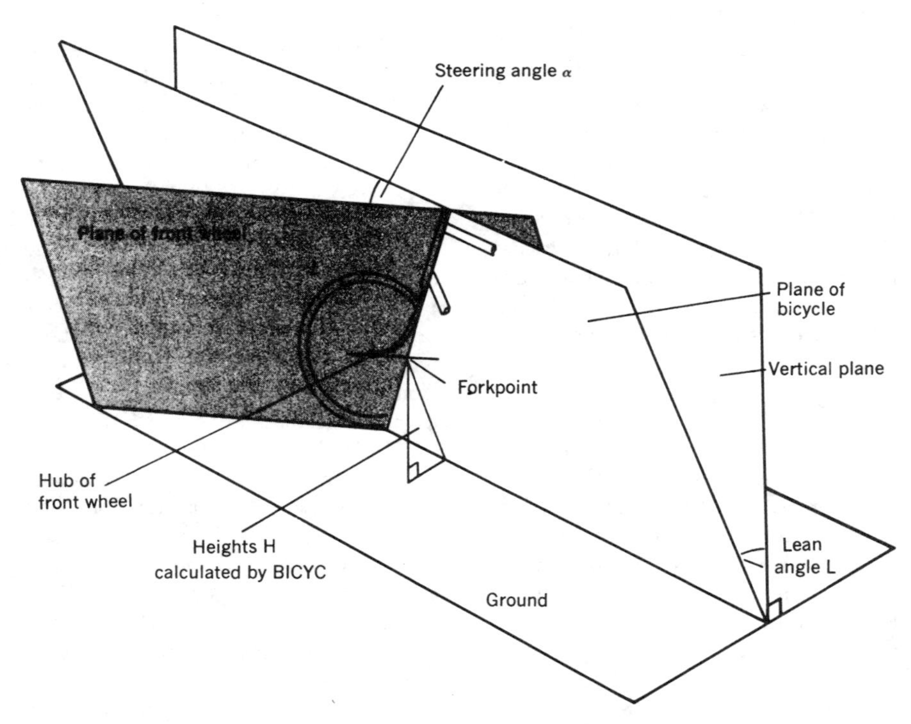

A TRICKY TRIGONOMETRICAL PROBLEM. We need to know *H*, the height of the forkpoint from the ground for a leaning bicycle. Subroutine BICYC calculates both the vertical height and the height in the plane of the bike. —FIG. 3

Computerized bicycles

It turns out that defining the height of the fork point of a bicycle in terms of the steering geometry and angles of lean and of steer (figure 3) is a remarkably tricky little problem. In fact I gave it up after a few attempts and instead wrote a Fortran subroutine, "BICYC," that solved the simultaneous trigonometrical equations iteratively and generated all the required dimensions for me. Armed with BICYC, I could now create all sorts of mad bicycles on the computer and put them through their steer-and-lean paces. The first few runs were most encouraging; they showed that with normal bicycle geometry, tilting the frame did indeed ensure that the center of gravity had its minimal elevation with the wheel twisted into the tilt. This had the makings of a really good theory. I hoped to prove that, for the observed steering geometry, the steering angle for minimal center-of-gravity height increased with the angle of lean by just the factor needed to provide perfect centrifugal stability, and that was why all bicycles have more or less the same steering geometry. As for the strange behavior of URB III, awkward to ride but incredibly stable if riderless, perhaps BICYC would provide a clue.

But further calculations shattered my hopes. Even with the bicycle dead upright, the forkpoint fell as the wheel turned out of plane (thus neatly disproving the contention of reference 7 that a bicycle tends to run true because its center of gravity rises with any turn out of plane), and the minimal height occurred at an absurdly large steering angle, 60 deg. Even worse, as the bike tilted, this minimum occurred at angles nearer and nearer the straight-ahead position (figure 4) until at 40 deg of tilt the most stable position was only 10 deg out of plane (these values are all for a typical observed steering geometry). Clearly the tilting wheel never reaches its minimal-energy position, and the minimum can not be significant for determining the stability of the bicycle.

I looked instead at the slope of the height versus steering-angle curve at zero steering angle, because this slope is proportional to the twisting torque on the front wheel of a tilted bike. Then, if H is the height of the forkpoint, the torque varies as $-dH/d\alpha$ at small values of α, the steering angle.

The curves in figure 4 show clearly that $dH/d\alpha$ varies linearly with lean angle L for small angles of lean. The more the bike leans, the bigger is the twisting torque, as required. The constant of proportionality for this relationship is $d^2H/d\alpha dL$, and the sign convention I adopted implies that a bicycle is stable if this parameter is negative. That is, for stability the forkpoint falls as the wheel turns into the lean when the bike is tilted.

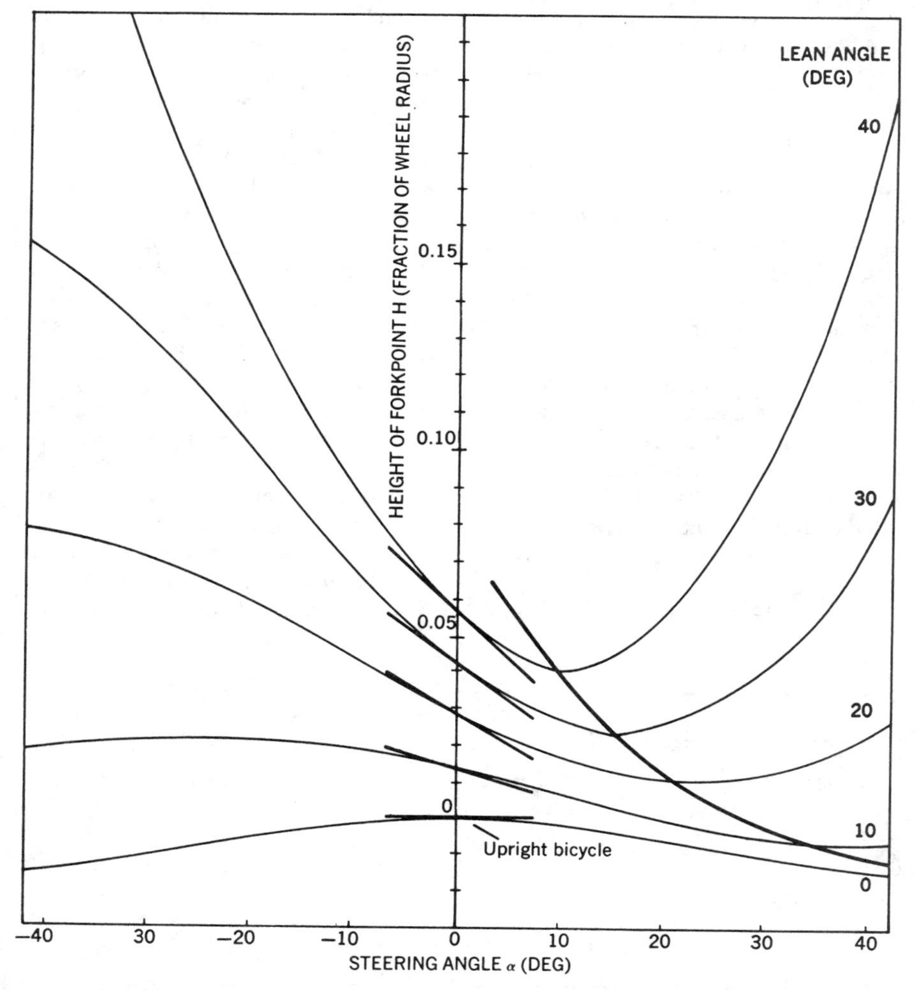

COMPUTERIZED BICYCLES. These data, from BICYC output, show that the minimal height of the forkpoint occurs nearer to the straight-ahead position for greater angles of lean. Note also that $dH/d\alpha$ varies linearly with lean angle L for small L. Curves, computed for typical steering geometry (20-deg fork angle, 0.2 radii front projection), are vertically staggered for clarity. —FIG. 4

I therefore computed $d^2H/d\alpha dL$ for a wide range of steering geometries, and drew lines of constant stability on a diagram connecting the two parameters of steering geometry—the angle of the front-fork steering axis and the projection of the wheel center ahead of this axis. I then plotted on my stability diagram all the bicycles I could find—ranging from many existing models to old high-wheeled "penny-farthings" to see if they supported the theory.

The results (figure 5) were immensely gratifying. All the bicycles I plotted have geometries that fall into the stable region. The older bikes are rather scattered but the modern ones are all near the onset of instability defined by the $d^2H/d\alpha dL = 0$ line. This is immediately understandable. A very stable control system responds sluggishly to perturbation, whereas one nearer to instability is more responsive; modern bicycle design has emphasized nimbleness and maneuverability. Best of all, URB III comes out much more stable than any commercial bike. This result explains both its wonderful self-righting properties and also why it is difficult to ride—it is too stable to be steered. An inert rider with no balancing reflexes and no preferred direction of travel would be happy on URB III, but its characteristics are too intense for easy control.

This mathematical exercise also made it plain that the center-of-gravity lowering torque is developed exactly as shown in figure 6, and is identical with that postulated in reference 6. But it does not vanish when the bicycle's lean is in equilibrium with centrifugal force, as therein supposed (BICYC calculated the height of the forkpoint in the plane of the bicycle—the "effective vertical"—to allow for this). It can only vanish when the contact point of the front wheel is intersected by the steering axis, which BICYC shows clearly is the condition for minimal height. There is thus an intimate connection between the "trail" of a bicycle, as defined in figure 6, and $d^2H/d\alpha dL$; in fact the $d^2H/d\alpha dL$ line in figure 4 coincides with the locus of zero trail.

Two further courses of action remained. First, I could make URB IV with a steering geometry well inside the unstable region, and second, I had to decide what force opposes the twisting torque on a bike's front wheel and prevents it reaching

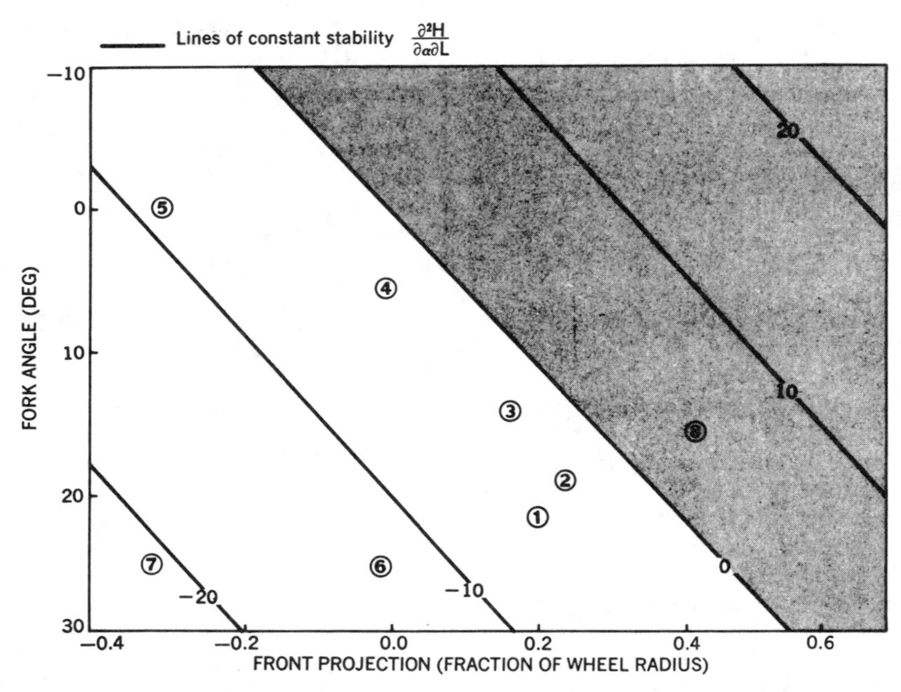

STABLE AND UNSTABLE BICYCLES. On this plot of fork angle versus front projection the $d^2H/d\alpha dL$ lines are lines of constant stability. Grey area shows the unstable region. Point 1 is a normal modern bicycle; 2 is a racing bike. 3 and 4 are high-wheelers (or "penny-farthings") from the 1870's. Point 5 is an 1887 Rudge machine, and 6 is a Lawson "Safety" of 1879. Point 7 is URB III, and point 8, the only unstable bicycle, is of course URB IV. —FIG. 5

BICYC's predicted minimal center-of-gravity position.

Self-centering

Let us consider the second point first; I was looking for some sort of self-centering in a bicycle's steering. Now this is well known in the case of four-wheeled vehicles: self-centering is built into all car steering systems, and various self-righting torques such as "pneumatic trail" are described by automobile engineers. Once again nasty variable frictional forces were rearing their ugly heads! But how could I check whether a bicycle wheel has self-centering? I examined a child's tricycle for this property, releasing it at speed and, running alongside it, giving the handlebars a blow. It certainly seemed to recover quickly and continue in a straight line, but unfortunately the tricycle (being free of the requirement of two-wheeled stability) has a different steering geometry.

So I made an experimental fixed-lean bicycle by fastening an extra "outrigger" wheel to the rear of the frame, converting it to an asymmetric tricycle. Adjustment of the outrigger anchorage could impose any angle of lean on the main frame. This machine was very interesting. Initially I gave it 15 deg of lean, and at rest the front wheel tilted to the 40-deg angle predicted by BICYC. When in motion, however, the wheel tended to straighten out, and the faster the bike was pushed the straighter did the front wheel become. Even if the machine was released at speed with the front wheel dead ahead it turned to the "equilibrium" angle for that speed and lean—another blow for gyro theory, for with the lean fixed there can be no precessional torque to turn the wheel. So clearly there *is* a self-centering force at work. It is unlikely to be pneumatic trail, for the equilibrium steering angle for given conditions appears unaltered by complete deflation of the front tire. Now I had encountered a very attractive form of self-centering action, not depending directly on variable frictional forces, while trying the naive experiment of pushing a bicycle *backwards*. Of course it collapsed at once because the two wheels travel in diverging directions. In forward travel the converse applies and the paths of the two wheels converge. So, if the front wheel runs naturally in the line of its own plane, the trailing frame and rear wheel will swing into line behind it along a tractrix, by straightforward geometry. To an observer on the bike, however, it will appear that self-centering is occurring (though it is the rest of the bike and not the front wheel that is swinging).

I modified my outrigger tricycle to hold the main frame as nearly upright as possible, so that it ran in a straight

SIDEWAYS FORCE on front tire produces a torque about the steering axis, so tending to lower the center of gravity of the bicycle. —FIG. 6

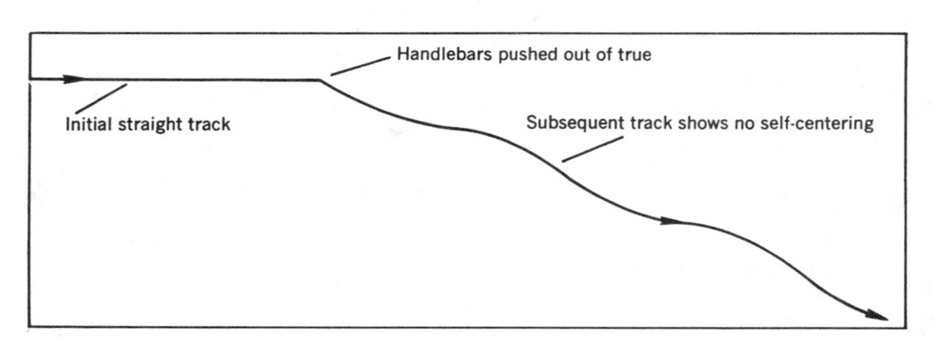

SELF-CENTERING? A bicycle with an "outrigger" third wheel to keep it upright was pushed and released riderless. At the point shown the handlebars were knocked out of true, resulting in a change of direction and no self-centering. The slight wave in the track resulted from oscillations in the framework. —FIG. 7

line. Then, first soaking the front wheel in water to leave a track, I pushed it up to speed, released it, and, running alongside, thumped the handlebars out of true. Looking at the bike, it seemed evident that the wheel swung back to dead ahead. But the track (figure 7) showed what I hoped to find—a sharp angle with no trace of directional recovery. The bicycle has only geometrical castor stability to provide its self-centering.

Success at last!

This test completed the ingredients for a more complete theory of the bicycle. In addition to the rider's skill and the gyroscopic forces, there are, acting on the front wheel, the center-of-gravity lowering torque (figure 6) and the castoring forces; the heavier the bicycle's load the more important these become. I have not yet formalized all these contributions into a mathematical theory of the bicycle, so perhaps there are surprises still in store; but at least all the principles have been experimentally checked.

I made URB IV by moving the front wheel of my bicycle just four inches ahead of its normal position, setting the system well into the unstable region. It was indeed very dodgy to ride, though not as impossible as I had hoped—perhaps my skill had increased in the course of this study. URB IV had negligible self-stability and crashed gratifyingly to the ground when released at speed.

It seems a lot of tortuous effort to produce in the end a machine of absolutely no utility whatsoever, but that sets me firmly in the mainstream of modern technology. At least I will have no intention of foisting the product onto a long-suffering public in the name of progress.

References

1. S. Timoshenko, D. H. Young, *Advanced Dynamics*, McGraw-Hill, New York (1948), page 239.
2. A. Gray, *A Treatise on Gyrostatics and Rotational Motion*, Dover, New York (1959), page 146.
3. J. P. den Hartog, *Mechanics*, Dover, New York (1961), page 328.
4. K. I. T. Richardson, *The Gyroscope Applied*, Hutchinson, London (1954), page 42.
5. R. H. Pearsall, Proc. Inst. Automobile Eng. **17**, 395 (1922).
6. R. A. Wilson-Jones, Proc. Inst. Mech. Eng. (Automobile division), 1951–52, page 191.
7. *Encyclopaedia Britannica* (1957 edition), entry under "Bicycle." □

The stability of bicycles

J. Lowell
Department of Physics, University of Manchester Institute of Science and Technology, Sackville Street, Manchester 1, United Kingdom

H. D. McKell
Department of Electrical Engineering and Electronics, University of Manchester Institute of Science and Technology, Sackville Street, Manchester 1, United Kingdom

(Received 7 November 1981; accepted for publication 15 February 1982)

We review previous ideas on why a bicycle is stable and analyze a somewhat simplified model. We show that a bicycle is almost, though not quite, self-stable; the most important parameter governing the stability is the "castor" of the front wheel, as was suggested by Jones [Phys. Today **23**, 34 (1970)]. The almost stable behavior explains why a bicycle is so easy to ride at speed.

I. INTRODUCTION

There does not seem to be any universally accepted answer to the question "how does a bicycle rider avoid falling over?" There are at least two aspects to the problem: first, we may ask what the rider does to counteract an incipient fall, and second, there is the possibility that a bicycle may be *self-stabilizing* at a sufficiently high speed, i.e., that the rider need do nothing to maintain his machine upright.

In this paper we shall be concerned mainly with the second question. We do not propose any new stability mechanisms; our purpose is to examine mechanisms previously suggested and to assess their relative importance with the help of measurements on an actual bicycle. We have tried to make the discussion accessible at undergraduate level, because we feel the problem may usefully be raised in a class on classical mechanics; many students are interested in cycling and will be strongly motivated to tackle the problem, which nicely illustrates features of motion in a circle, gyroscopes, oscillatory motion, and coupled differential equations.

We begin by reviewing previous work on the stability of bicycles. The most straightforward answer to the problem is that the rider simply "balances" the machine, i.e., that he moves his body and/or bicycle in such a way that the combined center of gravity always falls vertically above the line which joins the points of contact of the wheels and the ground. This explanation seems implausible, especially in view of the obvious fact that a stationary bicycle is difficult or impossible to balance whereas it is quite easy to balance a rapidly moving one. This fact is sometimes quoted as evidence that bicycles are gyroscopically stabilized, the rapidly rotating wheels acting in the same way as the gyroscope of a monorail car. Jones,[1] however, has cast doubt on all gyroscopic theories of stability with his experiments on a bicycle with a third wheel. This extra wheel was mounted clear of the ground and could be spun at any angular speed in either sense. Jones found that his modified bicycle could be ridden with equal ease, whatever the rate and direction of spin of the extra wheel; in other words, increasing or decreasing the total angular momentum made no difference to rideability. This is strong evidence that gyroscopic effects are not crucial to stability.[2]

Perhaps the most plausible account of how a bicycle stays upright is that given by (for example) Timoshenko and Young.[3] They point out that the centrifugal force developed when a bicycle travels a circular path can be used to overcome the couple due to gravity which tends to make the cycle fall. Thus, when a rider feels himself falling to the left (say), he turns the handlebars to the left so that the machine begins to travel in a circle. The resulting centrifugal force pushes the bicycle and rider upright and the handlebars can then be returned to the straight-ahead position. This accounts for the readily verified fact that it is virtually impossible to ride a bicycle if the handlebars cannot be turned (e.g., because the forks are lashed to the frame). To consider the matter quantitatively, we suppose that at some instant the bicycle is inclined at θ to the vertical, and is traveling in a direction that makes an angle η with some arbitrary reference direction (Fig. 1). If these angles are small the acceleration of the center of mass (CM in Fig. 1) in the direction normal to the frame is

$$h\ddot{\theta} + b\ddot{\eta} + v^2/R,$$

where R is the instantaneous radius of curvature of the path of the bicycle, and v the speed; h and b (and other quantities used below) are defined in Fig. 1. Since

$$R\dot{\eta} = v$$

and the component of the weight normal to the frame is $Mg \sin \theta \approx Mg\theta$ we have for the motion of the center of

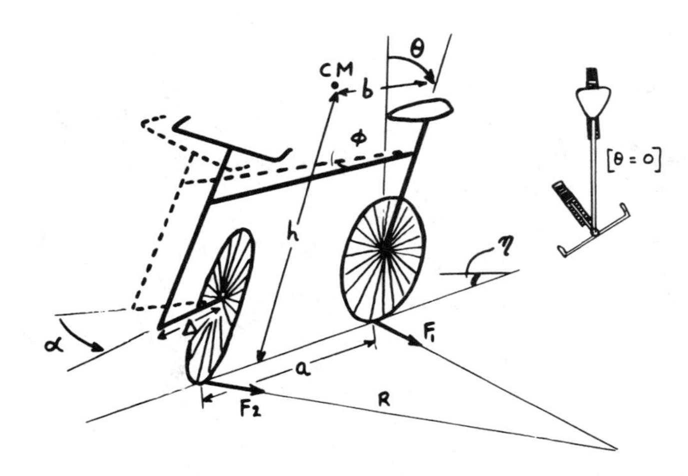

Fig. 1. Schematic sketch to show the parameters important in analyzing bicycle stability. CM is the center of mass of the bicycle with rider. The castor, Δ, is shown much exaggerated for clarity. R is the radius of curvature of the bicycle's path.

mass

$$g\theta = h\ddot{\theta} + b\frac{d}{dt}(v/R) + v^2/R.$$

Putting $R\alpha = a$ (Fig. 1), we find the equation

$$\ddot{\theta} + (bv/ha)\,\dot{\alpha} + (v^2/ha)\alpha - (g/h)\theta = 0. \quad (1)$$

Timoshenko and Young derive a much more complicated equation, because they do not restrict themselves to small values of θ and α; however, their equation reduces to Eq. (1) in the limit of small θ and α. This limit is adequate for considering stability under normal circumstances.

As Timoshenko and Young point out, it is clear from Eq. (1) that θ may be kept as small as may be desired, in principle, by suitably adjusting α. One interesting possibility is that the rider turns the handlebars by an amount proportional to his angle of leaning, i.e.,

$$\alpha = k\theta \quad (k = \text{const}).$$

Equation (1) then becomes

$$\ddot{\theta} + (kbv/ha)\dot{\theta} + [(kv^2/ha) - (g/h)]\theta = 0. \quad (2)$$

Evidently, this strategy ensures stability if $k > ga/v^2$, because Eq. (2) then describes a damped simple harmonic motion. However, there is no evidence to show that this is what a bicycle rider actually does; this question can only be resolved by experimental measurements, and we do not consider it further. Instead, we shall be concerned with the *self-stability* of a bicycle. Most cyclists would agree that, except at low speeds, it is hardly necessary to do anything to the bicycle—it seems to keep *itself* stable.[4] Assuming that Timoshenko and Young's is the correct explanation of bicycle stability, self-stability implies that a change in θ automatically causes a change in α which, through Eq. (1), tends to restore θ to its original value. Ways in which θ affects α are considered in Sec. II.

II. STABILIZATION MECHANISMS

It has been recognized for some time that the appreciable angular momentum of the front wheel must imply that it precesses (i.e., α increases) when the bicycle leans over (θ increases). If the wheel were rolling alone, this precession would ensure its stability if the speed were large enough, just as a rolling coin or a child's hoop is stable.[5] Grey[6] and Pearsall[7] regard this as the fundamental mechanism for stability of the bicycle; they treat the frame and rear wheel as complicating additions to the front wheel. To investigate the importance of this effect we need to know the rate of precession of the front wheel, which may be calculated as follows. Let I_z be the moment of inertia of the front wheel about the steering axis (roughly, a vertical diameter) and I its moment of inertia about its axis of rotation. Assuming that no couple is applied to the handlebars and that the front wheel can turn without friction, conservation of angular momentum about the steering axis gives

$$\frac{d}{dt}(I_z\dot{\alpha} - I\omega \sin\theta) = 0,$$

where ω, the angular speed of the front wheel, is vr if r is the wheel radius. Thus, if θ increases, the front wheel must precess, i.e., α must increase. For small θ,

$$\ddot{\alpha} = (Iv/I_z r)\dot{\theta}. \quad (3)$$

It is evident that the front wheel turns "into the fall," as in Fig. 1; thus the gyroscopic precession of the front wheel is in the direction necessary to arrest the fall.

Jones's[1] experiments with an extra wheel show that gyroscopic effects of the front wheel are probably not vital to the stability of bicycles; he therefore emphasized the importance of "castor" or "trail." In all bicycles, apparently, the steering axis intersects the ground in front of the point of contact between the ground and the front wheel; we shall see that this results in the front wheel turning when the cycle leans.[1,4] Figure 1 shows a bicycle with a grossly exaggerated castor (Δ); evidently, turning the handlebars through an angle α will cause the crossbar to turn through an angle $\phi \approx \alpha\Delta/a$, and the center of mass to move a distance $b\phi$ normal to the frame. If the cycle is leaning at an angle θ, the center of mass will fall, when the handlebars turn, through a distance $b\phi \sin\theta \approx (b\Delta/a)\alpha\theta$. Turning the handlebars through α, in the direction shown in Fig. 1, therefore lowers the potential energy by $(Mgb\Delta/a)\alpha\theta$, where M is the mass of the cycle (with rider). There is therefore a couple on the front wheel whose magnitude is $(Mgb\Delta/a)\theta$ and whose direction tends to increase α. Thus

$$I_z\ddot{\alpha} = (Mgb\Delta/a)\theta. \quad (4)$$

Again, the wheel turns into the fall, which tends to reduce θ; so the "Jones couple" is stabilizing (at least initially). The effect of castor is readily seen in a stationary bicycle—if the cycle is made to lean, the handlebars will turn into the direction of leaning. Jones[1] found that a bicycle with negative castor Δ was very difficult if not quite impossible to ride, so it seems probable that Δ and the associated couple are important factors in bicycle stability.

The castor Δ also causes a frictional couple which tends to reduce α. This couple is $F_2\Delta$, where (Fig. 1) F_2 is the sideways frictional force on the front wheel. The frictional forces F_1 and F_2 together provide the force necessary to maintain the bicycle on its circular path; therefore $F_2 + F_1 = Mv^2/R$. To prevent rotation about a vertical axis through the center of mass, F_2/F_1 must be equal to $b/(a-b)$. Thus

$$I_z\ddot{\alpha} = -(Mb\Delta v^2/a^2)\alpha. \quad (5)$$

Kirschner[4] has considered the Jones couple [Eq. (4)] and the frictional couple associated with the centrifugal force [Eq. (5)] in connection with the stability of a "*quasistatic*" bicycle: he shows that the potential energy (including a "centrifugal" term) decreases with θ for all values of θ if it is assumed that α always adjusts itself to its equilibrium value for the prevailing θ. He finds that under these circumstances there is no stability. However, as Kirschner notes,[4] the lack of quasistatic stability does not preclude *dynamic* stability; in Sec. III, we shall show that Eqs. (1) and (4) do imply that the bicycle has some degree of self-stability: the quasistatic assumption appears not to be valid.

It should be said that certain minor effects are not included in the above discussion. For example, the finite width of the tires must result in a frictional couple about the steering axis when the bicycle leans; we assume that the tires are so narrow that this can be neglected. Again, there will be centrifugal force acting on the center of mass of the front wheel tending to increase α. It is readily shown that this is negligible compared to the couple in Eq. (5), provided that the mass of the wheel is much less than that of the rest of the cycle plus the rider. Rewriting Eq. (1), and collecting the contributions to $\ddot{\alpha}$ gives

$$\ddot{\theta} - (g/h)[\theta - (v^2/ga)\alpha] + (bv/ha)\dot{\alpha} = 0, \quad (6)$$

$$\ddot{\alpha} - \mu[\theta - (v^2/ga)\alpha] - \gamma v\dot{\theta} = 0, \quad (7)$$

where $\mu = Mbg\Delta / I_z a$ and $\gamma = I / I_z r$.

We shall examine the behavior predicted by Eqs. (6) and (7) in Sec. III but it is useful first to discuss the magnitude of the various coefficients.

III. MAGNITUDES AND NUMERICAL ANALYSIS

Estimates and measurements of the relevant quantities were made for a real bicycle (Claude Butler "Majestic," 24½-in. frame) and are summarized in Table I. The distances a and b were measured by determining the distribution of weight between the two wheels, using two bathroom scales. The "Jones couple" μI_z was measured as follows: the front wheel was supported on a ballrace and a spring balance was used to find the couple required to keep the front wheel in the straight-ahead position while the rider leaned the bicycle at various angles, supporting himself as lightly as possible with fingertips against a wall. Plotting the couple against angle of lean enabled the frictional couple from the ballrace to be eliminated. This method gave a better estimate of μ than could be found by measurement of Δ, because of the rather complicated geometry of real bicycles and the small magnitude of Δ. The measured value of μ implies a value of Δ of about 2.5 cm, and direct inspection of the bicycle shows this to be reasonable. The values of Table I are given for rough guidance; no great accuracy was attempted, nor would it be useful, because most of the parameters will vary from one bicycle to another and many of them depend on the weight of the rider and on the riding position he adopts.

Analytic solutions of Eqs. (6) and (7) can be obtained, as we show in the Appendix. However, the analytic solutions are rather complicated and consequently unilluminating in algebraic form. We feel that an understanding of the details of the motion is best obtained by approximate solutions and semiquantitative discussion of the equations, backed up by graphical display of the exact solutions, and this approach is adopted below (except in the Appendix, where the exact analytic solution is outlined). In addition to the analytic solutions we have used a fourth-order Runge–Kutta method to obtain numerical solutions. These numerical solutions were in close agreement with the analytic solutions.

To examine stability, we assume that the bicycle is ini-

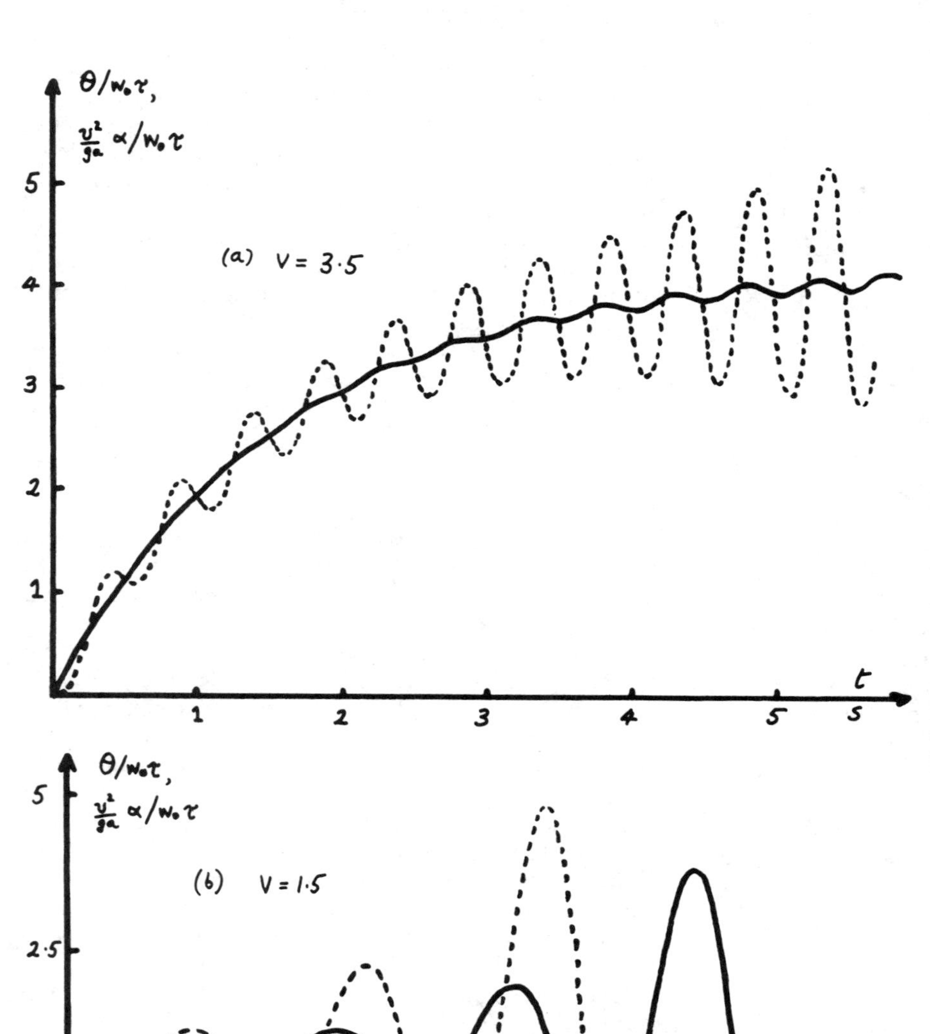

Fig. 2. Response of a bicycle to an impulse tending to make it fall over. The solid curve shows how the angle of lean (θ) depends on time and the dotted curve shows the time dependence of α, the angle that the front wheel makes with the plane of the frame. Both θ and α are measured in units of $W_0\tau$, where W_0 is the initial rate of change of θ and τ is the characteristic time $(h/g)^{1/2}$. Curves (a) are for a speed of 3.5 m/s and (b) for 1.5 m/s. Gyroscopic effects are neglected.

tially in equilibrium and upright ($\theta = \ddot{\alpha} = \dot{\theta} = \dot{\alpha} = \ddot{\theta} = \ddot{\alpha} = 0$) and study the effect of an impulse (e.g., a gust of wind) which imparts an initial rate of fall $\dot{\theta} = W_0$.

IV. SELF-STABILITY, NEGLECTING GYROSCOPIC EFFECTS

Although solutions of Eqs. (6) and (7) can be obtained without the need for simplifications, an intuitive feeling of how the bicycle behaves is best obtained by first seeing what happens in the absence of gyroscopic effects ($\gamma \to 0$). Examining the coefficients of the remaining terms (Table I), one is struck by the large magnitude of μ. This immediately suggests that $\theta - (v^2/ga)\alpha$ must always be small; for if $\theta - (v^2/ga)\alpha$ were much greater (say) than zero, $\ddot{\alpha}$ would be large and α would rapidly grow until $(v^2/ga)\alpha$ became almost equal to θ. Therefore we try as an initial approximation

$$\theta \approx (v^2/ga)\alpha. \tag{8}$$

Substituting for α in Eq. (6) gives

$$\ddot{\theta} + (gb/hv)\dot{\theta} = 0.$$

And if we take (Sec. II) $\theta = 0, \dot{\theta} = W_0$, at $t = 0$ the solution to this equation is

$$\theta = \nu^{-1} W_0 (1 - e^{-\nu t}), \tag{9}$$

where $\nu = gb/hv$.

Thus, if some perturbation causes the bicycle to start falling over at $t = 0$, the fall does not continue; instead, θ asymptotically approaches a steady value of $hv(W_0/gb)$, and α likewise approaches a value of $ha(W_0/bv)$. This means that the cycle responds by eventually following a circular path whose radius is $(bv/h)W_0$. The bicycle is stable in the sense that it does not fall over, though it does deviate from its linear path.

Figure 2 shows the results of the numerical solution of Eqs. (6) and (7) (with $\gamma = 0$) for $v = 1.5$ and 3.5 m/s. Broadly, the above conclusions are verified: if one takes the mean of the oscillations, $(v^2/ga)\alpha$ remains close to θ, and the curve of θ vs t is of the form predicted by Eq. (9). Numerically, we find from Table I that $hv/gb \approx 1.5$ s at $v = 3.5$ m/s and inspection of Fig. 2 shows that θ reaches 0.63 of its asymptotic value in a time very close to this, as required by Eq. (9).

However, the numerical solution (Fig. 2) shows that in reality an oscillation is superimposed on the solution Eq. (9); it is not difficult to see the origin of this oscillation. We can rewrite Eq. (7) (with $\gamma = 0$) as

$$\ddot{\alpha} + (\mu v^2/ga)\alpha = \mu\theta. \tag{10}$$

Suppose that the bicycle is initially stable with $\alpha = \theta = 0$ and that some perturbation causes α to deviate from zero. Ignoring for now the consequent change of θ, there will be a simple harmonic oscillation in α of frequency $(2\pi)^{-1}(\mu v^2/ga)^{1/2}$, i.e., ~ 2 Hz for $v = 3.5$ m/s. This oscillation would be steady if θ on the right-hand side of (10) were constant; but in fact it is not, because oscillation in α causes oscillation in θ, albeit with shifted phase [Eq. (6)]. Thus Eq. (10) represents, in fact, a driven harmonic oscillator, and the perturbation in α grows. In the specific example of Fig. 2, the "perturbation" starting the oscillations is a consequence of our initial conditions, $\dot{\alpha} = 0$, $\dot{\theta} \neq 0$ which imply $\theta \neq (v^2/ga)\alpha$ for $t \approx 0$. However, it is clear that any changes in θ and α which do not together satisfy Eq. (8) will lead to growing instability.

Table I. Parameters (for a real bicycle) used in the numerical analysis.

Quantity	Description	Magnitude
Measured quantities		
a	Wheelbase	1.0 m
b	Distance from real-wheel hub to CM	0.33 m
h	Height of CM	1.5 m
μI_z	Couple on front wheel/angle of lean	8 N m rad^{-1}
r	Wheel radius	0.33 m
m	Wheel mass	1.6 kg
M	Mass of cycle + rider	80 kg
Derived quantities		
$g/h\,(=\tau^{-2})$		6.5 s^{-2}
I	Polar M. of I. of front wheel	0.08 kg m^2
I_z	M. of I. of front wheel about diameter (includes allowance for handlebars)	0.06 kg m^2
μ		133 s^{-2}
γ		4.0 s^{-1}

To summarize, if gyroscopic effects are ignored, the bicycle is almost self-stable. A perturbation tending to push it over results, in first approximation, merely in the bicycle entering a curved path. However, the bicycle is *unstable* in the sense that oscillations in α tend to grow. In practice, the oscillatory instability would probably not matter; growth is very slow (Fig. 2) and it is possible that the oscillations would not be noticeable if the rider were anything other than completely inert; we shall discuss this point in Sec. V.

An interesting point that arises from this analysis is, why is μ so large? Cyclists like their bicycles to feel light and responsive, and this requires a small value of μ (so that the bicycle turns at a light pull on the handlebars). It is true that Eq. (9) depends on the assumption that μ is large, but it seems plausible that $\mu \gtrsim 10$ would be sufficient and this is, in fact, borne out by the exact solutions—we find that reducing μ has little effect until it is reduced below ~ 15. [Note that ν, and hence the asymptotic value of θ, is independent of μ according to Eq. (9)]. It is inconceivable that years of empirical optimization of bicycle design would unnecessarily leave μ so large. In fact, there is good reason for the large value, as will become clear in Sec. V.

V. GYROSCOPIC EFFECTS

We have seen that the solution of Eqs. (6) and (7) with $\gamma = 0$ can be divided into two parts—a slow approach of θ and α to steady values, which reflects the tendency of θ to stay close to $v^2(\alpha/ga)$, and a superimposed rapid oscillation which reflects the deviation from $\theta = v^2(\alpha/ga)$. When the gyroscopic term, $\gamma v\dot{\theta}$, is included in Eq. (7) the equations become less amenable to qualitative analysis. However, for small enough speeds the gyroscopic term is small and we can make progress by treating it as a perturbation. For the slow component of the solution [Eq. (9) in the case $\gamma = 0$] we can suppose $\ddot{\alpha}$ to be small and put $\theta - (v^2/ga)\alpha \approx \gamma v\dot{\theta}/\mu$ in Eq. (6). This leads to a solution of the form of (9) but with

$$\nu = \frac{gb}{hv}\left(1 + \frac{\gamma v^2}{\mu b}\right) \Big/ \left(1 + \frac{bg\gamma}{\mu h}\right),$$

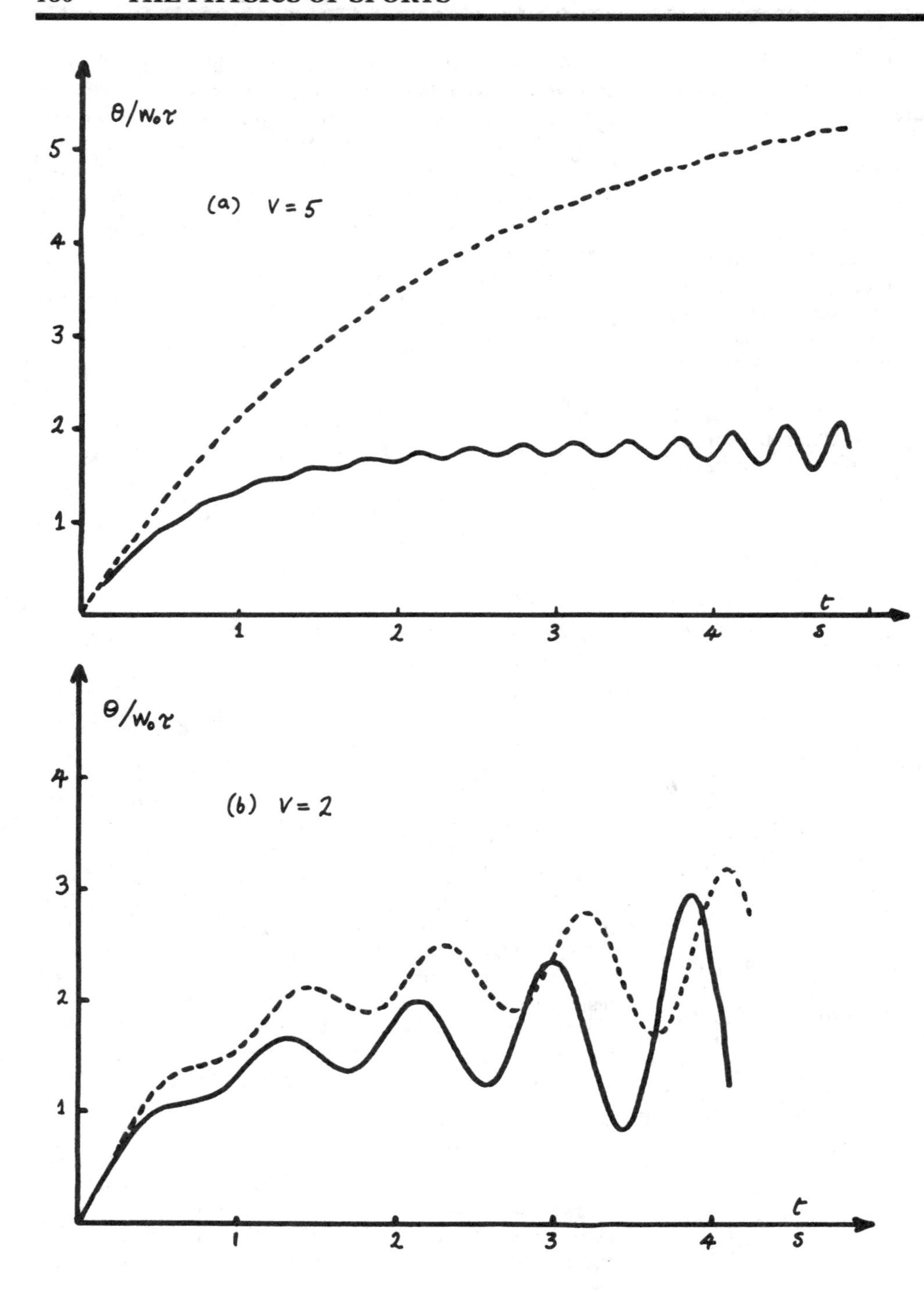

Fig. 3. These curves show the influence of gyroscopic action of the front wheel on bicycle stability. The dotted curves show the response that would result if gyroscopic effects were absent, and the solid curves show the response when they are included. Gyroscopic action is stabilizing in the sense that it results in a smaller (mean) value of θ, but destabilizing in the sense that it enhances the oscillatory instability. Curve (a) speed 5 m/s; (b) speed 2 m/s.

for $v > b(\sqrt{g/h})$ (~ 1 m/s), gyroscopic effects increase ν, so that the disturbance has a smaller effect. In this sense, the gyroscopic effect of the front wheel enhances stability.

The effect of the gyroscopic term on the oscillatory motion is not easily assessed by the perturbation argument. It is evident that oscillations in α will cause oscillations in θ of the same frequency but with shifted phase, so that $\gamma v\dot{\theta} = A\dot{\alpha}$ where A is in general complex. Equation (7) then becomes

$$\ddot{\alpha} + A\dot{\alpha} + (\mu v^2/ga)\alpha = \mu\theta,$$

which represents a driven oscillator, as before; but now there is a "damping" term due to the gyroscopic effect. This term will tend to damp or to enhance the oscillation in α, according as the real part of A is positive or negative. The full solution (Fig. 3) shows that the oscillations are enhanced, in fact markedly so at higher speeds. The reduction in the average value of θ for $t \rightarrow \infty$ is also apparent in Fig. 3.

To summarize, gyroscopic effects result in a smaller *mean* deflection of the bicycle, for a given perturbation, and they are in this sense stabilizing. However, they *enhance* the oscillatory instability, and Fig. 3 suggests that this second effect outweighs the former, i.e., that the gyroscopic action is a net nuisance. This appears to be the answer to the problem raised at the end of the last section, viz. why is μ so large? It seems that μ must be large to prevent the oscillatory instability becoming uncontrollably large. This is illustrated in Fig. 4, which shows θ vs t at 3.5 m/s for several values of μ (expressed as a multiple of the value for the actual bicycle). It is apparent that the μ chosen by the

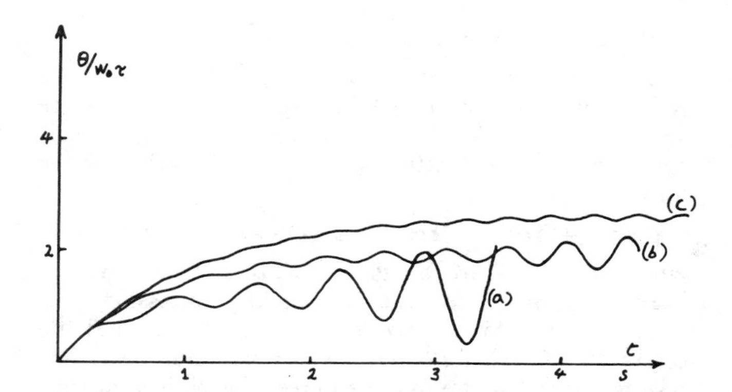

Fig. 4. This figure illustrates the importance of castor or trail, as measured by the parameter μ. Curve (b) is the predicted response of the bicycle studied (Table I) to an impulse pertubation. Curve (a) shows the predicted response, to the same perturbation, if the castor were halved ($\mu = 66.5$), and curve (c) is the prediction for twice the castor ($\mu = 266$). Speed is 5 m/s.

manufacturer is a reasonable compromise between the penalties of oscillatory instability at small μ and unnecessarily large "dc" response to perturbation (and heavy steering) for large μ.

VI. FRICTION

We have assumed so far that the front wheel steers without friction. This cannot, of course, be true and it might reasonably be supposed that the friction inevitably present in a real machine would damp the oscillating instability and make it unimportant. We have in fact found this to be so if the friction is "viscous," i.e., proportional to $\dot{\alpha}$. However, "viscous" friction is an implausible assumption, and "Coulomb" friction is much more probable; accordingly, we have carried out numerical solutions with Eq. (7) modified to

$$\ddot{\alpha} = 0 \quad \text{if } |-\gamma v\dot{\theta} + \mu(\theta - v^2/ga)| < F,$$

$$\ddot{\alpha} = \mu[\theta - (v^2/ga)\alpha] + \gamma v\dot{\theta} \pm F \quad \text{otherwise,}$$

with the sign of the constant F chosen so that $|\ddot{\alpha}|$ is always reduced. The results (Fig. 5) show that the oscillating instability is not removed by coulomb friction.

There is, however, another means whereby the oscillating instability may be rendered harmless, if it is slow growing. A bicycle rider does not sit still on his bicycle—probably it is psychologically almost impossible to do so. Rather, he will respond to perturbations by steering his bicycle back to a straight-line path. Every minor "correction" applied by the rider is in a sense a new perturbation which interrupts the oscillations with a random phase. Because of these constant corrections, we suppose, the oscillating instability is unable to grow. It remains true, however, that a large value of μ is an important feature of the bicycle—Fig. 4 shows that for $\mu \to 0$ the oscillatory instability would grow very rapidly.

It seems likely that a bicycle cannot be perfectly self-stable and that some intervention on the part of the rider is essential. However, a well-designed machine comes very close to stability—the oscillatory instability grows very slowly, except at rather low speeds. This is, presumably, why a bicycle is easy to ride—it is so close to being self-stable that the rider has plenty of time to make the rather small corrections required to keep his machine upright.

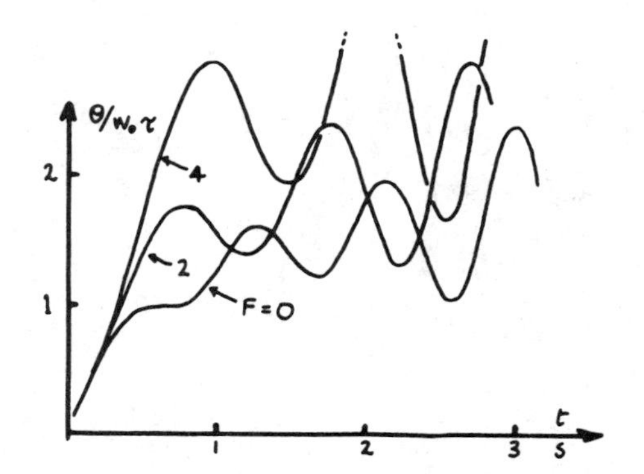

Fig. 5. Coulomb friction in the steering axis will not damp out the oscillatory instability. These curves, drawn for a speed of 2 m/s, show the response to an impulsive perturbation for three values of the friction parameter F (Sec. V).

VII. SUMMARY AND CONCLUSIONS

The most important factor in the self-stability of bicycles is the "castor" or "trail" of the front wheel, as was suggested by the experiments of Jones.[1] To first order, the bicycle responds to a perturbation (a push from the side) by following a curved path such that the "lean" caused by the perturbation is balanced by a centrifugal force. The disturbance is less pronounced at high speeds and for bicycles whose center of mass is well forward of the rear wheel. We find that the gyroscopic torque on the front wheel contributes to the stability in this sense.

Although the bicycle is stable in the sense that it does not fall over when pushed sideways, it is unstable against oscillation in the front wheel direction (α) and the resulting side-to-side oscillation of the machine as a whole. Gyroscopic effects make these oscillations grow much more rapidly than they would otherwise do, and Coulomb friction does not damp them out. It appears that the magnitude of "castor" required is governed by the need to keep the growth rate of this oscillatory instability manageably small. But for this, reasonable stability could be achieved with a much smaller castor.

Were it not for the oscillatory instability, riding a bicycle would be a simple matter of steering it to correct the changes in direction caused by perturbation. It seems likely that this is also true for a real bicycle, because the steering corrections are in effect new perturbations to the motion which force the oscillatory instability to begin again from small amplitude (the perturbations occur at random times and therefore for random phases of the oscillation). The fairly large castor of real bicycles ensures a slow growth of the oscillatory instability, so that the rider can be quite leisurely in his intervention. Figure 4 demonstrates that he would have to respond very quickly indeed if the castor were small, and shows why bicycles with small castor[1] are very difficult to ride.

ACKNOWLEDGMENTS

Catherine and Sarah Lowell assisted in the rather tricky measurement of the Jones couple μ.

APPENDIX

Equations (6) and (7) can be solved exactly by Laplace transforms.

The equations can be written

$$\ddot{\theta} = m(\theta - c\alpha) - b^{*}\dot{\alpha}, \quad (6')$$

$$\ddot{\alpha} = \mu(\theta - c\alpha) + d\dot{\theta}, \quad (7')$$

where $m = g/h$, $b^{*} = bv/ha$, $d = \gamma v$, $c = v^2/ga$.

If we denote the Laplace tansform of θ with respect to time by $\bar{\theta}$, i.e.,

$$\bar{\theta} = \int_0^{\infty} \theta e^{-st}\, dt,$$

then the Laplace transforms of Eqs. (6′) and (7′) are

$$s^2\bar{\theta} - s\theta(0) - \dot{\theta}(0) = m(\bar{\theta} - c\bar{\alpha}) - b^{*}[s\bar{\alpha} - \alpha(0)]$$

$$= m\bar{\theta} - (b^{*}s + mc)\bar{\alpha} + b^{*}\alpha(0), \quad (6'')$$

$$s^2\bar{\alpha} - s\alpha(0) - \dot{\alpha}(0) = \mu(\bar{\theta} - c\bar{\alpha}) + d\,[s\bar{\theta} - \theta(0)]$$

$$= -c\mu\bar{\alpha} + (\mu + sd)\bar{\theta} - d\theta(0), \quad (7'')$$

if $\bar{\alpha}$ from Eq. (7″) is substituted into Eq. (6″) we obtain, after some manipulation

$$\bar{\theta} = (As^3 + Bs^2 + Cs + D)/s(s^3 + ps + q), \quad (8')$$

where

$$A = \theta(0), \quad B = \dot{\theta}(0),$$

$$C = [(\mu c + b^{*}d)\theta(0) - mc\alpha(0) - b^{*}\dot{\alpha}(0)],$$

$$D = c[\mu b^{*}\alpha(0) + md\theta(0) + \mu\dot{\theta}(0) - m\dot{\alpha}(0)],$$

$$p = \mu c - m + b^{*}d, \quad q = b^{*}\mu + mcd.$$

Expression (8′) for $\bar{\theta}$ looks intractable, but it is possible to show (see for instance Tuma[8]) that the cubic factor in the denominator can be written

$$s^3 + ps + q = (s + u)[(s - u/2)^2 + w^2],$$

where u and w can be calculated from p and q (although the relationships are not simple). Finally, expression (8) can be split into partial fractions, each of which is the Laplace transform of a standard function and θ can be written in the form

$$\theta = W + Xe^{-ut} + Ye^{ut/2}\cos wt + Ze^{ut/2}\sin wt,$$

where W, X, Y, Z are functions of A, B, C, D, u, and w.

By the same methods a corresponding expression for α can be obtained. We have programmed a HP 9825 desk top calculator to evaluate θ and α as a function of time. The results (Figs. 2–4) are in close agreement with numerical solutions devised by the Runge–Kutta method. A complete analysis is available on request.

[1] D. E. H. Jones, Phys. Today **23**(4), 34 (April 1970).

[2] To avoid confusion, we note that gyroscopic effects are of possible importance for two distinct reasons. As well as causing the "monorail car" stabilizing effect discussed here, the rotating front wheel will precess when the cycle begins to fall, and this may influence the stability by causing the cycle to enter a circular path. This is considered further below.

[3] S. Timoshenko and D. H. Young, *Advanced Dynamics* (McGraw-Hill, New York, 1948).

[4] D. Kirschner, Am. J. Phys. **48**, 36 (1980).

[5] A. S. Ramsey, *Dynamics (Part II)* (Cambridge University, London, 1956).

[6] A. Grey, *A Treatise on Gyrostatics and Rotational Motion* (McMillan, London, 1918).

[7] R. H. Pearsall, Proc. Inst. Automobile Eng. **17**, 395 (1922).

[8] J. J. Tuma, *Engineering Mathematics Handbook*, 1st ed. (McGraw-Hill, New York, 1970), p. 7.

More bicycle physics

THOMAS B. GREENSLADE, JR.

A pervasive theme in *The Physics Teacher* is the application of basic physical principles to familiar situations. A number of recent articles and publications have used the common experience of riding a bicycle as motivation for examples in mechanics.[1] This note will present three examples which can be used to show students in introductory courses that they have the background to make and use models of real-life situations.

The broom seller

When I taught at the University of the West Indies in Kingston, Jamaica in 1972-73, the Rastafarian broom seller shown in Fig. 1 used to ride by my house every few days, selling small brooms for dusting, medium-sized brooms for sweeping, and long-handled brooms for knocking cobwebs out of the tall ceilings. He balanced the load of brooms atop his head, and I was always fascinated to see him pedal around a corner without dropping his bundle.

A first approximation to this situation would have the broom seller standing in one spot, and swinging the load of brooms through a right angle by turning his head. The model must make reasonable assumptions for the moment of inertia of the load and for the length of time necessary to make the turn, and then predict what maximum torque must be exerted to get the bundle of brooms around the corner. The torque can then be calculated in a second way by considering the interaction between the head and the bundle of brooms.

The bundle of brooms shown in Fig. 1 consists of brooms of various lengths, balanced at the center of mass. To make the calculation easy, let us assume that all of the brooms are of the same length L, and that there are N of them, each of mass M. Assuming further that the center of mass is at the geometrical center of the bundle, the load becomes the equivalent of a rod, and the moment of inertia I of the bundle through an axis perpendicular to the street and through the center of mass is

$$I = (1/12)NML^2$$

Typical data might be 5 of the long cobweb brooms, each 1 kg in mass and 3 m in length, which gives a moment of inertia of 3.75 kg-m^2.

The broom seller turns the corner slowly enough so that he does not tilt the bicycle, which means that the axis of rotation of the brooms always remains vertical. The right-angle turn is made in time $2t$. Assume that the load starts with zero angular velocity, and the angular velocity increases linearly up to time t, when it decreases linearly to zero again at time $2t$. A number of assumptions could be made here, but this one allows the use of the kinematic equation for rotation under constant angular acceleration α. The angle θ through which the bundle rotates, assuming zero initial angular velocity, is

$$\theta = (1/2)\alpha t^2$$

Let the angle be $\pi/4$ (half of a right angle) and the time for the half turn be 3.0 s; this relatively long time ensures a turn with negligible banking. The angular acceleration is 0.17 rad/s^2. The torque which must be exerted to swing the bundle around is the product of the moment of inertia and the angular acceleration, or 0.65 N-m.

Thomas B. Greenslade, Jr. *has done considerable long-distance solo bicycle touring. (Department of Physics, Kenyon College, Gambier, Ohio 43022)*

Reprinted from *The Physics Teacher* **21**, 360–363 (1983); © American Association of Physics Teachers.

Fig. 1. Jamaican broom seller. How can he turn a corner with brooms balanced on his head?

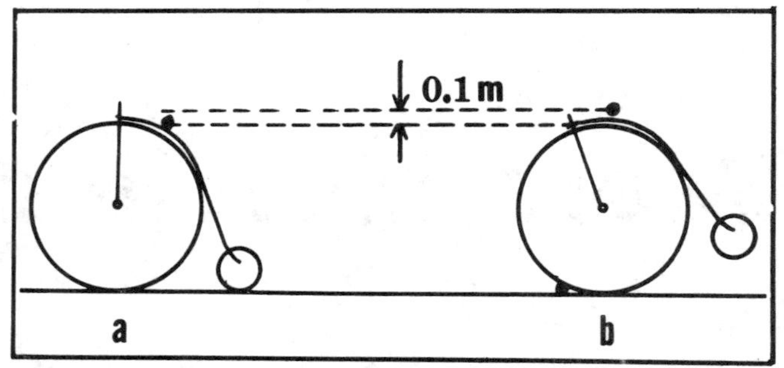

Fig. 2. (a) The location of the center of mass of the bicycle and its rider when in motion. (b) The location of the center of mass when the rider is just about to topple forward off the bicycle.

The torque may also be calculated if we make the assumption that the weight of the bundle is uniformly distributed over a distance, a, on top of the broom seller's knitted wool cap. Consequently, the normal force exerted on the bottom of the bundle by a length dR of the cap is

$$dF = (NMg)\,(dR/a)$$

and the maximum torque exerted by the frictional force μdF (μ is the coefficient of static friction between the cap and the wood of the brooms) is

$$dT = \mu(NMg/a)R\,dR$$

The torque exerted by the part of the cap between $R = 0$ and $R = +a/2$ is obtained by integrating over these limits, giving

$$T = \mu NMga/8$$

The total torque exerted by the entire cap is just twice this value. Let us assume a value for μ of 0.3, and take the value of a to be 0.15 m. The calculated value for the required torque is 0.55 N-m. The two values for the torque are in agreement within the wide range of assumptions which have been made.

The Penny-farthing

As an undergraduate, I restored a bicycle of late 1870's vintage, and learned how to ride it. The type of bicycle shown on our cover is called an Ordinary bicycle in the United States, and a Penny-farthing in the United Kingdom (from the way in which the combination of a large and a small wheel resembles the large 19th-century English penny placed next to a small farthing coin).

You mount this bicycle by standing with one foot on a small step to the rear of the backbone, and hopping along on the other foot until you get up enough speed to vault safely into the saddle. Once on, you had better find the pedals rapidly with your feet so that you can avoid being rapped sharply on the shin with a heavy pedal. Now that you are underway, lateral balance and steering are easy for anyone who can ride a modern bicycle.

Avoiding taking a header over the handlebars is not quite so easy. Even a small hole in the road, or a small stone on the surface, is enough to precipitate the unwary rider head first onto the road. I became curious about just how slowly one needed to travel forward to avoid the header. The following assumptions were made: Usually a large wheel will ride up and over a small irregularity in the pavement. As we shall see in the next paragraph, this form of bicycle is on the verge of instability because the center of mass needs to be raised only a small distance to reach a state of unstable equilibrium. The simplifying assumption is made that the consequence of hitting an obstruction is that the point of contact between the large wheel and the road stops moving forward, and the bicycle and rider, as a unit, tips forward about this point. An experienced rider will, of course, throw himself backward to prevent this from happening.

Figure 2(a) shows the outline of the wheels and the location of the center of mass. We assume that the center of mass of the bicycle and rider is located in the region of the hips, and that it is located 1.5 m above the point of contact of the large wheel to the road and 0.5 m behind this point. In Fig. 2(b) the bicycle is shown tipped up into a position of unstable equilibrium, with the center of mass directly above the one point of contact between the bicycle and the road. The center of mass has risen about 0.1 m. Assuming that the bicycle and rider ~~are~~ just poised, with no forward velocity, and ready to fall over in any direction, the increase in potential energy can be equated to the kinetic energy just before the machine started to tip up. If we assume that all of the kinetic energy is translational, then $\frac{1}{2}mv^2 = mgh$, and this corresponds to a velocity of 1.4 m/s, which is about 3.5 mi/hr – a walking pace. As I remember from my experiences as a sophomore, this seems to be a low, but reasonable value. Note that the calculation is independent of the mass of the system. This simple model implicitly assumes that the collision between the wheel and the obstacle is elastic.

A model for bicycle dynamics

This section is an exercise to see how much information can be obtained from a simple model of the forces acting on the bicycle and its rider. The model has been presented in somewhat different forms by Whitt and Wilson[2] and by DiLavore[1]; the object here is to give it an extended and unified treatment.

The model

There are two retarding forces acting on a bicycle and its rider. The first is a velocity-independent force usually called the rolling resistance. This can be reduced by using a narrow tire inflated to a high pressure. Whitt and Wilson suggest that this force is about 0.05 N per kilogram of total mass. Assume that the rider weighs 170 lb and the bicycle another 30 lb; this corresponds to a mass of about 90 kg. Therefore, let the constant retarding force of rolling friction be 4.5 N. The second is the force of air resistance on the bicycle which, at typical riding speeds, is proportional to the square of the speed of the bicycle. (Let us assume that there is no wind.) DiLavore states that at a speed of 8 km/hr, the force of air resistance is 1 N. This figure depends on the frontal area of the bicycle and its rider, among other factors. The total retarding force in newtons acting on the bicycle and rider when the speed is V km/hr is thus

$$F = 4.5 + 0.015\ V^2$$

and opposes the motion. In what follows, remember that the coefficients are typical values, and will be different when the bicycle-rider system is changed. From this model we can get information about the maximum speed which the rider can attain on the level and when ascending and descending hills of constant slope.

Maximum speed coasting down a hill

Assume that you are coasting down a long hill which has an angle of incline of θ with respect to the horizontal. The component of the gravitational force acting parallel to the pavement is $Mg \sin \theta$, where M is the total mass of the system. This force is equal and opposite to the retarding forces acting on the bicycle, giving

$$Mg \sin \theta = 4.5 + 0.015\ V^2$$

Three points should be noted. This equilibrium condition is reached asymptotically, and a rider with a larger mass reaches a higher terminal velocity than one with smaller mass, provided that they both have the same frontal area. The model fails for small speeds because the force of rolling friction must be zero at zero speed.

Figure 3 shows how the terminal velocity of the 90-kg bicycle and rider depends on the angle of the incline. The speeds may appear high for the seemingly small angles, but remember that a hill with an angle of 6.5 degrees, corresponding to the downhill speed of 50 mi/hr, has a rise of 1.0 m for every 8.8 m of horizontal advance. Few of us, except professional bicycle racers, would care to coast at such high speeds. We reduce the speed of the bicycle either by sitting upright on the bicycle (increasing the coefficient of the V-squared term by increasing the frontal area), or we put on the brakes (causing some of the gravitational potential energy to be transferred from the brake shoes and the rims to the air in the form of heat).

If we do want to increase the coasting speed, or the maximum speed on the level, bending over the handlebars to reduce the frontal area will reduce the coefficient of the V-squared term. Another approach is the use of partial or full fairings to reduce the drag forces.

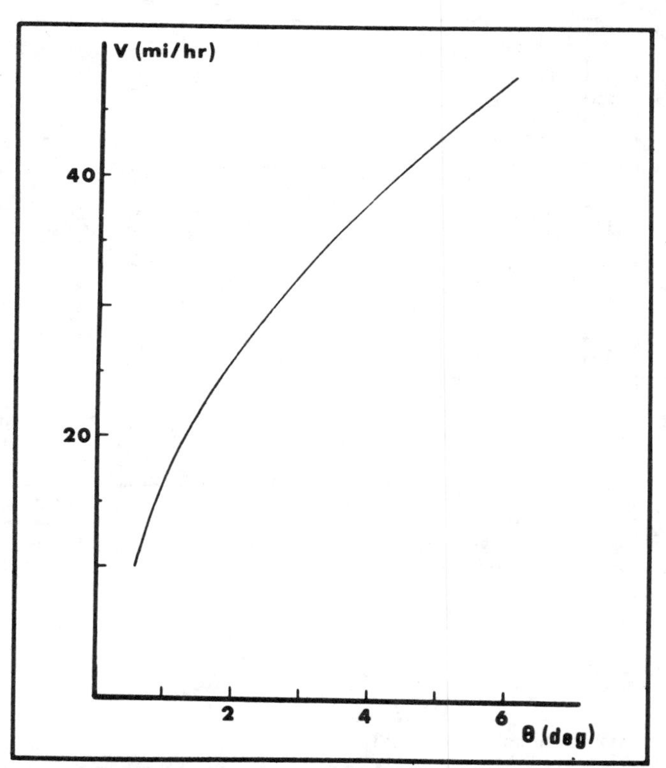

Fig. 3. The terminal velocity of the bicycle and rider as a function of the angle of the road with respect to the horizontal.

Maximum speed on the level

Assume that the rider is traveling with speed V on level ground with no wind. The power necessary to maintain this speed is $P = FV$, where F is the magnitude of the horizontal forward force which is exerted by the ground on the rear wheel at the point of contact with the ground. This force is equal and opposite to the retarding force of $4.5 + 0.015\ V^2$ N. The expression for the power expended by the rider is therefore

$$P = 4.5\ V + 0.015\ V^3$$

The unit of power here is the watt, and the speed is expressed in km/hr. In terms of speeds in mi/hr and power outputs in horsepower (these archaic units are used because we are accustomed to using them to evaluate vehicular performance), this equation is

$$P = 0.0038\ V + 0.0000048\ V^3$$

Figure 4 is a graph of the power output of the rider as a function of the constant speed which is maintained. The values for the power turn out to be surprisingly small! I find that a steady speed of 12 mi/hr on the level is a pace which I can keep up for an hour or more with no feeling of fatigue; the power expended here is only 0.053 hp, or about 40 W. This value is consistent with that given by Whitt and Wilson[3] for riders traveling at this speed. A reasonable *maximum* power output for a bicyclist is 0.50 hp, and this can be maintained for only a minute or two.

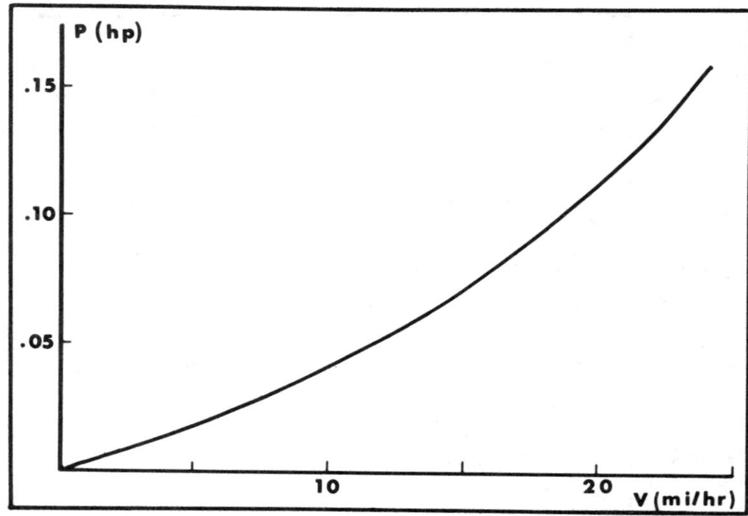

Fig. 4. The power output of the rider on the level as a function of the forward speed on the level.

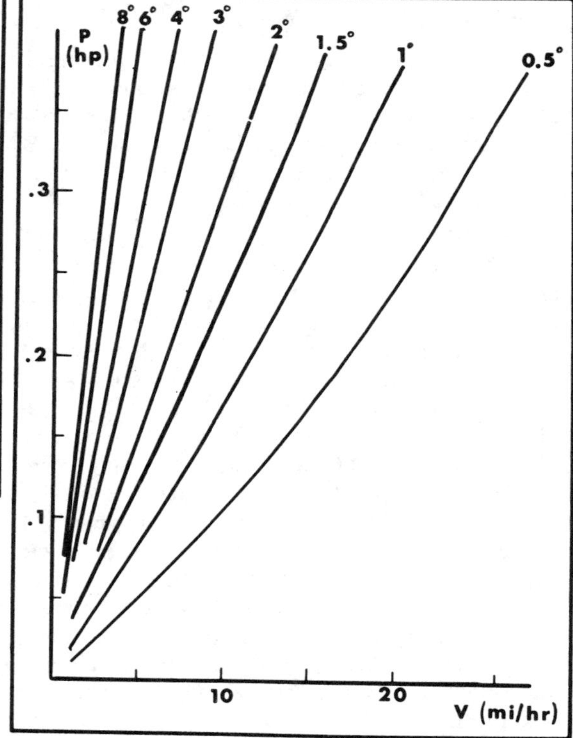

Fig. 5. Power versus speed curves for travel up hills of various angles.

Maximum speed up a hill

Now the rider is traveling up a hill which is inclined at an angle of θ degrees to the horizontal. A component of the gravitational force, $Mg \sin \theta$, must be added to the forces of air resistance and rolling friction. For the particular bicycle and rider (total mass 90 Kg) used as the example, the total force in newtons needed to keep the bicycle traveling up the hill at a constant speed V in km/hr is

$$F = 900 \sin \theta + 4.5 + 0.015 \text{ V}^2$$

and the power in watts which must be supplied is

$$P = (900 \sin \theta + 4.5)\text{V} + 0.015 \text{ V}^3$$

Most riders will have power outputs of 0.1 hp or less, and it can be seen in Fig. 5 that the lower portions of all of the curves are linear, indicating that the cubic term (due to air resistance) may be ignored. The equation therefore becomes

$$P = (900 \sin \theta + 4.5) \text{ V}$$

At an angle of 2 degrees, the first term in the parentheses is about seven times the second term, with the ratio getting larger for larger angles. Thus, as the angle of the hill increases, a greater fraction of the work done by the rider goes into raising the gravitational potential energy of the system, as we might expect.

Figure 5 gives a family of the resulting power versus speed curves for various angles; British units are used to allow comparisons with other curves. To see how fast you can ride up a given hill, use Fig. 4 to find your typical power output. The assumption is made that you will shift gears until you find one which will enable you to travel up the hill with the same power as you expend on the flat. Drawing a horizontal line on Fig. 6 then tells you what speeds you can expect on various hills. The middle-aged physics professor who can do 12 mi/hr on the level, reaches walking speed (3 mi/hr) on hills which have slopes of only one degree. By comparison, he can coast down this hill at a speed of 15.5 mi/hr. (Presumably, on climbing hills he will temporarily increase his power output; the increased rate at which oxygenated blood is pumped by the heart while hill climbing shows that this must be true.)

Conclusion

These three examples can be used as an addition to a regular class, or they can form the bases for homework problems or independent study. An interesting extension would be the calculation of the range of a battery-powered bicycle.

This material was first used as a single class which was presented to the Advanced Placement section taught by Mark A. Carle at University School in Cleveland, during the fall of 1981.

References

1. See, for example, Philip DiLavore, Phys. Teach. **19**, 194 (1981); Thomas B. Greenslade, Jr., Phys. Teach. **17**, 455 (1979); Richard Stepp, Phys. Teach. **14**, 439 (1976).
2. Frank R. Whitt and David G. Wilson, *Bicycling Science* (M.I.T. Press, Cambridge, MA, 1974) Chapters 2, 5, 6 and Appendix. This excellent reference uses British units primarily, with some translation into SI units.
3. Ref. 2, pp. 25.

More on Bicycle Physics

Greenslade's article "More bicycle physics" in the September issue 1983, certainly brought out nicely the basic physics of an exceedingly common and important device. There are, however, details which are not quite right. The expression for the expended power of the rider is not in the power of watt, but of 3.6 W.

$$\text{km/h} = \text{m}/(3.6\ \text{s}),$$

$$\text{Nkm/h} = \text{Nm}/(3.6\ \text{s}) = \text{W}/3.6.$$

Similar errors are found in the other power equations and figures.

Brynjolf Dokken, *Department of Physics, Teisen vider-gaende skole, Teisenvn, 5, Oslo 6, Norway.*

THE AUTHOR REPLIES

Brynjolf Dokken of Oslo, Norway has kindly drawn my attention to a numerical error in the third section of my article "More bicycle physics." [1] The power input P, necessary to keep the bicycle traveling at a constant velocity $\mathbf{v}$ by the application of a force $\mathbf{F}$ is calculated. In the article, an expression is given for F in newtons as a function of v in km/h. However, in the equation for the power, $P = \mathbf{F} \cdot \mathbf{v}$, I again express v in km/h. I should have expressed v in m/s to avoid the resulting erroneous factor of 3.6 (3.6 km/h = 1 m/s). Naturally, this affects the following equations and the vertical scales of the graphs. If SI units are used, the expression for F is $F = 4.5 + 0.20\,v^2$.

Moral: It is perfectly fine to express your final results in non-SI units to relate to everyday experience. But, when doing your basic calculations, rely exclusively on SI units!

Thomas B. Greenslade, Jr., *Department of Physics, Kenyon College, Gambier, Ohio 43022*

Reference

[1]Thomas B. Greenslade, Jr., "More bicycle physics," *Phys. Teach.* **21**, 360–363 (1983).

Bicycles in the Physics Lab

Robert G. Hunt
Johnson County Community College, 12345 College Boulevard, Overland Park, KS 66210

Riding bicycles is one of the most popular forms of physical activity in this country. It is enjoyed by people of all ages. Virtually every student of physics has learned to ride a bicycle, and many continue riding throughout their lives. Thus, in the search for lab experiences that are inherently interesting to students, using bicycles to illustrate physics seems a good choice. Students enjoy applying physics to familiar situations and are enthusiastic about learning more about why bicycles operate as they do.

I have developed a number of lab exercises aimed at developing understanding of principles of mechanics by using examples of bicycle operation. There is a wide range of difficulty in the exercises developed, from semi-quantitative discussions of Newton's laws for introductory physical science students to detailed applications of energy concepts for engineering physics students.

Newton's Laws

Students are familiar with the low rolling friction of bicycles, which provides a good example of the principle of inertia. Students are intuitively comfortable with the fact that bicycles tend to go in a straight line at constant speed and that a heavy load on a bicycle makes it harder to start or stop. These qualitative examples serve as good discussion topics to introduce the subject of Newton's laws to beginning physical science students.

Inertia can be quantified by measuring mass. Students can find their own mass by standing on a spring scale that is calibrated in kilograms. They can compare their mass to the mass of a bicycle by picking up a bicycle while standing on a scale and then subtracting their own mass (Fig. 1).

Fig. 1. A student finds her mass and the mass of a bicycle using a "bathroom scale" calibrated in kilograms.

To explore the quantitative relationships between forces and motion as related to bicycles, we purchased six inexpensive, mass-produced, "family-style" bicycles for students to use in experiments. (Another option is to allow students to bring their own bicycles into the lab, which might provide additional motivation.) For convenience in performing the measurements, each bicycle is mounted on a bicycle repair stand.[1] A 100-N force is applied to a spring balance. Care is taken that the force is applied in a direction parallel to the direction of motion of the pedal, that is, along a tangent to the path taken by the pedal. The resulting force at the outer edge of the rear wheel is measured with another spring balance hooked to a test tube clamp mounted on the tire (Fig. 2). Be careful to measure the force component that is tangent to the outer surface of the tire. Three or more different gears are used to show the range of forces that may be exerted on the ground by the rear wheel. With a 100-N force on the pedal, the force measured at the rear wheel will typically range from 14 to 35 N, depending on the gear used.

Using Newton's second law, students calculate the acceleration ($a = F/m$) they would achieve while riding on a horizontal surface, if friction were neglected. The

Reprinted from *The Physics Teacher* **27**, 160–165 (1989);

force they use is one fifth of their body weight. Research shows that this is a typical average force that people can comfortably exert on bicycle pedals for a long term.[2] The mass used is their own mass plus the mass of the bicycle. Of course in actual bicycle riding the force on the rear wheel is not constant, but varies with pedal position. For beginning students this point is not emphasized, but is carefully addressed for advanced students as described later in this note. Calculations for at least three different gears show that the range of accelerations possible for a particular bicycle might typically be from 0.2 to 0.6 m/s^2. This will depend on body mass, bicycle mass, and gearing. It is also interesting to have different students work on different styles of bicycles, such as 3-speed, 10-speed, 12-speed, BMX, and all-terrain. There is more similarity in gearing than students expect.

Equations of motion now can be introduced and used to predict the speeds that can be achieved. Neglecting friction, assuming that the acceleration is constant and that the bicycle starts from rest, speed is calculated from $v = at$ for two values of t: 10 s and 1 min.

Students can now consider another interesting aspect of bicycle riding. Will a heavier person accelerate at a higher or lower rate on a bicycle? A heavier person can exert more force on the pedal, which would tend to increase acceleration, but a heavier person also has greater inertia, which would tend to reduce the acceleration. The previous calculations can be repeated for a heavier person, perhaps one having twice the mass of the student who is doing the calculations. Students find that heavier people do accelerate at a slightly higher rate under the conditions specified in the problem.

At this point, concepts of frictional forces and terminal speed can be introduced. While riding, people accelerate to a speed at which they are physiologically comfortable. This is related to how fast the pedals are pumped, with a typical value of about one second per revolution of the pedals.[3,4] The rider then reduces the pedal force to maintain a constant speed. The force being produced by pushing down on the pedals at this point results in a propulsive force that equals frictional forces (mostly air friction), so that the net force along the line of travel is zero and constant speed results.[5]

The terminal speed will be different for different gears. The distance moved by a bicycle during one revolution of the pedals can be found directly by rolling a bicycle

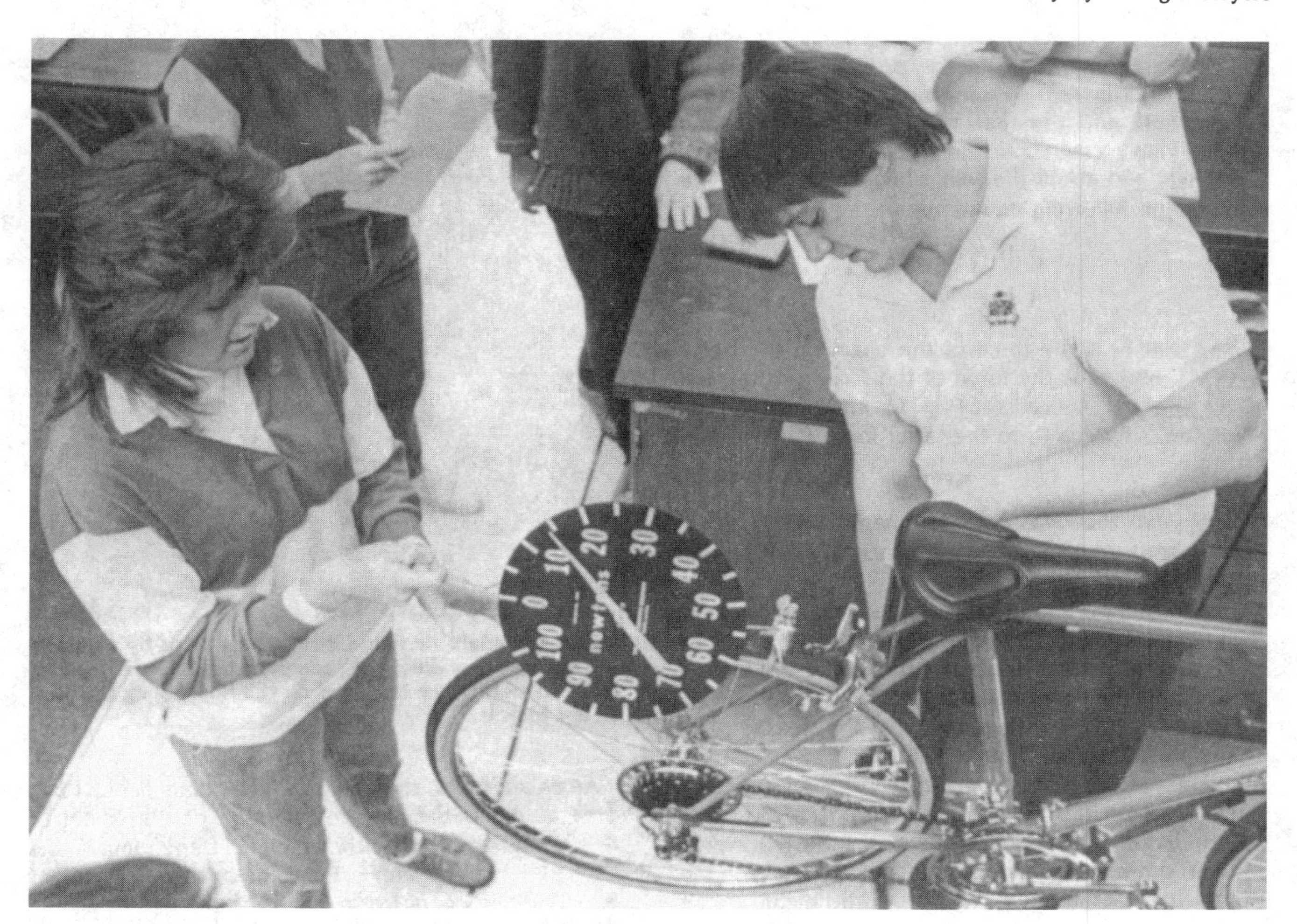

Fig. 2. One student exerts a 100-N force on a pedal in a direction tangential to pedal travel while another student measures the resulting tangential force at the rear tire.

across the floor while turning the pedals (Fig. 3). This distance, divided by the time, one second, gives an estimate of the speed of physiological comfort, or the terminal speed. Doing this for a range of gears illustrates the effects of cruising in different gears. Using a downward force of one-fifth body weight and other assumptions previously outlined, students find the typical values range from 3 to 8 m/s.

Bicycles also serve as good examples for Newton's third law. Action–reaction pairs abound on the bicycle power train. A large diagram is provided to students, together with a list of specific forces that they must indicate on the diagram. For example, students are instructed to draw an arrow on the diagram representing a force vector corresponding to the force of the foot on the pedal and to the force of the pedal on the foot. The students eventually see that the actual propelling force on the bicycle is the forward acting force of the ground on the rear wheel.

Statics and Dynamics

Introductory physics students can treat many of the above topics using fewer assumptions and more quantitative detail. One lab exercise we use examines the transfer of force from the foot on the pedal (F_1), through the sprockets and chain to the propelling force (F_4). Figure 4 is a diagram showing various forces and radii of sprockets and the rear wheel. Assuming a static condition (no acceleration), the sum of torques about the pedal axle and about the rear wheel axle is zero. This leads to the following equations:

$$F_1 r_1 = F_2 r_2 \quad (1)$$

$$F_3 r_3 = F_4 r_4. \quad (2)$$

Noting that F_2 is the force of the chain on the pedal sprocket, and F_3 is the force of the chain on the rear wheel sprocket, we can set $F_2 = F_3$, and find the ratio of the output force F_4 to the input force F_1:

$$F_4/F_1 = r_1 r_3 / r_2 r_4. \quad (3)$$

Students may then make measurements directly on the bicycle sprockets and calculate the ratio of F_4/F_1 for each gear.

This calculation requires that a decision be made to define the effective radius of a sprocket. The force of the chain on the sprocket is actually applied at some point between the tip of the gear tooth and its base. Observing a chain moving around the sprocket shows that the point of contact varies with chain tension and the magnitude of forces exerted. Thus, simply measuring the diameter of the sprocket from the tip of one tooth to the tip of the tooth directly across the diameter of the gear would result in some error. Taking a point halfway between the tip and the base (one-half of the tooth height) as the point from which measurements are taken gives good results.[6]

Fig. 3. A student finds the distance traveled by a bicycle in a selected gear during one complete revolution of a pedal. A metal meter tape is placed on the floor along the path of the bicycle.

As an alternate, students can now obtain actual force data at the pedal and at the edge of the wheel as described above and shown in Fig. 2. Depending on the time available, the ratio F_4/F_1 can be calculated for either all or selected gears. In our lab, comparison of these two methods of measurement shows good agreement; differences are usually five percent or less.

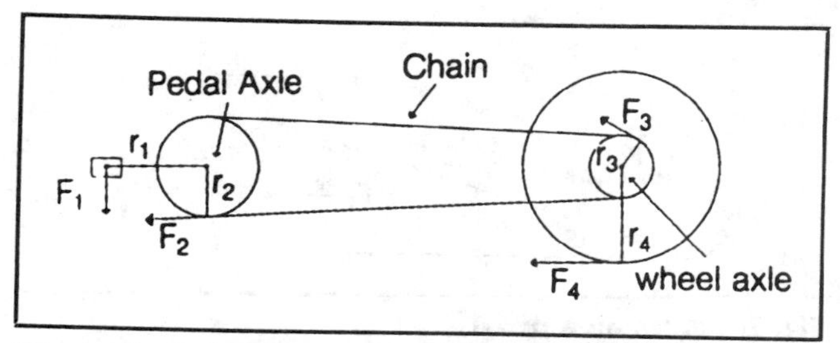

Fig. 4. The force exerted downward on a bicycle pedal (F_1) is transmitted first to the pedal sprocket, then to the chain (F_2), to the rear wheel sprocket (F_3), and finally to the ground (F_4).

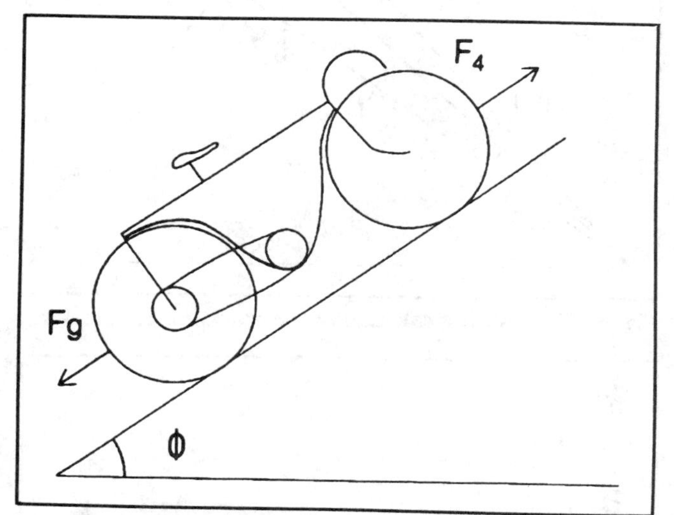

Fig. 5. A bicycle moving uphill experiences an up-slope force (F_4) caused by the force on the pedals, and a down-slope force (F_g) which is a component of the combined weight of the bicycle and rider.

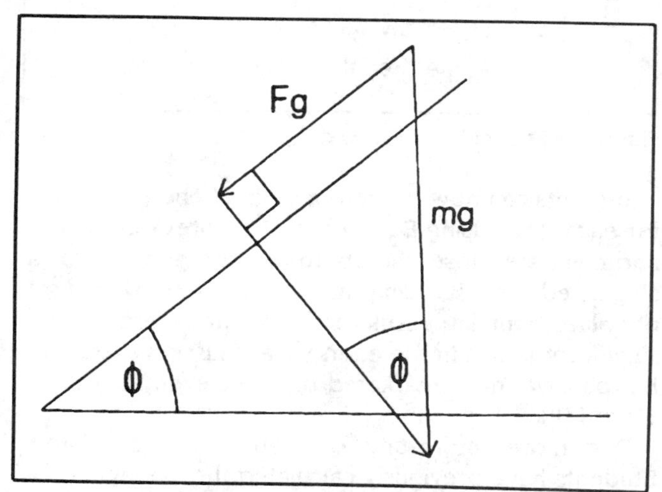

Fig. 6. The down-slope force (F_g) is a component of the combined weight (mg) of the bicycle and rider.

An interesting line to pursue at this point is the climbing ability of geared bicycles. As shown in Fig. 5, a bicycle moving up a slope at some angle θ is propelled uphill by the force of the ground on the rear wheel, F_4, and is pulled downhill by F_g, the gravitational force on the bicycle plus rider. The steepest hill that can be climbed is represented by the case where θ has the value leading to the condition $F_4 = F_g$, resulting in the net force = 0, and a constant speed uphill. An expression for θ under these conditions can be derived from the force diagram in Fig. 6. From that diagram, we can use the definition of the sine of an angle to write:

$$\sin\theta = F_g/mg = F_4/mg. \tag{4}$$

Theta is the maximum slope angle, and m is the total mass of bicycle and rider.

From the table of F_4/F_1 ratios previously constructed by students for the various gears, they can calculate F_4 once a value for F_1, the pedal force, is selected. For short periods of time, people riding up hill can exert as much as one half to two thirds of their body weight on the pedals.[7] For even shorter periods, they may exert their full body weight, or even exceed that value by pulling upward on the handle bars. However, the force exerted on the pedal which is transformed into locomotion depends on the pedal position, so that not all of the force exerted actually shows up as a force exerted on the rear wheel. Perhaps a reasonable value for F_1 for hill climbing is one half of body weight.

Students can now use that value of F_1 and the ratios of F_4/F_1 calculated previously to find F_4 for selected gears. Equation (4) can be used to construct a table of realistic values for maximum θ for various gears. Using these assumptions, typical values for the maximum slope attained by a 10-speed bicycle range from 17° in low gear to 6° in high gear. It is interesting for students to do a calculation for a heavier person. They will discover that a higher body weight will result in a slightly higher maximum θ.

Energy Relationships

Bicycles can be used to illustrate many of the work, energy, and power concepts found in beginning texts, and as a bonus provide unique examples from students' personal experiences. In one lab exercise, we examine the energy input to a bicycle by calculating the work done on a pedal as it moves through one-half revolution, from the top of its rotational path to its lowest point. This comprises the "power stroke" or "pedal thrust" as we push downward on the pedal. Referring to Fig. 7, this is the movement of the pedal crank from $\theta = +90°$ to $-90°$. A complication arises here because the applied force, F_1, can be assumed to be directed straight downward at all times while the pedal moves on a circular path. But energy is transferred only by the component of force in the direction of travel, as stated by the standard rendition of the energy transfer, or work equation:

$$W = F\cos\theta. \tag{5}$$

Here θ is the angle between the applied force and the direction of travel (Fig. 7).

As the bicycle pedal descends, θ varies. Thus, to find the total energy transferred to the bicycle during the one-half revolution requires summing energy contributions from small changes in θ. Figure 8 shows the pedal swinging through some small angle $\Delta\theta$, with the midpoint being represented by θ. The distance traveled by the pedal is the arc length subtended by $\Delta\theta$, or $r\Delta\theta$. The average force over this small angle is $F_1 \cos\theta$. The energy transferred to the pedal, from Eq. (5), is

$$\Delta W = (F_1 \cos\theta)(r\Delta\theta). \qquad (6)$$

Students can assume some value for F_1, such as one fifth of their body weight, and perform a series of calculations to find the total energy transferred to the bicycle during one-half revolution of the pedal. This involves choosing some interval for $\Delta\theta$, such as 10°, and then calculating a value for ΔW. For example, the first choice could be $\theta = 85°$, with $\Delta\theta = 90° - 80° = 10°$. Then using a measured value of r from a bicycle and a value for F_1, ΔW can be calculated. This procedure could be repeated over the full path. The sum of the values for ΔW is the work done, or the energy transferred for one-half revolution.[8]

Alternatively, students who can carry out integration will recognize that an easier way to arrive at the result is to evaluate the following integral, which is Eq. (6) translated into calculus:

$$W = F_1 r \int_{-\frac{\pi}{2}}^{+\frac{\pi}{2}} \cos\theta \, d\theta. \qquad (7)$$

Evaluation of this integral leads to the important relationship:

$$W = 2rF_1. \qquad (8)$$

This is the total energy transferred to the bicycle by a single pedal thrust. Students using the approximate numerical method should compare their results with Eq. (8).

Now that the energy input is known, it can be compared to the energy output. We already know how to find the output force F_4 from above. However, F_4 will vary from zero when the pedal angle is + 90°, to some maximum value when $\theta = 0°$, back to 0 at $\theta = -90°$. In fact, F_4 will exhibit a cosine variation as shown in Fig. 9. We now see that the values measured for F_4 in the previous experiments are the tangential forces and are actually the maximum values, F_{max}.

To simplify the calculation of energy output, it would be convenient to define a new value which would be a constant average force at the rear wheel. We will call this value F_{AV}. Because F_4 varies as $\cos\theta$, it has the value $F_{AV} = (2/\pi)\, F_{max}$, where F_{AV} is the average value over 90°. If x is the distance traveled while θ varies from +90° to −90°, the work output can be written as

$$W = F_{AV}\, x = (2/\pi)\,(F_{max})\, X. \qquad (9)$$

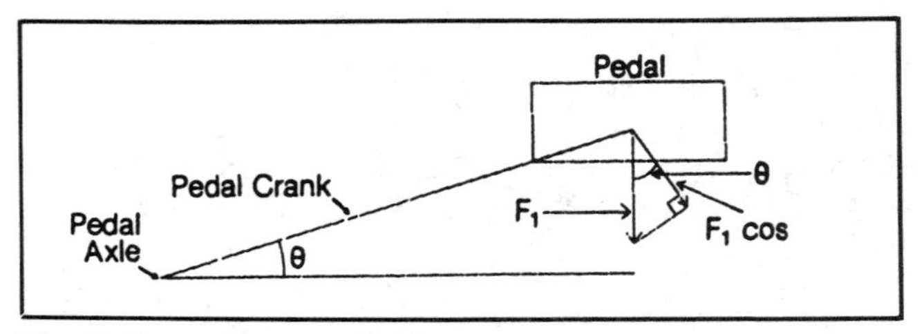

Fig. 7. Forces on a pedal.

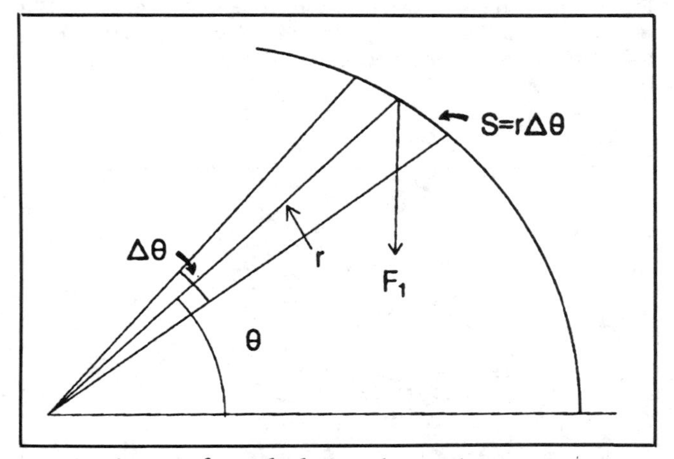

Fig. 8. Diagram for calculating the work increment ΔW.

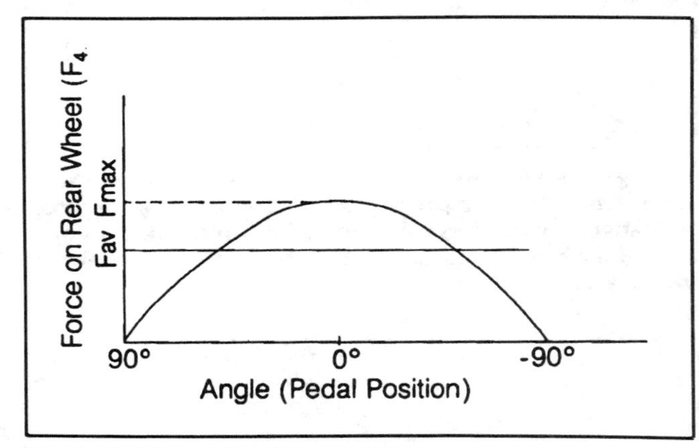

Fig. 9. Relationship of F_{AV} and F_{max}.

Students can now prepare a table of energy outputs for each gear, using Eq. (9) based on previous values, and compare those outputs to the energy input for a single pedal thrust, being careful to use a consistent set of values. Our lab results indicate that there can be significant losses in the energy transmission system of inexpensive, mass-marketed bicycles, even as high as 10 to 20%.

One more interesting calculation can be made. Students have previously calculated the average force at the rear wheel, F_{AV}, and have also calculated values of speed at selected times for various gears. We can now calculate power outputs using a standard form of the power equation, $P = Fv$, where the force would be F_{AV}. We have students construct a table showing power at various times after starting from rest for various gears, calculating v by using $v = at$ as before. This calculation is done for uphill slopes as well as for flat surfaces. The table is used to develop gear-shifting scenarios while traveling on the level ground or going up hills. People tend to adjust their pedaling speed so that they have a comfortable power output of about 100 to 200 W.[9] Thus, by consulting the table, students can see how a comfort level can be obtained on different slopes and at different speeds by appropriate use of gears. (Note that the power level of 100 to 200 W corresponds roughly to the comfort criterion of one second per pedal revolution mentioned above.) These results also show that for ordinary riding, there is substantial overlap in the gearing of a ten-speed bicycle; three to five speeds are probably sufficient for most people. However, mechanical access to certain gearing sequences on a ten-speed bicycle is inconvenient and leads to a practical need for redundancy.[10]

Conclusion

Bicycles can be a valuable tool in teaching many concepts of mechanics at all levels of complexity. The lab equipment is relatively inexpensive, and the activities lead to high levels of student interest. The few ideas presented here are just a sampling to indicate the potential of using bicycles. Additional ideas and much data on bicycle operation can be found in the references listed. *Bicycling Science* by Whitt and Wilson is a particularly valuable source of information on the technical details of bicycles and their operation.

I would be interested in hearing from readers about additional ways to use bicycles. If you contact me at the address given above, I would be happy to supply more details or send copies of the report forms we use in these labs.□

References

1. Bicycle repair stands were purchased from a local bicycle shop for $46 each.
2. Frank Rowland Whitt and David Gordon Wilson, *Bicycling Science*, 2nd ed. (MIT Press, 1982), see chap. 2.
3. Whitt and Wilson, chap. 2.
4. Thomas Greenslade, Jr., *Phys. Teach.* **17**, 455 (1979).
5. Whitt and Wilson, chap. 4.
6. Student calipers were purchased from Cenco, catalog #72610, $8.70.
7. Whitt and Wilson, chap. 2.
8. Philip DiLavore, *The Bicycle, A Module on Force, Work and Energy* (Distributed by AAPT, 1978).
9. Whitt and Wilson, chap. 2.
10. Greenslade, p. 455.

Chapter 7
THE HUMAN ELEMENT

CONTENTS

The first article in this chapter is "Newtonian mechanics and the human body: Some estimates of performance," by Herbert Lin. In it, the author uses introductory mechanics to estimate human performance levels for a number of athletic activities and then compares the estimates to empirical data for those same activities. The quantities estimated include maximum running speed, the speed for normal walking, the maximum range of the broad-jump (or long jump as it is now known), the maximum height of the pole vault, the maximum height of a vertical jump, the maximum speed of cycling, and the minimum time to run up a flight of stairs. The estimates provided agree quite well with actual performance records in the various events, despite the author's closing observation that "The oversimplified models used here are aimed at getting the right order of magnitude with minimal calculational effort."

"Locomotion: Energy cost of swimming, flying, and running," by Knut Schmidt-Nielsen is the second article in this chapter. Having been written by a physiologist primarily for other physiologists, this article has a different look and feel from most of the other articles included in this volume. First of all, there are virtually no equations; rather, linear regressions are used to organize empirical data graphically, and conclusions are developed primarily through carefully crafted text. Second, one notices that the scope of the article is quite broad, encompassing the energy cost of locomotion not only for humans but also for insects, birds, fish, and rodents.

Pleasantly, the biological science perspective of this article lends itself to insights of a type not normally found in physics articles, such as, for example, the "hydraulic" nature of earthworms and spiders:

> Other types of locomotion (i.e., other than swimming, flying, and running) exist, such as the hydraulic system of the earthworm, in which the body fluids in combination with the muscular body wall function as a 'skeleton.' Spiders lack extensor muscles in the limbs and use blood pressure to stretch their legs.

It is difficult to imagine how one can look at a spider in quite the same way after reading that last sentence.

The energy cost of locomotion in this article is defined as the amount of fuel needed to transport one unit of body weight over one unit of distance, and is measured in units of cal/g/km. Empirical energy cost data are then presented as a function of body weight on log-log plots in which the data points for similar types of locomotion are found to cluster about their respective regression lines.

So, for example, Schmidt-Nielsen shows that the energy cost of swimming for fish of different sizes decreases smoothly with body weight.[1] In addition, the author observes that water resistance drag is a complex function of shape, size, and speed, and, in particular, that:

> The importance of shape is expressed in what is commonly known as "streamlining," so well known from fish and whales, and regrettably absent in man.

This lack of appropriate streamlining contributes no doubt to the fact that the energy cost of swimming for a man is about 30 times as high as one would expect for a fish of his size! Man is clearly not one of nature's better swimmers.

In regard to flying locomotion, Schmidt-Nielsen concludes that:

> ...all flying animals fall amazingly close to one straight regression line, in spite of the vast differences in size, anatomy, flight pattern, speed, body temperature, and so on, between insects and birds.

If one extrapolates the regression line for flying locomotion to masses typical of humans, one discovers that the energy requirements for a flying man (or more precisely, for a bird the size and weight of a man) would be so great as to render such flight highly unlikely.[2]

In regard to running locomotion, the energy cost for human running clusters on the same regression line as that for other animals, a regression line that spans a 10 000-fold range of body weight. It would seem then that, unlike swimming or flying, in which human performance is either poor or nonexistent, running is one form of locomotion for which humans have achieved parity with other species.

One of the most interesting questions associated with running locomotion is the unresolved issue of whether "negative work," e.g., the gravitational potential energy made available when one runs *down* a hill, can be usefully recovered and stored in some fashion for later use by the runner. Although this question has been under active study since the 1920s, Schmidt-Nielsen points out that:

> The physiological role of negative work is still very much in dispute, both at the molecular level and in studies of the energetics of the whole organism.

The third article in this chapter is "Karate strikes," by Jearl D. Walker. In it, the author uses collision mechanics to analyze the forward karate punch. The analysis focuses on two main aspects of the blow, the deformation energy delivered to the target, as well as the impact force and resulting stress.

On the basis of his calculations, Walker concludes that a deformation energy of about 156 J can be delivered in something under 10 ms, with a corresponding maximum impact force of about 4900 N. He also shows that a force of this magnitude delivered to an opponent's head would result in a peak acceleration of about 90 g's, somewhat short of fatal but almost certainly capable of producing unconsciousness.

"The physics of karate," by S. R. Wilk, R. E. McNair,[3] and M. A. Feld, is the fourth article in this chapter and, despite some similarity in title, is really quite different in both method and scope from the previous article by Walker. While Walker's analysis employed a simple *static* model of impact force, Wilk, McNair, and Feld adopt a more sophisticated *dynamic* model. In addition, while Walker focused on the front foward punch, Wilk, McNair, and Feld considered a variety of punches and kicks. However, despite differences in approach, the two articles complement each other nicely.

Wilk, McNair, and Feld also point out a number of interesting biomechanical aspects of karate, noting, for example, that human bone is very strong indeed, having a modulus of rupture over 40 times larger than that of concrete! In addition, these authors also shed light on the apparent paradox contained in Newton's third law as it relates to karate: namely, that, since the force of the fist on the target must be equal and opposite to the force of the target on the fist, must not the damage to the fist always be equal to the damage to the target?

The resolution of this apparent paradox lies in the fact that bones and other objects are broken not by force, *per se*, but rather by the strain, i.e., deformation, induced by the *stress*, i.e., force per unit area, applied to them. As a result, successful karate techniques (at least as judged by the attacker) tend to be those in which the relative *placements* of the equal and opposite forces result in maximum strain on the target and minimum strain on the fist or foot of the attacker.

In analyzing the foot kick in the "knifed" position, for example, Wilk, McNair, and Feld note that

> A properly positioned foot can withstand forces several orders of magnitude larger than that needed to fracture a concrete slab!

"How can a downhill skier move faster than a sky diver?," by Angelo Armenti, Jr., is the fifth article in this chapter. In it, the author attempts to deal with the apparently paradoxical fact that the world record speed of 200 km/h (124 MPH) for downhill skiing is larger than the commonly cited terminal speed of 120 MPH for sky divers.

The resolution of this paradox centers on the opposite roles which air resistance is called upon to play in the two activities. For the skydiver, the object is to minimize the speed of descent so as to maximize the time of descent before the parachute must be deployed. For this reason, the skydiver adopts the familiar "spread-eagle" position in an effort to maximize his cross-sectional area. In this position, he doesn't have to move very fast (comparatively speaking) to make the air resistance force equal his weight, and hence achieve terminal velocity.

For the downhill skier, however, the object is to maximize the speed of descent. For this reason, the skier crouches in a "tuck" position so as to minimize his cross-sectional area. As a result, the skier must move relatively faster in order to make the air resistance force equal to the downhill component of his weight, resulting in terminal velocity.

The sixth article in this chapter is "Why do downhill racers prejump?," by Rob Hignell and Colin Terry. In it, the authors investigate the technique by which downhill skiers jump before reaching the edge of a steeper section of the slope. Hignell and Terry note:

> By prejumping before reaching the discontinuity of the slope, the skier attempts to land immediately at the beginning of, and traveling parallel to the steeper section of the run.

Their calculations show that the primary advantages of prejumping include significant reductions in (a) the time spent in the air, and (b) the landing forces encountered upon impact with the steeper portion of the slope.

The seventh and final article in this chapter, "Physics and a skiing record," by Albert A. Bartlett, describes an attempt at a Guiness world skiing record in which a man skied 234 000 vertical feet (more than eight times the elevation of the summit of Mt. Everest above sea level) in just under 24 h! Bartlett analyzes the kinematics and energetics of this feat and in so doing, provides insights into a host of good physics problems.

---

[1]Similar patterns (i.e., a decline in energy cost with body weight) exist for flying and running locomotion as well. It is necessary to recall, however, that the energy cost in question has units of cal/g/km. Hence, although larger animals have a lower unit energy cost (i.e., calories per gram of body weight), they nevertheless require more energy per kilometer than smaller animals simply because they have more units (i.e., more grams of body weight). Stated another way, the slopes of the regression lines are such that the body weight increase is always much larger than the corresponding energy rate decrease.

[2]Assuming a human body mass of 75 kg, the energy cost for flying extrapolates (Fig. 4 of Schmidt-Nielsen's article) to 0.6 cal/g/km, corresponding to 45 000 cal/km. Assuming for argument's sake a human flying speed of about 10 m/s (which is just below the maximum running speed for humans), it would take about 100 s (1.67 min) to travel a kilometer, and hence it would require about 27 kcal (the energy equivalent of about two teaspoons of sugar) for each minute of hypothetical human flight. Empirical studies on the power output of humans, however, indicate that an energy expenditure rate of this magnitude could only be sustained under "sprint" conditions, i.e., for 20 s or so. [See, for example, Henry A. Bent, "I: How much work can a person do?," *J. Chem. Educ.* **55**, 456 (1978).] For activities of 5-min duration or less, the maximum energy expenditure rate for humans falls to 16 kcal/min, and for activities of 2-h duration or so, the maximum energy expenditure rate drops to just 5 kcal/min. Clearly then, and quite aside from structural or aerodynamic considerations which could be expected to impose additional constraints, energy considerations alone suggest that, for creatures as heavy as humans, the air is not likely to offer a viable environment for locomotion.

[3]The holder of a black belt in karate, Ronald Erwin McNair, a distinguished African-American scientist and astronaut, was tragically killed along with six other crew members in the explosion of the Shuttle Challenger on January 28, 1986.

Newtonian mechanics and the human body: Some estimates of performance

Herbert Lin
Physics Department, Massachusetts Institute of Technology, Cambridge, Massachusetts 02139
(Received 9 June 1977; accepted 24 August 1977)

Non-trivial but elementary numerical examples of basic mechanics are lacking in most introductory textbooks. This paper gives several very simple examples which relate to numbers for which many students already have some intuition. In particular, we estimate the speeds of running, cycling, and walking, the length of a good broad-jump, the heights of a polēvault and a vertical high jump from rest, and the time for running up a flight of stairs. These estimates and the observed numbers are found to match rather well.

The purpose of this paper is to apply some very simple introductory mechanics to what people can do with their bodies. While not profound in themselves, these calculations are useful in demonstrating that physics does apply to real life and not just to massless ropes and frictionless planes. These results have the further advantage that many students already have a feel for the numbers involved in these calculations; by contrast, very few students have a feel for all the numbers required for even an elementary calculation of the stopping distance of a car or the momentum of a baseball. Finally, these calculations are useful for dispelling notions of super-human athletic performances. It should be noted that the underlying spirit of these order-of-magnitude calculations is that of "playing with the numbers to see if we can get the essential physics and still make them come out reasonably well."

When necessary, we will assume the following (invariant!) values: m is the mass of a typical person $\sim$50 kg; h is the height of a typical person $\sim$2 m; l is the length of a leg $\sim h/2 \sim 1$ m; $g = 10$ m/sec^2.

MAXIMUM SPEED OF RUNNING

To calculate the maximum speed of running V_{run}, we will assume that all of the work done by a runner's legs while accelerating to V_{run} goes into changing his/her kinetic energy, and that none of it goes into overcoming dissipative losses. (Of course, after the person reaches V_{run}, then dissipative losses are important, but since we are only concerned with how one gets up to V_{run}, and not what happens afterwards, we shall ignore these losses.)

The following points are relevant:

(1) The legs of almost all people can support at least their own weight, as evidenced by the fact that people are able to stand up. Most healthy people can support an additional dead load equal to their weight, but only with great effort, and thus we will say that the maximum force F available for accelerating the runner is on the order of αW, where α is about 1.5.

(2) The driving force supplied by the runner's legs can act on the runner over a length s on the order of the push-off length of a person's leg. This length is the difference between a fully extended leg and a bent one. If we assume that the foreleg and thigh are each of length $l/2$, and that when bent, they make a right angle, then we see that the push-off length s is

$$s = 2(l/2) - 2^{1/2}(l/2) \sim 1/3 \text{ m}$$

(3) When one is starting a run, one tries to push as near to horizontal as possible. We will assume that the vertical support needed to keep the body from falling comes from the support offered by the bones in the leg; as the body goes forward and starts to fall, the runner brings up a leg underneath the body and keeps it from falling. Thus we will assume that force of αW is available for propulsion.

(4) With each stride, the runner accelerates, and the kinetic energy of the body is increased. Maximum velocity is attained in several as we will set "several" equal to 10.

Putting the above together, and using the work-energy theorem, we write

$$\Delta KE = \begin{pmatrix} \text{number of} \\ \text{acceleration} \\ \text{periods } n \sim 10 \end{pmatrix} \times \begin{pmatrix} \text{force exerted} \\ \text{over one stride} \end{pmatrix} \times (\text{push-off length}).$$

This turns into

$$(1/2)m\, V_{\text{run}}^2 = nFs = n\alpha mgs,$$

or

$$V_{\text{run}} = (2n\alpha gs)^{1/2} \sim 10 \text{ m/sec}.$$

This analysis may seem unrealistic because it implies that by increasing n, a runner could reach arbitrary speeds. Of course, n is an *ad hoc* parameter determined from observation, so in this sense, the question is moot; n cannot be arbitrarily large, and in fact has a definite empirical range.

However, here is another argument. It states that the limiting speed is determined by the fact that the free leg is accelerated forward but then stops when it touches down, losing at that point the kinetic energy it gained during acceleration.

Carrying out this argument in the spirit of this paper, we assume that each leg accounts for about 1/8 of the total body mass. (My thigh is $\sim$40 cm in circumference, and my ankle $\sim$20 cm. The average circumference is then $\sim$30 cm, and so, assuming the leg to be a cylinder of circumference 30 cm and length 80 cm, we find a leg volume of about 6000 cm^3, or roughly 6 kg. My weight is $\sim$50 kg, and so $m_{\text{leg}}/m \sim 1/8$.)

We can then use the work-energy theorem again to write

$$KE_{\text{leg}} = (\text{Driving force supplied by one leg})(\text{push-off length}).$$

Reprinted from *American Journal of Physics* **46**, 15–18 (1978); © American Association of Physics Teachers.

To find KE_{leg}, we note that the bottom of the leg (the foot) is at rest relative to the ground. The top of the leg is moving at about V_{run}. If we assume that KE_{leg} is that of a straight, rigid rod rotating about an axis through one end, we see that

$$KE_{leg} = (1/2)I\omega^2 = (1/2)[(1/3)m_{leg}l^2](V_{run}/l)^2,$$

where ω is the angular velocity and I the moment of inertia each about the foot.

Putting these together, we get

$$(1/2)[(1/3)m_{leg}l^2](V_{run}/l)^2 = (1/2)\alpha mgs$$

which results in

$$V_{run} = [3\alpha gs(m/m_{leg})]^{1/2} = (24\ \alpha gs)^{1/2} \sim 11\ \text{m/sec}$$

which compares nicely to our earlier estimate of 10 m/sec.

If the rotational kinetic energy approach is too hard, we can also say that $KE_{leg} = (1/2)m_{leg}V_{leg}^2$, but realizing that V_{leg} is less than V_{run} because not all the leg is traveling at V_{run} (the bottom is at rest). Choosing $V_{leg} \sim V_{run}/2$ gives

$$KE_{leg} = (1/2)m_{leg}(V_{run}/2)^2$$

implying that

$$V_{run} = [4\alpha gs(m/m_{leg})]^{1/2} \sim 13\ \text{m/sec}.$$

A recent book[1] indicates that a substantial amount of energy is stored as elastic (spring) energy in the muscles and tendons when the moving foot strikes the ground, where "substantial" is about 60% of the energy supplied by muscle contraction. However, a more realistic number for the horizontal driving force would be less than αW we have assumed. We estimate these effects as having the same order of magnitude but opposite sign, and thank the powers that be for preserving our estimate.

SPEED OF WALKING

One walks by letting one leg swing forward freely, transferring weight to it, and repeating the process with the other leg. The "swing forward freely" is that of a simple pendulum (to first order), and thus we have a way of calculating the speed of walking V_{walk}.

From simple kinematics, we must have

$$V_{walk} \sim \text{(length of stride)/(time of stride)}.$$

However, if we assume the leg is a simple pendulum of length l, the time of stride must be just half the period of that pendulum. From any elementary textbook, the period T of a simple pendulum of length L is

$$T = 2\pi(L/g)^{1/2}.$$

Therefore, we can write

$$\text{time of stride} = (1/2)T = \pi(L/g)^{1/2}.$$

For L, we put in the effective length of the pendulum. The physical length of the pendulum is the length of the leg l, and so, assuming a uniform leg mass distribution, the effective length L is half the leg's length, so that we set $L \sim l/2 \sim 0.5$ m. Thus the stride time is about $\pi/(20)^{1/2} \sim 2/3$ sec. Hence,

$$V_{walk} = (\text{stride length} \sim 1\ \text{m})/(2/3\ \text{sec}) \sim 1.5\ \text{m/sec} \sim 3\ \text{mph}.$$

MAXIMUM RANGE OF BROADJUMPING

This is an application of projectile motion. From any textbook, we know that the range R of a point particle in projectile motion is $R = (V_0^2/g)\sin 2\theta$. If we assume that the broadjumper is a point particle (!), that the jumper's initial velocity upon takeoff is equal to V_{run}, and that the takeoff angle is the optimal 45°, we find for the length of a broad-jump R

$$R = (10\ \text{m/sec})^2/(10\ \text{m/sec}^2) \sim 10\ \text{m}.$$

In fact, the point broadjumper assumption is not bad, since the maximum length parameter associated with a person is a height of 2 m, which is much less than 10 m. This can be used as an example of a "self-consistent" theory; we make an assumption, carry out the calculation, and discover that our result is consistent with that assumption. We would have been in trouble if R had turned out to be 1 m.

MAXIMUM HEIGHT OF POLE-VAULTING

Pole-vaulting consists of running fast (building up maximum kinetic energy), sticking a pole into the ground (putting the kinetic energy into the potential energy of the bent pole), and allowing the pole to straighten out and fling the pole-vaulter over the bar. If we write H_{cg} for the height that the pole-vaulter's center of gravity rises, we can invoke energy conservation and write

$$mgH_{cg} = (1/2)mV_{run}^2,$$

which results in

$$H_{cg} = V_{run}^2/2g \sim 5\ \text{m}.$$

However, the height of the bar H_{pv} is measured from the ground, and the pole-vaulter's center of gravity starts out about 1 m off the ground. Therefore, since the pole-vaulter's center of gravity must pass very near the bar, we can write

$$H_{pv} = H_{cg} + 1\ \text{m} \sim 6\ \text{m}.$$

We realize that the pole-vaulter must carry a long, unwieldy pole which reduces the maximum running speed. On the other hand, the straightening of arms just before going over the bar adds an increment of height which we estimate makes up approximately for the loss due to reduced running speed. Thus we luck out once again.

MAXIMUM HEIGHT OF VERTICAL JUMP

The event needs a bit of explanation. A person makes a mark with chalk on the wall as high as (s)he can. (S)he then jumps as high as possible, making another mark on the wall. The difference in height between these two marks is the height of the jump, which we call H_{jump}. To calculate H_{jump}, we will again use conservation of energy. We write

(work done on human body)
= (change in gravitational potential energy).

The work done on the body is done by the ground exerting a force on the jumper through a certain crouch length. The maximum crouch is about equal to the length of one's legs, but no one starts a high jump from a fully squatting position, so let us assume that the crouch length in question is about the same as the push-off length for running.

The force which does the work on the body is once again the maximum force exertable by one's legs (which we assumed in calculating V_{run}). Thus we get

$$(\alpha mg)(\text{crouch length}) = mgH_{jump}$$

so that we get

$$H_{jump} \sim 1/2 \text{ m}.$$

MAXIMUM SPEED OF CYCLING

This problem is done in two parts. We first calculate the maximum power that a human can generate, and then equate this to the power dissipated by air resistance for a person riding a bike at a speed V_{bike}. (At high cycling speeds, air resistance is the dominant factor, and there is almost always at least one person in any class that knows this, as well as the ball-park maximum cycling speed.)

We can estimate the maximum power output P_{max} by taking the kinetic energy of running at V_{run} and dividing by the time t it takes to accelerate to that speed. This time is typically on the order of a few seconds, as it must be if the total time of a run is only 10 sec. Thus, letting t equal 2 sec,

$$P_{max} \sim (1/2)mV_{run}^2/t \sim 1.3 \text{ kW}.$$

To calculate the power dissipated by air resistance P_{dis}, we must first calculate the retarding force $F_{retarding}$.

When one rides at high speeds, one leaves behind a turbulent wake. A second rider riding in this wake feels very little air resistance; in this sense, the air stream has "disappeared" behind the first rider. Therefore, we will assume that the oncoming air stream has been deflected 90° in all directions perpendicular to its initial direction of motion, so that its final x component of momentum (in the reference frame of the bike) is zero.

In a time Δt, a mass of air $\Delta m_{air} = \rho A V_{bike} \Delta t$ strikes the cyclist, where A is the effective frontal area of the person on the bike {~0.5 m^2 if we assume height 1 m and width 0.5 m}, and ρ is the density of the air $\sim 1 \text{ kg/m}^3$. It carries x momentum $\Delta p = V_{bike}\Delta m_{air}$, which is lost due to the collision with the cyclist. Hence,

$$F_{retarding} = \Delta p/\Delta t = \rho A V_{bike}^2.$$

Thus $P_{dis} = F_{retarding}V_{bike}$ gives

$$P_{dis} = A\rho V_{bike}^3.$$

Equating P_{dis} to P_{max}, we get

$$V_{bike} = (P_{max}/A\rho)^{1/3} \sim 14 \text{ m/sec} \sim 30 \text{ mph}.$$

TIME TO RUN UP THE STAIRS

A typical flight of stairs might be about 5 m from landing to landing. To calculate the time t_{stairs} needed to run up one flight, we take the maximum power output P_{max} and equate this to the power needed to change the gravitational potential energy through a height of 5 m in a time t_{stairs}. Thus

$$P_{max} = mg\{5 \text{ m}\}/t_{stairs}.$$

This results in a value for t_{stairs} of 2 sec. This is low, but we must remember that some power must also go into going forward, and that also that one tends to exert less than full effort in running up stairs for fear of slipping.

Table I

Event	Observation	When	Ref.	Calculated
100 m dash	9.9 sec (10.1 m/sec)	1968 Mexico City	1	11 m/sec
Broad-jump	29′ 2″ (8.9 m)	1968 Mexico City	3	10 m
Vertical jump from rest	1–2 ft (30–60 cm)	1934 Columbia U.	4	0.5 m
Maximum cycling speed	1 km in 64.6 sec (15.5 m/sec)	1967 Mexico	5	14 m/sec
Pole-vault	18′ 6.5″ (5.7 m)	1975 Florida	6	6 m
leisurely stroll	3–4 mph (1.6–2 m/sec)	any Sunday afternoon		1.5 m/sec
mad dash up the stairs	3–4 sec	when late for class		2 sec

COMMENT AND DISCUSSION

Table I summarizes the results of the above calculations and compares the results to "observed data" in the form of world records[2-6] or other standards. We see that in most cases they compare rather nicely. Of course, we had no right to expect such good agreement, but at least they show that we are on the right track.

In no way does this paper pretend to be a substitute for more careful, detailed analysis; see, for example, Offenbacher[7] on jumping, Keller[8] on running, Magaria[1] on walking and running, and Dyson[9] on athletics in general. (In addition, I am advised that Ref. 10 contains material which deals with problems of the kind done here.) The oversimplified models used here are aimed at getting the right order of magnitude with minimal calculational effort. (This could be considered a form of "creative laziness"!) The fact that these results match the observed numbers is due largely to hindsight.

The approach used in this paper replaces a detailed analysis with a few *ad hoc* but nevertheless plausible parameters whose order of magnitude is experientially known to most students (and whose exact values can be adjusted on the basis of what seems reasonable).

I have found that students are quite interested in these estimates and find themselves challenged to do better than the estimates predict. Invariably, some student will claim to be able to jump higher than 1 m or climb a flight of stairs in less than 3 sec. This dispute can then lead to a student-led demonstration.

ACKNOWLEDGMENTS

My first contact with this style of "qualitive physics" was with Viki Weisskopf. His influence has never been far from me, and in reading this paper over, I believe it shows. An impressive demonstration of the utility and fun of this approach appears in Ref. 11.

[1]R. Margaria, *Biomechanics and Energetics of Muscular Exercise* (Oxford U.P., London, 1976), p. 94–100.

[2]N. McWhirter, *Guiness Book Of World Records* (Sterling, New York 1976), p. 648.

[3]Reference 2, p. 652.

[4]P. H. Gerrish, "A Dynamical Analysis of the Standing Vertical Jump," PhD thesis (Teacher's College, Columbia University, 1934).

[5]F. Whitt and D. Wilson, *Bicycling Science* (MIT, Cambridge, MA, 1974), p. 44.

[6]Reference 2, p. 652.

[7]E. F. Offenbacher, "Physics and the Vertical Jump," Am. J. Phys. **38**, 829 (1969).

[8]J. B. Keller, "A Theory of Competitive Running," Physics Today, p. 42, September 1973.

[9]G. Dyson, *The Mechanics of Athletics* (University of London, London, 1968).

[10]P. Davidovits, *Physics in Biology and Medicine* (Prentice-Hall, Englewood Cliffs, NJ 1975).

[11]V. F. Weisskopf, "Of Atoms, Mountains and Stars; A Study in Qualitative Physics," Science **187**, 605 (1975), Feb. 21 issue.

Locomotion: Energy Cost of Swimming, Flying, and Running

Knut Schmidt-Nielsen

The energy expended by animals as they swim through water, run on land, and fly in the air in the different kinds of locomotion is not immediately obvious to us. We know that a flying bird must continuously expend energy to keep from falling to the ground, and that in water many animals are neutrally buoyant and expend little effort to keep from sinking, but man has no experience in flying under his own power and he is a clumsy and ineffective swimmer. Walking and running we know more about, for man is our best and most cooperative experimental animal. Except for man and dog running animals have not received much attention, and studies of swimming and flying animals have only recently been carried out at a satisfactory level of success. In this article I shall compare the energy cost of these three kinds of locomotion.

Other types of locomotion exist, such as the hydraulic system of the earthworm, in which the body fluids in combination with the muscular body wall function as a "skeleton." Spiders lack extensor muscles in the limbs and use blood pressure to stretch their legs; the squid and octopus use jet propulsion in swimming. The mechanics of these types of locomotion have been well analyzed but their energy cost is completely unknown.

Differences in the physical qualities of the media in which animals move account for such structural adaptations as streamlining of the body in flying and swimming animals, and the use of levers for propulsion in running animals. The weight of most swimming animals is fully supported by the surrounding medium, but running and flying animals must support the full weight of their bodies. The running animal has a solid support, but the flying animal must support its weight against a fluid of low density and low viscosity. In contrast, when the aquatic animal swims through water, it meets the resistance of a medium of high viscosity and density; running and flying animals have the advantage of moving in a medium of low viscosity and low density.

The author is the James B. Duke professor of physiology in the department of zoology at Duke University, Durham, North Carolina 27706. This article was condensed from the text of an invited lecture given at the 25th International Congress of Physiology in Munich on 27 July 1971.

Comparison of Energy Costs

To compare the energy cost of activities as different as swimming, flying, and running, we need a suitable basis for comparison. There are several variables that must be taken into consideration. (i) The amount of fuel an animal uses in moving, which is expressed as an increase in the metabolic rate. (ii) The cost of moving obviously depends on the animal's size or weight; and the power available for moving is related to muscle mass, which is a function of size. Although aquatic animals do not support their weight, their size is of importance to the resistance they meet in moving. (iii) The distance over which an animal moves influences the energy cost. (iv) The speed at which the animal moves may also be considered. What is the most suitable basis for comparison?

To calculate the energy cost of locomotion, I have chosen to use the amount of fuel (expressed as calories) it takes to transport one unit of body weight (in grams) over one unit of distance (1 kilometer) (*1*). For the calculation of the cost of transportation in these units, we can use the ratio of the metabolic rate in calories per gram per hour, at a given speed in kilometers per hour, to that speed.

$$\frac{\text{Metabolic rate}}{\text{Speed}} = \frac{\text{cal g}^{-1}\text{ hr}^{-1}}{\text{km hr}^{-1}} = \text{cal g}^{-1}\text{ km}^{-1}$$

We thus find that time has conveniently been eliminated, and that the speed is of no direct consequence for comparing the metabolic cost of locomotion. The animals can therefore move within the range of speeds natural to them. This convenience is, as we shall see later, well justified for mammals, but is less justified for birds for which the cost of moving is more speed dependent than for mammals (*2*).

Swimming

An animal swimming through water meets with resistance, or drag; to propel itself the animal must supply a force which equals the drag. The drag on an aquatic animal is a complex function of shape, size, and speed. The importance of shape is expressed in what is commonly known as "streamlining," so well known from fish and whales, and regrettably absent in man. The size of the animal enters primarily as the surface area in contact with the water, for as water moves over a surface, the boundary layer dissipates energy. An important consideration is whether the flow in the boundary layer is laminar or turbulent. Important variables that influence the nature of flow in the boundary layer are speed and length, as expressed in the well-known Reynolds number. At high Reynolds numbers the drag is approximately proportional to the square of speed, but at very low Reynolds numbers the laws of fluid dynamics seem to change and drag becomes primarily a function of viscosity and is directly proportional to speed (Stokes' law).

A knowledge of elementary fluid dynamics, however, is not sufficient to define what animals really do; we need actual determinations of the energy expenditure. The energy cost of swimming salmon has been extensively studied by Brett (*3*). Brett's observations cover a wide range of sizes, from about 3 grams to 1.5 kilograms, and a variety of swimming speeds. From his data I have selected salmon of five size groups, swimming at three-fourths of their fatigue speed, and have plotted the calculated cost of transportation on logarithmic coordinates (Fig. 1). The points fall very close to a straight line, which thus expresses the energy cost of swimming for salmon within the size range studied.

Other investigators, studying other species of fish, have obtained results that give an energy cost of swimming virtually identical to that for salmon. Grayling (*Thymallus*) and whitefish (*Coregonus*), studied by Soviet investigators (*4*), are salmonid fishes. Thus it

Reprinted from *Science* **177**, 222–228 (1972);

is perhaps not surprising that the results are in accord with those obtained for the salmon, when they are recalculated to the same units. Similar results were obtained for the trout (*5*), another salmonid fish. It is interesting, however, that the energy cost of swimming for the goldfish (*6*), a carp, is so close to that of the trout that it could not be plotted as a separate point. Even the pinfish, *Lagodon rhomboides* (*7*), fits the curve for salmon, although its appearance and shape are quite different from the fast-swimming salmonids.

Even more interesting are the results obtained for the eel by Holmberg and Saunders (*8*). Their figures give a cost of transportation of 0.329 to 0.417 cal g^{-1} km^{-1}. Silver eel (mean weight, 248 grams) and green eel (mean weight 238 grams) gave similar results at swimming speeds ranging from 35 cm sec^{-1} to 65 cm sec^{-1}. Although the eel uses almost the entire body in propelling itself through water (anguilliform propulsion) and other fish use only the tail portion of the body (carangiform propulsion), the energy cost of swimming is similar.

It is worth noting that the investigators concerned with these different fish have worked independently and did not consider cost of transport in the terms used here. The close correspondence between their results therefore is striking.

To amuse my friends I have calculated the energy cost for 1 gram of mammalian sperm to travel 1 kilometer (*9*). The result is plotted in Fig. 1B, in which the points in the lower right hand corner and the regression line represent the data on salmon from Fig. 1A. A single bull sperm weighs 10^{-11} gram, and it takes a hundred thousand million sperm to make up the weight of the smallest salmon. The cost for this much sperm to swim 1 kilometer is nearly 10,000 times as high as for the one salmon. The coincidence between the sperm and the regression line for salmon should not be taken too seriously, however. The hydrodynamic considerations that govern the swimming of these two organisms are quite different; fish move at rather high Reynolds numbers, and the sperm at extremely low Reynolds numbers, of the order of 10^{-3}, where lift is negligible and viscous forces alone determine the drag.

The energy requirements of a swimming porpoise or dolphin have long been known among biologists as Gray's paradox. Dolphins can swim at high speeds; reliable determinations indicate that they readily swim at 15 knots, which is 28 kilometers per hour, and for short spurts they may reach 20 knots or 36 kilometers per hour (*10*). Gray (*11*) estimated the energy requirements of swimming dolphins by determining the drag on a model of the animal. He also calculated the drag from the area, the speed, and the drag coefficient, using a calculated Reynolds number of 1.6×10^7. Gray expressed the result in the following words: "If the resistance of an actively swimming dolphin is equal to that of a rigid model towed at the same speed, the muscles must be capable of generating energy at a rate at least seven times greater than that of other types of mammalian muscle" (*11*).

Gray's paradox is that the transition between laminar and turbulent flow takes place at Reynolds numbers of about 5×10^5, and at the speed dolphins swim, the calculated Reynolds number is more than 10^7. If the flow over the dolphin were laminar, however, the drag coefficient and thus the power requirement would be reduced to about one-ninth of the value estimated by Gray.

The oxygen consumption of swimming dolphins has not been determined. Let us instead compare the dolphin with the salmon and other fish, for which excellent information is available. Although we have no inherent right to extrapolate from the salmon to the dolphin, we can safely assume that both animals are highly adapted and effective swimmers, and extrapolating from a 1-kilogram salmon to a 90-kilogram dophin is not unreasonable. Gray's calculations were for the very high speed of 20 knots, but on the other hand, he did not include in his calculations that the efficiency of muscle in performing external work is not 100 percent, and the cost was therefore underestimated. Assuming an efficiency of 25 percent, I have obtained a metabolic rate of 18.1 kcal kg^{-1} hr^{-1}, which gives a cost of transport indicated by the upper open circle in Fig. 1C. (As a point of reference, the metabolic rate of man at rest is about 1 kcal kg^{-1} hr^{-1}, and during top athletic performances it is some 20 times as high.) If instead of turbulent flow we assume laminar flow, and if we assume that the dophin cruises at 10 knots, the power requirement is reduced accordingly. If we again assume 25 percent efficiency of the muscle in performing external work, we arrive at the lower open circle in Fig. 1C, which is located surprisingly close to the extended regression line for the salmon. In this case the extension of a regression line is better justified than it is in

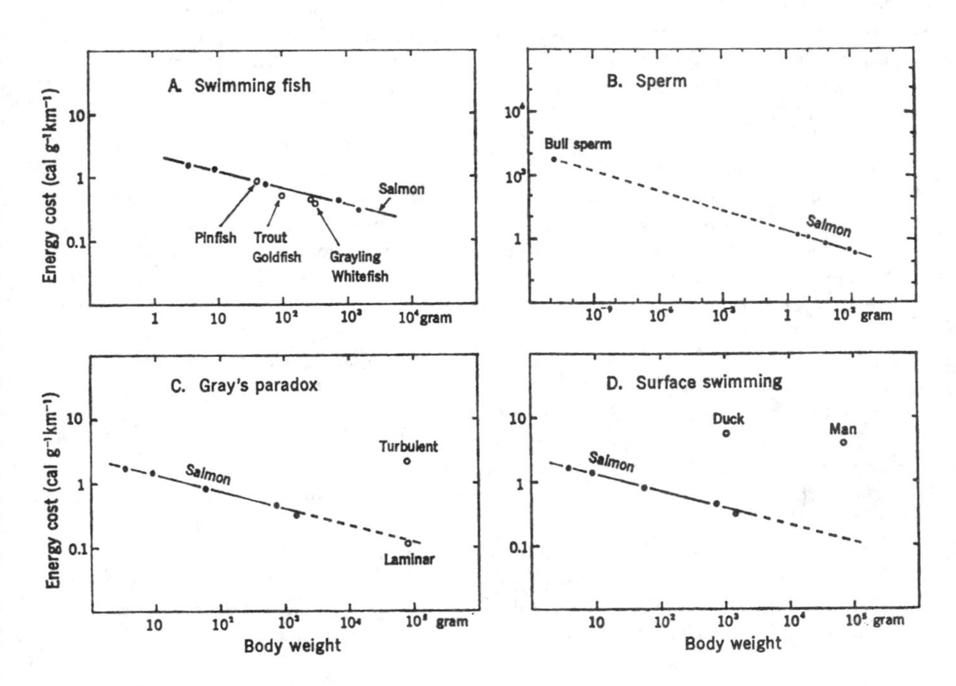

Fig. 1. Energy cost of swimming. (A) Cost of swimming for various fish relative to body size. Coordinates are logarithmic. Solid circles and regression line, represented in each graph, represent salmon with body weights of 3 to 1500 grams [data from Brett (*3*)]; open circles, other fish [data from (*4-8*)]. (B) Estimated energy cost for propulsion of bull sperm (*9*). Data for salmon and regression line are the same as in (A); the coordinate scales are different. (C) Estimated cost of swimming for dolphins, based on the assumption of turbulent flow or laminar flow of water over the body. Actual measurements on dolphins have not been made. (D) Cost of propulsion for two surface swimmers. Data for salmon are repeated from (A).

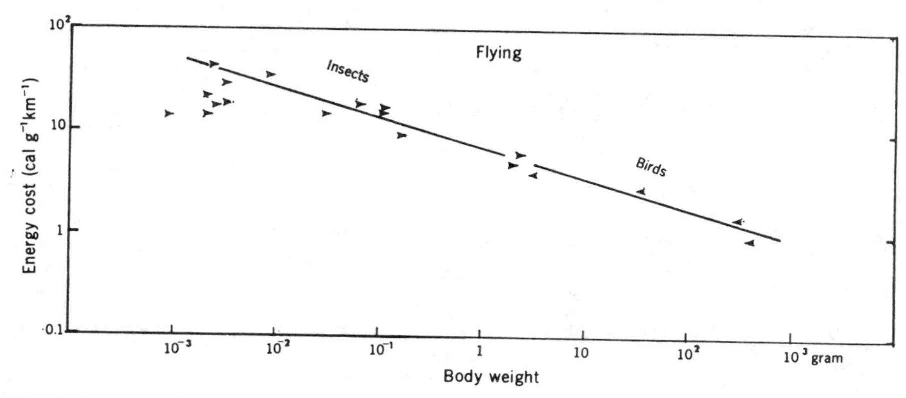

Fig. 2. Energy cost of flying relative to body size of insects (*23, 24*) and birds (*2, 16, 17, 22*). The regression line is based on birds and is arbitrarily extended to cover the size of insects.

the case of the sperm, because the fluid dynamics for salmon and dolphin should be similar. The calculated metabolic rate of the dolphin cruising at 10 knots, if we still assume laminar flow, is 2 kcal kg^{-1} hr^{-1}, about twice the resting metabolic rate of man. In other words, if the dolphin swims as effectively as a salmon, it should be able to cruise at 10 knots with little effort, and this is precisely the impression we gain when we observe these delightful animals from the deck of a ship.

The question of whether the flow can be laminar at this speed has not been resolved. The fluid dynamics of an oscillating body is not fully understood, but in popular terms we could say that as the tail undulates, perhaps it constantly moves out of the area of potential turbulence. Also, the thickest part of the dolphin's body is well behind the midpoint from snout to tail, and this aids in reducing a potentially turbulent boundary layer. An interesting but little-noticed observation was made some years ago by Steven (*12*). He noted that dolphins swimming at night in a phosphorescent sea produced a wake of two clean diverging lines of luminescence stretching behind the animal with very little luminescence, and therefore turbulence, in between. If two dolphins crossed each other's paths at close range, a definite pattern of crossing clean lines could be observed. These wakes indicate low turbulence, and are very different from the boiling mass of turbulence produced by a seal swimming at speed.

Before leaving the subject of swimming, I shall compare two surface swimmers to the salmon. The mallard duck, which Prange and I studied (*13*), swims at an energy cost which is nearly 20 times as high as that for a salmon of the same size (Fig. 1D). There are two important differences, the duck propels itself by paddling, and it swims on the surface where it sets up a wake of gravity waves. For a ship, a major cost of propulsion goes into surface waves, which contain both kinetic and potential energy, but how important this energy loss is for the duck we do not know.

The duck is a natural swimmer, but this cannot be said for man. The energy cost of swimming for man (*14*) is also high, and compared to the extrapolated cost of swimming for salmon, man struggles along at a cost which seems to be about 30 times as high as what we might expect for a fish of his size.

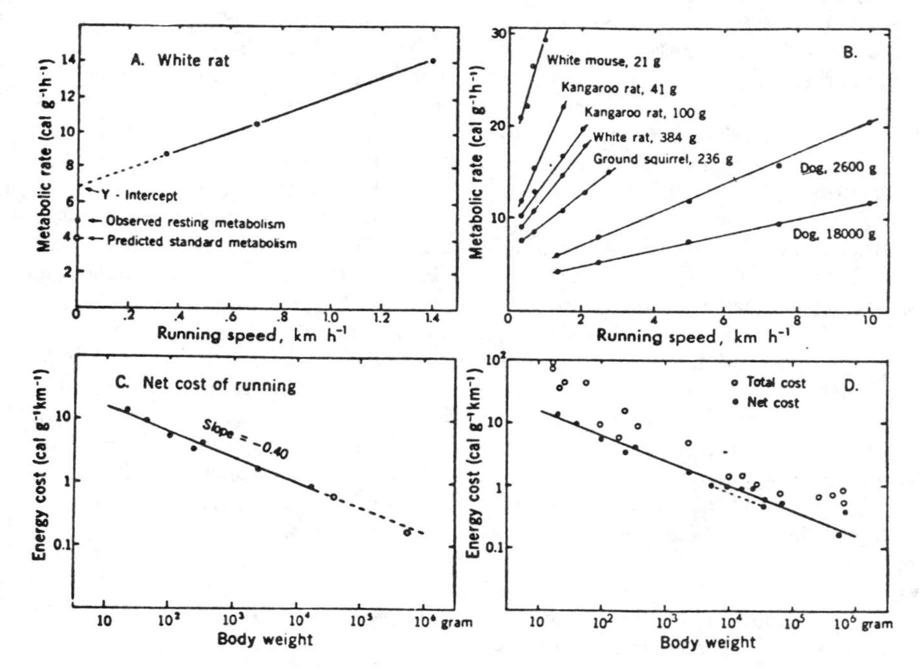

Fig. 3. Energy cost of running. (A and B) Metabolic rate (measured as oxygen consumption) of running animals relative to speed of running. Data for the white rat (*33*) shown in (A) are repeated in (B). Note the linear relationship between metabolic rate and speed, and that in (A) the regression line does not extrapolate to resting metabolic rate. (C) Net energy cost of running for mammals of different body size. Solid circles represent the slopes of each regression line in (B), and thus give the net cost of locomotion for the species in question. Open circles, sheep and horse (*33*). (D) Solid circles, net cost of running for mammals, same data as in (C), with additional data from (*39*). Open circles, total cost of locomotion [data from (*39, 40*)]. The dotted line represents an historical curiosity; it is the calculated regression line for all the data on dogs published by Slowtzoff in 1903 (*28*). This line has a slope of − 0.40, exactly the same value as that obtained in 1970 for animals ranging in size from 21 grams to 1.8 kilograms (*33*).

Flying

The medium in which flying animals move is much less dense and viscous than water, but flying animals must support the weight of their bodies. They must overcome drag, and this has several components. Friction drag is caused by the viscosity of the air in the boundary layer and depends on the shear rate. Friction drag occurs because the fluid in immediate contact with the surface of a moving body is dragged along with the body and establishes a velocity gradient in the boundary layer. What is meant by pressure drag can be described by imagining a flat plate moving perpendicularly to its plane through a fluid. It leaves a highly turbulent wake, and the work dissipated in this way represents the pressure drag. If we were to move the plate along its own plane, however, pressure drag would be negligible and friction drag would dominate. Induced drag can perhaps best be understood by reference to an airplane or a bird moving horizontally through the air. To move horizontally it needs lift which must equal the weight of the body. Lift is produced as air moves over the wing or airfoil, and the necessary force to move the wing in the direction of flight, to achieve this lift, is known as induced drag. Induced drag therefore represents the force necessary to support the body in the air.

The total drag is a complex function of air density and viscosity, and is related to area, shape, and speed of the flying animal. In spite of the complexities, the aerodynamics of bird flight has recently been successfully analyzed by Pennycuick (*15*).

Direct measurements of the energy cost of flying have been made for only a small number of birds, ranging from 3 to 300 grams (*2, 16, 17*), and quite recently for a medium-sized tropical bat (*18*). Table 1 shows that the metabolic rate during flight is quite high, but decreases with increasing body size. The bat, the only flying mammal that has been studied, falls within the range of the birds. Compared with the metabolic rate at rest, the increase in metabolic rate during flight for the budgerigar (parakeet) and the gull (*2, 17*) is about sevenfold, which is similar to the maximum increase obtained for running animals of similar size, such as mice and rats (*19*). Since the resting metabolic rate of birds in general is of the same magnitude as that of mammals of equal size, we can see that horizontal flight does not require extraordinary feats of power output (*20*). The apparently much higher increase factors for hummingbirds and bats are misleading because the resting metabolic rates for these animals are atypical; both hummingbirds and bats readily undergo torpor with considerable drops in body temperature and oxygen consumption. Therefore resting values are difficult to define accurately, and the increase factors are correspondingly uncertain.

Tucker's studies of flying birds are extraordinarily informative. By using a wind tunnel that could be tilted, Tucker determined the energy cost of flight at a variety of speeds for horizontal as well as for ascending and descending flight. The budgerigar had a definite minimum in the rate of oxygen consumption at a speed of 35 km hr$-^1$; when it flew faster or more slowly the rate of oxygen consumption increased (*2*). For the laughing gull (*17*) the cost of flying was far less dependent on the flying speed, however, and this difference points out the need for more extensive information for birds of various body sizes and different flight patterns. Most recently Bernstein *et al.* (*21*) have studied the fish crow, which has shorter, broader wings than gulls and in nature mostly uses continuous flapping flight while gulls use soaring as well as flapping flight. Despite these differences the energy cost of horizontal flight for the crow did not differ strikingly from the means reported for the gull (*17*).

Table 1. Metabolic rate during steady-state horizontal flight. (Information for hummingbird is for hovering flight.)

Flying animals	Body weight (g)	Metabolic rate (cal g^{-1} hr^{-1})		
		Rest	Flight	Increase
Birds				
Hummingbird (*16*)	3	14.3	204	14 ×*
Budgerigar (*2*)	35	15.8	105	6.7 ×
Gull (*17*)	300	7.3	54	7.4 ×
Bats				
Phyllostomis (*18*)	90	3.8	94	25 ×*

* Resting values for the hummingbird and bat are not well defined because these animals readily undergo torpor.

A particularly interesting aspect of the flying budgerigar was pointed out by Tucker (*2*). In analyzing the partial efficiency (the ratio of change in external work done to change in energy expenditure) for flight at different angles, he reached the conclusion that a flying budgerigar could ascend from and again descend to a given altitude and use no more energy than if it had spent the same time in horizontal flight at the same speeds. Furthermore, the budgerigar is a ground-feeding bird and does much of its ascending flight immediately after takeoff when it has not yet accelerated to normal flight speed. At these slow speeds the partial efficiency for ascent is greater than for descent, so that a combination of ascending and descending flight is more economical than horizontal flight over the same distance. These observations may explain why some birds seem to prefer an undulating flight pattern.

Oxygen consumption is technically far easier to determine for flying insects than for birds, and a larger number of such determinations is available. Information about birds and insects is combined in Fig. 2. In addition to the three birds in Table 1, a fourth (the point to the far right) refers to pigeons, in which metabolic turnover during flight was derived from an ingenious double-labeling technique with ^{2}H and ^{18}O (*22*). The bat is not included in the graph because the determinations were made on tethered animals that were flying in circles, and their speed therefore did not represent free or natural flight. The oxygen consumption of hummingbirds was determined for hovering flight; the calculation of cost of transport is based on an assumed flying speed of 50 km hr^{-1}.

The flying insects for which data are given in Fig. 2 cover a 1000-fold range of body weights, from a few milligrams to 3 grams (*23*). The two points to the right represent the desert locust and the cecropia moth, at the other end are mosquitoes that weigh less than 2 milligrams (*24*). The regression line is an arbitrary extension of the line for birds and some points for the smallest insects fall distinctly below this line. Some of these are based on older determinations that may be less reliable than more recent data. Whether or not we disregard these deviations, the overall conclusion is that all flying animals fall amazingly close to one straight regression line, in spite of the vast differences in size, anatomy, flight pattern, speed, body temperature, and so on, between insects and birds.

Running

More adequate physiological information is available for running than for swimming and flying. When an animal, including man, runs horizontally, he performs no useful external work. Energy loss due to frictional resistance against the ground is minimal, and work performed against air resistance is in most cases very small. Only when a man runs a sprint at top speed does the air resistance account

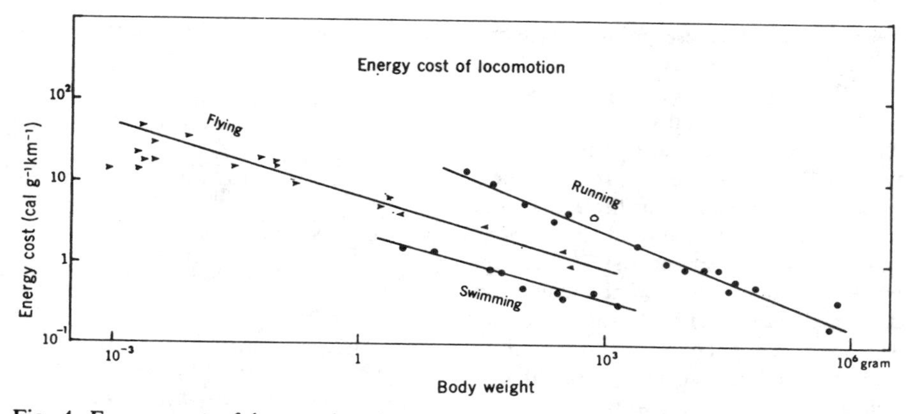

Fig. 4. Energy cost of locomotion for swimming, flying, and running animals, related to body size. Data are from Figs. 1A, 2, and 3D (net cost only). (The single open circle among the running animals refers to the cost of locomotion for the swimming duck.)

for more than a few percent of his total energy output (*25*); usually wind resistance is negligible compared to the common inaccuracies of measurements of oxygen consumption.

We know, however, that running fast requires a great deal of energy, and since it is not dissipated as work done on the environment, it must be dissipated internally. It can be divided into two categories, energy dissipated because of frictional and viscous resistances in joints and muscles, and energy used in the continuous accelerations and decelerations of the mass of the body and of the limbs. An exact analysis of these requirements is difficult because of geometrical complexities, such as the steadily changing angular momentum, the changing radius of curves followed by the center of mass of the different portions of the limbs, and so on. In spite of these difficulties an excellent kinematic analysis of running man was made by Wallace Fenn as early as 1930 (*26*). More recently, further analyses of the mechanical work of walking and running have been made at Margaria's laboratory in Milan (*27*). Such studies are essential for the analysis of the energy cost of locomotion in man, and studies extending to body shapes and sizes other than man's are much needed.

Running uphill, as we know, requires additional energy. A body that has been moved uphill has acquired potential energy, which is the product of its weight and the vertical distance through which it has moved. As we might predict, the cost of moving one unit of body weight uphill is similar for animals of widely different body sizes. This has been amply documented by many investigators; an early demonstration of this simple fact was given in 1903 by Slowtzoff (*28*), who studied dogs ranging in size from 5 to nearly 40 kilograms.

An animal moving downhill, on the other hand, loses potential energy, and there is some question of whether this energy can be usefully recovered by the organism. Conceptually, physiologists have attached the term "negative work" to this matter of events. To the physicist the term sounds appalling, for in actuality work cannot be negative; use of the term comes from the simple fact that work equals force times distance. For an animal moving downhill as opposed to uphill, the vertical distance is given a negative sign and the work therefore also gets a negative sign. The term is convenient and is likely to remain in use in physiology.

The question of whether any negative work can be recovered in the reversal of energy-yielding biochemical processes in the muscles has been seriously considered since the 1920's, when Fenn observed that lengthening of a muscle during contraction decreases the heat production in the muscle. In other words, to quote Fenn, "When the work done by the muscle is negative, the excess energy is also negative" (*29*).

An experiment as amusing as it is simple can be used to demonstrate that a given force can be exerted with less effort when the work is "negative." It is described in A. V. Hill's words as follows: "Two bicycles were arranged in opposition; one subject pedaled forward, the other resisted by back-pedaling. The speed had to be the same for both, and (apart from minor loss through friction) the forces exerted were the same. All the work done by one subject was absorbed by the other; there was no other significant resistance. The experiment was shown in 1952 at the Royal Society in London and was enthusiastically received, particularly because a young lady doing the negative work was able quickly, without much effort, to reduce a young man doing the positive work to exhaustion" (*30*).

Whether the biochemical events of muscular contraction actually can be reversed by stretching the muscle during contraction—that is, by performing work on the muscle, has not been finally resolved (*31*). The question is not without importance to locomotion, however. Let us consider the end of a single step of a running man. As the foot hits the ground, kinetic energy from the moving body is taken up by the partly contracted muscles of the leg. Part of the kinetic energy is stored as elastic energy in the contracted muscles, part of it seems to be recovered as the increased external work that the muscles can perform because of being stretched. The external work done by a muscle which has been stretched immediately before contraction has been found to be up to 2.5 times the work that the same muscle would perform during shortening from a state of isometric contraction (*32*), and only part of this increase can be accounted for by the elastic energy stored in the stretched muscle.

The physiological role of negative work is still very much in dispute, both at the molecular level and in studies of the energetics of the whole organism. I shall not attempt a further review of this particular field, instead I shall return to the energy expenditure of animals running horizontally.

The metabolic rate of a running animal increases with speed. For the white rat, as well as for other mammals, the relationship gives a straight line, as shown in Fig. 3A (*33*). It is now commonly agreed that within a wide range of speeds, the metabolic rate of a running man shows a similar straight line relationship (*34*). [For a walking man the relationship is distinctly nonlinear (*35*)]. The straight line in Fig. 3A does not extrapolate to the resting metabolic rate of the animal in question; the *Y*-intercept for zero running speed is higher than the value for the resting metabolism. However, the metabolic rate for a man who stands quietly on the treadmill instead of walking, or who is sitting passively on the bicycle ergometer (*36*), is close to the *Y*-intercept. The simplest interpretation of this observation is to consider that the difference between intercept and resting metabolism is due to the cost of maintaining the posture of locomotion, and the term "postural effect" seems acceptable.

To calculate the energy cost of running, we could use the total metabolic rate at a given speed, or we could subtract the resting (or standard) metabolic rate and say that the remainder goes into locomotion, or we could subtract the *Y*-intercept from the metabolic rate at a given speed and say that the excess energy is the cost of moving at that speed.

For my purposes this procedure is the simplest, for when the *Y*-intercept is subtracted, we obtain the metabolic cost of running as the slope of the line. Because the line is straight, the net cost of locomotion remains constant and does not change with the speed of running.

Determinations of the metabolic rate of animals of various body sizes in relation to the running speed shows that smaller animals have steeper regression lines than those of large animals (Fig. 3B). The slopes of these lines make it possible to compare animals running at quite different speeds, even though the maximum speed reached by the white mouse is less than the slowest speed of the dogs.

The slopes, calculated by dividing metabolic rate less intercept, by speed, indicate the cost of locomotion in the same units as used throughout this article. These slopes, the energy cost of moving 1 gram of animal over 1 kilometer, plotted against the body weight on logarithmic coordinates, yields a straight line as shown in Fig. 3C (*37*). The slope of the best-fitting straight regression line for these mammals is − 0.40. Thus the cost of locomotion, per unit weight, is lower for larger animals, but the meaning of the exact numerical value of the slope has so far not been analyzed successfully.

Additional available data on the cost of locomotion in mammals are compiled in Fig. 3D. The solid circles repeat the data from Fig. 3C, with additional points from other studies which permit a reasonably accurate calculation of the net cost of locomotion, obtained either by plotting the rate of metabolism against speed, or by subtracting an observed nonrunning metabolic rate. The open circles are for the total cost of running, obtained from the total oxygen consumption while the animal was moving. Since no intercept has been subtracted, these points fall above those for the net cost. Furthermore, the distribution of these points seems to be curved upward at both ends. This is probably because these animals were not running very fast, and their nonrunning metabolism therefore was a large fraction of the total. If the nonrunning metabolism or the *Y*-intercept were subtracted, the points would be displaced downward, toward the regression line for the net cost of running, and the curved distribution might disappear.

In Fig. 3D the energy cost of locomotion for the swimming duck is also inserted, in this case the total rather than the net cost. Is is interesting, although perhaps fortuitious, that a swimming duck moves across the surface of the water at a cost similar to that of a mammal of the same body size running on land.

The energy costs of swimming, flying, and running are compared in Fig. 4 [see also (*38*)]. The points used for running animals represent the net cost of running, but for flying animals, the total cost at the most economical speed has been used. Since the flight metabolism is a high multiple of the resting metabolism, subtraction of the latter would have only minor influence on the points. It is interesting that for a given body size, flying is a far cheaper way to move to a distant point than is running. This is not intuitively obvious to us, for man has no experience in flying under his own power and it seems that the mere cost of staying up in the air would be excessive. If we consider a migrating bird, however, things appear more reasonable. A migrating bird can fly nonstop for more than 1000 kilometers, but it is hard to imagine a small mammal, say a mouse, running 1000 kilometers without stopping to eat and drink. The energy cost of swimming appears to be lower again. Man, however, is so ill-adapted to moving in water that swimming costs him five or ten times as much energy as does running the same distance on land. If the dolphin had a similar energy expenditure while swimming, its ability to live in the ocean would indeed be a paradox.

References and Notes

1. This simple way of comparing the energy cost of locomotion is widely used for animals and man as well as machines. See, for example, T. Weis-Fogh, *Trans. Int. Congr. Entomol. 9th* **1**, 341 (1952); R. Margaria, *Int. Z. Angew. Physiol. Einschl. Arbeitsphysiol.* **25**, 339 (1968); V. A. Tucker, *Comp. Biochem. Physiol.* **34**, 841 (1970).
2. V. A. Tucker, *J. Exp. Biol.* **48**, 67 (1968).
3. J. R. Brett, *J. Fish. Res. Bd. Can.* **21**, 1183 (1964).
4. V. A. Matyukhin and A. J. Stolbow, in *Adaptatsiia Vodnykh Zhivotnykh* (Akademiia Nauk SSSR, Sibirskoe Otdelenie, Novosibirsk, USSR, 1970), pp. 149–152.
5. G. M. M. Rao, *Mar. Biol.* **8**, 205 (1971).
6. H. Smit, J. M. Amelink-Koutstall, J. Vijverberg, J. C. Von Vaupel-Klein, *Comp. Biochem. Physiol.* **39A**, 1 (1971).
7. D. E. Wohlschlag, J. N. Cameron, J. J. Cech, Jr., *Contrib. Mar. Sci.* **13**, 89 (1968).
8. B. Holmberg and R. L. Saunders, unpublished data.
9. This calculation is based on the estimated hydrodynamic cost of propulsion for sperm, its speed, and its oxygen consumption as given by R. Rikmenspoel, S. Sinton, and J. J. Janick [*J. Gen. Physiol.* **54**, 782 (1969)]; and by L. Rochschild [in *Mammalian Germ Cells* G. E. W. Wolstenholme, Ed. (Churchill, London, 1953), p. 122].
10. T. G. Lang and K. Pryor, *Science* **152**, 531 (1966).
11. J. Gray, *J. Exp. Biol.* **13**, 192 (1936).
12. G. A. Steven, *Sci. Progr.* **38**, 524 (1950).
13. H. D. Prange and K. Schmidt-Nielsen, *J. Exp. Biol.* **53**, 763 (1970).
14. K. L. Andersen, *Acta Chir. Scand.* **253** (Suppl.), 169 (1960).
15. C. J. Pennycuick, *J. Exp. Biol.* **49**, 527 (1968); *Ibis* **111**, 525 (1969).
16. R. C. Lasiewski, *Physiol. Zool.* **36**, 122 (1963); *Proc. Int. Ornithol. Congr. 13th* 1095 (1963).
17. V. A. Tucker, *Amer. J. Physiol.* **222**, 237 (1972).
18. S. P. Thomas and R. A. Suthers, *Fed. Proc.* **29**, 265 (1970).
19. L. Jansky, *Physiol. Bohemoslov.* **8**, 472 (1959); P. Pasquis and P. Dejours, *J. Physiol. Paris* **57**, 670 (1965); N. P. Segrem and J. S. Hart, *Can. J. Physiol. Pharmacol.* **45**, 531 (1967).
20. If the comparison is based on the standard metabolic rate of birds, as estimated from their body size [R. C. Lasiewski and W. R. Dawson, *Condor* **69**, 13 (1967)], the metabolic rate during flight gives a higher multiple. Also, the power output of birds flying faster or more slowly, or flying on an ascending path, is substantially higher. The maximum power output of birds therefore may exceed by several times the maximum power output of mammals of the same size.
21. M. H. Bernstein, S. P. Thomas, K. Schmidt-Nielsen, *Fed. Proc.* **31**, 325 (1972).
22. E. A. LeFebvre, *Auk* **81**, 403 (1964).
23. Points plotted for flying insects are compiled from the sources used by Tucker (*1*), with additional information added from J. L. Hanegan and J. E. Heath, *J. Exp. Biol.* **53**, 611 (1970); personal communication from J. E. Heath; and from J. K. Nayar and E. Van Handel (*24*).
24. J. K. Nayar and E. Van Handel, *J. Insect Physiol.* **17**, 471 (1971).
25. R. Margaria, *Int. Z. Angew. Physiol. Einschl. Arbeitsphysiol.* **25**, 339 (1968).
26. W. O. Fenn, *Amer. J. Physiol.* **93**, 433 (1930); *ibid.*, p. 447.
27. G. A. Cavagna, F. P. Saibene, R. Margaria, *J. Appl. Physiol.* **18**, 1 (1963); *ibid.* **19**, 249 (1964).
28. B. Slowtzoff, *Pfluegers Arch. Gesamte Physiol. Menschen Tierre* **95**, 158 (1903).
29. W. O. Fenn, *J. Physiol. London* **58**, 373 (1924).
30. A. V. Hill, *Science* **131**, 897 (1960).
31. R. E. Davies, *Essays Biochem.* **1**, 29 (1965); D. R. Wilkie, *Progr. Biophys. Mol. Biol.* **10**, 260 (1960).
32. G. A. Cavagna, B. Dusman, R. Margaria, *J. Appl. Physiol.* **24**, 21 (1968); G. A. Cavagna, *ibid.* **29**, 279 (1970).
33. C. R. Taylor, K. Schmidt-Nielsen, J. L. Raab, *Amer. J. Physiol.* **219**, 1104 (1970).
34. R. Margaria, P. Cerretelli, P. Aghemo, G. Sassi, *J. Appl. Physiol.* **18**, 367 (1963).
35. H. J. Ralston, *Int. Z. Angew. Physiol. Einschl. Arbeitsphysiol.* **17**, 277 (1958).
36. B. J. Whipp and K. Wasserman, *J. Appl. Physiol.* **26**, 644 (1969).
37. The regression line in Fig. 3C represents the net cost of locomotion, and not the total metabolic rate while running. There has been controversy about the correct way of calculating the energy cost of running, whether to use total metabolic rate at a given speed,

or to subtract the resting metabolism from the total and consider the remainder as the energy used in running, or, as I have done here, subtract value for the Y-intercept. The procedure must depend on the purpose we have in mind. I have discussed the net cost of locomotion—that is, the cost of locomotion per se. The total energy expenditure during locomotion is higher, but equally correct. For example, a migrating bird uses its fat reserves for flying, and the distance it can cover on the available fuel is determined by its total metabolic rate. Likewise, a grazing sheep must obtain enough food to cover, not only the energy cost of walking, but enough for its total metabolic needs. Therefore, there is no reason for controversy, so long as we have defined the terms in estimating the cost of locomotion.

38. A comparison with man-made machines shows that automobiles use about twice as much energy in locomotion, per unit weight, as a horse (about 0.5 cal g^{-1} km^{-1}). Big jet airplanes use about the same amount of energy as a horse per unit weight, but smaller planes use progressively more energy, and helicopters use nearly ten times as much as a horse. Ships use less energy, however, and an economical long distance freighter may, per unit weight, use almost as little as one-tenth the energy used in locomotion of a horse.
39. A. C. Barger, V. Richards, J. Metcalfe, B. Günther, *Amer. J. Physiol.* **184**, 613 (1956); S. Brody, *Bioenergetics and Growth* (Reinhold, New York, 1945); P. Cerretelli, J. Piiper, F. Mangili, B. Ricci, *J. Appl. Physiol.* **19**, 25 (1964); J. L. Clapperton, *Brit. J. Nutr.* **18**, 47 (1964); H. G. Knuttgen, *Acta Physiol. Scand.* **52**, 366 (1961); R. Margaria, P. Cerretelli, P. Aghemo, G. Sassi, *J. Appl. Physiol.* **18**, 367 (1963); R. Passmore and J. V. G. A. Durnin, *Physiol. Rev.* **35**, 801 (1955); B. Slowtzoff, *Pfluegers Arch. Gesamte Physiol. Menschen Tiere* **95**, 158 (1903); D. R. Young, R. Mosher, P. Erve, H. Spector, *J. Appl. Physiol.* **14**, 834 (1959); N. Zuntz and O. Hagemann, *Landwirt. Jahrb.* **27** (Suppl. 3), 1 (1898).
40. J. S. Hart, *Can. J. Zool.* **30**, 90 (1952); ——— and O. Heroux, *Can. J. Biochem. Physiol.* **33**, 428 (1955); J. S. Hart and L. Jansky, *ibid.* **41**, 629 (1963); L. Jansky, *Physiol. Bohemoslov.* **8**, 464 (1959); *ibid.*, p. 472; N. P. Segrem and J. S. Hart. *Can. J. Physiol. Pharmacol.* **45**, 531 (1967).
40. Supported by NIH research grant HE-02228 and research career award 1-K6-GM-21,522.

Karate strikes

Jearl D. Walker
Physics Department
Cleveland State University
Cleveland, Ohio 44115
(Received 21 November 1974; revised 24 March 1975)

Analysis of a forward karate punch is made as an example of collision mechanics. The energy lost to deformation of an opponent is evaluated, and the deformation is shown to be maximized if contact is made when the fist has the greatest speed. In karate fighting the maximum deformation is obtained by focusing a punch to terminate several centimeters inside an opponent's body. The average impact force is calculated and compared with that neeeded to break a human bone by stressing it beyond its ultimate bending stress value. Similar breaking forces needed for bricks and wooden boards are also computed. A brief description of breaking demonstrations for the physics classroom is made.

INTRODUCTION

In recent years, considerable popular attention has been directed towards the various Oriental forms of fighting, especially the several forms of karate. One of the peculiar features of karate is that the strikes to an opponent are designed to terminate several centimeters inside an opponent's body, in contrast to the follow-through of the wide swings employed by movie cowboys and most uninitiated street fighters. Here we shall examine the question[1] of whether there is any scientific basis for focusing a strike to terminate just inside the target body.

We shall also examine the forces involved in a strike to a bone of the opponent, for example, a forearm bone. In particular, we shall be interested in the feasibility of a karate expert breaking an opponent's bone much as they break boards, bricks, tiles, blocks of ice, etc., in exhibitions.

Though the martial arts are just now coming into widespread notice in the United States, they have been practiced for several thousand years in the East. Karate is believed to have begun about that long ago in China, spreading later to Okinawa, Japan, and Korea.[2,3] The principal characteristic of the resulting several styles of karate is the use of one's own body—hands, feet, elbows, head, and knees—as weapons and in defense. People in the West are often unaware of the large impact forces involved in such fighting. Even though a study of karate is immersed in violence, the fighting is graceful and appears to be well grounded scientifically.

DEFORMATION ENERGY

The primary point in striking an opponent is to maximize the deformation damage at the area of contact, and it is only rarely that moving the opponent's body as a whole is desired. In calculating the extent of damage, there appear to be two considerations, one being the amount of energy lost to deformation during the impact, and the other being the impact forces and stresses imposed. Let us first consider the energy lost to deformation.

If we treat the strike as a collision between two free bodies, one (M_1) initially at rest, and the other (M_2) initially moving at speed v, the energy ΔE lost to deformation and heat can be found from the conservations of total energy and linear momentum[4]:

$$\Delta E = \frac{(1-e^2)}{2}\frac{M_1 M_2}{M_1+M_2}v^2, \tag{1}$$

where e is the coefficient of restitution. That coefficient is the ratio of the velocity difference of the two bodies after the collision to that difference before the collision. A perfectly elastic collision has $e = 1$, and a totally inelastic collision has $e = 0$. In our development here, we shall neglect the heating of the two impacting masses and treat ΔE as being entirely lost to deformation.

The equation is used by Miller[5] and by Gray and Gray[6] in discussions of pile driving and forging. Miller questions whether elastic or inelastic collisions are desired in those activities, and in forging, whether a light or heavy hammer should be used. In forging, one wants to maximize the fraction of the initial energy lost to deformation, the fraction being

$$\frac{\Delta E}{M_2 v^2/2} = (1-e^2)\frac{M_1}{M_1+M_2}, \tag{2}$$

so that each strike is as efficient as possible. The fraction is largest for $e = 0$, that is, for totally inelastic collisions, as could have been guessed. The fraction is larger for smaller values of M_2, and thus a light hammer is more desirable than a heavy one. In pile driving, no deformation of the pile is sought, and one attempts to have elastic collisions where $e = 1$. The karate strike is similar to the forging considered by Miller. The karate strike is more complicated, however, because e is strongly dependent on where the opponent is hit. Even though hitting muscle and fat would involve a greater e, and hence less ΔE, than hitting a bone covered by a thin layer of skin, a strike to a high-e area may result in more pain to the opponent. Clearly, a strike to the groin may be more advantageous than one to the leg even though the e for the groin strike is greater. Another problem in the evaluation of any collision is that e may be dependent on v. Unlike Miller's forging, we are not so concerned with the efficient transfer of energy to deformation with each strike as we are with the absolute magnitude of the energy transfer. In forging, we can strike until the job is done; in karate we may have only one chance. We are therefore interested in (1) rather than (2).

The choice of M_2 for inclusion in (1) depends on not only the fighter's mass but also the type of strike employed. For example, the effective mass involved in a forward punch in which the attacker steps forward is greater than if he remains stationary. In our calculations here, however, we shall employ a typical value for the M_2 involved in a forward punch and not consider variations in M_2 due to stance or movement of the fighter's trunk.

Since (1) depends on v^2, the karate fighter attempts to maximize the speed of his strike. Let us consider a forward punch (Fig. 1) in which the attacker does not move his feet. The fist starts at the waist with the closed palm upward, and as the fist is hurled forward, it turns over until the closed palm is downward when the arm is fully extended. Karate blackbelts are said to complete this motion in about 0.2 sec with maximum speeds of about 7 m/sec occurring between 70% and 80% of the way through the motion.[7,8] This range of maximum speed is confirmed in Fig. 2, which is based on data taken from a high-speed movie of a forward punch by the author. In this paper we shall consider the 75% point as that of maximum speed.

Maximum energy transferred to deformation would occur if impact is made at that 75% point. Let us assume that M_2 is the mass of only the striking arm. Miller and Nelson[9] estimate that the arm is approximately 10% of a man's total mass. Taking $e = 0$, the total mass of both men to be 70 kg, and the maximum speed quoted above, we find from (1) that a maximum of 156 J can be transferred to deformation damage. In absence of a target, the karate expert is trained to terminate the punch when his arm is fully extended. If we assume constant acceleration and deceleration of his fist and if we used the time of

Fig. 1. Forward punch begins at the waist and terminates when arm is fully extended. Contact should be made before full extension.

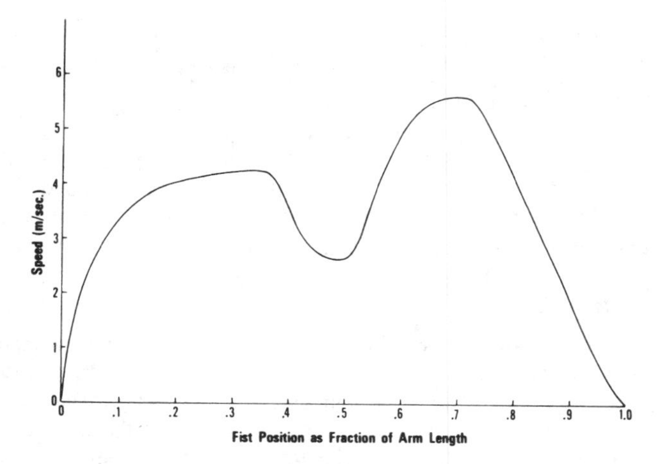

Fig. 2. Speed of fist versus position of fist measured as a fraction of the full arm extension. Data taken from high-speed movie of the author.

maximum fist speed given above, then the maximum speed is reached at 0.75 of the arm's full extension, or typically 10–14 cm from the stopping point. With only (1) considered, then, the maximum energy of about 156 J is delivered to the deformation of the opponent if the attacking fighter focuses his punch 10–14 cm inside the opponent's body. This way impact is made approximately when the fist has the greatest v.

Practically, however, the focus may be only half that distance inside the body since the fist should be fully turned to a palm-down orientation before impact is made to avoid any wrenching of the wrist. Focusing depths of 6 cm or more may be used in other forms of strikes which do not have that practical consideration. The backhand strike to the side involves rapidly pivoting the fist about the elbow while the elbow is rapidly brought from one's front to the side of the body. Contact should be made with the knuckles when the fist is traveling the fastest, and that will be several centimeters before the projected terminal point. In other words, the strike should again be focused several centimeters into the opponent's body.

Since (1) includes the square of the speed, we can see the great advantage of a karate fighter increasing the deformation energy by stepping forward with the punch or by striking when the opponent happens to lunge toward him. If both fighters quickly close, ΔE may even be doubled.

In street fighting there may be a continuation of the punch after maximum speed is obtained and even after contact is reached. Such follow-through is usually not made in karate for two reasons. One, if contact is made sometime during a wide swing, the attacker will probably jeopardize his balance. Two, if contact is made just as the follow-through begins, then the energy transfer during the follow-through results from pushing, and since pushing and displacement do not result in deformation damage, they are normally not worth the loss in the attacker's poise.

Introductory karate students often believe that the rotation of the fist in the forward punch described above significantly adds to the energy delivered to the opponent. If we treat the arm as a solid, uniform cylinder of radius $r = 3$ cm and use the arm mass and punching time from above, we find that the rotational energy E_r is

$$E_r = M_2 r^2 \omega^2/4 = 0.4\ \text{J}, \tag{3}$$

which is negligible compared to the 156 J calculated before. Similarly negligible results are obtained for punches in those styles of karate in which the fist is rotated immediately before contact rather than continuously throughout the punch. So, the fist rotation appears to be largely for the most comfortable use of the arm muscles.

IMPACT FORCE

The second aspect to be analyzed in evaluating the damage to the opponent is the impact force and the resulting stresses in the opponent's bones if bones are struck. Again, if we consider a forward punch with the arm as a free mass of about 7 kg, the maximum momentum of the punch is approximately 49 kg m/sec. Nakayama[10] reports that contact time appears to be 10 msec or less. If the fist comes to a complete stop within that time owing to the collision, then the average impact force can be as large as 4900 N. Since the collision occurs so quickly, it is approximately correct to neglect the muscular accelerations during the collision.

To evaluate roughly the effect of such an impact force, let us consider the resulting acceleration of the head if the strike is made there. Miller and Nelson[9] give the head's mass as typically 8% of a man's total mass. If we again assume that the total mass is 70 kg, the average impact force would give an average acceleration of about 875 m/sec^2, or about 89 g, to the head. This result is consistent with that of R. L. Le Fevre, as related to Benedek and Villars,[11] on the acceleration of a dummy's head when struck by a prizefighter. The peak acceleration of the dummy's head was 80 g. Such a blow, according to Benedek and Villars,[12] would be a factor of 2.5 times too low to be fatal to the average person but could result in unconsciousness. The blow from the prizefighter is spread over a rather large glove. The karate fighter impacts primarily only two knuckles, those of the first and second fingers, and the force can be considered to be applied at a single point. For our purposes here, would such a karate punch made directly to a bone, such as in the forearm, result in breakage?

To answer the question, we first must calculate the stress created by the impact force and then compare that value with the maximum value that can be withstood by a human bone. Our development follows Clements and Wilson[13] and Benedek and Villars.[14] If a static force F is applied to a material over an area A, then the stress σ on the material is F/A. If a straight bone is pulled at each end by a force F, then the bone will be elongated by an amount δl that is proportional to the force and the initial length L. The ratio $\delta l/L$ is the strain ϵ and is related to the stress by a constant E, the Young's modulus of elasticity:

$$\sigma = E\epsilon. \tag{4}$$

This linear relationship between stress and strain holds up to a point called the proportional limit. In our work here we shall continue to use (4) even up to the rupture point, but we shall partially compensate our error by eventually incorporating experimental stress data.

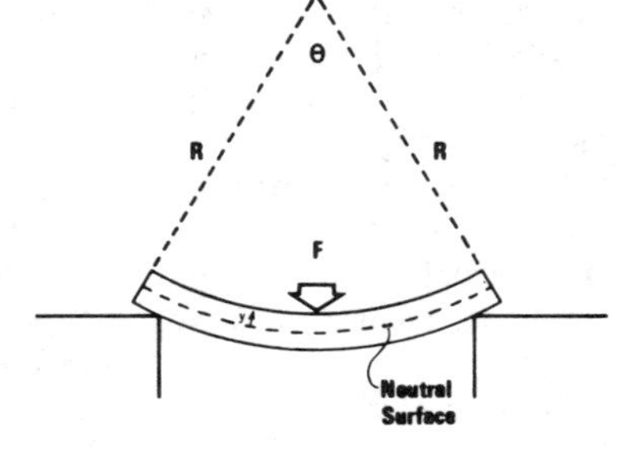

Fig. 3. Deflection of a supported bone is made by an applied force.

If a static force causes bending of a bone supported as shown in Fig. 3, then there will be compression along the inside of the curve and elongation along the outside curve. Somewhere inside the bone will be a surface called the neutral surface along which there is neither elongation nor compression. The magnitude of strain anywhere in the bone is expressed as a function of the distance y away from that neutral surface by

$$\epsilon = \delta l/L \simeq y\theta/R\theta = y/R, \tag{5}$$

where θ is the angle shown in Fig. 3 and R is the radius of curvature of the bent bone. The stress σ as a function of y follows from (4):

$$\sigma = Ey/R. \tag{6}$$

The external forces on the bone cause torques on the bone, and since we are considering a static case, all those torques must be in equilibrium. Consider the cross section of the bone on the near side of the imaginary cut indicated in Fig. 4. The stresses over the cross section cause a torque on the cross section about the axis OO. The bone does not turn about that axis, however, because there is an equal and opposite torque, $(F/2)(L/2)$, about OO due to the support force at the end of the bone. In relating the torque from the stresses on this cross section to the torque from the support force, we can eventually calculate a maximum force that can be endured by the

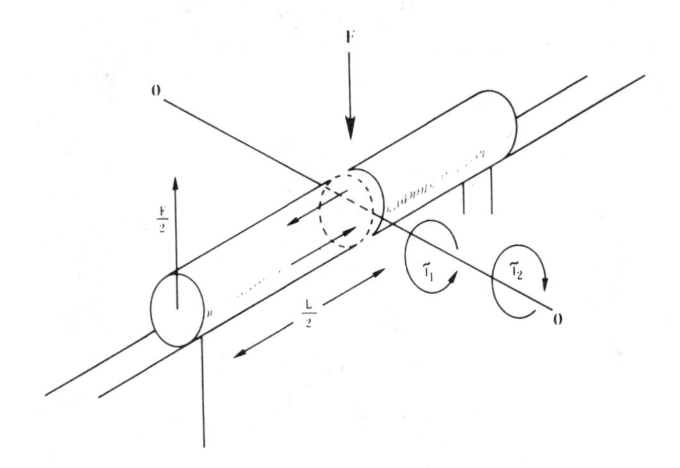

Fig. 4. The cross section on the near side of the imaginary cut through the bone experiences two torques about the axis OO. τ_1 is due to the stresses over the cross-sectional area indicated by the vectors. τ_2 results from the support force $F/2$.

bone without the stresses causing breakage.

The torque on the cross-sectional area element dA at a distance y from the axis OO due to stress is

$$d\tau = y\sigma\, dA. \qquad (7)$$

The total torque on the cross section about OO due to stress is

$$\tau = \int y\sigma\, dA.$$

By using (6), this expression is

$$\tau = (E/R) \int y^2\, dA = (E/R)I, \qquad (8)$$

where

$$I = \int y^2\, dA, \qquad (9)$$

and for equilibrium (Fig. 4), we must have

$$(F/2)(L/2) = \tau = (E/R)I. \qquad (10)$$

The maximum stress occurs at the bone surfaces along the inside and outside of the curve. If the stresses there exceed the ultimate bending stress σ_u, which is also called the modulus of rupture, then the bone ruptures. We approximate the distance between the neutral surface and the surface of the bone along the inside curve to be the radius a of the bone. The maximum force that can be endured follows from (10) and (6) as

$$F_{max} = 4IE/LR_{min} = 4I\sigma_u/La, \qquad (11)$$

since, by (6),

$$\sigma_u = Ea/R_{min}.$$

Benedek and Villars[15] give the ultimate bending stress from Yamada[16] for wet human long bones as being approximately 2×10^8 N/m^2. If the bone is assumed to be a uniform cylinder 1 cm in radius and 20 cm in length, then $I = \pi a^4/4$ and $F_{max} = 3142$ N. The center of the bone is deflected by less than 1 cm.

This result for F_{max} means that, if a bone is supported at both ends and if our static analysis is approximately equivalent to a situation in which the force is quickly applied to the bone, a karate fighter could easily rupture the bone with a punch. Of course, in a fight a bone such as an arm bone is unlikely to be broken. Good targets are the ribs and, if the opponent has mistakenly placed a foot firmly on the ground, the lower leg of that foot. Since the bone in the above calculation is deflected by less than 1 cm, the bending of a bone is in itself no reason for a fighter to focus a strike several centimeters into the body.

In karate demonstrations boards, bricks, tiles, blocks of ice, and other materials are broken rather than bones. Let us calculate the force needed to break a pine board of dimensions $1 \times 20 \times 30$ cm and a brick of dimensions $4 \times 10 \times 20$ cm. The symbol I from (9) for rectangular objects is $Wh^3/12$, where h is the height and W is the width. The ultimate bending stress is about 7×10^7 N/m^2 for pine[17] and can be about 6×10^6 N/m^2 for brick.[18] The board and the brick are each considered to be supported at the ends of their long sides and with the smallest dimension vertical (Fig. 5). The results of (10) for rupture are 3111 N for the pine and 3200 N for the brick, each of which is less than the average force of a punch by a karate blackbelt as calculated above.

Fig. 5. Horizontal boards broken by a hand chop.

At this point we can see why a karate demonstrator always chooses to break several thinner boards rather than a single board of the same total thickness. With the multiple boards, rupture can proceed successively through the boards, each such rupture involving a smaller value of h in the calculation above than if a single thick board were used.

We can also understand why a karate fighter accelerates his striking foot, fist, etc., before contact is made. The total elapsed time for a forward punch is at least 20 times longer than the contact time, which means that the force accelerating the striking fist is at least 20 times less

Fig. 6. Vertical board broken by a forward punch. Note that the punch terminates about 6 cm after contact with the board.

than the impact force and hence is too small to break bones, boards, and bricks. Significantly damaging strikes can be made even if the attacking fighter does not push against the ground during impact if the striking fist, foot, etc., has already received a large acceleration.

Karate demonstrations in a physics classroom by a blackbelt can bring the students to their feet in anticipation of the violence of the boards and bricks breaking. Board breaks are easier if the boards are horizontal and have solid supports on their ends (Fig. 5). The strike should be with the grain of the wood. If students hold the boards vertically (Fig. 6), several students are normally needed to hold the board firmly enough. A single student usually flinches and yields when the board is struck, and an embarrassing amount of pain to the striking hand can result. Bricks should also be supported horizontally. To protect the hand from the rough brick surface, bricks are often wrapped in a thin cloth.

The breaks are best accomplished by chopping with the side of the hand. If the fist is used, the wrist should be kept straight upon contact to avoid wrenching the wrist, and contact should be made on the knuckles of the first and second fingers. Unless the skin over the knuckles has previously been toughened, such punches will probably break the skin somewhat, and although any bleeding will greatly heighten the drama of the demonstration, punching should be avoided for this reason.

Proper and intelligent caution should always be exercised in any breaking demonstration and any simulated karate fighting, and unless a teacher holds a blackbelt, he should not attempt one of these demonstrations himself. With the popularity of karate increasing, there is a fair chance some physics students will hold blackbelts and can give the demonstration. If not, an instructor from a local karate school could be invited. Since many blackbelts have only an intuitive feel for the physics behind karate strikes, they may be very interested in cooperation just so they can finally see the scientific basis of their fighting styles.

ACKNOWLEDGMENTS

I would like to thank Frank Stanonik and Professor Karl J. Casper for many helpful comments on this paper and Ed Celebucki for posing in the photographs. This work was supported in part by an Instructional Assistance Grant from the Center for Effective Learning at the Cleveland State University.

[1]J. Walker, *The Flying Circus of Physics* (Wiley, New York, 1975), problem 2.10.

[2]M. Nakayama, *Dynamic Karate* (Kodansha, Palo Alto, CA, 1966), p. 13.

[3]H. Nishiyama and R. C. Brown, *Karate. The Art of "Empty Hand" Fighting* (Tuttle, Rutland, VT, 1960), pp. 16–17.

[4]B. C. McInnis and G. R. Webb, *Mechanics, Dynamics: The Motion of Solids* (Prentice-Hall, Englewood Cliffs, NJ, 1971), pp. 74–75.

[5]J. S. Miller, Am. J. Phys. **22**, 409 (1954).

[6]A. Gray and J. G. Gray, *A Treatise on Dynamics* (Macmillan, London, 1920), pp. 396–399.

[7]Reference 2, p. 298.

[8]J. A. Vos and R. A. Binkhorst, Nature (Lond.) **211**, 89 (1966).

[9]D. I. Miller and R. C. Nelson, *Biomechanics of Sport* (Lea and Febiger, Philadelphia, PA, 1973), p. 26.

[10]Reference 2, p. 299.

[11]G. B. Benedek and F. M. H. Villars, *Physics. With Illustrative Examples from Medicine and Biology* (Addison-Wesley, Reading, MA, 1974), Vol. 1, p. 4-74.

[12]Reference 11, p. 4-75.

[13]G. R. Clements and L. T. Wilson, *Analytical and Applied Mechanics* (McGraw-Hill, New York, 1951), pp. 363–378.

[14]Reference 11, pp. 3-59 to 3-92.

[15]Reference 11, p. 3-88.

[16]H. Yamada, *Strength of Biological Materials,* edited by F. G. Evans (Williams and Wilkins, Baltimore, MD, 1970).

[17]F. F. Wangaard, *The Mechanical Properties of Wood* (Wiley, New York, 1950), p. 23.

[18]H. C. Plummer, *Brick and Tile Engineering Handbook of Design* (Structural Clay Products Institute, Washington, DC, 1950), p. 87.

The physics of karate

S. R. Wilk,[a] R. E. McNair,[b] and M.S. Feld
Department of Physics and Spectroscopy Laboratory, Massachusetts Institute of Technology, Cambridge, Massachusetts 02139

(Received 20 July 1982; accepted for publication 28 September 1982)

This study analyzes physical aspects of a karate strike and its interaction with a target, examining some heretofore unexplored dynamical and biomechanical questions.

I. INTRODUCTION

In recent years, the martial arts has experienced a revitalization and achieved worldwide popularity. It is extraordinary that as early as 2000 years ago ancient Oriental priests, warriors, and physicians devised systems of weaponless self-defense that make nearly optimal use of the human body. Many of the stances and techniques employed are totally different from the combat methods of the Western world and are often counterintuitive. The ways in which animal movements were studied and incorporated into exercise forms, and the gradual infusion of militaristic and religious doctrines which led to karate, tae kwon do, tai chi, and other Asian fighting arts are well documented.[1]

Karate itself developed in Okinawa in the early 17th century when the Japanese conquered the island and confiscated all weapons. In order to continue their feudal practices, the Okinawans developed a system of combat based on the weaponless Chinese fighting methods. The Japanese word karate means Chinese hand or empty hand. Today these ancient techniques, along with the associated rituals and mental discipline, are practiced throughout the world. Though some variations have been introduced in modern styles and methods, the fundamentals of karate techniques remain unaltered and the development of these techniques follows the same general scenario as in the ancient systems.

The perfection of basic techniques is the prime objective of the karateka, the practitioner of the martial arts. To this end, the repetition of a variety of striking, kicking, and blocking techniques is the major constituent of a karate training session, with emphasis placed on proper breathing, balance, and focus. The development of these skills is further enhanced through the study of the "kata," each one a series of prescribed blocks and counterattacks against a number of imaginary attackers, and through free-style sparring (kumite), wherein the elements of timing and mental conditioning are incorporated.

The renewed interest in the martial arts has been stimulated, in part, by demonstrations in which martial arts practitioners perform physical and mental feats which often defy one's reasoning and intuition, thereby leaving an observer in awe. The destruction of formidable piles of wood and concrete by bare hands, feet, or other parts of the body (Fig. 1) is a dramatic manifestation of the inherent destructive capability of the human body. Although the breaking of objects is often the most memorable and sensational aspect of a karate demonstration, it is *not* an inherent part of karate. Breaking is neither an objective of the karateka nor an item in regular training; it is used primarily to gauge and demonstrate the effectiveness of various techniques. It is in this spirit that this article studies breaking as a measure of the effectiveness of the martial arts as a means of self-defense.

Recently, the underlying physical principles of the martial arts have begun to be explored through the study of karate strikes and the breaking of objects. Walker[2] has examined kinematical aspects of a karate punch, and he and Blum[3] have analyzed the breaking process, treating the interaction of hand and block in the static approximation. Cavanaugh and Landa,[4] Vos and Binkhorst,[5] and Tuinzing and Fischera[6] have measured different aspects of karate strikes through the use of force plates, stroboscopy, accelerometry, and electromiography. The present study further analyzes physical aspects of a karate strike and its interaction with a target, examining some heretofore unexplored dynamical and biomechanical questions.

An analysis of the deflection and subsequent fracture of a slab of material leads to expressions for the stored internal energy, time, and force required for breaking in terms of the bulk properties of the target. These quantities are measured by means of static loading, force plate, and acoustical techniques. Trajectories, velocities, and peak energies of karate techniques are also studied using two- and three-dimensional stroboscopic techniques. High-speed movies of the impact of the striking part of the body with the target are used to obtain detailed information about the deceleration and deformation of the fist during the strike, and hence the forces acting on it, and the onset of fracture in the target. The transfer of the kinetic energy of a strike into elastic energy in the target is explored both by simple mechanical approximations and a more quantitative dynamic model. The latter gives good agreement with the data and provides insight into the forces acting in the hand–forearm system. The ways in which the striking part of the body, properly positioned, protects itself from damage are also discussed.

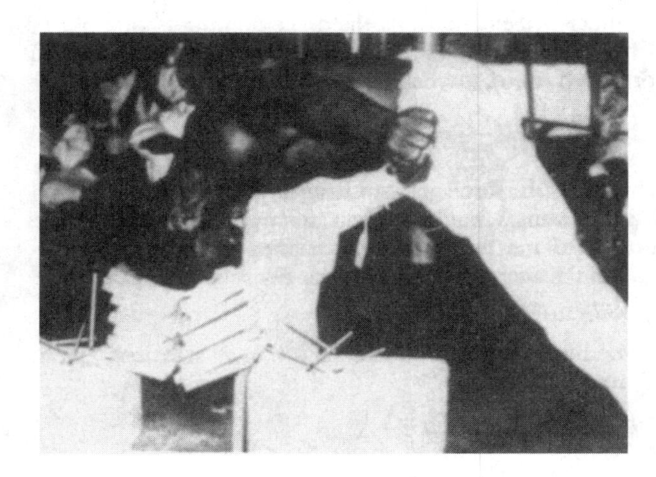

Fig. 1. Fracture of a stack of wooden boards using the forehead.

Reprinted from *American Journal of Physics* **51**, 783–790 (1983); © American Association of Physics Teachers.

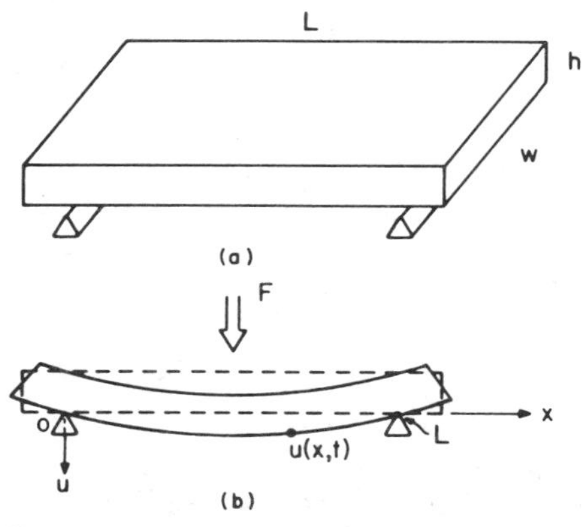

Fig. 2. Dimensions of blocks and their coordinates.

II. DYNAMICAL THEORY OF BREAKING

The blocks used in this study are rectangular bars of wood and concrete. Their dimensions l, w, h, and coordinates are defined in Fig. 2. x is the position along the length l. $u(x,t)$ is the deflection of the bottom surface from its equilibrium position.

The concrete bars are "patio blocks" made of type III cement and measuring $lwh = 40\times 19\times 4.0$ cm. The wooden blocks are pieces of dry white pine 1×12-in. shelving board measuring $28\times 15\times 1.9$ cm, cut so that the grain runs parallel to the width w. The patio blocks are dried in an oven for several hours to remove excess water, and thus improve the uniformity of the samples. In either case the blocks are supported at each end of their longest dimension, l, with their shortest dimension, h, vertical (Fig. 2). Two centimeters of overhang are allowed for at each end, reducing the effective length of the blocks by 4 cm. The mass of the blocks is 6.5 and 0.28 kg for concrete and wood, respectively.

The block is struck by the karateka at the top surface near the center of the span. For simplicity the force exerted by the striker is assumed to be evenly distributed along the line $x = l/2$, thus eliminating from consideration variations along w and reducing the problem to two dimensions. It is further assumed (and experiment confirms) that the deflection of the block is small compared to its overall dimensions.

Under these conditions the deflection $u(x,t)$ of a bar of density ρ and cross-sectional area $A = hw$ to an applied force is given, to good approximation, by the equation[7]

$$\rho A \frac{\partial^2 u}{\partial t^2} + EI \frac{\partial^4 u}{\partial x^4} = \mathscr{F}(x,t), \qquad (1)$$

where $\mathscr{F}$ is the force per unit length exerted on the bar and E is the Young's modulus of elasticity of the material. I, the moment of inertia for the cross section of the bar, depends only on its shape and dimensions. For a rectangular bar

$$I = h^3 w/12. \qquad (2)$$

For the case of free vibrations ($\mathscr{F} = 0$) solutions to Eq. (1) are of the form

$$u(x,t) = C \sin\left(\frac{n\pi x}{l}\right) \sin \omega_n t,$$

$$\omega_n = \left(\frac{n\pi}{l}\right)^2 \left(\frac{EI}{\rho A}\right)^{1/2}, \quad n = 1,2,\dots . \qquad (3)$$

If the impact occurs over a time interval $\gg 2\pi/\omega_2$ only the lowest mode ($n = 1$) will be appreciably excited, and the deflection is of the form

$$u(x,t) = \sin(\pi x/l) d(t), \qquad (4)$$

with d the time-dependent part of u. Equation (1) can then be reduced to a simple form by substituting Eq. (4), multiplying both sides by $\sin(\pi x/l)$, and integrating over x from 0 to l. Since the force is applied at $x = l/2$ the right-hand side of Eq. (1) is of the form $\mathscr{F}(x,t) = F(t)\delta(x - l/2)$. This yields

$$\ddot{d} + \omega^2 d = (2/\rho A l) F(t). \qquad (5)$$

Thus the motion of the block reduces to that of a harmonic oscillator of mass

$$m = \rho A l/2 \qquad (6a)$$

and spring constant

$$k = m\omega^2 = \frac{\pi^4 EI}{2l^3} = \frac{\pi^4}{24} \frac{h^3 w}{l^3} E. \qquad (6b)$$

The downward deflection of the block gives rise to stresses within it, the largest component, σ, being directed along l (x axis). This produces an elongation[8] along x, hence a strain, ϵ. The block fractures when, at some point x, σ reaches a critical value, σ_0, called the modulus of rupture. If the stresses along w and h are small compared to σ then the simple Hooke's law relation $\sigma = \epsilon E$ holds.[9] Using the well-known relationship between ϵ and u,[10]

$$\sigma = \epsilon E = Ey \frac{\partial^2 u}{\partial x^2}, \qquad (7)$$

where y is the distance of a point inside the block to the plane which bisects the block horizontally. Inserting Eq. (4) gives

$$\sigma = -(\pi/l)^2 Ey \sin(\pi x/l) d(t). \qquad (8)$$

The greatest stress occurs at $x = l/2$ and $y = \pm h/2$. Thus the fracture begins at the center of the surface of the block and rapidly propagates inward. In practice, the crack begins at the *lower* surface, since materials such as wood and concrete are stronger under compression than in stretching.[8,11]

The value of d at which fracture occurs may be obtained from Eq. (6):

$$d_0 = \frac{2l^2\sigma_0}{\pi^2 hE}. \qquad (9)$$

Using the value of the spring constant obtained in Eq. (6b), one can obtain expressions for the stored energy U_0 and the force F_0 exerted by the block at the point of fracture:

$$U_0 = \tfrac{1}{2} k d_0^2 = (\sigma_0^2/12E) hwl, \qquad (10)$$

$$F_0 = k d_0 = (\pi^2/12)(h^2 w/l)\sigma_0. \qquad (11)$$

These results are similar to those of Walker[2] and Blum,[3] obtained by assuming that the shape of the dynamically deformed block is the same as for static loading.[12]

In order to evaluate U_0 and F_0, values for the modulus of rupture and elasticity for our wood and concrete samples are needed. We have found it necessary to measure these parameters ourselves. In the case of concrete E and σ_0 vary considerably due to water content, age, and type of material.[11,13] As for wood, although tabulated values of the moduli for several different types are available (as in the USDA *Woods Handbook*), it appears that the parameters are measured for the wood stressed *along* the direction of the grain. They are therefore not applicable to our samples.

Table I. Breaking parameters for wood, concrete, and bone.

	Wood	Concrete	Bone
$E(10^8\ N/m^2)$	1.4 ± 0.5	28 ± 9	180[a]
$\sigma_0(10^6\ N/m^2)$	3.6 ± 1.0	4.5 ± 0.5	210[a]
d_0(mm)	16 ± 7	1.1 ± 0.4	11[b]
U_0(J)	5.3 ± 2.8	1.6 ± 0.6	14[b]
$F_0(10^2\ N)$	6.7 ± 1.9	31 ± 3.5	54[b]

[a] Representative values from Yadama, Ref. 15.
[b] Assumes a cylindrical bar of length 30 cm and radius 1 cm.[15]

The values of E and σ_0 were thus determined experimentally. A hydraulic testing machine was used to deflect the center of simply supported blocks identical to those used in the breaking studies. Graphs of the applied force f versus the peak deflection δ were made from which the moduli were determined making use of formulas of the type[14]

$$\sigma_0 = \tfrac{3}{2}(l/h^2 w) f, \quad (12a)$$

$$E = (l^3/48I\delta) f. \quad (12b)$$

The results are shown in Table I, along with the resulting values of d_0, U_0, and F_0.[15] For concrete our values for the moduli agree with the tabulated ones,[11,13] but for wood our values are considerably smaller.[16] For comparison the corresponding values for a long bone[17] are also given. As can be seen, the force needed to rupture such a bone is much greater than in the case of either wood or concrete, primarily because of the correspondingly larger value of σ_0.

As a check, the acoustical vibrations induced by lightly tapping the simply supported slabs with a pencil eraser were monitored by means of a small microphone. For the concrete blocks the lowest mode could be excited easily, yielding oscillations of period $\simeq 5$ ms which decayed in about 60 ms (Fig. 3). This is in reasonable agreement with the period of 6.6 ms $\pm 20\%$ predicted by Eq. (3). For the wood blocks a complex aperiodic acoustical pattern of peaks was observed, the largest ones being separated by 7–8 ms, as compared with the predicted value of 11 ms $\pm 20\%$. The nonsinusoidal behavior indicates that tapping unavoidably excites many modes, possibly including those when the slab responds as a plate rather than a bar,[7] and probably reflects the highly anisotropic elastic character of wood. The approximations made in the treatment above are better satisfied under karate strike excitation, where the duration of the impact is much longer so that the higher order modes are damped out during the interaction.

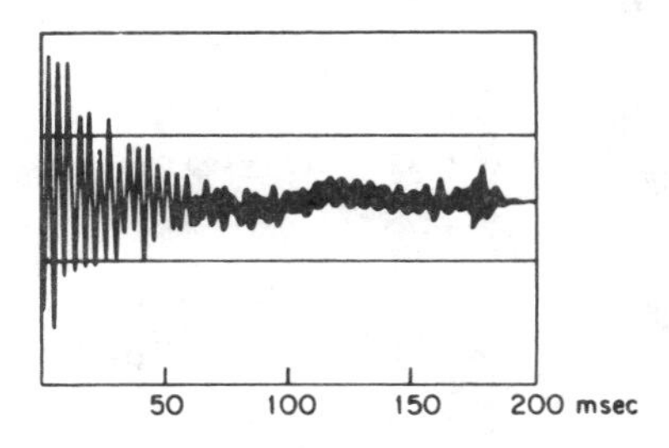

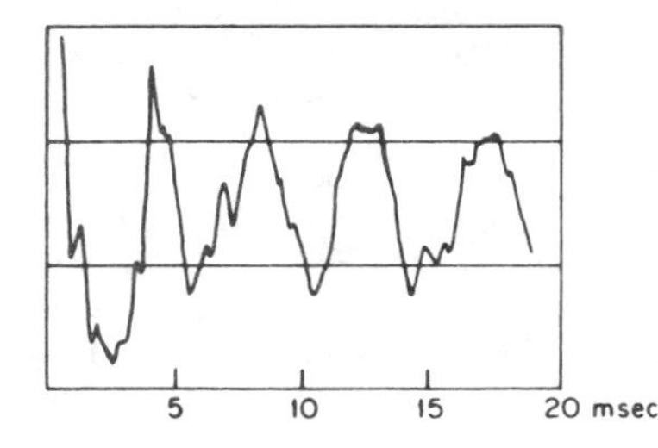

Fig. 3. Acoustical vibrations of a concrete slab, simply supported as in Fig. 2.

III. KINEMATICAL STUDIES OF KARATE TECHNIQUES

Since only a portion of the initial energy of the part of the body delivering the blow can be imparted to the target, the striker must have considerably more energy than U_0 if the block is to fracture. In order to determine the velocities, accelerations, etc. and the energies available to various techniques, strobe photographs were taken with an EEG model 553-11 Multiflash strobe lamp at 60 or 120 flashes/s. A variety of approaches was used: single exposures, acoustically triggered by the initial impact; multiple exposures; and sequential exposures using a rotating drum camera. In cases in which the motion was not in a plane, front and side views were recorded simultaneously by means of a large mirror placed to one side of the subject at 45° to the line of sight. The three-dimensional trajectory was then reconstructed using computer analysis,[18] from which peak velocities and other kinematical parameters could be determined.

As an example of the latter approach, Fig. 4(a) shows one frame of a drum camera sequence of a hammerfist strike (kensui-uchi), giving front and side views simultaneously. From this sequence the *net* speed of the hand as a function of time could be determined [Fig. 4(b)]. In this case a peak velocity of 10 m/s was reached at the point of impact. Other sequences with the same subject gave values as high as 14 m/s.

The results of the strobe studies are summarized in Table II. Peak speeds of 10–14 m/s were measured for hammer-

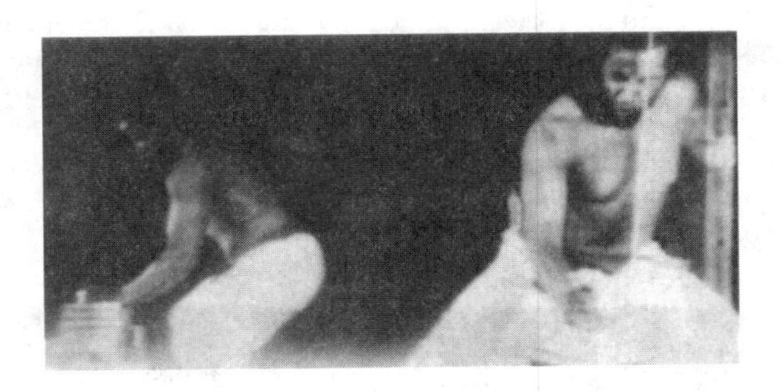

(a)

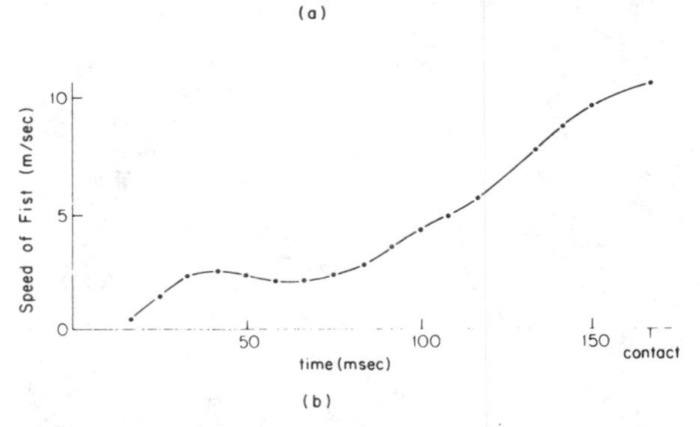

(b)

Fig. 4. Strobe photo analysis of hammerfist strike. (a) Single frame of drum camera sequence showing side and front views simultaneously. (b) Net speed of fist versus time, obtained from sequence.

Table II. Maximum speeds attained in various karate techniques, as measured from strobe photo studies of various subjects.

Technique	Maximum speed (m/s)
Front forward punch (seikan zuki)	5.7–9.8
Downward hammerfist strike (kentsui-uchi)	10–14
Downward knife hand strike (shuto-uchi)	10–14
Wheel kick	7.3–10
Roundhouse kick (mawashi-geri)	9.5–11
Back kick (ushiro-geri)	10.6–12
Front kick (mae-geri)	9.9–14.4
Side kick (yoko–geri)	9.9–14.4

fist and knife hand strikes (shuto-uchi), while somewhat smaller values were obtained for strikes in which the striking surface was the heel of the hand (shote-uchi). By way of comparison, forward punches (seikan-zuki) were found to reach speeds of 5.7–9.8 m/s, and various kicks 7.3–14.4 m/s.

Defensive techniques were also studied stroboscopically. Figure 5 shows a multiple strobe exposure of a downward block (gedan-barai), used to deflect an incoming blow to the lower portion of the body. As can be seen from the photo, maximum velocity is reached as the fist moves past the solar plexus region, well before the point of contact. This is different from offensive techniques, such as the hammerfist of Fig. 4 and the forward punch of Fig. 2 in Ref. 2, in which the peak velocity is reached at the point of contact. This feature is characteristic of blocks, whose purpose is to redirect an incoming technique and render it harmless rather than to meet it directly, a function which does not require developing large amounts of momentum.

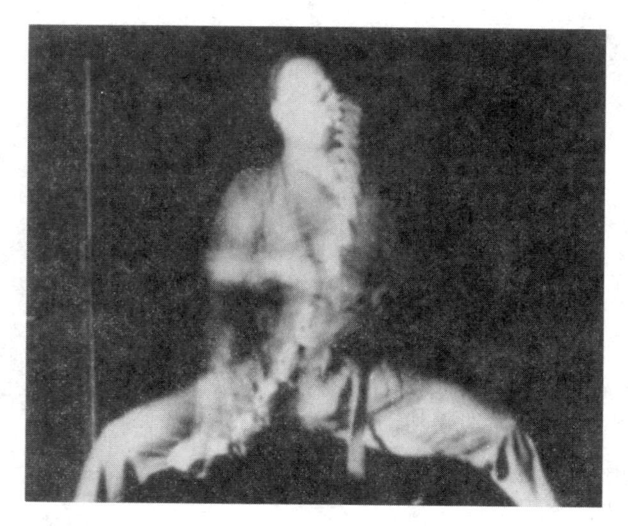

Fig. 5. Multiple strobe exposure of a downward block, 120 flashes/s.

The fist velocities measured in this study are comparable to those reported previously by other researchers. Vos and Binkhorst[5] measured velocities of 10–14 m/s in downward strikes, and Cavanaugh and Landa[4] reported angular velocities at the elbow of 24–29 rad/s. These agree with our measurements, 6–14 m/s and 25–27 rad/s, respectively. The time evolution of the forward punch recorded by Walker[2] is similar in shape and magnitude to those we measured.

IV. DETAILS OF THE IMPACT PROCESS

For a more detailed look at the collision between striker and target high-speed movies (1000 and 5000 frames/s) were taken. Figure 6 shows a sequence of frames of a hammerfist strike impinging on a concrete block. Successive frames are separated in time by 1 ms. These photos show the rapid deceleration of the fist on contact and its compression, and the subsequent initiation of the crack at the lower surface of the block and its rapid propagation upwards.

In analyzing such film strips the positions of the four marker dots on the fist were carefully recorded over a series of frames. The velocity and acceleration of each dot was then obtained by numerical differentiation. The results for the strike of Fig. 6 are shown in Fig. 7, in which is plotted the vertical position of the lower right dot and the corresponding velocity and acceleration. As the fist strikes the block it undergoes an extremely rapid deceleration, reaching a maximum value of 3500 m/s^2 for the lower right dot (Fig. 7) and up to 4000 m/s^2 for the other three. This brings the fist to a standstill, after which it continues downward. As seen in Fig. 7, the interaction lasts for 5 ms.

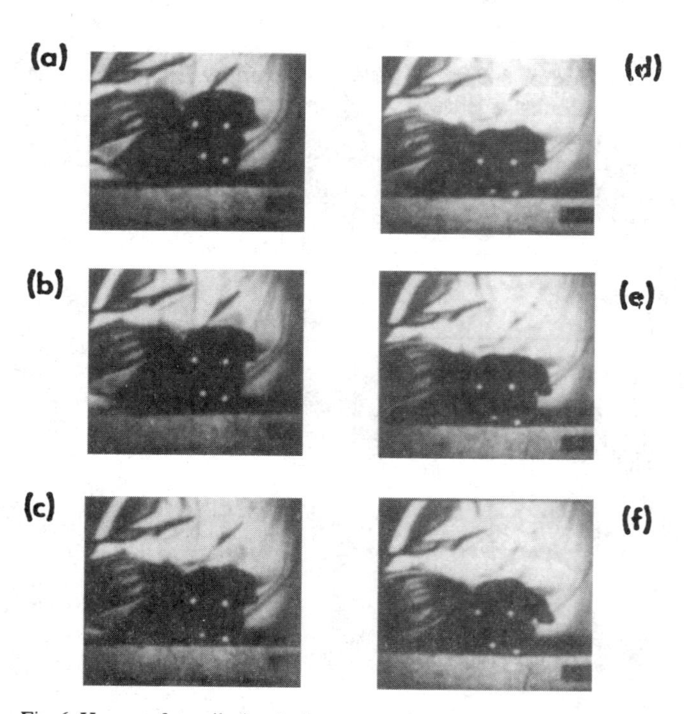

Fig. 6. Hammerfist strike impinging on a concrete patio block, high-speed movie film. Time between frames is 1 ms. The four dots are markers used as reference points. Although contact is made in frame (c), the crack does not develop until frame (e). Fracture occurs when the crack reaches the upper surface. Elapsed times for these frames are (a) 0, (b) 1, (c) 2, (d) 3, (e) 4, and (f) 5 ms.

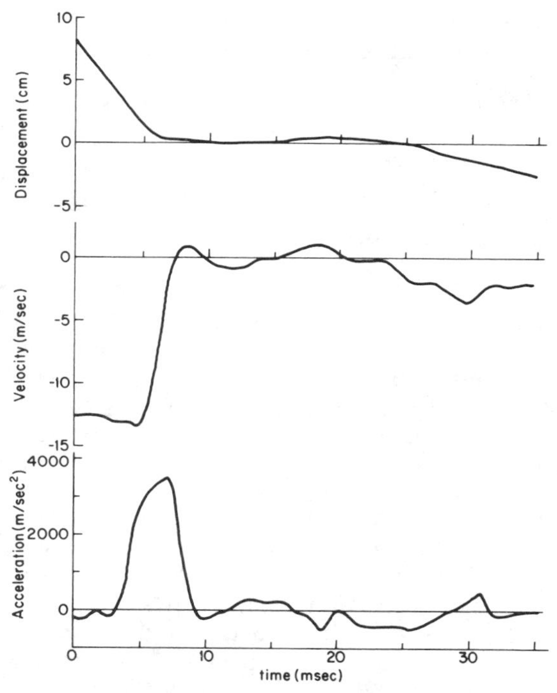

Fig. 7. Position, velocity, and acceleration of lower right-hand dot of fist from filmstrip of Fig. 6.

(a)

(b)

Fig. 8. Incomplete breaks of stacks of (a) wood and (b) concrete blocks. The technique in (a) is a hammerfist strike, and in (b) a palm–heel (shote) thrust.

This data may be used to estimate the peak force exerted on the fist during impact, given by the product of the hand mass and its acceleration. Applying the predictive formulas for the mass of body parts developed by Clauser *et al.*[19] to our subjects, we obtained an estimate of 0.7 kg for the mass of the fist. This yields peak forces in the range of 2400–2800 N.

Similar high-speed movies were also made using wood blocks, but decelerations when the fist strikes wood are very small and hard to measure.

Finally, as a check of the forces required to fracture the samples under actual conditions several blocks, supported at their ends on cinder blocks atop a force plate, were broken by karate strikes. The setup is essentially the same as that used by Vos and Binkhorst.[5] The force plate measures the force of the block pressing against the supports, hence the reaction force of the block (i.e., the product of its effective spring constant and displacement). The time dependence of the output signals, displayed on an oscilloscope, was about the same for both wood and concrete: The force rises rapidly over a period of 3–5 ms, after which the block either breaks or rebounds. For concrete, peak values as large as 2900 N were recorded when fracture occurred, with a mean peak value of 1900 N. The corresponding mean value for wood was 600 N. These are in reasonable agreement with the corresponding values of Table I. Interestingly enough, larger values were recorded when fracture did not occur, as large as 3600 N for concrete. Vos and Binkhorst[5] also observed larger peak values for unsuccessful breaks. This indicates (i) that the subject was capable of generating more than the required force for fracture, and (ii) that the force exerted by the striker is not the only important parameter—proper positioning of the hand or foot and its contact point along the length of the block are also critical.

The importance of making contact at the center of the target is graphically illustrated in the strobe photos of Fig. 8, which show incomplete breaks of stacks of (a) wood and (b) concrete blocks, separated by means of pencils at the ends. In each case all of the blocks were broken except for the bottom one, which remained intact. The subjects were capable of breaking the entire stack when the target was struck at the center, and failure to do so was attributed to the fact, evident in the photos, that contact occurred to the side of the target. This conclusion is further supported by the fracture pattern formed in successive blocks, which depicts the shock wave propagating towards the center of the stack. Photos such as these also reveal that the blocks fracture in succession, and that direct contact with the hand is not needed. This is particularly apparent in Fig. 8(a), where nine boards have been broken although the fist has only moved past the top two. Thus as a given block undergoes fracture it develops sufficient linear and angular momentum to fracture the block beneath it which, in turn, can fracture the next block. Huge stacks of material can be broken in this way.

V. COUPLING OF ENERGY INTO A TARGET

We seek a model that accounts for the behavior of the block and the hand during impact. As a specific example,

Table III(a). Initial velocities v_1 and kinetic energies U_1 needed to fracture slab. The uncertainties are due to variations among samples.

		Collision model					
		Elastic		Inelastic		Dynamic[20]	
		v_1(m/s)	U_1(J)	v_1(m/s)	U_1 (J)	v_1(m/s)	U_1 (J)
Slab material	Wood	5.2 ± 1.9	9.5 ± 5.0	4.3 ± 1.6	6.4 ± 3.4	6.1	12.3
	Concrete	2.8 ± 0.85	2.7 ± 1.0	5.0 ± 1.5	8.9 ± 3.3	10.6	37.1

the following discussion considers a downward hammerfist strike. In this technique the hand is tightly clenched and the block is struck with the bottom surface of the fist [Fig. 4(a)]. Consider first the rough approximation in which the interaction is assumed to be a simple mechanical collision—either elastic or inelastic. This approach necessitates choosing a value for the effective mass M of the hand–forearm system. High-speed films taken in side view indicate that in this technique the fist pivots at the wrist during impact, and that the resulting shock wave does not reach most of the forearm until after the block has fractured. Thus the appropriate value for M is the fist mass, ~0.7 kg.

The minimum initial kinetic energy, U_1, the hand must have in order to impart U_0 to the block is given by

inelastic collision: $U_1 = [(m + M)/M]U_0$ (13a)
(energy imparted to block-hand system),

elastic collision: $U_1 = [(m + M)^2/4mM]U_0$. (13b)

The effective mass m is 0.14 kg for wood and 3.2 kg for concrete [Eq. (6a)]. The resulting initial energies and velocities are given in Table III. The inelastic predictions for U_1 of 6.4 and 8.9 J for wood and concrete, respectively, appear to be reasonable (although perhaps too small), but the corresponding elastic values of 9.5 and 2.7 J do not, as they predict that the concrete targets are much easier to break than the wooden ones, in clear contradiction to experience. For the same reason both elastic and inelastic predictions using a hand–forearm mass of $M = 2$ kg must be rejected. These results therefore suggest that the interaction process is closer to inelastic than to elastic, with the effective striking mass being that of the fist alone, as anticipated. Elastic collisions are also ruled out by the high-speed movies, which show that the fist and block are in contact for several milliseconds, and thus appear to stick together.

Although the inelastic collision model is a useful first approximation, it cannot accurately describe the fist–block interaction process. For one thing, it assumes that the collision is completed before the block begins to be deflected, whereas the high-speed movies show that the fist and block are still interacting when fracture occurs. Another shortcoming of the inelastic model is that it gives no insight into the dynamics of the fracture process, nor about the internal forces acting on the hand–forearm system during the strike.

For a more accurate description of the impact process we used a dynamical model developed by Mishoe and Suggs[20] to describe the reaction of the hand to externally imposed vibrations. It is not *a priori* clear that this model, developed to describe the hand's response to vibrating machinery, can be extended to the much larger accelerations developed in a karate strike. Nevertheless, the results are encouraging.

This model (Table III) consists of three masses linked by springs and dampers. The first mass represents the skin, the second (M_1) the nearby hand tissue, and the third (M_2) the remainder of the hand mass (about 90%) and part of the forearm. The first mass is very small compared to the others and may safely be ignored in our application. Thus the hand is described by a pair of coupled linear differential equations of the harmonic oscillator type. The values of the relevant parameters are listed in Table IIIb.

In modeling the strike the hand oscillator system is incident on the block oscillator, described by Eq. (5), with a given initial velocity. Upon contact a system of three coupled oscillators is formed, as in the diagram of Table III. The coupled equations are solved by computer to obtain the deflection of the block as a function of time. In this way curves of maximum deflection as a function of the initial fist velocity are obtained (Fig. 9). The point at which the maximum deflection equals the critical deflection d_0, then gives the initial fist velocity required for fracture, v_1, from which the corresponding initial energy can be obtained (Table III). The relationship between initial velocity and maximum deflection is linear, and predicts U_1 values of 12.3 and 37.1 J for wood and concrete, respectively, somewhat larger than the inelastic values. The corresponding v_1

Table III(b). Parameters used in the dynamic model.

	Mass (kg)		Spring (10^4 N/m)		Damper (N s/m)	
Hand	M_1	0.0681	K_1	0.706	B_1	807
	M_2	0.595	K_2	1.16	B_2	275
Block						
Wood	m	0.14	k	4.34		
Concrete	m	3.2	k	295		

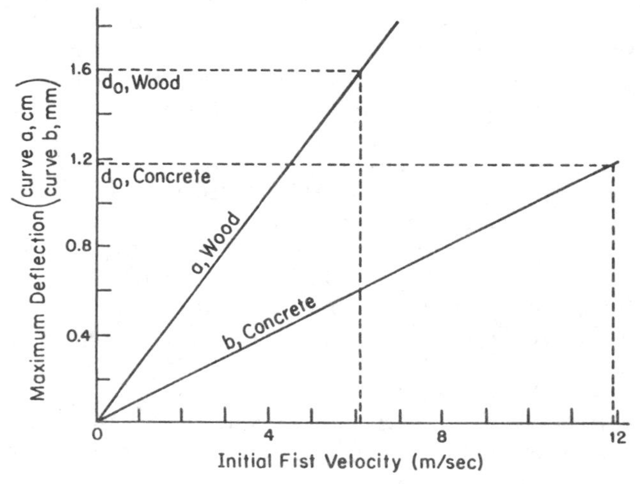

Fig. 9. Maximum deflection of block versus incident fist velocity, predicted by the dynamic model: (a) wood block; (b) concrete block.

values of 6.1 and 10.6 m/s agree with the common observation that beginners can break wooden boards but not concrete patio blocks: First velocities of 6 m/s are within range of a novice, whereas velocities in the 10 m/s range require training and practice.

Many of the details predicted by the dynamic model agree with the empirical data. For example, in the case of concrete the force on the hand rises rapidly, reaching peak values of up to 2200 N, and then falls quickly to zero. The block fractures in less than 5 ms, with peak deflections of about 1 mm. However, the predicted compression of the fist, somewhat under 2 cm, is too large. These predictions do not change greatly when the hand model is simplified by removing the last mass, M_1, and its associated spring and damper. Eliminating either the remaining spring (K_2) or damper (B_2), however, changes the results considerably and does not lead to reasonable values, even when the remaining spring or damping constant is varied over a wide range. It is interesting to note that in the limiting cases of either large spring constant or large damping constant the predicted value of U_1 approaches the inelastic value. However, this limit is not an accurate description of the actual situation.

These results suggest that with suitable modifications a model of the Mishoe–Suggs type can accurately describe the dynamics of the impact process, and provide useful information regarding the internal forces exerted on the hand and arm during the strike.

VI. BIOMECHANICAL ASPECTS

It is difficult to state a clear-cut criterion for the strength of a hand, foot, or other part of the body employed in a karate strike.[21] A hand or foot is a complex system of bones connected by visco-elastic tissue. As mentioned earlier, bone is a very strong material, its modulus of rupture being over 40 times larger than that of concrete (Table I). Thus a cylinder of bone 2 cm in diameter and 6 cm long, simply supported at the ends, can withstand in excess of 25 000 N of force exerted at the center.[15] This value represents a lower limit, since the bones in a striking hand or foot are neither simply supported nor impacted solely at the center. Under impact the bones can move and transmit part of the stress to the muscle and flesh in adjoining areas. Furthermore, some of the impact is absorbed by the skin and the muscles that lie between the point of impact and the bones.[22] In addition, much of the local stress is rapidly carried away to other parts of the body.

In the case of the hammerfist strike, for example, the adductor digiti minimi muscle protects the fifth metacarpal, the most vulnerable bone, from the blow. As the fist is tensed the adductor stiffens and thickens. The next line of defense is the set of tendons in the wrist, which cushions the blow as the fist is bent backwards. Finally, the energy transmitted to the arm is absorbed by muscles in the forearm and upper arm.

Proper positioning of the hand or foot is also critical. For example, in many techniques, such as the open hand strike or the side kick, contact is made with the edge of the hand (shuto position) or foot (knife foot). This concentrates the force in a small area of the target and reduces the likelihood of bending a bone to the point of fracture. Thus, if the striking part of the body is correctly oriented, the force required to break it is much greater than that needed to fracture the target.

To illustrate this point, consider breaking a concrete patio block with a side kick, where the foot is in the knifed position [Fig. 10(a)]. Bearing in mind the qualifications discussed above, a rough estimate of the relative strength of the foot as compared to the concrete slab can be made by considering the foot to be a rectangular block of bone of dimensions $l'w'h'$, the side of which strikes the central portion of the flat face of the slab [Fig. 10(b),(c)]. Equation (11) can then be used to compare the forces necessary to frac-

(a)

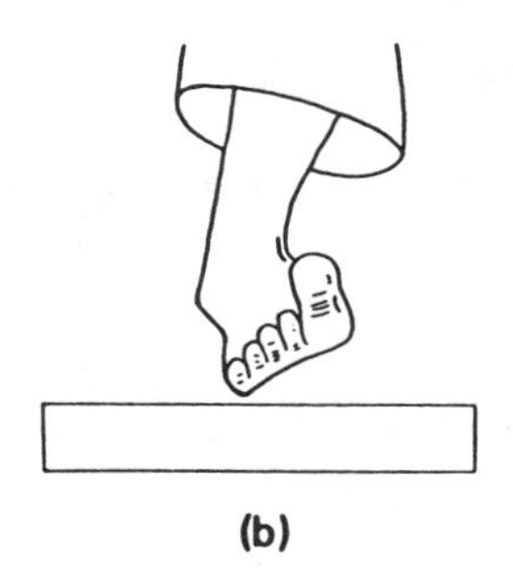

(b)

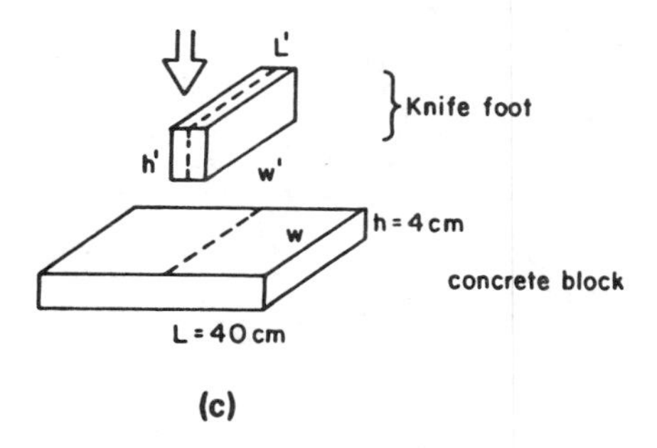

(c)

Fig. 10. Importance of orientation in striking a target. (a) Breaking a concrete patio block by means of a side kick. Note the knifed position of the foot. (b) Sketch of relative orientation of foot and block. (c) Model used in calculation. The knifed foot is approximated by a block of bone, oriented as shown.

ture the foot and the slab:

$$\frac{F'_0}{F_0} = \frac{h'^2 w'/l'}{h^2 w/l} \frac{\sigma'_0}{\sigma_0}, \tag{14}$$

where the primed quantities refer to the parameters of the foot. Since the foot is oriented so as to contact the face of the slab with its side surface [Fig. 10(c)], the primed dimensions l', w', h' should be identified as the thickness, length, and width of the foot, respectively. The length and thickness of the foot are comparable to the width and thickness of the concrete, but the width of the foot, typically about 8 cm, is only one-fifth the length of the block ($l = 40$ cm). Accordingly, we shall take $w' = w$, $l' = h$, $h' = l/5$. Since $h = 4$ cm, $l/h = 10$. From Table I, $\sigma'_0/\sigma_0 \simeq 40$. Accordingly,

$$\frac{F'_0}{F_0} \simeq \frac{1}{25}\left(\frac{l}{h}\right)^3 \frac{\sigma'_0}{\sigma_0} \simeq 2000. \tag{15}$$

Although this rough estimate is an oversimplification, the general conclusion is valid: A properly positioned foot can withstand forces several orders of magnitude larger than that needed to fracture a concrete slab!

ACKNOWLEDGMENTS

We are indebted to Charlie Miller of the MIT Strobe Lab for his invaluable aid in photographing the strikes, and to Ernie Cravalho, Woodie Flowers, and George Pratt of MIT, and Shelley Simons and Joseph Mansour of Boston's Children's Hospital for useful discussions and loan of equipment. Finally, we thank the members of the St. Paul A.M.E. Karate-Do for their patience and enthusiasm as willing subjects throughout the course of our research. This article is based in part on lecture–demonstrations presented at the 1976 and 1977 annual meetings of the American Association for the Advancement of Science. The authors gratefully acknowledge the encouragement and support of Rolf Sinclair, Secretary of AAAS Section B (Physics), who organized the sessions on "Science for the Naked Eye." This paper is dedicated to the memory of Kanwar Bidhi Chand.

[a] Present address: Department of Physics, University of Rochester, Rochester, NY 14627.

[b] Present address: NASA—Lyndon B. Johnson Space Center, Houston, TX 77058.

[1] An authoritative treatment is D. F. Draeger and R. W. Smith, *Asian Fighting Arts* (Kodansha International, Palo Alto, CA, 1969).

[2] J. D. Walker, Am. J. Phys. **43**, 845 (1975).

[3] H. Blum, Am. J. Phys. **45**, 61 (1977).

[4] P. R. Cavanaugh and J. Landa, Med. Sci. Sports **7**, 77 (1975).

[5] J. A. Vos and R. A. Binkhorst, Nature **211**, 89 (1966).

[6] K. Tuinzing and R. Fischera, Self-Defense World (March 1975).

[7] See, for example, P. M. Morse and K. U. Ingard, *Theoretical Acoustics* (McGraw-Hill, New York, 1968), Sec. 5.1; W. Goldsmith, *Impact: The Theory and Behavior of Colliding Solids* (Edward Arnold, London, 1960), Chap. III; and Ref. 10.

[8] Note that in the bending process the upper portion of the block is compressed, whereas the lower portion is stretched.

[9] The following analysis assumes a linear stress–strain relationship up to the point of rupture. Our static loading measurements show that this approximation is valid for wood, but that in concrete it breaks down as σ approaches σ_0. For concrete we obtained our value of E from the initial slope of the stress–strain curves, and σ_0 from the rupture point. See Ref. 18 for details.

[10] S. C. Crandall, N. C. Dahl, and T. S. Lardner, *An Introduction to the Mechanics of Solids*, 2nd ed. (McGraw-Hill, New York, 1972), pp. 423–24, 514.

[11] *Civil Engineering Handbook*, 4th ed., edited by L. C. Urquhart (McGraw-Hill, New York, 1959).

[12] The quasistatic analysis[2,3] leads to $F_0^{qs} = (8/\pi^2) F_0$ and $U_0^{qs} = (2/3) U_0$. {Blum,[3] using a geometrical approximation for the beam deflection [his Eq. (20)] instead of the static loading expression, obtains $U_0^B = (3/2) U_0$.}

[13] W. Spath, *Impact Testing on Materials* (Gordon and Breach, New York, 1961).

[14] Reference 10, Chaps. 7 and 8. The relationships are more complicated when the force is not applied exactly at $l/2$.

[15] These values were obtained using equations similar to Eqs. (9)–(11) but with $I = \pi a^4/4$, $a =$ radius of cylinder [cf. Eq. (2)], and substituting $y = a$ in Eq. (8) (rather than $h/2$). This gives $d_0 = l^2\sigma_0/\pi^2 aE$, $U_0 = (\sigma_0^2/16E)\pi a^2 l$, $F_0 = \pi^3 a^3 \sigma_0/4l$.

[16] See, for example, E. R. Parker, *Materials Data Book* (McGraw-Hill, New York, 1967).

[17] H. Yamada, *Strength of Biological Materials* (Krieger, Huntington, NY, 1970).

[18] S. R. Wilk, "A Biomechanical Study of a Karate Strike," S.B. thesis, MIT, June 1977 (unpublished).

[19] C. E. Clauser, J. T. McConville, and J. W. Young, "Weight, Volume, and Center of Mass of Segments of the Human Body," 1969 AMRL Technical Report, Yellow Springs, OH, Wright-Patterson Air Force Base (unpublished).

[20] J. W. Mishoe, "Dynamical Modeling of the Human Hand Using Driving Point Mechanical Impedance Techniques," Ph.D. thesis, North Carolina State University at Raleigh, 1974 (unpublished).

[21] A detailed discussion of the criteria under which a bone in the body will fracture is given in G. B. Benedek and F. M. H. Villars, *Physics with Illustrative Examples from Medicine and Biology* (Addison–Wesley, Reading, MA, 1974), Vol. 1, Chap. 3.

[22] R. M. Alexander, *Animal Mechanics* (University of Washington, Seattle, 1968); Chap. 4.

How can a downhill skier move faster than a sky diver?

Question:

Several years ago, *Ski Magazine* featured an article by the man who first broke the 200-km/h speed barrier for downhill skiers with a world record effort of 200.222 km/h (124.34 mph).[1] In that article, the speed of a skydiver at terminal velocity was quoted, for comparison purposes, as being 120 mph, prompting the question "How is it possible for a downhill skier to move faster than a skydiver?"

This question was asked by **Richard Parker**, *a physics student of* **Arthur Eisenkraft** *at Briarcliff High School, Briarcliff Manor, NY 10510. The answer given here was written by* **Angelo Armenti, Jr.** *who is Dean and Professor of Physics at Villanova University, Villanova, PA 19085. Dean Armenti has taught a course on the physics of sports and has written on the topic for* **The Physics Teacher**.

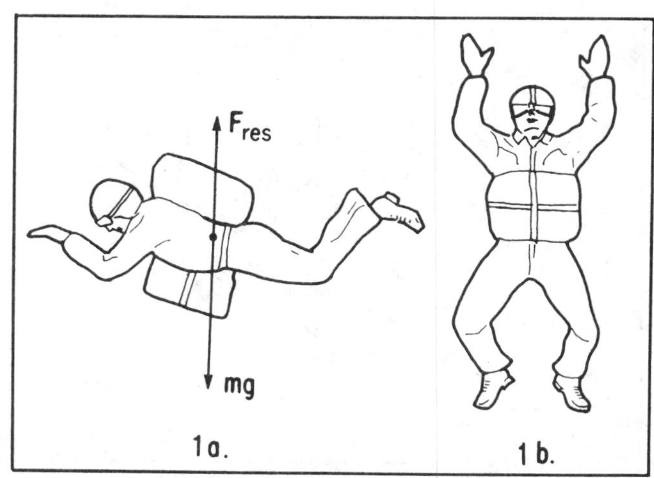

Fig. 1a. Side view of a skydiver in vertical fall. The forces acting on the skydiver are his weight $m\vec{g}$ and an air resistance force [Eq. (1)] proportional to the first power of his cross-sectional area A and the second power of his instantaneous speed v.

Fig. 1b. Front view of a skydiver in vertical fall. The skydiver is illustrated in the *spread eagle* position with cross-sectional area A.

Answer:

Qualitatively speaking, the skydiver assumes the familiar "spread-eagle" position as he falls and, consequently, presents to the air a relatively large area. As a result, he doesn't have to move very fast (comparatively speaking) to make the air resistance force equal his weight and hence achieve terminal velocity.

The speed skier crouches in a "tuck" position in order to present to the air the smallest possible area (typically less than *half* his area when upright). Consequently, the skier must move fairly fast to make the air resistance force equal the downhill component of his weight.[2]

Quantitatively speaking, it is necessary to consider the dynamics of each situation.

The skydiver:

Consider a skydiver of mass m and cross-sectional area A falling vertically with speed v as illustrated in Fig. 1a. The forces acting on the skydiver are assumed to be his weight $m\vec{g}$ and a resistive force of magnitude

$$F_{\text{res}} = -\kappa m A v^2 \quad (1)$$

in a direction opposite to the instantaneous velocity $\vec{v}$, where κ is a constant proportional to the *drag coefficient*.[3] From Newton's second law, we obtain for the skydiver

$$mg - \kappa m A v^2 = ma \quad (2)$$

where we have taken the downward direction to be positive. The left hand side of this equation is simply the net force on the skydiver as a function of his downward speed. At terminal speed, $a = 0$ and $v = v_T$ where

$$v_T \equiv \left(\frac{g}{\kappa A}\right)^{1/2} \quad (3)$$

The downhill skier

Consider a skier of mass m and cross-sectional area A' skiing down a slope of constant incline θ and coefficient of kinetic friction μ. The forces acting on the skier (Fig. 2a) are his weight $m\vec{g}$, a normal force $\vec{N}$, and a resistive force of magnitude

$$F'_{\text{res}} = -\mu mg \cos\theta - \kappa m A' v'^2 \quad (4)$$

in a direction opposite to the instantaneous velocity $\vec{v}'$. The first term in this equation represents the constant force of kinetic friction between skis and snow while the second term represents the variable force of air resistance encountered by the skier. As indicated by a comparison of Figs. 1b and 2b, the frontal area A' for the crouching skier is substantially smaller than the frontal area A of the skydiver.

From Newton's second law, we obtain for the skier

$$mg \sin\theta - \mu mg \cos\theta - \kappa m A' v'^2 = ma' \quad (5)$$

The left hand side of this equation is simply the net force on the skier as a function of his speed down the incline. At terminal speed, $a' = 0$ and $v' = v_T'$ where

$$v_T' \equiv \left(\frac{g'}{\kappa A'}\right)^{1/2} \quad (6)$$

and

$$g' \equiv g(\sin\theta - \mu\cos\theta) \quad (7)$$

is the effective acceleration due to gravity down the incline for the skier.

Discussion:

Using the definitions

$$\alpha \equiv \frac{A'}{A} \quad (8)$$

and

$$\beta \equiv \frac{v_T'}{v_T} \quad (9)$$

we see from Eq. (3) and (6) that the terminal velocity of the skydiver v_T will equal the terminal velocity of the skier v_T' whenever the right hand sides of these two equa-

Reprinted from *The Physics Teacher* **22**, 109–110 (1984); © American Association of Physics Teachers.

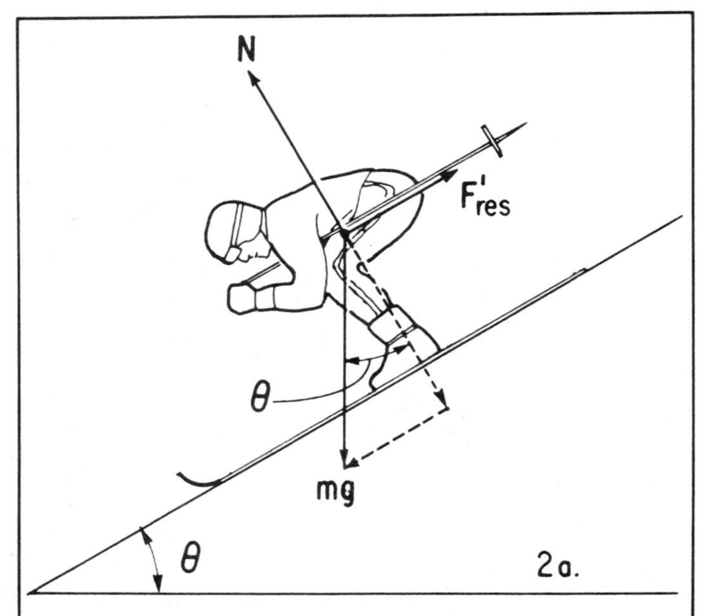

Fig. 2a. Side view of a downhill skier on a slope of constant incline θ. The forces acting on the skier are his weight $m\vec{g}$, a normal force $\vec{N}$, plus *two* resistance forces [Eq. (4)]. One resistance force is the constant force of kinetic friction between skis and snow and the other is the variable force of air resistance proportional to the first power of the cross-sectional area A' and the second power of the instantaneous speed v'.

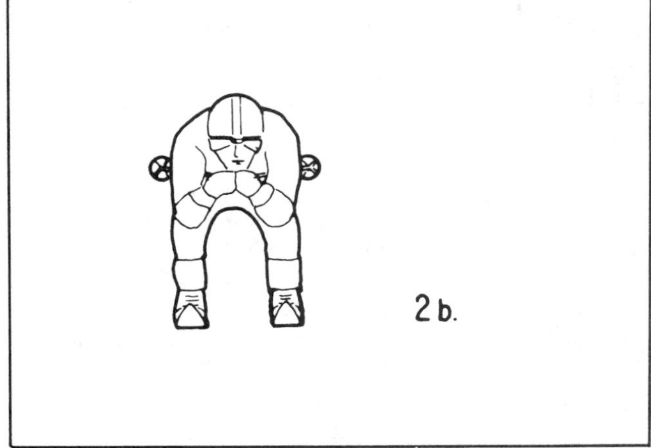

Fig. 2b. Front view of a downhill skier on a slope of constant incline θ. The downhill skier is illustrated in a *tuck* position with cross-sectional area $A' < A$.

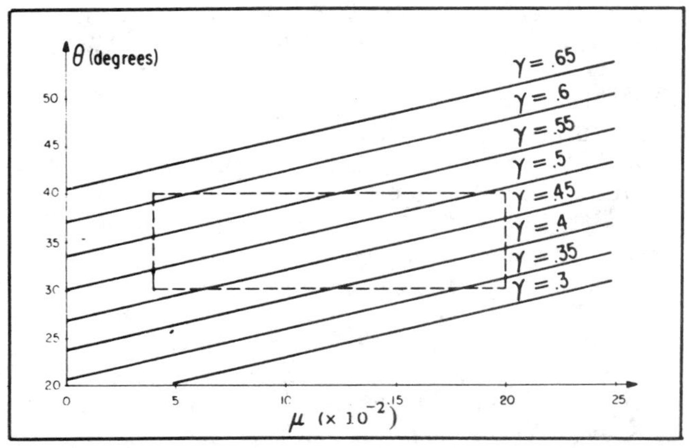

Fig. 3. θ versus μ for several different values of γ. The contours of slope versus coefficient of friction are shown for various values of γ [Eq. (11)]. The rectangular area depicted in the figure represents that portion of the θ, μ plane which is accessible to a real skier.

tions are equal, i.e., whenever

$$\left(\frac{g'}{\kappa A'}\right)^{1/2} = \beta\left(\frac{g}{\kappa A}\right)^{1/2} \quad (10)$$

with $\beta = 1$. Equation (10) may be written as

$$\sin\theta - \mu\cos\theta = \gamma \quad (11)$$

where $\gamma \equiv \alpha\beta^2$.

In Fig. 3 we plot Eq. (11) for several different values of γ. Typical values for the coefficient of kinetic friction between skis and snow range from $\mu = 0.04$ for dry snow at $0°$C to $\mu = 0.20$ for wet snow.[4] Typical values of slope at those places in the world where downhill ski records are attempted range between 30 and 40 degrees.[5] The rectangular area in Fig. 3 then indicates the ranges of slope θ and coefficient of friction μ available to a real skier. We see from Fig. 3 that, for a given ratio of areas α, there are many combinations of θ and μ which satisfy Eq. (11) and hence lead to equality of v_T' and βv_T. More important, a great many of these combinations fall inside the rectangle of physically accessible parameters thereby assuring that it is indeed possible (and in fact extremely likely) that a downhill skier would move as fast as ($\beta = 1$) and possibly even faster than ($\beta > 1$) a skydiver. The only thing needed to assure speeds for the skier equal to or greater than those of the skydiver is the right combination of slope θ, coefficient of friction μ, and fractional area α, i.e., any combination of θ, μ, and γ which satisfies Eq. (11) and falls inside the rectangle of Fig. 3.

As an example, if we assume a terminal velocity of $v_T = 53.65$ m/s (120 mph) for the skydiver, we find from Eq. (11) that on a slope with $\theta = 35°$ and $\mu = 0.09$, the skier will achieve the *same* terminal velocity as the skydiver ($\beta = 1$) provided only that $\alpha = 0.50$, i.e., provided that the skier crouches sufficiently to reduce his cross-sectional area A' to 50% of his upright area A. Furthermore, we can also determine from Eq. (11) that this same skier could achieve an even *greater* terminal velocity of $v_T' = 55.58$ m/s (which is the World Record of 124.23 mph) by simply decreasing his fractional area from $\alpha = 0.50$ to $\alpha = 0.466$.

References

1. Steve McKinney, "Getting Two," *Ski Magazine*, March, 1979, p. 33.
2. For an excellent discussion of the air resistance encountered by downhill skiers see A. E. Raine, "Aerodynamics of Skiing," *Science Journal*, March 1970, p. 26.
3. The force of air resistance on a moving object may be written as $F_{res} = -\frac{1}{2}C_D\,\rho\,v^2 A$, where v is the speed, A the cross-sectional area, ρ the density of air, and C_D is a dimensionless number called the *drag coefficient*. Upon comparison with Eq. (1), we see that $\kappa \equiv \rho C_D/2m$. For a discussion of the origin of the drag force see C. B. Daish, *The Physics of Ball Games* (The English Universities Press Ltd., London, 1972), p. 119. For a detailed description of the motion of an object under the influence of gravity plus a quadratic resistance force, see M. Peastrel, R. Lynch, and A. Armenti, "Terminal velocity of a shuttlecock in vertical fall," Am. J. Phys. 48, 511 (1980).
4. See for example *Handbook of Chemistry and Physics* (Chemical Rubber, Cleveland), p. 2170.
5. The author of reference 1 cites the course at Cervinia, Italy with its brief but spectacular 69-degree portion just before the speed trap. However, the world record was broken at Portillo, Chile on a course that begins with a 40-degree slope which gradually flattens out.

Why do downhill racers pre-jump?

Rob Hignell
Ryde School, Isle of Wight
Colin Terry
University of Wales, Bangor, Gwynedd, Great Britain

In a recent issue of *The Physics Teacher*, Armenti gave an interesting analysis of the motion of a downhill skier and compared it to the motion of a sky diver.[1] In this article we use elementary physics to show why, during a downhill race, skiers employ a technique known as "pre-jumping." Shortly before reaching the edge of a steeper section of the slope, the skier quickly gets up from the crouch position and pulls the legs upward. The result of this maneuver is that the skis leave the ground before reaching the steeper part of the slope. Why do skiers use this technique? We can get an answer to this question by analyzing a downhill course. For example, the ladies' downhill course at Val d'Isere in the French Alps has a vertical fall of 600 m over a length of about 2300 m. So the average angle of slope θ of this course is given by

$$\theta = \sin^{-1} \frac{600 \text{ m}}{2300 \text{ m}} = 15°$$

Suppose the skier comes to a place where the slope of the ski run changes abruptly from 15° to 20°. Figure 1 shows a skier approaching this part of the course. The Val d'Isere course is completed in about 85 s so the average speed $\overline{v}$ of the skier is

$$\overline{v} = \frac{2300 \text{ m}}{85 \text{ s}} = 27 \text{ m/s}$$

Suppose that in this case the skier does not pre-jump but maintains a steady position as she approaches the steeper part of the slope at a speed of 27 m/s. If she is in the air for a time t and lands at point A in Fig. 1, then if air resistance is ignored we have,

$$x = (27 \text{ m/s}) (\cos 15°)\, t$$

$$y = (27 \text{ m/s}) (\sin 15°)\, t + \frac{1}{2} (9.8 \text{ m/s}^2)\, t^2$$

We also know from Fig. 1 that

$$y = x \tan 20°$$

So

$$(27 \text{ m/s}) (\sin 15°)\, t + \frac{1}{2} (9.8 \text{ m/s}^2)\, t^2$$

$$= (27 \text{ m/s}) (\cos 15°) (\tan 20°)\, t$$

Thus

$$t = 0.51 \text{ s}$$

$$x = 13 \text{ m}$$

and

$$y = 4.8 \text{ m}$$

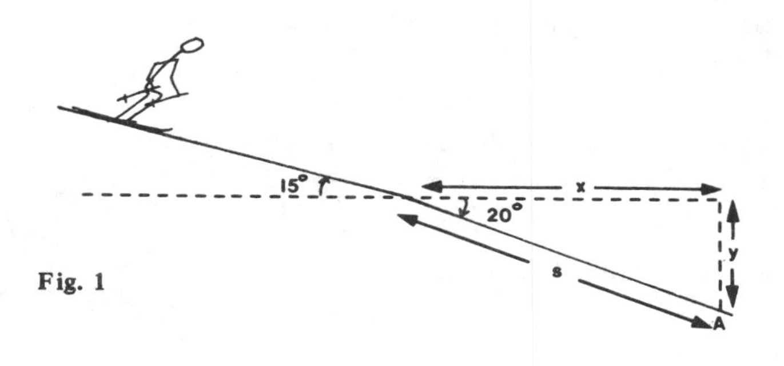

Fig. 1

This means that the distance s measured along the slope is

$$s = \sqrt{(13 \text{ m})^2 + (4.8 \text{ m})^2} = 14 \text{ m}$$

Immediately before landing the vertical component v_y of the skier's velocity is

$$v_y = (27 \text{ m/s}) \sin 15° + (9.8 \text{ m/s}^2)\, 0.51 \text{ s}$$
$$= 12 \text{ m/s}$$

The vertical component v_y' of the velocity after landing is

$$v_y' = v_x \tan 20° = 9.5 \text{ m/s}$$

So on landing, the change of vertical velocity Δv_y of the skier is 2.5 m/s. To calculate the force associated with this change of velocity we need to estimate the time over which the velocity change takes place. We suppose that the skier absorbs the impact by bending her knees, thereby lowering her center of gravity a distance of, say, 0.5 m. The average vertical velocity during the landing process is 1/2(9.5 + 12) m/s, so the time taken is

$$t = \frac{0.5 \text{ m}}{1/2\, (9.5 + 12) \text{ m/s}} = 0.04 \text{ s}$$

If the mass m of the skier is 50 kg, the vertical force F exerted on the skier's legs is

$$F = \frac{m\, \Delta v_y}{t} = \frac{(50 \text{ kg}) (2.5 \text{ m/s})}{0.04 \text{ s}} = 3 \text{ kN}$$

or about six times the skier's body weight.

A downhill racer does not want to risk being thrown off balance by such a landing force and in addition she

Reprinted from *The Physics Teacher* **23**, 487–488 (1985); © American Association of Physics Teachers.

wishes to minimize the time she travels out of control in the air. This is where the technique of pre-jumping comes in. By pre-jumping before reaching the discontinuity of the slope, the skier attempts to land immediately at the beginning of, and traveling parallel to the steeper section of the run (Fig. 2).

At the point of landing, a vertical component of velocity equal to (27 m/s) (cos 15°) (tan 20°) = 9.5 m/s is required. But at takeoff the vertical velocity is (27 m/s) sin 15° = 7.0 m/s. So the skier needs to increase her vertical velocity by 2.5 m/s using gravity. This takes a time t where

$$t = \frac{2.5 \text{ m/s}}{9.8 \text{ m/s}^2} = 0.26 \text{ s}$$

In order to land at the start of the steeper section of the slope the skier must take off at a horizontal distance x from the slope discontinuity where

$$x = (27 \text{ m/s}) (\cos 15°) (0.26 \text{ s}) = 6.8 \text{ m}$$

This corresponds to a distance s along the slope of

$$s = \frac{6.8 \text{ m}}{\cos 15°} = 7.0 \text{ m}$$

and to a vertical distance of

$$y = (6.8 \text{ m}) \tan 15° = 1.8 \text{ m}$$

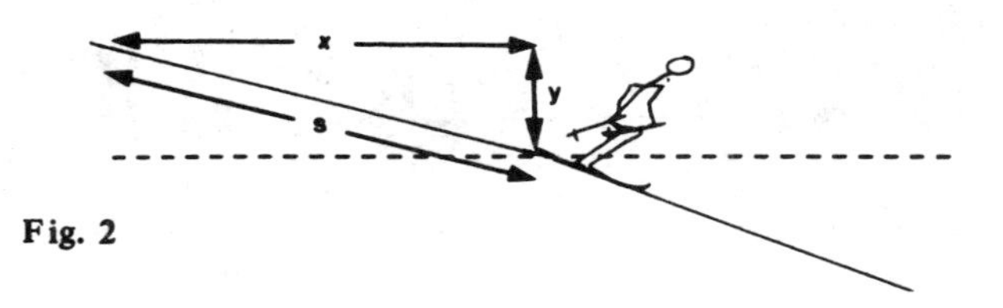

Fig. 2

But if the skier is in flight for 0.26 s she will fall a distance y' under gravity which is given by

$$y' = (27 \text{ m/s})(\sin 15°)(0.26 \text{ s}) + \frac{1}{2}(9.8 \text{ m/s}^2)(0.26 \text{ s})^2 = 2.1 \text{m}$$

So the skier must pull the skis upward a distance of (2.1 − 1.8) m = 0.3 m when she is at a point 7.0 m from the slope discontinuity. This is accomplished by standing up shortly before this point and pulling the legs up into a crouch position. When the skier pre-jumps in this way, she stays in better control. The skis are out of contact with the snow for 7.0 m compared with 14 m without pre-jumping. In addition, the landing force on the skis is reduced.

Reference

1. Angelo Armenti, Jr., Phys. Teach. 22, 109 (1984).

Physics and a Skiing Record

By Albert A. Bartlett

News stories often provide the ingredients of good physics problems. Consider this brief item from the sports page:[1]

> "Aspen Skier Sets Possible World Record."
> Ed McCaffrey of Aspen skied 234,000 vertical feet in a 24-hour period ending Friday morning in Keystone, to set what he hoped would be a Guinness world record.

Keystone, one of the major Colorado ski areas in the Rocky Mountains, is about 110 km west of Denver and has all the facilities of a major ski resort, including some lighted slopes for night skiing. Many good physics problems can be constructed from McCaffrey's skiing adventure.

McCaffrey made 100 rides up on the gondola and 100 runs down the mountain in 10 minutes less than 24 hours (85,800 s). The length of the gondola ride is 2829 m, and the average length of the downhill run on skis is 3.35×10^3 m. The elevations of the top and the base of the run are 3548 m and 2835 m, so the elevation loss of each run was 713 m. During the daylight hours (0800 to 1630), the gondola operated at a speed of 5.08 m/s, and the time for the ride was 557 s. During the night (1630 to 0800), the gondola speed was 4.00 m/s, and the ride time was 707 s. The mass of the skier plus equipment was 84 kg. The skier was in motion continuously; his only opportunity to rest was during the gondola rides. The skier took off his skis when he boarded the gondola and put on freshly waxed skis when he got off the gondola at the top of the mountain (Fig. 1).

Gross Calculations

The total distance traveled by the skier in the 100 round trips is

$$100\,(2829 + 3.35 \times 10^3) = 6.18 \times 10^5 \text{ m}$$

This is 618 km, for an average speed of 25.9 km/h or 7.20 m/s.

The total vertical distance traveled by the skier in 100 downhill trips is 7.13×10^4 m, which is more than eight times the elevation of the summit of Mt. Everest above sea level (8848 m).

The skier's loss of gravitational potential energy (PE) in one trip down the mountain is $mgh = 84 \times 9.80 \times 713 = 5.87 \times 10^5$ J, while the total loss of gravitational PE in 100 runs is 5.87×10^7 J. At the end of the 24 hours, the skier was at the same elevation as he was at the start. His net change in PE was thus zero. The gondola did 5.87×10^7 J of work on him, and he then lost this energy completely in skiing downhill.

The rate of loss of PE in the skiing—averaged over the 85,800 s—is $5.87 \times 10^7/85{,}800 = 684$ W. This must also be the average rate at which the gondola did work on the skier in getting him up the mountain. The rate at which the gondola did work (averaged over the actual time of a daytime uphill trip) is 5.87×10^5 J/557 s $= 1.054 \times 10^3$ W. This figure for the night trips is 5.87×10^5 J/707 s $= 830$ W.

The total work done on the skier by the gondola may be expressed in kilowatt-hours (3.6×10^6 J = 1 kWh) to give 16.3 kWh. If electrical energy costs $0.10/kWh, the cost of

Al Bartlett *received his B.A degree from Colgate University and his Ph.D. in nuclear physics from Harvard. He joined the faculty of the University of Colorado (Boulder, CO 80309-0390) in 1950 and retired at the end of 1987. In 1978, he was President of AAPT. He continues to be interested in applications of elementary physics.*

Reprinted from *The Physics Teacher* **28**, 72–76 (1990); © American Association of Physics Teachers.

Fig. 1. Ed McCaffrey (left) and C.J. Mueller (right) have just stepped off the gondola at the summit of Keystone Mountain (Colorado) and are putting on freshly waxed skis to start a run down the mountain. Photo courtesy of Keystone.

the work the gondola did on the skier is $1.63, much less than the cost of a lift ticket!

The effects of the variation of g with elevation are estimated in Appendix A. Calculations of the amount of coal or uranium needed to provide this energy are given in Appendix B.

How much snow at 0°C would melt to water at 0°C if all the lost gravitational PE were converted to heat and used to melt snow? The latent heat of fusion of water is 3.336×10^5 J/kg. The mass of water melted would then be 5.87×10^7 J/3.336×10^5 J/kg = 176 kg or about 7.4 kg/h.

As a very hypothetical problem, we could ask what would happen to the skier if all the heat equivalent of the gravitational PE were applied to the skier. Let us make the very approximate assumption that the thermal properties of the skier are the same as the thermal properties of an equal mass of water (84 kg). The specific heat of water is 4.18×10^3 J/kg C°, and the latent heat of vaporization of water at 100°C is 2.258×10^6 J/kg. The PE loss in a single run down the mountain, converted to heat and applied to 84 kg of water, would produce a temperature rise of $5.87 \times 10^5/(84 \times 4.18 \times 10^3) = 1.67$ C°. Thus the energy of 100 runs would be more than enough to cause 84 kg of water to boil. To raise 84 kg of water from the normal body temperature of 37°C to 100°C requires $4.18 \times 10^3 \times 84 \times (100 - 37) = 2.21 \times 10^7$ J. The PE available in 100 runs down the mountain is 5.87×10^7 J. There are thus $(5.87 - 2.21) \times 10^7 = 3.65 \times 10^7$ J available to vaporize water. The mass of water vaporized is thus $3.65 \times 10^7/2.258 \times 10^6 = 16.2$ kg.

At mountain elevations, water boils at temperatures lower than 100°C, and the latent heat of vaporization increases as the vaporization temperature decreases. Estimates of the effect of these are given in Appendix C.

Kinematic Calculations

The average incline of the gondola path is found by noting that the gondola rises 713 m in traveling 2829 m. Thus θ_g = Arcsin(713/2829) = 14.6 °. In the same way, the average incline of the ski path is θ_s = Arcsin(713/3.35 $\times 10^3$) = 12.3°.

When the gondola is moving with constant velocity, it exerts a force $mg = 84 \times 9.80 = 8.2 \times 10^2$ N on the skier. The share of the tension in the gondola support cable necessary to move the skier up the mountain at a constant velocity is $mg \sin \theta_g = 207$ N.

If the skier moved without friction on this constant "average" slope, his acceleration would be $g \sin \theta_s = 2.09$ m/s^2. The skier's velocity at the foot of the slope would be $\sqrt{2gh} = 118$ m/s, and the time required to make the run would be $t = 118/2.09 = 57$ s. The skier does not move with constant acceleration.

If it is assumed that the skier moves with constant speed down the "average" slope, the frictional force $\mu mg \cos \theta_s$ is equal to the accelerating force $mg \sin \theta_s$ and so the coefficient of friction $\mu = \tan \theta_s = 0.218$. This value is much higher than the tabulated values of 0.03 to 0.14 for skis running straight on snow in various conditions of speed, temperature, and wetness. This can be understood by noting that the skier is constantly turning as he comes down the slope. During these turns the skis are sliding sideways as well as moving parallel to their length. This sliding increases the average coefficient of friction and thus provides a deceleration, which the skier uses to keep his velocity from becoming too large. In addition, at these skiing speeds, air friction provides a substantial retarding force. Thus our calculated average coefficient of friction seems to be reasonable.

The skier starts and ends his run at rest. Air friction and friction of his skis on the snow is the principal mechanism by which the gravitational PE is lost. The "average" coefficient of friction is then the tangent of the angle of the ski slope.

A More Challenging Problem

Estimate the average speed of the skier on the downhill runs. If the gondola's speed was the same night and day, the problem would be elementary. For example, if it ran all the time at the daytime speed of 5.08 m/s, the time for 100 rides would add to 5.57×10^4 s, and the remaining time for skiing would be 3.01×10^4 s or 301 s/run. The average skiing speed on each of the 100 runs would then be 11.1

m/s. If the gondola ran all the time at the night speed of 4.00 m/s, the total ride time would be 7.07×10^4 s, which would allow 151 s for each run down for an average speed of 22 m/s. Note the enormous increase in the average skiing speed that is made necessary if the gondola speed is reduced and if we maintain the requirement that 100 round trips be completed in 85,800 s.

Our problem is made interesting by the fact that the gondola operates at the higher speed about one-third of the time and at the lower speed for two-thirds of the time. Before we solve this problem, let us define the following quantities.

N_D = number of gondola rides (day)
N_N = number of gondola rides (night)
T_D = gondola time up (day)
T_N = gondola time up (night)
T_S = average time to ski down
T_T = total time for the 100 complete round trips.

Our goal is to find the values of the three unknowns, T_S, N_D, and N_N. We will need three equations.

Our first equation represents the fact that the total number of round trips (RT) was 100.

$$N_D + N_N = 100 \qquad (1)$$

The day RT time is $(T_D + T_S)$ and the night RT time is $(T_N + T_S)$. Our second equation represents the fact that the sum of all the times is T_T.

$$N_D (T_D + T_S) + N_N (T_N + T_S) = T_T \qquad (2)$$

The number of day RT's in 8.5 hours of daytime operation is

$$N_D = (8.5 \times 3600)/(T_D + T_S)$$

and the number of night RT's in (15.5 hours − 10 min) is

$$N_N = (15.3 \times 3600)/(T_N + T_S)$$

Thus

$$\frac{N_D}{N_N} = \frac{8.5}{15.3} \frac{(T_N + T_S)}{(T_D + T_S)}$$

Our third equation is then

$$\frac{N_D}{N_N} = R \frac{(T_N + T_S)}{(T_D + T_S)} \qquad (3)$$

where $R = 8.5/15.3 = 0.554$. These three equations can be solved for N_D, N_N, and T_S. Doing this yields a quadratic equation for T_S of the form

$$AT_S^2 + BT_S + C = 0 \qquad (4)$$

where the coefficients are

$$A = N_T (1 + R) \qquad (5)$$

$$B = (1 + R) [N_T(T_D + T_N) - T_T] \qquad (6)$$

$$C = [(1 + R)(N_T T_D T_N) - (T_T)(T_D + RT_N)] \qquad (7)$$

Solving these three simultaneous equations gives the values $T_S = 211$ s, $N_D = 39.9$ runs, and $N_N = 60.1$ runs. In order to ski down the slope in time T_S, the skier's average speed is 15.9 m/s or 57 km/h. One can see that this is a high speed for a person on skis (Fig. 2). The rate at which the skier loses gravitational PE, averaged over the actual downhill skiing time of 211 s, is 5.87×10^5 J/211 s $= 2.78 \times 10^3$ W.

The consistency of these results may be checked by noting that the total time spent riding the gondola was 40 rides at 557 s plus 60 rides at 707 s or a total of 64,700 s. This left only 21,100 s to ski for an average of 211 s/run. The skier spent roughly 75 percent of his time riding the lift and only 25 percent of his time skiing. This indicates that the speed of the gondola is a major barrier for those who wish to try to break this record. For example, if the skier could continue to ski down in 211 s per run, and if the gondola ran all the time at the daytime speed, the skier could complete another 12 trips—which is another "Everest."

Fig. 2. Ed McCaffrey skiing down the mountain in his record-breaking series of runs. At the instant this photo was taken, both skis are in the air and the frictional force he is experiencing is due solely to air resistance. Photo courtesy of Keystone.

Uncertainties

The two largest uncertainties in these calculations are thought to be on the order of ± 1 percent in the total mass of the skier (plus equipment) and in the total distance traveled on skis. These give rise to uncertainties of the order of ± 2 percent or more in most of the calculated quantities and to an uncertainty of ± 5 percent in the final average skiing speed.

Conclusion

A single item from the sports page of the newspaper and a few additional data can provide a range of interesting and challenging problems. ♦

Appendix A

Would the decrease in the free-fall acceleration g with increasing altitude have an appreciable effect on any of the calculated results? Since g varies as r^{-2}, where r is the distance from the center of the Earth, a one percent increase in r will give a two percent decrease in g. The tabulated value of g at Denver (alt 1638 m) is 9.79609 m/s^2. The mean radius of the Earth is 6.371 $\times 10^6$ m, so r (Denver) is 6.373 $\times 10^6$ m. The mid elevation of the ski run is 1553 m higher than Denver. Thus $\Delta r/r$ = 1553/6.373 $\times 10^6$ = 2.44 $\times 10^{-4}$. So the fractional change in g is a decrease of 5 $\times 10^{-4}$. This is negligible compared with other uncertainties.

Appendix B

Let us estimate the quantity of fuel required to generate 5.87 $\times 10^7$ J of energy needed to lift the skier 100 times up the mountain.

If coal is the fuel used to generate the electrical energy, the amount of coal required is

$$\frac{5.87 \times 10^7 \text{ J}}{2.9 \times 10^7 \text{ J/kg} \times 0.34 \times 0.9 \times 0.8}$$

which gives 8.3 kg. The heat of combustion of coal[2] is 2.9 $\times 10^7$ J/kg. The assumed value of the thermodynamic efficiency of the steam plant[3] is 0.34; the assumed efficiency of transmission of the electrical energy is 0.9; and the assumed efficiency with which the electrical energy is converted to mechanical energy is 0.8.

Here we see that the consumption of 8 kg of coal produces 16 kWh of electrical energy. It is a rough rule of thumb that about half a kilogram of coal is required to produce one kWh of useful energy for the consumer.

Suppose that the electrical energy comes from the fission of atoms of U^{235} in a nuclear power plant. Assume that the energy released per fission is 200 MeV,[4] (3.20 $\times 10^{-11}$ J). Assume that the efficiencies of generation, transmission, and utilization are the same as those used in the coal calculation. Then

$$\frac{5.87 \times 10^7 \text{ J}}{3.20 \times 10^{-11} \text{ J/f} \times 0.34 \times 0.9 \times 0.8}$$

gives 7.48 $\times 10^{18}$ fissions. This is the number of atoms of U^{235} that are "consumed."

Let us continue our calculation in order to relate it to normal uranium, which is mostly U^{238}. The fraction of the atoms of normal uranium that are U^{235} is f = 0.007196.[5] Assume that normal uranium consists only of the two isotopes U^{235} and U^{238}. Then the number of U^{238} atoms that will accompany the calculated number of U^{235} atoms is

$$7.48 \times 10^{18} (1 - f)/f$$

which is 1.03 $\times 10^{21}$ atoms. The masses of these numbers of atoms can be calculated. Note that 1.0 $\times 10^{-3}$ kg of atoms of atomic mass 1.0 AMU would contain Avogadro's Number of atoms. Thus 1 AMU = 1.661 $\times 10^{-27}$ kg.[6] The atomic masses in AMU[7] are 235.04394 and 238.05082. The masses of the two isotopes are then

$$7.48 \times 10^{18} \times 235.04394 \times 1.661 \times 10^{-27} = 2.92 \times 10^{-6} \text{ kg of } U^{235}$$

and

$$1.03 \times 10^{21} \times 238.05082 \times 1.661 \times 10^{-27} = 4.03 \times 10^{-4} \text{ kg of } U^{238}$$

The total mass of normal uranium needed is the sum of these or 4.06 $\times 10^{-4}$ kg.

Appendix C

A more realistic solution to the problem of what would happen to the skier if the heat equivalent of the gravitational PE were applied to the skier involves

a) interesting physics
b) looking up data in a handbook
c) interpolating within the handbook data

These three things may no longer be stressed as they once were in introductory physics.

Originally, we calculated the result of applying 5.87 $\times 10^7$ J of heat to 84 kg of water at 37°C and at the sea-level pressure of one atmosphere. We found that the water was heated to 100°C and that 16.2 kg of water was then vaporized. We now ask what happens if the experiment is repeated at the elevation of 3191 m, which is approximately the midpoint of the ski run? This involves four steps. We must

1) Find the normal atmospheric pressure at this elevation.
2) Find the boiling temperature of water at this elevation. That is, at what temperature is the vapor pressure of water equal to the atmospheric pressure at this elevation?
3) Calculate the joules of heat energy needed to raise the temperature of the water from 37°C to this lower boiling temperature and then calculate the joules that are left over to vaporize the boiling water.
4) Look up the latent heat of vaporization of water at this reduced temperature and use it to calculate the kilograms of water that are vaporized. Very often the handbooks give data in the old c.g.s. system. At some point these must be converted to SI to complete the calculations.

Here are the calculations. From the table of the "Variation of Temperature, Pressure, and Density of the Atmosphere with Elevation,"[8] we find for water

3000 m	519.7 mm Hg
4000 m	455.9 mm Hg

We interpolate linearly to estimate the atmospheric pressure at 3191 m.

3191 m	507.5 mm Hg

In a table of "Vapor Pressure of Water below 100°C,[9] we look for the temperature at which the vapor pressure of water is 507.5 mm Hg. We find

89.0°C	506.1 mm Hg
89.2°C	510.0 mm Hg

A linear interpolation gives 89.1°C as the temperature at which water has the vapor pressure of 507.5 mm Hg. Thus we estimate the boiling temperature of water at this elevation to be 89.1°C. (We should note that the data of this table are the coordinates of the liquid-vapor line of the triple-point diagram for water.)

The joules of heat energy needed to raise 84 kg of water from 37.0°C to 89.1°C can be calculated

$$4.18 \times 10^3 \times 84 \times (89.1 - 37.0) = 1.83 \times 10^7 J$$

This is subtracted from the available 5.87×10^7 J to leave 4.04×10^7 J available to vaporize water.

We now go to a table of the "Properties of Saturated Steam."[10]

90°C	544.9 (15°)cal/g
89°C	545.6 (15°)cal/g

A linear interpolation gives 545.5 cal/g as the latent heat of vaporization of water at 89.1°C. This must be multiplied by 4.185×10^3 to convert it to J/kg. This gives 2.283×10^6 J/kg. The quantity of water vaporized is $4.04 \times 10^7 / 2.283 \times 10^6 = 17.7$ kg. At sea level we found that 16.2 kg of water were vaporized. At the mountain elevation the quantity is 17.7 kg. This is a nine percent increase.

Acknowledgements

I am greatly indebted to Barbara Keller, public relations coordinator at Keystone, for supplying the photographs and many of the data that are necessary for the calculations. This skiing event was part of a competition to benefit the Jimmie Heuga Center.

References

1. Boulder *Daily Camera*, Dec. 5, 1987.
2. R.H. Romer, *Energy Facts and Figures* (Sprint Street Press, Amherst, MA, 1985).
3. *Physics Vade Mecum*, edited by H.L. Anderson (American Institute of Physics, New York, 1981), p. 175.
4. H.A. Enge, *Introduction to Nuclear Physics* (Addison-Wesley, Reading, MA, 1966), p. 440.
5. C.M. Lederer, J.M. Hollander, and I. Perlman, *Table of Isotopes*, 6th ed. (John Wiley & Sons, New York, 1967), p. 142.
6. Ref. 3, p. 4.
7. Ref. 2, p. 21.
8. *Handbook of Chemistry and Physics*, 26th ed. (Chemical Rubber Publishing Co., Cleveland, OH, 1944), p. 2462.
9. *Handbook of Chemistry and Physics*, 44th ed. (Chemical Rubber Publishing Co., Cleveland, OH, 1962), p. 2422.
10. Ref. 9, p. 2548.

Chapter 8
FLUIDS AND SPORTS

CONTENTS

The first article in this chapter is "Aerodynamic effects on discus flight," by Cliff Frohlich. In it the author reviews the available literature on the physics of the discus,[1] and analyzes its flight under the combined influence of gravity, aerodynamic drag, and aerodynamic lift. He then calculates differences in the horizontal range of a well-thrown discus caused by changes in wind velocity, altitude, air temperature, gravity, and release velocity.

As to relative magnitudes, Frohlich notes that:

> a discus can travel: (i) 8.2 m farther against a 10-m/s wind than with such a wind; (ii) 0.13 m farther at 0 °C than at + 40 °C; (iii) 0.19 m farther with no wind at the elevation of Rome, Italy than at the elevation of Mexico City, Mexico; and (iv) 0.34 m farther at the equator than at the poles.

The largest of these, the calculated effect of the wind on the flight of a discus, is notable for two reasons: (1) the reported differential (8.2 m) is more than 10% of a world class discus throw, and (2) unlike most other track and field events, the discus throw is one event in which the athlete is helped rather than hindered by a head wind!

This ironic state of affairs stems from the fact that, while drag and lift are both increased by the presence of a head wind, the increased lift more than makes up for the increased drag, thereby resulting in additional range, at least for head wind velocities of 20 m/s or less. Also, due to aerodynamic factors, the launch angle for maximum range for a discus is about 35° as compared to the familiar 45° for projectiles in a vacuum.

Finally, in a theme familiar from Paul Kirkpatrick's 1944 article, "Bad physics in athletic measurements" (see Chap. 1), Frohlich notes that:

> Considering the importance of the aerodynamic forces affecting discus flight, it is surprising that record books make no distinctions between marks recorded in still air and wind-aided marks.

When, back in 1944, Kirkpatrick lamented the measurement procedures for the discus throw, his concern was based not on aerodynamics but on the fact that Coriolis effects were not taken into account. These effects, due to the rotation of the Earth and resulting from differences in launch direction (e.g., East vs West), lead to rather minuscule range differentials (at least as compared to 8.2 m) on the order of 5/16 of an inch for a throw of 200 ft, a difference of only one-hundredth of one percent.

The main point of Kirkpatrick's article was that Coriolis and other effects should be included as corrections to the athletic measurement process simply because their magnitudes, although typically small in absolute terms, nevertheless exceeded the "ostensible precision" of the recorded measurements. By showing that aerodynamic effects can produce range differentials a thousand times larger than those associated with Coriolis effects, Frohlich's analysis raises Kirkpatrick's original concerns about the adequacy of athletic measurement procedures to a new level.

While the first article in this chapter deals with the drag and lift on a discus due to a *single* fluid, air, the second article, "The study of sailing yachts," by H. C. Herreshoff and J. N. Newman, deals with the drag and lift on a sailboat due to *two* fluids, air and water.

As the authors note, the challenge of the naval architect is:

> to design a wind-driven craft that will move through both water and air with high efficiency.

And while that challenge has been aided by both experimental towing tanks and the rich history of empirical design gains flowing from centuries of "practical experience, shrewd observation and intuition" on the part of sailors around the world,[2] it has been hindered too by theoretical complexities resulting from the combined action of two dissimilar fluids, air and water, on the sails and hull of a single entity.

In discussing the fundamental differences between "running before the wind" (i.e., sailing in the same direction as the wind) and "sailing to windward" (i.e., in a direction typically within 20° to 40° of directly into the wind), the authors note that:

> In running before the wind the sails act simply as an aerodynamic drag to the wind; it follows that the boat speed will be less than the wind speed. The only relevant hydrodynamic force exerted by the water on the hull is a drag, which obviously should be minimized for maximum speed. In sailing to windward, on the other hand, the sails and hull both act as wings: lifting surfaces that meet the oncoming wind and water at various angles of attack and develop not only adverse aerodynamic and hydrodynamic drag forces but also favorable lift forces.

Also of note is the authors' assertion that:

> In principle there is no limit to the boat's speed in sailing to windward, particularly because the relative wind speed is increased by the boat's own speed.[3]

Although the truth of this last statement has been well documented, an understanding of why it should be so requires analysis of the forces acting on a sailboat, an exercise which is the subject of the next article in this chapter, and a topic which is discussed below in some detail.

Herreshoff and Newman make a strong analogy between the *fuselage* of an airplane and the *hull* of a sailboat:

> The mechanism by which a hull generates a horizontal lift is identical with that producing vertical lift in an airplane. The main body of a sailing-yacht hull, like the fuselage of an airplane, is in itself an inefficient lifting surface; lift is provided by a thin keel or retractable centerboard, which functions as a wing.

One of the key physical aspects of sailing to windward involves the balance between aerodynamic and hydrodynamic forces and turning moments (torques) on the boat. The force of the wind on the sails produces a substantial heeling moment which must be balanced by an equal and opposite restoring moment provided by the boat's ballast system.[4]

Towing tanks have been used in the search for efficient sailing hulls for more than 60 yr, principally as laboratories for measuring the forces on scale models. The usual strategy is to measure the actual forces on a scale model, and then to infer what the corresponding forces would be on a full-scale vessel. The benefit of this strategy is that it permits evaluation of different hull designs without the expense of constructing a full-scale vessel for each design. Its success, however, depends on the ability to accurately scale the forces, a task of some challenge.

The drag force on any hull is known to arise from two different sources, friction and wavemaking.[5] Friction resistance occurs because a moving hull sets into motion adjacent layers of water which were initially at rest. This requires the hull to exert a force on the water in those layers (Newton's second law) and this, in turn, requires the water in those layers to exert an equal and opposite force (frictional resistance) on the hull (Newton's third law).

Wavemaking resistance arises from the fact that the submerged portions of a moving hull create pressure changes below the surface that, when combined with the force of gravity, result in an irregular shape for the surface of the water, i.e., water waves. In effect, wavemaking resistance results directly from the work which the moving hull must do against gravity in producing these waves.

The scaling challenge referred to above occurs because, as the authors' note,

> When a hull is scaled down to model size, there is no way to scale down in equal degree the frictional effects and wave effects.

The reason for this is that frictional effects are scaled according to the *Reynolds* number which, for a given fluid, is proportional to vL, the product of boat speed and hull length. In order for the scaling to work, the full-scale vessel and the model must have the *same* Reynolds number. Hence, proper scaling requires that the speed of the model being tested be inversely proportional to its length.

The effects of wavemaking resistance, however, scale according to the *Froude* number which, for a given value of the acceleration due to gravity, is proportional to $v/L^{1/2}$. In order for the full-scale vessel and the model to have the same Froude number, insuring that the wave effects will scale properly, the speed of the model must be proportional to the *square root* of the model length, a requirement which is clearly incompatible with that for frictional scaling.

Although approximation methods have been developed for dealing with this situation, the authors report instead on a novel experiment in which a full-scale sail boat was tested in a very large towing tank, thereby obviating any need for scaling considerations. The authors report that the results obtained in this way were "substantially the same" as predictions based on scale models with similar hulls.

The third article in this chapter, "The physics of sailing," by Peter Rye, nicely complements the previous article by Herreshoff and Newman, first by focusing on the balance of aerodynamic and hydrodynamic forces and moments and second, by revealing the secret of sailing "close to the wind."

The ability of a sailboat to sail close to the direction of the oncoming wind is a desirable trait. Most boats can do so comfortably to within about 40° or so, while certain types of boats are capable of coming as close as 20° or less. By means of a simple force diagram analysis, Rye derives an important theorem which states that the *minimum angle for sailing to windward is precisely equal to the sum of the sail drag angle and the hull drag angle.* In effect, as sail drag and hull drag are reduced, the ability to sail close to the wind is increased.

The importance of this theorem can be seen in sailboat racing where great effort goes into designing hulls and sails having minimal drag. Since sailboats cannot sail directly into the wind, the "windward leg" of a race is negotiated by means of the familiar zig-zag path associated with "tacking" or, as it is also known, "beating to windward." Clearly then, reducing sail and hull drag produces two distinct benefits for sailboat racing: first, lowering the drag *increases the speed* of the boat and second, the associated reduction in the minimum angle for sailing to windward *decreases the distance* to be sailed by allowing the various zigs and zags to lie closer to the direction of the oncoming wind.

In regards to sailing "off the wind," the author describes two ways to compensate for large mainsail turning moments without resorting to the rudder and the unwanted drag which accompanies its use. On large boats, a spinnaker sail is set which generates both a counterbalancing turning moment as well as additional wind force and sailing speed due to added sail area. For small boats, rather than attempt to counterbalance the mainsail turning moment, the strategy is to render the turning moment zero by moving the centerboard until the mainsail force acts directly through it. In effect, the "lever arm" of the mainsail force is eliminated thereby causing the turning moment to vanish as well.

The fourth article in this chapter is "Drag force exerted by water on the human body," by A. Bellemans. In it the author draws attention to some direct measurements (by Clarys and Jiskoot) of the drag force on the human body as it moves through water. The actual experiment measured the force needed to tow male subjects through the water at speeds between 1.5 and 1.9 m/s. On the basis of those measurements, Bellemans estimates a value of 0.7 for CD, the drag coefficient for the human body in water. He estimates also that the minimum power demand associated with swimming at 1.7 m/s on the surface of the water is about 150 W.

The author notes, however, that the *total* power demand for swimming at this speed was previously deduced, from oxygen consumption measurements on swimmers, to be approximately 300 W, the difference being ascribed by Bellemans to the intense muscular effort of the limbs needed to propel a swimmer through the water at these speeds.

The fifth and final article in this chapter, "The physics of swimming," by M. R. Kent, focuses attention on theoretical and experimental aspects of the breaststroke. This stroke is particularly interesting because the instantaneous speed of the swimmer varies drastically in the course of each cycle, from a peak of about 1.95 m/s during the arm-propulsion phase, to a minimum of 0.0 m/s (!) when the legs are drawn forward in preparation for their propulsive phase, and back to about 1.85 m/s during the leg-propulsive phase at cycle's end, for an average velocity of about 1.41 m/s.

Kent's theoretical model is based on a propulsive force P and two resistive forces, L and R, where L is an active (dynamical) resistance force associated with limb movements, and R is a passive resistance force associated solely with water effects such as skin friction drag and wavemaking.

His experiment, on the other hand, involves towing a swimmer at a constant speed sufficiently large to insure that the towing line never goes slack, even as the person being towed executes powerful swimming strokes. [This condition may be summarized by the inequality $T > 0$, where T is the instantaneous tension in the towing line.] In the course of the experiment, the instantaneous position of the swimmer is filmed while, simultaneously, the instantaneous tension in the towing line is measured and recorded.

The cinematic record allows direct computation of the instantaneous position, velocity, and acceleration of the swimmer. In addition to these kinematic variables, the instantaneous tension in the towing line allows calculation of important dynamic variables such as water resistance and the propulsive power generated by the swimmer during a single explosive breaststroke cycle.

The author calculates a propulsive power output of 311 W for the breaststroke, a value comparable to that suggested by oxygen consumption data for swimmers, as cited in the previous article by Bellemans. In addition, a comparison of the velocity and resistance profiles suggests that, unlike passive resistance which is proportional to the square of the swimmer's velocity, the total resistance is roughly proportional to the first power of the swimmers velocity.

---

[1]The discus is one of the oldest sport implements in history and derives its name from the Greek word "diskos," the literal meaning of which is "thing for throwing." The early versions were made of stone or bronze, typically bore engraved inscriptions, and varied considerably in size and weight, although the familiar shape—circular, with a greater thickness in the middle than at the edges—appears to have remained fairly constant over time. See, for example, E. Norman Gardiner, *Athletics of the Ancient World* (Oxford-Clarendon, London, 1930), Chap. XI.

[2]History confirms that the development of the sailing ship was a truly global achievement, with the earliest

sailing ships appearing in Egypt around 4000 B.C. For a fascinating account of its evolution since that time see, for example, Romola and R. C. Anderson, *The Sailing-Ship: Six Thousand Years of History* (McBride, New York, 1947).

[3]Iceboats, for example, with their very flat sails (low sail drag) and smooth metal runners (low surface drag) are capable of speeds larger than the winds that propel them. In fact, such craft have been recorded at speeds well over 200 km/h! [See, for example, Paul Jackson, "What limits the speed of a sailboat?," *Phys. Teach.* **18**, 224 (1980).] To a lesser extent, catamaran-type sailboats achieve very high speeds by putting (and keeping) very little hull surface in the water. Sailboats with "normal" hulls, on the other hand, encounter large hydrodynamic resistance forces that substantially limit their speed when sailing to windward.

[4]In large boats, the ballast is normally provided by a heavy lead weight attached near the bottom of the keel. In small boats, ballast is normally provided by the weight of the crew, whose location is shifted as needed to provide the required restoring moment.

[5]As the authors point out, frictional effects account for all of the drag at low speeds and about half of the total drag at high speeds. For a thorough discussion of friction and wavemaking resistance see, for example, K. C. Barnaby, *Basic Naval Architecture*, 2nd ed. (Hutchinsons, London, 1954), Chap. X.

Aerodynamic effects on discus flight

Cliff Frohlich
University of Texas, Institute for Geophysics, P.O. Box 7456, Austin, Texas 78712
(Received 30 October 1980; accepted 22 January 1981)

Skilled discus throwers claim that a properly thrown discus will travel several meters farther if it is thrown against the wind, than if it is thrown along the direction of the wind. Numerical calculations confirm these claims for winds of up to about 20 m/sec and show that the extra distance is caused by the higher lift and drag forces acting on a discus that is thrown against the wind. Aerodynamic considerations influence numerous aspects of discus throwing, but these have not been dicussed in the scientific literature. In addition to reviewing the available literature, the present article calculates the effect on distance thrown caused by changes in wind velocity, altitude, air temperature, gravity, and release velocity. Some sample results are that a discus can travel: (i) 8.2 m farther against a 10-m/sec wind than with such a wind; (ii) 0.13 m farther at 0 °C than at +40 °C; (iii) 0.19 m farther with no wind at the elevation of Rome, Italy than at the elevation of Mexico City, Mexico; and (v) 0.34 m farther at the equator than at the poles.

INTRODUCTION

Wind drag is an important factor affecting performance in a number of individual sports, including bicycle racing, track running (sprinting), and long jumping. Generally, for best performance it is advantageous to be moving in the same direction as the wind. In an attempt to nullify this advantage, records are disallowed in certain track and field events if there is too large a component of wind velocity along the direction of the run or jump.

Discus flight is also influenced by wind, but unlike most other track and field events, *discus throwers can throw significantly farther if the wind blows against the direction of the throw than if there is no wind or if the wind blows in the same direction as the throw.* When thrown properly, a discus is an airfoil, and the aerodynamic lift more than compensates for the loss of performance due to drag. Discus enthusiasts have been aware of this paradoxical result for many years. Almost 50 years ago Taylor[1] measured drag and lift coefficients for a discus, calculated a few trajectories, and recommended that record performances be "adjusted" for the effect of the winds. His recommendation was not instituted, and to this day discus records are allowed under any wind conditions.

Suprisingly enough, there are apparently no physics textbooks or articles in scientific journals that discuss the aerodynamics of discus flight. In fact, most of the investigations of the aerodynamics of discus flight have been reported in exceedingly obscure places. For example, the most widely quoted numerical calculation of the effect of air on discus flight is the unpublished work of Cooper *et al.*,[2] who performed their analysis as a class project for an engineering course at Purdue. Several important studies, including the best discussion on the effects of discus rotation on discus flight[3] appeared in *Discobulus*, which was a mimeographed newsletter in the 1950s for a club of British discus enthusiasts. The most comprehensive work on discus aerodynamics is available only in Russian.[4] Measurements of the drag and lift coefficients for a discus at various angles of attack have been published in a physical education journal by Ganslen.[5] Several other authors have presented summaries of portions of the above work, including Lockwood,[6] Dyson,[7] and Hay.[8]

Many basic questions about the effect of physical variables on discus flight are not addressed at all by any of the previous work. As an airfoil, how will the discus be affected by changes in air density, Earth gravity, and its own mass and shape? Will a discus perform better when thrown at high altitudes and high air temperatures or at low altitudes and colder air temperatures? How much further will a discus travel if thrown at the Earth's poles than at its equator? If they are released at the same velocity, which will travel further, a men's discus or the smaller and lighter women's discus (see Table I).

FACTORS INFLUENCING DISCUS FLIGHT

Definitions of basic variables

When in flight, a discus is affected only by the forces of gravity, aerodynamic drag, and aerodynamic lift (see Fig. 1). If there is wind with nonzero velocity $\mathbf{v}_w$ then the aerodynamic drag will not act along a direction opposing the velocity $\mathbf{v}_d$ of the discus, but rather it will act along the direction of the relative velocity $\mathbf{v}_{\rm rel}$ (see Fig. 1) where

$$\mathbf{v}_{\rm rel} = \mathbf{v}_d - \mathbf{v}_w. \qquad (1)$$

The magnitude of the drag and lift forces are usually represented in terms of the dimensionless drag and lift coefficients c_d and c_L:

$$F_{\rm drag} = \tfrac{1}{2} c_d \rho A v_{\rm rel}^2; \quad F_{\rm lift} = \tfrac{1}{2} c_L \rho A v_{\rm rel}^2, \qquad (2)$$

where ρ is the density of the air and A is the maximum

Reprinted from *American Journal of Physics* **49**, 1125–1132 (1981); © American Association of Physics Teachers.

Table I. Characteristics of the discus and frisbee.

	Mass (kg)	Diameter (mm)	Thickness (mm)	M_{eff} (kg)	Current world record [d] (m)
Men's discus [a]	2.0	221	46	2.0	71.16
High school discus [b]	1.616	211	41.28	1.77	63.86
Women's discus [a]	1.0	182	39	1.47	70.72
Wham-O regular frisbee [c]	0.087	227	31	0.08	

[a]Reference 9.
[b]Reference 10.
[c]S. Johnson, *Frisbee* (Workman, New York, 1975).
[d]B. Nelson, Track & Field News **32**(12), 16 (1980).

cross-sectional area of the discus. From Eq. (2), the acceleration due to aerodynamic forces is

$$a = \tfrac{1}{2}\rho A v_{rel}^2 (c_d^2 + c_L^2)^{1/2}/M. \quad (3)$$

c_d and c_L depend strongly on the attack angle ψ, which is the angle between the plane of the discus and the direction of $\mathbf{v}_{rel}$. They may also depend weakly on $\mathbf{v}_{rel}$, ρ, and the discus rotational velocity $\boldsymbol{\omega}$. For simplicity, in the numerical calculations presented in this paper, it has been assumed that c_d and c_L are independent of v_{rel}, ρ, and $\boldsymbol{\omega}$, and that the rotation vector $\boldsymbol{\omega}$ is perpendicular to the plane of the discus.

Equations of motion

The equations of motion are particularly simple if $\mathbf{v}_w$, $\mathbf{v}_d$, and the normal to the plane of the discus all lie within a vertical plane, i.e., if the discus is thrown either with or against the wind, and so that the discus does not lean to the right or the left. In this case,

$$\ddot{x} = -\tfrac{1}{2}(\rho A v_{rel}^2/M)(c_d \cos\beta + c_L \sin\beta),$$
$$\ddot{y} = -g + \tfrac{1}{2}(\rho A v_{rel}^2/M)(c_L \cos\beta - c_d \sin\beta), \quad (4)$$

where β is the angle between the horizontal plane and v_{rel}. In addition if the lift and drag forces apply no torques to the discus, then the angle α between the plane of the discus and the horizontal remains constant throughout the flight. In this case, the path of the discus is determined completely by Eq. (4) if the initial conditions are known. The initial conditions include the initial release velocity v_{d_0}, the wind velocity v_w, the release angle R (the angle between the horizontal plane and the initial velocity vector), the discus inclination angle α (see Fig. 1), and the release height y_0.

These equations can be solved if the initial conditions are known. In particular, at time T after the discus has been released with initial velocity

$$\mathbf{v}_{rel_0} = \mathbf{v}_{d_0} - \mathbf{v}_w:$$

$$\dot{x} - v_w = v_{rel_x} = v_{rel_{x_0}} - \tfrac{1}{2}(\rho A/M) \times \int_0^T v_{rel}^2 (c_d \cos\beta + c_L \sin\beta) dt',$$

$$\dot{y} = v_{rel_y} = -gT + v_{rel_{y_0}} + \tfrac{1}{2}(\rho A/M) \times \int_0^T v_{rel}^2 (c_L \cos\beta - c_d \sin\beta) dt'.$$

Thus

$$x = v_w T + v_{rel_{x_0}} T - \tfrac{1}{2}(\rho A/M) \times \int_0^T \int_0^t v_{rel}^2 (c_d \cos\beta + c_L \sin\beta) dt' dt, \quad (5)$$

$$y = -\tfrac{1}{2}gT^2 + v_{rel_{y_0}} T + \tfrac{1}{2}(\rho A/M) \times \int_0^T \int_0^t v_{rel}^2 (c_L \cos\beta - c_d \sin\beta) dt' dt.$$

In this form, the solution neatly separates the effects of gravity g and wind v_w from the aerodynamic forces depending on v_{rel}, but it is not particularly useful for numerical calculations.

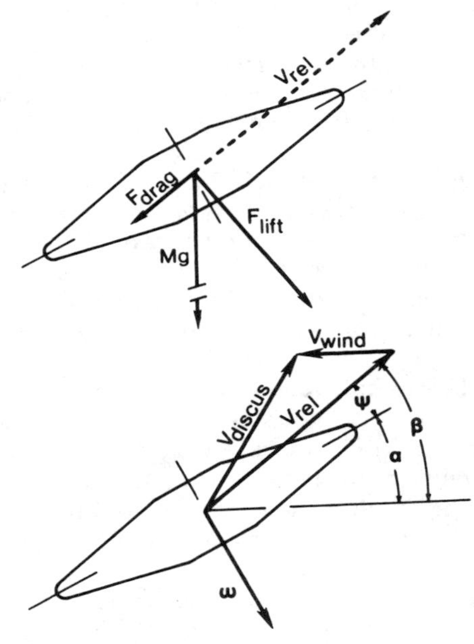

Fig. 1. Top: forces affecting discus flight include gravitational forces and aerodynamic forces. By definition the components of aerodynamic forces along the direction of the oncoming air are called drag forces, and the components perpendicular to the direction of the oncoming air are called lift forces. In this figure, the velocity of the discus relative to the air is such that the air strikes the upper surface of the discus, and so the direction of the lift is downward. Bottom: important variables controlling aerodynamic forces on the discus. The relative velocity $\mathbf{v}_{rel}$ depends on the actual discus velocity $\mathbf{v}_{discus}$ and the wind velocity $\mathbf{v}_{wind}$. The drag and lift coefficients depend strongly on the attack angle ψ. If the aerodynamic forces apply no torques to the discus, then the angle α between the plane of the discus and the horizontal does not change during flight. If torques do exist, their effect on the discus depends on its angular velocity.

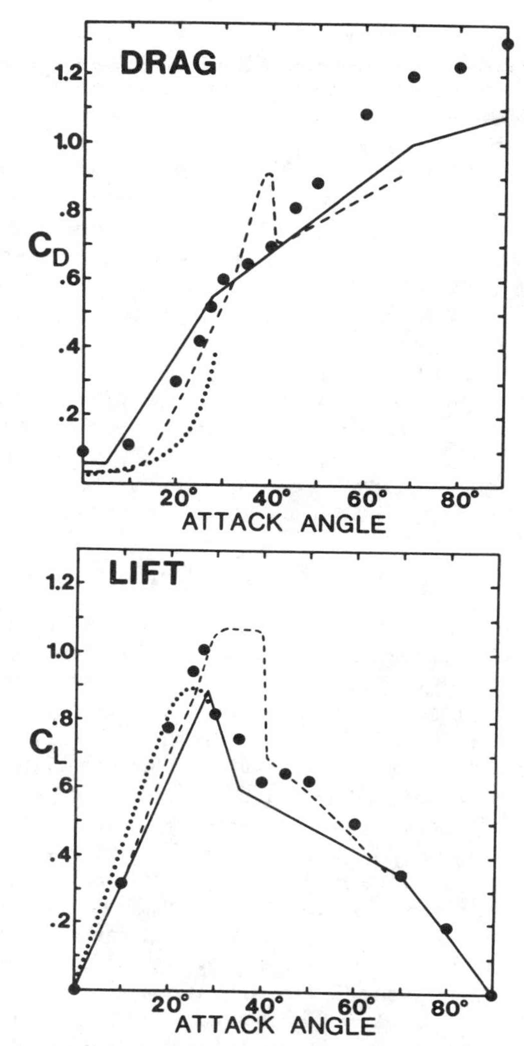

Fig. 2. Drag and lift coefficients for the discus. Solid circles are the measurements of Ganslen[5,8] for a relative air velocity of 24.4 m/sec. The solid line represents the values used in the numerical calculations of Cooper *et al.*[2] and in this paper. The dotted lines are the results of Kentzer and Hromas[3] for a relative velocity of 30.5 m/sec and a rotation of 2.5 rev/sec. The dashed lines are the coefficients reported by Tutevich.[4]

In practice, it is easier to calculate discus trajectories by making use of Eq. (4) and integrating over small time increments Δt. Since the drag and lift coefficients are nearly piecewise linear with the attack angle ψ (Fig. 2), numerical integration is improved by making use of $\dddot{x}$ and $\dddot{y}$. Since

$$\beta = \tan^{-1}(v_{\text{rel}_y}/v_{\text{rel}_x}),$$

$$\dddot{x} = (\rho A/2M)\{2(v_{\text{rel}_x}\ddot{x} + v_{\text{rel}_y}\ddot{y})(-c_d \cos\beta - c_L \sin\beta) + (v_{\text{rel}_x}\ddot{y} - v_{\text{rel}_y}\ddot{x})[(\partial C_d/\partial\psi - c_L)\cos\beta + (\partial C_L/\partial\psi - c_d)\sin\beta]\},$$

$$\dddot{y} = (\rho A/2M)\{2(v_{\text{rel}_x}\ddot{x} + v_{\text{rel}_y}\ddot{y})(c_L \cos\beta - c_d \sin\beta) + (v_{\text{rel}_x}\ddot{y} - v_{\text{rel}_y}\ddot{x})[(-\partial C_L/\partial\psi - c_d)\cos\beta + (\partial C_d/\partial\psi - c_L)\sin\beta]\}.$$

Thus over each time increment Δt,

$$\Delta x = \dot{x}\Delta t + \tfrac{1}{2}\ddot{x}\Delta t^2 + \tfrac{1}{6}\dddot{x}\Delta t^3,$$

$$\Delta y = \dot{y}\Delta t + \tfrac{1}{2}\ddot{y}\Delta t^2 + \tfrac{1}{6}\dddot{y}\Delta t^3.$$

The author has written a FORTRAN program using this method to calculate trajectories for the discus, and tests showed that accurate results could be obtained with time increments of 0.1 sec. With the 0.1-sec increments, calculated distances for typical throws varied less than 0.1 m from distances calculated using an increment of 0.01 sec.

Effect of rotation of the discus

The most important effect of the discus' rotation is to stabilize its orientation during flight. Upon release the discus rotates at about 7 rps.[1] In the absence of applied torques the discus' initial orientation is preserved throughout its flight.

At the moment of release, the discus thrower will usually attempt to orient the discus so as to maximize the lift forces and minimize the drag forces while the discus is traveling upward and outward. Most investigators agree that the optimum strategy in still air is to release the discus so that its inclination angle α is about 5° to 10° less than the release angle R.[2,5] Although this results in negative lift during the very beginning of flight, it allows for a minimum of drag and optimum average lift throughout the upward part of its flight.

In fact, the torques applied by aerodynamic forces during flight are not negligible, although they are quite small. If one carefully observes the orientation of a discus in flight, one notices that for a right-handed thrower the left side of the discus tilts (rolls) gradually downward about 10° during the latter portion of its flight.[3] As explained in Fig. 3, this is because the aerodynamic lift forces F_A are appar-

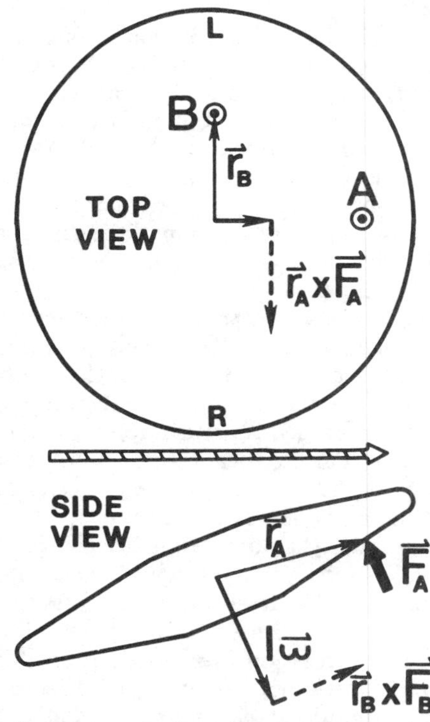

Fig. 3. One effect of air on the discus is to create torques that change the orientation of the axis of rotation of a discus. For a discus tilted with front slightly upwards and moving to the right the largest torques are caused because the lift forces are larger on the forward half of the discus than the rear half. Thus the lift force F_A acts at point A and creates a torque $\mathbf{r}_A \times \mathbf{F}_A$ that causes the angular momentum vector $I\omega$ to move in space. This causes the right edge of the discus to tilt slowly upwards during flight. Similarly, because of the discus rotation, the relative velocity is greater on the left side of the discus than the right, causing a net upwards force F_B to act on the left side of the discus. This force creates a torque $\mathbf{r}_B \times \mathbf{F}_B$, causing the forward edge of the discus to tilt progressively upwards during flight.

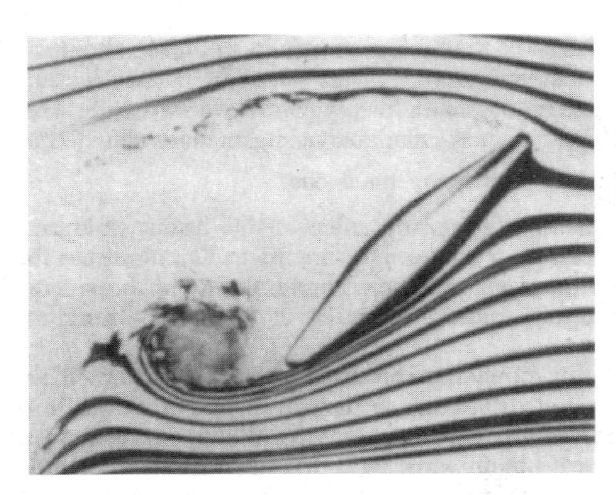

Fig. 4. Wind tunnel experiments demonstrate that the abrupt decrease in lift that occurs as the attack angle approaches 30° is accompanied by the formation of a turbulent wake. This wake is clearly visible in this wind tunnel photo of a discus at an attack angle ψ of 45°. Smoke has been introduced into the airstream so that the streamlines can be followed. This photo is reproduced with the permission of Richard Ganslen, who produced it when he was at Purdue University.

ently larger on the forward half of the discus, creating a torque vector pointing to the right. Since the angular momentum vector points mostly downward, and since torque equals the rate of change of angular momentum, the torque to the right causes the left side of the discus to tilt downwards.

Similarly, since the discus' rotation causes the relative air velocity to be slightly higher on the left side of the discus, the aerodynamic forces effectively are applied slightly to the left of the center of mass (Fig. 3). These forces create a torque vector pointing forward, causing the front edge of the discus to tilt (pitch) upward about $1\frac{1}{2}$°/sec during the course of its flight. Presumably aerodynamic forces during flight also slow the rate of rotation of the discus about its own axis, however, this effect has not been measured and cannot be very large.

Although the mass and shape of the discus are quite strictly controlled by the rules governing the sport,[9,10] there are no restrictions on the moment of inertia or the distribution of mass within the discus. Recently, most discus throwers have tended to prefer implements where most of the mass is concentrated near the rim. Such a discus has a large moment of inertia that resists changes in the discus orientation caused by aerodynamic torques. Of course, because of its symmetry, a nonrotating discus would also experience smaller aerodynamic torques than a rotating discus. This has caused at least one expert to suggest that a discus be constructed with a hollow mercury-filled rim in order to reduce the rotation after release (Ruudi Toomsalu, personal communication). However, experience suggests that a nonrotating discus lacks stability and will wobble and flutter, and thus the liabilities of a fluid-rim discus seem to be outweighed by the benefits.

Review of experimental measurements of c_d, c_L, and pitching moments

All reported investigations of the drag and lift coefficients for the discus agree that the drag coefficient increases steadily as the attack angle ψ increases from 0° to 90° (see Fig. 2).[2–5] In addition, they find that the lift is zero when $\psi = 0°$, it increases quickly to a maximum value at about $\psi = 30°$, and then drops quite sharply and approaches 0 as ψ approaches 90°. By studying the behavior of smoke streams passing a discus suspended in a wind tunnel (Fig. 4), Ganslen[5] was able to show that the abrupt decrease in lift at about 30° coincides with the formation of a turbulent wake behind the discus. Of the reported measurements of c_d and c_L, those of Ganslen[5] are the most reliable, and he is the only investigator reporting measurements of drag and lift coefficients over a range of air velocities. The basic form of the coefficients does not vary as the velocity changes, although the value of the drag and lift coefficients decreased by about 30% as the velocity of the wind increased from 21 to 30 m/sec.

The only published measurements of rotational torques on a rotating discus are those of Kentzer and Hromas,[3] who reported only the pitching moments. They found that pitching moment varies in a fashion similar to the lifting force, which apparently provides the force causing the moment.

RESULTS OF NUMERICAL CALCULATIONS

If reliable measurements existed at all attack angles for drag and lift coefficients, pitching and rolling moments, and discus rotation slowing torques, it would be possible to write a computer program to calculate the discus trajectory if eight initial conditions were known: the discus initial rotation speed and orientation vector (three variables), the initial release velocity (two variables if the azimuth of the throw is along the x axis), the release height (one variable) and the wind velocity (two variables, if horizontal winds only are assumed). In practice it is not practical to search for optimum combinations of eight variables, especially since the rotation moments have not been measured. For this reason, a number of simplifying assumptions will be made in order to simplify the calculations.

In addition, in all the calculations it has been assumed that wind velocity is independent of time and of height above the ground, and that the discus rotates only about an axis perpendicular to its faces (no "wobble"). Except as noted, the drag and lift coefficients of Cooper *et al.*[2] are used (see Table II) since they are the simplest and exhibit the same basic form as the other measurements. It will be assumed that pitching, rolling, and rotation slowing moments are zero, that the wind direction is parallel to the direction of the throw, and that the plane of the discus intersects the plane of the ground along a line perpendicular to the direction of the throw (i.e., the discus tilts neither to the right or the left). With these assumptions, Eq. (4) applies, and only five variables remain, R, α, v_{d_0}, v_w, and y_0.

Aerodynamic forces are definitely an important factor in discus flight. For example, the ratio a/g of aerodynamic to

Table II. Drag and lift coefficients of Cooper *et al.* (Ref. 2) used for numerical calculations in the present paper. Linear interpolation may be used to find coefficients between table entries.

Angle	Drag C_d	Angle	Lift C_L
0°	0.06	0°	0.00
5°	0.06	28°	0.875
30°	0.54	35°	0.60
70°	1.00	70°	0.35
90°	1.07	90°	0.00

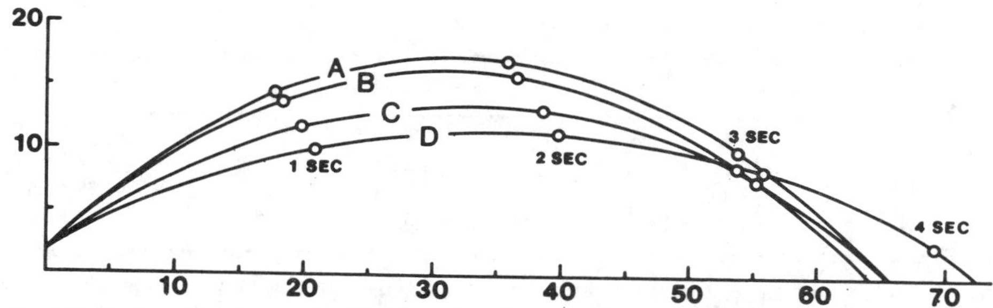

Fig. 5. Trajectory of a discus thrown with optimum release angle R and inclination angle α under various wind conditions, including: A—no air, discus thrown in a vacuum; B—a 10-m/sec wind in the direction of the throw; C—no wind; and D—a 10-m/sec wind against the thrower. Open circles show the point on the trajectory at each second after the time of release. Note that at these wind velocities, the longest distances are obtained when throwing against the wind with a low, flat trajectory. The release velocity is 25 m/sec, $A = 0.038$ m^2, $M = 2.0$ kg, $\rho = 1.29$ kg/m^3, $g = 9.8$ m/sec^2, release height $y_0 = 1.8$ m, and the drag and lift coefficients are those of Cooper *et al.*[2]

gravitational forces during discus flight is [see Eq. (3)]:

$$a/g = (\rho A v_{\text{rel}}^2/2Mg)(c_d^2 + c_L^2)^{1/2}. \tag{6}$$

Typically, this ratio will approach or exceed unity during some part of the discus flight. In this section we will consider the effects on distance thrown of variations in each of the parameters which contribute to this ratio.

Effect of v_w

Champion discus throwers claim that longer throws can be made throwing against fairly stiff winds than with the wind or with no wind[11] and my calculations confirm this result (see Fig. 5). For moderate winds, the worst possible situation is to throw with a wind whose velocity is 7.5 m/sec (Fig. 6). A discus will travel more than 6 m further if thrown against a 7.5-m/sec wind than if thrown with the wind. For winds to about 20 m/sec, a properly thrown discus will always travel further against the wind than with it. Thus under all conditions under which a discus competition could conceivably be held, discus throwers desiring record performances are correct in their preference for throwing in the face of stiff wind.

However, it is clear that this cannot be true for very high winds. This can be shown with an approximate calculation. Supposing the discus is thrown with an attack angle of zero (so that the drag is a minimum) and suppose it is released with a release angle of zero (so that its rate of horizontal progress is at a maximum). If the wind speed is high enough, the discus will stop its forward progress and travel back towards the thrower before it has traveled 60 m. In particular, if the attack angle and the release angle are zero, the horizontal acceleration against the direction of the throw is then at least [see Eq. (4)]:

$$\frac{dv_d}{dt} = -\frac{1}{2}\frac{\rho A c_d}{M}(v_d - v_w)^2.$$

This is integrable, with

$$v_d = v_w - 1/(K - Xt),$$

where K is a constant of integration that equals $-1/(v_{d_0} - v_w)$ if the initial horizontal velocity of the discus is v_{d_0}, $X = \frac{1}{2}\rho A c_d/M$ and t is time. The wind will decelerate the discus and when $v_d = 0$ the discus will reach its maximum distance $X_{\max}$. A straightforward calculation reveals that

$$X_{\max} = \frac{1}{X}\left(\frac{-1}{1 - v_w/v_{d_0}} + \ln\left|1 - \frac{v_{d_0}}{v_w}\right|\right).$$

Of course, this is an overestimate of the distance thrown, since the assumed drag force was the minimum possible force. However, for $c_d = 0.06$, $v_w = 3v_{d_0}$, and ρ, A, and M as in Fig. 5, one finds that $X_{\max}$ is 51 m. This confirms that optimum throws cannot be made against hurricane force winds.

For moderate winds (less than about 20 m/sec) a discus thrown with the wind must be thrown with a different strategy than a discus thrown against the wind. Against the wind, the discus inclination α at the moment of release should be about 10° to 15° less than the release angle R, so that during most of the flight the drag will be a minimum and the lift a maximum. As the wind velocity increases, the release angle R decreases, so that the trajectory becomes flatter, and the discus will not hang in the air too long and "catch" the wind at the end of the throw. In contrast, a discus thrown with the wind (Fig. 5) should be lofted higher in the air (R is larger), and in fact as the wind velocity reaches about 20 m/sec longer throws will be obtained if the discus is turned over so that the face of the discus catches the wind like a sail.

Although the longest throws occur against stiff winds, it is easier to obtain near-optimum performances when throwing with the wind. To obtain a throw within 1 m of the optimum when throwing against a 10-m/sec wind, the

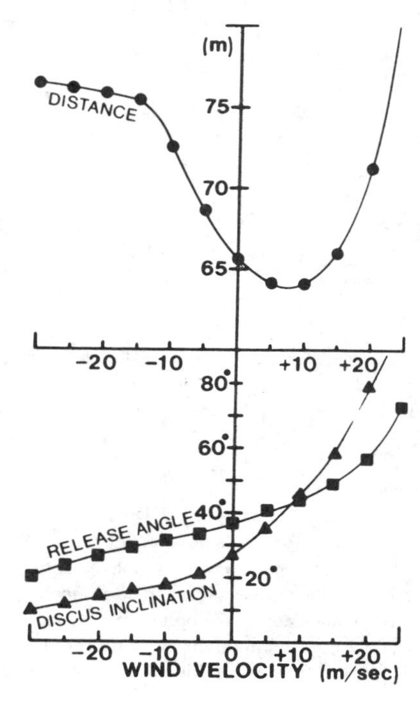

Fig. 6. Top: longest possible throw at various wind velocities. The solid line is the distance interpolated between the calculated points (solid circles). Note that the worst possible conditions to obtain long throws is to throw with a wind of 7.5 m/sec, and that if the wind velocity is less than about 20 m/sec, longer throws can always be obtained by throwing against the wind (negative wind velocities). ρ, g, A, M, y_0, c_d and c_L as in Fig. 5. Bottom: the release angle R (squares) and discus inclination α (triangles) that produce the longest throws at each wind velocity.

Table III. Approximate effect of variations in environmental conditions on maximum distance thrown with a men's discus released at 25 m/sec.

	Variation	0-m/sec wind	− 10-m/sec wind
Temperature	+ 10°C	− 3.4 cm	− 22.6 cm
Atmospheric pressure	+ 10 mm Hg	+ 1.2 cm	+ 8.1 cm
Elevation	+ 1 km	− 8.5 cm	− 55.8 cm
Gravity	+ 1 cm/sec^2	− 6.5 cm	− 7.5 cm

athlete must control the release angle R to within about $\pm 5°$ and the discus inclination α to within about $\pm 3°$ (Fig. 7). However, when throwing with a 10-m/sec wind, he can come within 1 m of the optimum by controlling these angles to within only $\pm 6°$ and $\pm 15°$. In competition, throwing against the wind would seem to favor experienced discus throwers, since more control is needed to obtain optimum performances.

The measured drag and lift coefficients are probably not accurate for relative velocities above about 40 m/sec, and so it is not meaningful to perform trajectory calculations for winds with velocities above about 20 m/sec. However, it is interesting to note that for optimum throws both with and against the wind at 20 m/sec the discus inclination is nearly parallel to the relative velocity vector at the point on the trajectory when the relative velocity (and thus drag) is a maximum. For throws against the wind, this occurs at the moment of release, and thereafter drag reduces the relative velocity. For throws with the wind, this occurs at the end of the discus flight. The following wind lifts the discus and increases its potential energy, and then the relative speed increases as the discus falls and loses potential energy.

Effect of ρ, *A*, and *M*

From Eq. (3), the effect of aerodynamic forces is proportional to the quantity $\rho A/M$. The discus area A and mass M are fixed by the rules governing the sport (Table I), but the air density ρ can vary nearly 50% between a high-temperature, high-altitude site and a low-temperature, low-altitude site (Table III). Although an athlete can do nothing to vary ρ to his favor during a particular competition, he may select competitions with favorable altitude and temperature if he desires to throw record distances.

A convenient quantity for calculating the effect of changing ρ, A, and M on distance is the effective mass, defined by

$$M_{\text{eff}} = M(\rho_{\text{STP}}/\rho)(A_{\text{men's}}/A), \tag{7}$$

where $\rho_{\text{STP}} = 1.29$ kg/m^3 and $A_{\text{men's}} = 0.038$ m^2. From Eq. (4), it is clear that for given initial release conditions two discuses with the same effective mass will possess similar trajectories. Thus the effective mass not only allows a simple comparison of the effect of changing air density, but also allows us to compare trajectories for implements with different surface area A and mass M.

The calculations show that changes in the effective mass do affect the distance of the throw (Fig. 8). The increase in distance due to decreasing the effective mass is not highly significant when there is no wind, but is about six times larger when the discus is thrown into a 10-m/sec wind. Figure 8 and Table III show that with no wind, changes in altitude and temperature can affect distance thrown by less than a meter, but with a wind the effect can be as large as several meters. Similarly, a women's discus will travel farther than a men's discus if both have the same release velocity. Both men and women attempting to obtain record per-

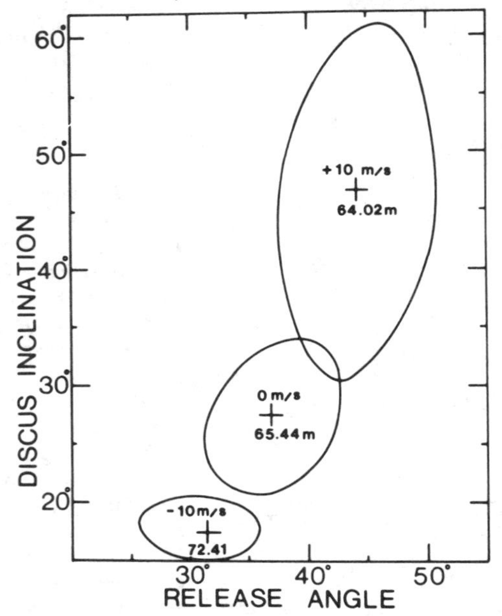

Fig. 7. Effect of changes in release angle and angle of inclination on distance thrown at various wind velocities. Plotted points (+) show optimum release angle and discus inclination for winds of 10 m/sec for and against the thrower, and for no wind. The solid line surrounding the optimum throw outlines the region containing release angle and discus inclinations that result in a throw within one meter of the optimum distance. Note that it is much easier to produce near optimum throws when throwing with the wind. ρ, g, A, M, y_0, c_d, and c_L are as in Fig. 5.

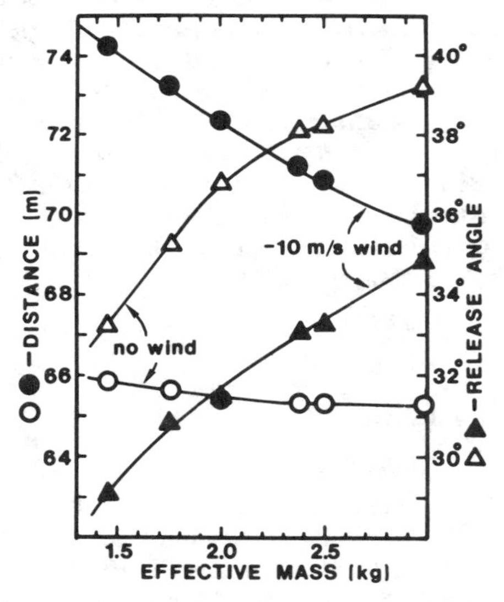

Fig. 8. Calculated optimum distance thrown (circles) and optimum release angle (triangles) versus effective mass (defined in text) for no wind (open symbols) and a 10-m/sec wind against the thrower (filled symbols). A discus with smaller effective mass is influenced more by aerodynamic forces than a discus with larger effective mass, and thus will travel somewhat further. For these calculations $v_{d_0} = 25$ m/sec, and g, y_0, c_d, and c_L are as in Fig. 5.

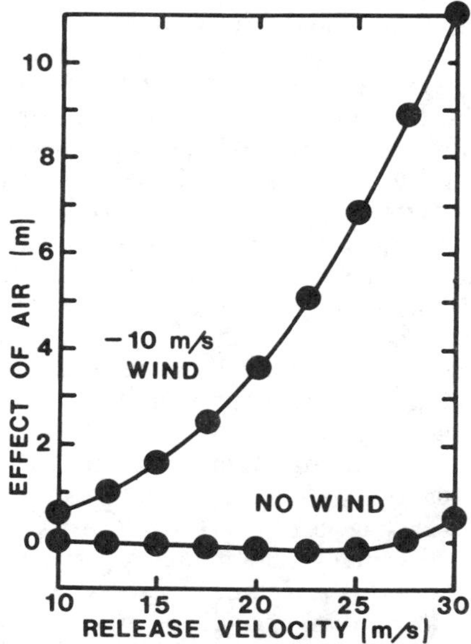

Fig. 9. Difference between optimum distance thrown in air and in a vacuum. In a vacuum, the optimum distance D is given by the equation $D = (v_{d_0}^2/g)(1 + 2gh/v_{d_0}^2)^{1/2}$ (see Lichtenburg and Wills[13]). Distances in air are calculated assuming ρ, A, y_0, M, c_d, and c_L as in Fig. 5.

formances should throw at low altitudes and low temperatures.

From the athlete's standpoint, the most significant result of decreasing the effective mass is that the release angle must be decreased in order to obtain the optimum distance. Thus for optimum performance women discus throwers should release the discus at an angle several degrees smaller than indicated in Fig. 7, and men throwing at high temperatures and altitudes should release it at an angle several degrees larger than indicated in Fig. 7.

Effect of v_{d_0}

To improve performance the single most important variable that a discus thrower can affect by training is his release velocity v_{d_0}. It is for this reason that training for discus throwers usually consists primarily of weight lifting to increase strength and agility drills to improve body quickness.[12] It is also for this reason that tall, strong athletes with long arms are most likely to be successful as discus throwers.

Aerodynamic forces become more important as the release velocity increases (Fig. 9). Up to about 25 m/sec with no wind, the optimum distance thrown is nearly the same whether the discus is thrown in a vacuum or in air, although the angle of release that should be used to obtain the optimum throw is quite different (see Lichtenberg and Wills[13] for a discussion of optimum release angles for projectiles in a vacuum). One can actually throw a discus further in the presence of air than in a vacuum if the release velocity is higher than 25 m/sec, or if one is throwing against a stiff wind (Fig. 9).

Other factors: effect of gravity g, release height y_0, and drag and lift coefficients c_d and c_L

In practice, neither gravity nor release height affect discus performance as significantly as the factors discussed above. Variations in g have little effect on discus flight, because g varies less than about 0.5% (or about 5 cm/sec^2) over the surface of the Earth. An increase in gravity of this magnitude may be expected to reduce the distance that a discus is thrown less than 0.34 m (see Fig. 10 and Table III).

The author has performed no explicit calculations concerning the effect of changing the release height y_0. A crude estimate of the effect of changing y_0 is that it would be proportional to cot θ, where θ is the angle between the discus trajectory and the Earth's surface at the point the discus strikes the ground, i.e.,

$$\Delta_{\text{distance}} = \Delta y_0 \cot \theta.$$

From Fig. 4 it appears that a change in y_0 of 1 m would change the distance thrown by about 2 m. However, in practice it is very unlikely that a discus thrower could maintain top form and change y_0 by more than a small fraction of a meter.

As reported earlier (Fig. 2), the measured drag and lift coefficients all have nearly the same general form, although at a given angle of attack the actual values vary somewhat. Since c_d and c_L only affect the aerodynamic forces on a discus, it is clear from Eqs. (3) and (4) and Fig. 8 that a proportional increase in drag and lift coefficients would have exactly the same effect on discus flight as a proportional increase in air density or a proportional decrease in the effective mass. Although a proportional increase in c_d and c_L will increase the distance thrown against winds of 10 m/sec (Fig. 8), increasing c_d while holding c_L constant will actually decrease the optimum distance thrown. In fact, the author's calculations found that doubling c_d at all angles of attack decreased the optimum distance thrown by 15.8 m, and halving c_d increased the optimum distance by 19.4 m.

DISCUSSION AND CONCLUSIONS

Aerodynamic forces are significant factor in discus flight, approaching the magnitude of gravitational forces under certain conditions. This allows a trained thrower to throw a discus several meters further against a wind than would be possible if no aerodynamic forces or no wind were present. It is because of the importance of aerodynamic factors that the discus must be thrown with a release angle of about 30° to 40° instead of the textbook value of about 45°. In addition, for maximum distance it must be oriented so as to take maximum advantage of the effects of drag and

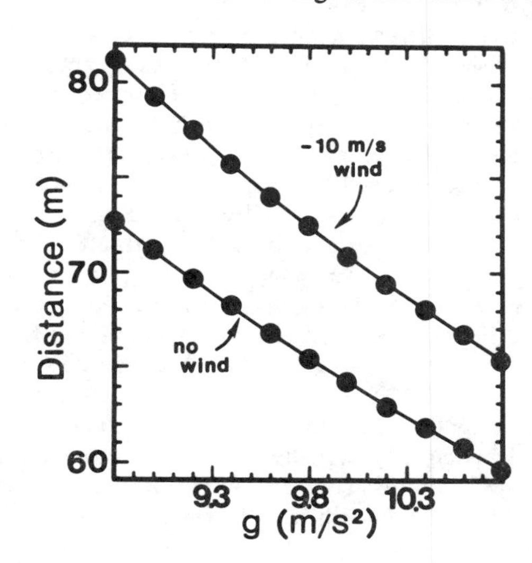

Fig. 10. Effect of changes in the Earth's gravity on the optimum distance thrown. ρ, A, y_0, M, c_d, and c_L are as in Fig. 5.

Fig. 11. Wham-O regular frisbee, and a Gill High School wood center discus. Because a discus and a frisbee are of similar size and shape, but different mass, they experience similar aerodynamic forces but vastly different gravitational forces.

lift forces during different portions of the flight.

As an airfoil, the discus is remarkably similar to its less massive cousin the frisbee (see Fig. 11 and Table I). In fact, in high winds a discus must be thrown just like a frisbee would be thrown under normal conditions, with a low release angle and a flat inclination. This is because the ratio of aerodynamic to gravitational forces is of similar magnitude. In particular, for a frisbee when $\psi = 10°$ and if $v_{d_0} = 5$ m/sec,

$$a/g = 0.184,$$

and for a discus if $v_{d_0} = 30$ m/sec, as might be encountered when throwing into a moderate wind,

$$a/g = 0.274.$$

Apparently the mass and shape of the discus place it in a regime where increasing the aerodynamic forces on it will increase the distance thrown. In particular, the numerical calculations presented in this paper indicate that for a given release velocity longer distance will be obtained with higher air densities, lower temperatures, lower altitudes, lighter discuses, larger discus areas, and stronger winds against the direction of the thrower. However, this trend cannot continue for extreme aerodynamic conditions, as indicated by the calculation showing that a discus must travel less far when thrown against exceptionally high winds. Experience suggests that a frisbee is probably in a regime where increasing the aerodynamic forces on it will decrease the distance thrown.

Considering the importance of the aerodynamic forces affecting discus flight, it is surprising that record books make no distinctions between marks recorded in still air and wind-aided marks. Apparently the present paper is the first work in English to quantitatively estimate the effect on distance thrown caused by changes in temperature, altitude, discus size, and gravity. It appears that a 10-m/sec wind is a more important factor in discus than in sprints,[14] the long jump or triple jump, or in the shot put.[13] Because they affect air density, temperature and altitude are also an important factor affecting discus results. Differences in air density alone would cause a discus released at 25 m/sec on a cold day (− 10 °C) in Moscow (elevation 120 m) to travel approximately 0.36 m further than an identical discus released in Mexico City (elevation 2239 m) in warm temperatures (+ 40 °C). Similarly, because it is lighter, a women's discus will experience relatively larger aerodynamic forces than a men's discus under otherwise identical conditions and thus can be thrown further if it has the same release velocity.

It is possible that head winds possessing a component blowing from the (right-handed) discus thrower's right side would permit even longer throws than those obtained with direct headwinds. This would explain why many champion throwers prefer to throw in "quartering winds." Under these conditions, the thrower would want to orient the discus with its highest point slightly to the right of straight ahead. Because the relative velocity would be smaller, both drag and lift would be reduced, but longer distances might be obtained since some of the drag forces would act perpendicular to the direction of flight. The question of whether longer distances would actually be attained has not been attacked quantitatively in the present study, but is the focus of an ongoing study.

Perhaps the most outstanding problem that remains to be studied quantitatively is the effect of rotation of the discus on its flight. It is possible to throw a discus in the wind in such a fashion that the aerodynamic torques cause it to pitch forward during flight, thereby allowing the discus to maintain a near-optimum angle of attack throughout a greater portion of the entire trajectory? Although it is unlikely that this would improve performance, it would be relatively easy to extend the author's numerical calculations to include the effects of rolling and pitching torques on the discus. However, no reliable wind tunnel measurements of these torques are available, and apparently no wind tunnel measurements of drag and lift coefficients have been performed in the last decade.

ACKNOWLEDGMENTS

Several individuals were kind enough to review an earlier draft of this manuscript, including David Caughy of Cornell University, Richard Ganslen of Denton, Texas, and Doug McCowan of the University of Texas. Barry Willis, of Great Britain, Ruudi Toomsalu of Estonia, and Richard Ganslen provided source materials that were unavailable in libraries in the United States. I am also grateful for conversations with Bob Cowles of Cornell University. This work has the University of Texas, Institute for Geophysics contribution number 458.

[1]J. A. Taylor, Athletic J. **12**, 9 (1932).
[2]L. Cooper, D. Dalzell, and E. Silverman, Flight of the Discus (unpublished, 1959). Available in 1980 from Purdue University Library.
[3]C. P. Kentzer and L. A. Hromas, Discobulus **4**, 1 (1958). Available in 1979 from Barry Willis, 14 Blueridge Avenue, Brookmans Park, Hatfield, Herts, Great Britain.
[4]V. N. Tutevich, *Teoria Sportivnykh Metanii* (in Russian, Moscow, 1969).
[5]R. V. Ganslen, Athletic J. **44**, 50 (1964).
[6]H. H. Lockwood, in *Athletics*, edited by G. F. D. Pierson (Nelson, Edinburgh, 1963).
[7]G. Dyson, *The Mechanics of Athletics*, 7th ed. (Holmes and Meier, New York, 1977).
[8]J. G. Hay, *The Biomechanics of Sports Techniques* (Prentice-Hall, Englewood Cliffs, NJ, 1978).
[9]H. Rico and J. Jackson, *Official AAU Track and Field Rules 1977* (Amateur Athletic Union of the U.S., Indianapolis, IN, 1977).
[10]*The Official Track and Field Guide 1979* (National Collegiate Athletic Association, Shawnee Mission, KS, 1978).
[11]K. Stone, Track & Field News **30**(6), 10 (1978).
[12]W. Paish, *Discus Throwing*, 4th ed. (British Amateur Athletic Board, London, 1976).
[13]D. B. Lichtenburg and J. G. Wills, Am. J. Phys. **46**, 546 (1978).
[14]C. R. Kyle, Ergonomics **22**, 387 (1979); L. Pugh, J. Physiol. **213**, 255 (1971).

The Study of Sailing Yachts

The design of their hulls is assisted by experiments with models. Now tests of the full-scale hull of a racing yacht in a towing tank are aiding the evaluation of results obtained from work with models

by Halsey C. Herreshoff and J. N. Newman

The modern sailing yacht has evolved from one of man's oldest efforts to harness natural forces for the purpose of locomotion. It is also a unique vehicle in that its performance depends on fluid flow in two mediums: water and air. For centuries the design of sailing vessels proceeded on the basis of practical experience, shrewd observation and intuition. Nonetheless, the successful design of fast sailing vessels played an important part in the fate of nations and the founding of empires.

Now that the sailing vessel is used almost solely for recreation and racing competition one can appreciate anew the complexity of the problem of designing a wind-driven craft that will move through both water and air with high efficiency. Nations that have no difficulty making the most advanced technological devices have discovered that building a superior sailing yacht is a task of considerable subtlety.

The principal means of reducing the unknown factors in sailing-yacht design is to test scale models of yacht hulls by towing them in an experimental tank. This procedure can be checked against the ultimate performance of the full-scale vessel at sea. Such confirmation is never entirely satisfactory, however, because of the many variables, such as wind and waves, that must be taken into account. In the case of a sailing yacht this is particularly true because the craft's ultimate performance is so largely controlled by the conditions of wind and sea, the skill of the crew and the quality of the sails.

Recently a full-sized sailing yacht has been tested in a towing tank, allowing for the first time a direct comparison of the yacht's behavior under controlled conditions with that of its scale model. The experiment was made possible through the interest and support of the Society of Naval Architects and Marine Engineers and the availability of large towing-tank facilities at the Navy's David Taylor Model Basin near Washington.

A sailing yacht must be designed to perform with maximum speed and seaworthiness over a wide range of conditions. The wind can blow from any direction with strength ranging from calms to storms, and the sea conditions are equally variable. Thus the boat must be fast not only when it is sailing downwind (when its course coincides with the wind direction) but also when it is sailing on a "reach," or at right angles to the wind [*see illustration on page 63*]. In the latter condition thrust from the sails is accompanied by a large sidewise force that tends to heel the boat over. Above all, a racing yacht must be able to perform well sailing upwind, when its course is at an angle of perhaps 40 degrees to the direction of the wind and when a delicate balance of circumstances determines the small difference between racing success and failure. It is not uncommon for well-matched yachts, such as those that compete for the America's Cup, to sail a four-mile leg of a course against the wind with a variation in elapsed time of only a few seconds, which amounts to a difference in speed of only a few parts in several thousand.

Indeed, an important stimulus to research in yacht performance is the prospect of continued challenges to U.S. possession of the America's Cup by Britain, Australia and probably other countries. Since 1851 U.S. Cup defenders have successfully met all challengers, but a close match is in prospect for September, 1967, when newly designed U.S. and Australian 12-meter boats will compete in the waters off Newport, R.I.

In discussing the mechanics of sailing we shall consider the two extreme conditions of operation: when a yacht is running before the wind, that is, sailing with the wind at its stern, and when it is sailing to windward, or against the wind. The mechanics of intermediate heading angles are essentially combinations of these two extremes. In running before the wind the sails act simply as an aerodynamic drag to the wind; it follows that the boat speed will be less than the wind speed. The only relevant hydrodynamic force exerted by the water on the hull is a drag, which obviously should be minimized for maximum speed. In sailing to windward, on the other hand, the sails and hull both act as wings: lifting surfaces that meet the oncoming wind and water at various angles of attack and develop not only adverse aerodynamic and hydrodynamic drag forces but also favorable lift forces.

In principle there is no limit to the boat's speed in sailing to windward, particularly because the relative wind speed is increased by the boat's own speed. In practice, however, the sails are most efficient and the boat's speed is greatest when the course is approximately at right angles to the wind; in that condition the forward component of the lift force on the sails is at a maximum. In sailing to windward there is an optimum heading angle with respect to the wind, generally 40 to 45 degrees. Accordingly for a boat to proceed in a direction opposite to the wind it is necessary to sail a zigzag course, which is termed "beating to windward." In sailing before the wind the hull is essentially upright and moving straight ahead, whereas in sailing to windward the large sidewise component of sail force causes the boat to heel on its longitudinal axis. In order for the hull to

Reprinted from *Scientific American* **215(2)**, 60–68 (1966); © Scientific American, Inc.

YACHT *ANTIOPE* undergoes tests in the towing tank of the U.S. Navy's David Taylor Model Basin near Washington. Above is a general view of the hull in the 1,700-foot tank; below, a closer view of a test in progress. In the close-up actual sailing conditions have been simulated in two ways: the hull has been given a "heel" to starboard partly by means of weights in a box on the platform at midships, and the bow has been offset toward the camera to simulate yaw angle. Several sensing devices are attached to the hull.

develop an equal and opposite sidewise force it must move through the water with a small angle of attack, or yaw angle, defined as a rotation around the hull's vertical axis. This angle is analogous to the angle of attack of the wind acting on the sails. The yaw angle is usually so small as to escape notice; it is generally less than three degrees, and in highly developed boats it is as small as one degree. Nevertheless, it is of fundamental importance to the entire process of sailing to windward [*see illustration below*].

The mechanism by which a hull generates a horizontal lift is identical with that producing vertical lift in an airplane. The main body of a sailing-yacht hull, like the fuselage of an airplane, is in itself an inefficient lifting surface; lift is provided by a thin keel or retractable centerboard, which functions as a wing.

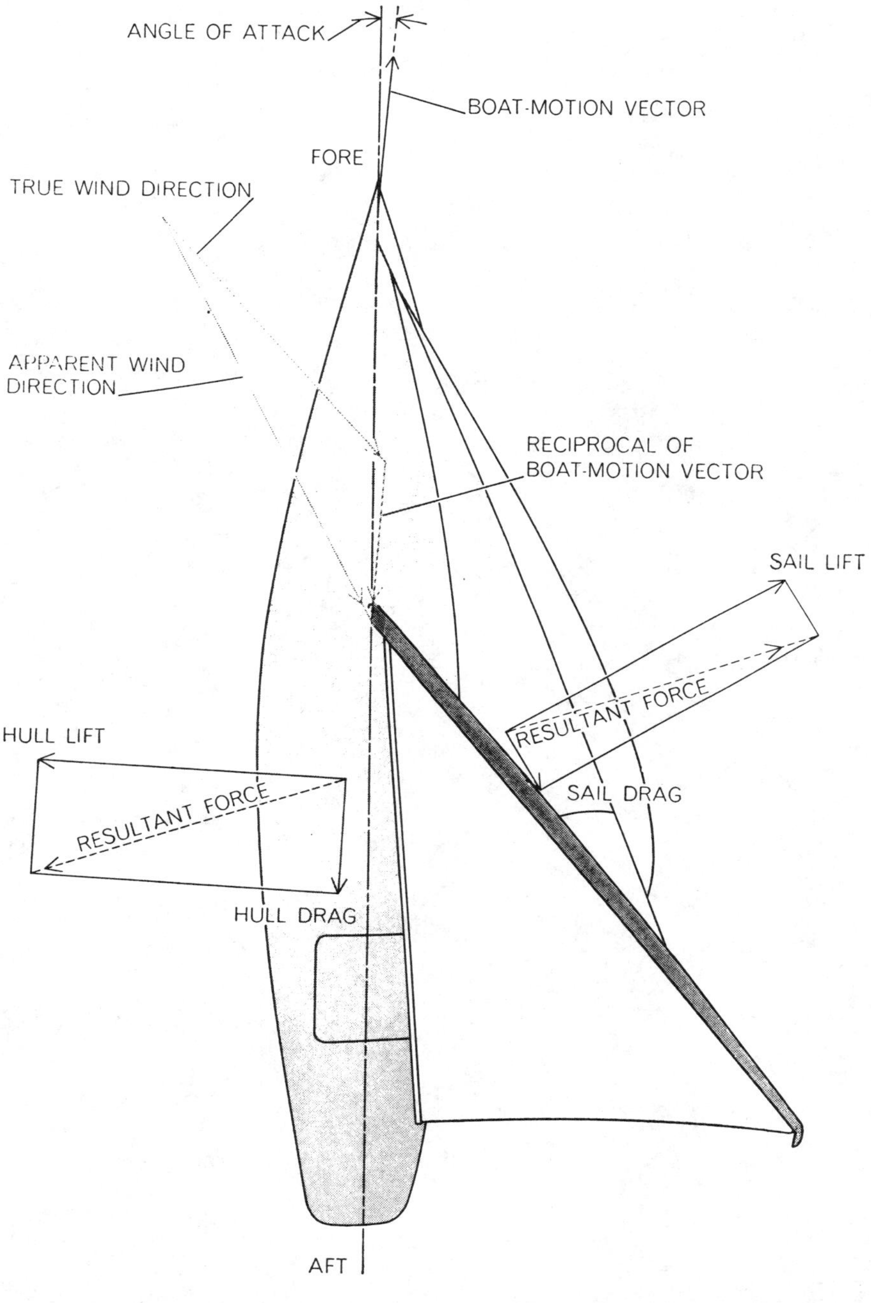

Also of fundamental importance is the equilibrium of hydrodynamic and aerodynamic moments: the turning influences exerted by water and wind. The aerodynamic heeling moment, arising from the sidewise force acting on the sails, is substantial. In order to resist this moment the hull must have a large restoring moment. This is provided in many large sailing yachts by a heavy lead ballast at the bottom of the keel and in small, light craft by a shifting of the crew's weight, which acts as "live ballast." The yaw, or vertical, moments are also significant; if they are unbalanced, the boat will tend to turn rather than sail a straight course. Small turning moments around the vertical axis are inevitable because the centers of hydrodynamic and aerodynamic pressure cannot be predicted with certainty and are subject to variation with sailing conditions. In general small angular adjustments of the rudder are required to maintain a steady course. It is necessary, however, to have the boat well balanced to obviate the need for large rudder angles, which induce additional drag forces. Moreover, the sign (plus or minus) of the turning moment, or the direction of the corrective rudder angle, is of crucial importance. It is desirable to design the hull and sails to produce a small turning moment in the direction that makes the bow head into the wind. This requires a small compensating rudder angle that increases the sidewise force on the hull—a force pushing the hull to windward—by giving an effective camber, or favorable curvature, to the combination of keel and rudder. A good sailor appreciates the importance of this balance and adjusts the fore-and-aft position of his mast and sails accordingly.

The improvement of sailboat hulls requires a quantitative study of the hydrodynamic forces and moments acting on the hull at various speeds and attitudes. Yacht design has advanced to the point where extremely small differences between yachts are important. Improving the average speed of an America's Cup contender by .01 knot would be a significant achievement; .1 knot would be a major breakthrough. Since no satisfactory theoretical predictions are available, hull shapes can be evaluated only by the observation of

FORCES ON SAILING YACHT are produced by the flow of air past the sails and the flow of water around the hull, including the effect of keel and rudder. The lift and drag produced by the sails are conventionally represented as being respectively perpendicular to and parallel to the apparent wind direction. Similarly, the lift and drag produced by the hull are shown as being perpendicular to and parallel to the direction of boat motion. When the yacht is sailing at constant speed, the resultant forces produced by the sails and hull are equal and opposite. In this diagram the yacht is sailing to windward, or within about 45 degrees of the true wind direction. Because of the boat's forward motion the angle between the boat's course and the apparent wind is only about 30 degrees. The hull meets the oncoming water at a small angle of attack, necessary to produce a lift force on the hull.

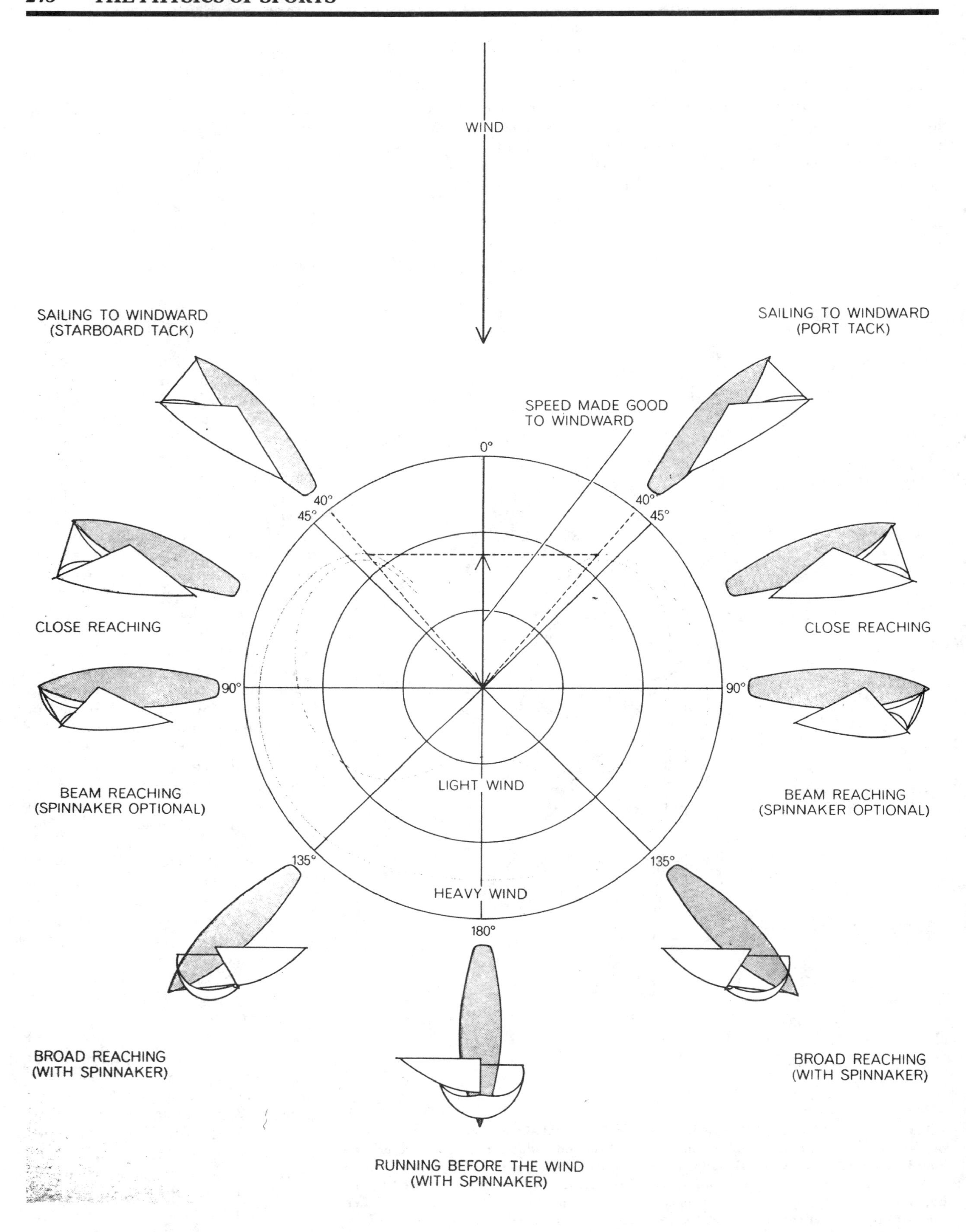

SAILING CONDITIONS depend on the intensity of the wind and the heading the boat has in relation to the wind. An angle of 40 to 45 degrees to the left or right of the direction from which the wind is coming is as close a heading as a sailboat can achieve efficiently when sailing to windward. Typical speeds for all possible headings are indicated by the two curves (*color*); the farther a point on a curve from the center of the circle, the higher the speed. The spinnaker is the large, ballooning sail at the bow.

actual sailing performance or by the direct measurement of the relevant forces in a towing tank.

The first intensive use of a towing tank for sailing craft was the pioneering effort of K. S. M. Davidson at the Stevens Institute of Technology in the 1930's. Davidson's tank studies played a central role in the development of the highly successful America's Cup yacht *Ranger,* which in 1937 defeated the British challenger *Endeavour II* in four straight races. Since that time model tests have been a key factor in developing the America's Cup contenders and many other racing yachts as well. Most of the test facilities used for this purpose are on the order of 100 feet long, eight feet wide and four feet deep, and they are equipped with carriages that span the tank and move along accurately aligned rails with controlled speeds of a few feet per second. The model hulls, usually about five feet long, are connected to the towing carriage through precision linkages and dynamometers that restrict the yaw and heel angles to specified values. The linkages must be designed so as not to restrain the vertical position and attitude of the hull, which change slightly with speed because of dynamic pressure effects. The gauges for measurement of drag and sidewise forces must be extremely sensitive; on this scale the forces and moments are very small and nearly infinitesimal differences between models are being sought.

ANTIOPE RACING sails to windward in an international race. The wind is about 40 degrees off the bow; the sails are trimmed almost flat with a small angle of attack to the wind.

Even assuming a perfect experimental regime, towing-tank studies present the investigator with a fundamental problem: There is no way to duplicate in miniature and in proper scale all the dynamical relations that affect the full-sized hull of a ship or yacht. In essence the problem is that the motion of the water moving along the hull is governed not only by the frictional properties of water but also by wave effects, which depend on gravity. When a hull is scaled down to model size, there is no way to scale down in equal degree the frictional effects and the wave effects. The frictional effects are scaled according to the Reynolds number (named for the British engineer Osborne Reynolds), which states that such effects can be scaled accurately only if the velocity of a body moving through a fluid is inversely proportional to its length (assuming that the viscosity of the fluid is held constant). The Reynolds number is more precisely defined as the product of the body's speed and length, divided by the kinematic viscosity coefficient of the fluid. This nondimensional parameter must have the same value for the full-scale vessel and its model.

Wave effects, on the other hand, are scaled according to the Froude number (named for another British engineer, William Froude), which states that the wave drag can be scaled only if the velocity of a hull is proportional to the square root of the length of the hull (assuming that the acceleration of gravity is held constant). The Froude number is defined as the ratio of the hull speed to the square root of the product of the hull length and gravitational acceleration.

Thus if one wants to carry out an experiment with ship models in which the dynamical forces are in proper scale, one must be able to adjust either the viscosity of the fluid or the force of gravity. Altering gravity is clearly impractical, and there is no fluid whose viscosity corresponds to that needed to represent water for models of practical size. Therefore one is forced to choose between test measurements that preserve the Reynolds number by varying the speed of the model inversely with the length of the model hull or that preserve the Froude number by varying the speed directly with the square root of the hull length.

Late in the 19th century Froude devised a practical procedure for separating the hull drag, or hull resistance, into two distinct components, one frictional and one residual. This procedure has made it possible to carry out towing-tank studies in spite of what appear to be the conflicting requirements set by the Reynolds number and the Froude number. Although the two components in Froude's method are not completely independent, they can be regarded as such for engineering purposes.

The frictional component of drag is derived from the transmission of viscous shearing forces that act in a boundary layer of fluid adjacent to the surface of the hull. This component contributes nearly all the drag at low speeds and about half of the total at the highest speeds of conventional ships and yachts. The residual component of drag is composed primarily of the wave resistance, which is the force required to transfer energy to the familiar system of waves that forms behind the hull of a moving ship. There are smaller secondary components, such as the drag produced if the flow of water over the hull becomes "separated" (that is, no longer flows smoothly). Other secondary components include the normal pressure forces that originate with the viscosity of water and the induced drag associated with the lift on a yawed hull. These components are arbitrarily included in the residual component. For a given hull shape and attitude the residual component of resistance is considered to depend solely on the Froude number, whereas the frictional component is assumed to depend solely on the Reynolds number.

Before this approach can be applied one must know how to separate the total drag into its frictional and residual components. Extensive research on flat plates and ship hulls has led to a useful technique for predicting the frictional component; this component is regarded as being simply proportional to the submerged surface area of a ship or its model and independent of geometrical form. Accordingly models are run at a speed set by the Froude scaling, so that the waves generated by the model are geometrically similar to those of the real ship. Frictional resistance is calculated from empirical formulas and subtracted from the total model drag to provide the residual component. This component is then multiplied by the cube of the scale ratio (yacht length divided by model length) to yield the residual component for the full-scale hull. To

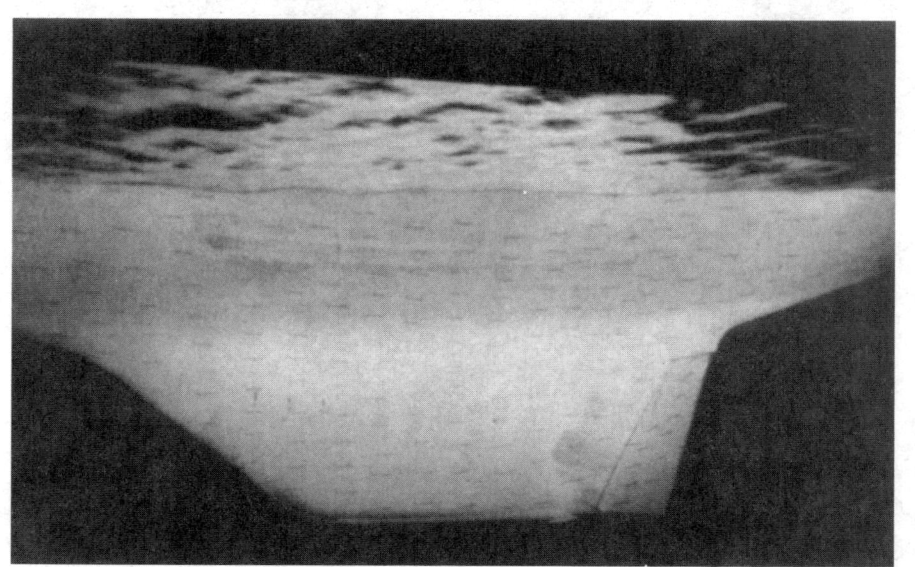

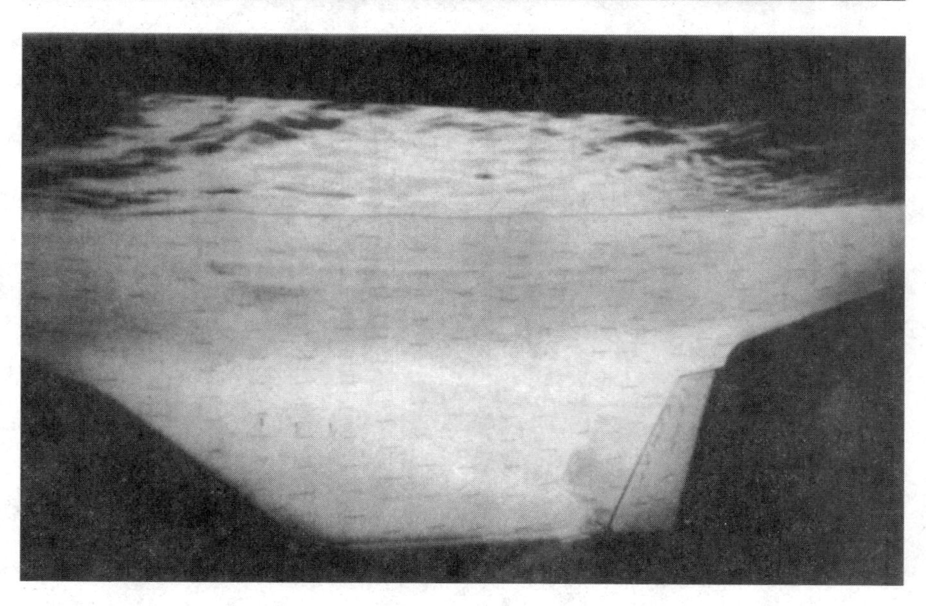

RUDDER ANGLE affects drag, as shown by strips of nylon yarn attached to the hull of the *Antiope*. The hull is in the David Taylor Model Basin's flow tank, in which the water moves but the vessel does not. At top the rudder angle is zero and the strips indicate an even flow. At middle a slight stalling effect appears with a rudder angle of five degrees; at bottom it is intensified with a 17-degree angle. Visible portion of hull is below waterline.

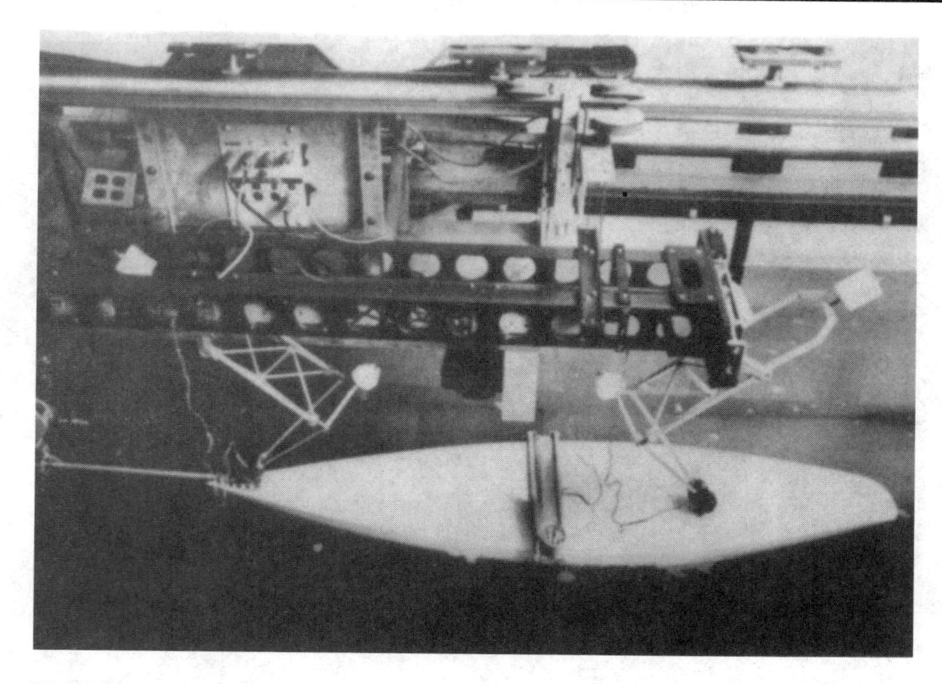

SCALE MODEL of the *Antiope* is fitted for tests in the ship-model towing tank of the Department of Naval Architecture and Marine Engineering at the Massachusetts Institute of Technology. Forces and moments are measured by means of the sensing devices mounted fore and aft; device at midships imparts heel. The model is about one-sixth the size of *Antiope*.

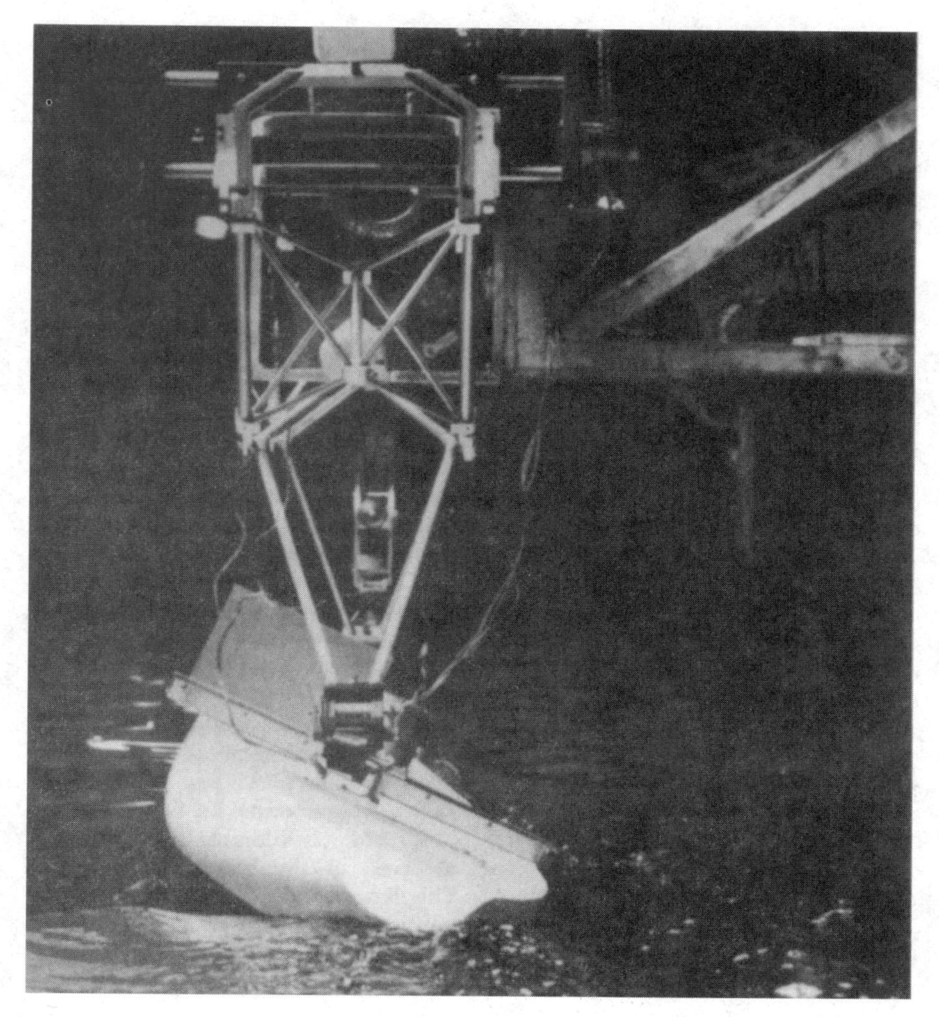

TEST RUN is conducted with the model of the *Antiope* in the M.I.T. tank. A typical test consists of 90 runs covering all the practical sailing attitudes and speeds. The tests provide a basis for predicting how the full-scale vessel will perform under various conditions.

this is added the frictional drag of the full-scale hull, calculated as in the case of the model but for the full-scale Reynolds number, to obtain a predicted value for the total resistance.

Unfortunately there is some uncertainty about the calculation of frictional drag. This is particularly true of yachts, because their hull forms differ so greatly from those of the conventional ship hulls for which the empirical formulations of frictional drag were derived. Moreover, since the frictional drag constitutes the largest portion of the total drag produced by a towed model, an accurate prediction is necessary for the overall Froude method to be valid. The scaling of the lift, or sidewise force, due to yaw is also a source of potential trouble, but this is usually assumed to scale down strictly according to Froude's law for wave effects, with frictional effects being of secondary importance.

Such was the state of the art when two years ago the Society of Naval Architects and Marine Engineers formed Panel H-13 (Sailing Yachts) as part of its technical and research program. This group, consisting primarily of yacht designers and towing-tank operators, immediately focused its attention on the scale-effect problem and possible experiments that might provide a correlation between full-scale and model results. It had been amply demonstrated that there was scant hope of obtaining reliable dynamic measurements on a real boat at sea or even in protected waters. The alternative was clear: to find some way to study a full-scale boat in a large towing tank. Through a chain of fortunate events this seemingly unlikely project quickly materialized. The Navy's large experimental facility at the David Taylor Model Basin is directed by charter to perform commercial testing services for the American maritime industry. The proposed experiment met the requirements of the charter.

The towing tank selected for the experiment is about 1,700 feet long, 51 feet wide and 20 feet deep. Normally this facility is used for towing scale models of ships and submarines. It is big enough, however, to accommodate the hull of a full-sized sailing yacht up to about 30 feet in length. Although larger hulls could be placed in the tank, they would be subject to wall effects (for example wave reflections from the sides of the tank) that might influence the measurements.

After considering various possibilities it was decided that a yacht from the highly competitive International 5.5-Meter Class would make a fine test subject. Designations such as "5.5-meter" and "12-meter" are derived not from the actual length of the boats but from international rules based on an empirical formula. The formula includes length, sail area, displacement and other factors to yield a figure, expressed as length, that is assumed to be a fair measure of the speed of the boat. Unlike "one design" yachts, each of which has a standard hull form and sail plan, the "meter" classes permit variations of design that lead to the development of new hull shapes and sail combinations. An analogous development process contributes to the improvement of ocean racing yachts.

One of the leading designers in the 5.5-meter class, A. E. Luders, offered his own boat the *Antiope* for the proposed tank tests. Her length at the waterline is 23 feet; her length overall, 31 feet. Financial support for the tests was obtained not only from the Society of Naval Architects and Marine Engineers but also from interested yachtsmen and naval architects. In November of last year the *Antiope* was shipped to the David Taylor Model Basin, where she was fitted onto the towing carriage by means of specially prepared linkages and equipped with gauges to record the three components of force at each end of the boat [*see illustrations on page 60*]. Yaw angle was adjusted by setting the bow off-center; heel angle was varied by moving 400 pounds of lead ballast along a 14-foot aluminum platform that extended from one side of the hull.

SAIL TEST is conducted in the flow-visualization wind tunnel at the U.S. Naval Experimental Station in Philadelphia. This is a 1:37-scale model of a 12-meter boat. The jib, or forward sail, is made of Mylar, the mainsail of solid plastic. Jets of smoke demonstrate that the flow of air begins to deflect well forward of the leading edges of the two sails.

Testing then proceeded for a week, both day and night, with runs spaced at half-hour intervals to ensure calm conditions of the tank water. Speed, heel and rudder angle were varied systematically, and some tests were duplicated with small brass studs fastened near the leading edge of the keel to stimulate turbulence. These tests produced 100 sets of data, which were automatically converted to digital form and reduced on a computer to give for each test condition the magnitude of the drag and sidewise forces and the center of the sidewise force.

The results obtained with the *Antiope* are substantially the same as predictions based on testing small-scale models of similar hulls in the 5.5-meter class, and a preliminary comparison shows no sign of drastic scale effects on either the resistance or the sidewise force. Now in progress is the step of providing a detailed comparison with exact scale models of the *Antiope*, which are being tested under identical conditions in the smaller towing tanks at the Massachusetts Institute of Technology, the Stevens Institute of Technology and other laboratories in this country and abroad. The resulting exact comparison will undoubtedly lead to detailed refinements in future test procedures.

While the *Antiope* was at the David Taylor Model Basin she was also studied in another large test facility: a flow-observation channel. In this facility the hull is held stationary while the water is made to flow at specified velocities, in a manner analogous to the testing of airplanes in a wind tunnel. Strips of yarn were fastened to the hull to show the flow pattern, which was observed and photographed through underwater windows in the sides and bottom of the channel [*see illustration on page 65*]. The flow studies revealed a rather regular, unseparated flow over most of the hull for the practical range of sailing conditions. Significant separation of flow did not occur until the yaw angle of the hull was increased to 10 degrees, far beyond the angles normally encountered in sailing. At less extreme yaw angles, characteristic of normal sailing conditions, considerable cross flow was observed at the bottom of the keel. This flow is indicative of the "tip vortex" associated with all lifting surfaces. Such a vortex is sometimes visible at the wing tip of an airplane when the reduced pressure at the core of the vortex causes the condensation of water vapor.

Parallel to the research on yacht hulls,

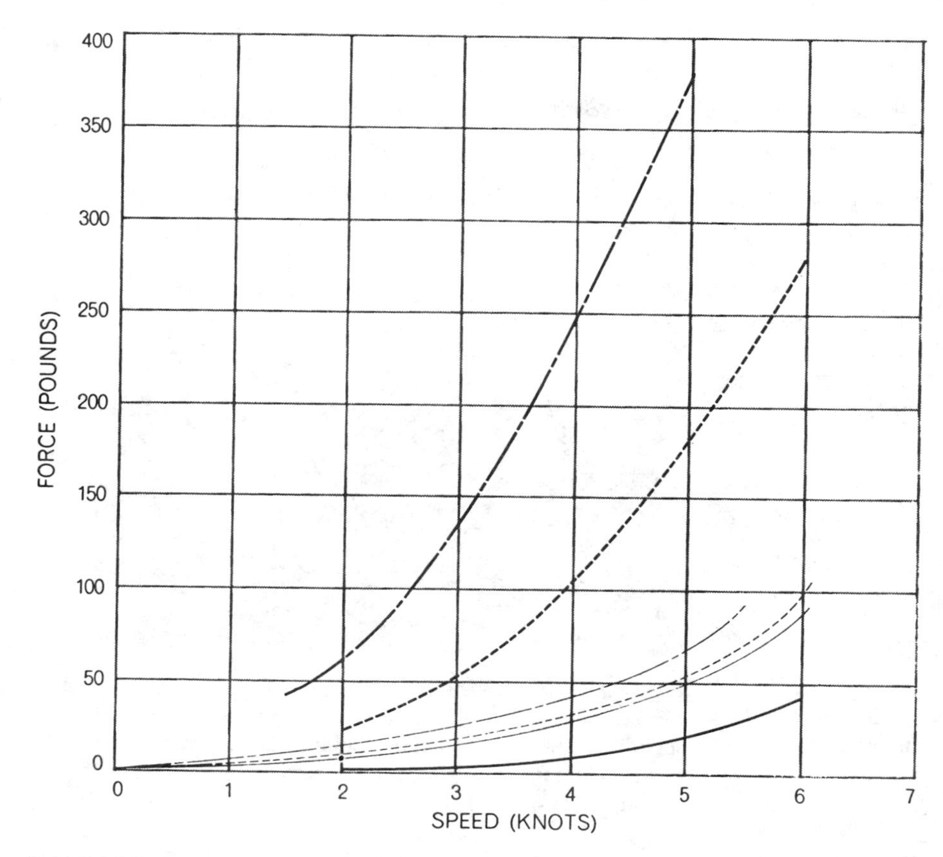

DATA FROM TESTS of *Antiope*'s hull in the model basin include relation of drag to speed (*color*) and side force to speed (*black*). Each set of curves shows, from bottom, yaw angles of zero, 2.65 and 6.45 degrees; heel angles vary from zero to 30 degrees. Small side force occurs even with no yaw angle because of asymmetry of hull resulting from heel angle.

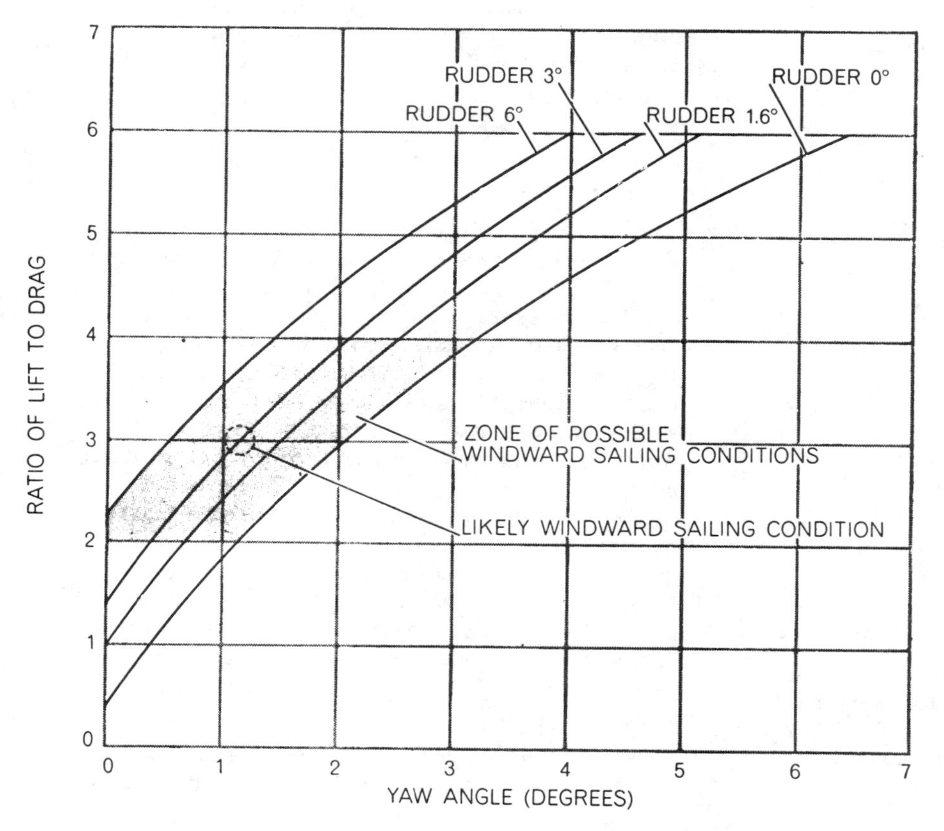

LIFT-TO-DRAG RATIO of *Antiope*'s hull is plotted against yaw angle for windward sailing at a speed of five knots and various rudder angles. Color shows possible and likely windward sailing conditions as derived from lift-to-drag ratio of sails on a 5.5-meter yacht.

experiments and analyses are under way in the study of sails. The scaling of force measurements on sails in a wind tunnel is simpler than the scaling of hydrodynamic forces on hulls; wave effects are no longer a factor and Reynolds scaling is valid. In the case of the sails, however, there is the additional problem that the shape is not constant: it varies with wind conditions because of the elasticity of the sail fabric and the distortion of the rigging that holds the sails. Moreover, the number of variables governing the choice of sails is large. Tests of jib, mainsail and hull combinations are now under way at the M.I.T. Wright Brothers Wind Tunnel and at the U.S. Naval Experimental Station in Philadelphia [*see illustration on preceding page*]. As in the towing-tank tests, dynamometers are used to measure the aerodynamic forces and moments on the sails. The test sails are made of aluminum or of a rigid fiberglass-reinforced plastic in order to maintain control over their shape.

Although it is useful for design and research to separate the treatment of the hull from that of the sails, the two are closely interrelated. A sail that is optimum for one boat is not optimum for boats of a different class or type. As an extreme example, an iceboat is capable of high speeds, and it can exert through its runner blades a high sidewise force with practically no drag. Under these circumstances flat sails with small angles of attack are appropriate; such sails produce rather small driving forces but have a high ratio of lift to drag. In contrast, a 12-meter sailing-yacht hull is subject to a considerable hydrodynamic resistance to forward motion and can develop a sidewise force of a large (but limited) extent. The sail design used in an iceboat would provide insufficient forward thrust for a 12-meter yacht, which needs a more cambered sail set at a larger angle of attack in order to produce the necessary drive.

The tremendous popularity of recreational boating and in particular the recent upsurge of interest in sailing vessels have stimulated renewed interest in yacht development. Yacht designers have long been aware that much research is needed to fill the existing gaps in information about the dynamics of sailing craft. It is hoped that model experiments and full-scale correlation, such as the *Antiope* tests, will lead to a better understanding of the fundamental mechanics of sailing-yacht performance.

THE PHYSICS OF SAILING

PETER RYE, Lectorer, School of Physics and Geosciences, Western Australia Institute of Technology

Summary

Few physicists know much of sailing theory, a failing well paralleled by a general lack of familiarity with the formalism of physics on the part of most yachtsmen. Despite this observation, discussions between avid sailors are well sprinkled with terms such as "lift", "balance" and "turning moment". This leads to conclusions that sailing is very much a physicist's sport, and that many yachtsmen know more physics than they realise . . .

The Balance of Forces and Moments

Yachtsmen place great store on the idea of a yacht sailing in "balance". This is known to be the state in which a yacht's performance potential is greatest. The two "balance" states correspond to the two concepts of the balance of moments and forces.

1. Sailing to Windard

Nowhere is the balance of forces more important than when a yacht is trying to sail as directly upwind as possible. Some simple trigonometry shows a very important theorem, relating how close a boat can sail to the wind to its hull and sail characteristics.

Consider Figure 1. β is the angle at which the wind blows across the yacht's bow; while this is not the actual angle between the course and the wind (being affected by the boat's velocity as well as the wind velocity), it is closely related to it. Due to the action of the yacht's centreboard or keel, the net force of the hull on the water is not directly backward, but close to a right angle to its course. The importance of hull drag is expressed by the term δ, called the drag angle of the hull. Similarly, the flow of air over the sail generates both lift (at right angles to the flow) and drag. The ratio of these defines the sail drag angle, γ.

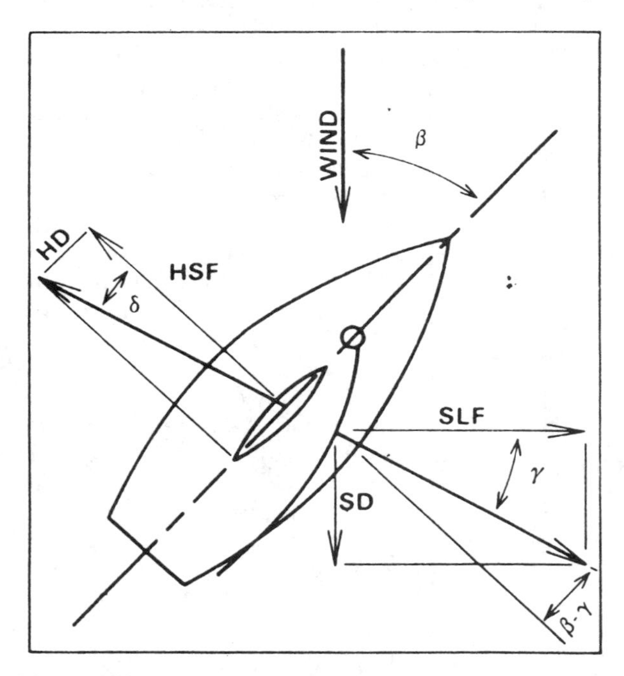

Figure 1. The balance of forces for a yacht sailing to windward. HSF denotes the hull (plus centreboard or keel) side force, HD the hull drag, SLF the sail lift force and SD the sail drag. The angle between the boat's course and the wind is β, δ is the drag angle of the hull and γ is that of the sail.

Figure 2. Short, bluff hulls such as that of the Heron shown here are limited by a large hull drag angle, and gain correspondingly less benefit from reduction of sail drag angle. They therefore use fuller sails (as may be seen by the curvature of the sail seams).

How important are these two parameters? By equating forward and side forces, we reach a very simple and useful result. If D is the drag, and L the lift (i.e., crosswind) force,

foward forces	$D \sin \delta$	$= L \sin (\beta - \gamma)$
side forces	$D \cos \delta$	$= L \cos (\beta - \gamma)$
therefore	$\tan \delta$	$= \tan (\beta - \gamma)$
i.e.,	β	$= \gamma + \delta$

At all times, the angle the yacht makes to the wind is the sum of the drag angles of hull and sail. To produce a "close-winded" yacht, such as a twelve metre, we must pay equal attention to the streamlining of both. However, if we are dealing with a short, bluff design such as the Heron sailing dinghy (Figure 2), there is a point at which we lose benefit from flattening the sails. Flat sails interact only weakly with the wind — their drag is reduced markedly, but their lift is also reduced so the power with which they drive the yacht also falls.

For this reason, sailmakers use the rule-of-thumb of "full hulls, full sails". The above results also explain why yachtsmen spend so much time working on the finish of their yachts' hulls — even going to the extent of graphite paints. Each small decrease in drag affects not only speed, but also the actual distance the yacht must cover in reaching that all important "windard mark".

2. "Off the Wind"

While the drag-angle law holds on all points of sailing, a yacht's helmsman has other problems to consider once his

Reprinted from *The Australian Physicist*, 27–29 (March, 1979); © The Australian Institute of Physics.

boat has reached the windward mark. He must then "bear off" to a course directed at about 135° from the wind direction. To gain maximum power on this leg of the course, he lets his mainsail swing out to make an angle of close to 90° to the direction of the wind. However, the point of action of the lift force on this sail moves out with it, until it may be a metre or two from the centreline. The effect is much like that of a twin-engined aircraft, forced to fly on one engine. If he is on, for example, a course with the wind from behind his right shoulder, his yacht experiences a strong clockwise moment. To prevent the yacht turning back into the wind, he can do any of three things. The most obvious correction is to turn the rudder, to the left in the above case. This solution — used by all yachtsmen in times of emergency — has an undesirable side effect, because the rudder, like the hull or the sail, creates drag as well as side force and slows the yacht down. In general, a rudder is not large enough to efficiently generate the required lift, and the drag penalty is correspondingly larger.

The other alternatives are much more constructive. The most common is to use the spinnaker, the large colourful sail set by many classes on downwind or crosswind courses. Not only does this sail increase sail area (and hence speed) considerably, but it also provides a lift force to the right of the centre of the yacht. The resulting anticlockwise moment largely cancels the effect of the mainsail. (Figure 3.)

In small classes such as the N.S.14 or the Heron (Figure 2), which do not set spinnakers, another approach is used. The turning moment is set to zero, not by providing an opposing moment, but by shifting the pivot point. The main source of side force is the centreboard, and this is pivoted backward until its centre coincides with the line of action. (Figure 3.)

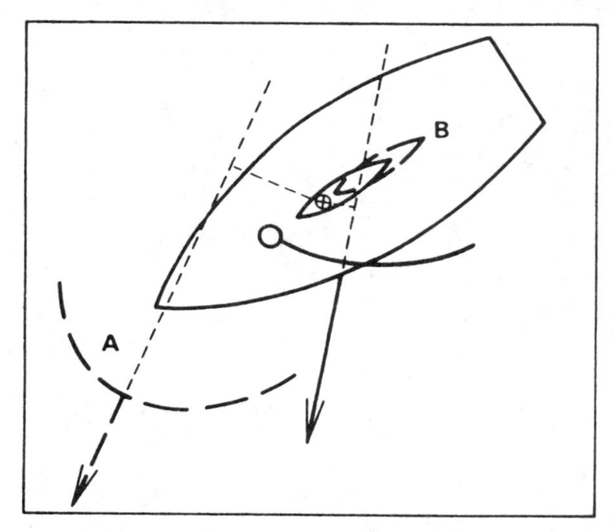

Figure 3. The turning moment of the mainsail may be compensated by either a counterbalancing moment from a spinnaker (A) or by movement of the fulcrum (the centreboard, B).

The Physics of Match Racing

When two yachtsmen are involved in a direct, boat-for-boat contest (such as the America's Cup) tactics become very important. With only two boats in the race, the result can be changed not only by making one's own boat go faster, but also by slowing the other down.

An obvious example is "covering", in which one yacht sails upwind of the other and spoils its wind. A much more subtle case involves so-called "safe leeward position", in which the disadvantaged yacht is more or less to windward.

Figure 4 shows the principle of this tactic. The yacht shown by the continuous line is sailing north-east in a northerly wind; the lift force on the sails acts roughly eastward (but for their drag angle). A certain Mr. Newton, a while back, said something about this situation that should be familiar to us all; the yacht exerts a westward force on the wind. Consequently, within the neighborhood of the yacht (effectively, a boat length or two) the wind direction is significantly to the east of north.

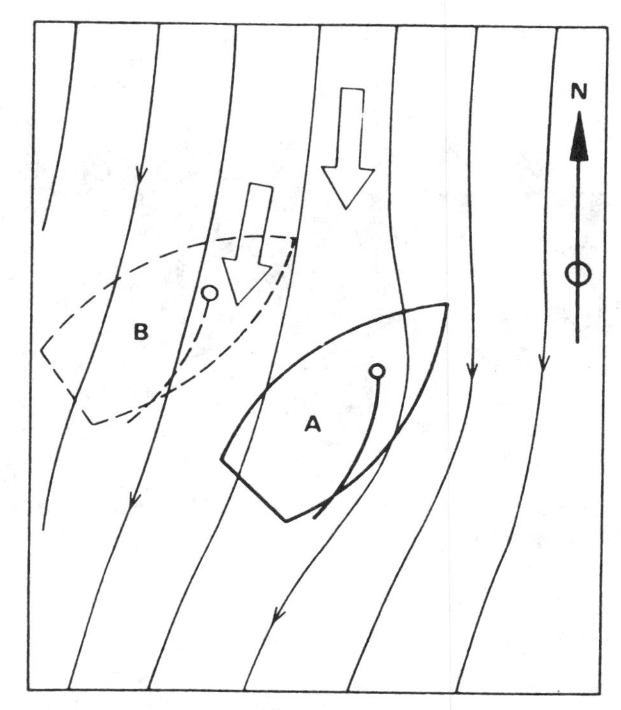

Figure 4. The "safe leeward position". Yacht A turns the wind direction, in the region immediately upwind, to a more north-easterly direction, so that an approaching yacht (B) would not be able to sail as close to the northerly course.

Even ahead of the yacht in question, another yacht (such as the one shown by the broken line in Figure 4) experiences this northeasterly tendency. As a result it can not sail on a course as close to the wind direction, and is disadvantaged.

If the helmsman of this second yacht is not careful, he may even be assisting the yacht in the "safe leeward position". The part of the wind which turns around his yacht to his right takes a more northwesterly direction. If this wind extends to the former yacht, it will be able to sail a correspondingly more northerly course.

Physics, Yacht Design and Safety

In the quest for the ultimate in performance, yacht designers have to decide on a trade-off between safety and speed. The main aim of many restrictions in a yacht class's specifications is the prevention of the worst excesses. However, some are inherent in the chosen type of yacht — for example, no dinghy sailor is competent unless he knows how to right a capsized yacht!

Other problems involve more explicit physics. Dinghies capsize, but catamarans nosedive. In fresh winds, the lift force on the sails of a downwind-sailing catamaran produces a moment that tends to turn its bow downward. This moment is opposed by the couple of the combined weight of yacht and yachtsman, with the distance of the centre of gravity from the "centre of buoyancy". In an attempt to reduce drag, catamaran designers reduce the beam at the forward end of the hull. This places the centre of buoyancy further aft, reducing the "righting moment" — with the occasional result shown in Figure 5.

A more recent arrival on the scene is the "windsurfer" style of craft, essentially a surfboard with a sail. The "yachtsman" holds the sail upright while standing on the board. This is fine in light winds, but in any significant blow, he must lean himself and the sail to windward to balance the sail's turning moment.

Figure 5. Catamaran suffering an imbalance of heeling and righting moments.

The unfortunate board sailor now begins to learn about the concept of unstable equilibrium. No wind is perfectly constant. In the case of a normal dinghy, with its sails heeled to leeward (by, say, an angle θ), the occasional drop in wind strength results in a drop in θ. Since the righting moment due to the weight of yacht and crew depends roughly on $\cos \theta$, while the "capsizing moment" of the sails depends on $\cos^2\theta$, the former increases more slowly than the latter, and for small wind fluctuations a new equilibrium value of θ results. But, due to the windsurfer's heel to windward, the righting moment of the "crew" increases as the capsizing moment of the sail decreases. An excessively large net righting moment is the end product — that is, a capsize to windward!

Conclusion

The above are only a few of the many ways in which physics may be applied to the sport of sailing. In a complete study of these connections, the basic mechanics explored here would be augmented by applications of fluid dynamics, meteorology and materials science.

It is little surprise that physicists take readily to sailing. The author has come across frequent examples — such as a recent trip to the United Kingdom, where three of the seven lecturing staff in his research group were avid yachtsmen. Any doubts of the connection were sealed when the author was fortunate to find a crewing position with the champion of his local yacht club. A conversation after the first race came to the point of discussing jobs. His reply was "I'm with Decca Radar. I'm a physicist".

Drag force exerted by water on the human body

A. Bellemans
Department of Physics, CP 223, Université Libre de Bruxelles, Brussels, Belgium
(Received 19 March 1980; accepted 18 April 1980)

One of the main contributions to the power demand in swimming arises from the drag force F_D exerted by water on the body. Our aim here is to call attention to some recent measurements of F_D for a *towed* subject, by Clarys and Jiskoot.[1] These data may be of interest for courses in physics or biomechanics addressed to students in physical education.

The experiments were carried in a dock, 200 m long and 4 m wide, the swimmer being towed at controlled speed by an electrically driven carriage. The reported values of F_D, for various positions and for velocities ranging from 1.5 to 1.9 m/sec, correspond to an average over a selected sample of 43 male subjects, all good swimmers. The Reynolds number is close to 10^6 at such speeds (highly turbulent flow) and one may expect the drag coefficient C_D, involved in[2]

$$F_D = (1/2)C_D \rho S v^2, \quad (1)$$

to be almost constant in this region (here ρ is the specific mass of water, S the frontal area of the body, and v its velocity). Indeed, when the reported values of F_D are plotted versus v^2, they fall reasonably well on straight lines passing through the origin of coordinates. See, e.g., Fig. 1. Least-squares adjustments of Eq. (1) on the experimental data lead to the results of Table I for $F_D/v^2 \equiv (1/2)C_D \rho S$. The determination of C_D requires the knowledge of S for which Clarys and Jiskoot provide no information. Further it is only when the body is fully immersed in water that C_D has a simple meaning. Assigning to S the plausible value of 0.1 m^2, one gets $C_D \simeq 0.7$ for the prone position, 60 cm below the water surface. Such a value for C_D compares very well with those of a sphere (~0.47), a nonstreamlined vehicle (0.6–0.8) and a racing cyclist (~0.9).

Table I. Least-squares estimates of F_D/v^2 from the data of Clarys and Jiskoot (Tables 1 and 2).[1]

	F_D/v^2 (N sec^2/m^2)	C_D (assuming S = 0.1 m^2)
Prone position, 60 cm below surface (18°C)	36.6 ± 1.0	0.73
Prone position, 60 cm below surface (24°C)	34.0 ± 0.6	0.68
Prone position at the surface (18°C)	30.2 ± 0.3	
Prone position at the surface (24°C)	28.5 ± 0.3	
Side (45°) position, at the surface (18°C)	29.4 ± 0.4	

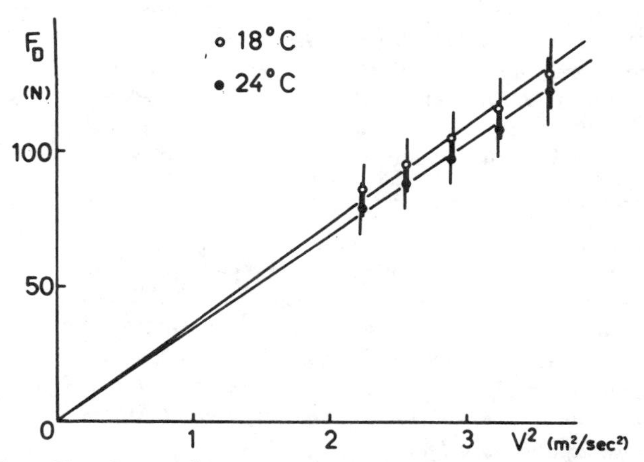

Fig. 1. Plot of F_D vs v^2 for a towed subject, 60 cm below water surface, at 18 and 24° C. The data (transformed to SI units) are taken from Table 2 of Clarys and Jiskoot.[1] Straight lines correspond to least-squares adjustments. Note that the vertical segments associated to the experimental points merely reflect the dispersion of the population of 43 subjects (rather than experimental uncertainties).

Given the values of Table I, it appears that the power required for swimming on the surface at a speed of 1.7 m/sec (i.e., approximately 100 m in 60 sec) cannot be lower than ~(30 N sec^2/m^2)(1.7 m/sec)^3, i.e., ~150 W. In the actual case, approximately twice this value is needed,[3] as the swimmer must himself insure his progression by moving his limbs adequately.

It may be interesting to note that the drag forces reported by Clarys and Jiskoot seem to decrease by nearly seven percents when the temperature varies from 18 to 24°C. Although C_D is a function of the Reynolds number *Re*, it is known to change extremely slowly in the neighborhood of $Re = 10^5$–10^6, so that the increase of 15 percent of *Re*, related to the drop of the water viscosity on going from 18°C (1.05 mN sec/m^2) to 24°C (0.91 mN sec/m^2) is insignificant. Assuming the change in C_D to be real, it could possibly be traced to the fact that the human body does not behave rigidly but will, to some extent, adapt its shape spontaneously in order to reduce the drag force; as the water temperature increases, the thermal stress of the subject is reduced and, by loosing some stiffness, its body will shape itself better in the flow.

[1] J.-P. Clarys and J. Jiskoot, Sciences et Motricité 1, 71 (1978).
[2] R. P. Feynman, R. B. Leighton, and M. Sands, *Lectures on Physics* (Addison-Wesley, London, 1964), Vol. 2, Chap. 41.
[3] The total power demand may be deduced from the oxygen consumption of the swimmer; see, e.g., P. O. Astrand and K. Rodahl, *Textbook of Work Physiology* (McGraw-Hill, New York, 1970), Chap. 16.

Reprinted from *American Journal of Physics* **49**, 367–368 (1981); © American Association of Physics Teachers.

The physics of swimming

M R Kent

Like all physical skills swimming is subject to the laws of mechanics. Until fairly recently, however, the biomechanical study of this crossdisciplinary activity has been largely left alone by the two specialists who would be primarily involved, the physicist and the physical educationalist. The reasons for this are probably that the physicist considered established coaches to be getting the best from their swimmers through well tested techniques, and the physical educationalist considered that although the more rigorous scientific approach of the physicist might produce some small improvement, the few technical data available were too difficult to interpret into meaningful modifications of style. Since the early 1970s, however, a greater crossfertilisation of ideas has taken place and the sports scientist has emerged. Recent television documentary programmes have shown how it is possible to analyse human performance and thus identify areas where for example greater strength, a slightly different angle or a slightly different degree of twist might produce that all important extra few millimetres in distance or that reduction in time of a few milliseconds. In man's search for excellence this is important.

The study of the performance of a sprinting swimmer in biomechanical terms makes it possible to build up a model of the stroke under investigation which may then be improved upon. The problem has a number of features which are not present for the dry land athlete. Water creates interesting mechanical effects due to friction, buoyancy, eddy formation and wave making for the swimmer, and optical and possibly electrical problems for the researcher. Other hidden dangers exist for both swimmer and researcher in that mains electricity is not welcome in poolside measuring instruments and all equipment must be sealed against damp. Once these problems have been overcome, however, the scientist can start his work.

If a continuous record in time is made of the instantaneous velocity V of the centre of mass of a swimmer an interesting series of regular curves emerge which although similar are never exactly repeated (Myashita 1971, Kent and Atha 1975, Bober and Czebanski 1975, Holmer 1979). The distances covered in consecutive cycles of the stroke are thus slightly different and this can be shown by simply calculating the area enclosed beneath the curve, between the start and end of the stroke cycle. A most important further implication is that there must be one 'best' stroke cycle and that if this can be repeated just once more by the swimmer without introducing adverse effects on any other stroke then a faster overall speed will result.

Theoretical model

In any attempt to evaluate the mechanical work and power of a swimmer the main obstacle to overcome is the determination of the force being applied to the water at every instant during the stroke cycle. This force is continually altering as a direct result of postural changes made by the swimmer. Further, this 'force' is not simply a propulsive force, since even though the main motion of a particular limb may be directed backward relative to the swimmer, much of the limb may be moving forwards with the body relative to the water. This implies that, in terms of retarding forces, a dynamic component must be added to what otherwise would be largely passive water resistance forces. As a result of the complex nature of the force generated by the swimmer it is not easy to separate the unseen work which has to be done in overcoming fluctuating water resistance from that which can be assessed from observed increases in speed. It is essential therefore that detailed information is obtained about the time course of total resistance throughout the stroke cycle for the purpose of analysis.

If it is assumed that propulsive forces generated by the swimmer are solely responsible for any acceleration but that oppositely directed forces produced by the propelling limbs exist at times of both acceleration and deceleration, then four conditions of motion for the moving swimmer may be considered.

Condition 1: If $P>L$ and $(P-L)>R$, where P represents the propulsive force generated by the swimmer, L the active retarding force created by 'propulsive' movements and R the water resistance created by skin friction, eddy formation and wave

---

Malcolm Raymond Kent *is the head of physics at Kelly College, Tanistock, Devon. After gaining a MSc at Loughborough University he became assistant master at Loughborough Grammar School, and then served five years in the Royal Navy as an instructor lieutenant commander. He has published a number of papers, mostly on biomechanics in swimming.*

Reprinted from *Physics Education* **15**, 275–279 (1980); © The Institute of Physics.

making and roughly proportional to velocity squared for prone swimmers, then *acceleration* is produced such that

$$P_A = (P-L) - R$$

where P_A represents the accelerating force. This condition will exist throughout the major part of any propulsive movement in any stroke where P is more than sufficient to cancel out the sum of L and R.

Condition 2: If $P>L$ but $(P-L)<R$ *deceleration* is produced. This will be the case towards the end of a so called 'propulsive' phase when as true propulsion dies away the high velocity of the swimmer produces very large water resistance forces.

Condition 3: If $L>P$ and $(L-P)>R$ *deceleration* is produced. This situation will only exist at low velocities when R is small but it is interesting to note that if R is allowed to be zero in the above condition, that is when velocity is zero, then since $L>P$ the swimmer will go backwards. Some breaststroke learner swimmers experience this from time to time, particularly when drawing the legs forward prior to the leg drive.

Condition 4: If $L>P$ and $(L-P)<R$ *deceleration* is produced. This will be the case at high velocity when R is large whilst the swimmer is again repositioning limbs ready for a power phase.

If it is assumed that all major propulsive actions are considered to exist as set out in condition 1 then during a freely performed swim when $P>L$ and $(P-L)>R$

$$P_A = (P-L) - R.$$

Therefore $P_A = P - (L+R)$

$$= M\left(\frac{dV}{dt}\right)_t$$

where V is the instantaneous velocity at time moment t and M is the mass of the swimmer. Then

$$M\left(\frac{dV}{dt}\right)_t = P - (L+R). \qquad (1)$$

Thus, to evaluate the propulsive force generated by the swimmer, M, $(dV/dt)_t$ and $(L+R)$ must be known. Although evaluation of the first two quantities is straightforward the calculation of the $(L+R)$ term is more complex. One solution is to actually tow the swimmer through water at a series of velocities which are high enough to ensure that when the swimmer performs a normal stroke during the tow, the towing line does not slacken. If the tension in the towing line is continually monitored and the posture of the swimmer is simultaneously recorded on cine film, it is possible to build up a family of tension variation curves each recorded at a different constant velocity. By means of these the continually changing $(L+R)$ component may be identified.

Furthermore, this component may be directly related to the equivalent intra-cycle moment during free swimming by ensuring that the free swim too, from which the intra-cycle instantaneous velocity is secured, is also recorded on cine film. With this information and assuming that during towing the force that would normally have produced acceleration in the freely performed stroke is instrumental only in reducing line tension T we obtain

$$T = (L+R) - P_A$$

at any instant. Considering a specific moment t during the freely performed stroke at which time in that particular posture the swimmer would be moving with velocity V, we obtain

$$T_t = (L+R) - M\left(\frac{dV}{dt}\right)_t$$

where T_t is the tension recorded at the same moment t, in the same posture, whilst the swimmer is pulled through the water at the same velocity V. Therefore

$$(L+R) = T_t + M\left(\frac{dV}{dt}\right)_t$$

and substituting for $(L+R)$ in equation (1) gives

$$M\left(\frac{dV}{dt}\right)_t = P - T_t - M\left(\frac{dV}{dt}\right)_t$$

$$P = T_t - 2M\left(\frac{dV}{dt}\right)_t. \qquad (2)$$

The above equation may thus be evaluated whenever the acceleration of the swimmer is positive.

Having secured information on the posture, propulsive force and instantaneous velocity of the swimmer it is possible to calculate intra-cycle work output, cycle work output, intra-cycle acceleration and, of course, compare these with one another, between swimmers and with the 'ideal' stroke sequence in order to identify those features of performance which require modification. Although the technique described may be applied to any stroke sequence the results which follow illustrate the intra-cycle biomechanical variations in a British Olympic breaststroke swimmer.

Biomechanical profiles

Figure 1 shows the intra-cycle velocity-time relationship of a single stroke cycle for this top class swimmer which was secured using equipment and methods reported elsewhere (Kent and Atha 1975). The timebase used originates from a clearly identifiable feature of the breaststroke cycle, that of zero velocity when the legs are drawn forward in preparation for their propulsive movement. This feature is unique to this stroke and may be used as a most suitable mid-stroke synchronisation point. The body configurations shown are copied from cine film synchronised to the velocity recording device and provide information for the coaching process. As can be seen, the velocities range from zero to 1.95 m s^{-1} and 1.85 m s^{-1} in the arm and leg propulsion phases respectively.

Figure 1 also shows the instantaneous acceleration of the swimmer calculated directly as the derivative of the velocity trace just described. From this curve it can be seen that the acceleration produced by the leg propulsion phase at 9.25 m s^{-2} is much greater than that produced by the arms at 1.00 m s^{-2}. Even this massive acceleration however fails to produce a higher velocity than that obtained during the arm propulsion phase because of the much lower (zero) starting velocity. The values of both acceleration and velocity throughout the stroke cycle are shown in table 1.

Figure 2 shows a typical family of tension variation traces as a function of time. These traces are required to obtain the term T_t to evaluate propulsive force as shown in equation (2). As stated previously, these curves are obtained by pulling the swimmer through the water at a number of prearranged and constant velocities whilst the tension in the towing line is continuously monitored and synchronised to cine film for a single stroke cycle (figure 3). From these curves it is possible to read off T_t by interpolation, knowing (a) the intra-cycle stroke posture of the swimmer at a given instant and (b) the instantaneous velocity at that instant from figure 1.

As can be seen, the tension in the towing line falls initially as the arms make their propulsive contribution, and under normal circumstances this reduction in tension would have produced acceleration. As the propulsive effort of the arms dies away a massive peak in resistance is observed as the legs are drawn forward ready for their drive. Although this massive force may look important it must be remembered that the largest forces are recorded at the highest towing velocities but at this stage of the stroke the swimmer has come momentarily to rest. As a result of this interpolation the calculated results produce the resistance profile

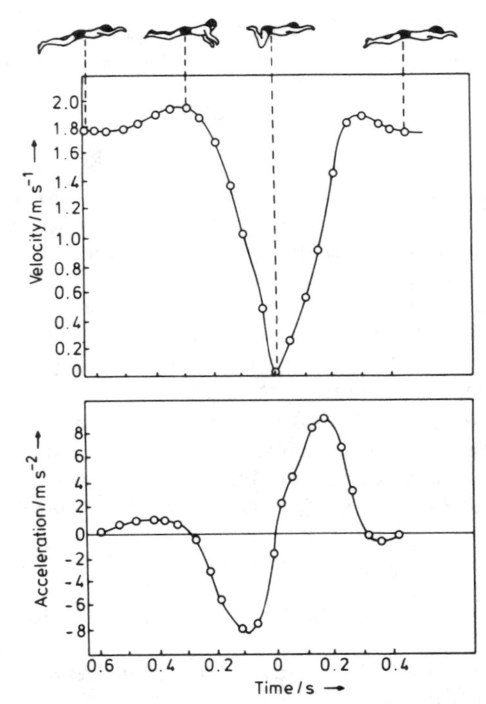

Figure 1 Velocity and acceleration changes during the breaststroke cycle

Figure 2 Tension variations in towing line for a top class breaststroke swimmer. Numbers on curves show velocities in m s^{-1}

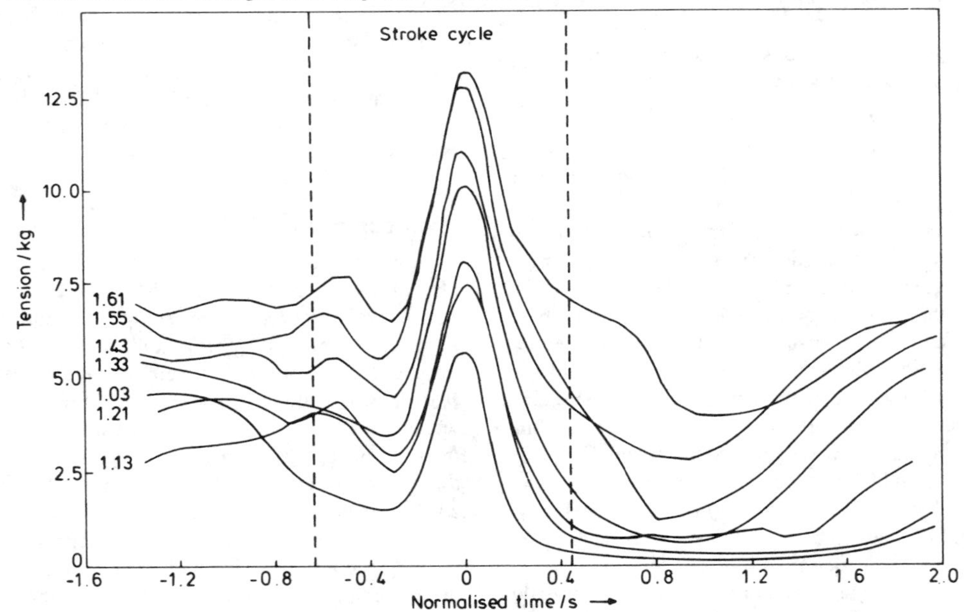

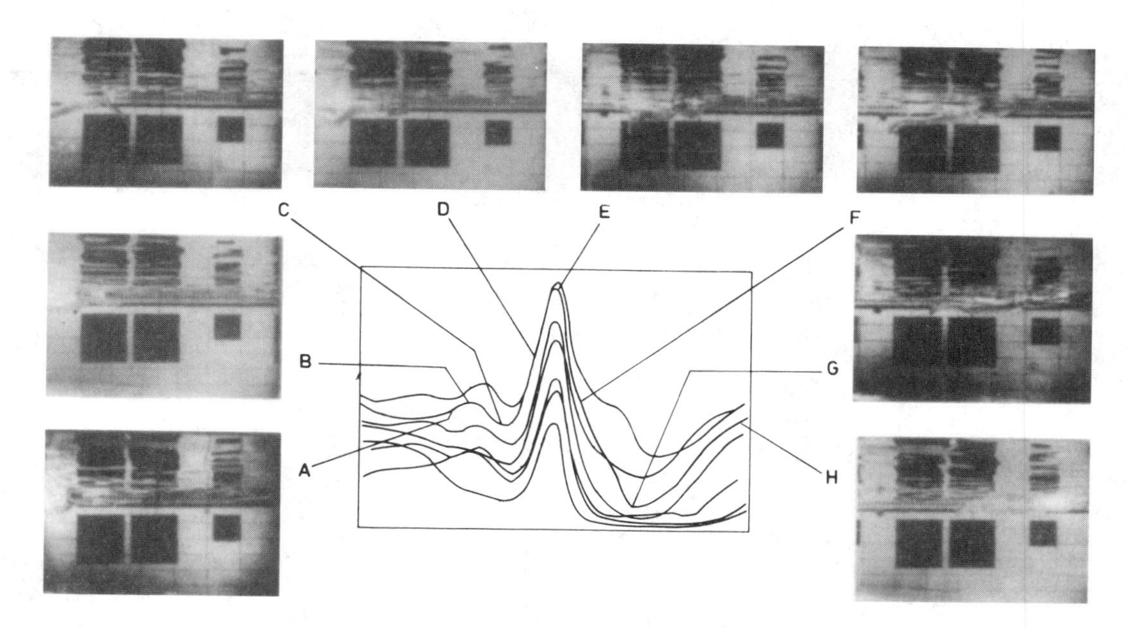

Figure 3 Relationship between posture and towing line tension for a top class breaststroke swimmer moving at a constant velocity of 1.55 m s^{-1}

shown in figure 4, the actual results being recorded in table 1. Here it may be observed that there is similarity between the resistance trace and the velocity trace shown in figure 1. Indeed it is possible to suggest that the intra-cycle resistance of the active breaststroke swimmer is more closely related to velocity than to the square of velocity, as is clearly the case for inactive, towed swimmers.

Figure 4 Water resistance and propulsive power changes during a breaststroke cycle

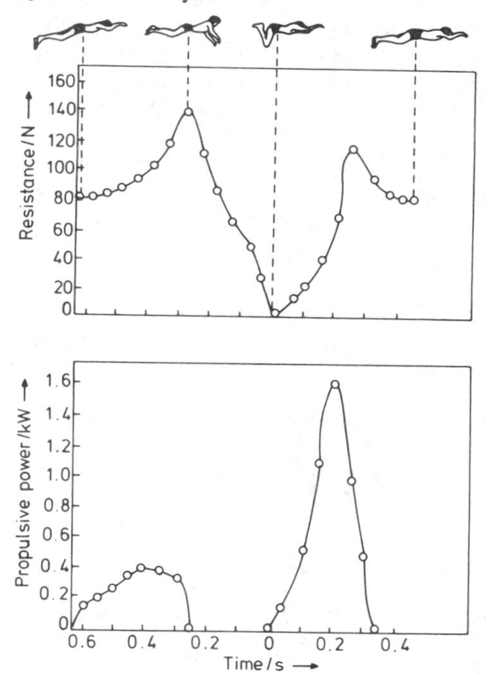

Figure 4 also shows the computed propulsive power for the swimmer calculated from equation (2) and using data shown in table 1. As stated previously it is possible to calculate propulsive force and hence propulsive power whenever the swimmer is accelerating, and this type of analysis clearly shows the relative contribution made by the arms and legs. Further, it also shows the massive counter-productive effect produced by the positioning of the legs prior to their propulsive drive.

It is clear that a physical analysis of swimming at the intra-cycle level as described enables features of performance to be identified which may not be observed in any other way. Such an analysis, together with the accompanying cine film record, gives valuable insight into the way in which man handles an aquatic environment. The results described above relate to a highly skilled performer and show that instantaneous propulsive powers of the order of 1.6 kW are generated when swimming at an average speed of 1.41 m s^{-1}. Tests for the whole performance of swimmers based on oxygen uptake measurements show that, at an average speed of 0.45 m s^{-1}, 475 W are generated (Cameron and Skofronick 1978). Such measurements contain little information relating to performance at the intra-cycle level although they are well established and thus serve as useful comparisons.

Finally, it is informative to perform one other

Table 1 *Mean computed intra-cycle biomechanical characteristics of a top class breaststroke swimmer. Mass of swimmer M = 60.8 kg. N denotes negative acceleration*

Time moment/s	Velocity/m s^{-1}	Acceleration/m s^{-2}	Resistance/N	Propulsive force/N	Propulsive power/W
−0.64	1.76	0.00	83	0	0
−0.60	1.76	0.15	82	100	176
−0.55	1.78	0.30	85	121	215
−0.50	1.80	0.60	97	170	306
−0.45	1.83	1.00	95	217	397
−0.40	1.90	0.95	103	219	415
−0.35	1.94	0.65	108	187	363
−0.30	1.95	0.55	136	203	396
−0.25	1.88	−2.55	118	N	N
−0.20	1.70	−4.30	84	N	N
−0.15	1.49	−6.15	73	N	N
−0.10	1.09	−8.15	45	N	N
−0.05	0.70	−7.50	31	N	N
0.00	0.00	0.00	0	0	0
0.05	0.29	3.65	11	455	132
0.10	0.54	7.40	19	918	496
0.15	0.96	9.25	26	1150	1104
0.20	1.51	8.15	76	1066	1610
0.25	1.82	3.75	114	570	1038
0.30	1.85	0.40	93	221	409
0.35	1.80	−0.95	87	N	N
0.40	1.77	−0.60	85	N	N
0.44	1.76	0.00	83	N	N

simple calculation for the performance of the top class swimmer whose results have been used to illustrate the physics of swimming. Given that the area enclosed by the propulsive power curve in figure 4 gives the total work output in that cycle and that the area enclosed by the velocity curve in figure 1 gives the total distance travelled, it is possible to calculate the propulsive power output for the stroke cycle described. In this case

Area enclosed by propulsive power curve = 345.6 J
Cycle time = 1.11 s
Therefore propulsive power output = 311 W
Mean velocity, from the velocity-time curve = 1.41 m s^{-1}.

It will be noted that this propulsive power output is less than the 475 W quoted by Cameron and Skofronick (1978) for a swimmer travelling much more slowly, but it must be remembered that the 311 W calculated represents propulsive power based on measurements of propulsion; it does not therefore, include any power used in maintaining body functions, or buoyancy or in making counter-productive limb movements.

In contrast to the propulsive power generated by this swimmer whose velocity changes continually throughout the stroke cycle, it is informative to calculate the theoretical power required by the same swimmer if it were possible for a constant velocity of 1.41 m s^{-1} to be maintained. Given that the drag on a prone swimmer is well accepted as being given by DV^2, where D is the drag constant for the prone swimmer and V is the velocity of the swimmer then

$$\text{Power} = DgV^3.$$

For the swimmer described, D = 2.76 kg s^2 m^{-2}, V = 1.41 m s^{-1}, and thus power = 76 W.

As can be appreciated, the 76 W required to maintain a steady 1.41 m s^{-1} for this swimmer is much less than the 311 W required to maintain an average velocity of 1.41 m s^{-1} where velocity changes by as much as 1.94 m s^{-1} in each cycle. In so far as it is impractical to coach a breaststroke swimmer or indeed any top class swimmer to maintain a constant velocity, it is reasonable to argue that, if the velocity fluctuation could be minimised, then the energy saved could be redirected into increasing overall performance. It is certain that with modern techniques of analysis such as the one described here, in which physics plays the major part, this will be done.

References

Bober T and Czebanski B 1975 *Proc. 2nd Int. Symp. Biomechanics in Swimming* (Baltimore: University Park Press) pp188–93

Cameron J R and Skofronick J G 1978 *Medical Physics* (New York: Wiley) p93

Holmer I 1979 *Swimming III* International Series on Sports Sciences Vol 8 (Baltimore: University Park Press) pp118–24

Kent M R and Atha J 1975 *Proc. 2nd Int. Symp. Biomechanics in Swimming* (Baltimore: University Park Press) pp125–9

Myashita M 1971 *Proc. 1st Int. Symp. Biomechanics in Swimming* (Brussels: Universite libre de Bruxelles) pp53–8

Chapter 9
SPORTS ASSORTMENT

CONTENTS

This chapter contains 12 articles in all, two on football, two on sports probabilities, two on drag racing, one on weight distribution in sailboats, two on rowing, and one each on badminton, fly casting, and springboard diving.

Football

"The physics of kicking a football," by Peter J. Brancazio, is the first article in this chapter. In it the author begins by presenting a statistical analysis of the range D and "hang time" T of hundreds of NFL kickoffs and punts. He then constructs a theoretical model for the trajectories in which the two key observables, D and T, are connected to the familiar initial conditions θ_0 and v_0 (angle of launch and initial velocity), through a parameter K which characterizes the degree of aerodynamic drag encountered by the football.

Since the drag force on an object depends on the area presented to the direction of its motion, the wobbling and tumbling motions typical of a kicked football lead to drag forces that vary substantially with the instantaneous orientation of the ball, unlike spherical projectiles for which the projection area and drag force tend to be constant along the entire trajectory. Due to the varying nature of the drag force on a tumbling football, it becomes necessary to choose an average value for the "drag factor" K in order establish the connection described above.

Once an appropriate average value of K has been selected, however, Brancazio's framework makes it possible to infer the launch angle and initial velocity for a kicked football from knowledge of its range and hang time, as well as to predict the range and hang time for a kicked football from given values of θ_0 and v_0.

The second article in this chapter (also by Brancazio) asks and answers the fascinating question, "Why does a football keep its axis pointing along its trajectory?" The trajectory in question is most familiar from the graceful flight of a well-thrown long pass so often captured in elegant slow motion by the cameras of NFL Films.

Brancazio's treatment of this problem, focusing on the central roles played by spin, angular momentum, and drag-induced torque, is equally elegant. As pointed out in the conclusion of his article, the key to the perfect spiral is a high degree of spin together with an alignment of the initial spin axis and velocity vector of the ball.

Probability and Sports

The third article in this chapter "Probability, statistics, and the World Series of baseball," by Kenneth S. Krane, compares the historical frequency of seven-, six-, five-, and four-game World Series with the expected frequency for each case, on the assumption that the teams are evenly matched.

Binary statistics are used and the agreement between actual and expected outcomes is reasonably good overall, and quite spectacular for one statistic in particular—the mean length of the series. In the words of the author,

> Excellent agreement is obtained between the expected mean length of the series (5.81 games) compared with the actual mean length ($441/76 = 5.80$ games).

The fourth article in this chapter is entitled "Exercise in probability and statistics, or the probability of winning at tennis," by Gaston Fischer. In it, the author defines p, x, s, and M as the probabilities of winning, respectively, a point, a game, a set, or a match in tennis. For simplicity's sake, he assumes that all probabilities are averages over both serving and receiving situations. He then sets out to find the relationships that exist between the four probabilities.

On the basis of binomial statistics and the following assumptions: (a) that a tie-break system is used when set scores reach 6–6 and, (b) that probabilities for winning tie-break games are identical to those for normal games, Fischer concludes that

> As is to be expected, a small advantage in the probability of winning a point leads to advantages which are amplified by large factors: 2.5 for games, 7.1 for sets with tie-break at 6–6, 10.6 for matches to the best of three sets, and 13.3 for matches to the best of five sets.

Drag Racing

The fifth article in this chapter is "On the physics of drag racing," by Geoffrey T. Fox. In it, the author describes a drag race as "an acceleration contest from a standing start over a distance of ¼ mile," and defines ET, the elapsed time from start to finish, and MPH, the speed of the vehicle at the end of the track, as dependent variables.

To first order, Fox limits the independent variables to the weight of the car and the power of the engine and, more specifically, to the car's power to weight ratio. This single-parameter model uses a constant power approximation to provide not only an excellent conceptual understanding of drag racing, but some very interesting predictions as well.

For example, Fox shows that MPH, the speed of a vehicle at the end of a ¼ mile race, should be proportional to the cube root of the power to weight ratio, with a constant of proportionality $K_{\text{theory}} = 280$. Fox's formula is identical to that of a well-known empirical formula for drag racing, the only difference being the constant $K_{\text{exp}} = 225$, which is only 80% of the theoretical value.

When one cubes the ratio of these two numbers, however, the result indicates a 50% loss in theoretical engine power, a result which Fox attributes to factors such as, e.g., wind and rolling resistance, traction limitations, and drive train inefficiencies. One of the largest sources of power loss between the engine and the rear wheels is that required to provide the large rotational energies stored in rapidly spinning engine, transmission, and drive train parts.

In second order, Fox includes the coefficient of friction between tires and track as an additional independent variable. He then divides the acceleration contest into two parts: a constant force portion at the beginning of the race, and a constant power portion as soon as the velocity is large enough so that the power is no longer sufficient to spin the tires (recall that power = force × velocity). In the first part of the race, performance is limited by the available traction while, during the second part of the race, it is limited by available *power*.

The sixth article in this chapter, "Constant power equations of motion," by Roger Stephenson, is a pedagogically useful companion piece to the previous article by Fox. In it, the author gives a scientific definition for "Zip," a word often used in the vernacular to describe a street car capable of noteworthy acceleration—and the exhilaration known to accompany it. Not surprisingly, Zip turns out to be closely related to Fox's power to weight ratio.

Stephenson notes that while every physics student is exposed to numerous examples of motion with constant force, very few seem to be familiar with examples involving constant power. In an effort to rectify this situation, the author provides useful graphical comparisons in which the time profiles for distance, velocity, and acceleration (which is *not* constant in the constant power case) are compared. Stephenson also discusses the importance of vehicle weight distribution, a characteristic which itself changes with acceleration.

Weight Distribution in Sailboats

The seventh article in this chapter, "Weight distribution: on its effect—or otherwise—on racing dinghy speed," by P. F. Hinrichsen, offers a fascinating glimpse at the interface between physics and the Olympic sport of sailboat racing in the Flying Dutchman class.

Hinrichsen describes a sequence of events in which (a) the need for a weight distribution rule is proposed, (b) the key physical quantity thought to confer advantage is identified, (c) techniques for rapid field measurement of that quantity are adopted, and (d) racing outcomes are assessed to determine whether significant advantage can be linked statistically to the quantity suspected of providing it.

Questions about the possible need for a weight distribution rule[1] arose from the following concern:

> Two identical boats, even as far as all-up weight and the balance point, may have significantly different boat speeds when sailing in waves if one boat has its weight distributed evenly while the other has very light bows and stern and a heavy center section.

The key physical quantity thought to be associated with a possible advantage was the "pitching moment," i.e., the moment of inertia about the boat's "athwartships axis," a horizontal axis perpendicular to both the vertical and fore-and-aft axes of the boat.

An oscillation-type test previously devised by Gilbert Lamboley was adopted for measuring the pitching moments of the various boats. This test was then administered on a voluntary basis to competitors at the 1976 Olympic Games.

However, when total moment of inertia was plotted against finishing time for each of the competitors, the author concluded that the data "gives no evidence of a trend." It was this somewhat disappointing observation, no doubt, that prompted Hinrichson to add the words "or otherwise" to the subtitle of his article.

Rowing

The eighth article in this chapter, "Rowing: A similarity analysis" by Thomas A. McMahon, begins with a question:

> Why should larger boats containing many oarsmen go faster than smaller ones containing fewer oarsmen?

In response to this question, McMahon develops an elegant model for calculating the relative speeds of geometrically similar racing shells with different numbers of oarsmen. The model predicts that the speed of a racing shell should be proportional to the number of oarsmen raised to the one-ninth power, a relationship which fits the empirical data for world class competition very well.

The author also probes the question of why eight-oared shells rowed by heavyweight crews should consistently beat equivalently advantaged lightweight crews, by margins typically on the order of five percent. McMahon not only answers this question, he also suggests a simple strategy, based on his model, for closing the gap:

> The remarkable conclusion becomes that, if the light-weight shell were shortened by 1.1 m in length and made 2.4 cm narrower in beam from its present dimensions, the lightweight crew could keep up with the heavyweights.

The ninth article in this chapter is "Why is it harder to paddle a canoe in shallow water?" by Angelo Armenti, Jr. In it, the author notes that the total resistance encountered by a hull moving on the water arises from two distinct sources, viscosity effects (skin friction drag) and gravity effects (wave-making resistance), the same two factors discussed in some detail previously in relation to the resistance associated with sailboat hulls (see Chap. 8).

As the author notes, each of these effects varies with water depth and, as a result, the total resistance encountered by a moving hull is quite sensitive to variations in depth. The article concludes by noting that the observed increase in total resistance with decreasing depth results precisely from the combined action of the skin friction drag and wave-making resistance effects cited above.

Badminton

The tenth article in this chapter, "Terminal velocity of a shuttlecock in vertical fall," by Mark Peastrel, Rosemary Lynch, and Angelo Armenti, Jr., describes a simple experiment in which the terminal velocity of a badminton shuttlecock was measured. The times required for the shuttlecock to fall given distances up to almost ten meters were measured in an open stairwell. The data were then compared to the predictions of several models.

The best data fit was obtained with a model which assumed a resistive force quadratic in the instantaneous speed of the falling object. This model was then used to determine the shuttlecock's terminal velocity, 6.80 m/s, 99% of which was reached in just 1.84 s, after falling 9.2 m.

Fly Casting

The eleventh article in this chapter, "The physics of fly casting " by John M. Robson, explores the physics of the process by which anglers retrieve and redeploy a 100-mg "fly" on the end of a 10-m line using only a 3-m rod and a flick of the wrist.

The author begins with an idealized analysis that neglects air resistance but captures the key aspects of that portion of the cast that begins with the line instantaneously at rest and fully extended behind the angler. When the rod is brought forward quickly, an amount of work is done on the line which concentrates an equal amount of kinetic energy in the entire length of line plus fly.

When the rod is abruptly stopped, however, the line doubles over with the fly end continuing forward with increasing speed. The reason for the increase is that, when the rod is abruptly stopped, a small but growing portion of the line closest to the rod tip stops also, causing the essentially constant kinetic energy to be concentrated in a smaller and smaller mass, viz., the mass corresponding to the decreasing portion of the line which is still moving forward.

In the balance of the article, Robson improves on the simple model described above by including air resistance effects in a more sophisticated analytical model, and he concludes with a detailed computer model which shows excellent agreement between theory and practice.

Springboard Diving

The twelfth and final article in this chapter, "Do springboard divers violate angular momentum conservation?" by Cliff Frohlich, begins by answering the title question firmly in the negative, and then proceeds to analyze a number of apparent paradoxes that "have confused both physicists and coaches for some time."

After analyzing the physical characteristics of dives and divers, Frohlich focuses on the precise manner in which somersaults and twists are initiated. Throughout the article, the author offers excellent insights into these paradoxes as he works toward the observation that:

> performers can and do execute torque-free somersaults, but only in such a manner that angular momentum is conserved.

---

[1]As noted by Hinrichsen, Flying Dutchman boats were already subject to a rule specifying the boat's *minimum total weight*, a "handicap" rule requiring lighter competitors to carry additional weight to make up any differences. Such handicap systems are common in competitive sports, e.g., thoroughbred horse racing, rowing, and automobile racing to name a few, and are designed to give advantages or compensations to different contestants so as to equalize the chances of winning. A minimum weight rule in amateur rowing, however, provides that the coxswain "weigh in" several hours before a race, a delay that has encouraged a few underweight, unethical, but very innovative coxswain's to improve the chances for winning by drinking a gallon or more of water prior to weigh in. Horse racing eliminates such time-delay loopholes by weighing the winning jockey within seconds of the race, and before the results are posted.

The Physics of Kicking a Football

PETER J. BRANCAZIO

The game of American football is filled with illustrations of the principles of physics. Typically, a substantial fraction of the students in any introductory physics course follow college or professional football rather closely and are reasonably familiar with the strategies and techniques of play. An enlightened physics teacher can take effective advantage of this interest by using game situations to clarify some of the concepts of physics. For example, the shifty movements of a good running back can serve to demonstrate the concept of acceleration, and various blocking and tackling situations can be used as dramatic examples of the theory of two-body collisions. It is also instructive to consider the football as a projectile. The trajectory of a football pass or kick illustrates not only the principles of projectile motion, but also some basic concepts of aerodynamics.

The "Kicker's Dilemma"

Let us take a physicist's view of the problems involved in kicking a football. Whether the ball is being kicked from a tee or punted (i.e., kicked in midair as it falls onto the kicker's foot), the kicker's usual purpose is to send the ball as far downfield as possible. At the same time, he is also trying to keep the ball in flight for as long as possible, so that his teammates can get downfield to tackle the kick receiver and prevent a substantial runback. In the language of physics, the kicker would like to maximize both the horizontal range *and* the time of flight or 'hang time' of the kick. However, an examination of the equations of projectile motion quickly reveals that this is not possible. If air resistance is neglected, the maximum horizontal range, R, and hang time, T, are given by the equations:

$$R = \frac{v_0^2}{g} \sin 2\theta_0 \quad \text{and} \quad T = \frac{2v_0 \sin\theta_0}{g}$$

where v_0 and θ_0 are the launching speed and launching angle, respectively. It is evident from the above equations that R is maximized for a given v_0 by launching a projectile at a 45° angle, whereas T is maximized when the launching angle is 90°.

Obviously, it is very poor strategy to launch a kick vertically upward just to maximize its hang time - but on the other hand, a kick launched at a 45° angle has a hang time 29.3% shorter than the maximum for the given launching speed. Thus the "kicker's dilemma" is that he must choose a launching angle, somewhere between the extremes of 45° and 90°, that provides the best strategic combination of distance and hang time. How much distance is the kicker willing to sacrifice to gain hang time (or vice versa) in each given kicking situation? Under these circumstances, what is the optimum launching angle?

Since there is no analytic solution to this problem, and since football kickers themselves have arrived at a solution through the process of trial and error, we must therefore observe the kickers in action in order to discover how they deal with this dilemma. It is difficult to measure the launching angle directly; such an approach calls for high-speed cinematography under artificial conditions. However, it is possible, at least in principle, to calculate the launching angle from measurements of the trajectory of the kick. In particular, the horizontal range and hang time can be readily obtained from televised football games, especially if one can videotape them.

To obtain data for this study, I examined the videotapes of 26 regular National Football League games played between September and November of 1984. For every kickoff and punt observed, I recorded the horizontal distance in yards (measured from the point where the kicker's foot hit the ball to the point where the ball was caught or hit the ground) and the hang time. The latter was measured with a digital stopwatch; three viewings of a kick were usually sufficient to give a value accurate to 0.1 sec. The distances could be measured to an accuracy of one yard. It should be noted that the distances thus recorded are quite

Peter J. Brancazio *is an associate professor of physics at Brooklyn College, CUNY, the author of* **SportScience** *(Simon & Schuster, 1984), and a notoriously poor punter. He gave up his Sunday afternoons and Monday nights for several months in order to obtain the data for this article. Department of Physics, Brooklyn College, CUNY, Brooklyn, NY 11210).*

Reprinted from *The Physics Teacher* **23**, 403–407 (1985); © American Association of Physics Teachers.

different from those reported by the game announcers since football rules specify that the length of a punt is to be measured from the line of scrimmage to the point where the ball is fielded or comes to a stop. Thus if a punt is not caught and is allowed to roll or bounce after it hits the ground, the distance that it travels in this manner is added to the distance the ball travels in flight. (So that the numerical results will be recognizable to football fans, I will be using U.S. customary units throughout this article.)

I obtained data for a total of 206 kickoffs (by 21 different kickers) and 238 punts (by 24 different punters). Kickoffs and punts were grouped separately because they are produced under quite different conditions. In a kickoff, the ball is kicked from a tee, and the kicker has the luxury of an unimpeded running start to begin the kick. A punter kicks the ball from a standing start while opponents are rushing at him from all sides. These conditions dictate different kicking strategies, techniques, and skills - so that almost without exception, every football team assigns the jobs of kicking and punting to two separate individuals.

These differences are clearly reflected in the statistical results. Kickoffs show much less variability than punts. They exhibit a relatively narrow spread of hang times; 74% of the kickoffs recorded had hang times between 3.8 and 4.3 sec. By comparison, 77% of the punts had hang times between 3.6 and 4.7 sec. Moreover, 88% of the kickoffs traveled 55 yards or more, whereas punting distances were distributed widely (one standard deviation about the mean ranged from 43 to 58 yards). In general, the results show that the average kickoff travels about 12 yards farther than the average punt, whereas hang times for punts average about 0.2 sec. longer. The statistical summary shown in Table I clearly indicates that kickoffs and punts must differ significantly in both launching speed and launching angle.

Table I. Statistical Summary of Kickoffs and Punts

(Data from 26 NFL games, Sept. to Nov. 1984)

	Kickoffs	Punts
Total Number of Kicks Recorded:	206	238
Hang Time (sec):		
Mean	3.8	4.1
Median	4.0	4.2
Longest	4.6 (D = 55 yds)	5.3 (D = 51 yds)
Distance (yds):		
Mean	63	50
Median	63	50
Longest	76 (T = 4.0 sec)	71 (T = 4.1 sec)
Best Kick Observed:	D = 73 yds, T = 4.4 sec	D = 70 yds, T = 5.0 sec

The Aerodynamics of the Football

It would be a simple matter to calculate the launching speed and angle from the distance and hang time for any kick if air resistance could be neglected. Unfortunately, this is decidedly not the case - indeed, the unusual aerodynamic shape of the football plays a significant role in determining the strategies and techniques of kicking (and passing) the ball.

For the range of speeds typical of football play, the aerodynamic drag force F_D is given by the equation:

$$F_D = \tfrac{1}{2} \rho C_D A v^2 \qquad (1)$$

Where ρ is the air density, C_D is the drag coefficient, A is the cross-sectional area of the projectile normal to the trajectory, and v is the speed of the projectile relative to the air. The major difficulty in applying this equation to the football is that both A and C_D depend upon the orientation of the long axis of the ball to its flight path - and, in fact, they change substantially over the trajectory if the football turns or tumbles in flight (as it often does).

An American football has a roughly ellipsoidal shape. It is 11.1 in. (28.2 cm) in length along its major axis and has a maximum diameter of 6.8 in. (17.3 cm). Thus the cross-sectional area varies from a minimum of 0.25 ft^2 (when the ball is traveling nose-first) to a maximum of 0.41 ft^2 (when it is turned broadside to its path). There do not seem to be any published measurements of the drag coefficient, C_D, for a football in its various orientations. Measurements of C_D for an ellipsoid[1] project a value of about 0.1 for a nose-first orientation, whereas a value of 0.6 seems about reasonable for a football traveling broadside at typical speeds. Thus the product $C_D A$ ranges from a minimum of 0.025 ft^2 to a maximum of 0.25 ft^2. In other words, the drag force on a football varies by a factor of 10 over the range of all possible orientations. The calculated magnitude of the drag force on a football traveling through air at 40°F at a speed of 50 mph (typical of a fairly ordinary pass or kick) thereby ranges from 0.15 lb to 1.50 lb. Since a football weighs about 0.91 lb, it is clear that aerodynamic forces have a significant effect on the trajectory of a football.

It is also quite evident that the orientation of the football in flight has a sizable influence on the nature of its trajectory - a fact that is amply reflected in the preferred throwing and kicking techniques. To attain distance and accuracy, it is essential that the ball be launched in a nose-first (minimum-drag) orientation. This requires that the ball be gripped on its side when it is thrown. By contrast, other sports projectiles (a baseball, for example) are gripped and pushed from behind so that the line of action of the accelerating force can act through the center of gravity of the ball. By holding a football on its side, the thrower cannot deliver the greatest accelerating force to the ball. However, this grip allows him to impart considerable spin to the ball about its long axis. This spin effectively turns the football into a gyroscope, and the law of conservation of angular momentum is called upon to maintain the nose-first orientation of of the ball in flight. My own observations of slow-motion videotapings of football passes indicate that a spin rate of 10 rev/sec is typical for a well-thrown forward pass. A football thrown in this manner is said to be "spiralled."

Almost any average athlete can learn to throw a decent spiral pass. However, *kicking* a spiral is another matter. Punters attempt to give a ball a spin by hitting the underside of the ball with the top of their foot, while angling their leg from right to left (for a right-footed kicker). Under the game conditions, they are successful in spiralling a punt roughly 60% of the time. Place kickers adopt a different strategy, however. for it is extremely difficult to "kick a spiral" when the football is positioned on a tee. Instead, they deliberately kick the ball so that the point of impact of their foot is below the center of the football. This causes the ball to tumble end-over-end as it travels through the air. While this does not minimize the aerodynamic drag, it does give the ball gyroscopic stability and a more predictable flight.

Calculation of Launch Parameters

The calculation of range and hang time for a projectile experiencing aerodynamic drag can be easily performed on a personal computer. The equations of motion for such a projectile cannot be made to yield an analytic solution, but they can be solved numerically by stepwise integration using standard methods.[2,3] The problem with applying this procedure to the trajectory of the football is - as previously noted - that the drag force varies with the orientation of the football. The best one can do under these circumstances is to adopt some average value for the drag force.

In performing trajectory calculations, it is customary to write the drag force in the form $F_D = mKv^2$, where m is the mass of the projectile and K is the drag factor. By comparison with Eq. (1) we see that $K = \rho AC_D/2m$. Taking into account the preceding estimates of the range of values for A and C_D, we find that K ranges from 0.001 ft^{-1} (football traveling nose-first) to 0.010 ft^{-1} (football traveling broadside). Theoretically, K ought to take on its minimum value for a perfectly spiralled football. For an end-over-end kick, the drag factor is presumably varying rapidly between the two extreme values, suggesting the use of some intermediate value of K. For a badly executed, non-spiralling pass or kick, the value of K is expected to be near the top of its range.

The effects of different amounts of aerodynamic drag on the range and hang time are shown in Fig. 1. The curves show more specifically the nature of the "kicker's dilemma" with respect to the best choice of launching angle. In air, maximum horizontal range is attained at a launch angle lying somewhere between 40° and 45°. For higher launching angles up to 60°, there is a clear and acceptable trade-off of distance for hang time. However, for angles greater than 60° the kicker gives up a great deal of distance for only a slight increase in hang time. It seems evident that the preferred launching angle should lie somewhere between 40° and 60°.

Given the wide range of values for the drag factor K, it is not possible to calculate the launching speed or angle for any given kick. In other words, the problem is undetermined: there are three unknown parameters—launching speed (v_o), launching angle (θ_o), and drag factor (K)—and only two "knowns" - i.e., the horizontal distance (D) and hang time (T). If, however, we select some appropriate value for K, then there will be a unique set of launching parameters (v_o, θ_o) corresponding to every distance-hang time combination.

The launching parameters for a given kick (for a specific value of K) can be most easily obtained graphically. Fig. 2 shows the D-T curves for selected values of v_o and θ_o at K = 0.003 ft^{-1}. The circles and crosses represent selected kicks and punts taken from the data in Table I. One can readily interpolate the corresponding launching parameters for any given kick from the graph. This procedure can then be repeated for different values of K to establish a range of possible launching parameters for a given kick.

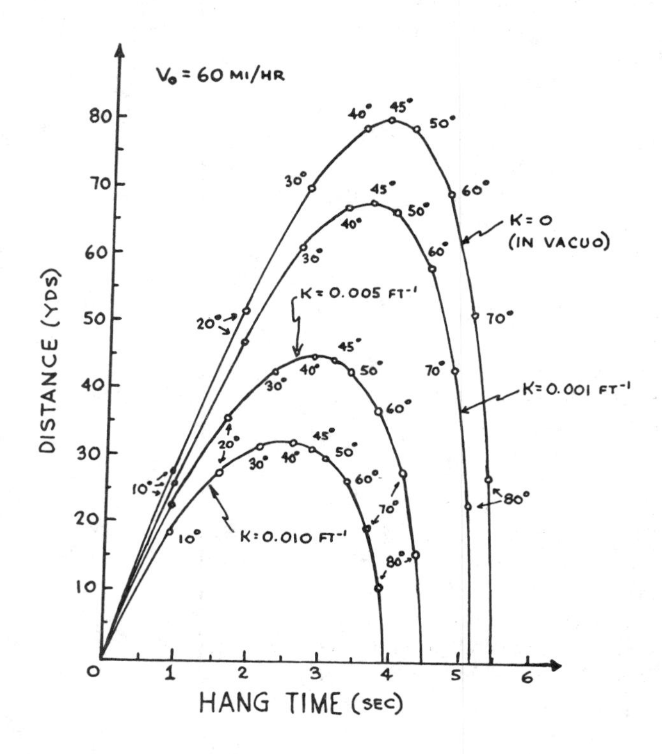

Fig. 1. Distance - hang time curves for a football launched at 60 mph as a function of launching angle and drag factor.

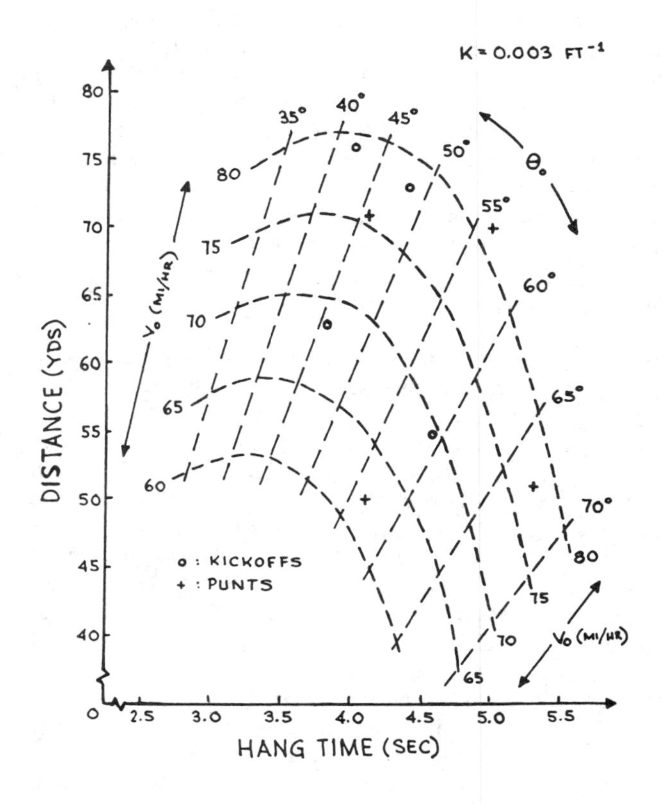

Fig. 2. Distance - hang time curves for K = 0.003 ft^{-1}. The launching speed (v_o) and launching angle (θ_o) for a given kick can be interpolated from this graph.

Results

For each of the chosen kicks or punts, the launching parameters v_o and θ_o have been evaluated for values of K ranging from 0.001 to 0.005 ft^{-1}. The results are presented in Table II. Values in the 0.005 to 0.010 range have not been listed, since these correspond to poorly oriented kicks that would require unattainably large launching speeds. The results for K = 0 (no air resistance) have also been tabulated for comparison.

This table reveals a rather interesting result—namely, that the range of possible launching angles for a given kick does not depend very strongly on K. It is seen that an increase in K by a factor of five produces at most a 17% increase (about 9°) in θ_o. Thus, for example, the launching angle for an average kickoff (D = 63 yds, T = 3.8 sec) must lie between 42° and 49°.

	PUNTS							
	Average Punt:		Longest Distance:		Longest Hang Time:		Best Overall Punt:	
	D = 50 yds T = 4.1 sec		D = 71 yds T = 4.1 sec		D = 51 yds T = 5.3 sec		D = 70 yds T = 5.0 sec	
K (ft^{-1})	v_o	θ_o	v_o	θ_o	v_o	θ_o	v_o	θ_o
0	51	61°	57	52°	61	71°	62	62°
0.001	55	60°	62	50°	66	70°	67	60°
0.002	58	58°	68	48°	71	68°	74	58°
0.003	62	56°	76	46°	77	67°	82	56°
0.004	68	55°	84	44°	84	65°	92	54°
0.005	71	53°	95	42°	92	64°	102	52°
Estimate	$v_o = 60$		$v_o = 70$		$v_o = 70$		$v_o = 70$	
of	$\theta_o = 57°$		$\theta_o = 47°$		$\theta_o = 69°$		$\theta_o = 59°$	
Parameters	K = 0.0025		K = 0.002		K = 0.002		K = 0.0015	

v_o = launching speed in mph; θ_o = launching angle; K = drag factor

Table II. Estimated Launching Parameters - Kickoffs and Punts

This range can be further narrowed by considering the corresponding values of the launching speed v_o. The sports research literature contains reports of actual measurements of the launching speed of a kicked football.[4,5] Typically, these are obtained from cinematographic studies, in a laboratory setting, using one or more kickers from the college football team. Generally, the top launching speeds reported do not exceed 60 mph. On this basis, we might expect the best professional punters to attain launching speeds around 70 mph. On kickoffs, the kicker augments his launching speed by running for about 10 yards prior to the kick. It is reasonable to expect that this maneuver will add about 10 mph to the kicker's stationary launching speed.

With these restrictions on the maximum launching speed in mind, I have made what I consider to be reasonable estimates of the launching parameters (v_o, θ_o, K) for each of the kicks and punts listed in Table II. On this basis, we can now come to some conclusions about kicking strategy. Assuming that the average well-executed professional kickoff is launched at a speed of about 70 mph, we find that the average kick is being launched at an angle of about 45°. (It also appears that the end-over-end motion of the football in flight creates an effective drag factor of K = 0.003 ft^{-1}). Longer-than-average kickoffs are evidently the result of higher-than-average launching speeds. It seems clear then, that kickoff specialists have adopted the strategy of kicking to maximize distance, by launching the kickoffs at about a 45° angle. One exception to this rule is presented in Table II by the kickoff with the longest observed hang time (4.6 sec.), which was probably launched (inadvertently?) at about a 60° angle.

Punters, however, are more willing to sacrifice distance to gain hang time. If we assume that punting launching speeds fall in the 60 to 70 mph range for well-executed punts, we see that the effective value of K is about 0.002 ft^{-1}. Accordingly, the typical launching angle for a punt is between 55° and 60°. Some punters may on occasion drop below 50° to gain more distance, or go as high as 70° to gain more hang time. In general, it turns out that by launching a punt at a 55° angle instead of 45°, a punter sacrifices about five yards of distance in exchange for an extra 0.55 sec. of hang time.

Finally, it is of interest to compare the estimated launching parameters with those calculated on the assumption that air resistance can be neglected. Bartlett has described a "take-home experiment" in which students are asked to record distances and hang times for kicks and punts, and then to calculate v_o and θ_o using the standard projectile motion equations.[6,7] This is, of course, a very admirable and creative way to familiarize students with simple measurement procedures as well as the fundamentals of kinematics. It should be noted, however, that if air resistance is ignored the calculated launching speeds will be undervalued by about 10 to 20 mph, while the calculated launching angles will be consistently too large by about 5°.

Conclusion

This study demonstrates that it is possible to construct a consistent model of the trajectory of a football kick, using the laws of projectile motion and basic aerodynamics. This model is able to determine within a fairly narrow range the launching angles used for kickoffs and punts. We have seen that kickers and punters have developed distinctly different strategies: kickoffs are launched at angles near 45° to maximize their distance, whereas punts are typically launched at angles closer to 60°, thereby trading distance to gain additional hang time.

Are these the best strategies to use? Generally, the most significant statistical measure of success ought to be the *net* yardage gained on the kick or punt, i.e., the length of the kick minus the runback. My own statistics show that on kickoffs, net yardage is in-

	KICKOFFS							
	Average Kickoff: D = 63 yds T = 3.8 sec		Longest Distance: D = 76 yds T = 4.0 sec		Longest Hang Time: D = 55 yds T = 4.6 sec		Best Overall Kick: D = 73 yds T = 4.4 sec	
K (ft^{-1})	v_o	θ_o	v_o	θ_o	v_o	θ_o	v_o	θ_o
0	54	51°	59	48°	56	64°	59	55°
0.001	58	49°	64	46°	60	62°	65	53°
0.002	63	47°	72	44°	65	61°	71	51°
0.003	68	45°	79	42°	70	59°	78	48°
0.004	75	44°	89	40°	75	58°	88	46°
0.005	83	42°	101	38°	82	56°	99	44°
Best Estimate of Parameters	$v_o = 70$, $\theta_o = 45°$, K = 0.003		$v_o = 80$, $\theta_o = 42°$, K = 0.003		$v_o = 70$, $\theta_o = 60°$, K = 0.003		$v_o = 80$, $\theta_o = 48°$, K = 0.003	

Table II. Estimated Launching Parameters - Kickoffs and Punts (Continued)

deed maximized by kicking for distance (specifically, by trying to put the ball deep into the end zone). With regard to punts, the statistics show a strong correlation between hang time and net yardage - indicating that it is effective to launch steeply-angled punts. Any further considerations, however, are best left to more knowledgeable football analysts.

References

1. H. Rouse, *Elementary Mechanics of Fluids* (John Wiley & Sons, New York, (1946).
2. P. J. Brancazio, *Phys. Teach.* **23**, 20 (1985).
3. C. Frohlich, *Am. J. Physics.* **52**, 325 (1984).
4. J. Kermond & S. Konz, *Res. Q.* (AAHPER) **49**, 71 (1978).
5. D. I. Miller, *Res. Q.* (AAHPER) **51**, 219 (1980).
6. A. Bartlett, *Phys. Teach.* **22**, 386 (1984).
7. T. R. Sandin, *Phys. Tech.* **23**, 6 (1985).

Why Does a Football Keep its Axis Pointing Along its Trajectory?

Question:

(Really—Questions Editors Ask) Why does a football thrown with a spin keep its axis pointing along its trajectory? It's most obvious in a long, soft pass. Even with a wobbly pass the ball's axis wobbles around the line of its trajectory.

This question was actually raised by the former editors of TPT.

Answer:

The answer was prepared by **Professor Peter J. Brancazio,** *Department of Physics, Brooklyn College of CUNY, Brooklyn, NY 11210. He is the author of the recently published article, "The Physics of Kicking a Football" (see Phys.Teach., 23, 403 1985).*

A football travels best when it is "spiralled" - that is, thrown or kicked with a substantial spin about its long axis. A spinning football possesses angular momentum and thus obeys the fundamental principle of conservation of angular momentum: *if there is no net external torque on an object, its angular momentum remains constant.* Since angular momentum is a vector quantity, it must have a constant direction in space as well as a constant magnitude when there is no net torque.

If the angular momentum of a football is conserved along its trajectory, then the axis of the ball should keep a constant orientation with respect to the ground, as shown in Fig. 1a. Footballs do sometimes exhibit this behavior, but more often the spin axis is seen to change direction so that it remains aligned with the tangent to the trajectory, as shown in Fig. 1b. The spin axis is said to "track" the trajectory. How can this behavior be explained?

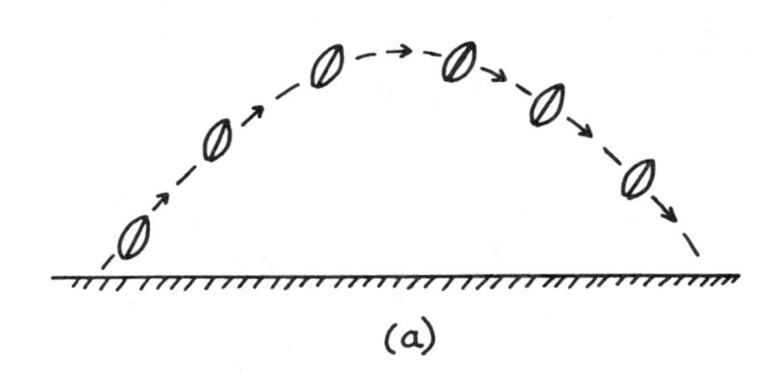

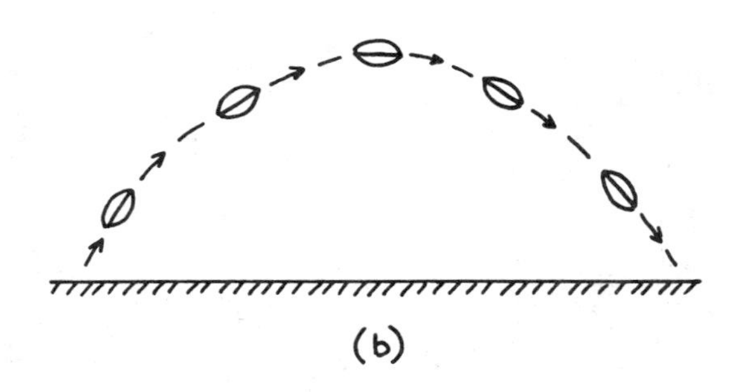

Fig. 1. Orientation of the spin axis of a football in flight. Angular momentum is conserved in (a), but not in (b).

Obviously, the angular momentum of the football must be changing in response to a torque. In fact, the aerodynamic drag acting on the football as it moves through the air can produce a torque about its center of mass. The force of the air pressure on the leading surface of the football can be resolved into a single force acting at a point known as the center of pressure. As shown in Fig. 2a, the line of action of the drag force F_D intersects the spin axis of the football at distance d from the center of mass, thereby producing a torque $N = F_D d \sin \theta$. The location of the center of pressure, the strength of F_D, and the size of d all depend on the air speed of the football and the inclination of the long axis,[1] and all of these quantities change as the football proceeds along its trajectory. For certain orientations (e.g., $\theta = 0°$ or $90°$), the line of action of F_D passes through the center of mass, so $d = 0$ and there is no torque at all.

A spinning football subject to a torque is very similar to a spinning top or gyroscope. Figure 2 shows that their torques take the same mathematical form. In general, the effect of such a torque on any spinning object is to cause the object to undergo precession; in the case of a top or gyroscope, its spin axis will sweep out a cone about a vertical line passing through its point of support.

It should be noted that a spinning football in flight can exhibit two kinds of precession. The first is *gyroscopic precession,* caused by the torque due to aerodynamic drag. The second is *torque-free*

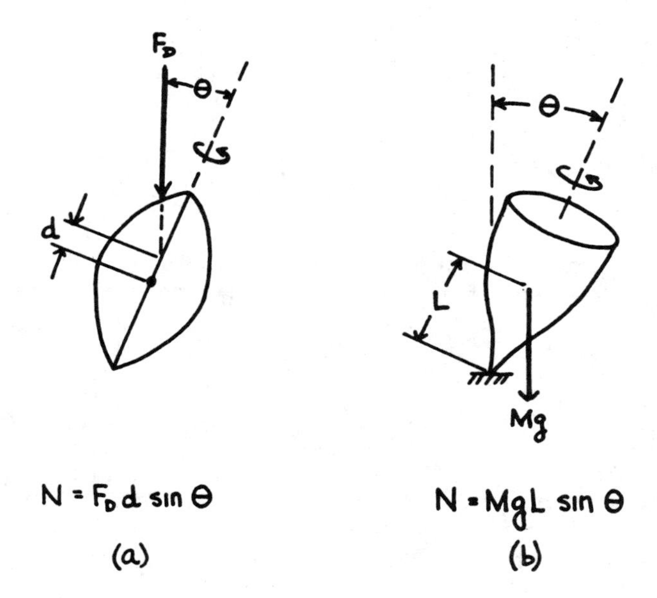

Fig. 2. Torques on a football and on a spinning top. (Football diagram has been rotated so that the trajectory is vertical to simplify the comparison.)

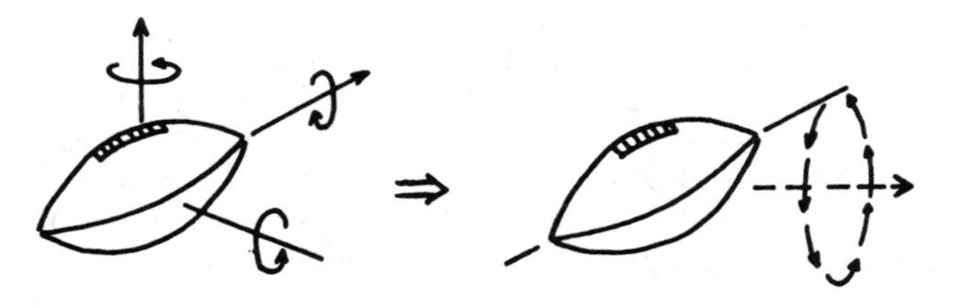

Fig. 3. Torque-free precession. Spins about more than one principal axis combined to produce a precession of "wobble."

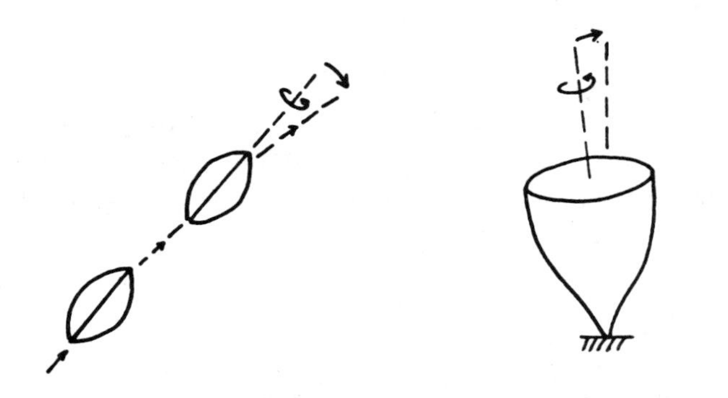

Fig. 4. A football with its spin axis initially aligned with its trajectory turns its axis to follow the curvature of its trajectory, just as a sleeping top rights itself when disturbed.

precession, which (as the name indicates) is independent of any applied force or torques. Torque-free precession occurs when a football is given a spin about more than one axis of rotation. Normally, a football is thrown with a lot of spin about its long axis. However, if it is also given a spin about either or both of its transverse axes, the two spins will combine to produce a precessional motion (see Fig. 3). The result is the "wobbly spiral" familiar to anyone who has tried to throw a football. (To avoid throwing a wobbly spiral, the football should be released with the wrist locked; a turn of the wrist downward or sideways will send the ball off with a spin about a transverse axis.)

To simplify matters, we will assume that the football is thrown with a spin about its long axis only. Hence any precession or other change in the orientation of the spin axis is attributable to aerodynamic torque, which in turn depends on the air speed and orientation of the axis of the football. In some cases, the aerodynamic torque is negligibly small, so that the spin axis remains fixed (as in Fig. la). Sometimes the torque will suddenly increase in flight (generally as the ball descends from the peak of its trajectory) and the football will begin a gyroscopic precession in response.

When a football turns its axis to keep it aligned with its trajectory as in Fig. lb, it represents a special case of gyroscopic motion. This behavior turns out to be similar to that of a "sleeping top" - a rapidly spinning top that is launched with its axis aligned vertically. In this configuration, the top is especially stable. The axis of the top remains vertical and does not precess; if the top is disturbed slightly it tends to return to the vertical.

When a football is launched with its axis aligned with its trajectory, there is no initial torque (note the similarity to the sleeping top). However, the trajectory of the football is curved so that as the football proceeds, a small angle begins to open between the spin axis and the tangent line to the trajectory (Fig. 4). Since the aerodynamic drag acts along this line (the drag is always opposite to the velocity vector), a small torque is created. The football responds to this disturbance in the same way that a sleeping top acts when disturbed slightly; namely, the axis of the football turns so as to reduce the angle (and the torque) back to zero. In this manner the axis of the football "tracks" its trajectory. The requirements are that:

1. the football is launched initially with its axis aligned with its trajectory, and
2. the football is thrown with sufficiently large spin.[2] If these conditions are met, the football should track its trajectory even if it is released with a small wobble; the latter is a torque-free precession that proceeds independently of the gyroscopic action of the football.

References

1. The aerodynamic drag on a football can be represented by the equation $F_D = \frac{1}{2}C_D A \varrho v^2$, where C_D is the drag coefficient, A is the cross-sectional area perpendicular to the air flow, ϱ is the air density, and v is the air speed of the football. Both C_D and A depend on the orientation of the football; they are smallest when the football is travelling nose first ($\theta = 0°$) and largest when the football is travelling broadside ($\theta = 90°$).
2. The specific mathematical condition for stability is

$$\omega_z = \frac{(4NI_{xy})^{1/2}}{I_z}$$

where ω_z is the angular velocity about the long axis, N is the aerodynamic torque, and I_{xy} and I_z are the principle moments of inertia of the football about its long and transverse axes, respectively.

Probability, statistics, and the World Series of baseball

Kenneth S. Krane
Physics Department, Orgeon State University, Corvallis, Oregon 97331
(Received 24 March 1980; accepted 21 July 1980)

In a recent article in this Journal,[1] Fischer has described an exercise in probability and statistics, in which he computed the probabilities of occurrence of various set scores in a tennis match based on the assumed probability of winning a game or an individual point. Fischer suggests some problems based on that computation which might be suitable for classroom work. I would like to offer in this note a similar calculation which offers both less mathematical rigor and perhaps more direct interest to introductory students. Indeed, I have used it in conjunction with our freshman and sophomore physics courses with great success. Not only does this problem stimulate student interest, but it also provides a simple example of "binary statistics" (a series of experiments, each of which has only two possible outcomes), in which all possible results can be tabulated and enumerated in a relatively short time.

The problem involves the calculation of the possible outcomes of the World Series in baseball, in which the champions of each of two leagues play a series of games, the winner being the first team to win four games. (Student interest in this topic is especially enhanced if the discussion takes place in early October.)

In the discussion the students usually make the assumption that either team has a 50% chance of winning any individual game. In class we explore this assumption (our "model") and discuss ways of testing it by comparison with "experiment." (Here the so-called scientific method is brought into the discussion.) Finally, after calculating various outcomes, the students are sent to the library to do the experiment by checking records of past World Series games. What is most remarkable and surprising to students and instructor is the good agreement between "theory" and experiment.

In the years from 1903 to 1979, there were 76 World Series contests (none was held in 1904). The total number of individual games played was 441.[2] The first test of the "model" is to verify that the chance of either team winning any individual game is 50%. Since different teams are generally involved in the contest each year, a criterion must be selected for dividing the teams into two groups. The logical criterion (always chosen by the students) is that of the two leagues represented by the competing teams. Applying such a criterion, we find that one league, the National League, was won 199, or 45%, of the 441 individual encounters. We generally conclude that this is sufficiently close to 50% to proceed with the "experiment."

One useful and interesting calculation to make is to consider the probability of each of the eventual outcomes of the Series. Most students will approach this by simply tabulating all of the possible outcomes ($AAAA$, $AAABA$, etc.), while the most clever and intuitive will attempt an explicit calculation of the probabilities. Table I shows the calculated results and a comparison with the actual outcomes. Although variations exist between theory and experiment, the general trend is reproduced. Excellent agreement is obtained between the expected mean length of the series (5.81 games) compared with the actual mean length (441/76 = 5.80 games).

Table I. Outcome of World Series contests.

Game score	Expected probability	Actual probability
4–0	1/8 = 12.5%	13/76 = 17%
4–1	1/4 = 25%	18/76 = 24%
4–2	5/16 = 31.25%	16/76 = 21%
4–3	5/16 = 31.25%	29/76 = 38%

Two other types of calculations which can be done have been relevant to the series outcomes for the past two years. The object is to calculate the probability that a team will be behind in games, either 2–0 or 3–1, and then emerge as the eventual victor in the series. A team with a 3–1 deficit (expected to occur 50% of the time, compared with the actual 43%) must win the remaining three games, which it will do one time in eight. The overall probability for a team first to achieve a 3–1 deficit and then to win the remaining three games is 1/16 = 6%. In actuality this has occurred 4 times (4/76 = 5%). A team losing 2–0 (expected probability of 50%, compared with actual 43%) will eventually emerge victorious, either 2–4 or 3–4, with a probability of 3/32 or 9%. This situation has actually occurred 7 times out of 76, which is likewise 9%!

The present problem is morphologically similar to the problems of calculating the outcomes of coin tosses or the spatial distribution of a small number of molecules in a box with two connected compartments, but this approach to "binary statistics" seems to be more successful in generating student interest in such calculations.

[1] Gaston Fischer, Am. J. Phys. **48**, 14 (1980).

[2] In four years (1903, 1919–1921) the series winner was the first team to win five, rather than four, games. For consistency, the results from these years have been analyzed as if the first team to win four games were the series victor.

Reprinted from *American Journal of Physics* **49**, 696–697 (1981); © American Association of Physics Teachers.

Exercise in probability and statistics, or the probability of winning at tennis

Gaston Fischer
58, rue de l'Observatoire, CH-2000 Neuchâtel, Switzerland
(Received 5 April 1979; accepted 24 July 1979)

The relationships between the probabilities p, x, s, and M, of winning, respectively, a point, a game, a set, or a match have been derived. The calculations are carried out under the assumption that these probabilities are averages. For example, x represents an average probability of winning a game when serving and receiving, and the same value of x is assumed to hold also for tie-break games. The formulas derived are for sets played with a tie-break game at the level of 6–6, as well as for the traditional rule requiring an advantage of two games to win a set. Matches to the best of three and five sets are considered. As is to be expected, a small advantage in the probability p of winning a point leads to advantages which are amplified by large factors : 2.5 for games, 7.1 for sets with tie-break at 6–6, 10.6 for matches to the best of three sets, and 13.3 for matches to the best of five sets. When sets are decided according to the traditional rule, the last three factors become, respectively, 7.4, 11.1, and 13.8. The theoretical calculations are compared with real and synthetic tennis scores and good agreement is found. The scatter of the data is seen to obey the predictions of a normal distribution. Some classroom problems are suggested at the end.

In competition tennis a match is usually played to the best of three or five sets. This means that the first player who wins two or three sets is declared the winner of the match. The probability M that a given player A will win in a match with player B is of course a function of the probability S that player A has of winning a single set against that same player B. In what follows we propose to calculate M when S is known.

Since the probability of winning a match is the result of a cumulative process of winning individual sets, it is clear that if player A is better than player B, A's chances of winning a match are greater than his chances of winning a single set. Labeling B's chances of winning a match or a set, respectively, as N and T, the above statement is written

$$M/N > S/T. \tag{1}$$

By definition the sums of probabilities M and N or S and T must total 100%, or in other words,

$$M + N = S + T = 1. \tag{2}$$

Quite similarly there is a relation between the probability S that player A will win a set and the probability x that he will win an individual game. Here again relations like (1) and (2) above exist. If y describes the probability that player B has of winning an individual game, then

$$S/T > x/y, \tag{3}$$

and

$$x + y = 1. \tag{4}$$

Let us begin with the derivation of the relation $S(x)$ between these two probabilities of winning a game and winning a set.

I. PROBABILITY OF WINNING A SET

According to the rules of tennis a player is declared the winner of a set if he scores six games before his opponent succeeds in scoring more than four games. If during play the score reaches a 5-5 tie, the set proceeds following a variety of rules. Up to a few years ago play was continued until one of the players secured a lead of two games, thereby winning the set. On occasion this led to very long-lasting sets, so today the practice of the tie-break game is adopted almost everywhere, although with several variants. When a set is to be decided by tie-break, play proceeds in the traditional fashion, but only till a score of 6-6, 7-7, 8-8, etc., is reached; at this point a single tie-break game, during which service is shared by both players, decides the set. In many recent tournaments the tie-break game has been played already at the level of 6-6. Our calculation will be carried out under this assumption, but could easily be modified to suit other conditions. The traditional case, however, will also be considered.

In general the probability of winning or losing a game is strongly dependent on whether a player is serving or receiving. In most instances serving is considered a decisive advantage. But since in the average set both players serve as often as they receive, we shall ignore that difference. Thus x represents something like the arithmetical mean of the probabilities that player A has of winning a game when serving and receiving. There is little doubt about the soundness of this assumption. It would be quite feasible to introduce different probabilities x_1 and x_2 for player A and y_1 and y_2 for player B, but it would then also become nec-

Reprinted from *American Journal of Physics* **48**, 14–19 (1980); © American Association of Physics Teachers.

essary to treat separately the sets in which player A starts serving and those where it is player B who starts. For the same reasons of simplicity we shall assume that the probabilities x and y also apply to the tie-break games, but as an example will suggest, this second assumption does not stand on as firm ground as the first one.

To begin the calculation of $S(x)$ we note that player A can win a set by different scores, starting with 6-0. The likelihood of A's winning by this result is obviously x^6. This can also be looked upon as the probability that B will not win a single game before A totals six games. Next we consider the probability that B wins one game and only one game, while A accumulates six games. Since a 6-1 score can arise in only six different ways (viz. *baaaaaa, abaaaaa, aabaaaa, aaabaaa, aaaabaa, aaaaaba*, but excluding *aaaaaab*), and since for each of these events the probability is x^6y, the likelihood of a 6-1 score is $6x^6y$. In similar manner the 6-2, 6-3, and 6-4 final scores can be calculated to occur with the probabilities listed in Table I (see Appendix). The probability of A winning over B by any score from 6-0 to 6-4 is simply the sum of the individual probabilities corresponding to these scores.

We consider now what may happen after a score of 5-5 has been reached. From 5-5, A can win either by scoring two successive games, thus reaching 7-5, or by going through 6-6 and subsequently winning the tie-break game. Let us ask, then, about the likelihood of a 5-5 intermediate score. This is calculated in the Appendix to be $252x^5y^5$. With this result it follows that A has $252x^6y^5$ chances to reach a 6-5 score, and thus $252x^7y^5$ chances to win by 7-5. The probability of a 6-6 intermediate result reached via 6-5 is $252x^6y^6$. But since a 6-6 score can also be obtained by going first through an intermediate result of 5-6, the chances of a 6-6 score total $504x^6y^6$. The probability of A's winning the set by tie-break is therefore $504x^7y^6$. Summing all the probabilities A has of winning the set yields

$$S(x) = x^6 + 6x^6y + 21x^6y^2 + 56x^6y^3 + 126x^6y^4 + 252x^7y^5 + 504x^7y^6. \tag{5}$$

As shown in the Appendix, when the traditional rule requiring an advantage of two games to win a set holds, $S(x)$ becomes:

$$S(x) = x^6 + 6x^6y + 21x^6y^2 + 56x^6y^3 + \frac{126x^6y^4}{1 - 2xy}. \tag{6}$$

Table I. Probability of a given set score (cf. Appendix).

Set score	Probability with tie break at 6-6	Probability with two game advantage
6-0	x^6	x^6
6-1	$6x^6y$	$6x^6y$
6-2	$21x^6y^2$	$21x^6y^2$
6-3	$56x^6y^3$	$56x^6y^3$
6-4	$126x^6y^4$	$126x^6y^4$
7-5	$252x^7y^5$	$252x^7x^5$
7-6	$504x^7y^6$	—
8-6	—	$504x^8y^6$
9-7	—	$1008x^9y^7$
...	—	...

Because of Eq. (4) both Eqs. (5) and (6) are truly functions of probability x only.

Clearly, if A does not win, B will win, and the probability of this occurring is, by symmetry with Eqs. (5) and (6):

$$T(y) = y^6 + 6y^6x + 21y^6x^2 + 56y^6x^3 + 126y^6x^4 + 252y^7x^5 + 504y^7x^6, \tag{7}$$

and

$$T(y) = y^6 + 6y^6x + 21y^6x^2 + 56y^6x^3 + \frac{126y^6x^4}{1 - 2yx}. \tag{8}$$

With Eq. (4) it can be verified that condition (3) is satisfied equally well by the pairs of Eqs. (5) and (7), and (6) and (8), i.e., that

$$S(x) + T(y) = 1. \tag{9}$$

Calculating $S(x)$ with Eqs. (5) and (6) we obtain the results given in Table II and plotted in Fig. 1. It is evident from these data that for values of x around 0.5 the slope of $S(x)$ is 2.83, i.e., much larger than unity. For example, if out of 100 games A usually wins 51 and B only 49, we may say that A has a 2% edge in games over B. His edge in sets will then be 5.66%, or nearly 6%. Evidently, when x is below 0.2 or above 0.8, which means that one of the players in general wins only one out of every five games played, his chances ever to win a set are down to 0.0091, or less than one out of 100 sets.

II. PROBABILITY OF WINNING A MATCH

Let us return now to the probability M that player A may win a match in terms of his probability of winning a set. What we want to derive, therefore, is the relation $M(S)$. By the same sort of reasoning as before we find that for a match decided in two winning sets, i.e., to the best of three sets,

Table II. Probability of winning a set. x is the probability of winning a game; S_7 is the probability of winning a set with tie break at 6-6; S_∞ is the probability of winning a traditional set (advantage of 2).

x	Percent edge in games	S_7	Percent edge in sets	S_∞	Percent edge in sets
0.30	−40	0.0696	− 86.1	0.0633	− 87.3
0.35	−30	0.1382	− 72.4	0.1295	− 74.1
0.40	−20	0.2369	− 52.6	0.2280	− 54.4
0.45	−10	0.3611	− 27.8	0.3554	− 28.9
0.46	− 8	0.3881	− 22.4	0.3835	− 23.3
0.47	− 6	0.4157	− 16.9	0.4121	− 17.6
0.48	− 4	0.4436	− 11.3	0.4411	− 11.8
0.49	− 2	0.4717	− 5.7	0.4705	− 5.9
0.495	− 1	0.4859	− 2.83	0.4852	− 3.0
0.5	0	0.5	0	0.5	0
0.505	+ 1	0.5141	+ 2.83	0.5148	+ 3.0
0.51	+ 2	0.5283	+ 5.7	0.5295	+ 5.9
0.55	+10	0.6389	+ 27.8	0.6446	+ 28.9
0.60	+20	0.7631	+ 52.6	0.7720	+ 54.4
0.70	+40	0.9304	+ 86.1	0.9367	+ 87.3
0.80	+60	0.9909	+ 98.2	0.9921	+ 98.4
0.90	+80	0.9998	+100.0	0.9998	+100.0
0.95	+90	1.0000	+100.0	1.0000	+100.0

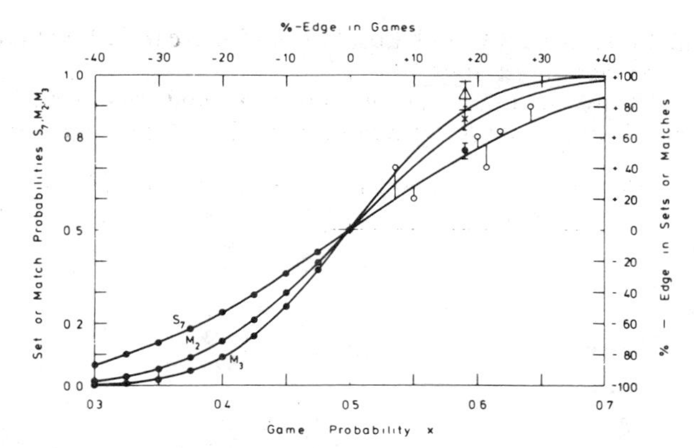

Fig. 1. Graphical representations of the probabilities S_7, M_2, and M_3 of winning, respectively, a single set with tie break at 6-6, or matches to the best of three or five sets, as a function of the probability x of winning an individual game. Open circles give S_7 values derived from real data (groups of 10 or 11 sets). Points with standard deviation bars are values of S_7, M_2, and M_3 calculated from a sample of 61 real sets. The closed circles with very small scatter on the left-hand portions of the curves are derived from large numbers of synthetic sets and matches (10 000 matches to the best of five sets, about 16 000 matches to the best of three sets, and 30 000 to 40 000 individual sets). The right-hand scale gives the percent edge in sets or matches, whereas the top scale gives the percent edge in games. The percent edge is so defined that a 2% edge in games, for example, means winning 51 out of 100 games on the average. Note the central symmetry of all the curves with respect to the center of the graph, which means, for example, that $T_7(y) = S_7(1-x) = 1 - S_7(x)$.

$$M_2(S) = S^2 + 2S^2T = S^2 + 2S^2(1-S) = S^2(3-2S). \quad (10)$$

For a match played to three winning sets, i.e., to the best of five sets, we find

$$M_3(S) = S^3 + 3S^3T + 6S^3T^2 = S^3(10 - 15S + 6S^2). \quad (11)$$

From these equations it is easy to see that a small percent edge in sets is multiplied by 1.5 for matches in 3 sets and by 1.875 for matches in five sets.

In Fig. 1 and Table III we combine Eqs. (10) and (11) with Eq. (5), obtaining thus the relation between the probability x of winning an individual game and the probabilities $M_2(x)$ and $M_3(x)$ of winning a match. For two players of nearly equal strength, i.e., x close to 0.5, we see that a small percent edge in games leads to an edge in matches over four times greater for matches in three sets, and more than five times greater if the match is played in five sets.

III. PROBABILITY OF WINNING IN TERMS OF POINTS

The basic scoring unit in tennis is the point. While point scores are expressed in a rather odd scale (15, 30, 40, 45, etc.) the actual point count is such that a player is declared the winner of a game if he totals four points with an advantage of at least two points over his opponent. When a point score of 3-3 is reached play continues until one of the players secures for himself a lead of two points and is then declared winner of the game. Calling p and q the points

Table III. Probabilities M_2 and M_3 of winning a match played to the best of three and to the best of five sets, with tie-break games at 6-6, in terms of the probability x of winning a game.

x	Percent edge in games	M_2	Percent edge in matches	M_3	Percent edge in matches
0.30	−40	0.0139	− 97.2	0.0030	− 99.4
0.35	−30	0.0520	− 89.7	0.0212	− 95.8
0.40	−20	0.1417	− 71.7	0.0902	− 82.0
0.45	−10	0.2970	− 40.6	0.2526	− 49.5
0.46	− 8	0.3350	− 33.0	0.2971	− 40.6
0.47	− 6	0.3747	− 25.1	0.3448	− 31.0
0.48	− 4	0.4157	− 16.9	0.3951	− 21.0
0.49	− 2	0.4576	− 8.5	0.4471	− 10.6
0.495	− 1	0.4788	− 4.24	0.4735	− 5.3
0.5	0	0.5	0	0.5	0
0.505	+ 1	0.5212	+ 4.24	0.5265	+ 5.3
0.51	+ 2	0.5424	+ 8.5	0.5529	+ 10.6
0.55	+10	0.7030	+ 40.6	0.7474	+ 49.5
0.60	+20	0.8583	+ 71.7	0.9098	+ 82.0
0.70	+40	0.9861	+ 97.2	0.9970	+ 99.4
0.80	+60	0.9998	+ 99.95	1.0000	+100.0
0.85	+70	1.0000	+100.0	1.0000	+100.0

totaled by players A and B, the probability x that A will win a game is given by

$$x(p) = p^4 + 4p^4q + 10p^4q^2/(1-2pq). \quad (12)$$

For $p = 0.5$ the slope of $x(p)$ is 2.5, as seen in Table IV and Fig. 2. This means that for nearly equal players a given point percent edge leads to a game percent edge 2.5 times greater.

In terms of the probability p of winning an individual point, an advantage of 1% will usually result in an advantage of 10.61% when playing a match in three sets, and in an advantage of 13.27% when playing a match to the best of five sets. When the difference in strength between the two players becomes larger than a few percent in points, the match soon becomes a "no contest." The weaker player has no chance ever to win a match. In competition tennis this has led quite naturally to the establishment of a rather fine gradation in the evaluation of the strength of players. This finding also explains why the fun in tennis is often very dependent on a close matching of the strength of the players.

IV. COMPARISON WITH PRACTICAL EXAMPLES

We have analyzed 61 set scores, established by two players A and B who regularly play together. Only the set results were recorded, not the point scores. The 61 sets were tabulated in the order in which they were played. In general three or four sets were played at a time, without regard to any match result. For the sake of comparison with our calculations, however, the 61 set results were grouped in their original order as if they represented match results, either to the best of three or to the best of five sets.

First the 61 sets were arranged in five groups of 10 sets and one group of 11 sets. For each group x and S were calculated and plotted as open circles in Fig. 1. These circles are seen to scatter around the $S(x)$ curve. All points are within 0.1 of the curve on the S scale, in accordance with

Table IV. Probabilities M_2 and M_3 of winning a match in 3 or 5 sets, with tie breaks at 6-6, in terms of the probability p of winning an individual point.

p	Percent edge in points	$x(p)$	$S_7(p)$	M_2	Percent edge in matches	M_3	Percent edge in matches
0.30	−40	0.0992	0.0002	0.0000	−100.0	0.0000	−100.0
0.35	−30	0.1704	0.0038	0.0000	− 99.99	0.0000	−100.0
0.40	−20	0.2643	0.0380	0.0042	− 99.2	0.0005	− 99.9
0.45	−10	0.3769	0.1876	0.0924	− 81.5	0.0488	− 90.2
0.46	− 8	0.4010	0.2390	0.1441	− 71.2	0.0923	− 81.5
0.47	− 6	0.4254	0.2973	0.2127	− 57.5	0.1596	− 68.1
0.48	− 4	0.4501	0.3614	0.2974	− 40.5	0.2531	− 49.4
0.49	− 2	0.4750	0.4296	0.3951	− 21.0	0.3698	− 26.05
0.495	− 1	0.4875	0.4647	0.4471	− 10.6	0.4340	− 13.2
0.499	− 0.2	0.4975	0.4929	0.4894	− 2.12	0.4868	− 2.65
0.5	0	0.5	0.5	0.5	0	0.5	0
0.501	+ 0.2	0.5025	0.5071	0.5106	+ 2.12	0.5133	+ 2.65
0.505	+ 1	0.5125	0.5353	0.5529	+ 10.6	0.5660	+ 13.2
0.51	+ 2	0.5250	0.5704	0.6049	+ 21.0	0.6302	+ 26.05
0.55	+10	0.6231	0.8124	0.9076	+ 81.5	0.9512	+ 90.2
0.60	+20	0.7357	0.9620	0.9958	+ 99.2	0.9995	+ 99.9
0.70	+40	0.9008	0.9998	1.0000	+100.0	1.0000	+100.0

a normal distribution which predicts for the average width of the scatter, the so-called standard deviation within which range 68% of the data will usually be found, a value given by

$$S = (nST)^{1/2}/n = [nS(1 - S)]^{1/2}/n, \tag{13}$$

where $n = 10$ or 11 is the number of sets. With an average S of 0.75 we find

$$\Delta S = 0.14 \tag{14}$$

Note that for a constant n value ΔS is largest when S approaches 0.5 and becomes smaller when S goes toward 0 or 1. For $S = 0.5$ we get $\Delta S = 0.16$ and for $S = 0.9$ we find $\Delta S = 0.09$, in good agreement with the behavior of the data displayed in Fig. 1. But Fig. 1 also makes it quite clear that $n = 10$ is too small a number of sets to permit a valid comparison with the theory. If then we look at all 61 sets, we find $x = 0.591$, $S = 0.754$, and $\Delta S = 0.055$, and this is plotted as a closed circle with standard deviation bars in Fig. 1. The example now comes much closer to the theoretical curve, but 61 sets still seem far too few to warrant the claim that theory and example agree perfectly well. For S between 0.3 and 0.7 over 2000 sets would be needed to bring ΔS down to about 0.01. The smallness of our sample is even more evident if we look at the M_2 and M_3 match values. The small number of matches leads to large standard deviations, so that no very definite conclusions can be drawn. All that can be said from the study of our sample of real data, is that theory and example agree within the rather large standard deviations allowed by a normal probability distribution when the sample is small.

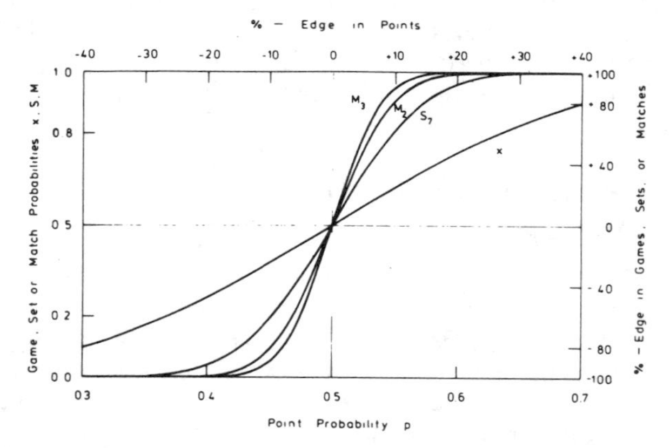

Fig. 2. Graphical representation of the probabilities x, S_7, M_2, and M_3 of winning, respectively, a game, a set with tie break at 6-6, or matches to the best of three or five sets, as a function of the probability p of winning an individual point. The right-hand scale gives the percent edges in games, sets, or matches, whereas the top scale gives the percent edge in points. The percent edge is so defined that a 2% edge in sets, for example, means winning 51 out of 100 sets on the average. Note the central symmetry of all the curves with respect to the center of the graph, which means, for example, that $N_2(q) = M_2(1\text{-}p) = 1 - M_2(p)$.

As it would be difficult to find a real sample of sufficient size, synthetic samples were produced by computer, by means of a program of random number generation. 10 000 synthetic matches to the best of five sets were thus played, for various game probabilities. This usually yielded about 16 000 matches to the best of three sets and about 30 000 to 40 000 individual sets. Now the standard deviations are down to less than

$$\Delta S = 0.003, \quad \Delta M_2 = 0.004, \quad \Delta M_3 = 0.005, \tag{15}$$

and it is clearly evident in Fig. 1 that this synthetic example agrees perfectly with the probability curves we have derived.

A somewhat striking feature of the real example is that the values of S, M_2, and M_3 are all significantly larger than predicted by the probability calculations. Could this be a consequence of our assumption that tie-break games are subject to the same average probability x as other games? Trying to answer this question we plot in Fig. 3 the various possible scores of a set against their predicted probability of occurrence for $x = 0.5$ and $x = 0.591$, together with the real and the synthetic example data. Again there is a very wide scatter in the real data, indicating that the sample studied is much too small. The synthetic data also appear more scattered than in Fig. 1. This is a consequence of the

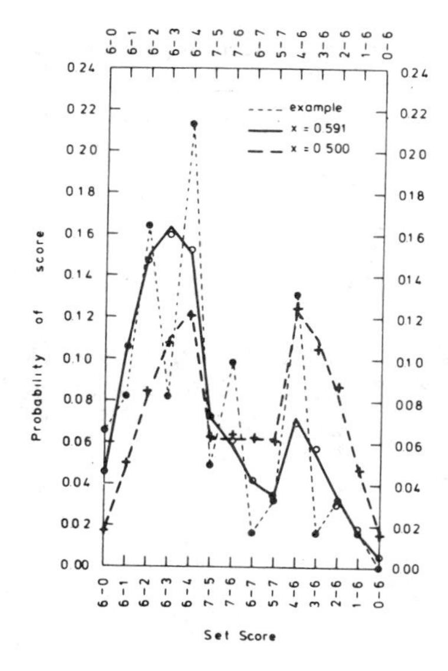

Fig. 3. Representation of the probability distribution of the various possible set scores, with tie break at 6-6; for two values of the probability x of winning a game: $x = 0.5$ and 0.591. Closed circles are from a real sample of 61 sets, for which $x = 0.591$. Crosses ($x = 0.5$) and open circles ($x = 0.591$) are derived from samples of about 40 000 synthetic sets.

expansion of the ordinate scale (compare with Fig. 1), but is not in violation with a normal distribution. Returning to the real data, we cannot fail to notice that certain scores occur much more frequently than predicted (6-4, 7-6, and 4-6), whereas others occur much less frequently than predicted by the calculation (6-3, 6-7, and 3-6). The high frequency of the two 6-4 and 4-6 scores suggests that both players succeed in improving their playing once they lead by 5-4. Consequently the 5-5 intermediate scores are less frequent, and therefore also the final 7-5 and 5-7 scores. But somewhat surprisingly, therefore, the 6-6 intermediate scores, the number of which is given by all the tie-break games, are about as numerous in the example (7) as predicted by the theory (6.13). Now if the tie-break games were decided by the same average probability of 0.591, there should be about four 7-6 outcomes and three 6-7 outcomes. But in the example there have been six 7-6 results and only one 6-7 result. So it appears that player A was much more successful in winning tie-break games than in winning regular games. It should be stressed, however, that even this highly odd distribution of the 7-6 and 6-7 scores remains within the scattering range tolerated by a normal distribution and could, therefore, be explained as a mere chance result. It remains true, nevertheless, that the excess values of S, M_2 and M_3 in the real example can be accounted for entirely by the excess of 7-6 over 6-7 scores.

In conclusion it can be said that the data from our two examples agree well with the probability relations derived. It should be mentioned that these relations are not restricted to scores obtained by two players where a constant probability relationship (x constant in time, for example) may be assumed. Any large random collection of set or match scores should satisfy the relations that we have derived. While scores created synthetically by a computer are seen to follow these relations very closely, it would be interesting to verify if the same holds true for a large catalog of real scores picked at random, or if observations similar to those we have made with our real sample could be confirmed, in particular the rather frequent incidence of 6-4 and 4-6 scores and the concomitant scarcity of 7-5 and 5-7 results.

V. DIDACTIC SUGGESTIONS

Several problems suitable as classroom exercises have been touched upon in this paper, e.g., (i) Derive any of the probability relationships (5-8) or (10-12). (ii) Prove that $S(x) + T(y) = M(S) + N(T) = x(p) + y(q) = 1$. (iii) Many textbooks on mathematics only sketch the passage from the binomial to the Gaussian or normal distribution (cf. for example "Mathematical Methods of Physics," by J. Mathews and R. L. Walker, Benjamin, New York, 1970). Carry out that passage in detail. (iv) Even though the passage from the binomial to the normal distribution involves many approximations (e.g., Stirling's approximation and the assumption that $\xi = m - np \ll np$, such that higher-order terms in ξ are then neglected), the normal distribution is in general taken to hold for all values of ξ, especially since the integral from $-\infty$ to $+\infty$ of $P(\xi)$ is unity, just as the sum over all m of $P(m)$. Comment on this observation. (v) Each of the various terms of Eq. (5) represents the probability of a particular set score. Find the standard deviation for these probabilities. (vi) Write a computer program, based on a subroutine of random number generation, to create synthetic tennis scores. How would you verify that the subroutine produces truly random numbers?

APPENDIX

Calculation of S(x)

The probability of occurrence of a score of 6-1 is given by the number of ways this score can arise. If a symbolizes a win by player A and b a win by player B, it is easy to see that any arrangement of six a's and one b is a possible score, except the arrangement terminating with b. Since there are six different arrangements possible, each with a probability of x^6y, the total probability of a 6-1 score is $6x^6y$.

To get the probability of a 6-2 score we look first at all the possible arrangements of a 6-1 score. We then have to insert a second b in any of the seven allowed positions, i.e., excluding the right-hand extremity. Since there were six positions for the first b, there are seven for the second b. There are therefore 42 arrangements. But every arrangement occurs twice because it is not possible to distinguish between the first and second b (interchanging the two b's does not yield a new arrangement). The correct probability is then

$$(6 \cdot 7/2)x^6y^2 = 21x^6y^2. \qquad \text{(A1)}$$

Similarly, the probabilities of a 6-3 and of a 6-4 score are given by

$$(6 \cdot 7 \cdot 8/2 \cdot 3)x^6y^3 = 56x^6y^3, \text{ for 6-3,} \qquad \text{(A2)}$$

$$(6 \cdot 7 \cdot 8 \cdot 9/2 \cdot 3 \cdot 4)x^6y^4 = 126x^6y^4, \text{ for 6-4.} \qquad \text{(A3)}$$

In the two last equations the denominator again expresses the number of indistinguishable arrangements obtained through permutation of the b's alone. The coefficients of (A1) to (A3) are in fact simply binomial coefficients, e.g.,

$$(6 \cdot 7 \cdot 8 \cdot 9/1 \cdot 2 \cdot 3 \cdot 4) = [9!/5! \cdot (9-5)!] = \binom{9}{5}. \qquad \text{(A4)}$$

The probability of occurrence of a 5-5 intermediate score is calculated in similar manner, except that in this case the last game can be won by B as well as by A. This probability is therefore

$$(6 \cdot 7 \cdot 8 \cdot 9 \cdot 10/2 \cdot 3 \cdot 4 \cdot 5)x^5y^5 = 252x^5y^5. \quad \text{(A5)}$$

This result is necessary to evaluate the chances of final scores like 7-5 and 7-6 occurring, where the second alternative involves a tie-break game.

Result (A5) is also necessary to calculate the probability that player A has of winning a set played according to the traditional rule requiring an advantage of two games. Since the probability of a 5-5 score is $252x^5y^5$, the probability of a final score of 7-5 is $252x^7y^5$. The probability of a 6-6 intermediate score is $504x^6y^6$, as this score can be reached equally well through 6-5 and 5-6. An 8-6 final score has thus the probability $504x^8y^6$. Continuation of this reasoning leads to the following expression for the probability that A will win by scores of 6-4, 7-5, 8-6, 9-7, etc.,

$$126x^6y^4[1 + 2xy + (2xy)^2 + (2xy)^3 + (2xy)^4 + \cdots] = 126x^6y^4/(1 - 2xy). \quad \text{(A6)}$$

Each of the terms on the left-hand side of (A6) represents the probability that a given high score will occur. For players of exactly the same strength a score of 8-6 is therefore only half as likely as a 7-5 score, 9-7 in turn is half as likely as 8-6, and so on.

On the Physics of Drag Racing

GEOFFREY T. FOX
Department of Physics
University of Santa Clara
Santa Clara, California 95053
(Received 10 July 1972)

The sport of drag racing can be very usefully examined in an introductory mechanics course. The concepts of energy, momentum, acceleration, velocity, torque, and power are all involved in the analysis of the problem. Many students find the problem to be very contemporary and close to their own experience. High interest among the students usually results in increased understanding of the physical principles involved.

INTRODUCTION

As undergraduate physics majors my brother, Robert C. Fox, Jr., and I spent our leisure time building and racing vehicles in drag raicng competition. We found our knowledge of physics very useful in that realm, enabling us to understand relationships involved in building and driving a winning car. Entering graduate school, I was forced financially to discontinue this colorful sport. Instead, I contented myself with continuation of theoretical studies and compilation of those made previously. As an assistant professor of physics, I have given seminars on the subject which have been extremely well received. I have also used particular topics as special credit problems in both my introductory physics course for engineers and scientists and in an upper division course in intermediate mechanics. Especially students in mechanics have been stimulated by the consideration of this topic.

I. WHAT IS A DRAG RACE?

Drag racing is an extremely popular spectator/participant sport in the United States which is rapidly growing. Interest among persons in their teens and early twenties is particularly high. If one omits horse racing, which I believe is popular only because of the gambling ingredient, then automobile racing is *the* largest spectator sport in the country. A drag race is the simplest of all forms of auto racing. The course is straight and therefore the problems are essentially one- or at most two-dimensional. The race is an acceleration contest from a standing start over a distance of $\frac{1}{4}$ mile. Two variables are measured for each competitor, they are ET and MPH. ET stands for elapsed time and is just the time it takes from crossing the starting line until crossing the finish line. MPH is essentially the terminal speed at the end of the 1320 ft, measured through a 132 ft time trap on either side of the finish line. Both measurements are done electronically with the aid of photocells.

The motor vehicles that compete in drag racing are many and varied. They range from everyday street-driven vehicles which have ET's of 20 sec and MPH of 70, through hopped up street legal cars turning 120 mph in 11.00 sec on up to the all out dragsters running over 220 mph in slightly more than 6 sec.

II. AN EXAMINATION OF THE VARIABLES

In trying to develop any theory it is always wise to start from an examination of the variables. In drag racing there are really just two dependent variables, ET and MPH. The goal of our theory is to predict or explain these dependent variables as a function of the independent variables. The key difficulty is that the list of independent variables is quite long, some of the key ones are listed: vehicle weight, engine horsepower (also the details of the horsepower curve), location of the center of gravity (both horizontal and vertical), coefficient of friction of tires on road (this in turn is a function of tire size, tire pressure, normal force, and surface conditions to mention a

Reprinted from *American Journal of Physics* **41**, 311–313 (1973);

few), air drag (depends basically on frontal area and drag coefficient), gearing (both rear axle and transmission), shifting techniques, moment of inertia of rotating parts, driver skill, etc.

III. THE FIRST ORDER THEORY OR THE CONSTANT POWER APPROXIMATION (CPA)

In trying to develop a first order theory it would be useful to examine the basic rules which determine the breakdown of classes in drag racing. The key parameter here is the ratio of weight to engine displacement. Engine displacement, is of course, closely related to power, hence our first order theory could be expected to use power to weight ratio as the independent variable. In the early 1960's an engineer, Roger Huntington, developed an empirical law relating power to weight ratio to MPH. His rule is $\text{MPH} = K_{\text{exp}}(\text{Power/Weight})^{1/3}$. The empirical K_{exp} is 225 where power is in horsepower, weight in pounds, and MPH in miles per hour.[1] No such simple relationship has been found for ET however.

Huntington's empirical rule has a theoretical basis which I discovered in 1964. It is just the result to be expected in the Constant Power Approximation (CPA). This is easily developed in the following manner,

$$\Delta \text{ Energy} = \text{Work}$$

$$\tfrac{1}{2}mv^2 = \int_0^t P\,dt.$$

Assuming constant power one gets

$$\tfrac{1}{2}mv^2 = Pt \quad \text{or} \quad v = (2Pt/m)^{1/2}. \tag{1}$$

Since

$$x = \int_0^t v\,dt = \frac{2}{3}\left(\frac{2P}{m}\right)^{1/2} t^{3/2}$$

we get

$$t = [(3/2)(m/2P)^{1/2}x]^{2/3}. \tag{2}$$

Now we eliminate time from Eqs. (1) and (2) giving

$$v = (3xP/m)^{1/3}$$

which is of the same form as the empirical law except $K_{\text{theory}} = 270$, using the same units as previously.

Although the discrepancy between K_{theory} and K_{exp} doesn't appear to be large, if one cubes it one finds that about 50% of the theoretical power is wasted. How can one account for this? There are a number of factors which can account for power loss. Some are: power losses in the drive train, shifting losses (no power is transmitted during shifting), average engine horsepower is less than peak engine horsepower (in the rpm range used), wind and rolling resistance, full power can't be transmitted initially due to traction limitations (traction), effective weight (due to Moment of Inertia and Rotational Energy).

Of these the last two are the most important. Consider the last first. $\text{Energy} = \tfrac{1}{2}Mv^2 + \tfrac{1}{2}I\omega^2 = \tfrac{1}{2}M_{\text{eff}}v^2$ and $M_{\text{eff}} = M + I(\omega/v)^2$. And we must consider "$I$" due to both those items rotating at axle speed and also those rotating at engine speed. The latter includes clutch, flywheel, crankshaft, etc. Notice that ω/v which is related to gearing comes in as a squared term, thus for a typical vehicle like our own 1955 Chevy 339 in.[3] (total weight 3700 lbs) the $I(\omega/v)^2$ term accounts for the following additional weight in each gear: 1st gear, 1500 lbs; 2nd gear, 800 lbs; 3rd gear, 500 lbs; 4th gear, 250 lbs.

Obviously this is a very important correction factor, but still there is the question: "Why does the first order theory work so well?" Or phrased another way, "Why doesn't MPH depend very strongly on traction whereas ET is extremely sensitive to this effect?" To answer this question let us proceed to the second order theory.

IV. THE SECOND ORDER THEORY

In this theory we divide up the run into two portions: the traction limited portion, and the power limited portion. For the traction limited portion of the run we have

$$F = \text{const.} = \mu_e Mg$$

in which μ_e is related to the coefficient of friction through geometrical factors and includes "weight transfer" effects. In practice μ_e varies between 0.5 and 2.0. Thus

$$\tfrac{1}{2}Mv^2 = Fx = \mu_e Mgx$$

giving

$$v_0 = (2gx_0\mu_e)^{1/2}.$$

For the rest of the run, we use the constant power assumption giving

$$\tfrac{1}{2}M(v^2 - v_0^2) = P(t - t_0)$$

which can be solved for v and then integrated to get position and one gets:

$$v = [(2\mu_e g x_0)^{3/2} + 3P/M(x - x_0)]^{1/3}.$$

Now we can eliminate x_0 from the equation because this is just the point at which power is no longer sufficient to spin the tires, or

$$Fv_0 = P = \mu_e M g v_0$$

giving

$$x_0 = P^2/2M^2\mu_e^3 g^3.$$

Finally giving

$$v = (3x/R)^{1/3}[1 - (1/6x\mu_e^3 g^3 R^2)]^{1/3}$$
$$\approx (3x/R)^{1/3}[1 - (1/18x\mu_e^3 g^3 R^2)]$$

in which $R = M/P$ is the ratio of mass to power. The one in brackets corresponds to the first order theory; the $-(1/18x\mu_e^3 g^3 R^2)$ term is the correction due to traction limitations. From this equation one can see that changes in coefficient of friction have only slight effects on terminal speed. Even for trap speeds as high as 140 mph the correction term is only a few percent.

V. OTHER TOPICS OF INTEREST

Among other effects that can be discussed are power shifting and "wheelies." In power shifting the rpm of the engine is not synchronized to the speed of the engine. Instead, the accelerator is floored during the shifting process. In this manner the considerable rotational energy stored in the flywheel, crankshaft etc., is not all lost, rather some of it is transformed into translational energy of the vehicle. The actual amount lost can be calculated as soon as one realizes what kind of problem is involved. Quite simply it is one of collision. The flywheel collides with the clutch—angular momentum is thus conserved. The excess energy is dissipated as heat caused by slippage against friction during the collision.

Wheelies are the ultimate manifestation of what is called "weight transfer." The normal forces exerted by the road on the front and rear tires depends not only upon total vehicle weight, and location of the center of gravity (CG) in the horizontal plane, but also upon location of the CG in the vertical plane and the acceleration of the vehicle. A wheelie occurs when the acceleration is so great that the front wheels lift off the road. The problem is simply solved using deAlembert's principle and taking torques about an axis passing through the point of contact of the rear wheels with the pavement. The "wheelie criterion" is thus found to be:

$$a/g = [\alpha(WB/h)]$$

where a is acceleration, g the acceleration due to gravity, WB the wheelbase, h the height of the CG, and α the horizontal distance between the rear axle and the CG.

The influence of shifting time, air and rolling drag, etc., can be easily brought into the problem via a digital computer. The problem can be straightforwardly solved by numerical integration of $F = ma$. The effects are found to be small. Interestingly enough, the first shift has more effect on ET while the last shift has more effect on MPH.

CONCLUSION

Drag racing has been shown to be uniquely well suited to analysis in an introductory physics course. This is largely due to the fact that it is a one dimensional problem. In the analysis many of the basic concepts of mechanics are applied in a context which is particularly interesting to many students. The relevance of the example therefore helps them to appreciate the most basic discipline of physics—mechanics.

[1] Private communication.

Constant power equations of motion

Roger Stephenson
Physics Department, University of Wisconsin at Platteville, Platteville, Wisconsin 53818

(Received 13 July 1981; accepted for publication 29 January 1982)

Correct solutions for the common problems associated with bodies accelerating under the application of constant power are generally missing from physics textbooks. The kinematic equations for this motion are not overly difficult to derive and bear strong similarities to their familiar constant force counterparts. The concept of "zip" is here defined as the square root of the power to mass ratio, and is shown to be a useful parameter of this kind of motion. Several specific problems are solved and the results compared with available real data.

I. INTRODUCTION

Most physics texts discuss the motion of automobiles because they are an important part of the student's experience of acceleration, velocity, distance, power, etc. Yet quite often the equations used are completely inappropriate.[1] In particular, the assumption that the force acting on the car remains constant is realistic only during braking

Reprinted from *American Journal of Physics* **50**, 1150–1155 (1982);

and at low speeds when the car is able to spin its wheels.[2]

The correct equations are more complicated than the constant force equations, but students appreciate being able to solve practical problems and getting answers that correspond well to their experience. This can be accomplished without analyzing air drag or friction. Simply getting the form of the equations for frictionless motion correct will greatly improve our predictions.

An automobile does not experience a constant force because its engine is not capable of delivering a constant torque to the drive wheels over a range of speeds. The engine can produce a constant torque at a fixed rpm, but in order for the car to accelerate, either the engine rpm must change or the gear ratio in the drive train that links the engine to the drive wheels must change. In cars that are currently mass produced, a combination of the two is used to produce acceleration.

To maximize acceleration, race drivers try to change gears in such a way as to keep the engine rpm as near as possible to the value that produces maximum power. With a continuously variable transmission, such as one developed at Physics Apparatus Research, Inc., it is possible for a driver to keep engine rpm (and thus power and torque at the engine) constant during an acceleration run. However, even then, the torque at the drive wheels is not constant. The gear ratio change that produces a speed increase also produces a proportional torque decrease. The power, however, is constant under these conditions and, ignoring friction, the power at the drive wheels will equal the power at the engine. This will also apply rather well for most automatic transmissions. The so-called torque converters they contain provide continuously variable ratios even though they waste some power.

II. STANDING START EQUATIONS FOR CONSTANT POWER MOTION

Constant power P at the drive wheels will result in a uniform increase in kinetic energy for the vehicle.[3] This will allow us to derive equations for the distance, velocity, and acceleration (x, v, and a) as functions of the time t for a vehicle of mass m. These equations are derived for the general case in Appendix A. However, since these will be too complicated for some students, let us examine the standing start case where x and v equal zero at $t = 0$. The kinetic energy will build up according to

$$(1/2)mv^2 = Pt. \tag{1}$$

Therefore the velocity must be given by

$$v = (2Pt/m)^{1/2}. \tag{2}$$

The square root of the ratio of power to mass occurs so often in these equations that we will assign it a symbol, Z, and, following a well-established tradition in physics, we will choose a vague common word and give it a precise meaning. Zip:

$$Z = (P/m)^{1/2}. \tag{3}$$

Since the dimensions of zip are length/time$^{3/2}$, it follows from dimensional analysis that distance, velocity, and acceleration will each be found by multiplying zip by time raised to the appropriate power and by some dimensionless quantity. Alternatively, if force is constant, zip will be a variable and will be found by multiplying acceleration by the squre root of time. In either case, zip will be measured in units such as ft/s$^{1.5}$ or m/s$^{1.5}$.

Table I. Comparison of equations of motion from a standing start.

Quantity	Value with constant power	Value with constant force
x	$(1/3)\,Z\,(2t)^{3/2}$	$(1/2)\,(F/m)\,t^2$
v	$Z\,(2t)^{1/2}$	$(F/m)\,t$
a	$Z\,(2t)^{-1/2}$	F/m

The value of zip as a parameter of motion is illustrated in the equation below in which distance, velocity, and acceleration are all proportional to zip. During a drag race between two trucks, the one that has 7% more zip will always be 7% farther along, be going 7% faster, and be accelerating at a 7% greater rate.

We now rewrite Eq. (2) as

$$v = Z(2t)^{1/2}. \tag{4}$$

Differentiating Eq. (3) yields

$$a = Z(2t)^{-1/2}. \tag{5}$$

Setting $v = dx/dt$ in Eq. (2), separating variables, and integrating[4] yields

$$x = (Z/3)(2t)^{3/2}. \tag{6}$$

These, then are the standing start equations of motion for a body to which constant power is applied. They are not much more complicated and probably no more difficult to remember then their constant force counterparts to which they are compared in Table I.

A graph of distance versus time for constant force (CF), constant power (CP), and constant velocity (CV) motions will show how constant power motion compares with the familiar kinds of motion. In Fig. 1 we see that ${}_{CP}x$ with its dependence on $t^{1.5}$ lies intermediate between ${}_{CV}x$ and ${}_{CF}x$ with their dependence on $t^{1.0}$ and $t^{2.0}$, respectively. While we cannot compare actual values on this graph, we can see how the motions differ in their functional dependence on time.

Figure 2 is similarly useful in that it shows how velocity increases with the application of constant power. We have all noticed on our speedometers that during acceleration, velocity builds very quickly at first and very slowly later. Many students will mistakenly assign this effect to air drag instead of recognizing it as a natural consequence of constant power motion. Their physics text probably contains the equation $\mathbf{F}\cdot\mathbf{v} = P$, but the usual application is for the computation of power which is regarded as a variable. In fact, physics texts generally[5] do not solve motion problems using the concept of constant power.

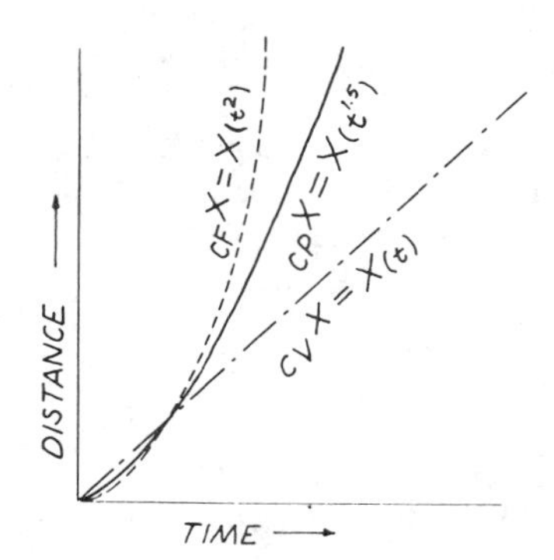

Fig. 1. Distance traveled by a vehicle with constant force (CF); constant power (CP); constant velocity (CV).

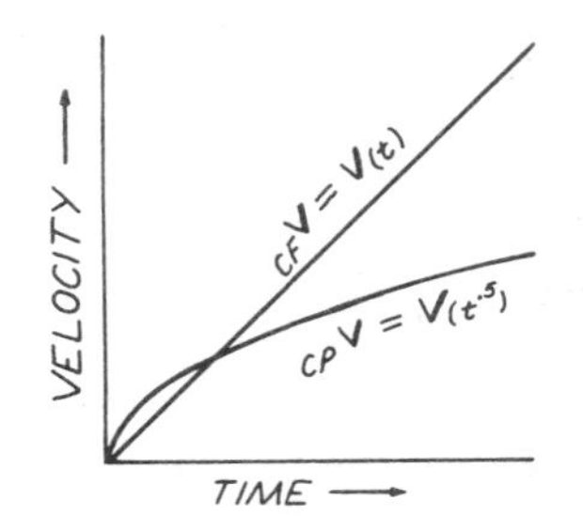

Fig. 2. Velocity of a vehicle with constant force (CF): constant power (CP).

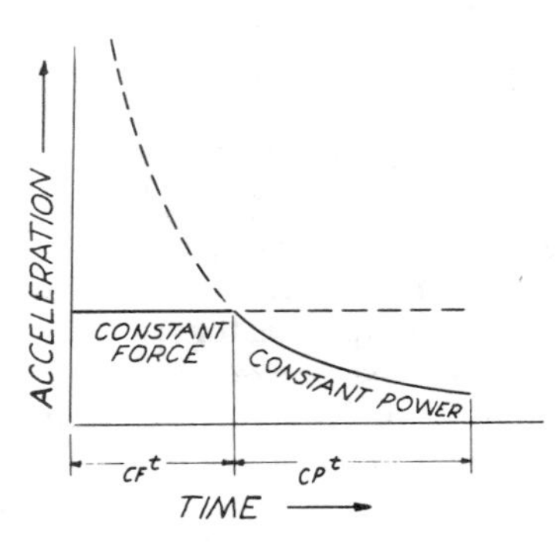

Fig. 3. Acceleration of a vehicle that undergoes constant force motion (limited by tire slip) until its velocity becomes so great that its engine has insufficient power to produce that much force. Thereafter, the vehicle experiences constant power motion.

III. SOLVING PROBLEMS INVOLVING CONSTANT POWER MOTION

The motion of many vehicles can be well described using just the equations of Table I. Generally[6] any vehicles such as bicycles, trucks, and economy cars having a low zip value can be regarded as having constant power motion. Such problems involve a simple substitution into the appropriate constant power equation and even high school students can handle the 3/2 exponent involved.

We can also use the equations of Table I to solve some nonstanding start problems. The time required for a car to accelerate from 50 to 60 mph can be found as the difference between the 0→50 and 0→60 times. All that we assume is that the motion be constant power motion between 50 and 60 mph. Therefore we can use this technique for cars that have a considerable zip.

There is one region where constant power equations do not apply and that is for small velocities where they predict large forces—larger in fact than can be provided by rubber tires. Thus the practicable acceleration run for a vehicle with sufficient zip to spin its wheels really consists of a constant force section followed by a constant power section. This is because the force that accelerates such a vehicle cannot exceed a certain fraction, k, of the vehicle's weight. If all of a vehicle's weight rests upon its drive wheels, as is the case for a four-wheel-drive vehicle, k will be equal to the coefficient for friction μ. A typical value for μ for hard pavement and rubber tires is 0.67 although this can vary considerably with conditions.[7]

However, most cars are driven by either the front wheels only or by the rear wheels only. This complicates the determination of k, not only because the weight distribution becomes important but also because acceleration effectively changes the weight distribution. This problem is treated in some detail in Appendix B where it is shown that, for a rear-wheel-drive car that is *not capable of lifting its front wheels*, k is given by

$$k = \mu j/(L - \mu h), \tag{7}$$

where j, L, and h are shown in Fig. 4. It is interesting that even the height h of the center of mass will affect k. Substituting some typical values into Eq. (7) yields a k value of about 0.31.

The change from constant force motion to constant power motion happens at some velocity U when $F_{\text{CF}} = F_{\text{CP}}$. Expressing each force at velocity U we have

$$kmg = P/U, \tag{8}$$

which can be solved for the change velocity U:

$$U = P/kmg = Z^2/kg. \tag{9}$$

At any velocity greater than U, the force provided by the engine is less than kmg and constant power equations apply. It is interesting that a car with velocity greater than U cannot spin its wheels. The solid line in Fig. 3 illustrates the accleration throughout such a run.

Equation (9) suggests that any vehicle should be able to spin its wheels at very low velocities, yet experience shows us that vehicles with little zip cannot. Such vehicles have such a low value for U that with any practical drive train gear ratio, the vehicle's velocity exceeds U before the engine can attain the rpm necessary for full power. Thus the actual acceleration curve for such a vehicle stays below the solid line in Fig. 3 at least until U is attained. However, it is this same group of low zip vehicles for which the constant power equations may be used without correction. The reason is that U is so low for these vehicles that the extra time required to reach it (even by a motion a little slower than constant force motion) is very small.

For a vehicle that can spin its wheels, the total time spent accelerating will be the sum of the times spent in each type of motion:

$$t_{0\text{–}V} = {}_{\text{CF}}t_{0\to U} + {}_{\text{CP}}t_{U\to v}. \tag{10}$$

Since the constant power part of this run does not begin with a standing start, we must either use the equations of Appendix A or we must once again express the time for a change between two velocities as the difference between the times required to reach each velocity from a standing start. Following the latter course, we alter Eq. (10) and reduce it as follows:

$$t_{0\to v} = {}_{\text{CF}}t_{0\to U} + {}_{\text{CP}}t_{0\to v} - {}_{\text{CP}}t_{0\to U}, \tag{11}$$

$$t_{0\to v} = U/kg + v^2/2Z - Z^2/2k^2g^2, \tag{12}$$

$$t_{0\to v} = v^2/2Z^2 + Z^2/2k^2g^2. \tag{13}$$

The first term of this expression is the time required to reach v with constant power motion. The second term is the additional time required because of wheel slip and amounts to half the time required to reach U under constant force kmg.

If we wish to find the time required to reach distance x in a vehicle able to spin its wheels for part of the trip, we must find the distance d at which it changes from constant force motion to constant power motion. Since d is reached under constant force, we write

$$d = U^2/2kg. \tag{14}$$

Substituting the value for U from Eq. (9) we find

$$d = Z^4/2k^3g^3. \tag{15}$$

Rewriting Eq. (11) for distance rather than velocity we get

$$t_{0\text{–}x} = {}_{\text{CF}}t_{0\to d} + {}_{\text{CP}}t_{0\to x} - {}_{\text{CP}}t_{0\to d}, \tag{16}$$

$$t_{0\to x} = (2d/kg)^{1/2} + (1/2)(3x/Z)^{2/3} - (1/2)(3d/Z)^{2/3}. \tag{17}$$

Substituting for d from Eq. (15) yields

$$t_{0\to x} = (Z/kg)^2 + (1/2)(3x/Z)^{2/3} - (1/2)(3/2)^{2/3}(Z/kg)^2, \tag{18}$$

$$t_{0\to x} \approx (1/2)(3x/Z)^{2/3} + 0.345(Z/kg)^2. \tag{19}$$

Once again the solution is in the form of the constant power solution plus additional time caused by wheel slip.

We have now developed the equations necessary to solve some of the common acceleration problems. The three most often quoted performance figures are the 0 to 60 mph time, the standing start quarter-mile time, and the velocity at the end of the quarter mile. Let us consider the performance of an automobile having a k value of 0.31, a mass of 100 slugs, and a 100-hp engine.[8]

The first thing we must do is calculate the zip for this hypothetical car:

$$Z = (P/m)^{1/2} = (55\,000 \text{ ft lb}/100 \text{ slug s})^{1/2} = 23.5 \text{ ft/s}^{1.5}.$$

Next we can substitute Z, k, and g into Eq. (13) to get

$$t_{0\to 60 \text{ mph}} = 7.0 \text{ s} + 2.8 \text{ s}.$$

The 2.8-s correction to the 7.0-s constant power prediction is certainly not negligible and justifies the trouble we took to evaluate it.

The standing start quarter-mile time is found by substituting Z, k, and g into Eq. (19) to get

$$t_{0\text{–}1/4 \text{ min}} = 15.3 \text{ s} + 1.9 \text{ s}.$$

Once again, the correction for wheel slip proves significant—nearly 2 s.[9]

Finally, we can find the velocity at the end of the quarter mile by substituting the 17.2-s quarter-mile time into Eq. (13), along with Z, k, and g, and solving for v:

$$v = 126 \text{ ft/s}.$$

If we had regarded the quarter-mile run as being entirely constant power motion, we would have obtained a result only 3% higher.[10] To understand why this is so, one may combine Eqs. (19) and (13) to obtain a complicated expression for v in which it is the *difference* between the two corrections for wheel slip that forms the new correction.

Table II shows a comparison of our predictions[11] with some data published in an automative magazine.[12] Several facts are made apparent by the table. First, although Lincolns are much more massive and powerful than Volvos, they will have the same performance if they have the same zip. Secondly, we notice that cars with nearly the same zip as our hypothetical car fall noticeably short of its predicted performance. Similarly, we notice that a car having nearly the performance of our hypothetical car claims a greater zip. Apparently, and not surprisingly, cars squander about 13% of their zip combating friction and air drag,[13] building up rotational energy,[14] and by not maintaining their engines at the correct rpm for maximum power. Since a 13% loss of zip corresponds to a 24% loss in power, discounting manufacturer's power specifications by 1/4 leads to fairly accurate performance predictions for many cars. The 1/4 loss factor does not negate the validity of the equations because the resulting predictions work equally well for a variety of distances, velocities, and times.

The author has found that all of the material in the body of this article can be taught with good comprehension in one hour in a calculus-based engineering physics course, and that problem solving using constant power equations can be taught in noncalculus college physics and even in high school physics with good effect.

Table II. Comparison of performance and zip of real cars with those of hypothetical 100-hp, 100-slug car that is assumed to make an acceleration run composed of constant force and constant power sections.

	Lincoln Mark V	Volvo GL	Hypo. CF→CP	Volvo VTEV/1
$t_{0\to 60 \text{ mph}}$	11.6 s	12.0 s	9.8 s	9.5 s
$t_{0\to 1/4 \text{ mi}}$	18.5 s	18.5 s	17.2 s	17.2 s
$v_{\text{end of } 1/4 \text{ mi}}$	76.1 mph	73.8 mph	85.8 mph	81.2 mph
zip (ft/$s^{1.5}$)	24.3	24.4	23.5	27.0

APPENDIX A

The acceleration a, velocity v, and position x for a body of mass m can be found at any time t provided we know that the body has initial conditions v_0 and x_0 and that it is increasing its kinetic energy at a constant rate, P. We begin with the expression for power as the rate of doing work W:

$$dW/dt = P, \tag{A1}$$

however, $W = \mathbf{F}\cdot d\mathbf{x}$, and therefore

$$\mathbf{F}\cdot d\mathbf{x}/dt = P. \tag{A2}$$

Recognizing the expression for velocity we now write

$$\mathbf{F}\cdot\mathbf{v} = P, \tag{A3}$$

and, invoking Newton's second law, we obtain

$$m\mathbf{a}\cdot\mathbf{v} = P. \tag{A4}$$

We now introduce the quantity "zip" which combines the constant mass and constant power in a way that simplifies subsequently derived equations. Zip:

$$Z = (P/m)^{1/2}. \tag{A5}$$

Thus Eq. (A4) becomes

$$\mathbf{a}\cdot\mathbf{v} = Z^2 \tag{A6}$$

and since $\mathbf{a} = d\mathbf{v}/dt$, we can write

$$v\,dv = Z^2\,dt. \tag{A7}$$

Integrating both sides of this equation yields

$$v^2/2 = Z^2 t + C_1. \tag{A8}$$

Evaluating the integration constant C_1 at $t = 0$, we find that $C_1 = v_0^2/2$ which substitutes in Eq. (A8) to yield

$$v^2/2 = Z^2 t + v_0^2/2. \tag{A9}$$

Solving for velocity we find

$$v = Z(2t + v_0^2/Z^2)^{1/2}. \tag{A10}$$

In order to find x we must first substitute dx/dt for v in Eq. (A10) and separate variables to get

$$dx = Z(2t + v_0^2/Z^2)^{1/2}\,dt. \tag{A11}$$

However, the differential of the quantity inside the parentheses is 2 dt so we must multiply and divide the right-hand side of the equation by 2 to obtain an expression we can integrate:

$$\int dx = (Z/2)\int (2t + v_0^2/Z^2)^{1/2} 2\, dt \tag{A12}$$

upon integration yields

$$x = (Z/3)(2t + v_0^2/Z^2)^{3/2} + C_2. \tag{A13}$$

Evaluating the integration constant C_2 at $t = 0$, we find that $C_2 = x_0 - (Z/3)(v_0/Z)^3$ which substitutes into Eq. (A13) to yield

$$x = (Z/3)[(2t + v_0^2/Z^2)^{3/2} - (v_0/Z)^3] + x_0. \tag{A14}$$

To find acceleration as a function of time, we note that Eq. (A6) shows acceleration to be a function of velocity written

$$a = Z^2/v, \tag{A15}$$

and from Eq. (A10) we have velocity as a function of time which lets us write

$$a = Z(2t + v_0^2/Z^2)^{-1/2}. \tag{A16}$$

Equations (A16), (A10), and (A14) are solutions for a, v, and x, respectively, in terms of Z and the original conditions v_0 and x_0 and, of course, time. For the special case of constant power motion from, a standing start where $v_0 = 0$ and $x_0 = 0$ at $t = 0$, these equations simplify greatly and become

$$a = Z(2t)^{-1/2}, \tag{A17}$$

$$v = Z(2t)^{1/2}, \tag{A18}$$

$$x = (Z/3)(2t)^{3/2}. \tag{A19}$$

Note that for constant power motion from a standing start, the acceleration, velocity, and position are all directly proportional to zip at all times.

APPENDIX B

Figure 4 shows the forces acting on a rear-wheel-drive car during maximum acceleration. It is assumed that this is one of the many cars that is capable of spining its wheels and is incapable of lifting its front end from the ground. This latter assumption fits virtually every production car and truck but not every motorcycle. Since the car has no vertical acceleration, we may write

$$\sum F_y = A + B - mg = 0. \tag{B1}$$

We may also invoke the second condition for equilibrium. But, since the car has linear acceleration, we must sum torques about the center of mass so that no inertial force in our accelerated reference frame will contribute a torque to the equation:

$$\sum \tau_{CM} = Bj + \mu Ah - A(L - j) = 0. \tag{B2}$$

We should pause here to consider the conditions under which Eq. (B1) is true. The assumption that there is no vertical component of acceleration means that the front wheel remains in contact with the road. Thus

$$B > 0. \tag{B3}$$

Equation (B2) shows us that if B is positive,

$$L - j > \mu h. \tag{B4}$$

This relationship among the three dimensions and μ must hold to guarantee that the front wheels will remain in contact with the road and that Eq. (B1) will apply. A car that is very tall or has very sticky tires or has its center of mass near the rear axle may fail to meet this criterion.

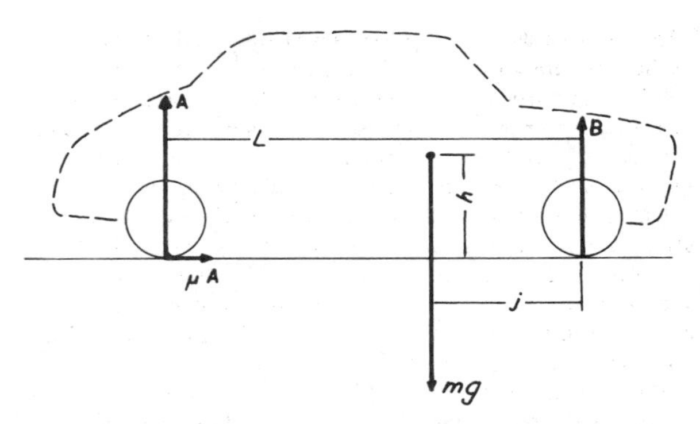

Fig. 4. Forces acting upon a car of mass m and wheelbase L that is experiencing maximum acceleration. The accelerating force is μA; where μ is the coefficient of friction, and A is the verticle force with which the earth supports the driving wheels (considered as one wheel). The center of mass is a height h above the road and a distance j behind the front axle. The force B with which the earth supports the front wheels is a function of acceleration as is A.

For ordinary rear-wheel-drive cars we may solve Eq. (B1) for B and substitute in Eq. (B2) to get

$$(mg - A)j + \mu Ah - A(L - j) = 0, \tag{B5}$$

$$A = jmg/(L - \mu h). \tag{B6}$$

We can now express the horizontal force F_x as a function of knowable parameters:

$$F_x = uA = [\mu j/(L - \mu h)]mg, \tag{B7}$$

$$F_x = kmg. \tag{B8}$$

Where we express F_x as a fraction, k, of the weight such that

$$k = \mu j/(L - \mu h), \tag{B9}$$

provided, of course, that the relationship (B4) is satisfied.

A similar derivation for a front-wheel-drive car yields an equation that is the same as Eq. (B9) except that the denominator becomes $L + \mu h$. Once again, j represents the horizontal distance from the center of mass to the undriven axle. We can see that a high center of mass would diminish the maximum driving force for a front-wheel-drive car and increase it for a rear-wheel-drive car.

Of course, to evaluate k we must find μ. Published experimental values of μ for a variety of road surfaces are available.[15] Alternatively, one may solve for μ by setting the work done in stopping equal to the kinetic energy.[16] 45 ft is a typical stopping distance listed for cars moving at 30 mph. This yields a μ value of 0.67. Substituting this value into Eq. (B10) along with some hypothetical dimensions ($L = 10$ ft, $j = 4$ ft, and $h = 2$ ft) yields a typical k value of 0.31.

[1]For example: D. Haliday and R. Resnick, *Fundamentals of Physics* (Wiley, New York, 1974) contains four kinematic problems on pp. 41 and 42 and a dynamic problem on p. 88 that specify a constant acceleration assumption. Further, a problem on p. 108a invites the student to treat power as a variable.

[2]If the car does spin its wheels, the force is determined by the coefficient of sliding friction. Since the coefficient of static friction is larger, the prudent driver can achieve greater acceleration by not quite spinning the wheels. This topic is treated more fully by R. C. Smith, Am. J. Phys. **46** (8), 858–859 (1978).

[3]The only use of the constant power approximation the author could find in the literature was done by G. T. Fox, Am. J. Phys. **41** (3), 311–313 (1973). In this article, Fox integrates $P\,dt$ to find energy and solves for velocity. He then integrates $v\,dt$ to find distance in an equation he does not number. Fox does not mention the constant of integration in either instance and does not differentiate v to obtain the acceleration. The focus of his article is the prediction of the conventional results for the standing start quarter-mile drag races, and Fox gets there by the shortest route. He solves the distance expression for time (the elapsed time result), and substitutes this into the velocity expression to have velocity as a function of distance (the terminal velocity result). Fox's pioneering work has not found its way into the physics curriculum. Perhaps the simple elegance of the equations of Table I of the present work will find wider acceptance. Fox also recognizes the necessity of dividing the acceleration runs of powerful cars into traction limited (constant force) and power limited (constant power) portions. Once again, he is interested in drag race results and therefore expresses velocity as a function of distance rather than time. Surprisingly, he does not solve for the elapsed time of the composite run. His effective friction term u_e is explicity developed as k in Appendix B of the present work. Although Fox did not recognize the unifying concept of zip, he did begin writing R for the mass to power ratio to simplify equations toward the end of the article.

[4]The constant of integration is zero since $x = 0$ at $t = 0$.

[5]The author was unable to find any that do until the the 1981 edition of Ref. 1 was published. It contains one problem.

[6]Vehicles for which $Z \approx 9$, such as large trucks, can be regarded as having only constant power motion. Assuming that they can spin their wheels makes only a 1% addition to the $t_{0\to 60\,\mathrm{mph}}$ prediction. More pessimistic assumptions will not add much more.

[7]J. S. Baker, *Traffic Accident Investigation* (The Traffic Institute, Northwestern University, Chicago, 1975), p. 210.

[8]The choice of engineering units for these problems is made to facilitate the use of and comparison with available real data.

[9]At first it might seem surprising that ${}_{\mathrm{CP}}t_{0\to d} \neq {}_{\mathrm{CP}}t_{0\to U}$. However, d and U are actually achieved under constant force acceleration and indeed ${}_{\mathrm{CF}}t_{0\to d} = {}_{\mathrm{CF}}t_{0\to U}$.

[10]Reference 3 notes the effect also.

[11]Performance prediction is attempted by writers in the automotive field. However, they tend to be content with any equation that works. The curve fitting done by Colin Campbell, *The Sports Car* (Chapman and Hall, London, 1978), pp. 211–222, is a good example.

[12]Motor Trend **32** (4), 34–48 (1980).

[13]G. I. Rochlin, Am. J. Phys. **44** (10), 1010–1011 (1976).

[14]J. L. Koffman, Automobile Eng. **45** (12), (1955).

[15]Reference 7.

[16]H. Erlichson, Am. J. Phys. **45** (8), 769 (1977).

Weight Distribution

P. F. Hinrichsen on its effect — or otherwise — on racing dinghy speed

Most racing helmsmen are very conscious of the effects of excess weight on boat speed, and this is especially true of those who sail fast planing dinghies like the Flying Dutchman. Whenever a new fitting is considered its usefulness must pay for its weight and the one it replaces is removed immediately. Although it has been known to people such as Twelve Metre-designers and those in the Finn class, it has not been generally recognised that the distribution of weight also can have an effect on boat speed. Two identical boats, even as far as all-up weight and the balance point, may have significantly different boat speeds when sailing in waves if one boat has its weight distributed evenly while the other has very light bows and stern and a heavy centre section. The pitching motion of these boats as waves try to rotate the boat periodically about a horizontal athwartships axis (Fig. 1) can be quite different. The pitching response of the boat determines how the bows hit the waves and this affects the resistance to forward motion and the more energy is wasted on pitching, the less is available for forward motion.

An example of the effect of weight distribution is the resistance of a dumbbell (two equal masses on the ends of a bar) to a twisting motion which is affected not only by the total mass, but is increased as the distance between the two masses increases. This is why tightrope walkers carry a long bar to steady themselves. Most sailors have experienced this when carrying a mast. It is much more difficult to turn around holding the mast than it is when carrying a tool kit of the same weight, and the fact that it is equally difficult to stop turning once the mast is swinging, proves that this is not windage but rather is because the average distance of the mass of the mast from its centre is large and this is what makes it difficult to change its rotation. Another good example is the reduction in rolling when the helmsman and crew sit out on opposite sides of a dinghy sailing downwind. When crew weight is spread far apart, it is more difficult to make the boat roll. Similarly, it is more difficult to make a boat with heavy ends start pitching and also more difficult to stop it, once it is pitching.

Recently, modern technology has made it possible to build planing dinghies consistently below the minimum weight specified by the class rules. About half the Flying Dutchman class at the 1976 Olympics carried corrector weights; one as much as seven kilograms. The gold medal boat carried a large piece of wood, and the Canadian FD was 15 kilograms underweight when first measured in 1975. Carbon fibre and Kevlar make it possible to build exceedingly strong and light production boats and builders now have a choice as to where to put the extra weight. The tendency to centralise this weight could produce faster boats which would defeat the object of one-design racing, and if this tendency were taken to extremes, the ends ▶

The Swedish FD suspended for the Lamboley Test. The two brackets hook under the gunwale and pivot on the crossbeam so that the boat is free to oscillate in the pitching mode. The spirit level on the foredeck is used to level the boat before measuring the position of the balance point, and the period of oscillation.

Fig. 1. The motions of a sailing boat. Surge, sway and heave are the linear oscillations along the fore-and-aft, athwartships and vertical axes respectively. Roll, yaw and pitch are the rotational oscillations around these axes. The co-ordinate system used to calculate the centre of gravity and moment of inertia are the position 'l' and height 'h' referred to an origin on the centre line at the intersection of the deckline and the transom. Only the pitching is affected by the fore-and-aft weight distribution.

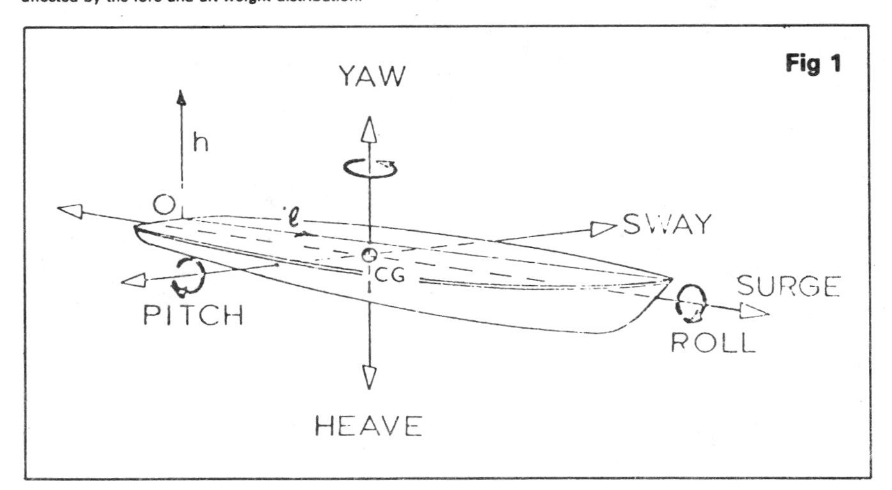

Reprinted from *Yachts and Yachting*, 347–351 (Feb. 3, 1978); © Yachting Press Limited.

Weight Distribution

could become so light and weak that the boats might become unseaworthy. So there seemed to be good reasons for considering rules to prevent the ends from being lightened excessively and the pros and cons of such a weight distribution rule were discussed at an informal meeting of the Flying Dutchman class held during the 1975 world championship in Buffalo, New York. Many sailors were convinced that such a rule was necessary, but it was also clear that weight distribution data, and a technique for measuring it under regatta conditions, would be needed before such a rule could be considered. The 1976 Olympic Regatta was an ideal opportunity to obtain such data, and some preliminary tests showed that the Lamboley test used by the Finn class would be adequate. The plan went ahead and at the 1976 Olympics each competitor was asked if he would allow a Lamboley test to be performed at the conclusion of the official measurement. Practically everyone co-operated enthusiastically and many competitors also made their tune-up boats available. Even the Queen displayed some interest in the Lamboley test of an FD during her visit to Kingston.

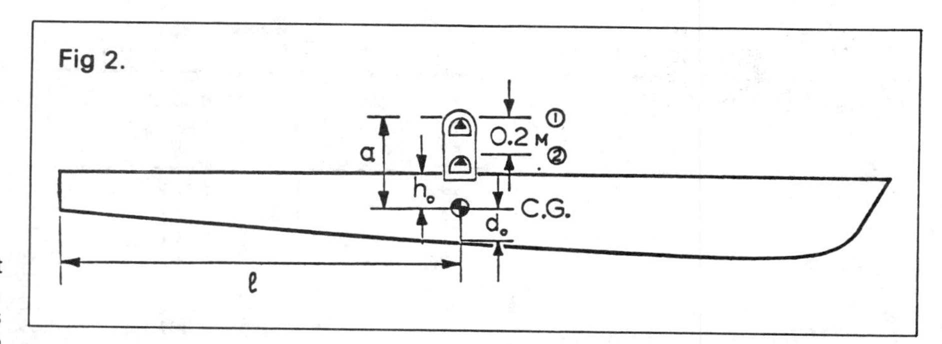

Fig 1. For the Lamboley Test the hull was suspended first from pivot '1' and then from pivot '2', which is exactly 200 mm below pivot 1, and the oscillation period T_1 and T_2 were measured with an electronic timer. To determine the fore and aft position of the 'centre of gravity' the distance 'l' was measured with the deck horizontal. The height of the CG was calculated from 'a ' and the distance ($d_0 - a$) ie from pivot 1 to the keel through the centreboard slot. As a check the distance from pivot 2 to the deckline was also measured.

In 1971 Gilbert Lamboley introduced an oscillation test to measure weight distribution of Finn hulls and could obtain reliable data in less than ten minutes, under regatta conditions. The weight distribution is measured in terms of the radius of gyration, or gyradius, which is the half-length of a dumbbell, of the same mass as the boat, and which has the same inertia to being turned (ie the same moment of inertia). A boat with heavy ends has a large gyradius. The location of the centre of gravity is also found by the test. This method of measuring, and hence controlling, the weight distribution has the great advantage that it avoids complex scantling rules or destructive tests such as drilling holes to measure skin thickness. For the test the hull is suspended from a horizontal athwartships pivot (Fig. 2) and when it is displaced from the rest position and released it oscillates back and forth, with a motion similar to the pitching motion induced by waves. The time it takes for the boat to complete one oscillation depends on the weight distribution, and by measuring the periods about two different pivots both the gyradius and the height of the centre of gravity can be calculated.

The results showed that the hulls with a bulkhead across the boat in front of the mast had significantly lower gyradii and also had their centres of gravity further aft. With the exception of the French boat, with its balsa wood deck and thin double bottom, there was no marked difference between wood and glassfibre hulls, though what effect the use of Kevlar will have remains to be seen. The pitching response depends on the moment of inertia which is the product of the mass and the gyradius squared. So a low gyradius does not necessarily mean it is a good hull, unless the weight is also down at the minimum allowed. For example, if you compare a new FD hull, (especially if it has no fittings) with an old boat, the new hull may have a disappointingly large gyradius, while the old one has a low gyradius. However, when you have added the fittings, which mostly go in the centre of the boat, the gyradius will decrease even though the moment of inertia increases. Conversely on old boat with heavy ends may take us so much extra weight in the centre section that, although it has an enormous moment of inertia, it has a relatively low gyradius.

Fig 2. The total moment of inertia I_T of the boat, helmsman and crew is shown for different fore and aft positions of the helmsman and crew. 'l' is the distance from the transom. The curve marked 'crew' shows the variation in I_T as the crew position is changed, assuming the helmsman remains just behind the genoa fairleads at $l = 1.8$ metres. The helmsman curve assumes the crew to remains at $l = 2.5$ metres and shows that changes in the helmsman's position have a larger effect than changes in the crew's position. To reduce I_T the helmsman should sit as far forward as possible. The solid parts of the curve emphasise the practical region, and the right hand scale shows the percentage change from the average value.

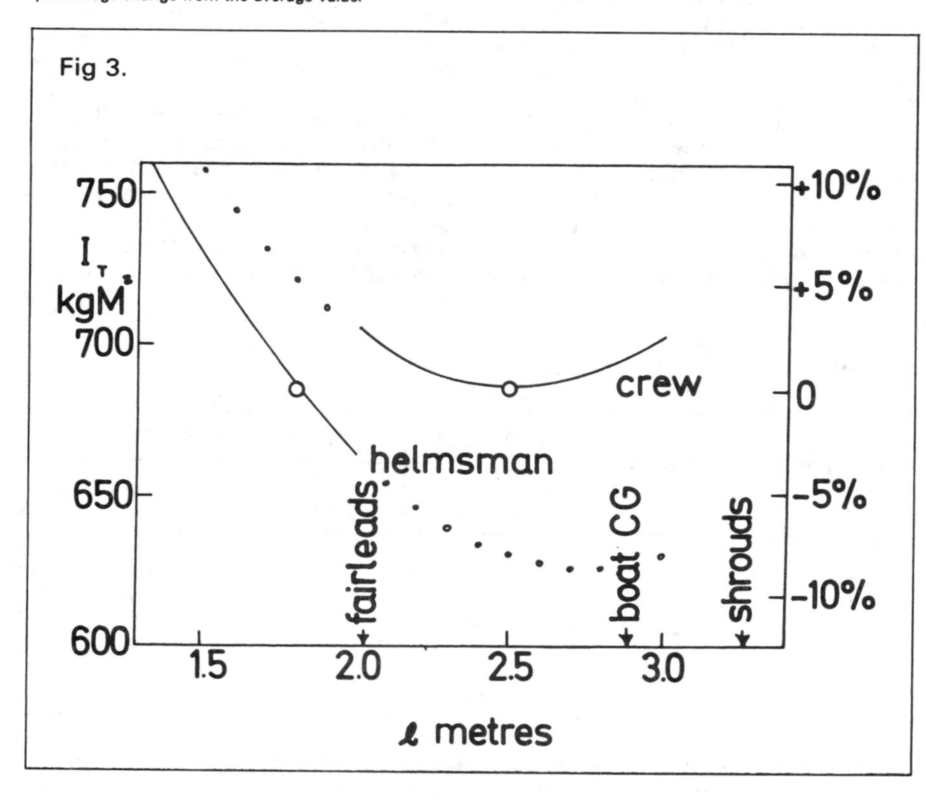

Table 1

COMPONENT	MASS m kg	MASS %	GYRADIUS k_0 Metres	CENTRE OF GRAVITY h Metres	CENTRE OF GRAVITY l Metres	MOMENT OF INERTIA i kg Metres²	MOMENT OF INERTIA %
HULL	125.1	74.2	1.526	−0.300	2.897	308.9	50.3
MAST	14.1	8.4	2.068	2.996	3.667	189.3	30.8
BOOM	3.7	2.2	0.821	0.765	2.219	5.9	1.0
MAINSAIL	3.2	1.9	1.650	2.933	2.692	35.0	5.7
GENOA	2.6	1.5	1.424	1.750	3.715	14.4	2.3
RUDDER & TILLER	5.1	3.0	0.545	−0.265	−0.050	46.0	7.5
CENTREBOARD	6.6	3.9	0.387	−0.880	3.000	7.1	1.2
ANCHOR, PADDLES, ETC.	8.1	4.8	0.904	−0.220	3.200	8.1	1.3
COMPLETE BOAT	168.5	100.0		0.074	2.885	614.6	100.0
MAXIMUM BOAT	184.3	+ 9.4		0.109	2.773	723.2	+ 17.7
MINIMUM BOAT	165.0	− 2.1		−0.017	2.957	499.1	− 18.8
CLOTHED HELMSMAN	90.0	53.4	0.130	0.150	1.800	46.4	7.5
CLOTHED CREW	97.0	57.6	0.130	0.150	2.500	1.8	0.3
TOTAL	355.5	211.0		0.114	2.505	687.3	111.8

The input data for an 'average Flying Dutchman' and the calculated contributions to the moment of inertia of the 'complete boat', taken as 100%. The results of similar calculations using maximum and minimum values are also quoted. The last three lines show the effect of the helmsman and crew on the total moment of inertia.

The response of the boat to waves depends on the total moment of inertia of the hull, mast, crew, and so on just as it is the total weight which is important. However the calculation of the total moment of inertia is not simple, because the contribution of each component depends on its shape and location as well as its weight, so the gyradius of each component has to be measured separately; see results in Table 1. An FD hull accounts for only about 50% of the moment of inertia of the complete boat, the mast contributes 30% and the rudder, sails and other gear a further 20%. In short, variations in the weight, position and gyradius of the mast and components can change the total moment of inertia by about as much as the variations in the hull.

For our calculations the spinnaker, pole, anchor, paddles, sheets, and other gear were assumed to be in typical positions for windward sailing. The symbol k_0 is the gyradius of each component about its own centre of gravity and I is the moment of inertia about the centre of gravity of the boat; h and l are the CG height above the deckline and distance forward of the transom. The calculation was repeated using the worst values, maximum boat, and the best values, minimum boat, for each component. These calculations used extreme values, which may not be compatible with boat tuning (eg the mainsail and boom were lowered to minimum height) and therefore the variation within a typical FD fleet would be much less than that shown. Note that the increase for the maximum boat is mainly due to the increased weight of the components, the gyradius having only changed from 1.91 to 1.98 metres. However this is not so for the minimum boat. Most of the boats were close to minimum weight so the improvement comes from concentration of the weight: 10% from the hull, 5% from the mast, and smaller amounts from lighter sails and rudder. The Canadian boat, when equipped with its original Elvström mast, had a moment of inertia of 540kg M² which is comparable to the minimum boat.

The mast accounts for 30% of the total moment of inertia, so the effect of different mast sections was investigated, see Table 2. The mast gyradii did not differ significantly; however, substantial improvements can be achieved by reducing both the weight and the CG height of the mast to the minimum values. The 8% contribution from the mainsail and genoa is equal to that from the rudder and is much larger than that from either the centreboard or the boom. Many FDs had heavy rudders, but their added strength probably offset any weight

Table 2
Mast Gyradius Measurements

MAST	WEIGHT m kg	TOTAL LENGTH l Metres	CG HEIGHT h Metres	GYRADIUS k Metres	MOMENT OF INERTIA i kg Metres²	MOMENT OF INERTIA i_T kg Metres²
b)						
Elvström MK II	12.8	7.63	2.95	2.13	57.8	172
	12.8	7.60	3.00	2.15	59.0	178
Z-Spar	14.4	7.55	2.98	2.07	61.6	192
Proctor F	13.3	7.57	3.12	2.11	59.2	191
Proctor B	12.0	7.73	2.98	2.25	61.0	172
Fogh	13.8	7.61	3.02	2.06	58.3	187

a) i_T is the contribution of the mast to the total moment of inertia of an "average boat", and includes the effect of the CG Height.
b) Data was recorded for fully rigged masts, including magic boxes, and with different step to deckline lengths. Thus the variations are not solely due to the mast sections.

Weight Distribution

The basic idea is that the helmsman, crew and boat centres of gravity should be as close together as possible — photo Black.

distribution disadvantage. Measurements on a complete boat confirmed the calculated results and also showed that such measurements are impractical under regatta conditions. The windage of the mast requires that measurements be made in absolutely still air: not the best conditions for sailing.

Table 1 shows that the helmsman and crew together contribute about 53% of the all-up weight of a Flying Dutchman. The distribution of this crew weight affects the total moment of inertia, and one might expect that changes in crew position and weight would be more significant than any differences between individual hulls. Fig. 3 shows the calculated effect of the crew position on the total moment of inertia and it is seen to be surprisingly small. The crew were assumed to be rigid bodies (some helmsmen have been known to believe this about their crews and vice versa). However a good crew naturally moves so as to absorb some of the irregularities in the motion of the boat, hence their weight does not have exactly the same motion as the boat. However, the major effect of the crew is due to their fore and aft position, and this does not change significantly from wave to wave. Thus the calculations should correctly reproduce the trend of the effect of the helmsman and crew on the weight distribution.

Fig. 4. The hull gyradius k is plotted versus the distance l_{CM} from the transom to the hull CG, for the FD hulls present at the 1976 Olympics. The diagonal curves are labelled with the total moment of interia I_T which these hulls would have if equipped with an 'average' crew, mast, and equipment. Note that boats which do not carry the double bottom forward of the mast have low values of both k and l_{CM}.

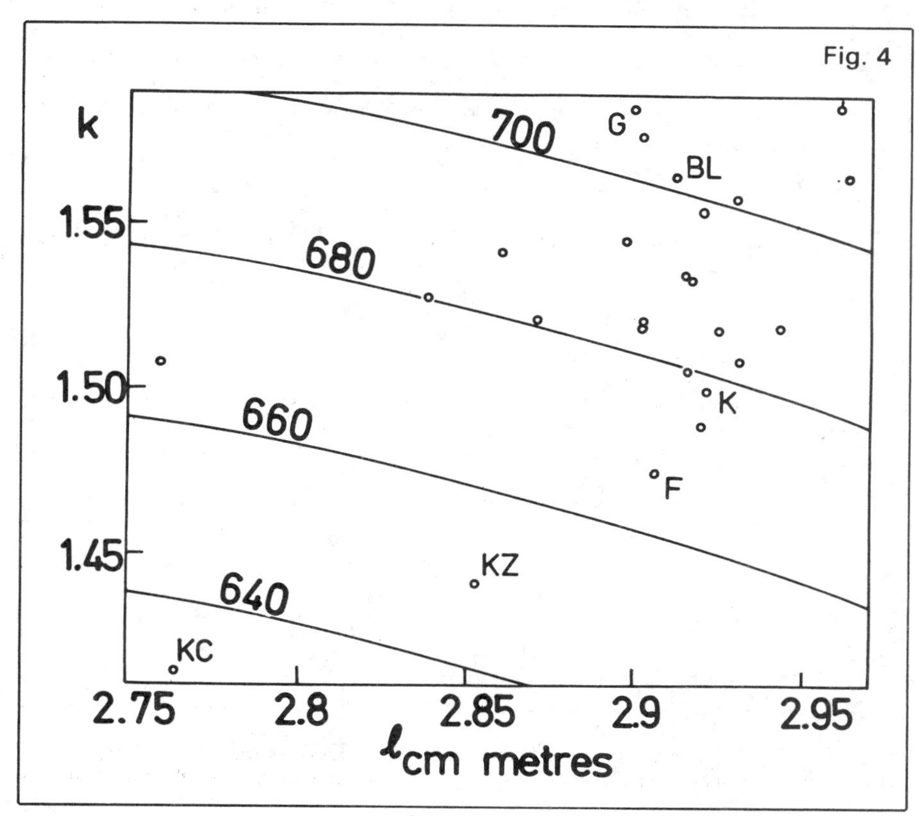

The crew is assumed to be flat out on the trapeze, the helmsman to be hiking out, and both are assumed to be wearing 16kg of clothing. Because the crew is very close to the centre of gravity of the combined boat plus helmsman his contribution to the moment of inertia is very small and, as shown in Fig. 3, even quite a large change in his position has only a minor effect on the weight distribution. However, the helmsman, because he is farther from the CG of the combined boat plus crew, has a far greater effect. Clearly it is advantageous for an FD helmsman to be both as far forward and as light as possible. (A 470 helmsman plus a Tempest crew should be ideal.)

Just as it is advantageous for the helmsman to be as far forward as possible, it is also advantageous for the centre of gravity of the boat to be as far aft as possible, and the basic idea is that the helmsman, crew, and boat centres of gravity should be as close together as possible. The total moment of inertia can be reduced by reducing the hull gyradius, or by moving the hull centre of gravity aft. For example, a hull that has an average gyradius but with the CG aft will, when the crew are added, have a smaller total moment of inertia than a hull that has a smaller gyradius but the CG farther forward, see Fig. 4. Variations in crew, mast, and equipment can produce up to an 11 per cent change in the total moment of inertia, while the positions of crew and helmsman can produce 5% differences, compared to a 17% variation from the hulls. What this really means is that although the hull still is the most important single factor, the combined effect of optimising all the other components can be as large as that from the hull.

As a test of the effects of the gyradius on the subsequent racing performance the 1976 Olympics had the great advantage that details of the wind speed, sea state and finishing times, as well as the final places, were available for each race. The quality and consistency of the helmsmanship and equipment in an Olympic Regatta should reduce the performance variations from these factors, so that if the gyradius does have a major effect we might hope to see a statistically significant trend in the results. The disadvantages of the 1976 Olympic Regatta as a test were that there were relatively few boats, and there were only seven races, one of which was essentially meaningless. Thus the wind and wave conditions might not have included those that induce resonant pitching in an FD. Also remember that even a definite correlation is not conclusive evidence: it may just mean that all the best helmsmen had heard that it might help them to reduce weight in the ends and had taken no chances, while the rest had not. The measurers awaited the Olympic results with great interest: four of the top contenders had boats with low gyradii, while two others had boats with relatively large gyradii, and to avoid any one-upmanship or bias, the results of the gyradius measurements were not released to the competitors until after the racing was over.

To make a long story short, the gold medal winning hull had the largest gyradius, the silver medal hull was lower than average, the bronze medal hull had the third largest gyradius and the hull with substantially the lowest gyradius was fourth. Data for the final placings of the rest of the fleet show the same scatter as the first four boats and gives no evidence of a trend, (see Fig. 5) nor did an analysis done in terms of finishing times. Analysing the total moment of inertia for the complete boat did not affect these conclusions, for practically every boat in the fleet used the Z Spar mast, and similar equipment. The one exception was the gold medal winner who, together with the Swiss boat, used an Elvström Mk II mast. It has been suggested that this made a major difference to the weight distribution but as can be seen in Fig. 4 their total moment of inertia was still significantly above the average. In contrast to the results for the Flying Dutchman class at the 1976 Olympics, there are some results from the 1972 US Finn Trials which show some correlation with the finishing order. Much more data, recorded under controlled conditions, will be required before a scientific case for the effect of the gyradius on the boat speed of a planing sailing dinghy can be established.

The Lamboley method of measuring weight distribution was used to measure Flying Dutchman hulls, masts, and components, and the results allow the total moment of inertia to be calculated reliably. The effect of the helmsman and crew on the weight distribution is much smaller than might be anticipated from their large contribution to the overall weight. The hull does make the most important contribution to the weight distribution, but the fore and aft position of the centre of gravity is almost as important as the gyradius. Of the other components, only the mast makes a more important contribution than the helmsman. The Flying Dutchman hulls found to have significantly the lowest gyradius and weight well aft were also very seaworthy and of a construction the class should encourage, so at the present time the introduction of a rule to control weight distribution of the Flying Dutchman does not seem warranted, but future developments could change this.

For the scientifically-inclined dinghy sailor there are two hobbies: sailing and attempting to analyse how a sailing dinghy should work. Success in the former depends to a large extent on not letting oneself become confused by the latter. Analysing the motion of a sailing boat is so complex that many simplifying assumptions must always be made, so that the effects of single components can be isolated. These assumptions are usually made so that the analysis can be completed rather than because they apply to all situations encountered during a race. Progress must therefore be made by a judicious mixture of theory and practice, and it may be suggested still that assiduous practice will give better results than devotion to theory. A more detailed account of these measurements and calculations is available from the author.

Fig. 5a. The total moment of inertia I_T of the Flying Dutchman dinghies at the 1976 Olympics is plotted versus their finishing times in the light air (wind 9 knots, sea state 1ft) 3rd race. The dashed line is a statistical fit, and has a correlation coefficient r = 0.2. If the last four boats are ignored it would almost be believable.

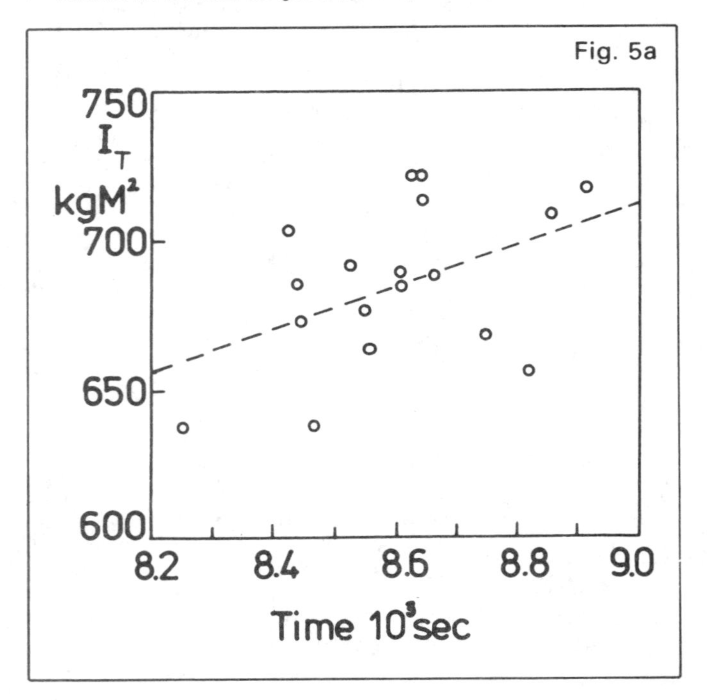

Fig. 5b. The total moment of inertia I_T of the Flying Dutchmen is plotted against their average finishing times for the 2nd, 4th, 6th and 7th races at the 1976 Olympics, ie the races with a 'sea state' of 3ft to 4ft waves. The dotted line is a statistical best fit but is essentially meaningless as the scatter is very large. (r = 0.04).

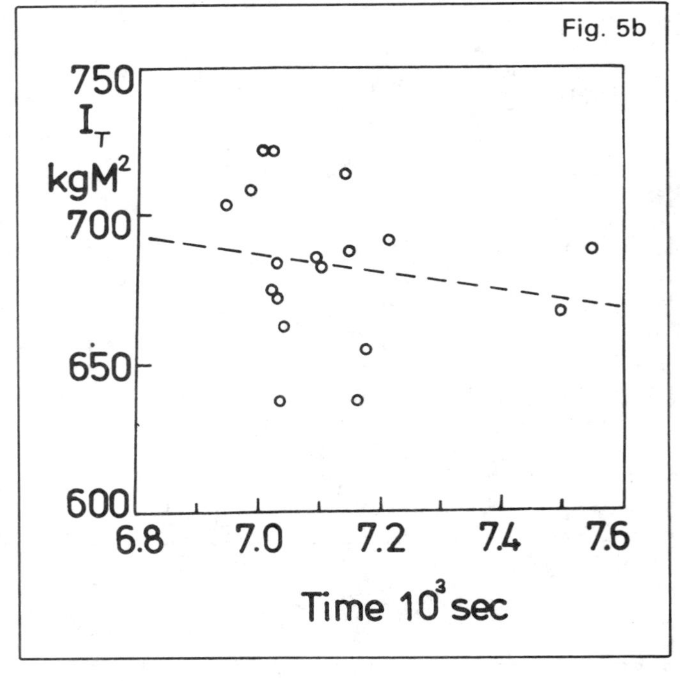

Rowing: A Similarity Analysis

Abstract. *The theory of models is employed for calculating the speed of geometrically similar racing shells. The theoretical prediction, that shells should have a speed proportional to the number of oarsmen raised to the 1/9 power, is in excellent agreement with observations. A significant implication for the proportions of lightweight and heavyweight eight-oared shells, indicating how both might attain the same speed, is developed.*

Why should larger boats containing many oarsmen go faster than smaller ones containing fewer oarsmen? The task of predicting the performance of a single man as he engages in athletic tasks such as running, jumping, flying, or, as in this case, rowing is in general difficult, since a specific knowledge is required of how chemical energy is converted to work and of how that work is utilized against external forces to produce motion. The theory of geometrically similar models makes possible enormous simplifications, provided the objective is changed from a prediction of absolute performance of a particular boat and crew to a comparison of the speeds of geometrically similar boats containing different numbers of oarsmen.

Consider the schematic representation of a shell in Fig. 1. The fundamental postulate of the following analysis will be that competitive shells are geometrically similar. The validity of this assumption is revealed in Table 1, where the length l and beam b are compared for shells accommodating one, two, four, and eight oarsmen. Despite the variety of sizes among the boats, the slenderness ratio l/b is reasonably invariant of the length of the boat. Also reasonably invariant is the boat weight per oarsman, which will have important consequences.

Each oarsman in either the lightweight or heavyweight class is assumed indistinguishable from other members of his class in weight and capability for sustained power output. The principal force that hinders the motion of the boat through the water is assumed to be skin friction drag. The great l/b ratio and relatively low velocity of the boats suggest that wave drag is small, and, in fact, full-scale towing tank tests have shown that the resistance due to leeway and wave-making together constitute only 8 percent of the total drag at 20 km per hour, the Olympic target speed, for an eight-oared shell (*1*).

The assumptions may be summarized as follows.

1) There is geometric similarity between boats, and the draft, or, equivalently, the volume of water displaced by the boats, is similar.

2) The boat weight per oarsman is a constant, k.

3) Each oarsman contributes power P_0 and weight W_0.

4) Skin friction drag is the only hindering force.

A cross section of a given boat is schematically represented in Fig. 1, in which the cross-sectional area A under the water line is illustrated. The volume of water displaced by the whole boat, length l, is proportional to the total weight of the boat plus n oarsmen:

$$Al \propto (\text{weight of boat}) + (\text{weight of oarsmen}) \propto nW_0 + knW_0 \propto n$$

But the area A is proportional to the square of some characteristic length of the boat (that is, L^2), and the length is directly proportional to L. Therefore,

$$n \propto L^3$$

The total power required to move the boat at velocity v is proportional to the product of the drag force D and the velocity v:

$$vD \propto nP_0$$

The skin friction drag is proportional to the product of the wetted area $A^{1/2}l$ and the square of the velocity. Thus, D is proportional to L^2v^2. Since P_0 is independent of scale and L^2 is proportional to $n^{2/3}$ from above, the final result for the dependence of the velocity v on the number of oarsmen n becomes

$$v^3 \propto \frac{n}{L^2} \propto \frac{n}{n^{2/3}} = n^{1/3}$$

$$v \propto n^{1/9}$$

The last four columns of Table 1 present representative times for 2000 m for Olympic and World Championship crews of heavyweight oarsmen, who were rowing several types of boats. These times, drawn as triangles, circles, and crosses, are plotted against the number of oarsmen per boat on a log-log plot in Fig. 2. Excellent agreement is found between these data and the similarity analysis, shown as a solid line of slope $-1/9$ in the figure.

An intriguing application of the similarity rule derived above concerns the observation that, among crews in eight-oared shells, the heavyweight crew (average weight, 86 kg per man) consistently beats the equivalently advantaged lightweight crew (73 kg per man) by four or five lengths in 2000 m and is therefore 5 percent faster. The description of the shells themselves in Table 1 indicates that the length and weight of the eight-oared shells are the same, and the only difference between the two lies in the beam of the heavyweight boat, which is 2 percent wider than the lightweight boat. The beam of the heavyweight boat would have to be

Reprinted from *Science* **173**, 349–350 (1971);

Table 1. Shell dimensions and performances. Values in the last four columns represent: I, 1964 Olympics, Tokyo (2); II, 1968 Olympics, Mexico City (3); III, 1970 World Rowing Championships, St. Catharines, Ontario (4); IV, 1970 Lucerne International Championships, Switzerland (5). Shell dimensions from J. A. Garofalo, shell builder.

No. of oarsmen	Modifying description	Length, l (m)	Beam, b (m)	l/b	Boat weight per oarsman (kg)	Time for 2000 m (minutes)			
						I	II	III	IV
8	Heavyweight	18.28	0.610	30.0	14.7	5.87	5.92	5.82	5.73
8	Lightweight	18.28	0.598	30.6	14.7				
4	With cox	12.80	0.574	22.3	18.1				
4	Without cox	11.75	0.574	21.0	18.1	6.33	6.42	6.48	6.13
2	Double skull	9.76	0.381	25.6	13.6				
2	Pair-oared shell	9.76	0.356	27.4	13.6	6.87	6.92	6.95	6.77
1	Single skull	7.93	0.293	27.0	16.3	7.16	7.25	7.28	7.17

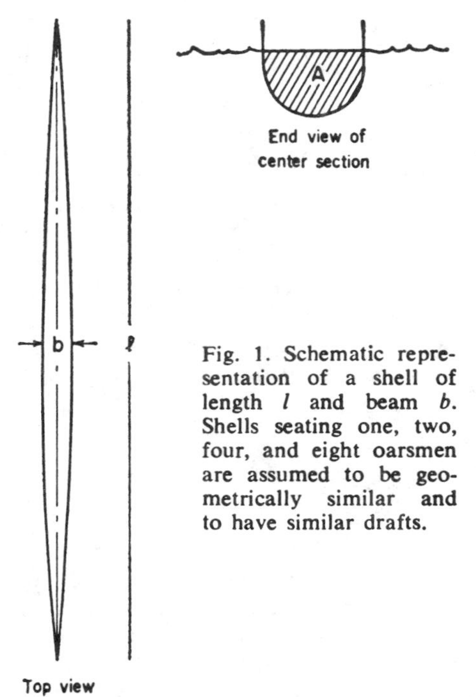

Fig. 1. Schematic representation of a shell of length l and beam b. Shells seating one, two, four, and eight oarsmen are assumed to be geometrically similar and to have similar drafts.

9 percent wider than the lightweight boat if the cross sections (A in Fig. 1) below the water were to be similar, so that a mild departure from similarity between light and heavy boats must be anticipated; that is, the heavy boat sits deeper in the water. Nevertheless, let us take the assumption that the wetted area of both light and heavy boats is the same. When this assumption is made, the speed of the boat is limited by the capacity of the crew for sustained power output. An argument rages as to the precise power of body weight that provides the best fit with data, but physical reasoning demonstrates that the factors limiting power output—that is, tensile strength of muscles, tendons, and bones, the rate of supply of oxygen admitted by the lungs and carried by the blood, and the rate of removal of heat from the working muscles—all are proportional to body surface area (thus, $W^{2/3}$) for geometrically similar animals. Then the power available to the heavyweight oarsmen is $K^{2/3}$ that of the lightweights, where K is the weight ratio, 86/73. Therefore,

$$\frac{v_{\text{heavyweight}}}{v_{\text{lightweight}}} \propto K^{2/9} \simeq (1.2)^{2/9} = 1.05$$

The heavyweights are thus predicted to be 5 percent faster than the lightweights, as is observed. Notice that if the assumption is taken instead that the length of both light and heavy boats are equal but that the below-water cross section A is proportional to displacement and thus to total weight, the result is

$$\frac{v_{\text{heavyweight}}}{v_{\text{lightweight}}} K^{1/18} \sim 1.01$$

In this case, the heavyweights beat the lightweights by only 1 percent, which is smaller than the observed margin.

What would happen if the lightweight shell were geometrically similar to the heavyweight but shorter by the ratio $K^{-1/3}$? In this case, the wetted area of the lightweight shell would be $K^{-2/3}$ the wetted area of the heavyweight shell, but since the power available would also be $K^{-2/3}$ that of the heavyweights, the two shells would have the same speed. The remarkable conclusion becomes that, if the lightweight shell were shortened by 1.1 m in length and made 2.4 cm narrower in beam from its present dimensions, the lightweight crew could keep up with the heavyweights. The practical validity of this results is certain to be checked, sooner or later, by a sufficiently enlightened lightweight crew.

THOMAS A. MCMAHON
Division of Engineering and Applied Physics, Harvard University, Cambridge, Massachusetts 02138

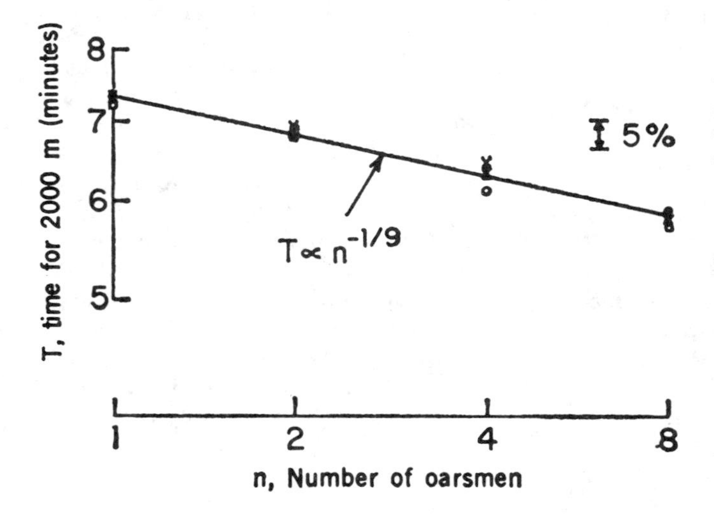

Fig. 2. Test of the theory of models result. The −1/9 power law (solid line) is compared with racing times for 2000 m, all in calm or near calm conditions. (Triangles) 1964 Olympics, Tokyo; (solid circles) 1968 Olympics, Mexico City; (crosses) 1970 World Rowing Championships, Ontario; (open circles) 1970 Lucerne International Championships.

References

1. J. F. Wellicome, "Report on Resistance Experiments Carried out on Three Racing Shells" [Ship Technical Memorandum 184, National Physical Laboratory (Great Britain), July 1967].
2. National Association of Amateur Oarsmen, *Rowing Guide* (Detroit, Mich., 1964), p. 46.
3. H. Parker (Harvard rowing coach), personal communication.
4. J. Frailey, *Oarsman* 2 (No. 5), 13 (November 1970).
5. ———, *Rowing* 15 (No. 19), 17 (August September 1970).

12 February 1971; revised 13 May 1971

Why is it harder to paddle a canoe in shallow water?

Question:

Our canoe club has noted that it's much more difficult to paddle in shallow water and maintain speed. In all cases the water has been deep enough to paddle in, and other factors such as wind, current and waves have been eliminated as the cause.

How can we explain the effect that depth of water has on the speed of Olympic-style racing canoes and kayaks?

This interesting and complex question was asked by **Ann Bowness,** *a student of physics teacher* **E. Locking** *of Chippewa Secondary School, North Bay, Ontario P1B 6G8. The answer was written by* **Angelo Armenti** *of Villanova University, Villanova, PA 19085. As well as being Dean of University College, Dr. Armenti is professor of physics and has supplied answers for several of our previous questions that involved the physics of sports.*

Answer:

Qualitatively speaking, the total resistance encountered by a hull moving on the water arises from two distinct sources, viz.,

a) viscosity effects, and

b) gravity effects.

As we shall see below, both types of resistance are affected by water depth and, as a result, the total resistance encountered in paddling a canoe or kayak is quite sensitive to the depth of water.

Viscosity effects

Viscosity effects give rise to "skin friction drag," a type of resistance which is characterized by a Reynolds number[1] and which is produced by the interaction between a hull and the surrounding water via the "wetted surface" of the hull. In effect, the moving hull sets into motion water which was previously at rest. By Newton's second and third laws of motion, the resulting drag force on the hull results from and is precisely equal to the rate of change of momentum of that water.

On the most elementary level, the hull drags along a thin layer of water in contact with the boat, and that layer (because water is viscous) attempts to drag along an adjoining layer, etc. The net result is the formation of a boundary layer[2] which attaches itself to the wetted surface of the moving hull in a manner similar to that illustrated by Fig. 1. The turbulent portion of the boundary layer gives rise to a *frictional wake* consisting of moving water behind and on either side of the hull. This wake represents energy which has been transferred to the water by the hull through viscous friction.

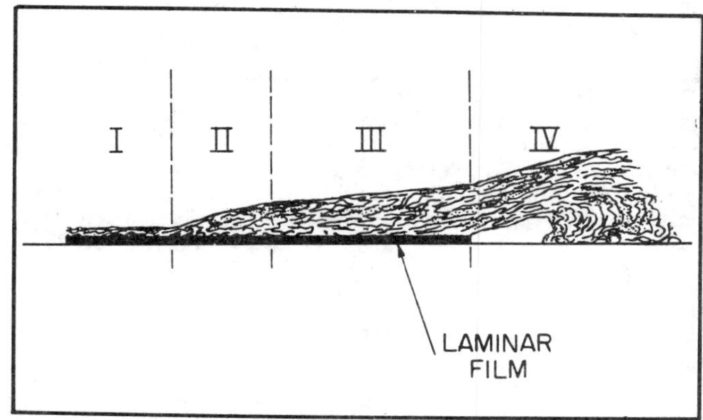

Fig. 1. Flow within the boundary layer. A typical boundary layer consists of four distinct regions as indicated in the figure: I-laminar; II-transitional; III-turbulent; IV-breakaway or separation. A laminar film underlies the first three regions. (Taken from K. C. Barnaby, Ref. 1.)

An empirical formula useful in predicting the magnitude of the skin friction drag on a hull was discovered by William Froude in 1872. According to his formula, the skin friction drag is given by

$$R = fSV^{n} \tag{1}$$

where R is the resistance in pounds, S is the wetted area of the hull in square feet, and V is the velocity of the hull in knots.[3] The values of the parameters n and f which gave the best fit to the experimental data were $n = 1.825$ and

$$f = 0.00871 + 0.053/(8.8 + L) \tag{2}$$

where L is the waterline length of the hull in feet.

Since skin friction drag increases with viscosity and viscosity, in turn, increases rapidly with decreasing temperature, skin friction drag decreases noticeably with increasing water temperature.[4]

Gravity effects

The motion of a hull through the water produces pressure changes *below the surface* which, together with the force of gravity, determines the instantaneous equilibrium contour at the (two-dimensional) free surface of the water. This irregular elevation of the free surface of the water (i.e., wave pattern) represents energy which has been diverted to wavemaking by the hull, essentially by doing work against gravity. Ironically, although the surface waves generated by a moving hull are eventually damped out by *viscosity*, wavemaking is nevertheless a *gravity* type resistance. Historically, *wavemaking resistance* came to be known as *residual* resistance since it was the resistance left over after skin friction drag had been accounted for. Wavemaking resistance is characterized by a Froude number in a manner similar to the way that viscous resistance is charac-

Reprinted from *The Physics Teacher* **23**, 310–313 (1985); © American Association of Physics Teachers.

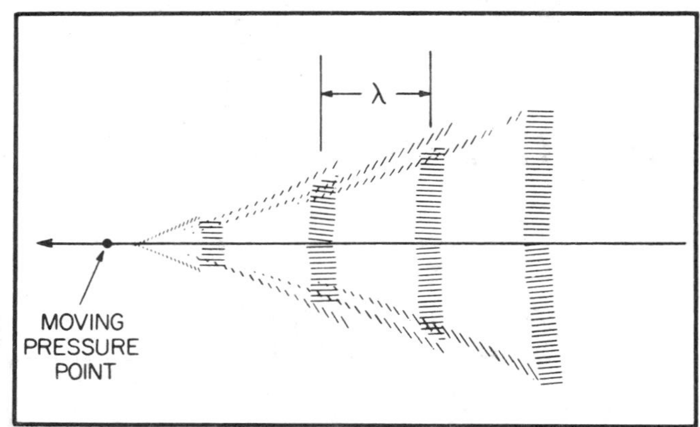

Fig. 2. Kelvin wave pattern for a moving pressure point. A moving pressure point produces transverse waves of wavelength λ that lie inside a V-shaped wake line. In deep water, $\lambda = 2\pi V^2/g$, where V is the velocity of the pressure point and g is the acceleration due to gravity. The wave pattern produced by a real hull may be approximated by the superposition of two such Kelvin patterns located respectively near the bow and stern of the hull. (Taken from J. F. Wellicome, Ref. 2.)

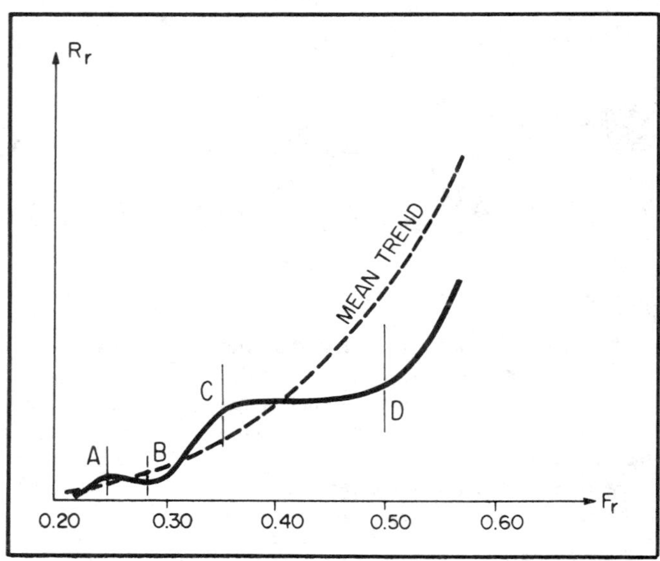

Fig. 3. Variation of wave resistance with Froude number. The total wave resistance is given by $R_w = \frac{1}{2}\rho SV^2R_r$ where ρ is the density of water, S is the surface area of the hull, V is the velocity of the hull, and R_r is the reduced resistance indicated in the figure. Resistance humps (at A and C) and resistance hollows (at B and D) are superimposed on a sharply upward mean trend of increasing resistance with Froude number. (Taken from J. F. Wellicome, Ref. 2.)

terized by a Reynolds number. The Froude number is given as

$$F_r = V/(gL)^{1/2}$$

where V is the velocity of the hull through the water, L is the waterline length of the hull, and g is the acceleration due to gravity.

Kelvin wave pattern

When a point of excess or reduced pressure moves beneath the surface of the water, a Kelvin wave pattern is formed.[5] As shown in Fig. 2, a Kelvin pattern is characterized by a) a diverging and approximately V-shaped wave system moving outwards from the line of motion of the pressure point, and b) a series of transverse waves constrained to lie inside the arms of the diverging wave pattern. In *deep* water, the distance between successive crests (i.e., the wavelength λ) of the transverse wave system is proportional to the square of the speed V of the pressure point. In fact,

$$\lambda = 2\pi V^2/g \tag{3}$$

The shallow water case is discussed below.

Although the wave pattern generated by a real hull appears extremely complicated, it may be represented fairly well by the superposition of *two* Kelvin wave patterns, one located slightly behind the bow and the other located slightly ahead of the stern. The net result of the superposition of the two Kelvin patterns depends on the speed of the hull and the distance d between the two pressure points, a distance which is approximately equal to the waterline length of the hull.[6] At certain speeds, the transverse wave crests from the bow system will add constructively with the transverse wave crests from the stern system, thereby producing a large (and hence, energetic) wave train behind the hull. At other speeds, crests from one system will add to troughs from the other system resulting in a very reduced wave train behind the hull. Obviously, the additional energy contained in large wave trains implies larger resistance forces for the hull while the reduced energy contained in smaller wave trains implies smaller resistance forces for the hull. Resistance "humps" occur when the distance between the two pressure points satisfies the condition for constructive interference, while resistance "hollows" occur when the distance between the two pressure points satisfies the condition for destructive interference.

The condition for constructive interference is

$$d = \lambda, 2\lambda, 3\lambda, \text{etc.} \tag{4}$$

while the condition for destructive interference is

$$d = (1/2)\lambda, (3/2)\lambda, (5/2)\lambda, \text{etc.} \tag{5}$$

Hence, from Ref. 6 and Eqs. (3), (4), and (5), we see that these humps and hollows correspond to fixed values of the Froude number $F_r = V/(gL)^{1/2}$. In particular, there is a resistance hump near $F_r = 0.35$ and a resistance hollow near $F_r = 0.50$. As indicated in Fig. 3, these humps and hollows are superimposed on a sharply upward trend of increasing resistance with Froude number.

In order to relate Fig. 3 to the case of a canoe or kayak, one needs to know the Froude number for the particular hull in question. For a kayak of length $L = 4.5$ m moving at a world class speed of V = 4.5 m/s, the Froude number comes out to about 0.68. Clearly, from the very definition of the Froude number, short, fast hulls will transfer a lot of energy into wavemaking. Long hulls, on the other hand, will have more wetted surface (and hence more skin friction drag) and relatively less wavemaking resistance.[7]

Shallow water effects

The theoretical relationship between the wavelength and velocity of water waves may be derived from a particular wave model of hydrodynamics.[8] The result is

$$v = [(g\lambda/2\pi)\tanh(2\pi d/\lambda)]^{1/2} \tag{6}$$

where v is the wave velocity, λ is the wavelength, and d is

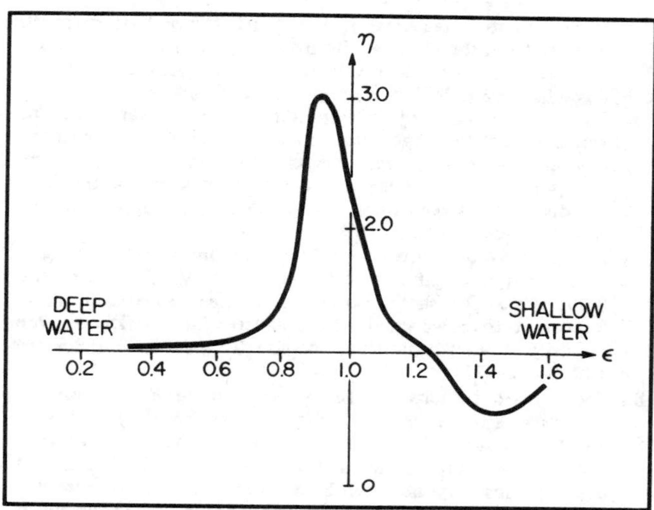

Fig. 4. Typical shallow water resistance variation for a motor launch. $\eta \equiv$ (Resistance in shallow water/Resistance in deep water) and $\epsilon \equiv V/(gd)^{1/2}$, where V is the velocity of the hull and d is the depth of the water. $\epsilon = 1$ corresponds to the case where the hull moves at a speed equal to the *critical wave speed*

$$v_c = (gd)^{1/2}$$

the maximum speed that water waves can have, regardless of wavelength, in water of depth d. Note that the maximum wave resistance occurs when the hull speed is about 90% of the critical wave speed. (Taken from J. F. Wellicome, Ref. 2.)

the depth of the water. Tanh x is the hyperbolic tangent, which is defined as:

$$\tanh x = \frac{e^x - e^{-x}}{e^x + e^{-x}}$$

For small x, a series expansion can be used.

The first two terms of the series are:

$$\tanh x \approx x - 1/3x^3 \ldots \qquad (\text{for } |x| < \tfrac{1}{2}\pi)$$

Therefore, in shallow water where $d/\lambda \ll 1$, $\tanh(2\pi\, d/\lambda) \approx (2\pi\, d/\lambda)$, and

$$v = (g\, d)^{1/2} \tag{7}$$

For deep water, on the other hand, $(d/\lambda) \gg 1$.

In this regime $\quad e^{-x} = e^{-(2\pi d/\lambda)} \ll e^x$.

Consequently, $\tanh(2\pi\, d/\lambda) \approx \frac{e^x}{e^x} = 1$, and

$$v = (g\lambda/2\pi)^{1/2} \tag{8}$$

which is exactly equivalent to Eq. (3) for the transverse waves in a Kelvin wave pattern for the case where the waves move at the same speed as the hull.[9]

From Eq. (7) we see that the maximum velocity of water waves in shallow water goes as the square root of the depth, independent of wavelength. Hence, as the water gets shallower and the relationship between λ and V changes, so too does the location of the humps and hollows of Fig. 3. As a result, the amount of wavemaking resistance varies drastically with depth, increasing to several times the deep water value as the water becomes shallower and the speed of the hull approaches the critical wave speed, and then decreasing to about one-half the deep water value as the now much slower waves can no longer keep pace with the hull, and rapidly cease to exist altogether.

Although there does not seem to be any data available on the effect of shallow water on the resistance encountered by a kayak or canoe, a typical curve of resistance for a motor launch is given in Fig. 4. One expects that the general trend indicated in Fig. 4 would also obtain in the case of a kayak. Assuming that to be the case, then, a kayak of length 4.5 m moving at 4.5 m/s would encounter the greatest wavemaking resistance for $\epsilon \equiv V/(gd)^{1/2} = 0.9$, which corresponds to a water depth of 2.55 m (8.4 ft). The minimum wavemaking resistance would occur near $\epsilon = 1.4$, which corresponds to a depth of 1.05 m (3.5 ft) and would again rise some 20% above this minimum value at $\epsilon = 1.6$, corresponding to a depth of 0.81 m (2.6 ft).

As can be seen from the graph, added wavemaking resistance (due to shallow water) begins to appear once the hull speed exceeds about one-half the critical wave speed $v_c \equiv (gd)^{1/2}$. Alternatively, one might say that the added wavemaking resistance appears once the depth of water falls below about four times the critical depth $d_c = V^2/g$. For our kayak, this depth is about 8.3 m (27 ft!).

The depth dependence of skin friction drag does not begin to manifest itself until the water becomes so shallow that the bottom begins to approach the edge of the boundary layer. When this happens, the boundary layer thickens very rapidly and skin friction drag increases very sharply. The onset of this added resistance occurs when the depth of the water falls below about two or three times the thickness of the boundary layer at the stern, which for our kayak is a depth of about two or three feet.

In summary, then, as one goes from deep water to shallow water, the wavemaking resistance on a moving hull first increases to several times the deep water value, and then falls to about half the deep water value, and finally begins to rise again, as illustrated in Fig. 4. The depth dependence of skin friction drag, on the other hand, does not manifest itself at all until the depth of the water decreases to the point where the boundary layer begins to approach the bottom, at which point the skin friction drag increases appreciably.

Finally, if the added resistance noted by the canoe club members occurred as their boats went from truly deep water (about 27 ft) to about 8.4 ft, such an increase would be the result of the added wavemaking resistance depicted by the sharply increasing leftmost portion of the curve (corresponding to $0.5 < \epsilon < 0.9$) in Fig. 4. If, on the other hand, the added resistance occurred as the water depth fell below about 3.5 ft, it is clear that this added resistance is due to *both* the 20% increase in wavemaking resistance encountered as the depth falls from 3.5 ft to 2.6 ft (corresponding to $1.4 < \epsilon < 1.6$), as well as the increase in skin friction drag due to the increasing proximity of the boundary layer to the bottom.

References

1. For the case of a hull moving through water, the Reynolds number is taken as $\mathcal{R} = LV/\nu$, where L is the waterline length of the hull, V is the velocity of the hull through the water, and ν is the kinematic viscosity of water. For an elegant general discussion of the Reynolds number see R. P. Feynman, R. B. Leighton, and M. Sands, *The Feynman Lectures on Physics* (Addison-Wesley, Reading, MA, 1964), Vol. II, Sec. 41-3. For a specific discussion of the role of the Reynolds number in the motion of hulls through water, see K. C. Barnaby, *Basic Naval Architecture*, 2nd ed. (Hutchinson's Technical Press, London, 1954), Ch. X.
2. See J. F. Wellicome, "Some Hydrodynamic Aspects of Row-

ing," in J. G. P. Williams and A. C. Scott, editors, *Rowing: A Scientific Approach* (A. S. Barnes & Co., New York, 1967), Ch. 2. The actual dimensions of a boundary layer are difficult to predict accurately. However, empirical evidence suggests that for hulls similar to those of kayaks, the boundary layer has approximately molecular thickness at the bow, increasing to several inches at the mid-point of the hull and to perhaps as much as a foot at the stern.

3. A speed of one knot is equal to one nautical mile (6080.2 ft) per hour. Hence, one knot equals 6080.2/5280 = 1.15 miles/hour = 1.69 f/s = 0.51 m/s.
4. The expression for f given in Eq. (2) is valid only for seawater at 59°F. For every 5°F increase (decrease) in water temperature, there is a 1.2% decrease (increase) in resistance. For fresh water, one must multiply the resistance calculated in Eq. (1) by 0.975. Also, skin friction drag varies with the roughness of the hull since the rougher the hull, the more likely that water from one layer will be transferred into another *within the boundary layer*. Hence, the rougher the hull surface, the *thicker* the boundary layer and the greater the resistance encountered by the hull. On the other hand, the injection of certain chemical additives (such as Guar Gum or Polyox) into the boundary layer can inhibit the growth of the boundary layer to the extent that the frictional drag is reduced by almost half! See Ref. 2, pp. 28 and 60. See also the "Search and Discovery" section of *Physics Today*, March 1978, p. 17.
5. The distinctive features of the Kelvin wave pattern result directly from the fact that the group velocity for these waves is precisely half their phase velocity. See F. S. Crawford, Am. J. Phys. 52, 782 (1984), and references cited therein.
6. The physical basis of this approximation is clear since one would expect the largest pressure differences to occur near the bow and the stern. While the precise location of the representative pressure points depends on the exact shape of the hull, the distance between these two points is typically 80% of the waterline length, i.e., $d \approx 0.8\, L$.
7. For the sake of comparison, the Froude number for an eight-oared shell of length $L = 18$ m and velocity V = 5.2 m/s is only 0.39. Hence for shells, skin friction drag tends to be more important than wavemaking resistance while for kayaks, the opposite appears to be true. See also, T. A. McMahon, Science 173, 349 (1971).
8. See Bernard LeMehaute, *An Introduction to Hydrodynamics and Water Waves* (Springer-Verlag, New York, 1976), p. 219.
9. In deep water, the longest wavelengths have the greatest speed. As a result, in the Pacific ocean which is both large and deep, storms very far from land generate various wavelengths which then reach land at different times. As every surfer knows, the long wavelength (good surfing) waves must be enjoyed immediately since their arrival will soon be followed by the short choppy waves unsuitable for surfing.

Terminal velocity of a shuttlecock in vertical fall[a)]

Mark Peastrel, Rosemary Lynch, and Angelo Armenti, Jr.
Department of Physics, Villanova University, Villanova, Pennsylvania 19085
(Received 11 June 1979; accepted 11 December 1979)

We have performed a straightforward vertical fall experiment for a case where the effects of air resistance are important and directly measurable. Using a commonly available badminton shuttlecock, a tape measure, and a millisecond timer, the times required for the shuttlecock to fall given distances (up to almost ten meters) were accurately measured. The experiment was performed in an open stairwell. The experimental data was compared to the predictions of several models. The best fit was obtained with the model which assumes a resistive force quadratic in the instantaneous speed of the falling object. This model was fitted to the experimental data enabling us to predict the terminal velocity of the shuttlecock (6.80 m/sec). The results indicate that, starting from rest, the vertically falling shuttlecock achieves 99% of its terminal velocity in 1.84 sec, after falling 9.2 m. The relative ease in collecting the data, as well as the excellent agreement with theory, make this an ideal experiment for use in physics courses at a variety of levels.

I. INTRODUCTION

In discussing the effects of air resistance on the motion of a vertically falling object, it is generally accepted that the resistive force is proportional to the first or second power of the instantaneous speed.[1] Further, it is generally recognized that the Reynolds number $\mathcal{R}$ plays a crucial role in determining whether the linear or the quadratic force law must be used. For $\mathcal{R} < 1$, linear resistance applies while for $1 < \mathcal{R} < 10^5$, quadratic resistance applies.[2] In this paper we report on a rather straightforward experiment in which the correct resistive-force law for a vertically falling badminton shuttlecock was determined quite conclusively.[3] The shuttlecock was chosen primarily because of its small mass (small gravitational force) and large area (large resistive force). One expects that these factors should result in velocities of fall low enough to permit accurate measurements using equipment readily available in the laboratory. The results of our experiment fully confirmed this expectation. The shuttlecock is ideal for studying the effects of air resistance in the laboratory.

II. EXPERIMENT

Using a commonly available badminton shuttlecock, a tape measure, and a millisecond timer, we were able to measure the times required for the shuttlecock to fall given distances. The experiment was performed in an open stairwell. Ten trials were made for each data point and the average times (given in Table I) were calculated. Standard deviations ranged from 0.008 sec for the longest time measured to 0.01 sec for the shortest time measured.[4] A schematic of the experimental apparatus used is given in Fig. 1.

For convenience in releasing the shuttlecock, a fine thread was attached to the inside of the shuttlecock enabling it to be supported before release by the pressure exerted on the string when the spring-loaded microswitch was pressed in the closed position. Opening the microswitch released the shuttlecock and simultaneously started the millisecond timer. After falling through a measured distance y, a collision between the falling shuttlecock and the cardboard target dislodged the flattened metal rod held by two bar magnets, breaking that circuit and stopping the millisecond timer. Specifically, the millisecond timer begins counting when the trigger voltage is about 1.5 V, and stops when the trigger is grounded. The two bar magnets and the rod form a switch. With the rod in place, the trigger voltage (across R_2) is approximately 1.5 V, thereby enabling the timer to run. With the shuttlecock suspended by its thread and the microswitch pressed in the closed position, the trigger will remain grounded (the timer stopped) until the switch is opened (the shuttlecock released, the timer running). When the falling shuttlecock strikes the target and dislodges the rod, the trigger voltage falls to zero, stopping the timer. The two bar magnets were held in place by a ring stand with insulated clamps. Moving the two magnets closer together resulted in a more sensitive apparatus. In an effort to minimize our timing errors, the distance between the two magnets was reduced until the slightest pressure on the target dislodged the rod.

Table I. Experimental data for a vertically falling shuttlecock.

Distance fallen y (m)	Time elapsed t (sec)
0.61	0.347
1.00	0.470
1.22	0.519
1.52	0.582
1.83	0.650
2.00	0.674
2.13	0.717
2.44	0.766
2.74	0.823
3.00	0.870
4.00	1.031
5.00	1.193
6.00	1.354
7.00	1.501
8.50	1.726
9.50	1.873

Reprinted from *American Journal of Physics* **48**, 511–513 (1980); © American Association of Physics Teachers.

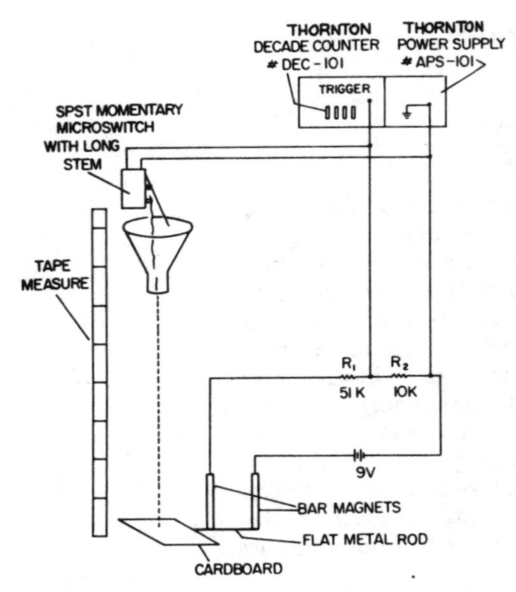

Fig. 1. Experimental apparatus. The bar magnets and the flat metal rod form a switch. With the microswitch closed and the rod in place, the millisecond timer will not run. When the microswitch is opened, releasing the shuttlecock, the timer starts counting and continues counting until the metal rod is dislodged by the collision between the falling shuttlecock and the cardboard target.

III. THEORY

Assuming a resistive force of the form

$$F_{res} = -kmv^2, \tag{1}$$

we obtain from Newton's second law

$$mg - kmv^2 = m\frac{dv}{dt}, \tag{2}$$

where we have taken the downward direction to be positive. Clearly, at terminal velocity the acceleration becomes zero and

$$v_T^2 = g/k. \tag{3}$$

If we assume that $v = 0$ at $t = 0$, we may integrate Eq. (2) to obtain[5]

$$t = \frac{v_T}{g}\tanh^{-1}\left(\frac{v}{v_T}\right). \tag{4}$$

Solving Eq. (4) for the velocity as a function of time, we get

$$v = v_T \tanh\left(\frac{gt}{v_T}\right). \tag{5}$$

Equation (5) may be integrated in turn to obtain the position with respect to time as follows,[5]

$$y = \frac{v_T^2}{g}\ln\left[\cosh\left(\frac{gt}{v_T}\right)\right], \tag{6}$$

where we have taken $y = 0$ at $t = 0$.

For the case of a quadratic resistance force, we see from Eqs. (1) and (3) that the terminal velocity is a parameter which must be specified for the particular object and medium in question. In Fig. 2, we plot the experimental data from Table I as well as the theoretical result, Eq. (6), using for the terminal velocity v_T the value which gave the best fit to the data. That value was determined from a least-

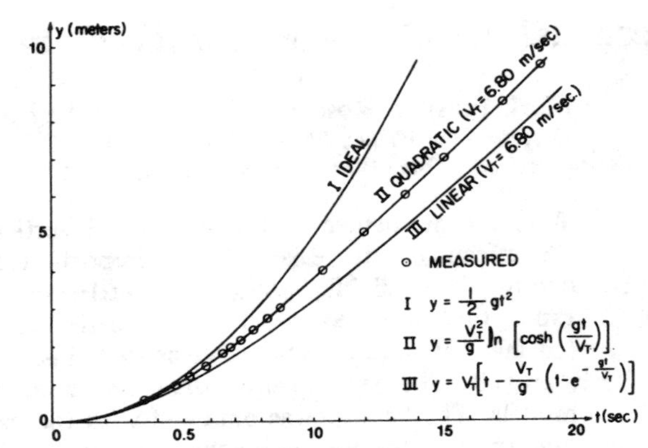

Fig. 2. Comparison of experiment and theory. I: A plot of distance fallen versus time elapsed for the ideal case of no resistance. II: The experimental data points plus a plot of distance fallen versus time elapsed for the case of a resistance force quadratic in the speed of the falling object. A least-squares analysis was done to determine the best fit to the experimental data. The terminal velocity which gave the best fit was found to be v_T = 6.80 m/sec. III: A plot of distance fallen versus time elapsed for the case of a resistance force linear in the speed of the falling object, using for the parameter v_T the value 6.80 m/sec.

squares analysis to be v_T = 6.80 m/sec. As can be seen from Fig. 2, the agreement between theory and experiment is most striking if one assumes a resistive force quadratic in the speed.[6] Also shown in this figure are the curves for the ideal case (no resistance) and the case of a resistive force linear in the instantaneous speed.[7] For this latter case, one assumes a resistive force of the form

$$F_{res} = -(mg/v_T)v, \tag{7}$$

which leads, after straightforward integration, to the results[5]

$$v = v_T(1 - e^{-gt/v_T}) \tag{8}$$

and

$$y = v_T[t - (v_T/g)(1 - e^{-gt/v_T})], \tag{9}$$

where once again we have taken $v = y = 0$ at $t = 0$.

In Fig. 3, we plot the instantaneous velocity of the shut-

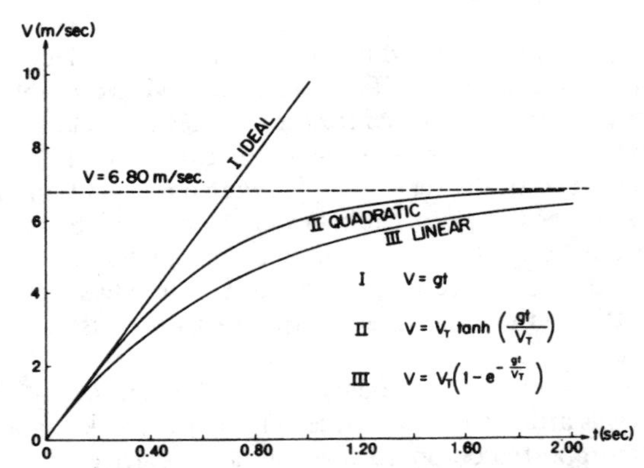

Fig. 3. Velocity versus time curves (theoretical) for vertical fall. I: A plot of instantaneous velocity versus time elapsed for the ideal case of no resistance ($v = gt$). II: A plot of instantaneous velocity versus time elapsed for the case of a resistance force quadratic in the speed of the falling object, Eq. (5), using for the parameter v_T the value 6.80 m/sec. III: A plot of instantaneous velocity versus time for the case of a resistance force linear in the speed of the falling object, Eq. (8), using for the terminal velocity the value v_T = 6.80 m/sec.

tlecock versus time as given by Eq. (5). Also shown are the curves for the ideal case as well as the case of a resistance force linear in the speed, Eq. (8). From Eqs. (5) and (6) we find that, according to the quadratic model, the vertically falling shuttlecock achieves 99% of its terminal velocity in 1.84 sec, after falling 9.2 m. As can be seen from Figs. 2 and 3, the effects of quadratic resistance are minimal for $t < 0.2$ sec (i.e., for the first 20 cm of fall), although these effects become substantial for $t \approx 1$ sec, resulting in a rapid approach to terminal velocity for $t > 1$ sec (i.e., $y > 4$ m). At first glance, the relative location of the curves for the linear and quadratic models in Figs. 2 and 3 might appear to be incorrect in that the quadratic model seems to be "closer" to the ideal case than the linear model. This somewhat paradoxical result stems from the fact that for a given terminal velocity v_T and a given instantaneous velocity v, the ratio of quadratic resistive force to linear resistive force is equal to the ratio of instantaneous velocity to terminal velocity, i.e.,

$$\frac{F_{\text{quad}}}{F_{\text{lin}}} = \frac{-(mg/v_T^2)v^2}{-(mg/v_T)v} = \frac{v}{v_T}. \tag{10}$$

For the case of vertical fall from rest, $v < v_T$ and hence $F_{\text{quad}} < F_{\text{lin}}$. Hence, in the regime $v < v_T$ the quadratic resistance force will always be less than the linear resistance force (for a *given* terminal velocity) and, as a result, the quadratic curves will always be closer to the ideal case than the linear curves.

IV. CONCLUSION

The experiment reported on in this paper shows quite conclusively that the correct resistive force law for the shuttlecock in vertical fall is the one which is quadratic in the instantaneous speed of the object [viz., Eq. (1)]. Our experiment also showed that the shuttlecock, although nonspherical, nevertheless satisfies the claimed connection between Reynolds number and appropriate force law stated in the Introduction.[8] We believe the most striking result obtained to be the excellent agreement between experiment and theory, specifically, the fact that 16 experimental data points could be fitted to the quadratic model with a correlation coefficient approaching unity. This agreement must be attributed, at least in part, to the inherent precision of the timing apparatus which we devised, an apparatus which is ideally suited to the experiment which we performed and which is readily adaptable to many other experiments.

In summary, the vertically falling shuttlecock experiment is a worthwhile one for physics students to perform. The apparatus is generally available, the data easy to collect (and surprisingly reproducible) but best of all, the agreement between theory and experiment is wonderfully reassuring!

[a]This paper was presented at the Second Annual Meeting of the South Eastern Pennsylvania Section of the AAPT, March 17, 1979.

[1]See, for example, Jerry B. Marion, *Classical Dynamics of Particles and Systems*, 2nd ed. (Academic, New York, 1970), p. 53.

[2]G. W. Parker, Am. J. Phys. **45**, 606 (1977), and references cited therein. For an elegant discussion of the role played by the Reynolds number, see R. P. Feynman, R. B. Leighton, and M. Sands, *The Feynman Lectures on Physics* (Addison-Wesley, Reading, MA, 1964), Vol. II, Sec. 41-3.

[3]Following the notation of Parker in Ref. 2, the Reynolds number for a sphere of radius r moving with velocity V in a fluid of density ρ and viscosity η is given by $\mathcal{R} = 2r\rho V/\eta$. Taking $r = 3$ cm and $V = 680$ cm/sec for the shuttlecock and $\eta/\rho = 0.15$ cm^2/sec for air, we calculate for our (admittedly nonspherical) shuttlecock a Reynolds number of $\mathcal{R} = 2.7 \times 10^4$. On the basis of this, one would expect that the quadratic rather than the linear force law would apply. This expectation is clearly borne out by the experimental results which we obtained.

[4]These numbers indicate timing errors of about 3% for the shortest times measured to about 0.5% for the longest times measured. As shown in Sec. III, it is the long-time data which determines the terminal velocity (See Ref. 7). Fortunately, it is this long-time data which we are able to measure most precisely. For a general discussion of the kind of error analysis suitable for our experiment (including least-squares principle and correlation), see H. T. Epstein, *Elementary Biophysics: Selected Topics* (Addison-Wesley, Reading, MA, 1963), Chap. 1.

[5]See, for example, R. S. Burington, *Handbook of Mathematical Tables and Functions* (Handbook, Sandusky, 1956).

[6]To appreciate just how striking the agreement is, consider the correlation coefficient defined by

$$\gamma = \{1 - [\sum(y_{\text{exp}} - y_{\text{th}})^2]/[\sum(y_{\text{exp}} - y_{\text{av}})^2\}^{1/2},$$

where y_{exp} are the experimentally determined distances for given times, y_{th} the theoretical values for corresponding times, and y_{av} is the numerical average of all the y_{exp}. Using the 16 experimental data points in Table I, and the theoretical prediction of the quadratic model [Eq. (6) with $v_T = 6.80$ m/sec], one calculates for the correlation coefficient the value $\gamma = 0.99996$! For a discussion of correlation coefficients and their relation to experimental data, see Ref. 4.

[7]It is interesting to note that if one considers just the first ten data points in Table I (i.e., $t < 1$ sec), it is possible to get a good fit using the linear model provided one takes for v_T an artificially high value of 13.4 m/sec. The apparent agreement is fortuitous, however, as seen by the fact that the last three data points in Fig. 2 fall on a straight line with slope (i.e., terminal velocity) equal to 6.80 m/sec. As Parker points out in Ref. 2, and as our shuttlecock experiment confirms, the linear resistance model [Eq. (7)] is of little relevance for objects having sizes and speeds of practical interest.

[8]As a check on whether sphericity (or the lack of it) would affect our results, we subsequently performed a vertical fall experiment using an ordinary ping-pong ball. Interestingly enough, the ping-pong ball also satisfied the quadratic resistive force law. Our nine data points were fitted to the quadratic model and yielded a terminal velocity of $v_T = 6.94$ m/sec with a correlation coefficient $\gamma = 0.995$. We did observe, however, that the motion of the ping-pong ball was somewhat erratic compared to that of the shuttlecock for comparable distances of fall. For example, when dropped from heights near 10 m, the ping-pong ball would often miss the target entirely, even though the release point was not varied from trial to trial. It appears likely that the "knuckleball effect" was responsible for this erratic sideways deflection of the ping-pong ball. For a discussion of this effect as it relates to nonrotating (or slowly rotating) baseballs, see R. G. Watts and E. Sawyer, Am. J. Phys. **43**, 960 (1975). For a description of a laboratory experiment on vertically falling (and nonrotating) ping-pong balls, see R. Terrell, Sports Illustrated **10**(26), 14 (1959).

The physics of fly casting

John M. Robson[a)]
Department of Physics, Sultan Qaboos University, Oman

(Received 2 March 1988; accepted for publication 24 March 1989)

The physics of the forward motion of the flyline during a fly cast is discussed, and a computer model of this motion is proposed which reproduces the main features of this part of the cast.

I. INTRODUCTION

In fly-fishing the fisherman uses a rod of about 3 m in length to cast the line out onto the water. The line is usually about 10 m long with a diameter of about 1 mm and a mass of about 10 g. An artificial fly is attached to the end of the line by a light thin length of monofilament called a leader. The mass of the leader is usually only about 300 mg and that of the fly about 100 mg. They are carried along without seriously affecting the motion of the line. However, the shape and length of the leader does affect the motion of the fly toward the end of the cast. In a complete cast, the angler raises the rod to lift the line off the water and to throw it out behind so that it straightens out roughly parallel to the surface. The rod is then brought forward so that the line unrolls in the air in front and settles down gently onto the water. The sequence is shown schematically in Fig. 1.

Many variations of this case are used by fishermen, but this is the basic line motion and is the one that will be discussed here. In fact, only the last part, the forward mo-

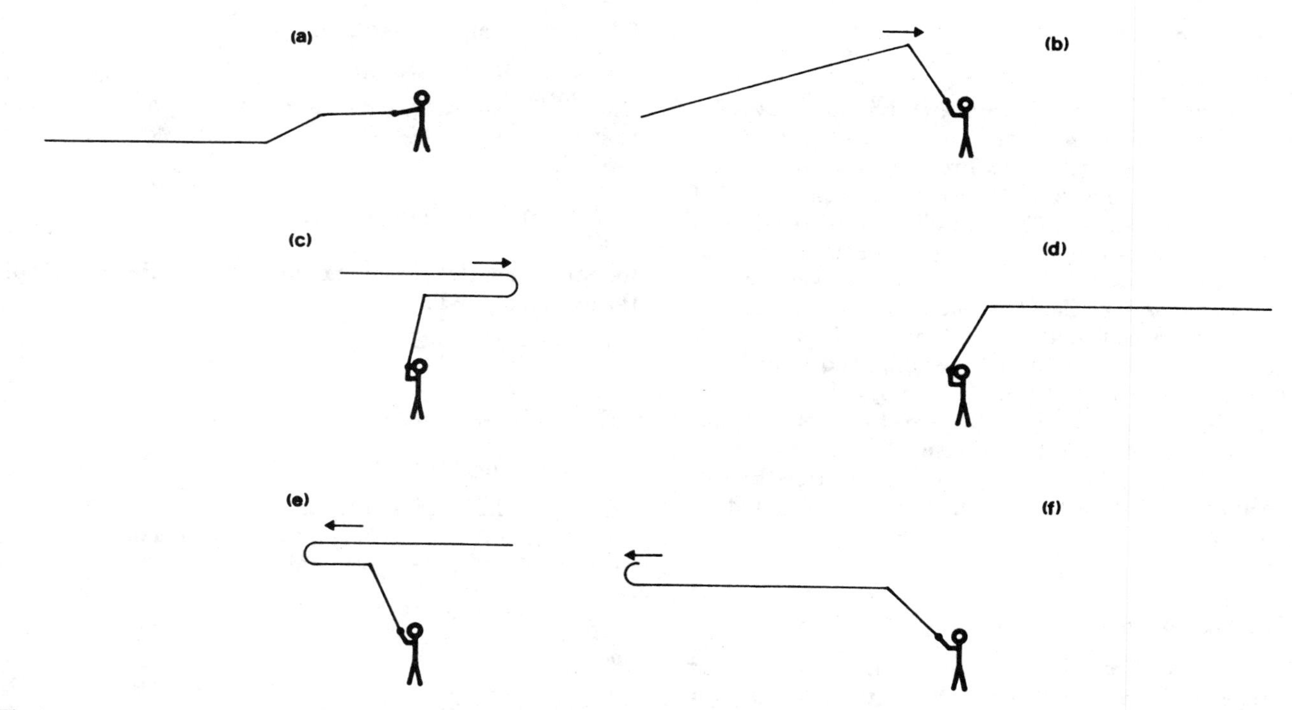

Fig. 1. A schematic representation of the motion of the rod and line during the casting of a fly. The sequence runs from (a) to (f).

tion of the line, will be treated in detail. It is the most significant part of the cast and embodies most of the physics associated with the line motion. First, the basic physics of the cast will be outlined in Sec. II; then a more detailed model will be used in Sec. III and will be compared to data taken from videos of actual casts.

The mechanics of this forward motion of the line and fly have been discussed recently by Spolek.[1] He develops the basic physics and uses it to calculate the velocity of the fly during the cast. He also shows how this depends critically on the taper of the line. The detailed model presented in this article emphasizes the actual shape of the line during its motion as well as the velocity of the fly.

II. OUTLINE OF THE PHYSICS OF THE FORWARD CAST

A. Basic theory without air resistance

The forward cast is assumed to start with the line fully extended to the rear as in Fig. 1(d) and instantaneously at rest. The rod is brought forward quickly to impart a forward velocity and kinetic energy to the line. It is then abruptly stopped so that the line doubles over with the fly end continuing forward as shown in Fig. 1(e) and (f). During this motion, the rod initially flexes backward; it then straightens out after the abrupt stop so that its stored potential energy is also given to the line.

To get a general idea of the physics associated with the motion following the abrupt stop, we will first assume that the rod does not flex and that the abrupt stop is instantaneous. With this simplification the rod would do no work on the line after the stop, though it may exert a force. Thus, if air resistance and gravity are neglected, no work is done after the stop, and consequently, the kinetic energy of the line remains constant. Nevertheless, since the rod can still exert a force, the momentum of the line can decrease.

If z is the length of the portion of line that is still moving and v is its velocity, the kinetic energy after the abrupt stop will be

$$T=\mu z v^2/2,$$

where μ is the mass per unit length of the line, which in this simple analysis is assumed to be uniform in diameter and density. Furthermore, the mass of the fly is neglected. Figure 2(a) illustrates the situation. Since T is constant, the velocity is related to the length of the line, which is still moving by

$$v=(2T/\mu z)^{1/2},$$

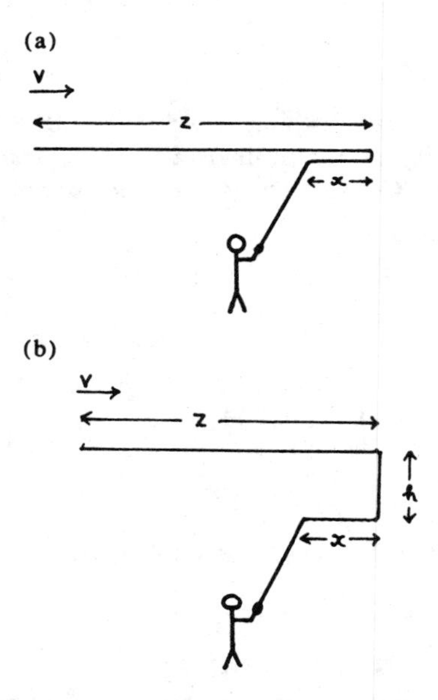

Fig. 2. The rod and line during the forward cast. In (a) the loop is very tight, whereas in the more realistic situation (b), the loop represents a significant fraction of the length of the line. The theory of these two cases is given in Sec. II.

and the momentum by

$$p = (2T\mu z)^{1/2}.$$

Hence, the velocity would increase as the line moves forward and in this idealized case would go to infinity with the momentum going to zero. The force accelerating the moving part of the line is provided by the tension in the line which, in turn, is opposed by the pull on the tip of the rod. However, since the rod tip is assumed to be stationary after its abrupt stop, this pull does no work and does not change the kinetic energy of the line. The pull increases as z decreases and, in fact, both the pull and its impulse go to infinity as z goes to zero in this oversimplified theory.

The horseman's whip has been analyzed in a similar manner by Rosenberg,[2] but differs in that the butt end, which corresponds to the rod tip end of the fishing line, is moved backward at a finite velocity. Thus work is done on the whip by the horseman in contrast to the situation described here.

B. Inclusion of air resistance

The general effect of air resistance can be seen from an extension of this simple theory. The shape of the line will now be approximated by the line in Fig. 2(b). If l is the total length of the line and v is the velocity of the fly,

$$z + h + x = l,$$

$$v = \dot{z} - \dot{x} = 2\dot{z}.$$

The kinetic energy of the line is

$$T = \mu z v^2/2 + \mu h \dot{x}^2/2$$

$$= \mu(4z + h)\dot{z}^2/2.$$

If time $t = 0$ refers to the moment when the rod is abruptly stopped, the kinetic energy at $t = 0$ is

$$T(0) = \mu(l - h)V^2/2,$$

where V is the maximum tip velocity of the rod just before it is stopped.

The force of air resistance on the loop part of the line of length h will be

$$Q = c_D r h \rho_a \dot{x}^2 = c_D r h \rho_a \dot{z}^2,$$

where c_D is the drag coefficient of the line of radius r, and ρ_a is the density of air. Skin friction is neglected in this simple analysis. Consequently, the work done by the line against air drag is

$$\int_0^x Q\, dx = -\int_0^t Q\dot{z}\, dt.$$

Hence

$$T(0) - T = -\int_0^t c_D r h \rho_a \dot{z}^3\, dt,$$

so that

$$\mu(4z + h)\dot{z}^2 = 2\int_0^t c_D r h \rho_a \dot{z}^3\, dt + T(0).$$

Differentiating this gives

$$(2\mu - c_D r h \rho_a)\dot{z}^2 + (4z + h)\mu\ddot{z} = 0.$$

This same equation can be obtained from the Lagrangian

$$L = T = 2\mu z \dot{z}^2 + \mu h \dot{z}^2/2.$$

If μ and h are assumed to be constant, this gives the following equation of motion:

$$2\mu\dot{z}^2 + (4z + h)\mu\ddot{z} = Q,$$

where Q is, as before, the force on the length h due to air resistance.

Since

$$\frac{d}{dt}(y^k\dot{y}) = y^{k-1}(k\dot{y}^2 + y\ddot{y}),$$

the solution of this equation of motion gives the velocity of the fly end of the line as

$$v = 2\dot{z} = c/(z + h/4)^k, \tag{1}$$

where

$$k = \tfrac{1}{2} - c_D r h \rho_a / 4\mu$$

$$= \tfrac{1}{2} - (h\rho_a / 4\pi r \rho_l)c_D,$$

where ρ_l is the density of the line.

When $x = 0$, $z = l - h$, and $v = V$, the maximum velocity of the tip of the rod just before it is abruptly stopped:

$$\therefore C = V(l - 3h/4)^k,$$

and

$$v = V[(4l - 3h)/(4z + h)]^k.$$

Consequently, the velocity of the fly will increase or decrease during the cast depending on whether k is positive or negative. The critical value of the loop size, h_c, is given by

$$h_c = 2\pi r \rho_l c_D / \rho_a.$$

For a typical line, $r \sim 0.7$ mm and $h_c \sim 200$ cm. Thus, if the loop exceeds about 2 m, the cast is likely to collapse in front of the angler.

The maximum velocity v_m of the fly will occur when $z = 0$, provided $h < h_c$:

$$v_m = V[(4l - 3h)/h]^k.$$

This can become very large if h is small.

In an actual cast, gravity and the bending of the rod are involved as well as air resistance. These inevitably complicate the detailed description of the line motion and make an extension of this simple theory rather difficult. Spolek,[1] however, has pursued this approach and has developed numerical solutions of the resulting equation. In this work an alternative model has been developed, and will be described in Sec. III, which enables reasonably precise calculatons to be made with the aid of a computer. Thus it also depends on numerical solutions, but provides an alternative insight into the motion of the line during the cast.

III. A MORE DETAILED MODEL

A. Introduction

In this section an attempt is made to extend the analysis further by including the fact that the line does not always travel in a neat straight line, but, like a flag in a wind, has undulations in its motion. Gravity, the detailed movement of the butt end of the fly rod by the caster, and the bending of the rod are also included. Furthermore, whereas Sec. II only treated the motion of the line after the rod had been stopped at the end of its forward motion, this model will treat the entire forward cast from its initiation at Fig. 1(d) until the line finally stops in front of the caster.

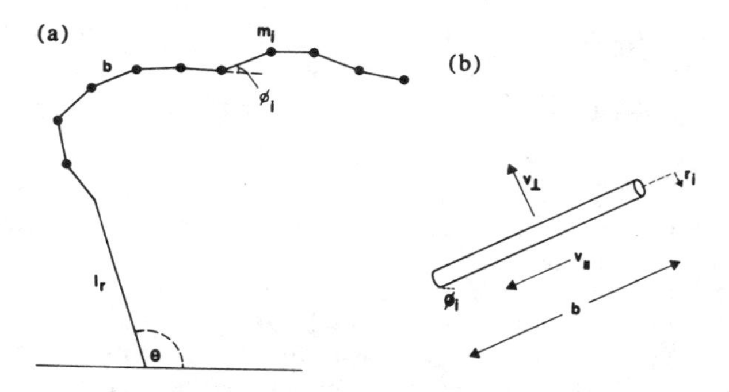

Fig. 3. (a) The model of the flyline used in most of these calculations. The line is represented by a number of masses joined to each other by massless stiff rods with frictionless pivots located at each mass. (b) To introduce air resistance, the massless rods are assumed to have dimensions equivalent to those of a flyline. $v_\perp$ and $v_\parallel$ are the components of the velocity perpendicular and parallel to the line.

B. The equations of motion of the line

It is fairly straightforward to set down the exact equations of motion of a continuous line subject to air resistance and gravity as it moves through the air. This has been done by Wood[3] who has used them to obtain approximate solutions for cables towed behind aircraft. These equations are hard to handle, however, and a more tractable representation has been used here. It is to represent the line by a number n of masses joined to each other by massless stiff rods as shown in Fig. 3(a). The pivots located at the masses are assumed to be frictionless, and the motion is restricted to two dimensions. An individual segment of the line, which will be referred to as a linelet, consists of a mass m_i joined to the preceding mass m_{i-1} by a massless rod of length b. The mass m_i is given by $\pi r_i^2 b\rho_l$, where r_i is the radius that a continuous line would have had at the position i; ρ_l is the density of the material of the continuous line. The angle that the ith linelet makes to the horizontal will be called ϕ_i.

The coordinates x_i and y_i of the ith mass m_i are

$$x_i = l_r \cos\theta + \sum_{k=1}^{i} b\cos\phi_k$$

and

$$y_i = l_r \sin\theta + \sum_{k=1}^{i} b\sin\phi_k,$$

so that the kinetic and potential energies of the line are given by

$$2T = \sum_{i=1}^{n} m_i \left[\left(l_r \sin\theta\dot\theta + \sum_{k=1}^{i} b\sin\phi_k\dot\phi_k\right)^2 + \left(l_r\cos\theta\dot\theta + \sum_{k=1}^{i} b\cos\phi_k\dot\phi_k\right)^2\right],$$

$$V = \sum_{i=1}^{n} m_i g\left(l_r\sin\theta + \sum_{k=1}^{i} b\sin\phi_k\right).$$

By forming the Lagrangian and using the n angles ϕ_i as the coordinates, the simultaneous equations relating these angles are found to be

$$\begin{matrix} a_{11}\ddot\phi_1 + a_{12}\ddot\phi_2 + \cdots + a_{1n}\ddot\phi_n + a_{1n+1} = 0 \\ \vdots \quad \vdots \quad \vdots \quad \vdots \quad \vdots \\ a_{n1}\ddot\phi_1 + a_{n2}\ddot\phi_2 + \cdots + a_{nn}\ddot\phi_n + a_{nn+1} = 0 \end{matrix} \qquad (2)$$

If we neglect air resistance and assume that the rod does not bend, the coefficients are

$$a_{ij} = B_{ij}\cos(\phi_i - \phi_j), \quad \text{for } j \leqslant n,$$

and

$$a_{in+1} = \sum_{k=1}^{n} B_{ik}\dot\phi_k^2 \sin(\phi_i - \phi_k) + Y_{in+1},$$

where

$$B_{ij} = \sum_{k=i}^{n} m_k, \quad \text{if } i \geqslant j, \quad \text{or}$$

$$\sum_{k=j}^{n} m_k, \quad \text{if } i < j,$$

and

$$Y_{in+1} = \sum_{k=i}^{n} m_k\left(\frac{l_r}{b}\left[\ddot\theta\cos(\theta-\phi_i) - \dot\theta^2\sin(\theta-\phi_i)\right] + (g/l_r)\cos\phi_i\right). \qquad (3)$$

The angle that the rod makes to the horizontal is included in Eq. (3) and θ, which throughout this work refers to the angle of the butt end of the rod as shown in Figs. 3 and 4. Various modifications must now be made to allow for air resistance and for the bending of the rod.

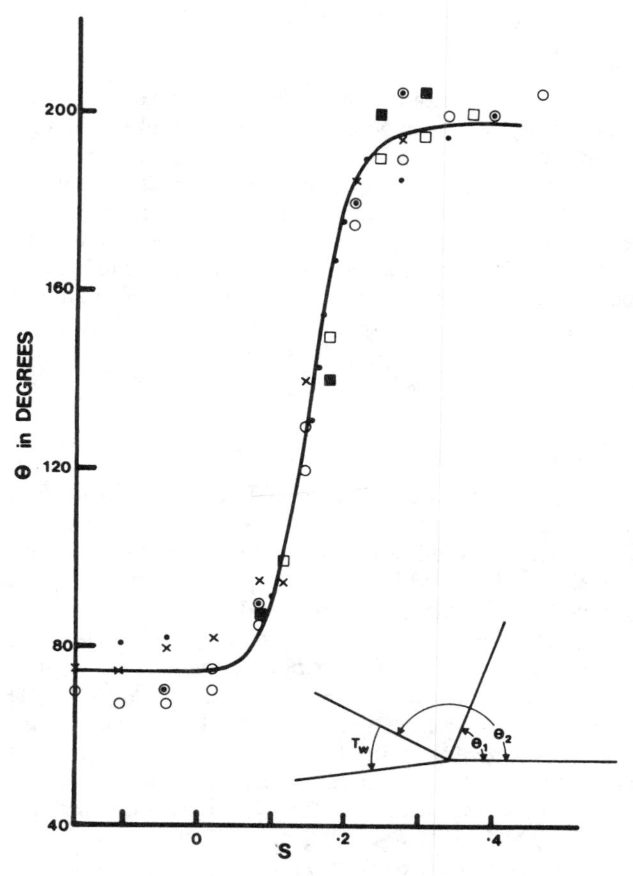

Fig. 4. The rod motion is described by θ_1 and θ_2, the intended start and finish of the cast, a rate of increase of angular acceleration α between θ_1 and θ_2, and a wrist time constant T_w governing the final stopping of the butt of the rod. The graph shows the data for several casts with the points indicating angles from film recordings.

In order to include air resistance, the pressure drag and the skin friction drag on a linelet were taken to be the drags on a cylinder of length b and radius r_i at the location of the ith linelet; see Fig. 3. If $v_\perp$ and $v_\parallel$ are the perpendicular and parallel components of the velocity of the linelet, the drags on the ith linelet are

$$F_{Di} = r_i b c_{Di} \rho_a v_{\perp i} |v_{\perp i}|,$$

and

$$F_{Si} = \pi r_i b c_{Si} \rho_a v_{\parallel i} |v_{\parallel i}|,$$

where c_{Di} is the drag coefficient, and c_{Si} is the skin friction coefficient for the ith linelet. When this is included, Eq. (3) is modified by replacing the last term, $m_k(g/l_r)\cos\phi_i$, by

$$m_k(g/l_r)\cos\phi_i - (Q_{Di} + Q_{Si})/l_r,$$

where

$$Q_{Di} = \sum_{k=i}^{n} F_{Dk}\cos(\phi_i - \phi_k)$$

and

$$Q_{Si} = \sum_{k=i}^{n} F_{Sk}\sin(\phi_i - \phi_k).$$

The drag coefficients depend on the velocity of the line, and the following approximation was used for c_D in most of the calculations:

$$c_D = 7.16, \quad \text{for } R < 1,$$

$$c_D = 7.16R^{-0.42}, \quad \text{for } 1 < R < 34,$$

$$c_D = 3.02R^{-0.165}, \quad \text{for } 34 < R < 1580,$$

$$c_D = 0.9, \quad \text{for } R > 1580,$$

where R is the Reynolds number for the linelet and numerically equals $13.3v_i r_i$, where v is in cm s^{-1} and r_i is in cm. This algorithm reproduces the handbook data given for a wire of cylindrical shape for Reynolds numbers between 1 and 5000. For the motion of a flyline, R is in this range for nearly all of the cast. The skin friction has a less significant effect on the line motion, and for simplicity the skin friction coefficient c_S was assumed to have the constant value of 0.005. A more detailed expression for this coefficient is given by Spolek,[1] which could be used in an extension of this work.

C. The motion of the butt end of the rod

Different casters obviously use different styles of casting, but a reasonable approximation to a normal forward cast with a light rod and line appears to be a linearly increasing acceleration followed by a rapid exponential decay of the angular velocity after the attempted stop. Thus a cast is described in these calculations by an initial butt angle θ_1, an attempted angle of stop θ_2, a rate of increase of angular acceleration α between these angles, and a stopping time constant T_w. These are shown in Fig. 4. Also shown are some experimental points for θ plotted against time obtained from a film record of some actual casts. The solid line is the computed position of the rod butt for $\alpha = 100\,000°/\text{s}^3$ and $T_w = 0.03$ s.

The expressions used for θ were

$$\theta = \theta_1 + \alpha t^3/6, \quad \text{for } \theta \leqslant \theta_2,$$

and

$$\theta = \theta_2 + \dot{\theta}_2 T_w\{1 - \exp[-(t - t_2)/T_w]\}, \quad \text{for } \theta > \theta_2,$$

where $\dot{\theta}_2$ and t_2 are values of $\dot{\theta}$ and t when $\theta = \theta_2$, and are given by

$$t_2 = [6(\theta_2 - \theta_1)/\alpha]^{1/3} \quad \text{and} \quad \dot{\theta}_2 = \alpha t_2^2/2.$$

D. The bending of the rod

Equation (3) must be further modified in order to take into account the bending of the rod. Fly rods are tapered so that the radius at the tip is noticeably less than at the butt; obviously, during the cast, they bend significantly. If the taper is small, the rod will bend over most of its length, whereas if there is a considerable taper, the bend will be localized more toward the tip as shown in Fig. 5. The bending of a tapered rod through a large angle can be analyzed by a method similar to that used by Timoshenko and Gere[4] for uniform rods and by Newman[5] for a flexible rope, but the theory does not lend itself to a simple analytical expression. Consequently, in this treatment of the line motion, the bending of the rod during the cast has been approximated as shown in Fig. 5. The rod of overall length a is assumed to bend only over a length c and into a circular arc over that length. Furthermore, the angle of deflection, shown as ψ in Fig. 5, is assumed to be proportional to the tension in the line at the rod tip. By itself, this would endow the rod with the properties of a simple harmonic oscillator with a natural frequency ω_0 related to the moment of inertia I of the bending tapered portion and the deflection per unit force d as follows:

$$\omega_0^2 = c^2/Id.$$

Measurements were made on a particular fly rod and suggested that values of $d = 0.0007$ cm dyn^{-1}, $c = 155$ cm, and $I = 1.2\times10^5$ g cm^2 would be appropriate. This gives $\omega_0 = 16.9$ s^{-1} or a natural period of 0.37 s. Many different rod designs are commercially available with periods as long as 1 s or as short as 0.1 s; this particular set of constants was used in these calculations.

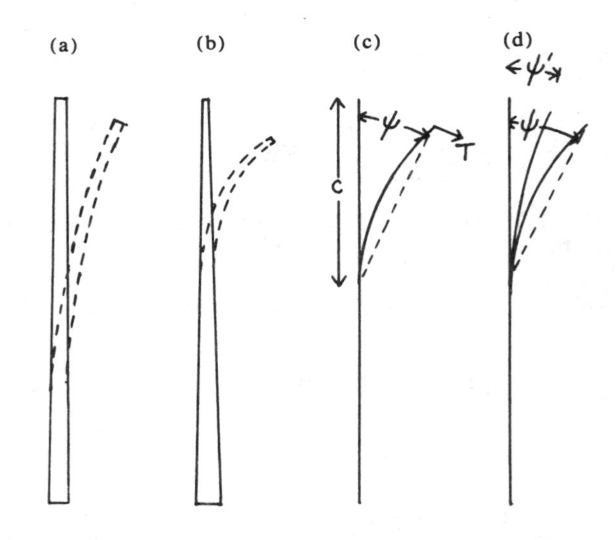

Fig. 5. A rod having a small taper is shown in (a) and bends over its entire length, whereas a rod with a larger taper only bends over the upper part as shown in (b). In this treatment the rod is assumed to bend into a circle over a length c as shown in (c). The linear angle of deflection ψ is assumed to be proportional to T, but the actual nonlinear deflection ψ' shown in (d) was calculated from ψ as described in the text.

The butt end of the rod usually consists of a cork handle glued around the thicker end of the tapered rod. The rod damping depends on how tightly this handle is grasped by the caster; usually the grip is light during the forward motion of the rod followed by a rapid tightening as the rod motion is stopped. In these calculations an underdamped coefficient of 0.5 was used for the forward motion with an overdamped coefficient of 3.0 when the caster grips the handle firmly as he tries abruptly to stop its motion.

As the rod is moved from θ_1 to θ_2, it is loaded by the line, and its natural period of oscillation is therefore increased. If the total mass of the line was m, it was assumed that the new angular frequency was given by

$$\omega^2 = c^2/d(I + c^2m).$$

For a typical cast this increased the period from 0.37 to 0.7 s. From these numbers it can be seen that the line does have a considerable effect on the response of the rod.

The deflection of the rod during the forward motion was calculated by a standard Laplace transform for an acceleration αt. This gives

$$\psi = \alpha[\omega t - \gamma + \exp(-\gamma\omega t/2) \times (\gamma \cos \omega_3 - f \sin \omega_3 t)]/\omega^3, \quad (4)$$

where γ is the damping coefficient, $\omega_3 = \omega(1 - \gamma^2/4)^{1/2}$, and $f = (2 - \gamma^2)/(4 - \gamma^2)^{1/2} \simeq 0.9$ for $\gamma = 0.5$.

When the forward motion of the butt stops at θ_2, it is assumed that the caster grips the handle tightly and the rod deflection dies away as if it was overdamped with a damping coefficient of 3.0.

The deflection of the rod is only linear up to about 15° and then becomes progressively more and more nonlinear. To allow for this, a deflection correction was applied using the empirical relation

$$\psi' = \psi \exp(-\beta\psi),$$

where ψ' is the actual deflection corresponding to the calculated linear deflection. A value of $\beta = 0.68$ gave reasonable agreement with the experimental deflections obtained for the particular rod and was used for most of the calculations.

Whereas the deflection during the forward motion between θ_1 and θ_2 is probably quite well represented by Eq. (4), the simple assumption of an overdamped decay after the motion is stopped is a considerable oversimplification. It was felt, however, that it was justifiable in this approximate treatment since the pull of the line is then roughly in the direction along the length of the rod rather than perpendicular to it.

E. The procedure used in the calculations

To obtain the motion of the line, the coordinates of the rod tip and of each linelet were determined at successive time intervals Δt from the start of the forward cast until t_m, the time when the line was fully extended in front of the caster. In order to avoid instabilities in the numerical solutions of the second-order differential equations, the ratio $b/\Delta t$, where b is the length of each linelet, must be kept greater than any velocities appearing in the motion of the line. These include lateral velocities analogous to those occurring in the flapping motion of flags in a wind. The length b was in general kept at 30 cm, which was the largest value that did not distort the shape of the line. A value of $\Delta t = 0.0001$ s then assured stable solutions and was used

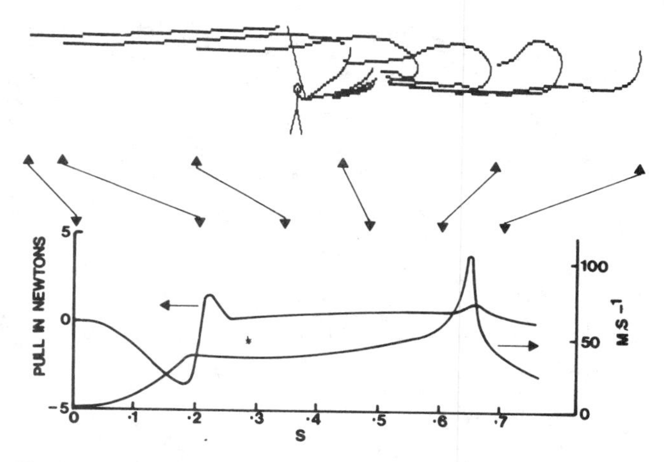

Fig. 6. The result of the calculation for a forward cast. The upper part shows the line at times indicated by the arrows that point at the fly end. The lower graph shows how the pull of the line on the rod and the velocity of the fly vary during the cast. Details of the particular cast are given in the text.

for most of the calculations. Usually, t_m was between 0.6 and 1.0 s so that up to 10 000 calculations were made for each cast. For each of these calculations, the angles θ and ψ were obtained, and the set of simultaneous equations shown as Eq. (2) were solved to give the n values of $\ddot{\phi}_i$. The values of ϕ_i and $\dot{\phi}_i$ from the preceding calculation were used in solving these equations. New values of ϕ_i and $\dot{\phi}_i$ were then computed and used to obtain new Cartesian coordinates of the linelets and as input for the next set of simultaneous equations. A typical cast took about 6 h on an IBM PC with a math coprocessor using IBM Professional FORTRAN.

Figure 6 shows the result of the calculations for a level line of length 900 cm and radius 0.74 mm with a leader tapering from 0.45 to 0.1 mm over a length of 180 cm. The fly was approximated by a cylinder of length 2 cm and radius 0.3 cm. The butt of the rod was moved from 75° to 155° with a rate of increase of acceleration of 50 000° s^{-3}, and the wrist stopping time constant was taken as 0.01 s. The graphs shown on the same figure indicate the pull of the line on the rod and the velocity of the fly plotted against time from the start of the forward cast. This would be regarded as quite a good cast by an angler in that the fly turned over smoothly toward the end of the cast and did not have an excessive velocity; there was only a slight pull on the rod toward the end of the cast, indicating that no excessive energy had been expended in the rod motion. Higher rates of acceleration of the rod led to a substantial pull on the rod as the line is unfurling at the end of the cast; this can be used to "shoot" extra line and extend the cast, but does not lead to a smooth presentation of the fly.

F. A test of the model

In order to test the model as a whole, some actual casts were recorded with a high-resolution video camera and compared to the computer predictions for similar casts. It was necessary to use a video camera with a fast response to avoid blur of the image on individual frames. Furthermore, a rather heavy line, a number 9WF, had to be used to produce an easily visible image, and consequently a rod was used that was rated for such a line; it was 3 m long. With

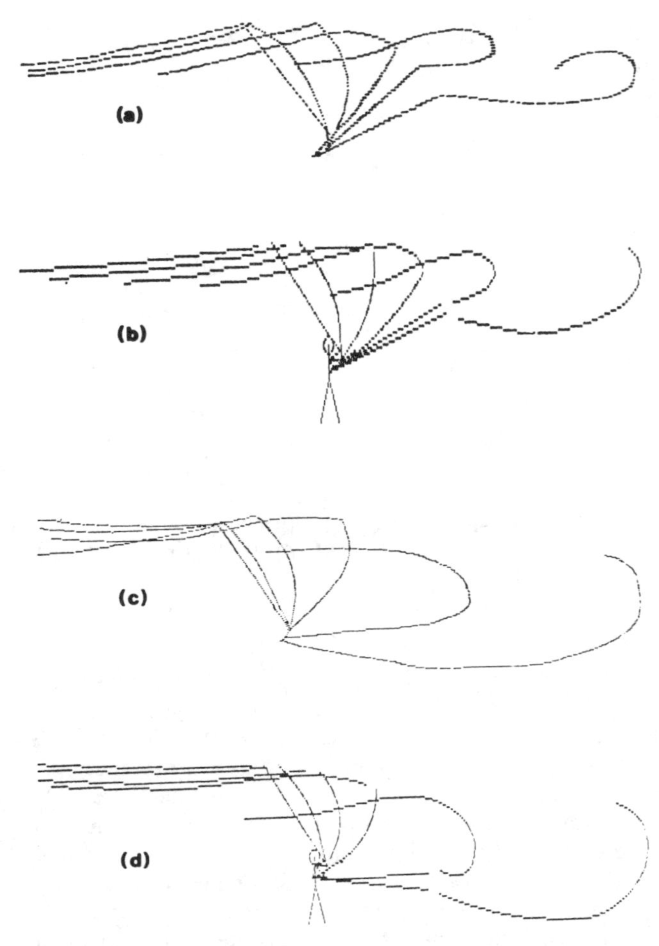

Fig. 7. Two comparisons between actual casts recorded by a video camera and computer predictions. (a) and (c) are the actual casts, and (b) and (d) are the computer simulations. For (a) and (b), the rod and line are shown at 0, 0.16, 0.32, 0.44, 0.64, and 1.0 s, whereas for (c) and (d) the times are 0, 0.12, 0.24, 0.36, 0.60, and 0.96 s. For details of the rod and elbow motions, refer to the text.

this heavy line and rod, it was found that the rod motion was represented better by a short constant acceleration followed by a uniform angular velocity; this was terminated by an exponential decay at the end of the cast. To improve the comparison, the program was also slightly modified to reproduce the downward motion of the elbow during the cast.

Figure 7 shows two comparisons between actual casts and their computer simulations. In the first cast a short line was used, and the caster was keeping the line airborne and not lowering onto the water; (a) shows the actual line and (b) the computer reproduction. In the second cast, the line was being lowered onto the water; (c) is the actual cast and (d) the computer reproduction. In view of the difficulty in reproducing the actual rod motions by the simple approximations used in the program, the agreement is considered to be quite reasonable.

IV. CONCLUSION

It has been shown that the motion during the forward cast of a flyline is approximated by a system of constant kinetic energy. Air resistance and the flexing of the fly rod modify this and introduce complications that prevent a simple analytical solution. However, a model has been presented that is amenable to solution by a small computer and that reproduces the main features of the forward motion of a flyline in reasonable agreement with those of actual casts. It has been used to look at the way in which fly casts depend on various characteristics of the rod and line and on different rod motions. These are in general agreement with the results of Spolek and actual experience, but are not reproduced here since they are of greater interest to anglers than to physicists.

ACKNOWLEDGMENTS

The author wishes to acknowledge with thanks the help given by the Centre for Educational Technology of Sultan Qaboos University in taking videos of various actual fly casts and providing equipment for their frame-by-frame playback. He also wishes to thank Dr. B. G. Newman for several helpful and informative discussions on the motion of flags and lines.

[a] Permanent address: Physics Department, McGill University, 3600 University Street, Montreal H3A 2T8, Canada.

[1] G. A. Spolek, "The Mechanics of flycasting: The flyline," Am. J. Phys. **54**, 832–836 (1986).

[2] R. M. Rosenberg, *Analytical Dynamics of Discreet Systems* (Plenum, New York, 1977), pp. 332–334.

[3] W. W. Wood, Aeronautics Research Laboratory Rep. A. 122, Australian Defence Scientific Service, P.O.B. 4331 Melbourne, Victoria 3001, Australia (1961).

[4] S. P. Timoshenko and J. M. Gere, *Theory of Elastic Stability* (McGraw-Hill, New York, 1961), pp. 76–82.

[5] B. G. Newman, "Shape of a towed boom of logs," Proc. R. Soc. London Ser. A **346**, 329–348 (1975).

Do springboard divers violate angular momentum conservation?

Cliff Frohlich
Marine Science Institute, University of Texas, 700 The Strand, Galveston, Texas 77550
(Received 15 November 1978; accepted 26 February 1979)

No. However, divers and trampolinists can perform somersaults and twists even though they have zero angular momentum at all times during the stunt. Also, if a diver is somersaulting in space and possesses angular momentum only about his somersaulting axis, he can make a single discrete change in the position of his arms which initiates continuous twisting motion even in the absence of any applied torque. These apparent paradoxes have confused both physicists and coaches for some time. The present paper attempts to reduce this confusion. It discusses several different methods that performers use to initiate somersaults and twists and presents concrete examples of each method. Wherever possible quantitative calculations are presented and evaluated using information about the moments of inertia, mass, etc., of "typical" performers.

I. INTRODUCTION

Divers, trampolinists, and gymnasts can perform rotations of their bodies of several different kinds. The first kind is the *somersault,* which is a rotation of the body about an axis along a line going from the performer's left side through his center of mass to his right side. In Fig. 1, the axis of rotation for somersaulting motion is marked with an "*S*" for a performer in several common diving positions. The second basic kind of rotation is the *twist,* which is a rotation of the body about an axis going from the performer's head through his center of mass to his feet. In Fig. 1, this axis is marked with a "*T*" for the diver in positions *A, B, C, D,* and *E.* Twists are usually performed with the body straight, and with no bend in the knees or at the waist. Only a tiny fraction of all stunts consist principally of a rotation about the third possible axis, which is along a line going from directly in front of the performer through his center of mass, and out through the center of his back. However, this paper will show that small rotations about this axis do play an important role in the initiation of some kinds of twisting stunts, particularly twisting somersaults.

Apparently very few physicists have observed springboard divers or trampolinists closely. Recently, all the graduate students, postdoctorates, and faculty in the Physics Department at Cornell University were given a questionnaire with specific multiple choice questions about the physical possibility of performing certain somersaulting and twisting stunts (see Fig. 2). One question asked if it is possible for a trampolinist with no angular momentum to execute a quarter somersault (a 90° rotation about his left-to-right axis). Another question asked if a somersaulting springboard diver could initiate twisting motion without any torques being applied to his body, that is, beginning the twist only *after* he is no longer in contact with the diving board. As we show later in this article, the answer is that trained divers can perform both of these motions without violating angular momentum conservation, Nevertheless, of the 59 physicists who responded to the questionnaire (see Table I), 34% incorrectly answered the first question, and 56% incorrectly answered the second—an astonishingly high proportion of misses for multiple choice questions.

These results illustrate the need for a paper discussing the basic physics of somersaulting and twisting stunts, particularly since there are at present no physics books or articles which treat these phenomena in any detail. Furthermore, divers, trampolinists, and gymnasts do provide excellent illustrations of certain peculiar features of the free rotation of rigid and semirigid bodies. Since more than 300 colleges and universities in the U.S. possess competitive swimming teams,[1] these phenomena can be observed by physicists and physics students on almost any campus.

Coaches and physical educators (as opposed to physics educators) also may find a review of the physics of diving to be useful. Several books written by or for coaches discuss somersaulting and twisting, and exhibit varying degrees of insight and/or confusion about the physics processes that occur.[2-5] Although the author of this paper is not qualified to write a "how-to" paper for performing stunts, the present work does provide the most precise account available on the physics of somersaulting and twisting stunts.

Historically, the earliest research relevant to these problems was not concerned with the motions of the human body, but instead with the twisting of cats, who land on their feet when dropped upside down from a height of a meter or so.[6-9] More recently, the possibility of space exploration has caused more quantitative work to be done.[10,11] Although Smith and Kane[10] were primarily concerned with motions of men in space, their paper also includes an excellent comprehensive review of much of the earlier work. In other related work, high-speed cameras have been used to make detailed observations of an athlete performing a back somersault with a twist[12] and a computer has been used to simulate the motions that occur in somersaults without twists.[13] However, neither of these studies of athletes discussed the physics associated with these stunts in any detail.

Because divers, coaches, astronauts, and physicists commonly use different terminology to describe motions of the human body, a quick summary of some synonymous terms is given here to minimize confusion. For example, the *left-to-right axis* is sometimes called the transverse axis, and somersaulting motion about this axis may be called "motion in the sagittal plane." Similarly, the *head-to-toe* axis is also called the longitudinal axis or the long axis, and twisting motion is said to take place "in the transverse plane." Finally the *front-to-back axis* is also called the

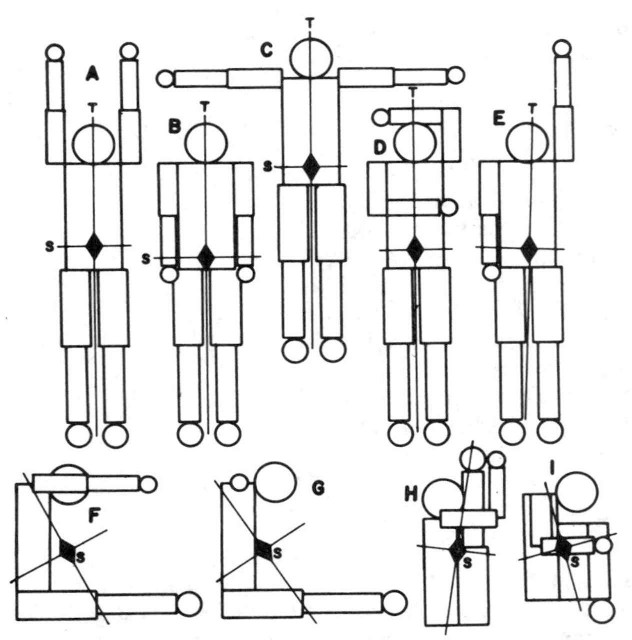

Fig. 1. Configuration of body segments of "typical" man in nine positions commonly used by divers and trampolinists. For each position, the solid diamond is at the center of mass, with the long axis of the diamond oriented along the principal axis associated with the principal moment of inertia I_1 in Table III. The axis of rotation for somersaulting dives is marked with an "S" and the axis of rotation for twisting is marked with a "T." The nine positions are; A: Layout throw; B: Layout somersault; C: Pretwist layout; D: Twist position; E: Twist throw; F: Loose pike; G: Pretwist pike; H: Tight pike; and I: Tuck.

medial axis, and rotations about this axis occur "in the frontal plane." When more than one term could be used to describe a stunt, in this paper the term which appears in italics in this section will be used.

Divers perform somersaulting rotations in several positions, for example, the *tuck* position (knees and waist bent, as in I in Fig. 1), the *pike* position (knees straight but waist bent, as in F, G, or H in Fig. 1), and the *layout* position (knees straight and waist not bent, A, B, and C in Fig. 1). When both somersaulting and twisting motions occur in the same dive, the rules place no restrictions on bending the knees and waist, and hence the dive is performed in the so-called *free* position (D and E in Fig. 1 are among the typical configurations assumed during twisting somersaults). Physicists commonly specify the direction of rotations using the right hand rule. When performing a *forward* somersault, a diver leaves the diving board facing the water and his rotation direction along his left-to-right axis is to his left. In a *backward* somersault he leaves the diving board with his back to the water and the direction of rotation along his left-to-right axis is to his right. For reference, Batterman[3] has published in paperback a collection of diving pictures

Table I. The fraction of respondents to the questionnaire who incorrectly answered the questions posed in Fig. 2. Eight diving coaches and 59 physicists responded to the questionnaire. The last column is the fraction of incorrect answers expected if the respondents had chosen their answers randomly.

	Coaches	Physicists	Random guess
Question 1	0.38	0.34	0.50
Question 2	0.13	0.56	0.67

which includes most of the dives commonly performed in collegiate competition.

II. PHYSICAL CHARACTERISTICS OF DIVES AND DIVERS

Some stunts are "possible" because no law of physics prevents their performance, but are "impossible" in practice because of the limitations of their performers. For example, ballistics tells us it is possible for a cow to jump over the moon, but we know that cows can not jump that high. In order to evaluate in terms of physics what occurs in diving and trampolining stunts, it is useful to have numerical estimates of certain characteristics of typical performers executing typical stunts.

A. Masses of body segments

The relative masses of body segments in Table II have been taken from a standard physical education text by Dyson,[2] and then rounded slightly for convenience in the calculations of the moments of inertia. The classic study of relative masses of body segments was done by Dempster,[14] and his results have been summarized elsewhere by several authors.[15] Dempster's source of data consisted of cadavers

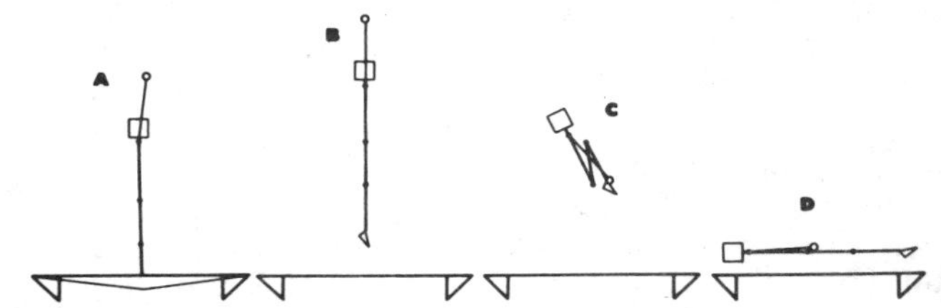

Fig. 2. *Question 1:* The stick-figure above is performing a simple trampoline stunt known as a "back-drop." In this stunt, he bounces, tucks, and lands on his back on the trampoline bed. *Question 1 in divers language:* Could a trained performer complete a back drop without any "throw" from the bed of the tramp? In other words, if the body of the trampolinist is not rotating at all at point "B" (after his feet have left the tramp bed) can be still do a back drop? *Question 1 in physicists' language:* If the trampoinist has zero angular momentum at "B" is it possible for him to land on his back as in "D"? Check one answer: □ Yes, a trained trampolinist could successfully complete the stunt as shown. □ No, even a well-trained individual could not do that stung if at "B" he had no rotating motion. *Question 2:* One dive that is listed in the AAU's *Official Diving Rules 1978* for a three meter springboard is a forward one and one half somersaults with three twists. Although this is a very difficult dive, there are probably more than a hundred divers in the United States who can do it. *Question 2 in divers' language:* Suppose the diver throws the somersault, but does not begin to throw the twists after his feet have left the board. Is it possible for him to do multiple twist dives like the triple twister even though he does not throw the twist until after he leaves the board? With practice, could a diver learn to throw twists in either direction, even though he did not begin the twist on the board? *Question 2 in physicists' language:* If the diver has initiated only somersaulting motion on the board, is it possible for him to initiate the twists after his feet have left the board? Could he choose to twist either to the left or to the right, depending on how he threw the twist after he left the board? Check one answer: □ Yes, a diver really can do a multiple twist dive such as a triple-twisting one and one-half somersault, and if properly trained he could choose the twist direction after his feet left the board. □ Yes, trained divers can do multiple twist dives where they initiate the twist after they leave the board. However, no matter how hard they train the direction of the twist is already determined before their feet leave the board. □ No, even a well-trained diver could only do a multiple twist dive if he initiated the twisting motion while his feet still touched the board.

Table II. Body segment masses and shapes for "typical" man with total mass of 75.64 kg and height 1.82 m.

Segment	Mass (Kg)	Description (cm)
Head	5.575	sphere: $r = 11$
Trunk	32.400	rectangular solid; 60 × 30 × 18
Upper arms (2)	2.356	cylinder; $r = 5$, $h = 30$
Forearms (2)	1.781	cylinder; $r = 4.5$, $h = 28$
Hands (2)	0.523	sphere; $r = 5$
Thighs (2)	8.650	cylinder; $r = 8$, $h = 43$
Lower legs (2)	4.086	cylinder; $r = 5.5$, $h = 43$
Feet (2)	1.436	sphere; $r = 7$

of eight subjects who died in middle or old age, and as such his results are not typical for divers and trampolinists. The relative masses reported in Table II are within the ranges reported by Dempster,[14] but on the average a lower percentage of the body weight is concentrated in the trunk, and a higher percentage in the limbs.

B. Principal moments of inertia

The moments of inertia reported in Table III are calculations of the author, as no detailed moment of inertia calculations for most diving body positions could be found in the literature. To make the calculations, the author divided the body into 14 segments and assigned each segment an approximate convenient shape which had an easily calculable moment of inertia (e.g., sphere for head, rectangular solid for trunk, cylinders for legs, etc.; see Table II). The principal dimensions of each segment were determined by measuring the appropriate dimensions of the author's own body with a meter stick. The remaining dimension (e.g., the radius of the leg cylinders) was chosen so that the conveniently shaped segment would have the density of water and the mass reported in Table II. Finally, a computer program calculated the center of mass and principal moments of inertia for typical positions used while somersaulting and twisting (Table III and Fig. 1).

For a few of the body positions it is possible to compare these results to other work. The calculations reported in Table III fall within the range of values calculated by Hanavan[11] for men of comparable height in body positions similar to B, C, F, and G in Fig. 1. Hanavan's[11] calculations extended the work of Santschi *et al.*[16] They measured the moments of inertia for 66 subjects in several body positions by the compound pendulum method, but had made no attempt to correct their measurements to obtain values for the moments along the principal axes. Rackham[4] reported that the ratio of moments for divers somersaulting about the left-right axis in the tuck, pike, and layout positions is approximately 1:2:4, also in substantial agreement with the results in Table III.

C. Stability

For a rigid body with three distinct moments of inertia, stable rotational motion is possible only about the principal axes associated with the largest and the smallest moments of inertia.[17] Rotation about the third principal axis is only metastable; the smallest perturbations of the rotation grow to become mysterious "wobbles" or twists.

When divers and trampolinists perform somersaults, are they rotating about a stable axis? Somersaults are rotations about the left-right axis, and Table III suggests that a performer somersaulting in a tuck or loose pike position is somersaulting about the axis associated with his largest moment of inertia, and thus the motion is stable. The moment of inertia for somersaulting in a tight pike position (Table III) is very close to the largest moment of inertia. Since few real divers can pike as tightly as the "typical man" in Fig. 1, probably the largest moment of inertia for piked real divers is associated with the somersaulting axis, and so this motion is stable. On the other hand, for a layout position, the largest moment of inertia is definitely not associated with the somersaulting axis, and so the somersaulting motion is not a stable one. The fact that the layout position is unstable for somersaulting may be partly responsible for the "side cast" or partial twist of the body which is sometimes observed as a diver enters the pool. This is a common problem for divers performing layout somersaults.

D. Time

Time is the fundamental constraint that limits all divers and trampolinists from performing more difficult stunts. A diver leaving a board is a projectile—if he simply drops from a 1-m board, 0.45 sec will elapse before he reaches the water. Similarly, 0.8 sec will elapse if he drops from a 3-m board. However, if he can use the board to raise his center of mass 2.0 m, he has about 1.5 sec to complete a dive from the 1-m board, and about 1.7 sec from a 3-m board. This additional 0.2 sec permits the performance of more complicated motions from the 3-m board motions with an additional twist or an additional somersault. For example, in American Athletic Union (AAU) sanctioned competition divers can perform forward one and one-half somersaults with four twists from the 3-m board, but with no more than three twists from the 1-m board.[18]

Clearly, a key element in any diver's technique is to achieve the maximum height possible so as to have sufficient time to complete complicated stunts. However, since this paper is concerned primarily with somersaulting and twisting, the physics of achieving height will not be discussed.

III. INITIATING SOMERSAULTS AND TWISTS

Somersaults and twists can be classified in more than one way. In this section, they are discussed in terms of the forces and torques acting to initiate the body rotation. In these terms, there are three types of somersaults, and three types of twists.

Table III. Principal moments of inertia for "typical" man in positions shown in Fig. 1. I_1 is the principal moment for the axis with direction closest to that of the man's spine (twist axis or head-to-toe axis), I_2 is the principal moment along a left-to-right axis (somersault axis), and I_3 is the principal moment along an axis closest to the front-to-back axis. Units are kg m^2. Letters beneath "Figure 1" column refer to appropriate diagram in Fig. 1.

Figure 1	Description	I_1	I_2	I_3
A	Layout throw	1.10	19.85	20.66
B	Layout sault	1.10	14.75	15.56
C	Pretwist layout	3.42	16.38	19.17
D	Twist position	1.06	16.65	17.24
E	Twist throw	1.08	17.41	18.20
F	Loose pike	4.83	10.45	7.53
G	Pretwist pike	5.54	10.09	10.42
H	Tight pike	1.75	5.89	6.05
I	Tuck	20.3	3.79	3.62

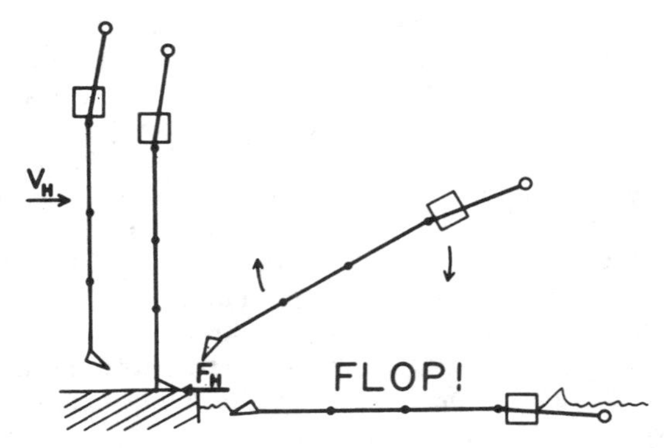

Fig. 3. Somersaulting can be initiated by horizontal forces even if the performer does not "throw" with his arms. In a "front header" from the side of the pool the diver runs and jumps into the air so that his body is moving forward with no rotational motion, but with only translational velocity (stick figure at far left). However, when his feet contact the pool deck, the pool deck exerts a horizontal impulsive force F_H on his feet, slowing his translational velocity and creating a torque that tends to initiate rotation. Although this dive can be observed at any neighborhood swimming pool, in competition the forward dive is generally not performed in this way.

A. Somersaults

(*i*) *Torque for somersault due to horizontal forces.* The most common way to initiate somersaulting motion is to apply torques to the body by applying horizontal forces to the feet. This can be done in more than one way. For example, if the diver's body is moving forward as he reaches the end of a diving board, the board applies horizontal forces to his feet to initiate a forward somersault (Fig. 3). Somersaulting motion will occur even if he never "throws" the somersault with his arms, shoulders, and head.

A much more controlled somersault is possible if the individual is allowed to "throw" with his upper body. As an example, the left picture in Fig. 4 shows how a forward somersault generally is initiated on a trampoline. "Throwing" the arms and head forward and down involves a rotation of the arms, which possess angular momentum. In free space, the lower body would rotate in the opposite direction (since angular momentum is conserved). However, since the trampolinists's feet are in contact with the trampoline, the trampoline exerts a force on the trampolinist which resists this opposite rotation, and provides the torque which begins the somersaulting rotation. The harder a performer "throws" with his upper body, the larger the horizontal force will be providing the torque, and the more angular momentum his body will have when the vertical forces provided by the trampoline cause his body to become airborne. The individual also can minimize or control the horizontal motion that results from these horizontal forces by making subtle adjustments in the position of his center of mass while his feet still touch the trampoline. This method of initiating somersaults—with torque provided by horizontal forces that counteract a rotation or "throw" of a portion of the body—is by far the most important method for trampolinists and divers.

In spite of the important role that horizontal forces play to initiate somersaulting, few divers or trampolinsts are aware that horizontal forces are being applied to their feet. Instead their attention is focused on how they "throw" with the upper body, and on their "balance" (the position of their center of mass). A brief calculation shows that this is because the magnitude of the horizontal forces that produce the torque is considerably smaller than the magnitude of the vertical forces that make the performer airborne. For example, consider a person performing a single somersault on the trampoline. If F_v and F_h are the vertical and horizontal forces acting for a time T on a performer's feet, then the initial takeoff velocity V_i and the initial rotational velocity ω_i are given by

$$V_i = F_v T/M \text{ and } \omega_i = F_h TL/I_{lay}, \quad (1)$$

where M is his mass, I_{lay} is his moment of inertia about his left-right axis, and L is the distance from his feet to his center of mass while the forces are being applied by the trampoline: To simplify the calculation, assume his body is rigid and ignore the angular momentum possessed by his upper body because of the "throw." If he tucks his body at the moment of takeoff, his rotational velocity will increase by a factor of I_{lay}/I_{tuck}, where I_{tuck} is his moment of inertia in the tuck position, and so

$$\omega = F_h TL/I_{tuck}. \quad (2)$$

When his center of mass is at the highest point, he will have completed one half somersault. The elapsed time T_e after takeoff will be

$$T_e = V_i/g = \pi/\omega \quad (3)$$

and his height H above the trampoline will be

$$H = (1/2)gT_e^2 = V_i\pi/2\omega. \quad (4)$$

Substituting in for V_i and ω from (1) and (2), one finds

$$F_h/F_v = \pi I_{tuck}/2MLH. \quad (5)$$

If $L = H = 1$ m, and I_{tuck} and M are as in Tables II and III, then $F_h/F_v = 0.08$.

(*ii*) *Torque for somersault due to vertical forces.* An individual can perform a somersault without any throw from the diving board or trampoline, even if only vertical forces act on his feet. In this case, his center of mass cannot be directly above his feet, and it is because he leans either forward or backward that the vertical forces produce a torque (see Fig. 5). This method of producing somersaulting motion is seldom employed on the trampoline, as it is difficult for the trampolinist to control the amount of lean, and

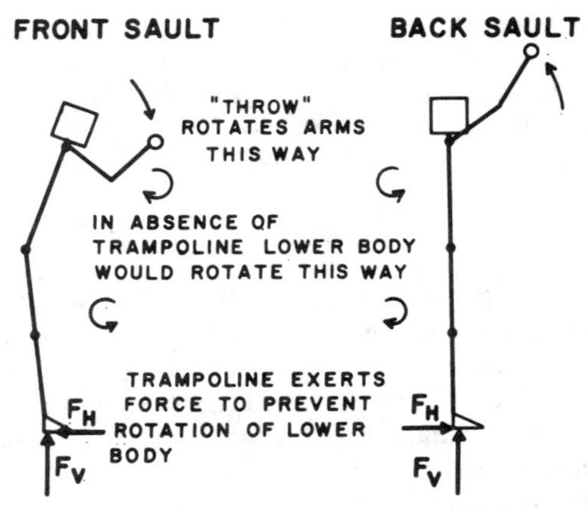

Fig. 4. Somersaulting can be initiated by a "throw" of the upper body while the performer's feet are still in contact with a trampoline or diving board. The throw creates horizontal forces, producing torques which initiate rotation. The direction of the rotation is the same as the direction of the throw.

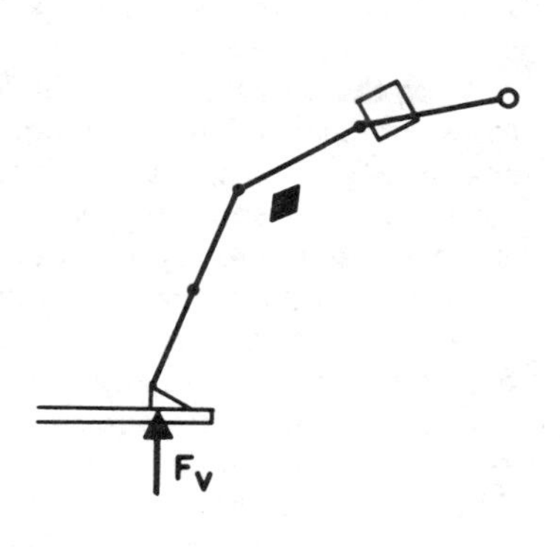

Fig. 5. Somersaulting can be initiated by vertical forces, even if the performer does not "throw" with his arms. In the picture the diving board exerts an upward force on the diver which creates a torque about his center of mass (solid diamond) causing him to rotate forward. In practice both front and back somersaults can be initiated in this manner if the diver "leans" as the board exerts a vertical force. The lean is exaggerated in the picture above.

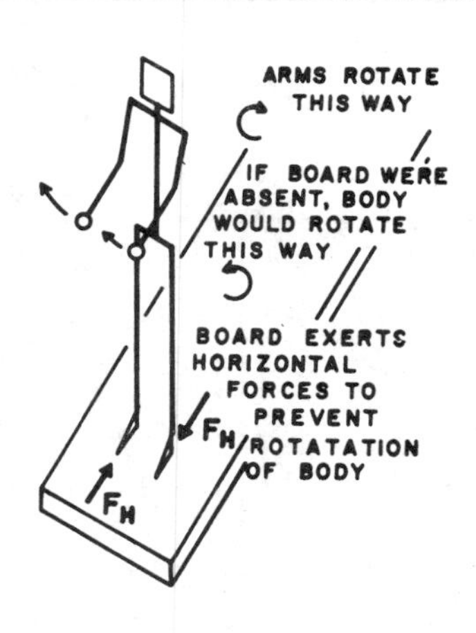

Fig. 7. Torque twist can be initiated by "throwing" the arms in the direction of the twist. If the performer's feet are in contact with the diving board, trampoline, or floor, then when the performer jumps into the air while throwing his arms in this manner, the horizontal forces create torques which cause him to twist.

a great deal of unwanted horizontal motion or "travel" often results. The lean also makes it more difficult to get enough height to perform difficult stunts.

Nevertheless, torques arising from vertical forces are important for a few kinds of stunts, particularly for divers performing multiple forward somersaults. For example, a slight lean is useful for divers performing forward two and one-half somersaults and forward three and one-half somersaults where a great deal of angular momentum is needed. In addition to the angular momentum resulting from the vertical forces, in practice the slight lean permits the diver to "throw" slightly earlier and slightly harder, which also helps to increase his angular momentum.

(*iii*) *Torque-free somersaults.* A limited amount of body rotation can be accomplished in the complete absence of any torque whatsoever. For example, if a trampolinist with legs and body straight "windmills" his arms forward about his shoulders, his entire body will tilt backwards. Because the moment of inertia of his arms is much smaller than the

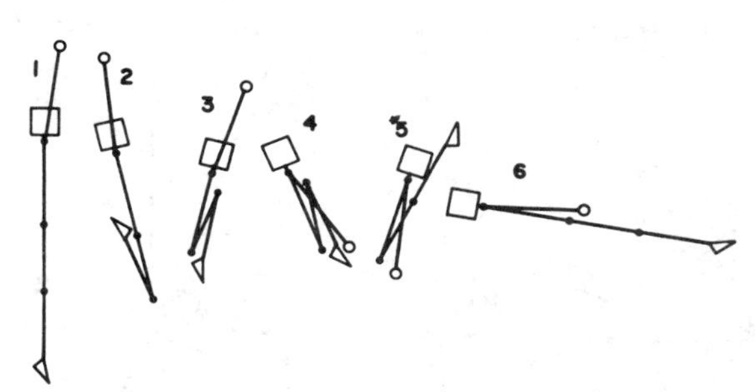

Fig. 6. Torque-free back quarter somersault ("back-drop"). If a "back drop" is executed in discrete steps as shown, the body will have rotated a total of 82° between the first and the sixth step, even though his angular momentum is exactly zero at all times. If the performer wished to finish the back drop with his body position exactly similar to his position in the first diagram but rotated 82°, after the sixth diagram he could raise his arms in the lateral plane of his body (along his sides). In principle a trampolinist could also rotate 82° forward by performing the steps shown in the reverse order, or 6-5-4-3-2-1. For each diagram, the amount of rotation shown was calculated as described in the Appendix, with the body consisting of two rigid pieces connected by a hinge. Each rigid piece is built from segments having the masses and intrinsic moments of inertia of the body parts listed in Table II. In practice, back drops are seldom done exactly as indicated in this figure. For example, usually steps two and three would be performed simultaneously, with the legs being tucked up in one motion. Also, in practice the neck and trunk are somewhat flexible, allowing a performer with zero angular momentum to rotate more than 82°.

moment of inertia of his body, a 360° rotation of his arms produces about a 20° rotation of his body. Thus, even though his angular momentum is exactly zero throughout the entire stunt, he has rotated or "somersaulted" 20°.

Figure 6 illustrates that it is possible to perform a torque-free somersault with nearly 90° rotation even without obviously "windmilling" the arms. Formulas used in the calculation of the amount of rotation drawn in Fig. 6 are derived in the Appendix. In practice, torque-free somersaults are not the essential part of any usual stunt, with the possible exception of the tuck back drop on the trampoline (a 90° back somersault, illustrated in Fig. 6). However, the same principles do play a small part in the execution of the ending of certain stunts, e.g., allowing a diver performing tucked somersaults a certain amount of control over his body's angle of entry into the water.

B. Twists

(*i*) *Torque twists.* Conceptually, the simplest way to initiate a twist is to apply a torque to the performer's body. For example, even nonathletic individuals standing on a hard surface generally can jump into the air and twist 180° or more by "pushing off" from the floor with their feet. Similarly, divers and trampolinists can initiate twists by pushing with their feet against the diving board or against the trampoline.

As with somersaults, the forces that initiate the rotation can be controlled most easily if the performer "throws" his arms in the direction of the twist *before* his feet lose contact with the floor. In the absence of forces, his lower body would rotate in the opposite direction from his arms. However, since the floor can apply forces to his feet these forces prevent this rotation in the opposite direction, and tend to cause his body to rotate in the same direction that he threw his arms (Fig. 7).

(*ii*) *Torque-free twists with angular momentum.* No torque whatsoever is applied to initiate twisting in the most important and most common method of initiating twists during somersaulting stunts. This mechanism can best be appreciated by carefully studying an example (Fig. 8). Suppose that a diver has initiated stable somersaulting motion and is in the layout position. Suppose further that he is no longer touching the diving board, that his body possesses considerable angular momentum about his left-to-right axis, and that his arms are extended to his sides as in Fig. 8(a) and in C in Fig. 1. Now the diver suddenly "throws" his right arm above his head and his left arm down to his side as in E in Fig. 1, moving his arms in the plane of

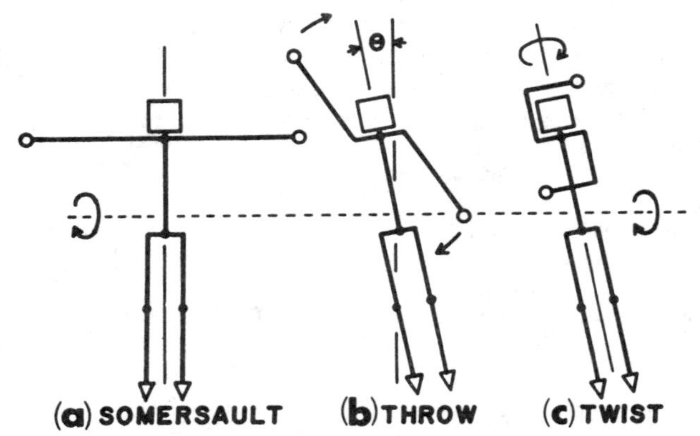

Fig. 8. Diver possessing angular momentum can initiate twisting even in the absence of any external torques. In (a) above, a somersaulting diver has angular momentum only about his left-right axis (dotted line) and has no twisting motion. At the instant pictured in (b) he sharply "throws" his left arm down and his right arm up laterally in the plane of his body. Because he has no angular momentum about his front-back axis (normal to the plane of the picture) as his arms rotate clockwise, his body rotates an angle θ counterclockwise. However, this causes his body to begin a continuous tumbling or twisting motion about his long axis in order to conserve angular momentum (see Fig. 9). In (c) this twisting motion as well as his somersaulting motion continue even when he is no longer "throwing" with his arms, but is in effect a rigid body with no external forces acting upon him.

his body as in Fig. 8(b). Viewed from the front, his two arms effectively are rotating clockwise, and so his lower body must react by rotating counterclockwise an angle θ about his front-to-back axis. Because of this rotation of θ of his body axes, his left-to-right body axis is no longer aligned with his angular momentum axis, and so to conserve angular momentum his body begins to tumble or twist about his head-to-toe axis. His body will continue to twist with considerable rotational speed even after his "throw" is complete, and even if he is holding his body rigidly.

The cause of this torque-free twisting is particularly clear if one considers the diver's moment of inertia tensor **I** in an inertial reference frame. Suppose the inertial reference frame is chosen so that just before the "throw" [Fig. 8(a)] the diver's principal axes are along the axes of the inertial frame, and thus **I** possesses no off-diagonal components. His angular momentum vector **H** and angular velocity vector ω also are parallel to one of the coordinate axes, e.g., call it the (0, 1, 0) axis. Immediately after the throw [Fig. 8(b)] the diver's principal axes are no longer along the axes of the inertial coordinate system, and thus **I** now possesses off-diagonal components. Because of these off-diagonal components, ω must possess components along the (1, 0, 0) axis and/or the (0, 0, 1) axis in order to ensure that the product $\mathbf{I}\cdot\boldsymbol{\omega} = \mathbf{H}$ remains fixed in space along the (0, 1, 0) axis.

With only a few assumptions one can derive an extremely simple expression for the rate of twisting per somersault for torque-free twists with angular momentum. First assume that the "throw" of the arms [occurring in Figure 8(b)] is fast enough so that no twist about the diver's head-to-toe axis occurs during the throw (e.g., in a time small enough so that a negligible amount of somersaulting takes place). Second, assume that after the "throw" is complete, two of the diver's principal moments of inertia are equal. Although this is not strictly true, in a layout position the two largest principal moments are close enough to one another (Table III) so that this assumption should give a good approximation of the twisting motion.[19]

With these assumptions, the diver is a rigid symmetrical top experiencing force-free motion which can be described completely in terms of simple trigonometric functions by solving Euler's equations of motion.[17] His rate of twisting is the angular velocity ω_t (total spin) about the body axis associated with his smallest principal moment of inertia (the long axis), and his rate of somersaulting is the rate of free precession ω_s that his angular velocity vector experiences about the angular momentum axis. In particular, if I_1 and I_2 are the two principal moments of inertia of the twisting diver, and if $\omega_1 = \omega_t$ and ω_2 are his rotational velocities about these principal axes, then Fig. 9 shows that

$$I_1\omega_t/I_2\omega_2 = \tan\theta \text{ and } \omega_2 = \omega_s \cos\theta. \tag{6}$$

Combining, we see that the rate of twisting per somersault is

$$\omega_t/\omega_s = I_2 \sin\theta/I_1. \tag{7}$$

This is the fundamental equation that governs the number of twists per somersault that a diver can perform if he initiates the twisting entirely after his feet have left the board. Note that the direction of the twists depends solely on whether the diver throws his arms clockwise or counterclockwise. Note also that the rate of twisting depends critically on the ratio of the principal moments of inertia (so that "skinny divers twist easier") and also on the angle θ that his body goes off axis when he "throws" his arms (Fig. 8). Practical ways of making this rate as large as possible will be discussed in the last section of this paper.

It is worth emphasizing that the kind of "throw" that initiates torque twists is quite different from the throw that initiates torque-free twists with angular momentum. For torque twists, the arms are thrown so that they effectively

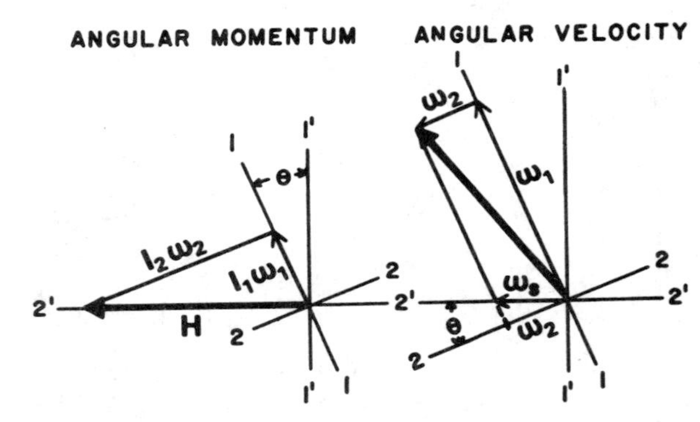

Fig. 9. Vector diagrams of the angular momentum vector and angular velocity vector in the inertial frame the instant after the twist is thrown. In both diagrams, the performer's long axis (twist axis) is along the 1′ axis before he "throws" his twist, but along the 1 axis after the twist begins. The performer is somersaulting about the 2′ axis, along which angular momentum **H** (heavy arrow in left diagram) is conserved. Before the twist is thrown, there is no component of angular momentum about the performer's long axis, and so the performer's angular velocity vector is along the 2′ axis, parallel to the angular momentum vector. After the twist is thrown, there is a component $I_1\omega_1$ of angular momentum along the 1 axis, and so twisting rotation begins. The vector components of angular momentum along the 1 and 2 axes satisfy $I_1\omega_1/I_2\omega_2 = \tan\theta$. The angular velocity vector is no longer along the 2′ axis. The diagram on the right shows that ω_s, the component of the angular velocity vector (heavy arrow) along the 2′ axis (somersaulting axis) is related to ω_2 by the relation $\omega_s \cos\theta = \omega_2$.

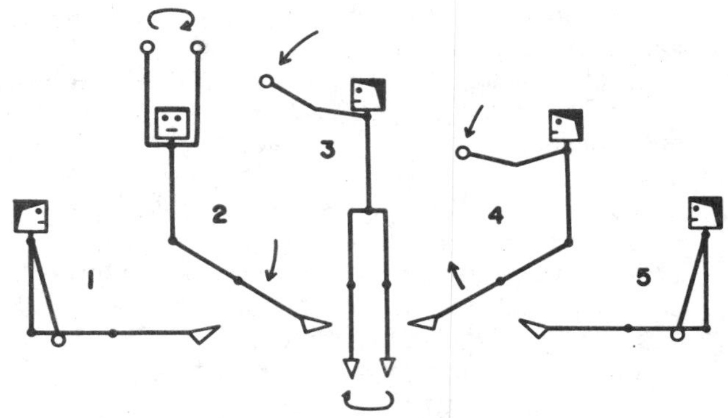

Fig. 10. "Swivel hips," a simple example of a trampoline stunt in which a performer can change the relative moments of inertia of his upper and lower body to accomplish a torque-free twist. The performer bounces from a sitting position (1), and begins to raise his arms over his head by bringing them forward and up as shown by the arrow. In (2), he rotates his upper body to turn his arms and shoulders as far around as possible, and he beings to drop his legs. Because his legs are piked and his arms are raised, the moment of inertia of his lower body is large relative to the amount of inertia of his upper body, and thus his legs experience little rotation. In (3) he brings his arms forward and down and "swivels" his hips beneath him while his arms are extended. In this step the moment of inertia of his lower body is smaller than that of his upper body. In (4) he brings his feet and legs up into a pike, and in (5) the stunt ends as it began, in a sitting position.

rotate about the body's head-to-toe axis, thus producing torques about this axis. For torque-free twists with angular momentum, the arms are thrown so that they rotate the body's principal axes relative to the angular momentum vector. In fact any body motion made during the dive will initiate twisting if the effect of the body motion is to reduce the angle between the head-to-toe axis and the angular momentum vector.

(*iii*) *Torque-free twists with zero angular momentum* or "cat twists." It is possible to rotate the body about the head-to-toe axis even though the body has no angular momentum and even though no external torques are applied. Like torque-free somersaults, these twists do not provide continuous twisting motion, but easily can produce rotations of 180° with only a few simple motions.

For example, if a cat is held upside down by its four legs and dropped without any initial rotation from a height greater than about half a meter, the cat will perform a sequence of motions that varies the relative moments of inertia of the front and rear portions of its body so as to cause it to land on its feet.[6,7] On the trampoline, a simple stunt that utilizes this type of twisting is called "swivel hips." One way of doing this stunt is described in Fig. 10.

It is also possible to perform torque-free twists without varying the relative moments of inertia of different portions of the body.[10] Figure 11 describes a simple body which demonstrates this. It consists of two identical plumb-bob shaped objects which are in contact along their conical ends. The three principal moments of inertia of each object are I, I, and J. The two objects roll without slipping on their conical ends with angular velocity ω, and simultaneously rotate with angular velocity Ω about the axis bb' (see Fig. 11). The angular momenta $\mathbf{H}_T$ and $\mathbf{H}_B$ of the two bodies are

$$\mathbf{H}_T = J\omega\mathbf{j}_T - J\Omega \cos\alpha\mathbf{j}_T + I\Omega \sin\alpha\mathbf{i}_T,$$
$$\mathbf{H}_B = J\omega\mathbf{j}_B - J\Omega \cos\alpha\mathbf{j}_B - I\Omega \sin\alpha\mathbf{i}_B, \qquad (8)$$

where $\mathbf{j}_T$, $\mathbf{j}_B$, $\mathbf{i}_T$, and $\mathbf{i}_B$ are unit vectors as in Fig. 11. To find the net angular momentum of the system, one adds $\mathbf{H}_T$ and $\mathbf{H}_B$ and resolves the components of $(\mathbf{j}_T + \mathbf{j}_B)$ and $(\mathbf{i}_T - \mathbf{i}_B)$ along the axis bb'. If the net angular momentum is zero, one finds that

$$\frac{\omega}{\Omega} = \frac{1 + \cos^2\alpha(J/I - 1)}{(J/I)\cos\alpha}. \qquad (9)$$

For example, if $I = J$ and $\alpha = 60°$, then $\omega/\Omega = 2$, and thus if the plumb-bob shaped objects roll once around on their conical ends, the entire body will rotate exactly 180° about the axis bb'.

This demonstrates that it is possible for a body to twist even though at all times during the twist: (i) the net angular momentum of the body is zero; (ii) no external torques are applied to the body; (iii) the principal moments of inertia of the body remain constant; (iv) the "shape" of the body does not change. Of course, twisting is possible only because the body is not truly a rigid body.

In principal a diver or trampolinist with good flexibility in the hip and waist region could perform a twist in a similar fashion. His arms and upper body would be analogous to body "T", in Fig. 11, and his legs would be like body "B." Kane and Scher[9] performed a detailed mechanical analysis on photos of a twisting cat, and concluded that this kind of mechanism was responsible for the twist. McDonald[8] has published drawings of photographs of a twisting man and has suggested that this kind of mechanism may be at least partly responsible for the twist. However, the present author does not agree that most torque-free twists with zero angular momentum are performed in this way by divers, trampolinists, or cats. In practice the performance of these twists is accompanied with considerable motion of the arms (or forelegs) relative to the upper body. In addition, during

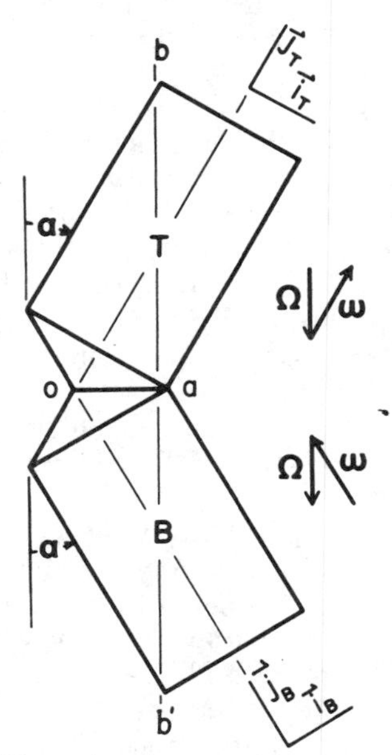

Fig. 11. Diagram illustrating that torque-free twisting can occur even though throughout the twisting process, the angular momentum is zero, the principal moments of inertia of the body are constant, and the shape of the body does not change. The body consists of two identical objects T and B, each shaped like a plumb bob (a cylinder with a cone on one end). If objects T and B each rotate with angular velocity ω about axes $\mathbf{j}_T$ and $\mathbf{j}_B$, respectively, then to conserve angular momentum the entire system must twist with angular velocity Ω about axis bb'. During the process, the concial ends roll without slipping and touch along the line Oa, which rotates in space with angular velocity Ω.

the twist the body generally is bent forward at the waist to a greater degree than to the back or to the sides. Both of these observations favor the hypothesis that the twist is initiated by varying the relative moments of inertia of body parts as in "swivel hips" (Fig. 10).

IV. OBSERVATIONS AND CONCLUSIONS

One of the main conclusions of this paper is that performers can and do execute torque-free somersaults, but only in such a manner that angular momentum is conserved. Uncertainty about this matter exists among divers and trampolinists as well as physicists (see Table I). For example, Rackham[4] says that if a diver jumps from the board with no angular momentum:

> No action on the part of the diver during a jump can cause him to rotate to enter the water headfirst. Whatever he does when he finally straightens for his entry he will be in the same position relative to the vertical as when he started the jump.

In fact, without bending his head or trunk, a diver could enter the water head first by twice tucking and untucking his body using the sequence of motion shown in Fig. 6.

As an experimental demonstration of this type of somersault, the author recently performed the following test on a trampoline to show that he could choose to rotate his body 90° after he was in the air. On each designated test jump, a friend watched the author's feet and after his feet had left the trampoline the friend either shouted "Back!" or remained silent. The author then attempted to perform a back one-quarter somersault (see Fig. 6) if "Back!" was called, or a straight jump (no somersault) if nothing was called. To prevent guessing the call, the author's friend determined the calls prior to each test jump by tossing a coin. Of the 20 test jumps that were attempted, the author performed the stunt as called by his friend 19 times.

In terms of physics, the torque-free somersault is closely related to the torque-free twist with zero angular momentum (cat twist). In both stunts motion in one portion of the body possesses angular momentum which must be countered by opposite motion of the remainder of the body. In both stunts rotation is not continuous—it ceases if the performer holds his body rigid. In practice, human performers are generally much more adept at torque-free twists than at torque-free somersaults. For example, Rackham[4] reports that a diver named Brian Phelps could hang by his hands from a ten meter platform, let go, and then do two 180° cat twists—one to the left and one to the right before entering the water.

Incidentally, using both cat twists and torque-free somersaults in combination would allow a hypothetical "spaceman" with zero angular momentum to position his body in any orientation in space that he chooses. This conclusion is in accord with NASA's own research on this question,[20] although the limb movements proposed in the NASA work generally are not used by divers and trampolinists. More often astronauts carry small gas guns which apply torques in order to effect body reorientations.

Another major conclusion of this paper is that when a performer does have angular momentum, "throwing" a twist can indeed produce a continuous rotating motion about his head-to-toe axis even if no torques are applied. Dyson[2] expressed uncertainty about this matter. He noted that

> . . . a gymnast somersaulting forward . . . can originate movements in the air which lead to displacement of mass about the main axis. In consequence, his body absorbs some angular momentum by twisting about the longitudinal axis, automatically reducing a tendency to somersault. Conversely, twisting can be "traded" for somersaulting. Thus ∴. . gymnasts, divers, etc. . . . sometimes appear to acquire angular momentum while free in space. However, other reputable students of this subject deny that a "trading" of angular momentum is possible. Here, then, is an important hypothesis requiring more solid experimental evidence.

Other recent authors accept the existence of this kind of twisting.[3,5,12] For example, Batterman[3] presents excellent pictures of a diver showing that his body is inclined to the vertical [as in Fig. 8(b)] while performing a back one and one-half somersault with two and one-half twists. A rather difficult dive which graphically illustrates the possibility of performing torque-free twists with angular momentum is the forward two and one-half somersault with two twists, which is performed from the 3-m springboard.[18] Generally, after leaving the board, the diver executes the first one and one-half somersaults, and only then "throws" for a twist to perform the last somersault with two twists before entering the water.

In practice, do performers really initiate twisting only after they have left the board? Recently the author viewed 64 frame/sec films of divers taken by Richard Gilbert at the 1972 Olympic Trials. For each twisting somersault, the author noted whether the diver dropped his arm or shoulder to initiate twisting before or after his feet left the board. These films revealed significant variations between different divers with respect to when they initiated twisting motion, and with respect to how fast they twisted. Although more than half of the twists were initiated with the feet still on the board for forward somersaults with twists, a significant fraction of the twists were initiated only after the diver was well away from the board. Twisting was more likely to be initiated away from the board for dives with only one or two twists. However, pure torque-free twisting was observed for a dive as complex as a forward one and one-half somersault with three twists.

If a diver initiates twisting after leaving the board, how fast can he twist while executing a forward twisting somersault? Equation (7) in Sec. III reveals that this will depend on the angle θ (see Fig. 9). Although the ratio I_1/I_2 will vary slightly from diver to diver, an individual diver has much more control over the $\sin\theta$ term in (7) than over the ratio of his principal moments of inertia. Exactly how he "throws" will determine how fast he twists.

For example, suppose the diver throws the twist from the layout position (Fig. 8), bringing one arm up over his head and the other arm down to his side. If we think of this as a rotation of his arms 90° relative to his body, Eq. (A6) from the Appendix tells us that if $I_1 = 2.31$ kg m^2 and if the hinge point of the arms is directly between the shoulders $\theta = 10.85°$. Thus from Eq. (7) $\omega_t/\omega_s = 3.0$ twists/somersault.

On the other hand, suppose he "throws" with his arms while he is still in a loose pike position (F in Fig. 1) and then straightens to a layout position to do the twists. Since his body's moment of inertia is smaller in the pike than in the layout (Table III), Eq. (A5) tells us that: $\theta = 19.96°$; and

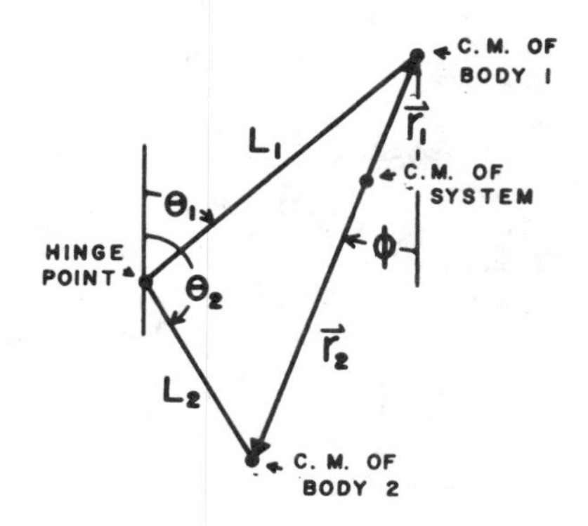

Fig. 12. Variables describing hinged diver. Many body motions of a diver can be interpreted as if the diver's body consisted of two rigid objects connected by a hinge. For example, as the diver pikes or "jackknifes" the hinge point is at his waist, one rigid body is his legs, and the other rigid body consists of his trunk, head and arms.

so $\omega_t/\omega_s = 5.5$ twists/somersault. Thus after he straightens to the layout position he can twist significantly faster because he threw the twist while still in the pike. In practice, motions of the diver's head and trunk as well as his arms will help him to increase the angle θ, and thus his rate of twisting.

The results above are in excellent quantitative accord with the observations of diving films. In the films of the 1972 Olympic Trials, divers performing multiple twisting somersaults were observed to twist at average rates varying from 3.8 twists/somersault to 6.0 twists/somersault. As mentioned previously, one diver performed a forward one and one-half somersault with three twists and clearly initiated the twists well after leaving the board. He was observed to initiate twisting in the pike position, and to complete one-half twist after somersaulting 185°, and two and one-half twists after somersaulting 315°, for an average rate of exactly 5.5 twists/somersault.[21]

ACKNOWLEDGMENTS

Several individuals besides the author made significant contributions to this work. Joe Burns, of the Department of Theoretical and Applied Mechanics at Cornell University, carefully reviewed an earlier draft of this manuscript, as did Charles Batterman, diving coach at the Massachusetts Institute of Technology. I am particularly indebted to Ted Grand, at the Oregon Primate Center, for critical discussions, and to Richard Gilbert, diving coach at Cornell University, for allowing me to view his film collection. Michael E. Fisher, of the Department of Chemistry, Physics, and Mathematics at Cornell, provided the analytical solution to Eq. (A4) in the Appendix after I provoked him by stating that it was not integrable. I also thank the members of the National Collegiate Athletic Association, (NCAA) Diving Rules Subcommittee who took the time to answer the diving questionnaire. Most of all, however, I am indebted to Ray Obermiller, diving coach at Grinnell College, who taught me almost everything I know about diving.

APPENDIX: THE DIVER AS A HINGED BODY

Suppose that a diver behaves mechanically like two rigid bodies which make angles θ_1 and θ_2 with the vertical and are attached to one another by a hinge so that the two bodies are constrained to move in a plane. We wish to calculate how θ_1 and θ_2, change as the diver varies the hinge angle $(\theta_2 - \theta_1)$. Call the masses of the two bodies M_1 and M_2, and the moments about their individual centers of mass I_1 and I_2, and the distance between the hinge point and their individual centers of mass L_1 and L_2 (Fig. 12). Let ϕ be the angle that the line between the centers of mass makes with the vertical.

If $\mathbf{r}_1$ and $\mathbf{r}_2$ are the position vectors of the centers of mass of the individual bodies in a coordinate system whose origin is at the diver's center of mass, then his angular momentum $\mathbf{H}$ of the system is

$$\mathbf{H} = I_1\dot{\theta}_1 + M_1(\mathbf{r}_1 \times \dot{\mathbf{r}}_1) + I_2\dot{\theta}_2 + M_2(\mathbf{r}_2 \times \dot{\mathbf{r}}_2).$$

This expression can be simplified, since $|\mathbf{r}_1 \times \mathbf{r}_1| = r_1{}^2\dot{\phi}$. In center of mass coordinates, if $\mathbf{r} = \mathbf{r}_2 - \mathbf{r}_1$, then

$$r_1 = [M_2/(M_1 + M_2)]r$$

and

$$r_2 = [M_1/(M_1 + M_2)]r.$$

Thus the angular momentum H becomes

$$\begin{aligned} H &= I_1\dot{\theta}_1 + M_1 r_1^2\dot{\phi} + I_2\dot{\theta}_2 + M_2 r_2^2\dot{\phi} \\ &= I_1\dot{\theta}_1 + I_2\dot{\theta}_2 + M_* r^2\dot{\phi}, \end{aligned} \tag{A1}$$

where M_* is the reduced mass, i.e.,

$$M_* = M_1M_2/(M_1 + M_2).$$

To eliminate ϕ and r from this equation, note that

$$\tan\phi = (L_1\sin\theta_1 - L_2\sin\theta_2)/(L_1\cos\theta_1 - L_2\cos\theta_2).$$

Differentiating, we find that

$$\begin{aligned} r^2\dot{\phi} = {} & [L_1^2 - L_1L_2\cos(\theta_2 - \theta_1)]\,\dot{\theta}_1 \\ & + [L_2^2 - L_1L_2\cos(\theta_2 - \theta_1)]\,\dot{\theta}_2, \end{aligned} \tag{A2}$$

where

$$r = |\mathbf{r}_2 - \mathbf{r}_1| = [L_1^2 - 2L_1L_2\cos(\theta_2 - \theta_1) + L_2^2]^{1/2}.$$

Substituting (A2) into (A1), we find that

$$\begin{aligned} H = {} & \{I_1 + M_*L_1[L_1 - L_2\cos(\theta_2 - \theta_1)]\}\,\dot{\theta}_1 \\ & + \{I_2 + M_*L_2[L_2 - L_1\cos(\theta_2 - \theta_1)]\}\,\dot{\theta}_2. \end{aligned} \tag{A3}$$

In particular, if $H = 0$,

$$\dot{\theta}_2 = -\frac{I_1 + M_*L_1[L_1 - L_2\cos(\theta_2 - \theta_1)]}{I_2 + M_*L_2[L_2 - L_1\cos(\theta_2 - \theta_1)]}\,\dot{\theta}_1. \tag{A4}$$

Surprisingly enough, this expression can be solved exactly in closed form.[22] If neither L_1 nor L_2 are zero, and if we define

$$A_1 = \frac{I_1 + M_*L_1^2}{M_*L_1L_2};\quad A_2 = \frac{I_2 + M_*L_2^2}{M_*L_1L_2},$$

$$a_+ = \frac{A_1 + A_2}{2};\quad a_- = \frac{A_1 - A_2}{2}$$

and if $h = (\theta_2 - \theta_1)$; then one can show easily that

$$\frac{d\theta_2}{dh} = \frac{1}{2}\left(1 + \frac{a_-}{a_+ - \cos(h)}\right).$$

This is integrable, and in fact if the hinge angle h changes from h_i to h_f, then the change in θ_2 is

$$\theta_{2_f} - \theta_{2_i} = \left\{\frac{h}{2} + \frac{a_-}{(a_+^2 - 1)^{1/2}} \times \tan^{-1}\left[\left(\frac{a_+ + 1}{a_- - 1}\right)^{1/2}\tan\left(\frac{h}{2}\right)\right]\right\}\Bigg|_{h_i}^{h_f}, \tag{A5}$$

where the indefinite integral (in braces) is evaluated at h_f and h_i.

Equation (A4) simplifies considerably if $L_1 = 0$, with the first body having its center of mass at the hinge point. In this case

$$\dot{\theta}_2 = [-I_1/(I_2 + M_* L_2^2)]\,\dot{\theta}_1.$$

If $\Delta\theta_2$ and Δh are the changes in θ_2 and h, respectively, then this becomes

$$\Delta\theta_2 = [-I_1/(I_1 + I_2 + M_* L_2^2)]\,\Delta h.$$

Since $M_* L_2^2 = M_1 r_1^2 + M_2 r_2^2$ this is just

$$\Delta\theta_2 = (-I_1/I_{\text{tot}})\,\Delta h, \qquad \text{(A6)}$$

where I_{tot} is the diver's total moment of inertia about his center of mass.

Equation (A4) is equivalent to Eq. 3.2.7 of Smith and Kane.[10] Smith and Kane[10] also derive expressions to describe more general motions of bodies having two and three segments.

[1]National Collegiate Athletic Association, *The Official Swimming Guide 1978* (National Collegiate Athletic Association, Shawnee Mission, KS, 1977).

[2]G. Dyson, *The Mechanics of Athletics*, 3rd ed. (University of London, London, 1964).

[3]C. Batterman, *The Techniques of Springboard Diving* (MIT, Cambridge, MA, 1968).

[4]G. Rackham, *Diving Complete* (Faber and Faber, London, 1975).

[5]J. Hay, *The Biomechanics of Sports Techniques*, 2nd ed. (Prentice Hall, Englewood Cliffs, NJ, 1978).

[6]Editor, Nature **51**, 80 (1894).

[7]D. McDonald, New Sci. **7**, 1647 (1960).

[8]D. McDonald, New Sci., **.J**, 501 (1961).

[9]T. R. Kane and M. P. Scher, Int. J. Solids Struct. **5**, 663 (1969).

[10]P. G. Smith and T. R. Kane, The Reorientation of a Human Being in Free Fall (Technical Report No. 171, Stanford University, U.S. Government Accession Number N67-31537, 1967) (unpublished).

[11]E. P. Hanavan, American Institute of Aeronautics and Astronautics Paper 65-498, 1 (1965).

[12]J. Broms, W. Duquet, and M. Hebbelinck, Med. Sport **8**, Suppl. Biomech. III, 429 (1973).

[13]D. I. Miller, Med. Sport **8**, 116 (1973).

[14]W. T. Dempster, Space Requirements of the Seated Operator (Wright Air Development Center WADC-TR-55-159, Ohio, 1955) (unpublished).

[15]M. Williams, and H. R. Lissner, *Biomechanics of Human Motion* (Sanders, Philadelphia, 1962).

[16]W. R. Santschii, J. DuBois, and C. Omoto, *Moments of Inertia and Centers of Mass of the Living Human Body* (Technical Documentary No. AMRC-TDR-63-36, Wright-patterson Air Force Base, Ohio, 1963) (unpublished).

[17]L. D. Landau and E. M. Lifshitz, *Mechanics* (Pergamon, Oxford, 1965).

[18]*Official Diving Rules 1978*, edited by J. Walker, (Amateur Athletic Union of the U.S., Indianapolis, IN, 1978).

[19]It is possible to obtain a closed solution for the diver's motion without this second assumption. See, e.g., Ref. 17.

[20]T. R. Kane and M. P. Scher, J. Biomech. **3**, 39 (1970).

[21]Perhaps the earliest recorded reference to a twisting somersault appeared in 1674 in *Grammatica Linguae Anglicanae* by John Wallis, a grammar published in Oxford, England:

When a Twister, a-twisting, will twist him a twist;
For the twisting of his twist, he three times doth intwist;
But, if one of the twists of the twist do untwist,
The twine that untwisteth, untwisteth the twist.
Untwirling the twine that untwisteth between,
He twirls, with his twister, the two in a twine;
Then, twice having twisted the twines of the twine,
He twisteth, the twine he has twined, in twain.
The twain that, in twining, before in the twine;
As twins were untwisted, he now doth untwine;
Twixt the twain intertwisting a twine more between,
He, twirling his twister, makes a twist of the twine.

[22]Michael E. Fischer (private communication).

Chapter 10
BIBLIOGRAPHY

CONTENTS

Articles

1. H.G. Magnus, "On the deviation of projectiles; and on a remarkable phenomenon of rotating bodies," *Memoirs of the Royal Academy*, Berlin (1852). English translation in *Taylor's Foreign Scientific Memoirs* (1853), p. 210.
2. Lord Rayleigh (John William Strutt), "On the irregular flight of a tennis-ball," *Messenger of Mathematics* **VII**, 14 (1877).
3. P. G. Tait, "On the path of a rotating spherical projectile," *Trans. R. Soc. Edinburgh* June/July (1893).
4. P. G. Tait, "On the path of a rotating spherical projectile. II," *Trans. R. Soc. Edinburgh* January (1896).
5. P. G. Tait, "Long driving," *Badminton Mag.* March (1896).
6. J. J. Thomson, "The dynamics of a golf ball," *Nature* **85**, 2147 (1910).
7. John W. Maccoll, "Aerodynamics of a spinning sphere," *J. R. Aeronaut. Soc.* **32**, 777 (1928).
8. C. N. Hickman, "Velocity and acceleration of arrows. Weight and efficiency of bows, as affected by backing of bow," *J. Franklin Inst.* **208**, 521 (1929).
9. G. J. Higgins, "The aerodynamics of an arrow," *J. Franklin Inst.* **210**, 805 (1930).
10. Paul Kirkpatrick, "Unscientific measurement in athletics," *Sci. Am.* **156**(4), 226 (1937).
11. Arthur E. Oxley, "Prehistoric airfoils," *Sci. Am.* **161**(2), 90 (1939).
12. Frank L. Verwiebe, "Does a baseball curve?," *Am. J. Phys.* **10**, 119 (1942).
13. Arthur Taber Jones, "Physics and bicycles," *Am J. Phys.* **10**, 332 (1942).
14. Paul E. Klopsteg, "Physics of bows and arrows," *Am. J. Phys.* **11**, 175 (1943).
15. Paul Kirkpatrick, "Bad physics in athletic measurements," *Am. J. Phys.* **12**, 7 (1944).
16. John M. Davies, "The aerodynamics of golf balls," *J. Appl. Phys.* **20**, 821 (1949).
17. Paul Kirkpatrick, "Notes on jumping," *Am. J. Phys.* **25**, 614 (1957).
18. D. Williams, "Drag force on a golf ball in flight and its practical significance," *Q. J. Mech. Appl. Math.* **XII**(3), 387 (1959).
19. Lyman J. Briggs, "Effect of spin and speed on the lateral deflection (curve) of a baseball; and the Magnus effect for smooth spheres," *Am. J. Phys.* **27**, 589 (1959).
20. Carl Selin, "An analysis of the aerodynamics of pitched baseballs," *Res. Q.* **30**(2), 232 (1959).
21. Roy Terrell, "Nobody hits it," *Sports Illustrated* **10**(26), 14 (1959).
22. A. V. Hill, "Production and absorption of work by muscle," *Science* **131**, 897 (1960).
23. J. H. Bayes and W. T. Scott, "Billiard-ball collision experiment," *Am. J. Phys.* **31**, 197 (1963).
24. Paul Kirkpatrick, "Batting the ball," *Am. J. Phys.* **31**, 606 (1963).
25. R. Margaria, P. Cerretelli, P. Aghemo, and G. Sassi, "Energy cost of running," *J. Appl. Physiol.* **18**, 367 (1963).
26. H. C. Herreshoff and J. N. Newman, "The study of sailing yachts," *Sci. Am.* **215**(2), 60 (1966).
27. J. D. Patterson, "Home run hitting," *Phys. Teach.* **5**, 167 (1967).
28. D. Williams, "The dynamics of the golf swing," *Q. J. Mech. Appl. Math.* **20**, 247 (1967).
29. R. D. Edge, "The surf skimmer," *Am. J. Phys.* **36**, 630 (1968).
30. Seville Chapman, "Catching a baseball," *Am. J. Phys.* **36**, 868 (1968).
31. Felix Hess, "The aerodynamics of boomerangs," *Sci. Am.* **219**(5), 124 (1968).
32. R. L. Garwin, "Kinematics of an ultraelastic rough ball," *Am. J. Phys.* **37**, 88 (1969).
33. Burton G. Shuster, "Ballistics of the modern-working recurve bow and arrow," *Am. J. Phys.* **37**, 364 (1969).
34. W.C. Marlow, "Comment on 'Ballistics of the modern-working recurve bow and arrow' [Burton G. Shuster, *Am. J. Phys.* **37**, 364 (1969)]," *Am. J. Phys.* **48**, 983 (1980).
35. Theodore Jorgensen, Jr., "On the dynamics of the swing of a golf club," *Am. J. Phys.* **38**, 644 (1970).
36. Elmer Offenbacher, "Physics and the vertical jump," *Am. J. Phys.* **38**, 829 (1970).
37. David E. H. Jones, "The stability of the bicycle," *Phys. Today* **23**(4), 34 (1970).
38. A. E. Raine, "Aerodynamics of skiing," *Sci. J.* **6**(3), 26 (1970).
39. Roger Bannister, "Towards the three-and-a-half minute mile," *Sci. J.* 36 (November 1970).
40. Jeffrey Lindemuth, "The effect of air resistance on falling balls," *Am. J. Phys.* **39**, 757 (1971).
41. Paul Nielsen, "Basketballs for vibration isolation," *Am. J. Phys.* **39**, 966 (1971).
42. G. Preston Burns, "Deflection of projectiles due to rotation of the earth," *Am. J. Phys.* **39**, 1329 (1971).
43. Thomas A. McMahon, "Rowing: a similarity analysis," *Science* **173**, 349 (1971).
44. Knut Schmidt-Nielsen, "Locomotion: energy cost of swimming, flying and running," *Science* **177**, 222 (1972).
45. J. I. Shonle and D. L. Nordich, "The physics of ski turns," *Phys. Teach.* **10**, 491 (1972).
46. R. Margaria, "The sources of muscular energy," *Sci. Am.* **226**(3), 88 (1972).
47. S. S. Wilson, "Bicycle technology," *Sci. Am.* **228**(3), 81 (1973).
48. Geoffrey T. Fox, "On the physics of drag racing," *Am. J. Phys.* **41**, 311 (1973).
49. J. D. Memory, "Kinematics problem for joggers," *Am. J. Phys.* **41**, 1205 (1973).
50. Ellen D. Yorke, "Energy cost and animal size," *Am. J. Phys.* **41**, 1286 (1973).
51. Joseph B. Keller, "A theory of competitive running," *Phys. Today* **26**(9), 42 (1973).
52. T. A. McMahon, "Size and shape in biology," *Science* **179**, 1201 (1973).
53. Albert Gold, "Energy expenditure in animal locomotion," *Science* **181**, 275 (1973).
54. J. Higbie, "The motorcycle as a gyroscope," *Am. J. Phys.* **42**, 701 (1974).
55. Arthur Hershman, "Animal locomotion as evidence for the universal constancy of muscle tension," *Am. J. Phys.* **42**, 778 (1974).
56. J. Richard Shanebrook, "Power to overcome air resistance at highway speeds," *Am. J. Phys.* **42**, 1028 (1974).
57. A. Armenti, Jr., "Physics and sport: a new course for nonscience majors," *Phys. Teach.* **12**, 349 (1974).
58. C. L. Stong, "Hang gliding, or sky surfing, with a high-performance low-speed wing," *Sci. Am.* **231**(6), 138 (1974).
59. G. D. Scott, "The swimmer's twin rainbows," *Am. J. Phys.* **43**, 460 (1975).
60. Allen L. King, "Project boomerang," *Am. J. Phys.* **43**, 770 (1975).
61. J. D. Walker, "Karate strikes," *Am. J. Phys.* **43**, 845 (1975).

62. Robert G. Watts and Eric Sawyer, "Aerodynamics of a knuckleball," *Am. J. Phys.* **43**, 960 (1975).
63. Robert Weaver, "Comment on 'aerodynamics of a knuckleball' [*Am. J. Phys.* **43**, 960 (1975)]," *Am. J. Phys.* **44**, 1215 (1976).
64. J. W. Halley and B. Eaton, "A course in the physics of human motion," *Am. J. Phys.* **43**, 1007 (1975).
65. C. L. Stong, "The ultimate in sailing is a rig without a hull," *Sci. Am.* **232**(3), 118 (1975).
66. T. H. Maugh, "Introducing the happy non-hooker," *Science* **187**, 941 (1975).
67. T. C. Soong, "The dynamics of the javelin throw," *J. Appl. Mech.* **42**, 257 (1975).
68. George W. Ficken, Jr., "Center of percussion demonstration," *Am. J. Phys.* **44**, 789 (1976).
69. F. X. Hart and C. A. Little, III, "Student investigation of models for the drag force," *Am. J. Phys.* **44**, 872 (1976).
70. Gene Rochlin, "Drag force on highway vehicles," *Am. J. Phys.* **44**, 1010 (1976).
71. C. H. Bachman, "Some observations on the process of walking," *Phys. Teach.* **14**, 360 (1976).
72. Joseph C. Baiera, "Physics of the softball throw," *Phys. Teach.* **14**, 367 (1976).
73. Richard Stepp, "The bicycle lab," *Phys. Teach.* **14**, 439 (1976).
74. T. C. Soong, "The dynamics of the discus throw," *J. Appl. Mech.* **43**, 531 (1976).
75. Henry W. Ryder, Harry Jay Carr, and Paul Herget, "Future performance in footracing," *Sci. Am.* **234**(6), 109 (1976).
76. Alan D. Bernstein, "Listening to the coefficient of restitution," *Am. J. Phys.* **45**, 41 (1977).
77. Haywood Blum, "Physics and the art of kicking and punching," *Am. J. Phys.* **45**, 61 (1977).
78. D. C. Hopkins and J. D. Patterson, "Bowling frames: paths of a bowling ball," *Am. J. Phys.* **45**, 263 (1977).
79. Richard A. Bartels, "Braking distance versus mass for automobiles," *Am. J. Phys.* **45**, 398 (1977).
80. G. W. Parker, "Projectile motion with air resistance quadratic in the speed," *Am. J. Phys.* **45**, 606 (1977).
81. Bernard L. Cohen, "Body weight as an application of energy conservation," *Am. J. Phys.* **45**, 867 (1977).
82. Herbert Lin, "Newtonian mechanics and the human body: Some estimates of performance," *Am. J. Phys.* **46**, 15 (1978).
83. J. Liesgang and A. R. Lee, "Dynamics of a bicycle: Nongyroscopic aspects," *Am. J. Phys.* **46**, 130 (1978).
84. D. B. Lichtenberg and J. G. Wills, "Maximizing the range of the shot put," *Am. J. Phys.* **46**, 546 (1978).
85. William Hooper, "Comment on 'Maximizing the range of the shot put'," *Am. J. Phys.* **47**, 748 (1979).
86. Richard C. Smith, "General physics and the automobile tire," *Am. J. Phys.* **46**, 858 (1978).
87. R. L. Armstrong, "Relative velocities and the runner," *Am. J. Phys.* **46**, 950 (1978).
88. Walter C. Connolly, "The physics of sport activities," *Phys. Teach.* **16**, 392 (1978).
89. Pierre Lafrance, "Ship hydrodynamics," *Phys. Today* **31**(6), 34 (1978).
90. John Jerome, "How high can a man jump?," *Quest* 15 (November 1978).
91. T. A. McMahon and P. R. Greene, "Fast running tracks," *Sci. Am.* **239**(6), 148 (1978).
92. Henry A. Bent, "How much work can a person do?," *J. Chem. Educ.* **55**, 456 (1978).
93. Henry A. Bent, "Caloric costs of mass transport," *J. Chem. Educ.* **55**, 526 (1978).
94. Henry A. Bent, "Heart work," *J. Chem. Educ.* **55**, 586 (1978).
95. Henry A. Bent, "Energy storage problems," *J. Chem. Educ.* **55**, 659 (1978).
96. Henry A. Bent, "Limiting reagents," *J. Chem. Educ.* **55**, 726 (1978).
97. Henry A. Bent, "Reactions for every occasion," *J. Chem. Educ.* **55**, 796 (1978).
98. P. F. Hinrichsen, "Weight distribution," *Yachts and Yachting* 347 (February 3, 1978).
99. Howell Peregrine, "Surfing brought to rest," *New Scientist* 8 (6 July 1978).
100. D. P. Whitmire and T. J. Alleman, "Effect of weight transfer on a vehicle's stopping distance," *Am. J. Phys.* **47**, 89 (1979).
101. H. Brody, "Physics of the tennis racket," *Am. J. Phys.* **47**, 482 (1979).
102. Cliff Frohlich, "Do springboard divers violate angular momentum conservation?," *Am. J. Phys.* **47**, 583 (1979).
103. J. D. Memory, "Kinematic strategy of self-preservation for the jogger," *Am. J. Phys.* **47**, 749 (1979).
104. B. L. Coulter and C. G. Adler, "Can a body pass a body falling through the air?," *Am. J. Phys.* **47**, 841 (1979).
105. P. Raychowdhury and J. Boyd, "Center of percussion," *Am. J. Phys.* **47**, 1088 (1979).
106. Peter Rye, "The physics of sailing," *The Australian Physicist* **27** (March 1979).
107. M. Hubbard, "Lateral dynamics and stability of the skateboard," *J. Appl. Mech.* **46**, 931 (1979).
108. R. L. Huston, C. Passerello, J. M. Winget, and J. Sears, "On the dynamics of a weighted bowling ball," *J. Appl. Mech.* **46**, 937 (1979).
109. M. Enterline and R. Lewin, "Science of an ancient art," *New Sci.* **8** (5 April 1979).
110. Richard Wilde, "Take a cool turn for pleasure," *New Sci.*, **936** (20/27 December 1979).
111. Jearl Walker, "Boomerangs!," *Sci. Am.* **240**(3), 162 (1979).
112. M. S. Feld, R. E. McNair, and S. E. Wilk, "The physics of karate," *Sci. Am.* **240**(4), 150 (1979).
113. Thomas B. Greenslade, Jr., "Exponential bicycle gearing," *Phys. Teach.* **17**, 455 (1979).
114. T. A. McMahon and P. R. Greene, "The influence of track compliance on running," *J. Biomech.* **12**, 893 (1979).
115. Gaston Fischer, "Exercise in probability and statistics, or the probability of winning at tennis," *Am. J. Phys.* **48**, 14 (1980).
116. Daniel Kirshner, "Some nonexplanations of bicycle stability," *Am. J. Phys.* **48**, 36 (1980).
117. J. D. Edmonds, Jr., "Speed and displacement relations for tire skids," *Am. J. Phys.* **48**, 253 (1980).
118. Harold Weinstock, "Thermodynamics of cooling a (live) human body," *Am. J. Phys.* **48**, 339 (1980).
119. M. Peastrel, R. Lynch, and A. Armenti, "Terminal velocity of a shuttlecock in vertical fall," *Am. J. Phys.* **48**, 511 (1980).
120. Paul Jackson, "What limits the speed of a sailboat," *Phys. Teach.* **18**, 224 (1980).
121. R. Mehta and D. Wood, "Aerodynamics of the cricket ball," *New Sci.* **442** (7 August 1980).
122. M.R. Kent, "The physics of swimming," *Phys. Educ.* **15**, 275 (1980).
123. Gideon Ariel, "Can computers win gold medals?," *New Sci.*, 282 (24 July 1980).
124. V. E. Neie and S. W. Ball, "The physics of human motion," J. Chem. Soc. Trans. **277** (May 1980).
125. J. Fagenbaum, "Electronics aids the athlete," *IEEE Spectrum* 36 (July 1980).
126. Cliff Frohlich, "The physics of somersaulting and twisting," *Sci. Am.* **242**(3), 154 (1980).
127. M. Hubbard, "Dynamics of the pole vault," *J. Biomech.* **13**, 965 (1980).
128. S. Mochon and T. A. McMahon, "Ballistic walking," *J. Biomech.* **13**, 49 (1980).
129. A. Bellemans, "Power demand in walking and pace optimization," *Am. J. Phys.* **49**, 25 (1981).
130. P. A. Smith, C. D. Spencer, and D. E. Jones, "Microcomputer listens to the coefficient of restitution," *Am. J. Phys.* **49**, 136 (1981).
131. I. Alexandrov and P. Lucht, "Physics of sprinting," *Am. J. Phys.* **49**, 254 (1981).
132. W. C. Marlow, "Bow and arrow dynamics," *Am. J. Phys.* **49**, 320 (1981).
133. Peter Brancazio, "Physics of basketball," *Am. J. Phys.* **49**, 356 (1981).
134. Peter Brancazio, "Erratum: Physics of basketball [*Am. J. Phys.* **49**, 356 (1981)]," *Am. J. Phys.* **50**, 567 (1982).
135. A. Bellemans, "Drag force exerted by water on the human body," *Am. J. Phys.* **49**, 367 (1981).

136. A. Tan and G. Miller, "Kinematics of the free throw in basketball," *Am. J. Phys.* **49**, 542 (1981).
137. P. Palffy-Muhoray and W. G. Unruh, "Turning quickly: A variational approach," *Am. J. Phys.* **49**, 685 (1981).
138. K. S. Krane, "Probability, statistics, and the World Series of baseball," *Am. J. Phys.* **49**, 696 (1981).
139. H. Brody, "Physics of the tennis racket II: The 'sweet' sport," *Am. J. Phys.* **49**, 816 (1981).
140. Cliff Frohlich, "Aerodynamic effects on discus flight," *Am. J. Phys.* **49**, 1125 (1981).
141. Philip DiLavore, "Why is it easier to ride a bicycle than to run the same distance?," *Phys. Teach.* **19**, 194 (1981).
142. Peter F. Hinrichsen, "Practical applications of the compound pendulum," *Phys. Teach.* **19**, 286 (1981).
143. A. Armenti, "Could an athlete run a 3-m radius 'loop-the-loop?'," *Phys. Teach.* **19**, 624 (1981).
144. J. A. Zufiria and J. R. Sanmartin, "Influence of air drag on the optimal hand launching of a small, round, projectile," *Am. J. Phys.* **50**, 59 (1982).
145. J. C. Lewis and H. Kiefte, "Simple experiment to illustrate the functional form of air resistance," *Am. J. Phys.* **50**, 145 (1982).
146. P. Palffy-Muhoray and D. Balzarini, "Maximizing the range of a shot put wihout calculus," *Am. J. Phys.* **50**, 181 (1982).
147. Frank S. Crawford, "Superball and time-reversal invariance," *Am. J. Phys.* **50**, 856 (1982).
148. Peter J. Brancazio, "Comments on Kinematics of the free throw in basketball," *Am. J. Phys.* **50**, 944 (1982).
149. A. Tan and K. Taylor, "Reply to Comments on 'Kinematics of the free throw in basketball'," *Am. J. Phys.* **50**, 945 (1982).
150. J. Lowell and H. D. McKell, "The stability of bicycles," *Am. J. Phys.* **50**, 1106 (1982).
151. Roger Stephenson, "Constant power equations of motion," *Am. J. Phys.* **50**, 1150 (1982).
152. G. J. van Ingen Shenau, "The influence of air friction on speed skating," *J. Biomech.* **15**, 449 (1982).
153. Alan L. Myers, "Thermodynamics of running," *Chem. Eng. Educ.* **18** (Winter 1982).
154. Chester R. Kyle, "Go with the flow: Aerodynamics and cycling," *Bicycling* **59** (May 1982).
155. P. A. Lightsey, "An analysis of the rotational stability of the layout back somersault," *Am. J. Phys.* **51**, 115 (1983).
156. H. Erlichson, "Maximum projectile range with drag and lift, with particular application to golf," *Am. J. Phys.* **51**, 357 (1983).
157. R. C. Nicklin, "Bikework," *Am. J. Phys.* **51**, 423 (1983).
158. S. K. Bose, "Maximizing the range of the shot put without calculus," *Am. J. Phys.* **51**, 458 (1983).
159. S. R. Wilk, R. E. McNair, and M. S. Feld, "The physics of karate," *Am. J. Phys.* **51**, 783 (1983).
160. Jearl Walker, "The physics of the follow, the draw and the masse (in billiards and pool)," *Sci. Am.* **249**(1), 124 (1983).
161. A. C. Gross, C. R. Kyle, and D. J. Malewicki, "The aerodynamics of human-powered land vehicles," *Sci. Am.* **249**(6), 142 (1983).
162. Jacques Thomas, "Why boomerangs boomerang (and killing sticks don't)," *New Sci.* **838** (22 September 1983).
163. Peter Brancazio, "The hardest blow of all," *New Sci.* **880** (22/29 December 1983).
164. A. J. Ward-Smith, "The influence of aerodynamic and biomechanical factors on long jump performance," *J. Biomech.* **16**, 655 (1983).
165. Thomas B. Greenslade, Jr., "More bicycle physics," *Phys. Teach.* **21**, 360 (1983).
166. Brynjolf Dokken, "More on bicycle physics," *Phys. Teach.* **22**, 138 (1984).
167. Thomas B. Greenslade, Jr., "The author replies," *Phys. Teach.* **22**, 138 (1984).
168. Larry Mallak, "Athletic optimization," *Ill. Technogr.* **99**, 10 (1983).
169. R. J. Shephard, "Sports science: Its prostitution and prospects," *Interdiscip. Sci. Rev.* **8**, 120 (1983).
170. J. Carey and M. Brunom, "Physics in the batter's box," *Newsweek* (19 September 1983).
171. J. L. Groppel, C. J. Dillman, and T. J. Lardner, "Derivation and validation of equations of motion to predict ball spin upon impact in tennis," *J. Sports Sci.* **1**, 111 (1983).
172. Cliff Frohlich, "Aerodynamic drag crisis and its possible effect on the flight of baseballs," *Am. J. Phys.* **52**, 325 (1984).
173. O. Helene, "On "waddling" and race walking," *Am. J. Phys.* **52**, 656 (1984).
174. Frank S. Crawford, "Elementary derivation of the wake pattern of a boat," *Am. J. Phys.* **52**, 782 (1984).
175. M. E. Brandan, M. Gutierrez, R. Labbe, and A. Menchaca-Rocha, "Measurement of the terminal velocity in air of a ping-pong ball using a time-to-amplitude converter in the millisecond range," *Am. J. Phys.* **52**, 890 (1984).
176. W. G. Unruh, "Instability in automobile braking," *Am. J. Phys.* **52**, 903 (1984).
177. A. Armenti, "How can a downhill skier move faster than a sky diver?," *Phys. Teach.* **22**, 109 (1984).
178. Albert A. Bartlett, "Television, football, and physics: Experiments in kinematics," *Phys. Teach.* **22**, 386 (1984).
179. H. Brody, "That's how the ball bounces," *Phys. Teach.* **22**, 494 (1984).
180. P. J. Brancazio, "Sir Isaac and the rising fastball," *Discover* **44** (July 1984).
181. P. J. Brancazio, "Getting a kick out of physics," *Discover* **64** (November 1984).
182. Lloyd Hunter, "The art and physics of soaring," *Phys. Today* **37**, 34 (1984).
183. L. P. Remizov, "Biomechanics of optimal flight in ski-jumping," *J. Biomech.* **17**, 161 (1984).
184. P. J. Brancazio, "Science and the game of baseball," *Sci. Dig.* 66, July 1984.
185. T. A. McMahon, "Mechanics of locomotion," *Int. J. Robotics Res.* **3**, 4 (Summer 1984).
186. J. Strnad, "Physics of long-distance running," *Am. J. Phys.* **53**, 371 (1985).
187. Kenneth S. Mendelson, "Why is ice so slippery?," *Am. J. Phys.* **53**, 393 (1985).
188. Herman Erlichson, "Is a baseball a sand-roughened sphere?," *Am. J. Phys.* **53**, 582 (1985).
189. Cliff Frohlich, "Comments on 'Is a baseball a sand-roughened sphere?'," *Am. J. Phys.* **53**, 583 (1985).
190. Cliff Frohlich, "Effect of wind and altitude on record performance in foot races, pole vault, and long jump," *Am. J. Phys.* **53**, 726 (1985).
191. J. Kwasnoski and R. Murphy, "Determining the aerodynamic drag coefficient of an automobile," *Am. J. Phys.* **53**, 776 (1985).
192. A. Tan, "The moment of inertia of a basketball," *Am. J. Phys.* **53**, 811 (1985).
193. P. J. Brancazio, "Looking into Chapman's homer: The physics of judging a fly ball," *Am. J. Phys.* **53**, 849 (1985).
194. A. F. Rex, "The effect of spin on the flight of batted baseballs," *Am. J. Phys.* **53**, 1073 (1985).
195. K. Voyenli and E. Eriksen, "On the motion of an ice hockey puck," *Am. J. Phys.* **53**, 1149 (1985).
196. P. J. Brancazio, "Trajectory of a fly ball," *Phys. Teach.* **23**, 20 (1985).
197. George A. Amann and Floyd T. Holt, "Karate demonstration," *Phys. Teach.* **23**, 40 (1985).
198. H. Brody, "The moment of inertia of a tennis racket," *Phys. Teach.* **23**, 213 (1985).
199. A. Eisenkraft, "The softball trajectory: An outdoor lab," *Phys. Teach.* **23**, 300 (1985).
200. A. Armenti, "Why is it harder to paddle a canoe in shallow water?," *Phys. Teach.* **23**, 310 (1985).
201. P. J. Brancazio, "The physics of kicking a football," *Phys. Teach.* **23**, 403 (1985).
202. R. Hignell and C. Terry, "Why do downhill racers pre-jump?," *Phys. Teach.* **23**, 487 (1985).
203. P. J. Brancazio, "Why does a football keep its axis pointing along its trajectory?," *Phys. Teach.* **23**, 571 (1985).
204. T. A. McMahon, "The role of compliance in mammalian running gaits," *J. Exp. Biol.* **115**, 263 (1985).
205. V. Foley, G. Palmer, and W. Soedel, "The crossbow," *Sci. Am.* **252**(1), 104 (1985).
206. Jearl Walker, "Fly casting illuminates the physics of fishing," *Sci. Am.* **253**(1), 122 (1985).
207. J. Witters and D. Duymelinck, "Rolling and sliding resistive

forces on balls moving on a flat surface," *Am. J. Phys.* **54**, 80 (1986).

208. Ernie McFarland, "How Olympic records depend on location," *Am. J. Phys.* **54**, 513 (1986).
209. Cliff Frohlich, "Resource letter PS-1: Physics of sports," *Am. J. Phys.* **54**, 590 (1986).
210. Joseph M. Zayas, "Experimental determination of the coefficient of drag of a tennis ball," *Am. J. Phys.* **54**, 622 (1986).
211. H. Brody, "The sweet spot of a baseball bat," *Am. J. Phys.* **54**, 640 (1986).
212. Graig A. Spolek, "The mechanics of flycasting: The flyline," *Am. J. Phys.* **54**, 832 (1986).
213. Martin H. Edwards, "Zero angular momentum turns," *Am. J. Phys.* **54**, 846 (1986).
214. A. Tan, C. H. Frick, and O. Castillo, "The fly ball trajectory: An older approach revisited," *Am. J. Phys.* **55**, 37 (1987).
215. Robert G. Watts and Ricardo Ferrer, "The lateral force on a spinning sphere: Aerodynamics of a curveball," *Am. J. Phys.* **55**, 40 (1987).
216. Peter J. Brancazio and Howard Brody, "How much air is in a basketball?," *Am. J. Phys.* **55**, 276 (1987).
217. Peter J. Brancazio, "Rigid-body dynamics of a football," *Am. J. Phys.* **55**, 415 (1987).
218. Albert A. Bartlett and Paul G. Hewitt, "Why the ski instructor says, 'lean forward!'," *Phys. Teach.* 25, 28 (1987).
219. Howard Brody, "Models of tennis racket impacts," *Int. J. Sports Biomech.* **3**, 293 (1987).
220. T. A. McMahon, G. Valiant, and E. C. Frederick, "Groucho running," *J. Appl. Phys.* **62**, 2326 (1987).
221. Antonin Štěpánek, "The aerodynamics of tennis balls: The topspin lob," *Am. J. Phys.* **56**, 138 (1988).
222. George C. Goldenbaum, "Equilibrium sailing velocities," *Am. J. Phys.* **56**, 209 (1988).
223. S. Lingard, "Note on the aerodynamics of a flyline," *Am. J. Phys.* **56**, 756 (1988).
224. R. Evan Wallace and Michael C. Schroeder, "Analysis of billard ball collisions in two dimensions," *Am. J. Phys.* **56**, 815 (1988).
225. George Y. Onoda, "Comment on 'Analysis of billard ball collisions in two dimensions'," by R. E. Wallace and M. C. Schroeder [*Am. J. Phys.* **56**, 815–819 (1988)]," *Am. J. Phys.* **57**, 476 (1989).
226. John J. McPhee and Gordon C. Andrews, "Effect of sidespin and wind on projectile trajectory, with particular application to golf," *Am. J. Phys.* **56**, 933 (1988).
227. Mark Carle, "Olympic wrestling and angular momentum," *Phys. Teach.* **26**, 92 (1988).
228. Chris Buethe and Dick Simon, "Physics at the Indy 500," *Phys. Teach.* **26**, 274 (1988).
229. Eugene E. Nalence, "Using automobile road test data," *Phys. Teach.* **26**, 278 (1988).
230. Richard A. D. Hewko, "The racing car turn," *Phys. Teach.* **26**, 436 (1988).
231. Jeffrey Kluger, "What's behind the home run boom?," *Discover* **78** (April 1988).
232. Robert G. Watts and Steven Baroni, "Baseball-bat collisions and the resulting trajectories of spinning balls," *Am. J. Phys.* **57**, 40 (1989).
233. Howard Brody, "An experiment to measure the density of air," *Phys. Teach.* **27**, 46 (1989).
234. Robert G. Hunt, "Bicycles in the physics lab," *Phys. Teach.* **27**, 160 (1989).
235. Bengt Magnusson and Bruce Tiemann, "The physics of juggling," *Phys. Teach.* **27**, 584 (1989).
236. Dexter Ford, "Physics and the motorcycle," *Motorcyclist* 82 (July 1989).
237. Roger Cox, "State-of-the-art sports equipment," *Vis a Vis* 39 (November 1989).
238. David T. Kagan, "The effects of coefficient of restitution variations on long fly balls," *Am. J. Phys.* **58**, 151 (1990).
239. John M. Robson, "The physics of fly casting," *Am. J. Phys.* **58**, 234 (1990).
240. M. G. Calkin, "The motion of an accelerating automobile," *Am. J. Phys.* **58**, 573 (1990).
241. Howard Brody, "Models of baseball bats," *Am. J. Phys.* **58**, 756 (1990).
242. Tom Southworth, "Potential energy analysis of a bow," *Phys. Teach.* **28**, 42 (1990).
243. Albert A. Bartlett, "Physics and a skiing record," *Phys. Teach.* **28**, 72 (1990).
244. Michael Hanson, "The flight of a boomerang," *Phys. Teach.* **28**, 142 (1990).
245. Eric Kincanon, "Juggling and the theorist," *Phys. Teach.* **28**, 221 (1990).
246. Howard Brody, "The tennis-ball bounce test," *Phys. Teach.* **28**, 407 (1990).
247. Mark M. Payne, "Electric fields and football fields," *Phys. Teach.* **28**, 563 (1990).
248. R. McN. Alexander, "Optimum take-off techniques for high and long jumps," *Philos. Trans. R. Soc. London Ser. B* **329**, 3 (1990).
249. Steven Ashley, "Baseball bats: Getting good wood (or aluminum) on the ball," *Mech. Eng.* **112**, 40 (October 1990).
250. G. Franke, W. Suhr, and F. Reiß, "An advanced model of bicycle dynamics," *Eur. J. Phys.* **11**, 116 (1990).
251. Wim Hennekam, "The speed of a cyclist," *Phys. Educ.* **25**, 141 (1990).
252. William M. MacDonald, "The physics of the drive in golf," *Am. J. Phys.* **59**, 213 (1991).
253. Peter J. Brancazio, "Why does a tennis racket have a 'sweet spot'?" *Cenco Phys. Sports Ser.*
254. Peter J. Brancazio, "Why do good jumpers seem to hang in the air?," *Cenco Phys. Sports Ser.*
255. Peter J. Brancazio, "How do karate experts break boards with their bare hands?," *Cenco Phys. Sports Ser.*
256. Peter J. Brancazio, "How do you balance a bicycle?," *Cenco Phys. Sports Ser.*
257. Peter J. Brancazio, "Why does a golf ball have dimples?," *Cenco Phys. Sports Ser.*
258. Peter J. Brancazio, "How does a boat sail into the wind?," *Cenco Phys. Sports Ser.*
259. Peter J. Brancazio, "How do you throw a curve ball?," *Cenco Phys. Sports Ser.*

BOLD denotes articles included in this volume.

Books

1. H. Nishiyama and R. C. Brown, *Karate. The Art of "Empty Hand" Fighting* (Tuttle, Rutland, VT, 1960).
2. Marian Williams and Herbert R. Lissner, *Biomechanics of Human Motion* (Saunders, Philadelphia, 1962).
3. J. W. Bunn, *Basketball Technique and Team Play* (Prentice-Hall, Englewood Cliffs, New Jersey, 1964).
4. H. Hobson, *Scientific Basketball* (Prentice-Hall, Englewood Cliffs, New Jersey, 1964).
5. C. A. Marchaj, *Sailing Theory and Practice* (Dodd Mead, New York, 1964).
6. Alan Watts, *Wind and Sailing Boats* (Adlard Coles, London, 1965).
7. *Rowing: A Scientific Approach*, edited by J. G. Williams and A. C. Scott (Barnes, New York, 1967).
8. A. Cochran and J. Stobbs, *The Search for the Perfect Swing* (Lippincott, New York, 1968).
9. H. F. Kay, *The Science of Yachts, Wind & Water* (deGraff, Tuckahoe, NY, 1971).
10. Stanley Plagenhoef, *Patterns of Human Motion* (Prentice-Hall, Englewood Cliffs, NJ, 1971).
11. W. Satty, *The Cosmic Bicycle* (Straight Arrow, San Francisco, 1971).
12. C. B. Daish, *The Physics of Ball Games* (English Universities, London, 1972).
13. D.I. Miller and R.C. Nelson, *Biomechanics of Sport* (Lea & Febiger, Philadelphia 1973).
14. *Mechanics and Sport*, AMD—Vol. 4, edited by J. L. Bleustein (American Society of Mechanical Engineers, New York, 1973).
15. G. B. Benedek and F. M. H. Villars, *Physics. With Illustrative Examples from Medicine and Biology* (Addison-Wesley, Reading, MA, 1974).

16. William G. Van Dorn, *Oceanography and Seamanship* (Dodd Mead, New York, 1974).
17. Edmond Bruce and Henry A. Morss, Jr., *Design for Fast Sailing* (Amateur Yacht Research Society, Newbury, England (1975).
18. R. Margaria, *Biomechanics and Energetics of Muscular Exercise* (Oxford University, London, 1976).
19. G. Dyson, *The Mechanics of Athletics* (7th ed.) (Holmes and Meier, New York, 1977).
20. A. Sharp, *Bicycles & Tricycles: An Elementary Treatise on their Design and Construction* (Reprint of 1896 ed.) (MIT, Cambridge, MA, 1977).
21. *Mechanics and Energetics of Animal Locomotion*, edited by R. M. Alexander and G. Goldspink (Halsted, New York, 1977).
22. C. A. Marchaj, *Aero-Hydrodynamics of Sailing* (Dodd Mead, New York, 1979).
23. Eadweard Muybridge, *Muybridge's Complete Human and Animal Locomotion* (Dover, New York, 1979).
24. Joseph Norwood, Jr., *High Speed Sailing* (Granada, New York, 1979).
25. John Jerome, *The Sweet Spot in Time* (Summit, New York, 1980).
26. F. R. Whitt and D. C. Wilson, *Bicycling Science* (2nd ed.) (MIT, Cambridge, MA, 1982).
27. J. Forester, *Bicycle Transportation* (MIT, Cambridge, MA, 1983).
28. John Howe, *Skiing Mechanics* (Poudre, Colorado, 1983).
29. *Newton at the Bat*, edited by Eric W. Schrier and William F. Allmann (Scribner's, New York, 1984).
30. Peter Brancazio, *Sport Science* (Simon and Schuster, New York, 1984).
31. M. Stewart Townend, *Mathematics in Sport* (Halsted, New York, 1984).
32. T. A. McMahon, *Muscles, Reflexes, and Locomotion* (Princeton University, Princeton, NJ, 1984).
33. John Forester, *Effective Cycling* (MIT, Cambridge, MA, 1984).
34. James G. Hay, *The Biomechanics of Sports Techniques* (3rd ed.) (Prentice-Hall, Englewood Clifs, NJ, 1985).
35. Lee Torrey, *Stretching the Limits* (Dodd Mead, New York, 1985).
36. Cliff Frohlich, *Selected Reprints "Physic of Sports"* (AAPT, College Park, MD, 1986).
37. Ross Garrett, *The Symmerty of Sailing* (Adlard Coles, London, 1987).
38. David Griffing, *The Dynamics of Sports* (3rd ed.) (Dalog, Oxford, OH, 1987).
39. Howard Brody, *Tennis Science for Tennis Players* (University of Pennsylvania, Philadelphia, 1987).
40. Ira Flatow, *Rainbows, Curve Balls and other Wonders of the Natural World Explained* (Harper & Row, New York, 1989).
41. Robert K. Adair, *The Physics of Baseball* (Harper & Row, New York, 1990).
42. R. G. Watts and A. T. Bahill, *Keep Your Eye On The Ball: The Science and Folklore of Baseball* (Freeman, New York, 1990).
43. Neville de Mestre, *The Mathematics of Projectiles in Sport* (Cambridge University, 1990).
44. A. Cochran, *Science and Golf*, 1st Annual Edinborough Scotland Conference (Chapman and Hall, London, 1990).

APPENDIX 1

FREE THROW (FT) FRACTION VS PLAYER HEIGHT IN FEET NBA DATA (1972–1973)

INDIVIDUAL STATISTICS (216 PLAYERS)		AVERAGE STATISTICS (TWO PLAYERS OR MORE)	
HEIGHT (ft)	CAREER FT FRACTION	HEIGHT (ft)	CAREER FT (AVERAGE)
5.75	0.854		
6.00	0.800		
6.00	0.776	6.00	0.788
6.08	0.791		
6.08	0.799		
6.08	0.850		
6.08	0.827		
6.08	0.719		
6.08	0.739		
6.08	0.781		
6.08	0.768		
6.08	0.564		
6.08	0.690	6.08	0.753
6.17	0.840		
6.17	0.682		
6.17	0.732		
6.17	0.713		
6.17	0.757		
6.17	0.772		
6.17	0.707		
6.17	0.779		
6.17	0.757		
6.17	0.800		
6.17	0.719		
6.17	0.739		
6.17	0.801		
6.17	0.804	6.17	0.757
6.21	0.825		
6.25	0.813		
6.25	0.721		
6.25	0.816		
6.25	0.784		
6.25	0.737		
6.25	0.622		
6.25	0.738		
6.25	0.804		
6.25	0.587		
6.25	0.863		
6.25	0.656		
6.25	0.753		
6.25	0.755		
6.25	0.832		
6.25	0.782		
6.25	0.850		
6.25	0.644		
6.25	0.778		
6.25	0.818		
6.25	0.812		
6.25	0.611		
6.25	0.764		
6.25	0.645		
6.25	0.803		
6.25	0.759		
6.25	0.825		
6.25	0.767		
6.25	0.854	6.25	0.757

FREE THROW (FT) FRACTION VS PLAYER HEIGHT IN FEET NBA DATA (1972–1973)

INDIVIDUAL STATISTICS (216 PLAYERS)		AVERAGE STATISTICS (TWO PLAYERS OR MORE)	
HEIGHT (ft)	CAREER FT FRACTION	HEIGHT (ft)	CAREER FT (AVERAGE)
6.29	0.733		
6.29	0.794		
6.29	0.667	6.29	0.731
6.33	0.711		
6.33	0.667		
6.33	0.760		
6.33	0.668		
6.33	0.786		
6.33	0.783		
6.33	0.808		
6.33	0.760		
6.33	0.626		
6.33	0.763		
6.33	0.750		
6.33	0.777		
6.33	0.779	6.33	0.741
6.38	0.731		
6.42	0.725		
6.42	0.550		
6.42	0.802		
6.42	0.750		
6.42	0.747		
6.42	0.763		
6.42	0.592		
6.42	0.833		
6.42	0.722		
6.42	0.747		
6.42	0.738		
6.42	0.833		
6.42	0.659		
6.42	0.781		
6.42	0.731		
6.42	0.722		
6.42	0.839		
6.42	0.738		
6.42	0.800		
6.42	0.762		
6.42	0.838		
6.42	0.800		
6.42	0.682		
6.42	0.747		
6.42	0.784		
6.42	0.816		
6.42	0.776		
6.42	0.802		
6.42	0.805		
6.42	0.780	6.42	0.755
6.46	0.803		
6.50	0.636		
6.50	0.692		
6.50	0.749		
6.50	0.712		
6.50	0.731		
6.50	0.541		
6.50	0.706		
6.50	0.740		
6.50	0.622		
6.50	0.549		
6.50	0.769		
6.50	0.810		
6.50	0.699		
6.50	0.694		
6.50	0.728		

FREE THROW (FT) FRACTION VS PLAYER HEIGHT IN FEET NBA DATA (1972–1973)

INDIVIDUAL STATISTICS (216 PLAYERS)		AVERAGE STATISTICS (TWO PLAYERS OR MORE)	
HEIGHT (ft)	CAREER FT FRACTION	HEIGHT (ft)	CAREER FT (AVERAGE)
6.50	0.719	6.50	0.694
6.54	0.838		
6.58	0.700		
6.58	0.761		
6.58	0.693		
6.58	0.771		
6.58	0.610		
6.58	0.730		
6.58	0.651		
6.58	0.724		
6.58	0.659		
6.58	0.715		
6.58	0.739		
6.58	0.794		
6.58	0.730		
6.58	0.695		
6.58	0.699		
6.58	0.712		
6.58	0.560		
6.58	0.626		
6.58	0.669		
6.58	0.721		
6.58	0.778		
6.58	0.500		
6.58	0.477		
6.58	0.636	6.58	0.681
6.63	0.715		
6.63	0.630		
6.63	0.750	6.63	0.698
6.67	0.785		
6.67	0.679		
6.67	0.794		
6.67	0.626		
6.67	0.469		
6.67	0.747		
6.67	0.643		
6.67	0.543		
6.67	0.798		
6.67	0.675		
6.67	0.700		
6.67	0.735		
6.67	0.766		
6.67	0.696		
6.67	0.681		
6.67	0.611		
6.67	0.222		
6.67	0.795	6.67	0.665
6.71	0.755		
6.75	0.500		
6.75	0.554		
6.75	0.736		
6.75	0.590		
6.75	0.685		
6.75	0.727		
6.75	0.322		
6.75	0.722		
6.75	0.710		
6.75	0.697		
6.75	0.796		
6.75	0.615		
6.75	0.700		
6.75	0.708		
6.75	0.689		

FREE THROW (FT) FRACTION VS PLAYER HEIGHT IN FEET NBA DATA (1972–1973)

INDIVIDUAL STATISTICS (216 PLAYERS)		AVERAGE STATISTICS (TWO PLAYERS OR MORE)	
HEIGHT (ft)	CAREER FT FRACTION	HEIGHT (ft)	CAREER FT (AVERAGE)
6.75	0.620		
6.75	0.587	6.75	0.645
6.79	0.657		
6.79	0.677	6.79	0.667
6.83	0.721		
6.83	0.694		
6.83	0.665		
6.83	0.624		
6.83	0.611		
6.83	0.747		
6.83	0.657		
6.83	0.737		
6.83	0.772		
6.83	0.676		
6.83	0.594		
6.83	0.500	6.83	0.667
6.92	0.750		
6.92	0.639		
6.92	0.722		
6.92	0.586		
6.92	0.633		
6.92	0.672	6.92	0.667
7.00	0.672		
7.00	0.800		
7.00	0.676		
7.00	0.769		
7.00	0.611	7.00	0.701
7.02	0.594		
7.08	0.511		
7.08	0.467		
7.08	0.534	7.08	0.504
7.17	0.677		
7.17	0.630	7.17	0.654

Source: Official National Basketball Association Guide—1972–73; Published by *The Sporting News*, St. Louis (1972).

APPENDIX 2

WORLD SERIES DATA: 1903–1990

		TEAM A	TEAM B				TEAM A	TEAM B	
YEAR	**GAMES**	**WINS**	**WINS**	**TIES**	**YEAR**	**GAMES**	**WINS**	**WINS**	**TIES**
1903	8	5	3	0	1947	7	4	3	0
1904	0	0	0	0	1948	6	4	2	0
1905	5	4	1	0	1949	5	4	1	0
1906	6	4	2	0	1950	4	4	0	0
1907	5	4	0	1	1951	6	4	2	0
1908	5	4	1	0	1952	7	4	3	0
1909	7	4	3	0	1953	6	4	2	0
1910	5	4	1	0	1954	4	4	0	0
1911	6	4	2	0	1955	7	4	3	0
1912	8	4	3	1	1956	7	4	3	0
1913	5	4	1	0	1957	7	4	3	0
1914	4	4	0	0	1958	7	4	3	0
1915	5	4	1	0	1959	6	4	2	0
1916	5	4	1	0	1960	7	4	3	0
1917	6	4	2	0	1961	5	4	1	0
1918	6	4	2	0	1962	7	4	3	0
1919	8	5	3	0	1963	4	4	0	0
1920	7	5	2	0	1964	7	4	3	0
1921	8	5	3	0	1965	7	4	3	0
1922	5	4	0	1	1966	4	4	0	0
1923	6	4	2	0	1967	7	4	3	0
1924	7	4	3	0	1968	7	4	3	0
1925	7	4	3	0	1969	5	4	1	0
1926	7	4	3	0	1970	5	4	1	0
1927	4	4	0	0	1971	7	4	3	0
1928	4	4	0	0	1972	7	4	3	0
1929	5	4	1	0	1973	7	4	3	0
1930	6	4	2	0	1974	5	4	1	0
1931	7	4	3	0	1975	7	4	3	0
1932	4	4	0	0	1976	4	4	0	0
1933	5	4	1	0	1977	6	4	2	0
1934	7	4	3	0	1978	6	4	2	0
1935	6	4	2	0	1979	7	4	3	0
1936	6	4	2	0	1980	6	4	2	0
1937	5	4	1	0	1981	6	4	2	0
1938	4	4	0	0	1982	7	4	3	0
1939	4	4	0	0	1983	5	4	1	0
1940	7	4	3	0	1984	5	4	1	0
1941	5	4	1	0	1985	7	4	3	0
1942	5	4	1	0	1986	7	4	3	0
1943	5	4	1	0	1987	7	4	3	0
1944	6	4	2	0	1988	5	4	1	0
1945	7	4	3	0	1989	4	4	0	0
1946	7	4	3	0	1990	4	4	0	0

Source: *The Baseball Encyclopedia* (Eighth Edition): The Complete and Official Record of Major League Baseball, Macmillan Publishing Company, New York (1990).

AUTHOR INDEX